Rolling of Advanced High Strength Steels

Theory, Simulation and Practice

Rolling of Advanced High Strength Steels

Theory, Simulation and Practice

Editors

Jingwei Zhao and **Zhengyi Jiang**

School of Mechanical, Materials and Mechatronic Engineering
University of Wollongong
NSW, Australia

CRC Press is an imprint of the
Taylor & Francis Group, an **informa** business
A SCIENCE PUBLISHERS BOOK

CRC Press
Taylor & Francis Group
6000 Broken Sound Parkway NW, Suite 300
Boca Raton, FL 33487-2742

First issued in paperback 2020

ISBN-13: 978-1-4987-3031-0 (hbk)
ISBN-13: 978-0-367-78211-5 (pbk)

Library of Congress Cataloging-in-Publication Data

Names: Zhao, Jingwei, 1980- editor. | Jiang, Zhengyi, 1962- editor.
Title: Rolling of advanced high strength steels : theory, simulation and practice / editors, Jingwei Zhao and Zhengyi Jiang, School of Mechanical Materials and Mechatronic Engineering, University of Wollongong, NSW, Australia.
Description: Boca Raton, FL : CRC Press, [2017] | Includes bibliographical references and index.
Identifiers: LCCN 2017004457| ISBN 9781498730310 (hardback) | ISBN 9781498730341 (e-book)
Subjects: LCSH: Steel, High strength. | Rolling (Metal-work)
Classification: LCC TA479.H43 R65 2017 | DDC 629.2/32--dc23
LC record available at https://lccn.loc.gov/2017004457

Visit the Taylor & Francis Web site at
http://www.taylorandfrancis.com

and the CRC Press Web site at
http://www.crcpress.com

Preface

Steel is one of the most widely used metals in our lives, with applications ranging from making fences, entrance gates, cooking and garden utensils, buildings to more specialist types of use in the building of skyscrapers, roads, bridges, automotive products and military hardware. The steel industry, however, is saddled with issues such as high energy consumption and high greenhouse gas emissions. To keep a long term and sustainable development, the steel industry, on the one hand, should continue to make the process innovation to reduce energy use and greenhouse gas emissions in production, and on the other hand, may pursue product innovation with the development of society such as to develop new high strength steels for the modern automotive industry for weight reductions given that the automotive industry is a key contributor to the national economy particularly for industrialised countries.

In today's automotive industry, one most direct and effective way of achieving the goals of conserving natural resources and preventing global warming, is by reducing the weight of automobiles and therefore lowering the rate of fuel consumption and greenhouse gas emissions. To achieve these objectives of the automotive sector, the steel industry has been developing new advanced high strength steels (AHSSs) whose unique metallurgical properties and processing methods enable the automotive industry to meet requirements of safety, efficiency, emissions, manufacturability, durability, and quality at a relatively low cost.

AHSSs for auto-making are primarily produced by rolling, plus heat treatment technologies if necessary. The term "AHSSs" refers to a group of high strength steels. Each grade of AHSSs possesses unique chemical compositions, microstructural features, mechanical properties, advantages, and challenges associated with its application. Due to the metallurgical complexity of AHSSs, it is impossible to roll all of the AHSS grades in a rolling mill with the same rolling technology. Each of AHSSs has unique applications in vehicles, and specified rolling technologies are required to produce high quality AHSS products where they might be the best employed to meet performance demands of the automotive parts. This has prompted and encouraged us to publish a scholarly book in the area of rolling of AHSSs with a purpose of providing readers with a valuable technical document that can be used in the research and development of AHSSs for automotive and other manufacturing industries.

This book focuses on the rolling of AHSSs which contains theory, simulation and practice, with contributors from USA, Germany, Poland, Italy, Spain, Austria, Australia, China, India and Iran. Topics covered in this book include introduction of AHSSs and their processes, steel rolling theory, modelling of temperature distribution and microstructural evolution in AHSSs rolling, thermomechanical processing of AHSSs, hot and cold strip rolling practice of the most popular AHSSs, and other AHSSs rolling-related technologies including chattering in rolling of AHSSs, hot dip galvanising of AHSSs, and control model of heat treatment process in the rolling of AHSSs.

Rolling of Advanced High Strength Steels: Theory, Simulation and Practice will be useful for both theoretical and applied research aimed at AHSSs rolling technologies, and will be a scientific and valuable literature for the metallurgists, engineers, materials scientists, academics and graduate students who are studying and working with AHSSs and their rolling technologies worldwide. We would like to thank all of the authors who have made significant contributions to this book.

Jingwei Zhao
Zhengyi Jiang
University of Wollongong, Australia

Contents

1

Advanced High Strength Steels and Their Processes

Minghui Cai[1,*] and *Hongshuang Di*[2]

ABSTRACT

Recent trends in the automotive industry towards the improved passenger safety and reduced weight have led to an interest in research, development and manufacturing of advanced high strength steels (AHSSs), in which both high strength and high ductility are achieved with reasonable cost of production. This chapter summarizes the history of AHSSs, the types of AHSSs currently being used and the advances in AHSSs that can be used for lightweight automotive structural components. It also discusses and compares the differences in typical chemistry, microstructure and properties between AHSSs and conventional high strength low alloy (HSLA) steels, in conjunction with disadvantages and advantages related to the products manufacturing and practical applications.

1.1 Definition and Types of Advanced High Strength Steels

Advanced high strength steels (AHSSs) are a family of steels that are stronger and have better formability or higher ductility than the conventional high strength steels (HSSs), which have been reported by Bouaziz et al. (2013),

[1] School of Materials Science and Engineering, Northeastern University, Shenyang 110819, China.

[2] State Key Laboratory of Rolling and Automation, Northeastern University, Shenyang 110819, China.

* Corresponding author: caimh@smm.neu.edu.cn

Keeler et al. (2014) and Kuziak et al. (2008). The AHSS family may be distinguished based on the strength levels that can roughly be defined: yield strength >300 MPa and ultimate tensile strength (UTS) >600 MPa. The principle difference in microstructures between conventional HSSs and AHSSs is that the former are single phase ferritic steels, in which ductility decreases with strength; whereas the later are primarily multi-phase steels that contain ferrite, martensite, bainite, and/or retained austenite for an improved combination of strength and ductility due to a higher strain hardening capability. In response to automotive demands for additional capabilities of AHSSs, materials scientists and technologists from the steel industry and academic institutions have been dedicating to search for new types of AHSSs or modification of existing types. Lee and Han (2015) summarized the representative types of AHSSs and the corresponding tensile properties using the well-known global formability diagram, Figure 1.1.

The 1st generation AHSSs involve-dual phase (DP) (Song et al. 2006), ferrite & bainite (FB) (Cai et al. 2011a), complex phase (CP) (Lee et al. 2008), hot press forming (HPF) (Karbasian and Tekkaya 2010), martensitic (MART) (Galindo-Nava and Rivera-Diaz-del-Castillo 2016) and transformation induced plasticity (TRIP) steels (Wang et al. 2014a), with an UTS of over 600 MPa, but has a relatively low product of strength and elongation (PSE) below 20 GPa%. The low ductility of 1st generation-AHSS is considered to be associated with the body-centered cubic (bcc) structured ferrite matrix. The 2nd generation AHSSs involve high manganese TRIP (Cai et al. 2011b), twinning induced plasticity (TWIP) (Grasel et al. 2000), shear band-induced

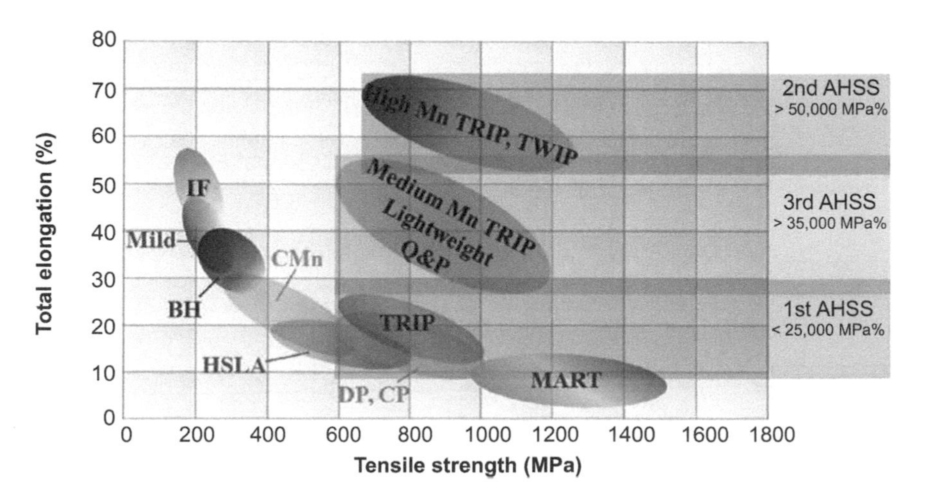

Figure 1.1 Relationship between tensile strength and total elongation of various advanced high strength steels (Lee and Han 2015).

plasticity (SBIP) (Gutierrez-Urrutia and Raabe 2012a) steels, exhibiting a remarkable combination of high strength and ductility, e.g., their PSE values exceeding 60 GPa, which is a result of TRIP/TWIP/shear banding effects from the face-centered cubic (fcc) structured austenitic matrix during plastic deformation. More recently, the 3rd generation AHSSs, which involve duplex medium Mn TRIP (Cai et al. 2014) and quenching and partitioning (Q&P) (Wang et al. 2014b) processed steels, have attracted more attention in the automotive industry recently, because they exhibit a good tradeoff between the material cost and mechanical properties.

1.2 The 1st Generation Advanced High Strength Steels

1.2.1 Dual Phase Steels

DP steels are currently one of the most widely used 1st generation AHSSs that contain the hard phase, known as islands of martensite with a volume fraction of 10–40%, which is embedded in the ductile ferritic matrix phase, as schematically represented by Keeler and Kimchi (2014) in Figure 1.2. They usually have a carbon content (typically 0.06–0.15 wt.%), which acts as the austenite stabilizer, strengthens the martensite and controls the phase distribution. The Mn content of 1.5–3 wt.% is used to stabilize the austenite and strengthen the ferritic matrix. Si promotes the ferritic transformation, while strengthening the ferrite. The Cr & Mo addition of about 0.4 wt.% retards the formation of both pearlite and bainite. Additionally, micro alloying elements, such as Nb and V are usually used as the precipitation and grain refinement strengtheners of the ductile ferrite, as reported by Schemmann et al. (2015).

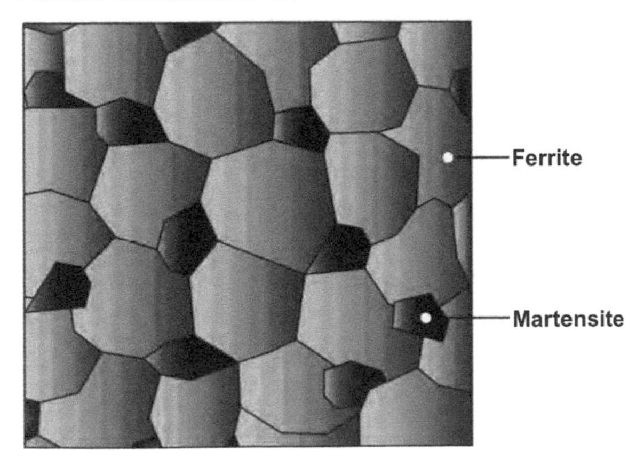

Figure 1.2 A schematic microstructure of dual phase steels (Keeler and Kimchi 2014).

The DP type of microstructure can be produced by intercritical annealing (Han et al. and Kang et al. 2013, Mazaheri et al. 2014) or thermo-mechanical treatment involving combination of hot rolling and water-quenching (Leon-Garcia et al. 2010, Speich et al. 1981 and Zhang et al. 2015). The former process involves holding in the austenite and ferrite intercritical regime between the ferritic transformation starting temperature (A_{c1}) and the ferritic transformation finishing temperature (A_{c3}), and subsequent quenching to transform the austenite into the martensite below the martensite starting temperature. This kind of treatment usually results in a coarse ferrite/martensite microstructure. In contrast, the latter process, e.g., controlled hot rolling, requires the optimal cooling schedules following hot rolling to obtain the microstructures of ferrite and martensite. The recently developed thermo-mechanical treatment by Zhang et al. (2015) results in the finer microstructures. Different intercritical annealing or thermo-mechanical treatments cause the differences in microstructural features, e.g., the volume fraction, morphology (size, aspect ratio, interconnectivity, etc.), and the grain size, which will significantly influence the overall mechanical behavior of DP steels.

As illustrated in Figure 1.3 (Bouaziz et al. 2013), the microstructure of DP steels provides a quite low yield ratio of about 0.5, together with high strength levels of 500 to 1200 MPa, which allows for a broad range of applications from crumple zone to body structure. In addition, DP steels also possess the following advantages when compared with the conventional HSS steels such as macroscopically homogeneous plastic flow and higher initial strain hardening behavior, a higher strain rate sensitivity and good resistance to fatigue crack propagation.

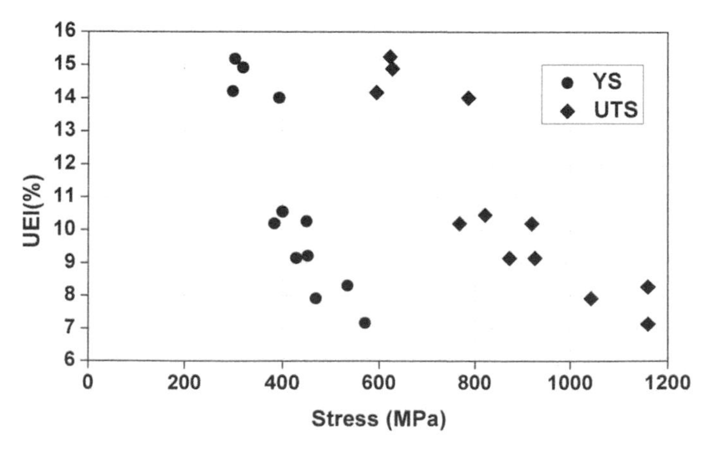

Figure 1.3 The combined properties of yield and tensile strengths and uniform elongation of DP steels (Bouaziz et al. 2013).

The particular microstructure combination of DP steels usually exhibits poor hole-expansion ratio due to the large difference in hardness between soft and hard phases, which means that these steels are not good candidates for applications that require high stretch-flange ability. The drawback, however, can be overcome by either adding Ti for strengthening the ferrite matrix, or developing microstructures with better combinations of strength and fracture strain, such as FB, CP or TRIP steels.

1.2.2 Ferrite-Bainite Steels

FB steels are also known as high hole-expansion (HHE) or stretch flangeable (SF) steels, as investigated by Cai et al. (2011a), Kumar et al. (2008a), Sudo et al. (1983) and Takahashi (2003), and the representative microstructure is composed of the soft ferrite and the hard bainite, as shown in Figure 1.4. In general, the carbon content in FB steels should be less than 0.1 wt.% for the improved formability and weldability, while a relatively high amount of Mn (1.0–1.5 wt.%) is added to refine ferrite grains and prevent the fine carbides from growing by lowering the eutectoid temperature (A_{r1}). Additionally, Si is often added to FB steels to promote the finer, equiaxed ferrite grains, inhibit the carbide precipitation in bainite and decrease the solubility of carbon in ferrite (Cai et al. 2009a, 2009b and 2011a).

Similarly, both heat treatment and thermo-mechanical control process can be used to produce the ferrite and bainite microstructure. However, the final microstructure can be quite different due to various manufacturing routes as well as the difference in chemical composition and initial microstructure, as summarized by Cai (2009c) in Table 1.1.

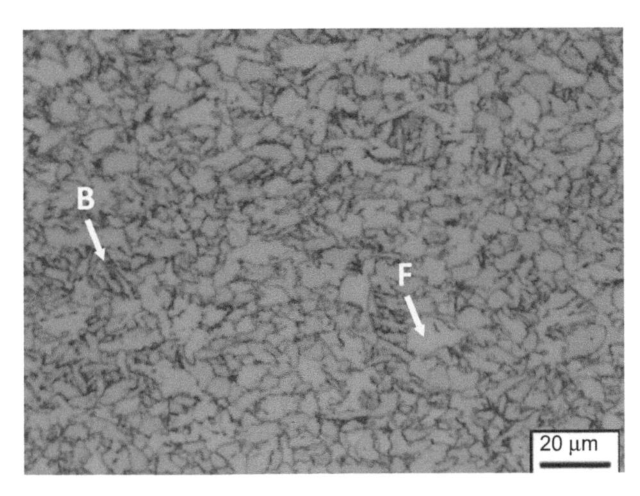

Figure 1.4 Representative microstructure of thermo-mechanical control processed FB steels, involving hot rolling, followed by multi-stage cooling routes in a Fe-0.08C-1.5Mn-0.5Si (wt.%) steel (Cai et al. 2011a).

Table 1.1 Comparison of chemical composition, microstructure and mechanical properties of FB steels fabricated by various thermo-mechanical processing routes (Cai et al. 2012).

Chemical composition (wt.%)	Methods	Shape of ferrite	V_B (%)	YS (MPa)	UTS (MPa)	δ (%)	HER (%)	Ref.
0.08C-0.49 Si-1.38 Mn-0.022 Nb	ST (1373 K, 20 min) + Soaking (773 K, 60 min + Quenching)	Polygonal	50~90	439~445	552~556	22~27	-	Kumar et al. 2008b
(0.074~0.085)C-0.15Si-1.5 Mn-0.04 Nb	Rolling in the austenite region + three-step controlled cooling	Polygonal	10~15	510~560	580~700	19~27	103~120	Cho et al. 1999
(0.15~0.20)C-1.5 Mn-0.5Si-(0~0.21)Cr-(0~0.13) V-(0~0.16) Mo	Two-step rolling in the austenite region + air cooling	Irregular	48~74	325~462	525~716	22~28	-	Podder et al. 2007
0.077C-1.35Si-0.97 Mn	Rolling in the austenite region + two-step controlled cooling	Polygonal	25	421	633	34	112	Cai 2009c, Cai et al. 2011a
0.077C-0.57Si-1.43 Mn		Quasi-polygonal	31	473	722	28	95	
		Polygonal	24	525	600	29	118	
0.079C-0.92Si-1.46 Mn		Polygonal	21	614	708	33	101	
0.075C-0.5Si-1.0 Mn-0.45Al-(0~0.018) Nb	Rolling in the austenite region + two-step controlled cooling	Polygonal	9~14	360~425	500~570	28~34	91~108	Wang et al. 2009

The combination of high strength and hole-expansion ratio is one of the most important advantageous features of FB steels due to the enhanced stretch capability when compared with conventional HSS and DP steels with the same levels of strength. Another important feature of such a unique mixture of ferrite and bainite is the potential for dynamic load bearing in an event such as car crash because of high energy absorption capacity, while it exhibits good fatigue properties. Thus, FB steels are good candidates for shock towers and control arms, as shown in Figure 1.5.

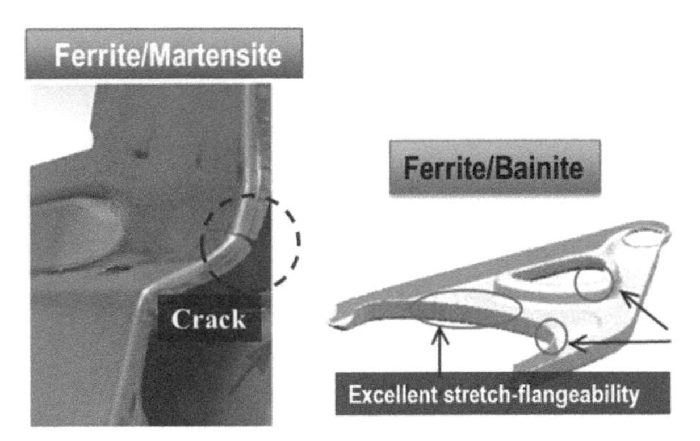

Figure 1.5 A comparison of stretch-flangeability between ferrite-martensite/ferrite dual-phase steels.

1.2.3 Complex Phase Steels

The microstructure of CP steels is typically characterized by small amounts of martensite within the ferrite/bainite matrix, which has been investigated by Zare and Ekrami (2011) and Dan et al. (2012). CP steels typically contain 0.1–0.25 wt.% C, 1.5–2.5 wt.% Mn and less than 1 wt.% Si with additional microalloying elements like Nb, V or Ti to cause further strengthening by fine carbon or nitrogen precipitates of Nb, V or Ti.

It should be noted that it is possible to form a multi-phase microstructure in the hot rolled condition, in which the multi-stage controlled cooling technology is usually utilized, as reported by Deardo (1995). The factors influencing the fractions and morphologies of different phases such as ferrite, bainite, martensite and retained austenite, involve alloy composition, finish temperature, cooling rate and coiling temperature, etc.

The complex phase type of microstructure in CP steels offers the following advantageous features over DP steels:

- High yield strength at equal tensile strengths of 800 MPa and greater;
- Good roll-forming, bending and hole-expansion behavior for high strength level;

- Achievement of very narrow bending radii;
- High energy absorption capability and fatigue strength;
- Good resistance spot weld ability.

Given their high energy absorption capacity and fatigue strength, these grades are particularly well suited for automotive safety components requiring good impact strength and for suspension system components. In addition, CP steels exhibit good roll-forming, bending and hole-expansion behaviors. Figure 1.6 compares the hole-expansion rates of DP and CP steels with tensile strength of 1000 MPa, indicating CP steels has exceptional stretch flangeability (ArcelorMittal 2009).

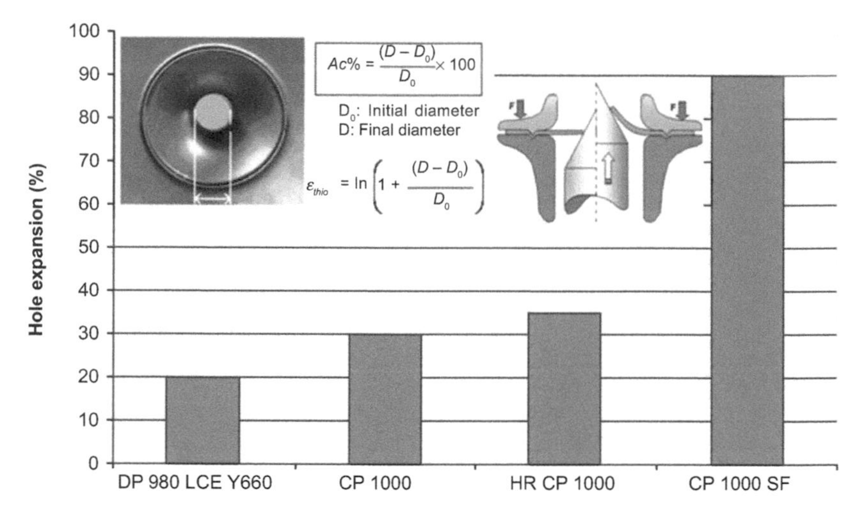

Figure 1.6 Relationship between hole-expansion and tensile strength of DP and CP steels (ArcelorMittal 2009).

1.2.4 Transformation-Induced Plasticity Steels

TRIP steels exhibit a multi-phase microstructure consisting of some amount of retained austenite (>5 vol.%) embedded in the soft ferrite matrix (50–60 vol.%), plus some hard phases like bainite and martensite, as schematically illustrated by Tang (2007) in Figure 1.7. In general, the total content of alloying elements is relatively low, about 3.5 wt.%, including 0.12–0.55 wt.% C, 0.20–2.5 wt.% Mn, and 0.40–1.8 wt.% Si. For the weldability and hardenability reasons, C and Mn contents in low alloy TRIP steels are usually 0.20–0.25 wt.% and around 1.5 wt.%, respectively. Although Si is believed to be an efficient alloying element to prevent against the formation of cementite, the partial replacement of 1 wt.% Si by 1 wt.% Al has been applied, taking into consideration the continuous

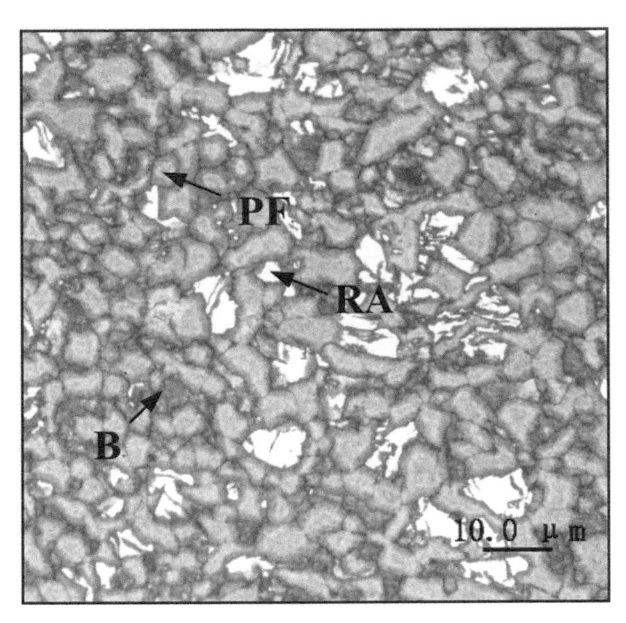

Figure 1.7 Tint-etched TRIP microstructure of 0.18C-1.38Mn-0.67Si-0.56Al-0.015Nb steel after deformation at 800°C, followed by holding at 400°C for 3 min (Tang 2007).

galvanizing of TRIP steel in practical production (Jacques, Meyer et al. and Pichler 1999, Jacques et al. 2001). In the recently developed TRIP steels, only partial replacement of Si by a limited amount of Al and 0.5–0.10 wt.% P has been proposed by Ding et al. (2011). The use of P is attempted to reduce the addition of Al, because P has a negative effect on the formation of cementite and has a strong solid solution strengthening effect. The role of micro alloying elements like Nb, Mo, Ti has been studied by Bleck et al. (2003) and Hashimoto et al. (2004) to further refine the microstructure and enhance the resulting properties of TRIP steels.

The manufacturing processes of TRIP steels usually involve several various stages: hot/cold rolling, intercritical treatment where austenite and ferrite are stable, followed by a holding in the banitic region where austenite is enriched in carbon and finally quenched to room temperature. As compared to DP steels, the cooling stage of TRIP steels after rolling is more complicated.

TRIP steels have high work hardening behavior, resulting in excellent formability and high stretch capability. Complex shapes are possible because TRIP steels exhibit good bendability and resist the onset of necking. Tang et al. (2010a and 2010b) demonstrated that TRIP steels have excellent bake-hardening capability. However, shear cracking at the interfaces between the ductile and hard phases also reduces the hole-expansion limits for edges of TRIP steels, as reported by the Auto/Steel Partnership (2010). It should be noted that poor resistance spot-welding behavior caused by alloying can be addressed by modifying welding cycles.

Automotive applications of TRIP steels include body structure and ancillary parts. With high energy absorption and strengthening under strain, they are often selected for components that require high crash energy management, such as cross members, longitudinal beams, A- and B-pillar reinforcements, sills and bumper reinforcements, as referred from the Steel Marker Development Institute (2011).

1.2.5 Martensitic Steels

The microstructure of martensitic steels is mostly lath martensite, with a small amount of very fine ferrite and/or bainite phases, as shown in Figure 1.8. Alloying elements such as Mn, Cr, Si, Mo, B, Ni, and V, etc. are usually used in various combinations with the proper C content, in order to increase the hardenability.

MART steels are produced by applying rapid quenching following hot rolling, annealing or post-forming heat treatment. Post-quench tempering is often utilized as well to improve the ductility and formability even at a relatively high strength level.

The martensitic microstructure is known for its extremely high strength, ranging from 900 to 1700 MPa, with relatively low elongation. Considering, MART steels have such high strength to weight ratio, they are often

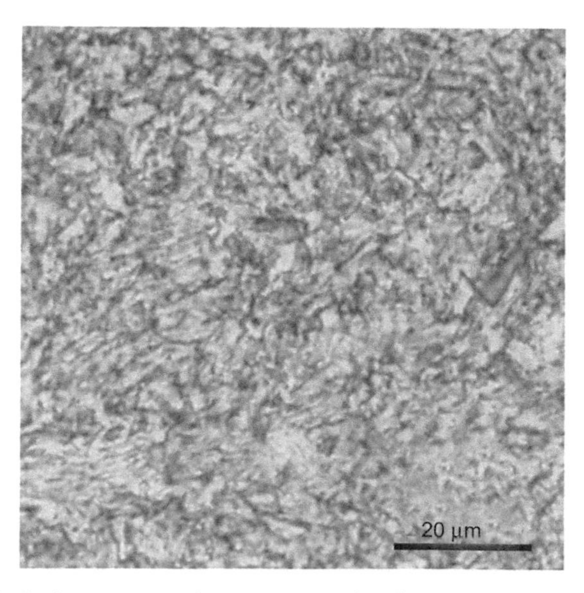

Figure 1.8 Optical microstructure of martensitic steel with a composition of 0.077C-1.43Mn-0.57Si (Cai 2009c).

recommended for bumper reinforcement and door intrusion beams, rocker panel inners and reinforcements, side sill and belt line reinforcements, springs, and slips.

1.3 The 2nd Generation AHSS

The 2nd generation-AHSS is typically austenitic steels at room temperature. The stacking fault energy (SFE) of the austenite (γ_{SFE}), which is closely related to the chemical composition and deformation temperature, plays a crucial role on determining the deformation modes, as schematically shown in Figure 1.9 by Wang et al. (2015) and Cai et al. (2014, 2016) reported that the low γ_{SFE} values of <20 mJ/m^2 favor the TRIP phenomenon related to the strain-induced either ε- or α′-martensitic transformation during tensile deformation. In the medium γ_{SFE} ranges of 20–40 mJ/m^2, no phase transformation occurs, and alternatively deformation twinning is instigated, together with the dislocation glide, as reported by Yang et al. (2006) and Tang et al. (2015). At high γ_{SFE} value of about 90 mJ/m^2, deformation is accommodated only via dislocation glide, as demonstrated by Ding et al. (2016). According to the γ_{SFE} differences, this group can be divided into high manganese TRIP, TWIP and SBIP steels. These micro structural features, such as ε- or α′-martensite plates, deformation twins and shear bands, cause a remarkable variety of work hardening behavior as they all act as effective obstacles for dislocation glide.

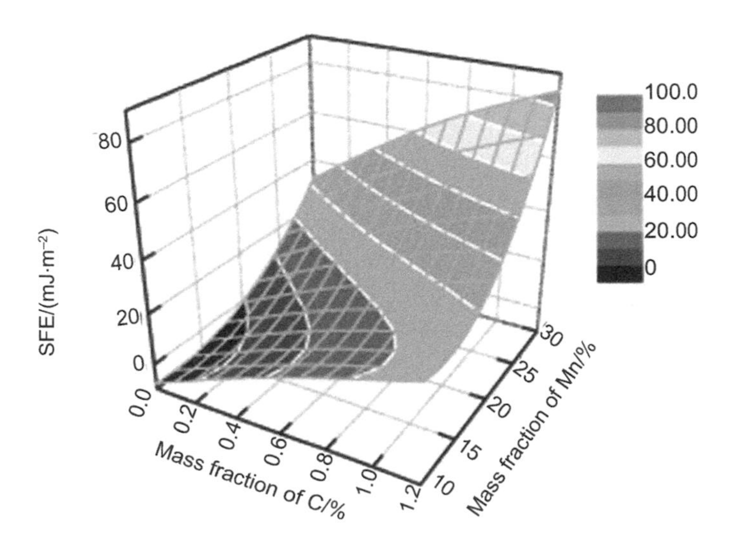

Figure 1.9 Dependence of stacking fault energy on alloying elements, C and Mn in Fe-Mn-C steels at 25°C (Wang et al. 2015).

1.3.1 High Manganese Twinning Induced Plasticity Steels

The microstructure of high manganese TWIP steels is mostly austenitic, with some amount of annealing twinning, as shown in Figure 1.10. Gutierrez-Urrutia and Raabe (2012b) and Pierce et al. (2015) reported that high manganese TWIP steels contain Mn content as high as 18–30 wt.%, because Mn is a crucial alloying element to remain the austenite based on the ternary system of Fe-Mn-Al and control the γ_{SFE} to the range of 20–40 $mJ{\cdot}m^{-2}$. Jeong et al. (2013), Lee et al. (2016) and Li et al. (2015a) demonstrated that silicon (<3 wt.%) improves the tensile strength of high manganese TWIP steels by solid solution strengthening, and increases the fracture strength, although it does not improve the ductility. The addition of Al (<3 wt.%) to HMn-TWIP steels increases the γ_{SFE} significantly and therefore stabilizes the austenite against phase transformations, which has been studied by Hamada et al. (2007), Jung et al. (2012) and Ma et al. (2015). Ghasri-Khouzani and McDeremid (2015) investigated that adding small amount of less than 1 wt.% C can improve the stability by inhibiting the formation of ε-martensite and strengthen the steels. Additionally, Reyes-Calderón et al. (2013), Park et al. (2014), Li et al. (2015b) and Mejia et al. (2015) have shown that these steels micro alloyed by Nb, V and Ti can control the grain growth under hot working conditions.

Zhang et al. (2011) focused on the hot rolled high manganese TWIP steels by a conventional production line. Meanwhile, Shen et al. (2013), Yanushkevich et al. (2016) and Vercammen et al. (2004) studied

Figure 1.10 Optical microstructure of Fe-26Mn-3Al-1C steel processed by solution treatment and quenching.

heat treatment of the cold rolled sheets to refine the austenite grain by recrystillization of the heavily deformed microstructure. However, the wide use of high manganese TWIP steels for various applications is limited due to technological problems related to poor casting, hot-working above 1150°C. A promising alternative approach to manufacture high manganese TWIP steels is twin-roll strip casting, which has been demonstrated by Daamen et al. (2011) and (2015) and Liu et al. (2007). This combined casting and forming process enables near-net shape production of these highly alloyed steel, as shown in Figure 1.11.

It is well known that the deformation twinning causes high and continuous work hardening behavior as the microstructure becomes finer and finer. The resultant twin boundaries act like grain boundaries, as fine as tens of nanometers and strengthen the steel. Thus, high manganese TWIP steels exhibit both high strength of 700–1200 MPa and extremely high ductility over 50%, even at relatively high strain rates, as investigated by Grasel et al. (2000). The high ductility together with high strength levels of high manganese TWIP steels can improve the crash resistance for car structural body parts such as A and B pillar, front and rear bumper beams, door impact beam, etc. (Bouaziz et al. 2011). The excellent formability enables deep and stretch formation of parts with complex shape at room temperature (Busch et al. 2014). Despite these advantages of high manganese TWIP austenitic steels, their practical application will probably be limited, because of high cost due to Mn above 20 wt.%, Al and Si alloying concept. In addition, some difficulties related to the aforementioned technical challenges, including poor casting and hot-working above 1150°C, corrosion resistance and Mn segregation, etc. (Haase et al. 2016).

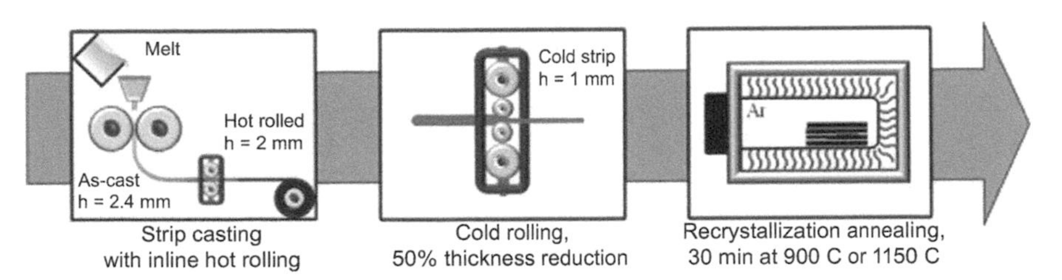

Figure 1.11 Schematic process line for production of HMn-TWIP steels, cold rolled strip via twin-roll strip casting at the IBF, RWTH Aachen University (Daamen et al. 2015).

1.3.2 High Manganese Transformation-Induced Plasticity Steels

High manganese TRIP steels contain a relatively lower Mn content, normally below 18 wt.%, corresponding to the lower γ_{SFE} value (<20 mJ/m^2), which has been reported by Shao et al. (2016), Figueiredo et al. and Xie et al. (2015). The mechanical stability of austenite significantly reduces with decreasing

Mn content. In this case, when external force is exerted, the microstructure evolves from the fcc austenite to the hexagonal ε-martensite or the bcc α′-martensite. In contrast, high manganese TRIP steels are much stronger than high manganese TWIP steels, with a moderate ductility of approximately 35%. However, this ductility is comparable to the conventional low-alloyed TRIP steels belonging to the 1st generation AHSSs, in which only a certain amount of the austenite is transformed to martensite, offering only 5% additional ductility in crash. The characteristics of high manganese TRIP steels, ductile yet strong, is the result of changes in the crystal lattice, as shown in Figure 1.12 (Total Materia 2007).

High manganese TRIP steels are particularly useful for side impact protection, because of high energy absorption capacity, which is caused by extremely high work hardening rates. This can prevent the side sections from collapsing too much and protects vehicle occupants from injury.

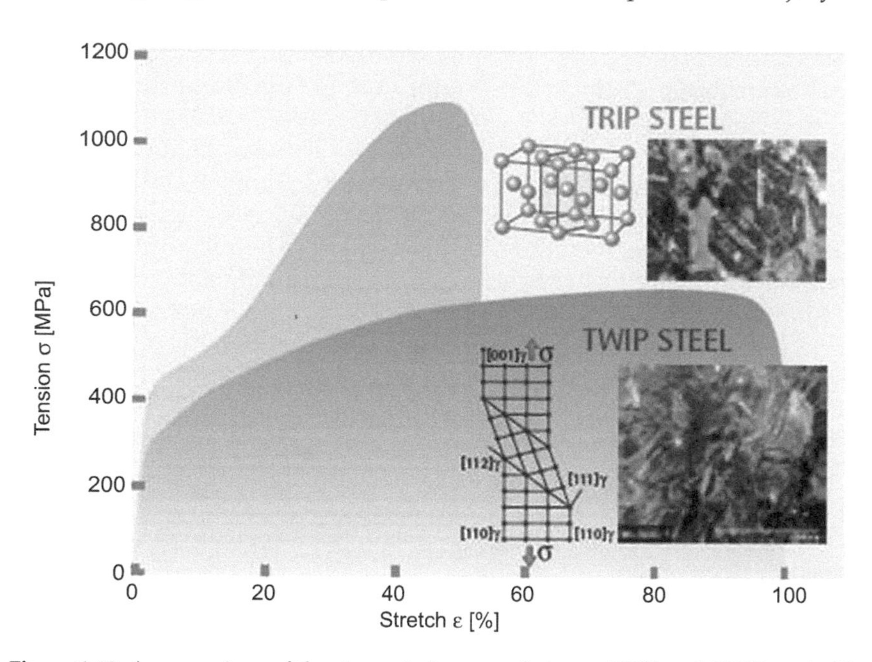

Figure 1.12 A comparison of the stress-strain curves between TRIP and TWIP steels (Total Materia 2007). TRIP steel can resist high stresses, while TWIP steel does not break until tensile elongation reaches approximately 90%.

1.3.3 Shear-Band Induced Plasticity Steels

SBIP steels with high contents of Mn and Al are also referred to as 'Triplex steels' or low density steels (Frommeyer and Brux 2006). This group can be distinguished into single austenite (Ding et al. 2016), austenite-based and ferrite-based structures (Yoo and Park 2008) based on the difference in alloying compositions, as reported by Raabe et al. and Yang et al. (2014). High manganese (15–30 wt.%) steels with an Al content of 2–12 wt.% and C content of 0.5–1.8 wt.%, have been considered as promising light weight structural steels (Gutierrez-Urrutia and Raabe 2013). Typically, low-density steels with a ferritic structure usually have compositions with <8 wt.% Mn, 5–8 wt.% Al and <0.3 wt.% C, while those with a complex structure consisting of austenite and ferrite can be generally synthesized by using compositions of 5–30 wt.% Mn, 3–10 wt.% Al and 0.1–0.7 wt.% C (Gutierrez-Urrutia and Raabe 2014). The addition of Al promotes the precipitation of nano sized $(FeMn)_3AlC$ carbides, which are called κ-carbides (Figure 1.13), contributing to the improvement of mechanical properties (Hwang et al. 1993).

In general, κ-carbides can form either through aging treatment or during rapid quenching for >10 wt.% Al, in conjunction with the rapid alloy prototyping (RAP) (Springer and Raabe 2012), which is proposed as an efficient tool to develop the Fe-Mn-Al-C low-density TWIP and κ-carbide hardened steels.

The Fe-Mn-Al-C system is a promising low density steel grade that offers outstanding combinations of high yield strength of 0.5–1.0 GP, high ultimate tensile strength of 1.0–1.5 GPa and total elongation of 30–80% as well as a reduced specific weight, i.e., 1.5% density reduction per 1 wt.%

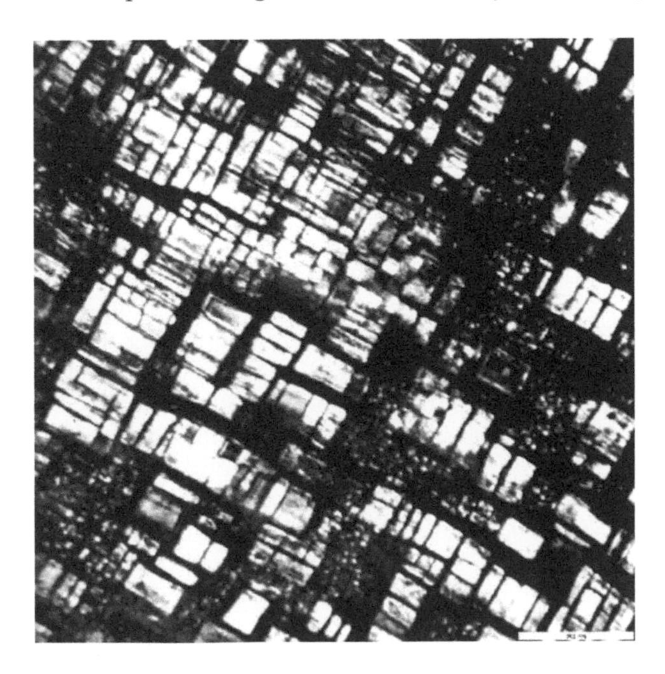

Figure 1.13 Transmission Electron Microscopy (TEM) analysis of κ-carbies in a Fe-30 Mn-1.2C-8 Al low density steel (Raabe et al. 2014).

Al addition. The research and development of advanced low density steels aims to structural applications in many industrial sectors, in particular in the automotive industry. These steel grades allow the manufacture of lightweight crash-resistance car body structures, leading to cars with safety and a considerable reduction in fuel consumption (Gutierrez-Urrutia and Raabe 2013 and Zuazo et al. 2014).

1.4 The 3rd Generation Advanced High Strength Steels

The 3rd generation AHSSs for automotive industry, which include medium Mn steel (Cai et al. 2014 and 2016), and quenching and partitioning processed steel (Wang et al. 2014a), is a further step in development of advanced high strength-ductility balance steel sheets. Steels in this category combine the advantages of multiphase features of 1st generation AHSS, and austenitic phase of 3rd generation AHSS that enables strain-induced martensitic transformation and mechanical twinning (Aydin et al. 2013, Grajcar et al. 2012, Lee and Han 2015). The main purpose of developing 3rd generation AHSSs is to balance the mechanical properties between 1st generation and 2nd generation AHSS at reasonable cost when compared to 1st generation AHSS.

1.4.1 Medium Mn Steels

Medium Mn steels are defined by Lee and Han (2015), as a kind of steel with Mn content ranging from approximately 3–10 wt.%. As a main austenite stabilizer, Mn increases the hardenability of steel, resulting in a dramatic decrease in the volume fraction of ferrite as a result of shifting the austenite to ferrite transformation region to the right according to the CCT diagrams (Tsukatani et al. 1991 and Hashimoto et al. 2004). As an important alloying element for lightweight steels, the addition of Al also allows a higher intercritical annealing temperature because it increases the equilibrium starting (Ae_1) and finishing (Ae_3) temperatures during reverse transformation, so as to shorten the annealing time within 2 min (Lee et al. 2011, Lee et al. 2013, Suh et al. 2010). Si concentration (<2 wt.%) is used to improve both the strength and ductility, because Si increases the hardenability and thermal stability of the reversed austenite, resulting in a relatively high fraction of retained austenite and subsequent active TRIP phenomenon (Furukawa et al. 1994). The micro alloying with Nb and Mo (Cai et al. 2014 and 2016), V and Ti (Lee et al. 2013 and Siciliano and Poliak 2005) is applied to obtained fine-grained/ultrafine-grained complex microstructures with a high volume fraction of retained austenite of optimal stability for continuous strain-induced martensitic transformation over a large strain regime during drawing, stretching and bending, etc. The microstructure of medium Mn steels is characterized by being multi-phase (e.g., ferrite, austenite, martensite), metastable, and multi-scale (so called as M^3) (Furukawa et al. 1994), as shown in Figure 1.14 (Cai et al. 2014).

The common production of medium Mn steels requires long-time intercritical annealing in the austenite and ferrite two-phase region, either directly after hot rolling of austenite or after hot rolling of austenite plus cold rolling of martensite, as literatured in Arlazarov et al. (2012), Cai et al. (2014 and 2016) and Gibbs et al. and Cao et al. (2011), respectively. The stability of the austenite is a function of the intercritical annealing temperature and the compositions such as Mn and C concentration in austenite, which can be optimized by thermo-calculations (Dong 2012). However, the dimensional variability of the cold-formed TRIP parts is too high due to high instantaneous plastic modulus at all strains; therefore, stamping companies are reluctant to this kind of TRIP steel. Alternatively, hot forming of medium Mn steels have been recently proposed as well to overcome the major challenges related to the spring back of cold-formed medium Mn steels (Cai et al. 2015).

Medium Mn steel exhibits an ultimate tensile strength of above 1000 MPa and a uniform elongation of more than 20%, so they are utilized to design lightweight automotive body structural components with enhanced crash resistance. For example, Bao steel has recently developed a series

Figure 1.14 Scanning Electron Microscopy (SEM) morphology of a Fe-6.5 Mn-0.17 C-1.1 Al-0.05 Nb-0.22 Mo steel processed by conventional hot rolling, cold rolling and intercritical annealing at 650°C for 30 min (Cai et al. 2014).

of medium Mn steel sheets including CR980, GI980 and GI1180 grades, which are suitable for A-pillar, B-pillar, anti-collision beams and door side reinforcements, etc.

1.4.2 Quenching & Partitioning Steels

The quenching and partitioning concept is proposed to produce the austenite-containing steels (Figure 1.15) (Speer et al. 2005), based on a new understanding of carbon partitioning hypothesized between martensite and austenite (Speer et al. 2003). QP steels usually have 0.2 wt.% C, 1–1.5 wt.% Si and 1–1.5 wt.% Mn (Wang and Speer 2013). It should be noted that alloying elements such as Si, Al or P (Speer et al. 2003) that can suppress carbide precipitation play a critical enabling role in the Q&P process. For example, Si suppresses cementite formation, or delays the transition from early-stage tempering (where ε or η carbides are present), to later-stage tempering (where θ-Fe_3C is present) (Krauss 1990 and Honeycombe and Bhadeshia 1995).

QP steels are produced via the Q&P process which consists of quenching and partitioning stages (Toji et al. 2014 and 2015). In the quenching stage, fully austenitized or intercritically annealed steels are quenched to temperatures below the martensite start temperature (M_s) but above the martensite finish temperature (M_f) to form a desirable fraction of martensite. Subsequently, the quenched steels are held at the same or higher temperatures than the quenching temperature for carbon partitioning. The thermal history and phase transformation process are schematically displayed in Figure 1.16 (Wang and Speer 2013).

The resultant microstructures of QP steels are mainly composed of the tempered martensite and the retained austenite, so that they exhibit

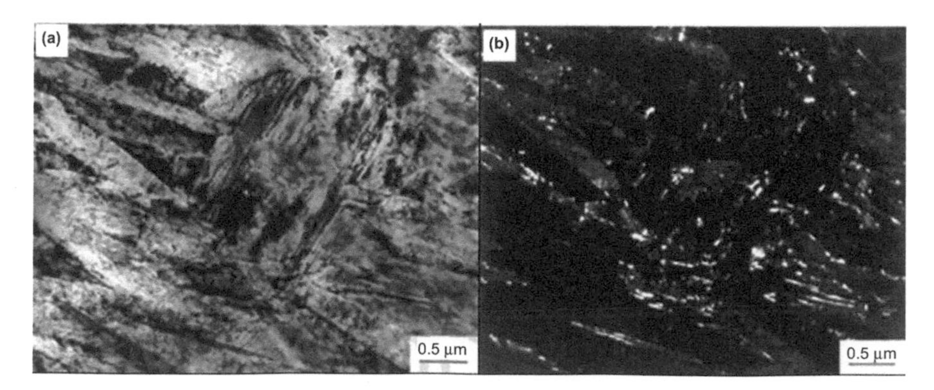

Figure 1.15 TEM bright field (a) and $(002)_\gamma$ dark field (b) images showing martensite and retained austenite after one-stage quenching and partitioning process before cooling to room temperature (Speer et al. 2005).

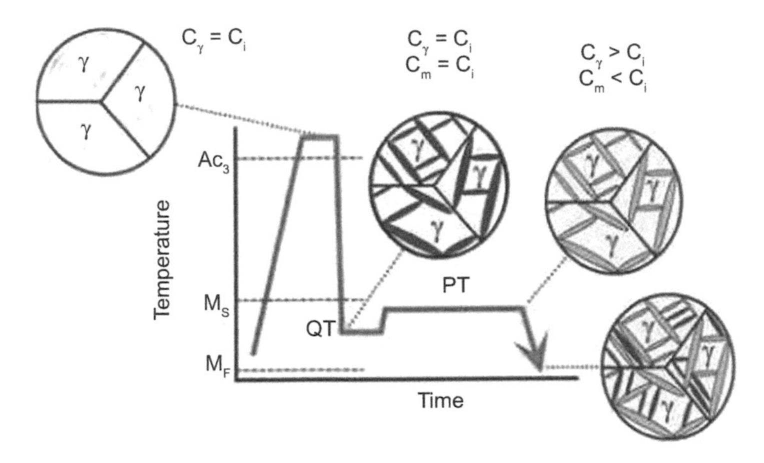

Figure 1.16 Schematic illustration of thermal profile and phase transformation behaviour of QP steels. QT-quenching temperature; PT-partitioning temperature (Wang and Speer 2013).

relatively higher strength levels when compared to the conventional 1G-TRIP steels, e.g., an ultrahigh strength of 1000–1400 MPa with adequate ductility of 10–20% (Speer et al. 2004). QP steels exhibit ultrahigh strength, good ductility and higher work hardening rates over conventional high strength steels, so they are well suited for automotive structural parts for its safety and weight reduction, such as cross members, longitudinal beams, B-pillar reinforcements, sills, and bumper reinforcements, which cannot be cold formed using conventional high strength steels with similar strength levels.

1.5 Summary

In the beginning of this chapter, we described the general definition of AHSSs based on their strength levels, and highlighted the significance of research activity dedicated to AHSSs, driven by the breakthrough combination of strength and ductility while maintaining low material and manufacturing costs.

- The existing inverse relationship between strength and ductility of low alloyed 1st generation AHSSs (e.g., DP, FB, CP, TRIP and MART steel) is related to the bcc-structured ferrite matrix, considering that their deformation is dominated by dislocation slip;
- The high alloyed 2nd generation AHSSs (e.g., high manganese TRIP or TWIP steel) exhibits an extra high ductility without sacrificing strength, which is a result of the fcc-structured austenitic matrix, considering that a continuous TRIP or TWIP effect occurs during plastic deformation;

- The more recently developed 3rd generation AHSSs (e.g., medium Mn or QP steel) exhibits a good tradeoff between the material cost and mechanical properties in comparison to 1st and 2nd generation AHSSs.

A review of the representative chemical compositions, microstructures, mechanical properties of AHSSs and their manufacturing processes, as well as examples of practical applications has been presented, revealing why the manufacturing of AHSSs is currently a topic of great practical interest and fundamental importance.

Acknowledgements

This research was supported by the Natural Science Foundation of China (Grant No. 51401050) and the Alfred Deakin Research Fellowship (Dr. Minghui Cai).

Keywords: Steel, Dual Phase, Complex Phase, TRIP, TWIP, SBIP, Quenching & Partitioning, Composition, Microstructure, Properties, Thermo-mechanical Processing, Strength, Ductility

References

Aydin, H., E. Essadiqi, I.H. Jung and S. Yue. 2013. Development of 3rd generation AHSS with medium Mn content alloying compositions. Mater. Sci. Eng. A 564: 501–508.

ArcelorMittal. 2009. http://automotive.arcelormittal.com/saturnus/sheets/C_EN.html.

Arlazarov, A., M. Gouné, O. Bouaziz, A. Hazotte, G. Petitgand and P. Barges. 2012. Evolution of microstructure and mechanical properties of medium Mn steels during double annealing. Mater. Sci. Eng. A 542: 31–39.

Auto/Steel Partnership. A Special Edition of AHSS Case Studies. Advanced High Strength Steel Applications: Design and Stamping Process Guidelines 2010.

Bleck, W., A. Frehn, E. Kechagias and K. Hulka. 2003. Control of microstructure in TRIP steels by niobium. Mater. Sci. Forum 426-432: 43–48.

Bouaziz, O., S. Allain, C.P. Scott, P. Cugy and D. Barbier. 2011. High manganese austenitic twinning induced plasticity steels: A review of the microstructure properties relationships. Curr. Opin. Solid State Mater. Sci. 15: 141–168.

Bouaziz, O., H. Zurob and M. Huang. 2013. Driving force and logic of development of advanced high strength steels for automotive applications. Int. Steel Res. 84: 937–947.

Busch, C., A. Hatscher, M. Otto, S. Huinink, M. Vucetic, C. Bonk, A. Bouguecha and B.A. Behrens. 2014. Properties and application of high-manganese TWIP steels in sheet metal forming. Proc. Eng. 81: 939–944.

Cai, M.H., H. Ding, J.S. Zhang, L. LI, X.B. Li and L.X. Du. 2009a. Transformation behavior of low carbon steels containing two different Si contents. Int. J. Iron Steel Res. 16: 55–60.

Cai, M.H., H. Ding, J.S. Zhang and L. Li. 2009b. Effect of silicon and prior deformation of austenite on isothermal transformation in low carbon steels. Acta Metall. Sin. (Engl. Lett.) 22: 100–109.

Cai, M.H. 2009c. Microstructural evolution and mechanical behavior of ferrite/bainite steels with high stretch flangeability, Ph.D. Thesis. North Eastern University.

Cai, M.H., H. Ding, Y.K. Lee, Z.Y. Tang and J.S. Zhang. 2011a. Effects of Si on microstructural evolution and mechanical properties of hot-rolled ferrite and bainite dual-phase steels. ISIJ Int. 51: 476–481.

Cai, M.H., H. Ding, Z.Y. Tang, H.Y. Li and Y.K. Lee. 2011b. Strain hardening behavior of high performance FBDP, TRIP and TWIP steels. Int. Steel Res. 82: 242–248.

Cai, M.H., H. Ding, J.S. Zhang and P.D. Hodgson. 2012. Fabrication, microstructure and mechanical properties of high performance ferrite/bainite steels. The Asia Steel International Conference, Beijing 1–6.

Cai, M.H., Z. Li, Q. Chao and P.D. Hodgson. 2014. A novel Mo and Nb microalloyed medium Mn TRIP steel with maximal ultimate strength and moderate ductility. Metall. Mater. Trans. A 45: 5624–5634.

Cai, M.H., P.D. Hodgson and B.F. Rolfe. 2015. Hot forming of medium Mn steel with TRIP effect, The 2nd International Conference on Hot Stamping of UHSS (ICHSU2015), Changsha 25–29.

Cai, M.H., Z. Li, Q. Chao and P.D. Hodgson. 2016. Dependence of deformation behavior on grain size and strain rate in an ultrahigh strength-ductile Mn-based TRIP alloy, Mater. Sci. Eng. A 653: 35–42.

Cao, W.Q., C. Wang, J. Shi, M.Q. Wang, W.J. Hui and H. Dong. 2011. Microstructure and mechanical properties of Fe-0.2C-5Mn steel processed by ART-annealing. Mater. Sci. Eng. A 528: 6661–6666.

Cho, Y.R., J.H. Chung, H.H. Ku and I.B. Kim. 1999. Effect of controlled cooling on the formability of TS 590 MPa grade hot-rolled high strength steels. Int. Met. Mater 5: 571–578.

Daamen, M., B. Wietbrock, S. Richter and G. Hirt. 2011. Strip casting of a high manganese steel (FeMn22C0.6) compared with a process chain consisting of ingot casting and hot forming. Int. Steel Res. 82: 70–75.

Daamen, M., C. Haase, J. Dierdorf, D.A. Molodov and G. Hirt. 2015. Twin-rolled strip casting: A competitive alternative for the production of high-manganese steels with advanced mechanical properties. Mater. Sci. Eng. A 627: 72–81.

Dan, W.J., Z.Q. Lin, S.H. Li and W.G. Zhang. 2012. Study on the mixture strain hardening of multi-phase steels. Mater. Sci. Eng. A 552: 1–8.

Deardo, A.J. 1995. Multi-phase microstructures and their properties in high strength low carbon steels. Int. ISIJ 35: 946–954.

Ding, H., D. Han, J. Zhang, Z.H. Cai, Z.Q. Wu and M.H. Cai. 2016. Tensile deformation behavior analysis of low density Fe-18Mn-10Al-xC steels. Mater. Sci. Eng. A 652: 69–76.

Ding, W., D. Tang, H.T. Jiang, B.F. Wang and Z.H. Gong. 2011. Mechanical properties of continuously annealed Si-free P-containing TRIP steel. Acta Metall. Sin. 47: 1022–1025.

Dong, H. 2012. High performance steels: Initiative and practice. Sci. China Technol. Sci. 55: 1774–1790.

Figueiredo, R.B., F.L. Sicupira, L.R.C. Malheiros, M. Lawasaki, D.B. Santos and T.G. Langdon. 2015. Formation of epsilon martensite by high-pressure torsion in a TRIP steel. Mater. Sci. Eng. A 625: 114–118.

Frommeyer, G. and U. Brux. 2006. Microstructure and mechanical properties of high-strength Fe-Mn-Al-C light-weight TRIPLEX steel. Steel Research Int. 9-10: 627–631.

Furukawa, T., H. Huang and O. Matsumura. 1994. Effects of carbon content on mechanical properties of 5% Mn steels exhibiting transformation induced plasticity. Mater. Sci. Technol. 10: 964–969.

Galindo-Nava, E.I. and P.E.J. Rivera-Diaz-del-Castillo. 2016. Understanding the factors controlling the hardness in martensitic steels. Script. Mater. 110: 96–100.

Ghasri-Khouzani, M. and J.R. McDeremid. 2015. Effect of carbon content on the mechanical properties and microstructural evolution of Fe-22Mn-C steels. Mater. Sci. Eng. A 621: 118–127.

Gibbs, P.J., E. De Moor, M.J. Merwin, B. Clausen, J.G. Speer and D.K. Matlock. 2011. Austenite stability effects on tensile behaviour of manganese-enriched-austenite transformation-induced plasticity steel. Metall. Mater. Trans. A 42: 3691–3702.

Grajcar, A., R. Kuziak and W. Zalecki. 2012. Third generation of AHSS with increased fraction of retained austenite for the automotive industry. Archi. Civil Mech. Eng. 12: 334–341.

Grasel, O., L. Kruger, G. Frommeyer and L.W. Meyer. 2000. High strength of Fe-Mn-(Al, Si) TRIP/TWIP steels development-properties-application. Int. J. Plast. 16: 1391–1409.

Gutierrez-Urrutia, I. and D. Raabe. 2012a. Dislocation and twin substructure evolution during strain hardening of an Fe-22 wt.%–0.6 wt.% C TWIP steel observed by electron channeling contrast imaging. Acta Mater. 59: 6449–6462.

Gutierrez-Urrutia, I. and D. Raabe. 2012b. Multi-stage strain hardening through dislocation substructure and twinning in a high strength and ductile weight-reduced Fe-Mn-Al-C steel. Acta Mater. 60: 5791–5802.

Gutierrez-Urrutia, I. and D. Raabe. 2013. Influence of Al content and precipitation state on the mechanical behavior of austenitic high-Mn low-density steels. Scripta Mater. 68: 343–347.

Gutierrez-Urrutia, I. and D. Raabe. 2014. High strength and ductile low density austenitic FeMnAlC steels: Simplex and alloys strengthened by nano scale ordered carbides. Mater. Sci. Technol. 30: 1099–1104.

Haase, C., T. Ingendahl, O. Guvenc, M. Bambach, W. Bleck, D.A. Molodov and L.A. Barrales-Mora. 2016. On the applicability of recovery-annealed twinning-induced plasticity steels: potential and limitations. Mater. Sci. Eng. A 649: 74–84.

Han, Q.H., Y.L. Kang, X.M. Zhao, N. Stanford and M.H. Cai. 2013. Suppression of Ms temperature by carbon partitioning from carbon supersaturated ferrite to metastable austenite during intercritical annealing. Mater. Design. 51: 409–414.

Hamada, A.S., L.P. Karjainen and M.C. Somani. 2007. The influence of aluminum on hot deformation and tensile properties of high-Mn TWIP steels. Mater. Sci. Eng. A, 467: 114–124.

Hashimoto, S., S. Ikeda, K.I. Sugimoto and S. Miyake. 2004. Effects of Nb and Mo addition to 0.2C-1.5Si-1.5 Mn steel on mechanical properties of hot rolled TRIP-aided steel sheets. Int. ISIJ. 44: 1590–1598.

Honeycombe, R.W.K. and H.K.D.H. Bhadeshia. 1995. Steels, Microstructure and Properties, London: Edward Arnold.

http://apac.totalmateria.com/page.aspx?ID=CheckArticle&LN=CN&site=kts&NM=207.

http://automotive.arcelormittal.com/saturnus/sheets/A2_EN.html.

Hwang, C.N., C.Y. Chao and T.F. Liu. 1993. Grain boundary precipitation of a Fe-8.0Al-31.5Mn-1.05C alloy. Scripta Mater. 28: 263–268.

Jacques, P. 1999. Bainite transformation of low carbon Mn-Si TRIP assisted multiphase steels: influence of Si content on cementite precipitation. Mater. Sci. Eng. A 273-275: 475–479.

Jacques, P.J., E. Girault, A. Mertens, B. Verlinden, J.V. Humbeeck and F. Delannay. 2001. The developments of cold-rolled TRIP-assisted multiphase steels: Al-alloyed TRIP assisted multiphase steels. ISIJ Int. 41: 1068–1074.

Jeong, K., J.E. Jin, Y.S. Jung, S. Kang and Y.K. Lee. 2013. The effects of Si on the mechanical twinning and strain hardening of Fe-18Mn-0.6C-twinning-induced plasticity steel. Acta Mater. 61: 3399–3410.

Jung, I., S.J. Lee and B.C. De Cooman. 2012. Influence of Al on internal friction spectrum of Fe-18Mn-0.6C twinning-induced plasticity steel. Scripta Mater. 66: 729–732.

Kang, Y.L., Q.H. Han, X.M. Zhao and M.H. Cai. 2013. Influence of nano particle reinforcements on the strengthening mechanisms of an ultrafine-grained dual phase steel containing titanium. Mater. Design 44: 331–339.

Karbasian, H. and A.E. Tekkaya. 2010. A review on hot stamping. J. Mater. Proc. Technol. 210: 2103–2118.

Keeler, S. and M. Kimchi. 2014. Advanced high-strength steels application guidelines version 5.0, The World Auto Steel.

Krauss, G. 1990. Steels: Heat Treatment and Processing Principles, Metals Park, OH: ASM International.

Kumar, A., S.B. Singh and K.K. Ray. 2008a. Short fatigue crack growth behavior in ferrite-bainite dual-phase steels. ISIJ Int. 48: 1285–1292.

Kumar, A., S.B. Singh and K.K. Ray. 2008b. Influence of bainite/martensite-content on the tensile properties of low carbon dual-phase steels. Mater. Sci. Eng. A 474: 270–282.

Kuziak, R., R. Kawalla and S. Waengler. 2008. Advanced high strength steels for automotive industry. Arch. Civil Mech. Eng. VIII 8: 103–117.

Lee, H., H.J. Koh, C.H. Seo and N.J. Kim. 2008. Microstructure and tensile properties of hot-rolled Fe-C-Mn-Si-Cu multiphase steel. Script. Mater. 59: 83–86.

Lee, Y.K. and J. Han. 2015. Current opinion in medium Mn steel. Mater. Sci. Technol. 31: 843–856.

Lee, S.M., I.J. Park, J.G. Jung and Y.K. Lee. 2016. The effect of Si on hydrogen embrittlement of Fe-18Mn-0.6C-xSi twinning-induced plasticity steels. Acta Mater. 103: 264–272.

Lee, S., S. J. Lee, S. Santhosh Kumar, K. Lee and B.C.D. Cooman. 2011. Localized deformation in multiphase, ultra-fine-grained 6 Pct Mn transformation-induced plasticity steel. Metall. Mater. Trans. A 42: 3638–3651.

Lee, S., Y. Estrin and B.C. De Cooman. 2013. Constitutive modelling of the mechanical properties of V-added medium manganese TRIP steel. Metall. Mater. Trans. A 44: 3136–3146.

Leon-Garcia, O., R.H. Petrov and L. Kestens. 2010. Effect of cooling rate on the damage micro mechanisms of DP steels. Mater. Sci. Forum 638–642: 3337–3342.

Li, D.J., Y.R. Feng, S.Y. Song, Q. Liu, F.Z. Ren and F.S. Shang. 2015a. Influence of silicon on the work hardening behavior and hot forming behavior of Fe-25 Mn wt%-(Si, Al) TWIP steel. J. Alloy Compound. 618: 768–775.

Li, D.J., Y.R. Feng, S.Y. Song, Q. Liu, Q. Bai, G. Wu, N. Lv and F.Z. Ren. 2015b. Influence of Nb-microalloying on microstructure and mechanical properties of Fe-25Mn-3Si-3Al TWIP steel. Mater. Design 84: 238–244.

Liu, Z.Y., Z.S. Lin, S.H. Wang, Y.Q. Qiu, X.H. Liu and G.D. Wang. 2007. Microstructure characterization of austenitic Fe–25Mn–22Cr–2Si–0.7N alloy processed by twin roll strip casting. Mater. Charact. 58: 974–979.

Ma, P.H., L.H. Qian, J.Y. Meng, S. Liu and F.C. Zhang. 2015. Influence of Al on the fatigue crack behavior of Fe-22Mn-Al-0.6C TWIP steels. Mater. Sci. Eng. A 645: 136–141.

Mazaheri, Y., A. Kermanpur and A. Najafizadeh. 2014. A novel route for development of ultrahigh strength dual phase steels. Mater. Sci. Eng. A 619: 1–11.

Mejia, I., A.E. Salas-Reyes, J. Calvo and J.M. Cabrera. 2015. Effect of Ti and B micro-additions on the hot ductility behavior of a high-Mn Fe-23Mn-1.5Al-1.3Si-0.5C TWIP steel. Mater. Sci. Eng. A 648: 311–329.

Meyer, M.D., D. Vanderschueren and B.C.D. Cooman. 1999. The influence of the substitution of Si by Al on the properties of cold rolled C-Mn-Si TRIP steels. Int. ISIJ 39: 813–822.

Park, I.J., S.Y. Jo, M.W. Kang, S.M. Lee and Y.K. Lee. 2014. The effect of Ti precipitates on hydrogen embrittlement of Fe-18Mn-0.6C-2Al-xTi twinning-induced plasticity steel. Corr. Sci. 89: 38–45.

Pichler, A. 1999. TRIP steels with reduced Si content. Steel Res. 70: 459–465.

Pierce, D.T., J.A. Jimenez, J. Bentley, D. Raabe and J.E. Wittig. 2015. The influence of stacking fault energy on the microstructural and strain-hardening evolution of Fe–Mn–Al–Si steels during tensile deformation. Acta Mater. 100: 178–190.

Podder, A.S., D. Bhattacharjee and R.K. Ray. 2007. Effect of martensite on the mechanical behavior of ferrite-bainite dual phase steels. ISIJ Int. 47: 1058–1064.

Raabe, D., H. Springer, I. Gutierrez-Urrutia, F. Roters, M. Bausch, J.B. Seol, M. Koyama, P.P. Choi and K. Tsuzaki. 2014. Alloy design, combinatorial synthesis, and microstructure-property relations for low-density Fe-Mn-Al-C austenitic steels, JOM, The Minerals, Metals & Materials Society.

Reyes-Calderón, F., I. Mejía, A. Boulaajaj and J.M. Cabrera. 2013. Effect of micro alloying elements (Nb, V and Ti) on the hot flow behavior of high-Mn austenitic twinning induced plasticity (TWIP) steel. Mater. Sci. Eng. A 560: 552–560.

Schemmann, L., S. Zaefferer, D. Raabe, F. Friedel and D. Mattissen. 2015. Alloying effects on microstructure formation of dual phase steels. Acta Mater. 95: 386–398.

Shao, C.W., P. Zhang, R. Liu, Z.J. Zhang, J.C. Pang and Z.F. Zhang. 2016. Low-cycle and extremely-low cycle fatigue behaviors of high-Mn austenitic TRIP/TWIP alloys: Property evaluation, damage mechanisms and life prediction. Acta Mater. 103: 781–795.

Shen, Y.F., C.H. Qiu, L. Wang, X. Sun, X.M. Zhao and L. Zuo. 2013. Effects of cold rolling on microstructure and mechanical properties of Fe-30Mn-3Si-4Al-0.093C TWIP steel. Mater. Sci. Eng. A 561: 329–337.

Siciliano, F. and E.I. Poliak. 2005. Modelling of the resistance to hot deformation and the effects of micro alloying in high-Al steels under industrial conditions. Mater. Sci. Forum. 500-501: 195–202.

Song, R., D. Ponge, D. Raabe, J.G. Speer and S.K. Matlock. 2006. Overview of processing, microstructure and mechanical properties of ultrafine grained bcc steels. Mater. Sci. Eng. A 441: 1–17.

Speer, J.G., D.K. Matlock, B.C De Cooman and J.G. 2003. Schroth. Carbon partitioning into austenite after martensitic transformation. Acta Mater. 51: 2611–2622.

Speer, J.G., D.V. Edmonds, F.C. Rizzo and D.K. Matlock. 2004. Partitioning of carbon from supersaturated plates of ferrite, with application to steel processing and fundamentals of the bainite transformation. Curr. Opin. Solid State Mater. Sci. 8: 219–237.

Speer, J.G., F.C.R. Assuncao, D.K. Matlock and D.V. Edmonds. 2005. The 'quenching and partitioning' process: background and recent process. Mater. Res. 8: 417–423.

Speich, G.R., V.A. Demarest and R.L. Miller. 1981. Formation of austenite during intercritical annealing of dual-phase steels. Metall. Trans. A 12: 1419–1428.

Springer, H. and D. Raabe. 2012. Rapid alloy prototyping: Compositional and thermos-mechanical high throughput bulk combinatorial design of structural materials based on the example of 30 Mn-1.2C-xAl triplex steels. Acta Mater. 60: 4950–4959.

Steel Marker Development Institute. 2011. AHSS 101: The Evolving Use of Advanced High-Strength Steels for Automotive Applications.

Sudo, M., S. I. Hashimoto and S. Kambe. 1983. Niobium bearing ferrite-bainite high strength hot-rolled sheet steel with improved formability. Int. ISIJ 23: 303–311.

Suh, D.W., S.J. Park, T.H. Lee, C.S. Oh and S.J. Kim. 2010. Influence of Al on the micro structural evolution and mechanical behaviour of low-carbon, manganese transformation-induced-plasticity steel. Metall. Mater. Trans. A 41: 397–408.

Takahashi, M. 2003. Development of high strength for automobiles. Nippon Steel Tech. Rep. 88: 2–7.

Tang, Z.Y. 2007. Development and research of low-silicon TRIP steel with niobium. Ph.D. Thesis, North Eastern University.

Tang, Z.Y., H. Ding. H. Ding, M.H. Cai and L.X. Du. 2010a. Effect of prestrain on microstructures and properties of Si-Al-Mn TRIP steel sheet with niobium. Int. J. Iron Steel Res. 17: 59–65.

Tang, Z.Y., H. Ding. H. Ding, M.H. Cai and L.X. Du. 2010b. Effect of baking process on microstructures and mechanical properties of low silicon TRIP steel sheet with niobium. Int. J. Iron Steel Res. 17: 68–74.

Tang, Z.Y., R.D.K. Misar, M. Ma, N. Zan, Z.Q. Wu and H. Ding. 2015. Deformation twinning and martensitic transformation and dynamic mechanical properties in Fe-0.07C-23Mn-3.1Si-2.8Al TRIP/TWIP steel. Mater. Sci. Eng. A 624: 186–192.

Tsukatani, I., S. Hashimoto and T. Inoue. 1991. Effects of silicon and manganese addition on mechanical properties of high-strength hot-rolled sheet steel containing retained austenite. Int. ISIJ. 31: 992–1000.

Toji, Y., H. Matsuda, M. Herbig, P.P. Choi and D. Raabe. 2014. Atomic-scale analysis of carbon partitioning between martensite and austenite by atom probe tomography and correlative transmission electron microscopy. Acta Mater. 65: 215–228.

Toji, Y., G. Miyamoto and D. Raabe. 2015. Carbon partitioning during quenching and partitioning heat treatment accompanied by carbide precipitation. Acta Mater. 86: 137–147.

Vercammen, S., B. Blanpain, B.C. De Cooman and P. Wollants. 2004. Cold rolling behavior of an austenitic Fe-30Mn-3Al-3Si TWIP steel: the importance of deformation twinning. Acta Mater. 52: 2005–2012.

Wang, C., H. Ding, M.H. Cai and B. Rolfe. 2014a. Multi-phase microstructure design of a novel high strength TRIP steel through experimental methodology. Mater. Sci. Eng. A 610: 436–444.

Wang, C., H. Ding, M.H. Cai and B. Rolfe. 2014b. Characterization of microstructures and tensile properties of TRIP-aided steels with different matrix microstrcuture. Mater. Sci. Eng. A 610: 65–75.

Wang, L. and J.G. Speer. 2013. Quenching and partitioning steel heat treatment. Metallogr. Microstruct. Anal. 2: 268–281.

Wang, W.W., H. Ding, Z.Y. Tang, Y. Shang and H. Ding. 2009. Hole-expansion properties of ferrite-bainite steels. J. Iron Steel Res. 21: 48–52.

Wang, Y.C., P. Lan, Y. Li and J.Q. Zhang. 2015. Effect of alloying elements on mechanical behavior of Fe-Mn-C TWIP steel. J. Mater. Eng. 43: 30–38.

Xie, P., C.L. Wu, Y. Chen, J.H. Chen, X.B. Yang, S.Y. Duan, N. Yan, X.A. Zhang and J.Y. Fang. 2015. A nano twinned surface layer generated by high strain-rate deformation in a TRIP steel. Mater. Design. 80: 144–151.

Yang, F.Q., R.B. Song, L.F. Zhang and C. Zhao. 2014. Hot deformation behavior of Fe-Mn-Al lightweight steel. Proc. Eng. 81: 456–461.

Yang, P., Q. Xie, L. Meng, H. Ding and Z. Tang. 2006. Dependence of deformation twinning on grain orientation in a high manganese steel. Scripta Mater. 55: 629–631.

Yanushkevich, Z., A. Belyakov, R. Kaibyshev, C. Haase and D.A. Molodov. 2016. Effect of cold rolling on recrystallization and tensile behavior of a high-Mn steel. Mater. Charact. 112: 180–187.

Yoo, J.D. and K. Park. 2008. Micro band-induced plasticity in a high Mn–Al–C light steel. Mater. Sci. Eng. A 496: 417–424.

Zare, A. and A. Ekrami. 2011. Influence of martensite volume fraction on tensile properties of triple phase ferrite-bainite-martensite steels. Mater. Sci. Eng. A 530: 440–445.

Zhang, J.C., H.S. Di, Y.G. Deng, S.C. Li and R.D.K. Misra. 2015. Microstructure and mechanical property relationship in an ultrahigh strength 980 MPa grade high-Al low-Si dual phase steel, Mater. Sci. Eng. A 645: 232–240.

Zhang, L., X.H. Liu and K.Y. Shu. 2011. Microstructure and mechanical properties of hot-rolled Fe-Mn-C-Si TWIP steel. Int. J. Iron Steel Res. 18: 45–58.

Zuazo, I., B. Hallsedt, B. Lindahl, M. Selleby, M. Soler, A. Etienne, A. Perlade, D. Hasenpouth, V. Massardier-Jourdan, S. Cazottes and X. Kleber. 2014. Low-density steels: Complex metallurgy for automotive applications. JOM. 66: 1747–1758.

2

A New Theory for Cold Flat Rolling

G. Echave Iriarte

ABSTRACT

Plastic Flow Theory is applied in this chapter to solve the mathematical problem of the cold flat rolling without considerations of spread. Symmetry considerations lead to the use of bipolar cylindrical coordinates. The Levy-Mises flow rule along with the Von Mises criterion and the momentum equations are used to obtain an analytical solution for the plastic stress field in the roll gap. It is found, that certain discrepancies in the fit of the roll pressure distribution using previous theories can be resolved, when the coupling between the structure of the plastic flow and the stress field is taken into consideration. The non-plastic transition at the neutral plane, the weird negative pressure regions along the middle plane of the strip, and the wavy roll pressure profiles with more than a single maximum can also be explained with the present theory. Because of the above mention reasons, the new theory is also helpful for cold flat rolling calculations of Advanced High Strength Steels (AHSSs).

Felipe IV-5-4°-B, San Sebastián (20011), Guipuzcoa (Spain).
Email: gechavei@yahoo.es

2.1 Introduction

A cold rolling theory is a set of mathematical expressions relating the main parameters of the process, which can be used by the mill manufacturers to design new rolling facilities, or by the operators for better handling of the process. And these expressions form the basis for the computer programs used to control the cold rolling mills.

Based on the Mathematical Theory of Plasticity, a new theory for cold flat rolling has been developed, relating the plastic velocity field, the strain rates, the flow rule and the flow criterion, the stress field and the roll pressure distribution, to the plastic properties of the material and the geometry of the process. Under the condition of incompressibility, an admissible velocity field and the corresponding strain rates are established, that form the basis for the calculation of the plastic stress field using the Levy–Mises flow rule, the Von Mises criterion, and the momentum equations.

Due to the high speed of the process in modern cold rolling mills, the adiabatic condition can be easily attained within the roll gap, establishing certain balance between strain hardening and thermal softening. In this case, the rigid perfectly plastic solution could make sense even in industrial practice. When the rolling speed is largely reduced, the thickness reduction is small or the material is very prone to harden, the strain hardening cannot be ignored. An isotropic hardening rule taking into account these facts is also proposed in the present chapter.

The theory predicts new effects inherent in the process and solve the discrepancies between the existing theories and the experimental facts, because the stress field has been successfully coupled with the kinematics of the plastic flow. The constant and progressive falls in pressure from the ends of the roll gap to the neutral point. The local fall around the neutral point and the corresponding rounding of the peak of pressure. The non-plastic transition and the possible existence of negative pressure regions along the middle plane of the strip. Even the vortical nature of the disturbed flow within the roll gap due to friction drag with the rolls, and the wavy roll pressure profiles with more than a single maximum.

Some important results from energetic considerations are obtained at the end of the present chapter, and to characterize the plastic flow in the roll gap, average measures of strain and strain rate are also given.

The new theory intends to be a full solution to the problem within the whole range of practical rolling speeds and thickness reductions, except for the case of thin foil and temper rolling, where plastic flow conditions and roll distortion differ greatly, and the elastic effects are of paramount importance. For the present case of Advanced High Strength Steels (AHSS), the theory provides not only the means to calculate the main magnitudes of the process: the roll pressure profile and the rolling loads and torque, but also the possibility to characterize in high speed rolling the dynamic response of this kind of materials.

2.2 Previous Theories on Cold Flat Rolling

Due to the importance of the process from an industrial perspective, plenty of attempts have been made in the last century to solve the problem of the "Roll Gap", beginning with the classical works conducted by Siebel (1925), Von Karman (1925), Nadai (1939), Tselikov (1939) and Bland and Ford (1948). All these theories have similar assumptions most of which have proven later to be erroneous, and they all have as their starting point a differential equation that represents the condition of equilibrium of an elementary vertical plane section of the strip between the rolls. The assumptions in which these theories rest are as follows:

- The plane vertical sections of the strip before rolling remain plane during rolling.
- The lateral spread can be ignored, and therefore, the material flows under plane strain. This condition is satisfied in more or less degree, if the thickness of the strip is much smaller than the width.
- The material under rolling is supposed to be incompressible, and the Von Mises criterion for the onset of plastic flow is considered.
- Slipping friction is supposed to occur between the strip and the rolls, and therefore, there is a point at which both have the same velocity (neutral point).
- The friction coefficient is supposed to be constant along the arc of contact with the rolls.
- The material of the strip is a rigid-perfectly plastic continuum, homogeneous and isotropic, and therefore, the strain hardening is not considered.
- The elastic deformation of the rolls is considered irrelevant. Nevertheless, for better accuracy and because the process is stationary in time, the flattened radius of the rolls is calculated using Hitchcock's equation (Hitchcock 1935).

The resultant differential equation of equilibrium is:

$$\frac{d}{dy}\left\{\frac{h}{2}(p-2k)\right\}-\frac{p}{2}\frac{dh}{dy}\pm\mu p=0 \qquad \mu\tan\phi\cong 0 \tag{2.1}$$

where h is the thickness of the strip at the position y from the exit of the roll gap, p is the vertical force acting on the corresponding plane section of width dy, μ is the friction coefficient along the arc of contact with the rolls, ϕ is the rolling angle measured from the exit of the roll gap and with centre in the rolls, and k is the yield stress in shear.

The theories by Von Karman, Nadai and Tselikov, differ in the assumptions made on Eq. (2.1). Nadai also considered the possibility of a friction force proportional to the relative velocity of slip between the rolls and the strip. The most important drawbacks of these theories come from some of the erroneous initial assumptions above, and the further simplifications used to solve the corresponding differential equation. Their solutions ignore the plastic flow.

A more refined theory by Orowan (1943), abandons the assumption of the plane sections and incorporates the static friction. The starting point of his theory is the solution by Prandt (1923), for the stress field of a compressed mass of plastic material between two rough parallel flat plates, and the corresponding extension by Nadai to plates at a small angle. He also assumes three regions for the contact between the rolls and the strip, the slip backward region, the slip forward region, and the static region. In the slipping regions Coulomb's friction law is assumed. When the friction stress attains the yield in shear of the strip, friction saturates due to the flow criterion and becomes static. The theory also provides the criteria to calculate the corresponding regions along the arc of contact with the rolls, and considers a strain hardening material.

The theory by Orowan gives better agreement with experimental data. However, it does not provide a solution to the rounding of the peak of the friction hill, and to the possible existence of negative pressure regions along the middle plane of the strip. Orowan's theory does not give any information on the plastic flow of the material between the rolls. The theory assumes from the outset the stress field of a different plastic problem (Prandt 1923).

2.3 Preliminaries on the New Theory

The approach of the new theory is to assume a Natural System of Coordinates, closely related to the symmetries of the problem, to obtain an incompressible velocity field and the corresponding strain rates. Once the kinematics of the plastic flow has been established, the Levy-Mises flow rule, the Von Mises criterion and the momentum equations are used to obtain, a self-consistent solution for the plastic stress field in the roll gap. The Coulomb friction condition with the rolls will allow the calculation of the zero order solution that helps to obtain the coupling between the structure of the plastic flow and the stress field. The theory follows, therefore, the most general kinematic approach of any problem in continuum mechanics.

2.3.1 Definition of the Process—The Plane Strain Hypothesis

Cold flat rolling is the process of plastic flow, in which a sheet or strip of metal below its recrystallization temperature is deformed passing through two opposite driving rolls to reduce its thickness. The term cold flat rolling applies to sheets of initial thickness up to eight millimetres approximately, and in which the radius of the rolls is at least one hundred times the final thickness of the strip. Because of the width of the rolled stock is significantly larger than the initial thickness, the lateral spread (less than 2%) can be ignored. Under these conditions, the problem reduces to one of plane strain except in narrow zones near the edges of the strip. Therefore, the assumption that the strip flows under plane strain is adopted in the present theory.

Due to the high pressures involved in the process the rolls are elastically deformed, and their increase in radius must be taken into account for a more accurate calculation of the roll pressure profile. Therefore, the radius of the rolls along the arc of contact with the strip is the flattened radius calculated with equations like the one by Hitchcock (1935).

Since the speed of the strip is increased passing through the rolls and its initial velocity is less than the rolls speed, there must be a point at which both strip and rolls move exactly with the same velocity. This is called the neutral point. Before the neutral point, the strip moves slower than the rolls and the friction force tends to draw the material into the roll gap. This is called the backward slip zone. Beyond the neutral point, in the forward slip zone, friction opposes the delivery of the strip which now moves faster than the rolls. Therefore, it is considered that the friction with the rolls is slipping and that the corresponding coefficient is constant along the arc of contact.[1] In a steady state situation, the strip adjusts its velocity with respect to the rolls in such a way, that the forces acting on the strip at the boundaries of the roll gap are in equilibrium. This last condition determines the position of the neutral point.

Front and back tensions are taken into consideration, because they are one of the most efficient ways of controlling the thickness of the rolled strip. Finally, unlike in the classical theories, the strain hardening can be incorporated into the present theory in a natural manner.

2.3.2 Bipolar Cylindrical Coordinates—The Roll Gap: Regions and Boundaries

Bipolar cylindrical coordinates have been chosen to solve the plastic problem in the roll gap because of obvious symmetry considerations. The arcs of contact with the rolls have a very simple equation $\eta = \pm\eta_0$, and the limits of the forming region are cylinders, $\xi = \xi_0$ at the entrance of the roll gap Σ_1^+, and $\xi = \pi$ at the exit of the roll gap Σ_2^+. Therefore, the boundaries of the roll gap are surfaces coordinates.

[1] This could be the case if boundary lubrication prevails because the influence of the sliding speed between the rolls and the strip is irrelevant. In the mixed regime, where most of the cold flat rolling is operated, significant variations of the friction coefficient along the arc of contact with the rolls can occur.

Bipolar cylindrical coordinates make necessary one obvious assumption regarding the nature of the plastic flow in the roll gap, and therefore, the plane sections hypothesis of the classical theories is ignored. Instead, the admissible velocity field for the kinematic approach to the problem is considered one-dimensional. Figure 2.1 is our geometric model for cold flat rolling.

Using the incompressibility condition before and after the roll gap, the following equation is obtained:

$$\oint_{\partial\Omega_2} \vec{v}d\vec{S} = 0 \rightarrow v_1 e_1 = v_2 e_2 \tag{2.2}$$

The relations between Cartesian and bipolar cylindrical coordinates are:

$$x_1 = \xi \qquad x_2 = \eta \qquad x_3 = z \tag{2.3}$$

$$x = \frac{a \sinh \eta}{(\cosh \eta - \cos \xi)} \qquad y = \frac{a \sin \xi}{(\cosh \eta - \cos \xi)} \qquad z = z \tag{2.4}$$

$$\pi \leq \xi < 2\pi \qquad -\infty < \eta < \infty \qquad -\infty < z < \infty \tag{2.5}$$

This system of coordinates is orthogonal and the components of the metric tensor are:

$$g_{11} = g_{22} = \frac{a}{(\cosh \eta - \cos \xi)} \qquad g_{33} = 1 \qquad g_{ij} = 0 \qquad \forall i \neq j \tag{2.6}$$

The equation for the power under front and back tensions is advanced at this stage. In order to perform the calculations, the material of the strip in the roll gap and its boundaries are considered as shown in Figure 2.1. Without loss of generality, only one unit in the *z* direction is taken into account.

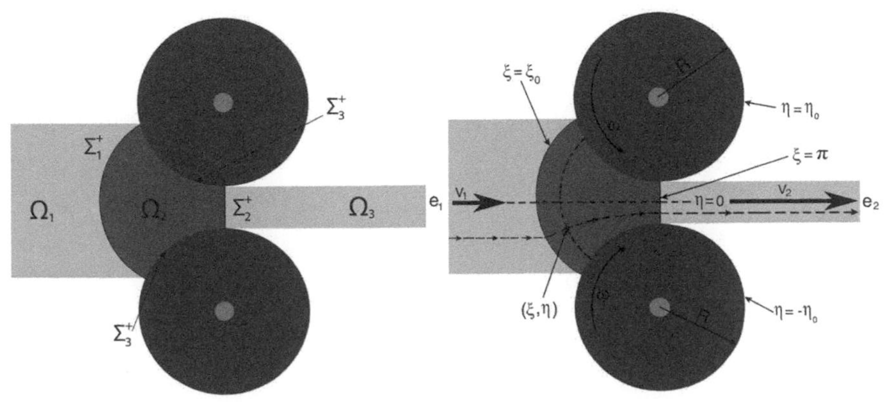

Figure 2.1 Geometric model for cold flat rolling.

$$\oint_{\partial\Omega_2} t_i v_i dS = \sigma_2 v_2 e_2 - \sigma_1 v_1 e_1 + \int_{\Sigma_3^+} \sigma_{\xi\eta}(\xi, \pm\eta_0) v_\xi(\xi, \pm\eta_0) dS_\eta \tag{2.7}$$

$$\oint_{\partial\Omega_2} t_i v_i dS = \sqrt{\frac{2}{3}} \int_{\Omega_2} \sigma_C \sqrt{D_{ij} D_{ij}} dV + \int_{\Sigma_1^+} \sigma_{\xi\eta}(\xi_0, \eta) \Delta v_\eta(\xi_0, \eta) dS_\xi \tag{2.8}$$

The first term on the right-hand side of Eq. (2.8) is the power of deformation, and the second term is the shear power at the entrance of the roll gap Σ_1^+ which is necessary for a sudden change in the velocity field. There is no redistribution of the velocity field at the exit of the roll gap because the velocity is already normal to the boundary Σ_2^+.

2.4 Kinematics of the Plastic Flow

2.4.1 Plastic Velocity Field

The incompressibility condition for the normal plastic flow in bipolar cylindrical coordinates is:

$$\text{div}\vec{v} = \frac{1}{g_\xi g_\eta g_z}\left[\frac{\partial}{\partial \xi}(g_\eta g_z v_\xi) + \frac{\partial}{\partial \eta}(g_\xi g_z v_\eta) + \frac{\partial}{\partial z}(g_\xi g_\eta v_z)\right] = 0 \tag{2.9}$$

In plane strain $v_z = 0$, and to start the calculations, from all the set of solutions of Eq. (2.9), the solution given by Eqs. (2.10) and (2.11) is considered. In this process, some modes of deformation are lost in the continuum, but this guess will allow to obtain an analytical solution in perturbation theory for the stress field, and in turn to approach the real structure of the plastic flow in the roll gap.

$$\frac{\partial}{\partial \xi}(g_\eta g_z v_\xi) = 0 \rightarrow \frac{\partial}{\partial \xi}\left[\frac{v_\xi(\xi, \eta)}{\cosh\eta - \cos\xi}\right] = 0 \tag{2.10}$$

$$\frac{\partial}{\partial \eta}(g_\xi g_z v_\eta) = 0 \rightarrow \frac{\partial}{\partial \eta}\left[\frac{v_\eta(\xi, \eta)}{\cosh\eta - \cos\xi}\right] = 0 \tag{2.11}$$

Integrating Eqs. (2.10) and (2.11), the following components are obtained:

$$v_\xi(\xi, \eta) = (\cosh\eta - \cos\xi) g(\eta) \tag{2.12}$$

$$v_\eta(\xi, \eta) = (\cosh\eta - \cos\xi) h(\xi) \tag{2.13}$$

Using the rigid-plastic assumption, the velocity of the strip at the exit of the roll gap must be v_2, therefore:

$$v_\xi(\pi,\eta)=(\cosh\eta+1)g(\eta)=-v_2=-v_1\frac{e_1}{e_2}\rightarrow g(\eta)=-v_1\left(\frac{e_1}{e_2}\right)\frac{1}{(\cosh\eta+1)} \quad (2.14)$$

$$v_\xi(\xi,\eta)=-v_1\left(\frac{e_1}{e_2}\right)\frac{(\cosh\eta-\cos\xi)}{(\cosh\eta+1)} \quad (2.15)$$

Because of symmetry, the component v_η must be null at the central plane $\eta = 0$:

$$v_\eta(\xi,0)=(1-\cos\xi)h(\xi)=0\rightarrow h(\xi)=0\rightarrow v_\eta(\xi,\eta)=0 \quad (2.16)$$

Therefore, the streamlines for the plastic flow are arcs of bipolar coordinate η constant. At the intersection of the central plane $\eta = 0$ with the entrance of the roll gap Σ_1^+, the velocity field must be equal to the input velocity of the strip, and therefore:

$$v_\xi(\xi_0^v,0)=-v_1=-v_1\left(\frac{e_1}{e_2}\right)\frac{(1-\cos\xi_0^v)}{2} \quad (2.17)$$

From the kinematic condition defined by Eq. (2.17), one of the points of the boundary Σ_1^+ as function of the thickness reduction is obtained:

$$\cos\xi_0^v = 1-(2e_2/e_1) = 2r-1 \quad (2.18)$$

If the position of the neutral point is given, the following condition must be fulfilled:

$$\left|v_\xi(\xi_N,\eta_0)\right|=v_1\left(\frac{e_1}{e_2}\right)\frac{(\cosh\eta_0-\cos\xi_N)}{(\cosh\eta_0+1)}=\omega R_0 \quad (2.19)$$

Therefore, the input and output velocities of the strip before and after rolling are:

$$v_1=\omega R_0\frac{(\cosh\eta_0+1)}{(\cosh\eta_0-\cos\xi_N)}\left(\frac{e_2}{e_1}\right) \quad (2.20)$$

$$v_2=\omega R_0\frac{(\cosh\eta_0+1)}{(\cosh\eta_0-\cos\xi_N)} \quad (2.21)$$

Finally, the velocity field as function of the angular velocity of the rolls is:

$$v_\xi(\xi,\eta) = -v_1\left(\frac{e_1}{e_2}\right)\frac{(\cosh\eta - \cos\xi)}{(\cosh\eta + 1)} = -\frac{(\cosh\eta - \cos\xi)(\cosh\eta_0 + 1)}{(\cosh\eta_0 - \cos\xi_N)(\cosh\eta + 1)}\omega R_0 \tag{2.22}$$

The plastic velocity field thus obtained, is not exactly the one of cold flat rolling. Nevertheless, it will be close enough to be considered as the starting point for the calculation of the strain rates. The functional form of the shear stress and of the difference in coordinate stresses will be obtained afterwards, using the Levy-Mises flow rule and the Von Mises flow criterion.

2.4.2 The Cold Flat Rolling Approximation

Before proceeding with the rest of the exposition and in order to simplify the calculations, what shall be called as the "cold flat rolling approximation" is going to be established. According to Eq. (2.16), the curves of constant η or η-isosurfaces are the streamlines for the plastic flow in the roll gap. They are non-concentric circles, whose centers lie along the x-axis, and approach the corresponding focus as the absolute value of η increase. The surface $\eta = 0$ has infinite radius and corresponds to the Cartesian y-axis. They have the following equation:

$$y^2 + (x - a\coth\eta)^2 = \left(\frac{a}{\sinh\eta}\right)^2 \qquad \forall\eta \in (-\infty,\infty) \tag{2.23}$$

The family of ξ-isosurfaces is orthogonal to the former, and are also non-concentric circles that intersect at the two focal points $x = \pm a$. The surface $\xi = \pi$ has infinite radius and corresponds to the Cartesian x-axis. They have the following equation:

$$x^2 + (y - a\cot\xi)^2 = \left(\frac{a}{\sin\xi}\right)^2 \qquad \forall\xi \in [\pi, 2\pi) \tag{2.24}$$

The arcs of contact with the rolls are streamlines, and according to Eq. (2.23) its equations are:

$$y^2 + (x \mp a\coth\eta_0)^2 = \left(\frac{a}{\sinh\eta_0}\right)^2 = R^2 \rightarrow x_0 = a\coth\eta_0 \tag{2.25}$$

Using the final thickness of the strip and Eq. (2.25), the following condition is obtained:

$$e_2 = 2(x_0 - R) = 2\left(a\coth\eta_0 - \frac{a}{\sinh\eta_0}\right) = 2R(\cosh\eta_0 - 1) \tag{2.26}$$

and from Eq. (2.26), the bipolar coordinates of the arcs of contact with the rolls and the focal distance are calculated:

$$\cosh\eta_0 = 1 + \frac{e_2}{2R} \tag{2.27}$$

$$a = R\sinh\eta_0 \rightarrow a = R\sqrt{\cosh^2\eta_0 - 1} = R\sqrt{\left(1 + \frac{e_2}{2R}\right)^2 - 1} \tag{2.28}$$

According to Eq. (2.4), the intersection of the entrance of the roll gap Σ_1^+ with the rolls is:

$$x\left(\xi_0^l, \eta_0\right) = \frac{a\sinh\eta_0}{\left(\cosh\eta_0 - \cos\xi_0^l\right)} = R\left(1 + \frac{e_2}{2R} - \cos\xi_0^l\right)^{-1}\left(\cosh^2\eta_0 - 1\right) = \frac{e_1}{2} \tag{2.29}$$

$$x\left(\xi_0^l, \eta_0\right) = R\left(1 + \frac{e_2}{2R} - \cos\xi_0^l\right)^{-1}\left[\frac{e_2}{R} + \left(\frac{e_2}{2R}\right)^2\right] = \frac{e_1}{2} \tag{2.30}$$

$$\cos\xi_0^l = 1 - \frac{2e_2}{e_1} + \frac{e_2}{2R}\left(1 - \frac{e_2}{e_1}\right) \geq 1 - \frac{2e_2}{e_1} = \cos\xi_0^v \rightarrow \xi_0^l \geq \xi_0^v \tag{2.31}$$

In the cold flat rolling approximation, $\frac{e_2}{2R} < 0.005$ and the following conditions are fulfilled:

$$\cosh\eta_0 = 1 + \frac{e_2}{2R} \cong 1 + \frac{\eta_0^2}{2} \rightarrow \eta_0 \cong \sqrt{\frac{e_2}{R}} \rightarrow 1 < \cosh\eta < 1.005 \rightarrow \cosh\eta = 1 \tag{2.32}$$

$$\sinh\eta_0 \cong \eta_0 = \sqrt{\frac{e_2}{R}} \rightarrow 0 < \sinh^2\eta < 0.01 \rightarrow \sinh^2\eta \cong \eta^2 \cong 0 \tag{2.33}$$

$$a = R\sinh\eta_0 \cong R\eta_0 = \sqrt{e_2 R} \tag{2.34}$$

$$\cos\xi_0^l = 1 - \frac{2e_2}{e_1} + \frac{e_2}{2R}\left(1 - \frac{e_2}{e_1}\right) \cong 1 - \frac{2e_2}{e_1} = \cos\xi_0^v \rightarrow \xi_0^l = \xi_0^v = \xi_0 \tag{2.35}$$

Equation (2.35) implies, that the ξ coordinate of three points of the entrance of the roll gap Σ_1^+ is the same at this level of approximation, and therefore, this boundary degenerates in a circle of constant bipolar coordinate: $\xi_0^l = \xi_0^v = \xi_0$.

The "cold flat rolling approximation" is the bipolar equivalent of the small angles of contact assumption of the classical theories, and is easily satisfied in all High Mills. For cluster mills with small diameter of the work rolls, the approximations given by Eqs. (2.32), (2.33), (2.34), and (2.35) hold depending on the final thickness of the strip and have to be taken with caution. It is important to note that this approximation unlike in the case of small angles of contact hypothesis, does not impose *a priori* any additional restriction on the reduction or the angle of contact ϕ_0 for a given roll radius and final thickness of the strip. This fact opens the possibility of analytical calculations in cluster mills where the angles of contact are in general much higher than the maximum of 5° allowed by the classical theories. This is yet another advantage of our election of the coordinate system.

2.4.3 Strain Rates and Strains—Equivalent Plastic Strain Rate

The components of the "strain rate" tensor D_{ij} in orthogonal curvilinear coordinates are:

$$D_{11} = \frac{1}{g_1}\frac{\partial v_1}{\partial x_1} + \frac{v_2}{g_1 g_2}\frac{\partial g_1}{\partial x_2} + \frac{v_3}{g_1 g_3}\frac{\partial g_1}{\partial x_3} \tag{2.36}$$

$$D_{22} = \frac{1}{g_2}\frac{\partial v_2}{\partial x_2} + \frac{v_1}{g_2 g_1}\frac{\partial g_2}{\partial x_1} + \frac{v_3}{g_2 g_3}\frac{\partial g_2}{\partial x_3} \tag{2.37}$$

$$D_{33} = \frac{1}{g_3}\frac{\partial v_3}{\partial x_3} + \frac{v_1}{g_3 g_1}\frac{\partial g_3}{\partial x_1} + \frac{v_2}{g_3 g_2}\frac{\partial g_3}{\partial x_2} \tag{2.38}$$

$$D_{12} = \frac{1}{2}\left[\frac{g_2}{g_1}\frac{\partial}{\partial x_1}\left(\frac{v_2}{g_2}\right) + \frac{g_1}{g_2}\frac{\partial}{\partial x_2}\left(\frac{v_1}{g_1}\right)\right] \tag{2.39}$$

$$D_{23} = \frac{1}{2}\left[\frac{g_3}{g_2}\frac{\partial}{\partial x_2}\left(\frac{v_3}{g_3}\right) + \frac{g_2}{g_3}\frac{\partial}{\partial x_3}\left(\frac{v_2}{g_2}\right)\right] \tag{2.40}$$

$$D_{13} = \frac{1}{2}\left[\frac{g_1}{g_3}\frac{\partial}{\partial x_3}\left(\frac{v_1}{g_1}\right) + \frac{g_3}{g_1}\frac{\partial}{\partial x_1}\left(\frac{v_3}{g_3}\right)\right] \tag{2.41}$$

In bipolar cylindrical coordinates and using the velocity field defined by Eq. (2.15), the components are:

$$D_{\xi\xi} = \frac{1}{g_\xi}\frac{\partial v_\xi}{\partial \xi} = -v_1\left(\frac{e_1}{e_2}\right)\left(\frac{\cosh\eta - \cos\xi}{a}\right)\frac{\sin\xi}{(\cosh\eta + 1)} \tag{2.42}$$

$$D_{\eta\eta} = \frac{v_\xi}{g_\eta g_\xi}\frac{\partial g_\eta}{\partial \xi} = -D_{\xi\xi} \tag{2.43}$$

$$D_{zz} = D_{\xi z} = D_{\eta z} = 0 \tag{2.44}$$

$$D_{\xi\eta} = \frac{1}{2}\left[\frac{g_\xi}{g_\eta}\frac{\partial}{\partial\eta}\left(\frac{v_\xi}{g_\xi}\right)\right] = \frac{1}{2}\frac{\partial}{\partial\eta}\left(\frac{v_\xi}{g_\xi}\right) = -\frac{1}{2}v_1\left(\frac{e_1}{e_2}\right)\frac{\partial}{\partial\eta}\frac{(\cosh\eta - \cos\xi)^2}{a(\cosh\eta + 1)} \quad (2.45)$$

$$D_{\xi\eta} = -v_1\left(\frac{e_1}{e_2}\right)\left(\frac{\cosh\eta - \cos\xi}{a}\right)\frac{\sinh\eta}{(\cosh\eta + 1)}\left[\frac{1}{2}\left(1 + \frac{\cos\xi + 1}{\cosh\eta + 1}\right)\right] \quad (2.46)$$

Taking the second order Taylor expansion of the last bracket in Eq. (2.46) and using the cold flat rolling approximation, the disturbing term can be utterly ignored. Therefore, as a further step in the kinematic approach to the problem, the following modified shear strain rate is considered instead:

$$\Psi(\xi,\eta) = \frac{1}{2}\left(1 + \frac{\cos\xi + 1}{\cosh\eta + 1}\right) \cong \left(\frac{3 + \cos\xi}{4}\right) - \left(\frac{1 + \cos\xi}{8}\right)\left(\frac{\eta^2}{2}\right) \cong \Psi^{(0)}(\xi) \quad (2.47)$$

$$D_{\xi\eta} = -v_1\left(\frac{e_1}{e_2}\right)\left(\frac{\cosh\eta - \cos\xi}{a}\right)\frac{\sinh\eta}{(\cosh\eta + 1)}\Psi(\xi) \quad (2.48)$$

The new function $\Psi(\xi)$ is considered as unknown, and therefore, new modes of deformation are open in the continuum making it kinematically less stiff, in order to accommodate at a later stage in the solution, the boundary condition of slipping friction with the rolls. Obviously, the condition of incompressibility is fulfilled by the strain rates above:

$$D_{ii} = D_{\xi\xi} + D_{\eta\eta} + D_{zz} = 0 \quad (2.49)$$

The following expression for the equivalent plastic strain rate is considered:

$$\bar{D} = \sqrt{\frac{2}{3}D_{ij}D_{ij}} = \sqrt{\frac{2}{3}\left(D_{\xi\xi}^2 + D_{\eta\eta}^2 + 2D_{\xi\eta}^2\right)} \quad (2.50)$$

$$\bar{D} = \frac{2}{\sqrt{3}}\left(\frac{v_1}{a}\right)\left(\frac{e_1}{e_2}\right)\left(\frac{\cosh\eta - \cos\xi}{\cosh\eta + 1}\right)\sqrt{\sin^2\xi + \Psi^2(\xi)\sinh^2\eta} \quad (2.51)$$

Integrating Eq. (2.51) in the cold flat rolling approximation and ignoring the shear strain rate, gives a measure of plastic strain.

$$\bar{D} = \frac{D\bar{\varepsilon}}{Dt} = \frac{\partial\bar{\varepsilon}}{\partial t} + \left\{\left(\frac{v_\xi}{g_\xi}\right)\partial_\xi\right\}\bar{\varepsilon} = \left(\frac{v_\xi}{g_\xi}\right)\partial_\xi\bar{\varepsilon} \quad (2.52)$$

$$\bar{D} = -\frac{v_1}{a}\left(\frac{e_1}{e_2}\right)\frac{(\cosh\eta - \cos\xi)^2}{(\cosh\eta + 1)}\partial_\xi\bar{\varepsilon} \quad (2.53)$$

$$\partial_{\xi}\bar{\varepsilon} = -\frac{2}{\sqrt{3}}\frac{\sqrt{\sin^2\xi + \Psi^2(\xi)\sinh^2\eta}}{(\cosh\eta - \cos\xi)} \cong \frac{2}{\sqrt{3}}\frac{\sin\xi}{(1-\cos\xi)} \tag{2.54}$$

$$\bar{\varepsilon} = \int \frac{2}{\sqrt{3}}\left(\frac{\sin\xi}{1-\cos\xi}\right)d\xi = \frac{2}{\sqrt{3}}\ln(1-\cos\xi) + C \tag{2.55}$$

Using the boundary condition $\bar{\varepsilon}(\xi_0) = 0$, the following equivalent plastic strain is obtained:

$$\bar{\varepsilon}(\xi) = \frac{2}{\sqrt{3}}\ln\left(\frac{1-\cos\xi}{1-\cos\xi_0}\right) \tag{2.56}$$

The components of the strain tensor can be calculated in a similar manner.

2.4.4 Average Measures of Strain and Strain Rate

In the cold flat rolling approximation and ignoring the influence of the shear strain rate in the flow structure, the expression for the equivalent plastic strain rate is:

$$\bar{D}(\xi) = -\frac{1}{\sqrt{3}}\left(\frac{v_1}{a}\right)\left(\frac{e_1}{e_2}\right)(1-\cos\xi)\sin\xi \tag{2.57}$$

The average measure of strain rate over the roll gap (Figure 2.2) is defined as follows:

$$\tilde{D} = \frac{1}{A}\int_{\Omega_2}\bar{D}(\xi)dA \cong \frac{1}{A}\int_{-\eta_0}^{\eta_0}\int_{\pi}^{\xi_0}\left[-\frac{a}{\sqrt{3}}v_1\left(\frac{e_1}{e_2}\right)\left(\frac{\sin\xi}{1-\cos\xi}\right)\right]d\xi d\eta \tag{2.58}$$

$$\tilde{D} = -\frac{1}{A}\int_{-\eta_0}^{\eta_0}\left[\frac{a}{\sqrt{3}}v_1\left(\frac{e_1}{e_2}\right)\right]\ln\left(\frac{1-\cos\xi_0}{2}\right)d\eta \tag{2.59}$$

$$\tilde{D} = -\frac{2}{\sqrt{3}}\frac{a\eta_0}{A}v_1\left(\frac{e_1}{e_2}\right)\ln\left(\frac{1-\cos\xi_0}{2}\right) \tag{2.60}$$

$$\tilde{D} = -\frac{2}{\sqrt{3}}\frac{v_1}{A}e_1\ln\left(\frac{e_2}{e_1}\right) \tag{2.61}$$

The area of the roll gap Ω_2 in the cold flat rolling approximation (Echave 2011a) is:

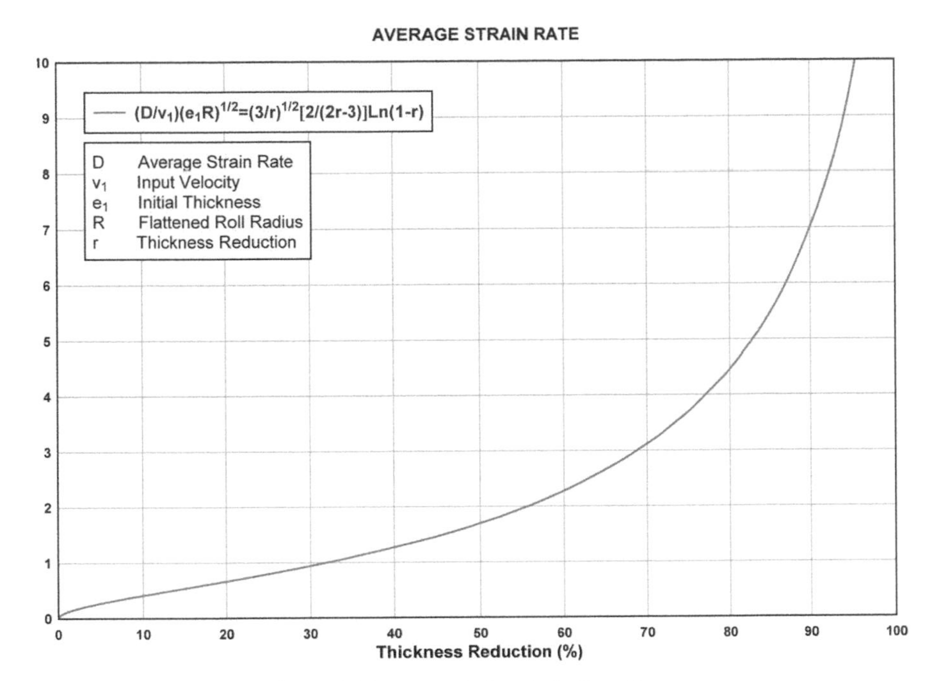

Figure 2.2 Average strain rate in the roll gap.

$$\mathrm{A} = \int_{\pi}^{\xi_0} \int_{-\eta_0}^{\eta_0} \frac{a^2}{(\cosh\eta - \cos\xi)^2} d\eta d\xi \cong \int_{-\eta_0}^{\eta_0} \int_{\pi}^{\xi_0} \frac{a^2}{(1-\cos\xi)^2} d\xi d\eta \tag{2.62}$$

$$\mathrm{A} = 2a^2\eta_0 \int_{\pi}^{\xi_0} \frac{d\xi}{(1-\cos\xi)^2} = -\frac{2}{3} a^2 \eta_0 \cot\left(\frac{\xi_0}{2}\right)\left[\frac{2-\cos\xi_0}{1-\cos\xi_0}\right] \tag{2.63}$$

$$\mathrm{A} = \frac{1}{3}\sqrt{R(e_1 - e_2)}\left[e_1 + 2e_2\right] = \frac{1}{3} e_1 \sqrt{e_1 R}(3-2r)\sqrt{r} \tag{2.64}$$

Using Eq. (2.64) in Eq. (2.61), the average strain rate over the roll gap is obtained:

$$\tilde{D} = -\frac{v_1}{\sqrt{R}}\sqrt{\frac{3}{e_1 - e_2}}\left(\frac{2e_1}{e_1 + 2e_2}\right)\ln\left(\frac{e_2}{e_1}\right) \tag{2.65}$$

$$\tilde{D} = \frac{v_1}{\sqrt{e_1 R}}\left(\frac{2}{2r-3}\right)\sqrt{\frac{3}{r}}\ln(1-r) \tag{2.66}$$

In the same manner, an average measure of strain (Figure 2.3) can be defined:

$$\tilde{\varepsilon} = \frac{1}{A}\int_{\Omega_2} \bar{\varepsilon}\, dA \cong \frac{1}{A}\int_{-\eta_0}^{\eta_0}\int_{\pi}^{\xi_0}\left[\frac{2}{\sqrt{3}}\ln\left(\frac{1-\cos\xi}{1-\cos\xi_0}\right)\right]\frac{a^2}{(1-\cos\xi)^2}d\xi\, d\eta \tag{2.67}$$

$$\tilde{\varepsilon} = \frac{1}{A}2a^2\eta_0\int_{\pi}^{\xi_0}\left[\frac{2}{\sqrt{3}}\ln\left(\frac{1-\cos\xi}{1-\cos\xi_0}\right)\right]\frac{d\xi}{(1-\cos\xi)^2} \tag{2.68}$$

Integrating by parts and using the integral of Eq. (2.63), the following expressions are obtained:

$$\tilde{\varepsilon} = \frac{4}{3\sqrt{3}}\left(\frac{a^2\eta_0}{A}\right)\int_{\pi}^{\xi_0}\cot\left(\frac{\xi}{2}\right)\sin\xi\left[\frac{2-\cos\xi}{(1-\cos\xi)^2}\right]d\xi \tag{2.69}$$

$$\tilde{\varepsilon} = \frac{4}{\sqrt{3}}\sqrt{\frac{1-r}{r}}\left[\frac{1-r}{3-2r}\right]\int_{\pi}^{\cos^{-1}(2r-1)}\cot\left(\frac{\xi}{2}\right)\sin\xi\left[\frac{2-\cos\xi}{(1-\cos\xi)^2}\right]d\xi \tag{2.70}$$

and performing the following change: $\xi = \cos^{-1}(2x-1)$, the average strain (Echave 2011a) is:

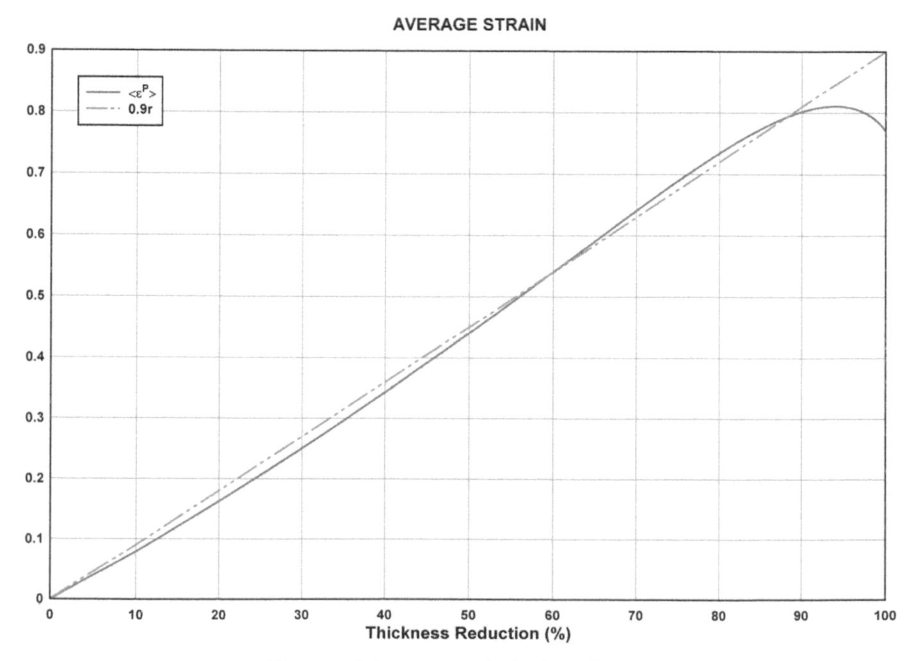

Figure 2.3 Average strain in the roll gap.

$$\tilde{\varepsilon} = \frac{2}{\sqrt{3}} \sqrt{\frac{1-r}{r}} \left[\frac{1-r}{3-2r} \right] \int_0^r \sqrt{\frac{x}{1-x}} \left[\frac{3-2x}{(1-x)^2} \right] dx \tag{2.71}$$

$$\tilde{\varepsilon} = \frac{4}{3\sqrt{3}} \left(\frac{6-5r}{3-2r} \right) - \frac{8}{\sqrt{3}} \sqrt{\frac{1-r}{r}} \left(\frac{1-r}{3-2r} \right) \sin^{-1} \sqrt{r} \cong 0.9r \tag{2.72}$$

This expression suggests that thickness reduction is a good measure of plastic strain.

2.5 Plastic Stress Field in the Rigid Perfectly Plastic Case

To compute the plastic stress field in the roll gap, the assumption of a rigid perfectly plastic Von Mises continuum along with the Levy-Mises flow rule as constitutive equation are used. This will allow a direct computation of the functional form of the shear stress and of the difference in coordinate stresses. The components of the stress tensor will be finally obtained integrating the equations of internal equilibrium along with the boundary conditions. The rigid perfectly plastic model works in practice, when the softening due to thermal activation of the dynamic recovery counterbalances the hardening due to strain and strain-rate. This effect is enhanced, when in high speed rolling the thermal field in the roll gap becomes adiabatic. The Levy-Mises equations in bipolar cylindrical coordinates under plane strain are:

$$D_{\xi\xi} = \frac{\dot{\lambda}}{3} \left[2\sigma_{\xi\xi} - \sigma_{\eta\eta} - \sigma_{zz} \right] \tag{2.73}$$

$$D_{\eta\eta} = \frac{\dot{\lambda}}{3} \left[2\sigma_{\eta\eta} - \sigma_{\xi\xi} - \sigma_{zz} \right] = -D_{\xi\xi} \tag{2.74}$$

$$D_{zz} = \frac{\dot{\lambda}}{3} \left[2\sigma_{zz} - \sigma_{\xi\xi} - \sigma_{\eta\eta} \right] = 0 \rightarrow \sigma_{zz} = \frac{1}{2} \left[\sigma_{\xi\xi} + \sigma_{\eta\eta} \right] \tag{2.75}$$

$$D_{\xi\eta} = \dot{\lambda} \sigma_{\xi\eta} \tag{2.76}$$

$$D_{\xi z} = \dot{\lambda} \sigma_{\xi z} = 0 \rightarrow \sigma_{\xi z} = 0 \tag{2.77}$$

$$D_{\eta z} = \dot{\lambda} \sigma_{\eta z} = 0 \rightarrow \sigma_{\eta z} = 0 \tag{2.78}$$

The Von Mises criterion under plane strain has the following expression:

$$\sigma_{\xi\eta}^2 + \frac{1}{4} \left[\sigma_{\xi\xi} - \sigma_{\eta\eta} \right]^2 = \frac{\sigma_C^2}{3} \tag{2.79}$$

Subtracting Eq. (2.74) from Eq. (2.73) and substituting the result along with Eq. (2.76) in Eq. (2.79), the parameter $\dot{\lambda}$ of the flow rule is obtained:

$$2D_{\xi\xi} = \dot{\lambda}\left[\sigma_{\xi\xi} - \sigma_{\eta\eta}\right] \tag{2.80}$$

$$\frac{D_{\xi\eta}^2}{\dot{\lambda}^2} + \frac{D_{\xi\xi}^2}{\dot{\lambda}^2} = \frac{\sigma_C^2}{3} \tag{2.81}$$

$$\frac{1}{\dot{\lambda}^2}\left[\left\{\left(\frac{v_1}{a}\right)\left(\frac{e_1}{e_2}\right)\left(\frac{\cosh\eta - \cos\xi}{\cosh\eta + 1}\right)\right\}^2 \left\{\sin^2\xi + \Psi^2(\xi)\sinh^2\eta\right\}\right] = \frac{\sigma_C^2}{3} \tag{2.82}$$

The shear stress and the difference in coordinate stresses are computed, using the parameter $\dot{\lambda}$ given in Eq. (2.82):

$$\sigma_{\xi\eta} = \frac{D_{\xi\eta}}{\dot{\lambda}} = -\frac{\sigma_C}{\sqrt{3}}\frac{\Psi(\xi)\sinh\eta}{\sqrt{\sin^2\xi + \Psi^2(\xi)\sinh^2\eta}} = \frac{\sigma_C}{\sqrt{3}}\frac{H'(\xi)\sinh\eta}{\sqrt{1 + H'^2(\xi)\sinh^2\eta}} \tag{2.83}$$

$$\sigma_{\xi\xi} - \sigma_{\eta\eta} = \frac{2D_{\xi\xi}}{\dot{\lambda}} = -\frac{2\sigma_C}{\sqrt{3}}\frac{\sin\xi}{\sqrt{\sin^2\xi + \Psi^2(\xi)\sinh^2\eta}} = \frac{2\sigma_C}{\sqrt{3}}\frac{1}{\sqrt{1 + H'^2(\xi)\sinh^2\eta}} \tag{2.84}$$

where the following functional change: $\Psi(\xi) = H'(\xi)\sin\xi$ has been made. The new function $H'(\xi)$ is very important in what follows, and it will be referred to as the "shear structure".

The average strain rates in cold flat rolling range: $1 \div 10^4$ s^{-1}. These conditions belong to both, the quasi-static strain rate regime ($\tilde{D} < \dot{\varepsilon}_t$) and to the dynamic strain rate regime ($\tilde{D} > \dot{\varepsilon}_t$), where $\dot{\varepsilon}_t \approx 10^2$ s^{-1} is the transition strain rate. Following the phenomenological model of Johnson and Cook (1983), the yield stress σ_C as function of the initial temperature T_0 of the strip and of the equivalent plastic strain rate $\dot{\varepsilon}_{\xi_0} = \bar{D}(\xi_0) \cong \tilde{D}$ at the entrance of the roll gap is:

$$\sigma_C = \sigma_Y\left\{1 + C\left(\dot{\varepsilon}_{\xi_0}, \dot{\varepsilon}_t, T_0\right)\ln\left[\dot{\varepsilon}_{\xi_0}/\dot{\varepsilon}_0\right]\right\} Z_\alpha(T_0) \quad \begin{aligned} C(\dot{\varepsilon} < \dot{\varepsilon}_t) &= C_Q(T) \\ C(\dot{\varepsilon} > \dot{\varepsilon}_t) &= C_D(T) \end{aligned} \tag{2.85}$$

where σ_Y is the quasi-static yield stress at the reference conditions ($\dot{\varepsilon}_0$, T_R), $Z_\alpha(T)$ is the thermal softening function defined by Eq. (2.201), $C(\dot{\varepsilon}, \dot{\varepsilon}_t, T)$ is the temperature-dependent strain rate sensitivity function of the material over the whole range of rolling conditions, and finally C_Q and C_D are its asymptotes at a given temperature.

2.5.1 Equations of Internal Equilibrium—Boundary Value Problem for the Stress Field in the Roll Gap

To calculate the unknown components of the stress tensor, the differential equations of internal equilibrium in the roll gap are used, ignoring volumetric and inertial forces:

$$\mathrm{div}\tilde{\sigma} = \rho\frac{D\vec{v}}{Dt} = \rho\left[\frac{\partial\vec{v}}{\partial t} + (\vec{v}\nabla)\vec{v}\right] \cong 0 \tag{2.86}$$

In orthogonal curvilinear coordinates the divergence of the stress tensor is:

$$(\mathrm{div}\tilde{\sigma})_i = \frac{1}{(g_1g_2g_3)g_i}\left\{\frac{\partial}{\partial x_i}(g_1g_2g_3\sigma_{ii}) - \frac{1}{2}\sum_{j=1}^{3}\frac{g_1g_2g_3}{g_j^2}\sigma_{jj}\frac{\partial g_j^2}{\partial x_i} + \sum_{j\neq i}\frac{\partial}{\partial x_j}\left(\frac{g_1g_2g_3g_i^2}{g_ig_j}\sigma_{ij}\right)\right\} \tag{2.87}$$

For the case of bipolar cylindrical, the equation for $i = z$ reduces to an identity under plane strain. The remaining two equations are:

$$\frac{\partial}{\partial\eta}\left[\frac{\sigma_{\eta\eta}}{(\cosh\eta-\cos\xi)^2}\right] - \frac{1}{2}(\sigma_{\xi\xi}+\sigma_{\eta\eta})\frac{\partial}{\partial\eta}\left[\frac{1}{(\cosh\eta-\cos\xi)^2}\right] + \frac{\partial}{\partial\xi}\left[\frac{\sigma_{\xi\eta}}{(\cosh\eta-\cos\xi)^2}\right] = 0 \tag{2.88}$$

$$\frac{\partial}{\partial\xi}\left[\frac{\sigma_{\xi\xi}}{(\cosh\eta-\cos\xi)^2}\right] - \frac{1}{2}(\sigma_{\xi\xi}+\sigma_{\eta\eta})\frac{\partial}{\partial\xi}\left[\frac{1}{(\cosh\eta-\cos\xi)^2}\right] + \frac{\partial}{\partial\eta}\left[\frac{\sigma_{\xi\eta}}{(\cosh\eta-\cos\xi)^2}\right] = 0 \tag{2.89}$$

Performing the derivative of the first bracket in Eqs. (2.88) and (2.89) and operating in order to obtain the difference in coordinate stresses, these equations take the following form:

$$\frac{1}{(\cosh\eta-\cos\xi)^2}\frac{\partial\sigma_{\eta\eta}}{\partial\eta} - \frac{1}{2}(\sigma_{\xi\xi}-\sigma_{\eta\eta})\frac{\partial}{\partial\eta}\left[\frac{1}{(\cosh\eta-\cos\xi)^2}\right] + \frac{\partial}{\partial\xi}\left[\frac{\sigma_{\xi\eta}}{(\cosh\eta-\cos\xi)^2}\right] = 0 \tag{2.90}$$

$$\frac{1}{(\cosh\eta-\cos\xi)^2}\frac{\partial\sigma_{\xi\xi}}{\partial\xi} + \frac{1}{2}(\sigma_{\xi\xi}-\sigma_{\eta\eta})\frac{\partial}{\partial\xi}\left[\frac{1}{(\cosh\eta-\cos\xi)^2}\right] + \frac{\partial}{\partial\eta}\left[\frac{\sigma_{\xi\eta}}{(\cosh\eta-\cos\xi)^2}\right] = 0 \tag{2.91}$$

The boundary conditions with the rolls according to the assumption of slipping friction are:

Backward slip region: $\sigma_{\xi\eta}(\xi, \eta_0) = \mu\sigma_{\eta\eta}(\xi, \eta_0) \quad \forall\xi\in(\xi_N, \xi_0]$ (2.92)

Forward slip region: $\sigma_{\xi\eta}(\xi, \eta_0) = -\mu\sigma_{\eta\eta}(\xi, \eta_0) \quad \forall\xi\in[\pi, \xi_N)$ (2.93)

The boundary condition of front tension at the exit of the roll gap Σ_2^+ is:

$$\int_{-\eta_0}^{\eta_0} \sigma_{\xi\xi}(\pi,\eta)\frac{a d\eta}{(\cosh\eta+1)} = \sigma_2 e_2 \tag{2.94}$$

The boundary condition of back tension at the entrance of the roll gap Σ_1^+ is:

$$\int_{-\eta_0}^{\eta_0}\left[\sigma_{\xi\xi}(\xi_0,\eta)\cos\alpha - \sigma_{\xi\eta}(\xi_0,\eta)\sin\alpha\right]\frac{a d\eta}{(\cosh\eta-\cos\xi_0)} = \sigma_1 e_1 \tag{2.95}$$

$$\sin\alpha = -\frac{x}{a}\sin\xi_0 = -\frac{\sinh\eta\sin\xi_0}{(\cosh\eta-\cos\xi_0)} \tag{2.96}$$

where α is the angle between the line perpendicular to the boundary Σ_1^+ at the corresponding point (ξ_0, η) and the negative Cartesian y-axis.

The boundary value problem defined by Eqs. (2.79), (2.90) and (2.91) along with the boundary conditions defined by Eqs. (2.92), (2.93), (2.94) and (2.95) not containing any reference to the structure of the plastic flow, make the rigid perfectly plastic problem statically determined or isostatic under plane strain.

2.5.2 The Self-Consistent Solution

Substituting Eqs. (2.83) and (2.84) in Eqs. (2.90) and (2.91), the following partial differential equations in the coordinate stresses and the shear structure are obtained:

$$\frac{\partial\sigma_{\eta\eta}}{\partial\eta} + \frac{2\sigma_C}{\sqrt{3}}\frac{\left[1-H'(\xi)\sin\xi\right]}{\sqrt{1+H'^2(\xi)\sinh^2\eta}}\left(\frac{\sinh\eta}{\cosh\eta-\cos\xi}\right) + \frac{\sigma_C}{\sqrt{3}}\frac{H''(\xi)\sinh\eta}{\left[1+H'^2(\xi)\sinh^2\eta\right]^{3/2}} = 0 \tag{2.97}$$

$$\frac{\partial\sigma_{\xi\xi}}{\partial\xi} - \frac{2\sigma_C}{\sqrt{3}}\frac{\sin\xi + H'(\xi)\sinh^2\eta}{\sqrt{1+H'^2(\xi)\sinh^2\eta}}\left(\frac{1}{\cosh\eta-\cos\xi}\right) + \frac{\sigma_C}{\sqrt{3}}\frac{H'(\xi)\cosh\eta}{\left[1+H'^2(\xi)\sinh^2\eta\right]^{3/2}} = 0 \tag{2.98}$$

The problem has been thus reduced to the two partial differential equations above and to Eqs. (2.83) and (2.84) for the shear stress and for the difference in coordinate stresses. A total of four equations and four unknowns: the three components of the stress and one shear structure.

It is evident that the only way out at this stage is numerical integration, but before approaching this point and in order to proceed within the framework of the analysis, further use is made of the cold flat rolling approximation. To simplify the calculations, the second order Taylor expansion is performed in Eqs. (2.97) and (2.98) (Echave 2011a).

$$\frac{\partial \sigma_{\eta\eta}}{\partial \eta}+\frac{2\sigma_C}{\sqrt{3}}\frac{\left[1-H'(\xi)\sin\xi\right]}{(1-\cos\xi)}\eta+\frac{\sigma_C}{\sqrt{3}}H''(\xi)\eta=0 \tag{2.99}$$

$$\frac{\partial \sigma_{\xi\xi}}{\partial \xi}-\frac{2\sigma_C}{\sqrt{3}}\left(\frac{\sin\xi}{1-\cos\xi}\right)+\frac{\sigma_C}{\sqrt{3}}H'(\xi)+\frac{\partial \sigma_{\xi\xi}^{(2,1)}}{\partial \xi}=0 \tag{2.100}$$

where the disturbing term in the Eq. (2.100) has the following expression:

$$\frac{\partial \sigma_{\xi\xi}^{(2,1)}}{\partial \xi}=\frac{\sigma_C}{\sqrt{3}}\left\{\frac{2\sin\xi}{(1-\cos\xi)^2}-H'(\xi)\left(\frac{3+\cos\xi}{1-\cos\xi}\right)+\left(\frac{2\sin\xi}{1-\cos\xi}\right)H'^2(\xi)-3H'^3(\xi)\right\}\left(\frac{\eta^2}{2}\right) \tag{2.101}$$

Performing the same expansion in Eqs. (2.83) and (2.84), the following is obtained:

$$\sigma_{\xi\eta}=\frac{\sigma_C}{\sqrt{3}}H'(\xi)\eta \tag{2.102}$$

$$\sigma_{\xi\xi}-\sigma_{\eta\eta}=\frac{2\sigma_C}{\sqrt{3}}\left[1-H'^2(\xi)\left(\frac{\eta^2}{2}\right)\right] \tag{2.103}$$

The stresses are computed integrating Eq. (2.99) and using Eq. (2.103) as follows:

$$\sigma_{\eta\eta}=\omega(\xi)-\left\{\frac{2\sigma_C}{\sqrt{3}}\frac{\left[1-H'(\xi)\sin\xi\right]}{(1-\cos\xi)}+\frac{\sigma_C}{\sqrt{3}}H''(\xi)\right\}\left(\frac{\eta^2}{2}\right) \tag{2.104}$$

$$\sigma_{\xi\xi}=\frac{2\sigma_C}{\sqrt{3}}+\omega(\xi)+\sigma_{\xi\xi}^{(2,2)} \tag{2.105}$$

$$\sigma_{\xi\xi}^{(2,2)}=-\frac{2\sigma_C}{\sqrt{3}}\left\{H'^2(\xi)+\frac{\left[1-H'(\xi)\sin\xi\right]}{(1-\cos\xi)}+\frac{1}{2}H''(\xi)\right\}\left(\frac{\eta^2}{2}\right) \tag{2.106}$$

From Eqs. (2.100), (2.105) and (2.106), the following partial derivatives are obtained:

$$\frac{\partial \sigma_{\xi\xi}}{\partial \eta}=\frac{\partial \sigma_{\xi\xi}^{(2,2)}}{\partial \eta}=-\frac{2\sigma_C}{\sqrt{3}}\left\{H'^2(\xi)+\frac{\left[1-H'(\xi)\sin\xi\right]}{(1-\cos\xi)}+\frac{1}{2}H''(\xi)\right\}\eta \tag{2.107}$$

$$\frac{\partial\sigma_{\xi\xi}}{\partial\xi}=\frac{2\sigma_C}{\sqrt{3}}\left(\frac{\sin\xi}{1-\cos\xi}\right)-\frac{\sigma_C}{\sqrt{3}}H'(\xi)-\frac{\partial\sigma_{\xi\xi}^{(2,1)}}{\partial\xi} \tag{2.108}$$

Due to the kinematic approach to the problem and the disturbing treatment above, Eqs. (2.107) and (2.108) are inconsistent. In order to find a consistent solution, the following system is considered instead:

$$\frac{\partial\sigma_{\xi\xi}}{\partial\eta}=\frac{\partial\sigma_{\xi\xi}^{(2,2)}}{\partial\eta}-\frac{\partial\sigma_{\xi\xi}^{(2,1)}}{\partial\eta}+\varphi'(\eta) \tag{2.109}$$

$$\frac{\partial\sigma_{\xi\xi}}{\partial\xi}=\frac{2\sigma_C}{\sqrt{3}}\left(\frac{\sin\xi}{1-\cos\xi}\right)-\frac{\sigma_C}{\sqrt{3}}H'(\xi)-\frac{\partial\sigma_{\xi\xi}^{(2,1)}}{\partial\xi}+\frac{\partial\sigma_{\xi\xi}^{(2,2)}}{\partial\xi} \tag{2.110}$$

Now the cross-derivatives are equal and the solution is straightforward:

$$\sigma_{\xi\xi}=\frac{2\sigma_C}{\sqrt{3}}\ln(1-\cos\xi)-\frac{\sigma_C}{\sqrt{3}}H(\xi)+\sigma_{\xi\xi}^{(2,2)}-\sigma_{\xi\xi}^{(2,1)}+\varphi(\eta) \tag{2.111}$$

Using Eq. (2.103) again, the component $\sigma_{\eta\eta}$ is obtained.

$$\sigma_{\eta\eta}=\sigma_{\xi\xi}-\frac{2\sigma_C}{\sqrt{3}}\left[1-H'^2(\xi)\left(\frac{\eta^2}{2}\right)\right] \tag{2.112}$$

The disturbing terms are computed integrating Eq. (2.101):

$$\sigma_{\xi\xi}^{(2,1)}=-\frac{2\sigma_C}{\sqrt{3}}\left\{\frac{1}{(1-\cos\xi)}+\frac{1}{2}\left[I(\xi)-2J(\xi)+3K(\xi)\right]\right\}\left(\frac{\eta^2}{2}\right) \tag{2.113}$$

$$\sigma_{\xi\xi}^{(2,2)}=-\frac{2\sigma_C}{\sqrt{3}}\left\{H'^2(\xi)+\frac{\left[1-H'(\xi)\sin\xi\right]}{(1-\cos\xi)}+\frac{1}{2}H''(\xi)\right\}\left(\frac{\eta^2}{2}\right) \tag{2.114}$$

$$\sigma_{\xi\xi}^{(2,2)}-\sigma_{\xi\xi}^{(2,1)}=-\frac{\sigma_C}{\sqrt{3}}\left\{\begin{matrix}2H'^2(\xi)-\dfrac{2\sin\xi}{(1-\cos\xi)}H'(\xi)\\+H''(\xi)-\left[I(\xi)-2J(\xi)+3K(\xi)\right]\end{matrix}\right\}\left(\frac{\eta^2}{2}\right) \tag{2.115}$$

where the following integrals are thus defined:

$$I(\xi)=\int H'(\xi)\left(\frac{3+\cos\xi}{1-\cos\xi}\right)d\xi \tag{2.116}$$

$$J(\xi)=\int H'^2(\xi)\left(\frac{\sin\xi}{1-\cos\xi}\right)d\xi \tag{2.117}$$

$$K(\xi) = \int H'^3(\xi)\, d\xi \tag{2.118}$$

The angle α between the input velocity v_1 and the corresponding streamline at Σ_1^+ is:

$$\sin\alpha = -\frac{\sinh\eta \sin\xi_0}{(\cosh\eta - \cos\xi_0)} \cong \frac{-\sin\xi_0}{(1-\cos\xi_0)}\eta = \sqrt{\frac{r}{1-r}}\eta \tag{2.119}$$

$$\cos\alpha \cong 1 - \frac{\sin^2\xi_0}{(1-\cos\xi_0)^2}\left(\frac{\eta^2}{2}\right) = 1 - \frac{r}{1-r}\left(\frac{\eta^2}{2}\right) \cong 1 \tag{2.120}$$

The corresponding boundary condition (Eq. 2.95) at the entrance of the roll gap is:

$$\int_{-\eta_0}^{\eta_0}\left[\sigma_{\xi\xi}(\xi_0,\eta) - \frac{\sigma_C}{\sqrt{3}}H'(\xi_0)\sqrt{\frac{r}{1-r}}\eta^2\right]\frac{\sqrt{e_2 R}}{(1-\cos\xi_0)}d\eta = \sigma_1 e_1 \tag{2.121}$$

$$\int_{-\eta_0}^{\eta_0}\sigma_{\xi\xi}(\xi_0,\eta)\frac{\sqrt{e_2 R}}{(1-\cos\xi_0)}d\eta = \sigma_1 e_1 + \frac{2\sigma_C}{\sqrt{3}}e_1 H'(\xi_0)\sqrt{\frac{r}{1-r}}\left(\frac{\eta_0^2}{6}\right) \cong \sigma_1 e_1 \tag{2.122}$$

The shear term in Eq. (2.122) can be ignored. Finally, the corresponding boundary condition (Eq. 2.94) at the exit of the roll gap is:

$$\int_{-\eta_0}^{\eta_0}\sigma_{\xi\xi}(\pi,\eta)\frac{\sqrt{e_2 R}}{2}d\eta = \sigma_2 e_2 \tag{2.123}$$

2.5.3 The Zero Order Solution

The stress field defined by Eq. (2.111) has one degree of freedom $\varphi(\eta)$. Ignoring disturbing terms and using the boundary condition of back tension defined by Eq. (2.122) at the entrance of the roll gap, the following expression for the stress is obtained:

$$\sigma_{\xi\xi}^{(0)}(\xi,\xi_0) = \sigma_1 + \frac{2\sigma_C}{\sqrt{3}}\ln\left(\frac{1-\cos\xi}{1-\cos\xi_0}\right) - \frac{\sigma_C}{\sqrt{3}}\left[H(\xi) - H(\xi_0)\right] \quad \forall\xi \in [\xi_N, \xi_0] \tag{2.124}$$

and using the boundary condition of front tension at the exit of the roll gap, the second branch of the stress is calculated:

$$\sigma_{\xi\xi}^{(0)}(\xi,\pi) = \sigma_2 + \frac{2\sigma_C}{\sqrt{3}}\ln\left(\frac{1-\cos\xi}{2}\right) - \frac{\sigma_C}{\sqrt{3}}\left[H(\xi) - H(\pi)\right] \quad \forall\xi \in [\pi, \xi_N] \tag{2.125}$$

The problem of cold flat rolling has been thus reduced, to a combined problem of forward and backward extrusion with the same velocity field, and source in the tensions and in the friction force with the rolls. Because the self-consistent solution for the stress field defined by Eq. (2.111) allows one constant of integration only, and due to the gap in the boundary condition of friction with the rolls, it is compelling to solve a field from the entrance of the roll gap to the neutral point, and another field from this point to the exit of the roll gap. At the neutral point, both solutions necessarily meet at a C^0 point or "peak". Therefore, the equation for the position of the neutral point is:

$$\sigma_{\eta\eta}^{(0)}(\xi_N,\xi_0)=\sigma_{\eta\eta}^{(0)}(\xi_N,\pi)\leftrightarrow\sigma_{\xi\xi}^{(0)}(\xi_N,\xi_0)=\sigma_{\xi\xi}^{(0)}(\xi_N,\pi) \tag{2.126}$$

Fulfilling the boundary condition of friction with the rolls, an ordinary differential equation for the shear structure is obtained. The problem from the entrance of the roll gap to the neutral point is considered first. In this case, the friction force with the rolls tends to draw the strip into the roll gap. Ignoring the disturbing term in Eq. (2.112) to calculate the roll pressure distribution, the following differential[2] equation is obtained:

$$\sigma_{\xi\eta}(\xi,\eta_0)=\mu\sigma_{\eta\eta}^{(0)}(\xi,\xi_0) \qquad \forall\xi\in(\xi_N,\xi_0] \tag{2.127}$$

$$\frac{\sigma_C}{\sqrt{3}}H'(\xi)\eta_0=\mu\left\{\sigma_1-\frac{2\sigma_C}{\sqrt{3}}+\frac{2\sigma_C}{\sqrt{3}}\ln\left(\frac{1-\cos\xi}{1-\cos\xi_0}\right)-\frac{\sigma_C}{\sqrt{3}}\left[H(\xi)-H(\xi_0)\right]\right\} \tag{2.128}$$

$$H'(\xi)-\vartheta\left\{\frac{\sigma_1}{k}-2+2\ln\left(\frac{1-\cos\xi}{1-\cos\xi_0}\right)-\left[H(\xi)-H(\xi_0)\right]\right\}=0 \tag{2.129}$$

$$F'(\xi)+\vartheta F(\xi)-2\vartheta\left\{\frac{\sigma_1}{2k}-1+\ln\left(\frac{1-\cos\xi}{1-\cos\xi_0}\right)\right\}=0 \qquad \vartheta=\mu/\eta_0 \tag{2.130}$$

where ϑ is the "rolling parameter". The problem is solved in the same manner from the neutral point to the exit of the roll gap. In this case, the friction drag with the rolls opposes the delivery of the rolled strip:

$$\sigma_{\xi\eta}(\xi,\eta_0)=-\mu\sigma_{\eta\eta}^{(0)}(\xi,\pi) \qquad \forall\xi\in[\pi,\xi_N) \tag{2.131}$$

[2] For the shear structure, Eq. (2.130) and other similar equations are strictly speaking integral not differential.

$$\frac{\sigma_C}{\sqrt{3}} H'(\xi)\eta_0 = -\mu\left\{\sigma_2 - \frac{2\sigma_C}{\sqrt{3}} + \frac{2\sigma_C}{\sqrt{3}}\ln\left(\frac{1-\cos\xi}{2}\right) - \frac{\sigma_C}{\sqrt{3}}\left[H(\xi)-H(\pi)\right]\right\} \tag{2.132}$$

$$H'(\xi) + \vartheta\left\{\frac{\sigma_2}{k} - 2 + 2\ln\left(\frac{1-\cos\xi}{2}\right) - \left[H(\xi)-H(\pi)\right]\right\} = 0 \tag{2.133}$$

$$F'(\xi) - \vartheta F(\xi) + 2\vartheta\left\{\frac{\sigma_2}{2k} - 1 + \ln\left(\frac{1-\cos\xi}{2}\right)\right\} = 0 \tag{2.134}$$

The solution of Eq. (2.130) is obtained, using the boundary condition $F(\xi_0) = 0$ at the entrance of the roll gap:

$$F(\xi) = 2\vartheta\exp[-\vartheta\xi]\int_{\xi_0}^{\xi}\exp[\vartheta\xi]\left\{\frac{\sigma_1}{2k} - 1 + \ln\left(\frac{1-\cos\xi}{1-\cos\xi_0}\right)\right\}d\xi \tag{2.135}$$

$$\begin{aligned} F(\xi) = {} & 2\left\{1-\exp\left[-\vartheta(\xi-\xi_0)\right]\right\}\left\{\frac{\sigma_1}{2k} - 1 - \ln(1-\cos\xi_0)\right\} \\ & +2\vartheta\exp[-\vartheta\xi]\int_{\xi_0}^{\xi}\exp[\vartheta\xi]\ln(1-\cos\xi)\,d\xi \end{aligned} \tag{2.136}$$

$$\begin{aligned} \frac{\sigma_{\eta\eta}^{(0)}}{2k}(\xi,\xi_0) = {} & \left\{\frac{\sigma_1}{2k} - 1 - \ln(1-\cos\xi_0)\right\}\exp\left[-\vartheta(\xi-\xi_0)\right] + \ln(1-\cos\xi) \\ & -\vartheta\exp[-\vartheta\xi]\int_{\xi_0}^{\xi}\exp[\vartheta\xi]\ln(1-\cos\xi)\,d\xi \end{aligned} \tag{2.137}$$

In the same manner, the solution of Eq. (2.134) for the exit branch of the stress field is obtained, using the boundary condition $F(\pi) = 0$ at the exit of the roll gap:

$$F(\xi) = -2\vartheta\exp[\vartheta\xi]\int_{\pi}^{\xi}\exp[-\vartheta\xi]\left\{\frac{\sigma_2}{2k} - 1 + \ln\left(\frac{1-\cos\xi}{2}\right)\right\}d\xi \tag{2.138}$$

$$\begin{aligned} \frac{\sigma_{\eta\eta}^{(0)}}{2k}(\xi,\pi) = {} & \left\{\frac{\sigma_2}{2k} - 1 - \ln 2\right\}\exp\left[\vartheta(\xi-\pi)\right] + \ln(1-\cos\xi) \\ & +\vartheta\exp[\vartheta\xi]\int_{\pi}^{\xi}\exp[-\vartheta\xi]\ln(1-\cos\xi)\,d\xi \end{aligned} \tag{2.139}$$

Using Eq. (2.127), the shear stress from the entrance of the roll gap to the neutral point is:

$$\begin{aligned} \frac{\sigma_{\xi\eta}^{(0)}}{2k}(\xi,\xi_0) = {} & \eta\vartheta\left\{\left[\frac{\sigma_1}{2k} - 1 - \ln(1-\cos\xi_0)\right]\exp\left[-\vartheta(\xi-\xi_0)\right] + \ln(1-\cos\xi)\right\} \\ & -\eta\vartheta^2\exp[-\vartheta\xi]\int_{\xi_0}^{\xi}\exp[\vartheta\xi]\ln(1-\cos\xi)\,d\xi \end{aligned} \tag{2.140}$$

and from the neutral point to the exit of the roll gap using equation (2.131) is:

$$\frac{\sigma_{\xi\eta}^{(0)}}{2k}(\xi,\pi) = \eta\vartheta\left\{\left[\frac{\sigma_2}{2k}-1-\ln 2\right]\exp\left[\vartheta(\xi-\pi)\right]+\ln(1-\cos\xi)\right\}$$
$$-\eta\vartheta^2\exp[\vartheta\xi]\int_{\pi}^{\xi}\exp[-\vartheta\xi]\ln(1-\cos\xi)\,d\xi \tag{2.141}$$

The shear stress is linear in the coordinate η, and therefore, according to Eq. (2.4) it is also linear in the Cartesian coordinate x, in perfect agreement with the solution to the problem of the parallel flat plates by Prandt (1923).

2.5.3.1 Neutral Point and Critical Friction Coefficient

To calculate the position of the neutral point using the zero order solution, the following transcendental equation coming from Eq. (2.126) has to be solved:

$$\left[\frac{\sigma_1}{2k}-1-\ln(1-\cos\xi_0)\right]\exp\left[-\vartheta(\xi_N-\xi_0)\right]-\left[\frac{\sigma_2}{2k}-1-\ln 2\right]\exp\left[\vartheta(\xi_N-\pi)\right]=$$
$$\vartheta\exp[\vartheta\xi_N]\int_{\pi}^{\xi_N}\exp[-\vartheta\xi]\ln(1-\cos\xi)\,d\xi+\vartheta\exp[-\vartheta\xi_N]\int_{\xi_0}^{\xi_N}\exp[\vartheta\xi]\ln(1-\cos\xi)\,d\xi \tag{2.142}$$

When the back tension increases, the neutral point moves to the exit of the roll gap. If the front tension increases, it will have the opposite result. For a given reduction and tensions, there is a lower bound or critical friction coefficient $\mu_{LB} = \eta_0\vartheta_{LB}$, for which the neutral point is situated at the exit of the roll gap Σ_2^+. Below this critical coefficient, the cold flat rolling becomes impossible because the forward slip zone disappears, there is not enough shear stress to pull the strip and the rolls begin to skid. From Eq. (2.142), the following equation is obtained:

$$\left[\frac{\sigma_2}{2k}-1-\ln 2\right]-\left[\frac{\sigma_1}{2k}-1-\ln(1-\cos\xi_0)\right]\exp\left[\vartheta_{LB}(\xi_0-\pi)\right]=$$
$$\vartheta_{LB}\exp[-\vartheta_{LB}\pi]\int_{\pi}^{\xi_0}\exp[\vartheta_{LB}\xi]\ln(1-\cos\xi)\,d\xi \tag{2.143}$$

2.5.3.2 Range of Validity of the Zero Order Solution

The influence of the shear stress on the structure of the plastic flow can be ignored, when the following condition is satisfied:

$$\sqrt{1+H'^2(\xi_N)\eta_0^2} \cong 1+\frac{1}{2}H'^2(\xi_N)\eta_0^2 \cong 1 \rightarrow \left|H'(\xi_N)\right| \ll \frac{\sqrt{2}}{\eta_0} \tag{2.144}$$

In this case, friction with the rolls is disturbing for the plastic flow. In practical mill operation the condition defined by Eq. (2.144) holds, and the average measures of strain and strain rate defined previously remain good measures.

With regard to the Von Mises criterion, the following conclusions are obtained:

$$\sigma_{\xi\eta}^{2}+\frac{1}{4}\left[\sigma_{\xi\xi}-\sigma_{\eta\eta}\right]^{2}=\frac{\sigma_{C}^{2}}{3}\rightarrow\left[\frac{\sigma_{C}}{\sqrt{3}}H'(\xi)\eta\right]^{2}+\frac{1}{4}\left[\frac{2\sigma_{C}}{\sqrt{3}}\right]^{2}=\frac{\sigma_{C}^{2}}{3} \tag{2.145}$$

$$\frac{\sigma_{C}^{2}}{3}H'^{2}(\xi)\eta^{2}+\frac{\sigma_{C}^{2}}{3}=\frac{\sigma_{C}^{2}}{3}\left[\mathrm{O}\left\{\mu^{2}(\eta/\eta_{0})^{2}\right\}+1\right]\cong\frac{\sigma_{C}^{2}}{3} \tag{2.146}$$

The Von Mises condition is fulfilled in the order of approximation, if the friction coefficient is small enough. In fact, the difference in coordinate stresses could have been established, using Eq. (2.145) and ignoring the shear stress. Therefore, the zero order solution can be obtained from the Von Mises criterion, and considerating equilibrium of forces using directly the Coulomb friction condition and ignoring all the details of the plastic flow in the roll gap (Von Karman 1925). In this way, the complex boundary value problem for the equations of internal equilibrium (Eqs. 2.90 and 2.91) and the flow criterion (Eq. 2.79) is replaced directly by an ordinary differential equation to be solved from both ends (entrance and exit) of the roll gap.

Although the solution could have been obtained ignoring the plastic flow altogether, it has been proved that it is compatible with the modified shear strain rate obtained from our kinematic approach to the problem. Therefore, the velocity field defined by Eq. (2.15) can be considered as a good first approximation to the real structure of the plastic flow in the roll gap.

The zero order solution as main representative of the stress state in the roll gap and as the basis for a disturbing solution does make sense (Echave 2011a) for the combination of parameters ($\vartheta\leq 1.2, \eta_0^2\cong 0.01$), for all the range of thickness reductions without tensions. Eventually, it can be extended to greater values of the rolling parameter $\vartheta\leq 1.5$, when the average error defined by Eq. (2.148) maintains its relevance, while the local error defined in Eq. (2.147) in fulfilling the Von Mises criterion next to the rolls and around the neutral point is ignored in the high range of thickness reductions. The largest value of the shear stress is obtained in rolling without tensions, and therefore, the range of validity of the solution increases with the application of front and back tensions.

$$\varepsilon_{\mathrm{VM}}^{\max}=\frac{\sigma_{\xi\eta}^{2}}{k^{2}}(\xi_{N},\eta_{0})=H'^{2}(\xi_{N})\eta_{0}^{2} \tag{2.147}$$

$$\tilde{\varepsilon}_{\mathrm{VM}}=\frac{1}{\mathrm{A}}\int_{\Omega_2}H'^{2}(\xi)\eta^{2}d\mathrm{A}=2\eta_{0}^{2}\sqrt{\frac{1-r}{r}}\left[\frac{1-r}{3-2r}\right]\int_{\pi}^{\xi_0}\left[\frac{H'(\xi)}{1-\cos\xi}\right]^{2}d\xi \tag{2.148}$$

On the other hand, the slipping friction condition along the full length of the arc of contact with the rolls breaks down, when due to the flow criterion, the following condition of saturation of the friction stress is fulfilled:

$$\mu\left|\sigma_{\eta\eta}\left(\xi,\eta_0\right)\right| \geq k \rightarrow \left|\sigma_{\xi\eta}\left(\xi,\eta_0\right)\right| = \frac{\sigma_C}{\sqrt{3}} = k \rightarrow \frac{1}{2k}\left|\sigma_{\xi\eta}\left(\xi,\eta_0\right)\right| = 0.5 \;\forall \xi \in \left[\pi,\xi_0\right] \tag{2.149}$$

Static friction cannot be incorporated into the theory, because Eq. (2.83) for the shear stress prevents the condition defined by Eq. (2.149) to be attained, and therefore, a new kinematic model for the plastic flow would be necessary. Furthermore, when the shear stress is large enough, Eqs. (2.102) and (2.103) do not hold near the rolls and around the neutral point, and therefore, the disturbing procedure to solve the plastic stress field in the roll gap breaks down.

Finally, the critical friction coefficient for the condition defined by Eq. (2.149) to be attained with the zero order solution at the neutral point in rolling without tensions is far greater than the corresponding μ_{LB} (Echave 2011a). Since the friction coefficient with present cold rolling lubricants move slightly above this lower bound, the slipping friction condition along the full length of the arc of contact with the rolls can be considered valid (Al-Salehi et al. 1973), and only for very high reductions, there is a necessity to incorporate static friction. The appointed values of μ_{LB} and its variation with the thickness reduction are consistent with the friction coefficient calculated using mixed hydrodynamic-boundary lubrication models for steel (Yamamoto et al. 2002) and aluminium (Le and Sutcliffe 2006), and with the values measured for μ_{LB} under tensions on experimental rolling mills (Tabary et al. 1996).

2.5.3.3 Comparison with the Theory by Von Karman

Figure 2.4 is the plot of roll pressure distribution as a function of the normalized rolling angle (Echave 2011a) using the zero order solution. Therefore, the exit of the roll gap is on the left and the entrance is on the right. The agreement with Von Karman theory in rolling with and without tensions is excellent. Only two parameters define the roll pressure distribution, the thickness reduction r and the rolling parameter ϑ. The corresponding shear structures are given in Figure 2.5.

In a nutshell, the zero order solution established using slipping friction along the full length of the arc of contact with the rolls is consistent with this assumption for all the range of practical thickness reductions and rolling

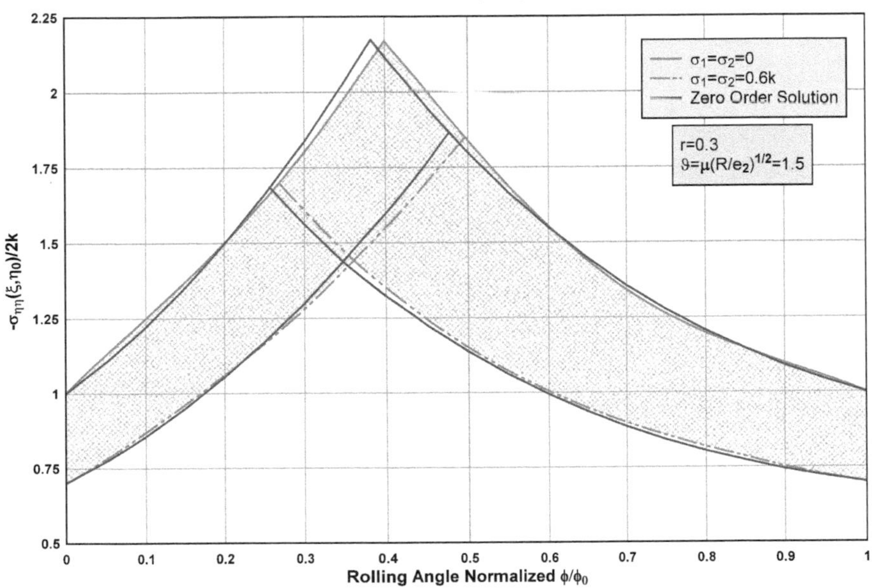

Figure 2.4 Zero order roll pressure distribution.

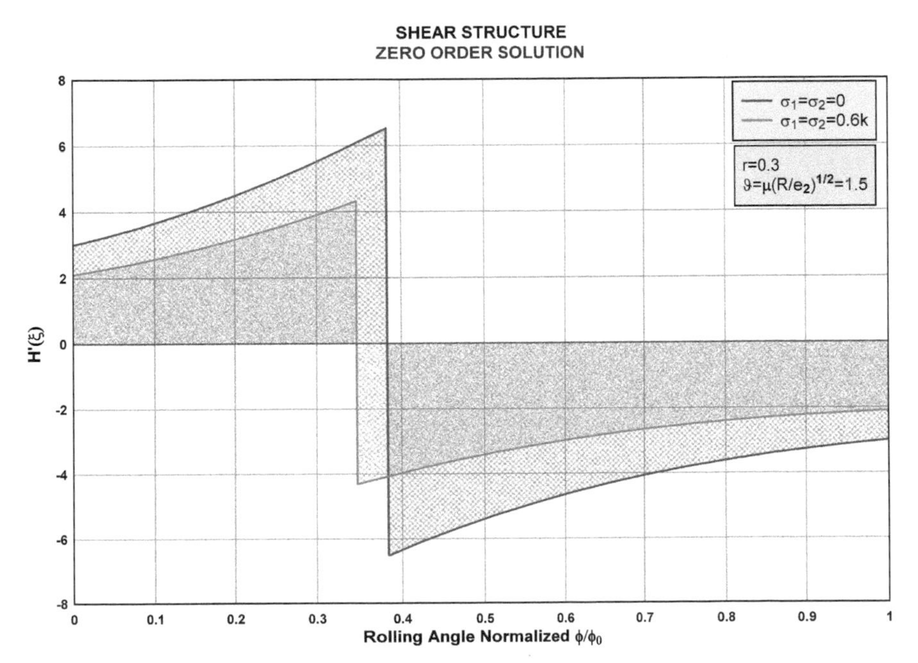

Figure 2.5 Shear structure.

parameters, and only in the very high range the static friction could be necessary around the neutral point. Therefore, in practical mill operation the solution holds, and another explanation for the fall in pressure, the non-plastic transition, and the rounding of the peak of the roll pressure profile must be found.

2.6 Disturbing Terms with the Zero Order Solution—First Order Solution

2.6.1 Falls in Pressure Due to the Dependence of the Coordinate Stresses on the Transversal Dimension

The next step will be to analyze how the disturbing terms modify the zero order solution. Instead of applying the boundary conditions with the rolls to calculate the shear structure to first order,[3] the usual disturbing procedure is used. Therefore, the shear structure in what follows is the corresponding to the zero order solution. The boundary condition defined by the Eq. (2.123) at the exit of the roll gap is considered first:

$$\sigma_2 = \frac{2\sigma_C}{\sqrt{3}}\ln 2 - \frac{\sigma_C}{\sqrt{3}}H(\pi) + \int_{-\eta_0}^{\eta_0}\left[\sigma_{\xi\xi}^{(2,2)}(\pi) - \sigma_{\xi\xi}^{(2,1)}(\pi) + \varphi(\eta)\right]\frac{\sqrt{e_2 R}}{2e_2}d\eta \quad (2.150)$$

Using Eq. (2.115), the disturbing terms are rewritten as follows:

$$\sigma_{\xi\xi}^{(2,2)} - \sigma_{\xi\xi}^{(2,1)} = -\frac{\sigma_C}{\sqrt{3}}\Theta(\xi)\left(\frac{\eta^2}{2}\right) \quad (2.151)$$

$$\Theta(\xi) = 2H'^2(\xi) - \frac{2\sin\xi}{(1-\cos\xi)}H'(\xi) + H''(\xi) - \left[I(\xi) - 2J(\xi) + 3K(\xi)\right] \quad (2.152)$$

$$\frac{1}{2\eta_0}\int_{-\eta_0}^{\eta_0}\varphi(\eta)d\eta = \bar{\varphi} = \sigma_2 - \frac{2\sigma_C}{\sqrt{3}}\ln 2 + \frac{\sigma_C}{\sqrt{3}}H(\pi) + \frac{\sigma_C}{\sqrt{3}}\Theta(\pi)\left(\frac{\eta_0^2}{6}\right) \quad (2.153)$$

Considering $\varphi(\eta)$ as an even polynomial of second order in η, its mean value is:

$$\frac{1}{2\eta_0}\int_{-\eta_0}^{\eta_0}\varphi(\eta)d\eta = \bar{\varphi} = \frac{1}{2\eta_0}\int_{-\eta_0}^{\eta_0}\left\{\mathrm{A} + \mathrm{B}\left(\frac{\eta^2}{2}\right)\right\}d\eta = \mathrm{A} + \mathrm{B}\left(\frac{\eta_0^2}{6}\right) \quad (2.154)$$

Comparing Eqs. (2.153) and (2.154), the stress field to first order is obtained:

[3] The corresponding integro-differential equation does not have analytical solution for the shear structure.

$$\sigma_{\xi\xi}^{(1)} = \sigma_2 + \frac{2\sigma_C}{\sqrt{3}} \ln\left(\frac{1-\cos\xi}{2}\right) - \frac{\sigma_C}{\sqrt{3}}\left[H(\xi)-H(\pi)\right] - \frac{\sigma_C}{\sqrt{3}}\left[\Theta(\xi)-\Theta(\pi)\right]\left(\frac{\eta^2}{2}\right) \tag{2.155}$$

$$\sigma_{\xi\xi}^{(1)} = \sigma_{\xi\xi}^{(0)} - \frac{\sigma_C}{\sqrt{3}}\left[\Theta(\xi)-\Theta(\pi)\right]\left(\frac{\eta^2}{2}\right) \tag{2.156}$$

where $\sigma_{\xi\xi}^{(0)}$ is the zero order solution. Using Eq. (2.112) the component $\sigma_{\eta\eta}$ is obtained:

$$\sigma_{\eta\eta}^{(1)} = \sigma_{\eta\eta}^{(0)} - \frac{\sigma_C}{\sqrt{3}}\left[\Theta(\xi)-\Theta(\pi)-2H'^2(\xi)\right]\left(\frac{\eta^2}{2}\right) \tag{2.157}$$

In the same manner, using Eq. (2.122) the entry branch of the stress field to first order is obtained. Ignoring the shear term in Eq. (2.122) coming from the equilibrium of forces on the boundary Σ_1^+, the final result is:

$$\sigma_{\xi\xi}^{(1)} = \sigma_{\xi\xi}^{(0)} - \frac{\sigma_C}{\sqrt{3}}\left[\Theta(\xi)-\Theta(\xi_0)\right]\left(\frac{\eta^2}{2}\right) \tag{2.158}$$

and the component $\sigma_{\eta\eta}$ using Eq. (2.112) has the following expression:

$$\sigma_{\eta\eta}^{(1)} = \sigma_{\eta\eta}^{(0)} - \frac{\sigma_C}{\sqrt{3}}\left[\Theta(\xi)-\Theta(\xi_0)-2H'^2(\xi)\right]\left(\frac{\eta^2}{2}\right) \tag{2.159}$$

Figure 2.6 is a comparison between the first order and the zero order solutions for the roll pressure distribution. The different disturbing terms in Eqs. (2.157) and (2.159) are now analysed.

2.6.1.1 Constant Falls in Pressure

$$\sigma_{\eta\eta}^{(1,1)}(\xi,\xi_0) = \frac{2\sigma_C}{\sqrt{3}} H'^2(\xi_0)\left(\frac{\eta^2}{2}\right) \geq 0 \tag{2.160}$$

$$\sigma_{\eta\eta}^{(1,1)}(\xi,\pi) = \frac{2\sigma_C}{\sqrt{3}} H'^2(\pi)\left(\frac{\eta^2}{2}\right) \geq 0 \tag{2.161}$$

These falls in pressure are constant because the effect of the disturbing term in Eq. (2.103) is shared by both coordinate stresses. It is evident that

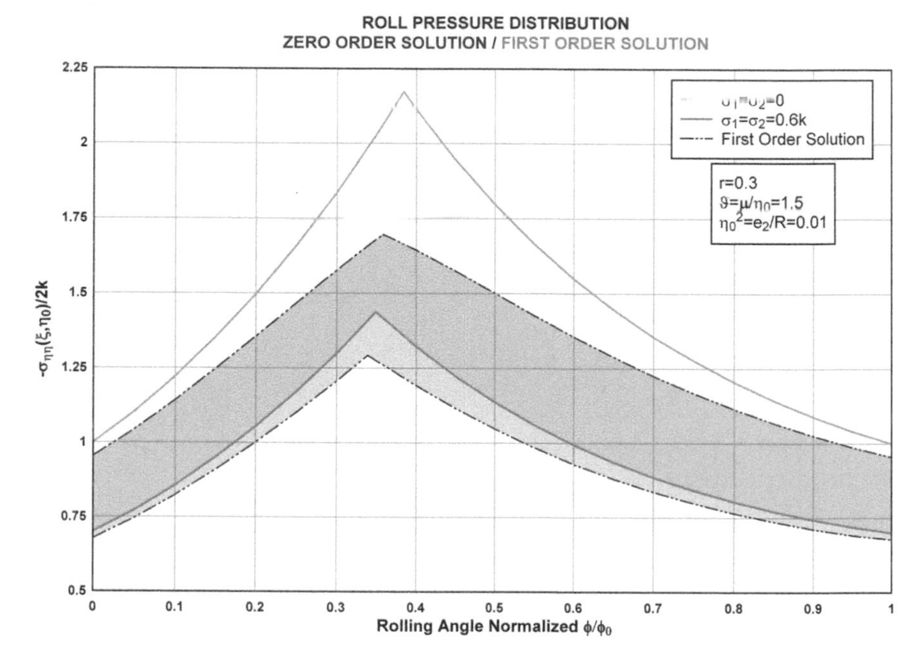

Figure 2.6 First order roll pressure distribution.

they are positive, and therefore, they imply the corresponding fall on each branch of the pressure profile.

2.6.1.2 Progressive Falls from the Ends of the Roll Gap to the Neutral Point

$$\sigma_{\eta\eta}^{(1,2)}(\xi,\xi_0)=\frac{\sigma_C}{\sqrt{3}}\left[I(\xi)-I(\xi_0)-2\{J(\xi)-J(\xi_0)\}+3\{K(\xi)-K(\xi_0)\}\right]\left(\frac{\eta^2}{2}\right)\geq 0 \tag{2.162}$$

$$\sigma_{\eta\eta}^{(1,2)}(\xi,\pi)=\frac{\sigma_C}{\sqrt{3}}\left[I(\xi)-I(\pi)-2\{J(\xi)-J(\pi)\}+3\{K(\xi)-K(\pi)\}\right]\left(\frac{\eta^2}{2}\right)\geq 0 \tag{2.163}$$

These terms are null at the corresponding end of the roll gap and increase from there up to the neutral point (Echave 2011a). The corresponding integrals are given by Eqs. (2.116), (2.117) and (2.118).

2.6.1.3 Localized Fall in Pressure around the Neutral Point

$$\sigma_{\eta\eta}^{(1,3)}(\xi,\xi_0) = -\frac{\sigma_C}{\sqrt{3}}\left[H''(\xi) - H''(\xi_0)\right]\left(\frac{\eta^2}{2}\right) \cong -\frac{\sigma_C}{\sqrt{3}}H''(\xi)\left(\frac{\eta^2}{2}\right) \geq 0 \quad \forall\xi \approx \xi_N \tag{2.164}$$

$$\sigma_{\eta\eta}^{(1,3)}(\xi,\pi) = -\frac{\sigma_C}{\sqrt{3}}\left[H''(\xi) - H''(\pi)\right]\left(\frac{\eta^2}{2}\right) \cong -\frac{\sigma_C}{\sqrt{3}}H''(\xi)\left(\frac{\eta^2}{2}\right) \geq 0 \quad \forall\xi \approx \xi_N \tag{2.165}$$

This term can be ignored except around the neutral surface, where the derivative of the shear structure can be very large and negative. It is the source of the additional fall and rounding of the peak of the roll pressure profile. The remaining disturbing term in Eq. (2.152) can be ignored.

Making Eqs. (2.157) and (2.159) equal with $\eta = \eta_0$, the new position $\xi_N^{(1)}$ of the neutral point is obtained, which due to the symmetric and disturbing nature of these falls in pressure, remains almost unaltered.

2.6.1.4 The Non-plastic Transition and the Possibility of Negative Pressure

In order to get some insight into what happens around the peak of pressure, the shear structure is considered at the neutral surface. Although this function is discontinuous at this point, its derivative in the smooth continuous limit can be taken as $H''(\xi_N) = -\infty$.

Therefore, according to Eqs. (2.164) and (2.165) the entire neutral surface $\xi = \xi_N$ is working at negative pressure or suction. The application of front and back tensions favors the appearance of the suction. The stress tensor at the neutral surface has, in this limiting case, the following components:

$$\sigma_{\xi\xi}(\xi_N) \cong \sigma_{\eta\eta}(\xi_N) \cong \sigma_{\eta\eta}^{(1,3)}(\xi_N) = -\frac{\sigma_C}{2\sqrt{3}}H''(\xi_N)\eta^2 \to +\infty \tag{2.166}$$

$$\sigma_{zz}(\xi_N) = \frac{1}{2}\left[\sigma_{\xi\xi}(\xi_N) + \sigma_{\eta\eta}(\xi_N)\right] \cong -\frac{\sigma_C}{2\sqrt{3}}H''(\xi_N)\eta^2 \to +\infty \tag{2.167}$$

$$\sigma_{\xi\eta}(\xi_N) = \sigma_{\xi z}(\xi_N) = \sigma_{\eta z}(\xi_N) = 0 \tag{2.168}$$

The stress state tends to be hydrostatic while remaining plastic. This is in itself a contradiction, since hydrostatic stresses cannot produce normal plastic flow, and therefore, this singular and fully degenerate stress state belongs to the ideal limiting case of the zero order solution. Along the middle transversal plane of the strip and due to symmetry the lateral spread is null, and at the intersection with the neutral surface, a non-plastic transition and negative pressure could begin to manifest under the right combination of the rolling parameters. The real shear structure is continuous at this surface, but its derivative can still be very large and negative, and the analysis above remains valid. The closer the conditions of rolling to the zero order solution, bigger is the possibility of negative pressure and stronger is the condition of non-plasticity. This could be the case in modern cold rolling mills, where the friction coefficient is maintained as low as possible. Negative pressure in the middle plane of the strip has been reported in previous work on cold flat rolling (Ch'ien and Ch'en 1952), and our simple analytical model for the plastic flow begins to shed light on this matter. First order corrections to the zero order solution must be taken into account in order to explain this weird phenomenon.

2.7 Zero Order Solution with Strain Hardening

As mentioned before, the rigid perfectly plastic solution could make sense even in industrial practice, when the softening due to the dynamic recovery counter balances the hardening due to strain and strain-rate. In most cases the hardening cannot be averted, and therefore, the following hardening function and the corresponding flow stress along the roll gap are considered:

$$\sigma_C(\xi) = \sigma_C\{1 + f(\xi)\} \tag{2.169}$$

Substituting Eq. (2.169) in Eqs. (2.83) and (2.84), the functional form of order zero for the shear stress and for the difference in coordinate stresses are obtained:

$$\sigma_{\xi\eta} = \frac{\sigma_C}{\sqrt{3}} H'(\xi)\{1 + f(\xi)\}\eta \tag{2.170}$$

$$\sigma_{\xi\xi} - \sigma_{\eta\eta} = \frac{2\sigma_C}{\sqrt{3}}\{1 + f(\xi)\} \tag{2.171}$$

Therefore, Eq. (2.91) of internal equilibrium to order zero takes the following form:

$$\frac{\partial \sigma_{\xi\xi}}{\partial \xi} - \frac{2\sigma_C}{\sqrt{3}}\left(\frac{\sin\xi}{1-\cos\xi}\right)\{1 + f(\xi)\} + \frac{\sigma_C}{\sqrt{3}} H'(\xi)\{1 + f(\xi)\} = 0 \tag{2.172}$$

Integrating by parts and using the boundary condition of back tension, the entry branch of the stress field is obtained:

$$\sigma_{\xi\xi}(\xi) = \sigma_1 + \frac{2\sigma_C}{\sqrt{3}}\{1+f(\xi)\}\ln(1-\cos\xi) - \frac{2\sigma_C}{\sqrt{3}}\{1+f(\xi_0)\}\ln(1-\cos\xi_0)$$
$$-\frac{\sigma_C}{\sqrt{3}}\int_{\xi_0}^{\xi} H'(\xi)\{1+f(\xi)\}d\xi - \frac{2\sigma_C}{\sqrt{3}}\int_{\xi_0}^{\xi} \ln(1-\cos\xi)f'(\xi)d\xi \tag{2.173}$$

Using Eq. (2.171) and the friction condition with the rolls from the entrance of the roll gap to the neutral point given by Eq. (2.127), the corresponding shear structure is calculated:

$$H'(\xi)\{1+f(\xi)\}$$
$$= \vartheta\begin{bmatrix} 2\{1+f(\xi)\}\left[\ln(1-\cos\xi)-1\right] - 2\{1+f(\xi_0)\}\ln(1-\cos\xi_0) \\ +\frac{\sigma_1}{k} - \int_{\xi_0}^{\xi} H'(\xi)\{1+f(\xi)\}d\xi - 2\int_{\xi_0}^{\xi}\ln(1-\cos\xi)f'(\xi)d\xi \end{bmatrix} \tag{2.174}$$

Defining $F(\xi) = \int_{\xi_0}^{\xi} H'(\xi)\{1+f(\xi)\}d\xi$, the following differential equation is obtained.

$$F'(\xi) + \vartheta F(\xi) - 2\vartheta\begin{bmatrix} \{1+f(\xi)\}\ln(1-\cos\xi) - \{1+f(\xi_0)\}\ln(1-\cos\xi_0) \\ +\left(\frac{\sigma_1}{2k}\right) - \{1+f(\xi)\} - \int_{\xi_0}^{\xi}\ln(1-\cos\xi)f'(\xi)d\xi \end{bmatrix} = 0 \tag{2.175}$$

where $F'(\xi) = H'(\xi)\{1+f(\xi)\}$ is the effective shear structure with strain hardening. The solution of Eq. (2.175) is:

$$F(\xi) = 2\vartheta\exp[-\vartheta\xi]\int_{\xi_0}^{\xi}\exp[\vartheta\xi]\left\{\begin{matrix} \left(\frac{\sigma_1}{2k}\right) - \{1+f(\xi)\} - \int_{\xi_0}^{\xi}\ln(1-\cos\xi)f'(\xi)d\xi + \\ \{1+f(\xi)\}\ln(1-\cos\xi) - \{1+f(\xi_0)\}\ln(1-\cos\xi_0) \end{matrix}\right\}d\xi \tag{2.176}$$

and performing the integration of the constant terms in the bracket, the following expression is obtained:

$$F(\xi) = 2\{1-\exp[-\vartheta(\xi-\xi_0)]\}\left\{\left(\frac{\sigma_1}{2k}\right) - 1 - \{1+f(\xi_0)\}\ln(1-\cos\xi_0)\right\}$$
$$-2\int_{\xi_0}^{\xi}\ln(1-\cos\xi)f'(\xi)d\xi \tag{2.177}$$
$$+2\exp[-\vartheta\xi]\int_{\xi_0}^{\xi}\exp[\vartheta\xi]\begin{bmatrix} \vartheta\left[\{1+f(\xi)\}\ln(1-\cos\xi) - f(\xi)\right] \\ +\ln(1-\cos\xi)f'(\xi) \end{bmatrix}d\xi$$

The entry branch of the roll pressure profile is calculated, by substituting Eq. (2.177) in Eq. (2.173) and using Eq. (2.171) for the difference in coordinate stresses. The final result is:

$$\begin{aligned}\frac{\sigma_{\eta\eta}}{2k}(\xi,\xi_0)=&\left[\left(\frac{\sigma_1}{2k}\right)-1-\{1+f(\xi_0)\}\ln(1-\cos\xi_0)\right]\exp[-\vartheta(\xi-\xi_0)]\\&+\{1+f(\xi)\}\ln(1-\cos\xi)-f(\xi)\\&-\exp[-\vartheta\xi]\int_{\xi_0}^{\xi}\vartheta\exp[\vartheta\xi]\ln(1-\cos\xi)d\xi\\&-\exp[-\vartheta\xi]\int_{\xi_0}^{\xi}\exp[\vartheta\xi][\vartheta f(\xi)+f'(\xi)]\ln(1-\cos\xi)d\xi\\&+\exp[-\vartheta\xi]\int_{\xi_0}^{\xi}\vartheta\exp[\vartheta\xi]f(\xi)d\xi\end{aligned}\tag{2.178}$$

The corresponding shear stress distribution is obtained using Eq. (2.127):

$$\frac{\sigma_{\xi\eta}}{2k}(\xi,\xi_0)=\frac{1}{2}F'(\xi)\eta=\eta\vartheta\frac{\sigma_{\eta\eta}}{2k}(\xi,\xi_0)\tag{2.179}$$

$$\begin{aligned}\frac{\sigma_{\xi\eta}}{2k}(\xi,\xi_0)=&\,\eta\vartheta\left[\left(\frac{\sigma_1}{2k}\right)-1-\{1+f(\xi_0)\}\ln(1-\cos\xi_0)\right]\exp[-\vartheta(\xi-\xi_0)]\\&+\eta\vartheta\left[\{1+f(\xi)\}\ln(1-\cos\xi)-f(\xi)\right]\\&-\eta\vartheta\exp[-\vartheta\xi]\int_{\xi_0}^{\xi}\exp[\vartheta\xi]\begin{bmatrix}\vartheta\left[\{1+f(\xi)\}\ln(1-\cos\xi)-f(\xi)\right]\\+\ln(1-\cos\xi)f'(\xi)\end{bmatrix}d\xi\end{aligned}\tag{2.180}$$

In the same manner, using the friction condition with the rolls from the neutral point to the exit of the roll gap defined by Eq. (2.131), the differential equation for the corresponding shear structure and its solution are:

$$F'(\xi)-\vartheta F(\xi)+2\vartheta\begin{bmatrix}\frac{\sigma_2}{2k}+\{1+f(\xi)\}\ln(1-\cos\xi)-\{1+f(\pi)\}\ln 2\\-\{1+f(\xi)\}-\int_{\pi}^{\xi}\ln(1-\cos\xi)f'(\xi)d\xi\end{bmatrix}=0\tag{2.181}$$

$$F(\xi)=-2\vartheta\exp[\vartheta\xi]\int_{\pi}^{\xi}\exp[-\vartheta\xi]\begin{Bmatrix}\left(\frac{\sigma_2}{2k}\right)-\{1+f(\xi)\}-\int_{\pi}^{\xi}\ln(1-\cos\xi)f'(\xi)d\xi\\+\{1+f(\xi)\}\ln(1-\cos\xi)-\{1+f(\pi)\}\ln 2\end{Bmatrix}d\xi\tag{2.182}$$

Finally, performing the integral in Eq. (2.182), the exit branch of the roll pressure profile and of the shear stress distribution are obtained:

$$\begin{aligned}F(\xi) = {} & 2\{1-\exp[\vartheta(\xi-\pi)]\}\left\{\left(\frac{\sigma_2}{2k}\right)-1-\{1+f(\pi)\}\ln 2\right\}\\ & -2\int_{\pi}^{\xi}\ln(1-\cos\xi)f'(\xi)d\xi\\ & -2\exp[\vartheta\xi]\int_{\pi}^{\xi}\exp[-\vartheta\xi]\begin{bmatrix}\vartheta[\{1+f(\xi)\}\ln(1-\cos\xi)-f(\xi)]\\ -\ln(1-\cos\xi)f'(\xi)\end{bmatrix}d\xi\end{aligned} \tag{2.183}$$

$$\begin{aligned}\frac{\sigma_{\eta\eta}}{2k}(\xi,\pi) = {} & \left[\left(\frac{\sigma_2}{2k}\right)-1-\{1+f(\pi)\}\ln 2\right]\exp[\vartheta(\xi-\pi)]\\ & +\{1+f(\xi)\}\ln(1-\cos\xi)-f(\xi)\\ & +\exp[\vartheta\xi]\int_{\pi}^{\xi}\vartheta\exp[-\vartheta\xi]\ln(1-\cos\xi)d\xi\\ & +\exp[\vartheta\xi]\int_{\pi}^{\xi}\exp[-\vartheta\xi][\vartheta f(\xi)-f'(\xi)]\ln(1-\cos\xi)d\xi\\ & -\exp[\vartheta\xi]\int_{\pi}^{\xi}\vartheta\exp[-\vartheta\xi]f(\xi)d\xi\end{aligned} \tag{2.184}$$

$$\begin{aligned}\frac{\sigma_{\xi\eta}}{2k}(\xi,\pi) = {} & -\eta\vartheta\left[\left(\frac{\sigma_2}{2k}\right)-1-\{1+f(\pi)\}\ln 2\right]\exp[\vartheta(\xi-\pi)]\\ & -\eta\vartheta[\{1+f(\xi)\}\ln(1-\cos\xi)-f(\xi)]\\ & -\eta\vartheta\exp[\vartheta\xi]\int_{\pi}^{\xi}\exp[-\vartheta\xi]\begin{bmatrix}\vartheta[\{1+f(\xi)\}\ln(1-\cos\xi)-f(\xi)]\\ -\ln(1-\cos\xi)f'(\xi)\end{bmatrix}d\xi\end{aligned} \tag{2.185}$$

2.7.1 Isotropic Hardening Rule—The Adiabatic Thermal Field

Consider the total power used in the roll gap, namely, the power injected as deformation by friction with the rolls, the power injected by the tensions and the heat dissipation by friction:

$$P^{TOTAL} = P^{rolls} + (\sigma_2-\sigma_1)v_1e_1 = P^{heat} + P^{fric} + (\sigma_2-\sigma_1)v_1e_1 \tag{2.186}$$

Substituting Eq. (2.283) in Eq. (2.186) and using the result given by Eq. (2.303), the following expression for the total power is obtained:

$$P^{TOTAL} = P^{heat} - \frac{2\sigma_C}{\sqrt{3}}v_1e_1\ln(1-r) + \frac{2}{3}kv_1e_1\vartheta\eta_0^2\left[1-\frac{\sigma_1}{2k}\right]\sqrt{\frac{r}{1-r}} \tag{2.187}$$

In the adiabatic regime of high speed rolling the heat generated by friction with the rolls can be ignored because it is a boundary generated heat, and the shear power term in the usual range of thickness reductions and rolling parameters is always disturbing. The important term from the thermal point of view is the bulk power of deformation. Therefore, it is assumed that this term gets fully[4] dissipated into heat as the deformation proceeds along the roll gap.

The equation of the energy in adiabatic regime (no heat flux $\bar{\Psi}$), without sources $\dot{q}$, and considering the strain hardening is:

$$\rho\frac{Du_T}{Dt}+\rho\frac{Du_D}{Dt}\cong\rho\frac{Du_T}{Dt}=\sigma_{ij}D_{ij}-\operatorname{div}\bar{\psi}+\rho\dot{q}=\sigma_{ij}D_{ij} \tag{2.188}$$

$$\sigma_{ij}D_{ij}=\sigma_{\xi\xi}D_{\xi\xi}+\sigma_{\eta\eta}D_{\eta\eta}+2\sigma_{\xi\eta}D_{\xi\eta}\cong\sigma_{\xi\xi}D_{\xi\xi}+\sigma_{\eta\eta}D_{\eta\eta} \tag{2.189}$$

$$\sigma_{\xi\eta}D_{\xi\eta}\cong-\frac{\sigma_C}{\sqrt{3}}v_1\left(\frac{e_1}{e_2}\right)\left(\frac{1-\cos\xi}{2a}\right)\sin\xi H'^2(\xi)\eta^2\propto H'^2(\xi)\eta^2\cong 0 \tag{2.190}$$

Therefore, the contribution of the shear to the dissipation function is ignored:

$$\rho c_V\frac{DT}{Dt}=-\left[\frac{2\sigma_C}{\sqrt{3}}\{1+f(\xi)\}+\sigma_{\eta\eta}\right]D_{\eta\eta}+\sigma_{\eta\eta}D_{\eta\eta}=-\frac{2\sigma_C}{\sqrt{3}}\{1+f(\xi)\}D_{\eta\eta} \tag{2.191}$$

$$\rho c_V\left(\frac{v_\xi}{g_\xi}\frac{\partial T}{\partial\xi}\right)=-\frac{2\sigma_C}{\sqrt{3}}\{1+f(\xi)\}D_{\eta\eta} \tag{2.192}$$

Considering Eq. (2.192) to order zero, the following ordinary differential equation for the thermal field in the roll gap is obtained:

$$\frac{d}{d\xi}T(\xi)=\frac{2}{\sqrt{3}}\left(\frac{\sigma_C}{\rho c_V}\right)\{1+f(\xi)\}\frac{\sin\xi}{(1-\cos\xi)} \tag{2.193}$$

Using the boundary condition at the entrance of the roll gap $T(\xi_0)=T_0$, the thermal field is:

$$T(\xi)=T_0+\frac{2}{\sqrt{3}}\left(\frac{\sigma_C}{\rho c_V}\right)\int_{\xi_0}^{\xi}\{1+f(\xi)\}\frac{\sin\xi}{(1-\cos\xi)}d\xi \tag{2.194}$$

[4] This is not completely true as some of it is stored in the material in the form of dislocations, its interactions, and other micro-structural phenomena that contribute to hardening. This energy is recovered in more or less degree in the different thermal cycles of annealing.

$$T(\xi) = T_0 + \left(\frac{\sigma_C}{\rho c_V}\right)\left\{\bar{\varepsilon}(\xi) + \frac{2}{\sqrt{3}}\left[f(\xi)\ln(1-\cos\xi) - \int_{\xi_0}^{\xi}\ln(1-\cos\xi)f'(\xi)d\xi\right]\right\} \tag{2.195}$$

where $T(\xi)$ must remain below the recrystallization temperature of the strip. Therefore, there is coupling between the thermal field and the strain hardening. It is evident, that the adiabatic isothermals are ξ-isosurfaces and that for a material rigid perfectly plastic, the adiabatic increment of temperature along the roll gap is proportional to the equivalent plastic strain. The time for a material particle to cross the roll gap is:

$$dt = \frac{dl}{v_\xi} = \left(\frac{g_\xi}{v_\xi}\right)d\xi = -2\frac{\sqrt{e_2 R}}{v_2}\frac{d\xi}{(1-\cos\xi)^2} \tag{2.196}$$

$$t_{RG} = 2\frac{\sqrt{e_2 R}}{v_2}\int_{\pi}^{\xi_0}\frac{d\xi}{(1-\cos\xi)^2} = -\frac{2}{3}\frac{\sqrt{e_2 R}}{v_2}\cot\left(\frac{\xi_0}{2}\right)\left[\frac{2-\cos\xi_0}{1-\cos\xi_0}\right] \tag{2.197}$$

$$t_{RG} = \frac{1}{3\eta_0}\left(\frac{e_2}{v_2}\right)\left[\frac{3-2r}{1-r}\right]\sqrt{\frac{r}{1-r}} \tag{2.198}$$

In high speed rolling and in the usual range of reductions, the time is of the order of $10^{-3} \div 10^{-4}$ seconds, and therefore, the assumption of an adiabatic heating is fully justified.

Based again on the rheological model of Johnson and Cook (1983), the following hardening function $f(\xi)$ in Eq. (2.169) is considered:

$$f(\xi) = \sigma_C^{-1} A\Phi(\bar{\varepsilon})\left\{1 + C(\bar{D}(\xi), \dot{\varepsilon}_t, \bar{T})\ln\left[1 + \bar{D}(\xi)/\dot{\varepsilon}_0\right]\right\} Z_\beta(\bar{T}) \tag{2.199}$$

where A is the strain hardening coefficient with dimensions of stress at the reference conditions ($\dot{\varepsilon}_0$, T_R), σ_C is the yield stress (Eq. 2.85), and $\bar{D}(\xi)$ is the equivalent plastic strain rate (Eq. 2.57). Because this function is null at the roll gap exit, a unit is added to the argument of the logarithm to avoid an irrelevant singularity. $\Phi(\bar{\varepsilon})$ is any physically acceptable function of the equivalent plastic strain. For a better fit in Figure 2.7, instead of the usual power law of the JC model, the following function that allows the hardening to saturate with the strain is considered:

$$\Phi(\bar{\varepsilon}) = \tanh\left[n\bar{\varepsilon}\right] \tag{2.200}$$

where n is a material constant. The function of thermal softening is defined as follows:

$$Z_\beta(\bar{T}) = 1 - \left[\frac{\bar{T}(\xi) - T_R}{T_M - T_R}\right]^\beta \qquad Z_\beta(\bar{T} \le T_R) = 1 \qquad \beta > 0 \tag{2.201}$$

where T_M is the melting temperature of the strip, and T_R is the reference temperature below which the thermal softening cancels. The exponent β measures the efficiency of the material of the strip to counteract the hardening due to strain and strain rate at a given temperature. In this way, and unlike in the JC model, the influence of the thermal softening on the yield stress (Eq. 2.85), and on the strain hardening (Eq. 2.199) is decoupled. It is an experimental fact that the effect of thermal softening is stronger on strain hardening than on yield stress, and therefore, the thermal softening exponents have to fulfil the following condition $\beta < \alpha$. The effective thermal field in Eqs. (2.199) and (2.201) is defined, depending on the rolling conditions, as an intermediate between the two limiting cases, the adiabatic and the isothermal:

$$\bar{T}(\xi) = T_0 + h(v_1/v_0)\left[T(\xi) - T_0\right] \tag{2.202}$$

where $T(\xi)$ is the adiabatic thermal field given by Eq. (2.195), and v_0 is the threshold velocity of rolling for the roll gap to become adiabatic under the given rolling conditions. In this way, the complex problem of heat transfer is replaced by the function h that controls the transition from an isothermal to an adiabatic roll gap, and that has to fulfil the following requirements:

$$h(v_1/v_0) \cong 0 \qquad \forall v_1 \ll v_0 \qquad\qquad h(v_1/v_0) = 1 \qquad \forall v_1 \geq v_0 \tag{2.203}$$

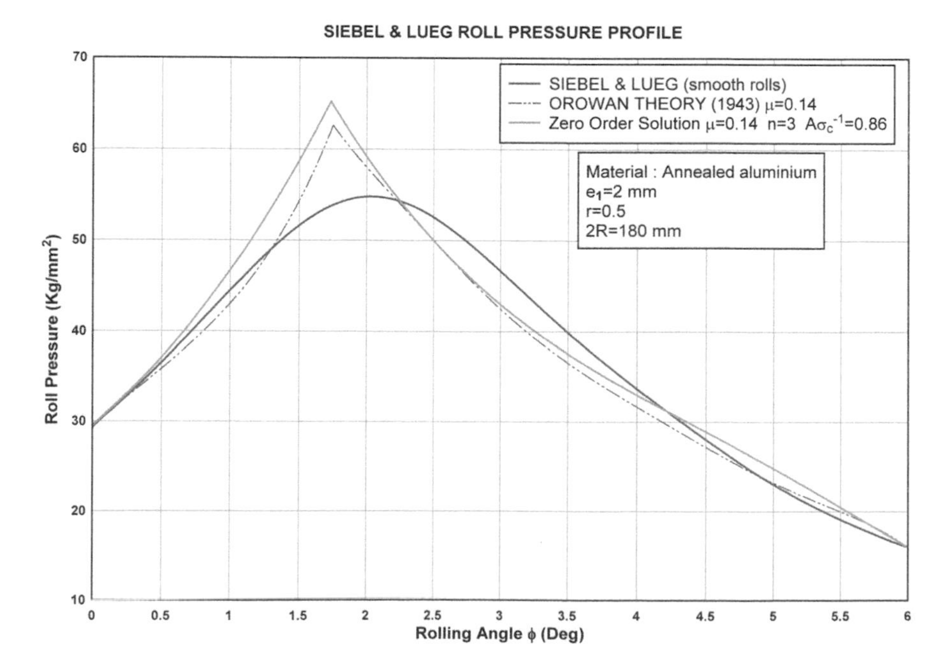

Figure 2.7 Siebel & Lueg experimental roll pressure distribution.

Finally, an explicit differential equation for the adiabatic thermal field is obtained using Eq. (2.199) in Eq. (2.193) with $h = 1$. Once its solution is found and using Eq. (2.199) with the value of h proper, the flow stress and the stress field are obtained without the need to iterate.

Figure 2.7 is the roll pressure distribution in rolling without tensions of annealed aluminium for the case of smooth rolls, obtained by Siebel and Lueg (1933) using the embedded pin-transducer technique. The rolling conditions are far from adiabatic ($h \cong 0$), and the temperature of the strip at the entrance of the roll gap is $\underline{T}_0 \approx T_R$, and therefore, the function of thermal softening takes the value $Z_\beta(\overline{T}) = 1$. The strain rate hardening at the rolling speeds of a laboratory mill can be ignored. The maximum and average errors in the fulfilment of the Von Mises criterion are:

$$\varepsilon_{\mathrm{VM}}^{\max} \cong 45\% \qquad \tilde{\varepsilon}_{\mathrm{VM}} \cong 5\% \tag{2.204}$$

Therefore, the zero order solution is beyond its range of validity around the neutral point due to a large violation of the Von Mises criterion in this region, when the influence of the shear structure in the difference between the coordinate stresses is ignored.

The dotted curve is the best fit using Orowan's theory with a combination of static and slipping friction. The fit in continuous is the zero order solution with the hardening law defined by Eq. (2.200) and with the same friction coefficient.

The zero order solution with slipping friction, produces a similar fit to Orowan's theory considering also static friction. No hardening law is able to predict the right position of the neutral point, indicating that this case goes beyond the disturbing expansion of our system of equations for the shear structure and the stresses, and therefore, the zero order solution as the main representative of the stress state within the roll gap does not make sense. Consequently, there is a need to find a non-disturbing numerical solution for the shear structure with strain hardening. This is the subject of the following section.

Due to the dependence of the hardening function on the equivalent plastic strain, the cold flat rolling under plane strain and with strain hardening is no longer isostatic, and therefore, it has to be considered as statically undetermined or hyper-static due to the rheology.

2.7.2 Non-disturbing Numerical Extension in the Cold Flat Rolling Approximation

When the condition defined by Eq. (2.144) breaks down, the disturbing expansion leading to Eqs. (2.99) and (2.100) is no longer valid, and the following boundary value problem coming from Eqs. (2.97) and (2.98) with strain hardening must be solved in the roll gap.

$$\frac{\partial \sigma_{\xi\xi}}{\partial \xi} - \frac{2\sigma_C}{\sqrt{3}} \left(\frac{\sin\xi + H'(\xi)\eta^2}{1-\cos\xi} \right) \frac{\{1+f(\xi)\}}{\sqrt{1+H'^2(\xi)\eta^2}} + \frac{\sigma_C}{\sqrt{3}} \frac{H'(\xi)\{1+f(\xi)\}}{\left[1+H'^2(\xi)\eta^2\right]^{3/2}} = 0 \tag{2.205}$$

$$\frac{\partial \sigma_{\eta\eta}}{\partial \eta} + \frac{2\sigma_C}{\sqrt{3}} \left(\frac{1-H'(\xi)\sin\xi}{1-\cos\xi} \right) \frac{\{1+f(\xi)\}\eta}{\sqrt{1+H'^2(\xi)\eta^2}} + \frac{\sigma_C}{\sqrt{3}} \frac{\partial}{\partial \xi} \left(\frac{H'(\xi)\{1+f(\xi)\}}{\sqrt{1+H'^2(\xi)\eta^2}} \eta \right) = 0 \tag{2.206}$$

$$\sigma_{\xi\eta} = \frac{\sigma_C}{\sqrt{3}} \frac{H'(\xi)\eta}{\sqrt{1+H'^2(\xi)\eta^2}} \{1+f(\xi)\} \tag{2.207}$$

$$\sigma_{\xi\xi} - \sigma_{\eta\eta} = \frac{2\sigma_C}{\sqrt{3}} \frac{1}{\sqrt{1+H'^2(\xi)\eta^2}} \{1+f(\xi)\} \tag{2.208}$$

The boundary conditions of slipping friction with the rolls, and the tensions at the entrance and at the exit of the roll gap defined by Eqs. (2.122) and (2.123) respectively, remain the same. This is an iterative boundary value problem to be solved numerically from both ends of the roll gap in a self-consistent manner, using the fact that Eq. (2.206) can be integrated in η. The procedure is similar to the disturbing case and the solution will have the corresponding "peak", where for a given value of η both branches of the stress field merge, defining in this way the neutral surface.

The iterative process starts using both branches of the shear structure to order zero. After the n-iteration and in order to obtain the n+1 shear structure, the corresponding friction condition along the arc of contact with the rolls has to be applied:

$$\sigma_{\xi\eta}(\xi,\eta_0) = \frac{\sigma_C}{\sqrt{3}} \frac{H'_{n+1}(\xi)\eta_0}{\sqrt{1+H'^2_{n+1}(\xi)\eta_0^2}} \{1+f(\xi)\} = \pm\mu\sigma_{\eta\eta}^{(n)}(\xi,\eta_0) \tag{2.209}$$

The iterative process stops and the solution can be considered valid, when the relative error in the shear structure between two successive iterations is satisfied.

$$\mathrm{E}(\xi) = \left| \frac{H'_{n+1}(\xi) - H'_n(\xi)}{H'_n(\xi)} \right| < M \qquad \forall \xi \in [\pi, \xi_0] \tag{2.210}$$

Finally, the following numerical procedures for the solution of the cold flat rolling can be defined. The general boundary value problem for the velocity and stress fields in the roll gap. The restricted boundary value problem for the stress field ignoring the plastic flow and using the analytical approach to the flow stress of the previous section. Filling the gap between them and the analytical solution, the above procedure considering strain hardening could be used for a further improvement in the fit of the Siebel & Lueg roll pressure profile.

2.8 The Disturbing Component of the Plastic Flow due to Friction Drag with the Rolls

2.8.1 Boundary Value Problem for the Detailed Structure of the Plastic Flow in Plane Strain

So far, an analytical solution for the plastic stress field in the roll gap was obtained for roll pressure calculations using the Levy-Mises flow rule, the Von Mises criterion, and the equations of internal equilibrium. The plastic stress field to order zero is a solution in complete agreement with the flow structure. Because the shear strain rate had to be modified to obtain a successful coupling with the plastic stresses, the initial velocity field given by Eq. (2.15) is only a good first approximation to the real structure of the plastic flow in the roll gap. A suitable integration of the modified shear strain rate reveals the details of the flow originated by friction with the rolls, as a perturbation of the initial plastic velocity field. To fulfil incompressibility and plane strain, a new transversal component of the velocity has to be considered.

Disturbing Eq. (2.15), the following expressions for the plastic velocity field and the strain rates are obtained:

$$v_\xi(\xi,\eta) = -v_1\left(\frac{e_1}{e_2}\right)\frac{(\cosh\eta - \cos\xi)}{(\cosh\eta + 1)} + \delta v_\xi(\xi,\eta) = v_\xi^{(0)}(\xi,\eta) + \delta v_\xi(\xi,\eta) \quad (2.211)$$

$$D_{\xi\xi} = \frac{1}{g_\xi}\frac{\partial v_\xi}{\partial \xi} + \frac{v_\eta}{g_\xi g_\eta}\frac{\partial g_\xi}{\partial \eta} = \frac{1}{g_\xi}\frac{\partial v_\xi^{(0)}}{\partial \xi} + \frac{1}{g_\xi}\frac{\partial \delta v_\xi}{\partial \xi} + \frac{v_\eta}{g_\xi g_\eta}\frac{\partial g_\xi}{\partial \eta} \quad (2.212)$$

$$D_{\eta\eta} = \frac{1}{g_\eta}\frac{\partial v_\eta}{\partial \eta} + \frac{v_\xi}{g_\eta g_\xi}\frac{\partial g_\eta}{\partial \xi} = \frac{1}{g_\eta}\frac{\partial v_\eta}{\partial \eta} + \frac{v_\xi^{(0)}}{g_\eta g_\xi}\frac{\partial g_\eta}{\partial \xi} + \frac{\delta v_\xi}{g_\eta g_\xi}\frac{\partial g_\eta}{\partial \xi} \quad (2.213)$$

$$D_{\xi\eta} = \frac{1}{2}\left[\frac{\partial}{\partial \xi}\left(\frac{v_\eta}{g_\eta}\right) + \frac{\partial}{\partial \eta}\left(\frac{v_\xi}{g_\xi}\right)\right] = -v_1\left(\frac{e_1}{e_2}\right)\left(\frac{\cosh\eta - \cos\xi}{a}\right)\frac{\sinh\eta}{(\cosh\eta + 1)}\Psi(\xi) \quad (2.214)$$

Therefore, the incompressibility condition is:

$$\frac{1}{g_\xi}\frac{\partial v_\xi^{(0)}}{\partial \xi} + \frac{1}{g_\xi}\frac{\partial \delta v_\xi}{\partial \xi} + \frac{v_\eta}{g_\xi g_\eta}\frac{\partial g_\xi}{\partial \eta} + \frac{1}{g_\eta}\frac{\partial v_\eta}{\partial \eta} + \frac{v_\xi^{(0)}}{g_\eta g_\xi}\frac{\partial g_\eta}{\partial \xi} + \frac{\delta v_\xi}{g_\eta g_\xi}\frac{\partial g_\eta}{\partial \xi} = 0 \quad (2.215)$$

Because the initial velocity field is already incompressible, the following condition applies:

$$D_{\xi\xi}^{(0)} + D_{\eta\eta}^{(0)} = \frac{1}{g_\xi}\frac{\partial v_\xi^{(0)}}{\partial \xi} + \frac{v_\xi^{(0)}}{g_\eta g_\xi}\frac{\partial g_\eta}{\partial \xi} = 0 \quad (2.216)$$

$$\frac{\partial \delta v_\xi}{\partial \xi}+\frac{v_\eta}{g_\xi}\frac{\partial g_\xi}{\partial \eta}+\frac{\partial v_\eta}{\partial \eta}+\frac{\delta v_\xi}{g_\xi}\frac{\partial g_\eta}{\partial \xi}=0 \tag{2.217}$$

Operating Eq. (2.217), the differential equation of incompressibility is obtained:

$$\frac{\partial \delta v_\xi}{\partial \xi}+\frac{\partial v_\eta}{\partial \eta}-\left(\frac{\sin \xi}{\cosh \eta-\cos \xi}\right)\delta v_\xi-\left(\frac{\sinh \eta}{\cosh \eta-\cos \xi}\right)v_\eta=0 \tag{2.218}$$

The differential equation for the modified shear strain rate is:

$$\frac{\partial}{\partial \xi}\left(\frac{v_\eta}{g_\eta}\right)+\frac{\partial}{\partial \eta}\left(\frac{v_\xi}{g_\xi}\right)=-2v_1\left(\frac{e_1}{e_2}\right)\left(\frac{\cosh \eta-\cos \xi}{a}\right)\frac{\sinh \eta}{(\cosh \eta+1)}\Psi(\xi) \tag{2.219}$$

$$\frac{\partial}{\partial \xi}\left(\frac{v_\eta}{g_\eta}\right)+\frac{\partial}{\partial \eta}\left(\frac{v_\xi}{g_\xi}\right)=\frac{1}{a}\frac{\partial}{\partial \xi}\left[(\cosh \eta-\cos \xi)v_\eta\right]+\frac{1}{a}\frac{\partial}{\partial \eta}\left[(\cosh \eta-\cos \xi)v_\xi\right] \tag{2.220}$$

In order to simplify the expressions, the disturbing fields are re-scaled as follows:

$$[\delta]v_{\eta[\xi]}(\xi,\eta)\leftrightarrow -v_1\left(\frac{e_1}{e_2}\right)[\delta]v_{\eta[\xi]}(\xi,\eta) \tag{2.221}$$

Operating Eq. (2.220) and using Eq. (2.211), the following expressions are calculated:

$$\frac{\partial}{\partial \eta}\left[(\cosh \eta-\cos \xi)v_\xi\right]=\frac{\partial}{\partial \eta}\frac{(\cosh \eta-\cos \xi)^2}{(\cosh \eta+1)}+\frac{\partial}{\partial \eta}\left[(\cosh \eta-\cos \xi)\delta v_\xi\right] \tag{2.222}$$

$$\frac{\partial}{\partial \eta}\frac{(\cosh \eta-\cos \xi)^2}{(\cosh \eta+1)}=2\sinh \eta\left(\frac{\cosh \eta-\cos \xi}{\cosh \eta+1}\right)\left[\frac{1}{2}\left(1+\frac{\cos \xi+1}{\cosh \eta+1}\right)\right] \tag{2.223}$$

$$\frac{1}{2}\left(1+\frac{\cos \xi+1}{\cosh \eta+1}\right)\cong\frac{1}{4}(3+\cos \xi)=\Psi^{(0)}(\xi) \tag{2.224}$$

Using Eqs. (2.219), (2.220), (2.222) and (2.224), the following equation is obtained:

$$\frac{\partial}{\partial \xi}\left[(\cosh \eta-\cos \xi)v_\eta\right]+\frac{\partial}{\partial \eta}\left[(\cosh \eta-\cos \xi)\delta v_\xi\right]=2\sinh \eta\left(\frac{\cosh \eta-\cos \xi}{\cosh \eta+1}\right)\Delta\Psi(\xi) \tag{2.225}$$

where $\Delta\Psi = \Psi - \Psi^{(0)}$, and $\Psi^{(0)}$ is the shear structure of the initial undisturbed velocity field, given by Eq. (2.47).

The final differential equation coming from the modified shear strain rate is:

$$\frac{\partial \delta v_\xi}{\partial \eta} + \frac{\partial v_\eta}{\partial \xi} + \left(\frac{\sinh \eta}{\cosh \eta - \cos \xi}\right)\delta v_\xi + \left(\frac{\sin \xi}{\cosh \eta - \cos \xi}\right)v_\eta = \frac{2\sinh \eta}{(\cosh \eta + 1)}\Delta\Psi \tag{2.226}$$

and the following boundary value problem in the roll gap Ω_2 is obtained:

$v_\eta(\xi, \eta_0) = 0$	No cavitations in the continuum.	(2.227)
$v_\eta(\xi, -\eta) = -v_\eta(\xi, \eta)$	Odd function due to symmetry considerations.	(2.228)
$\delta v_\xi(\pi, \eta) = \delta v_\xi(\xi_0, \eta) = 0$	Entrance and exit in the roll gap as a rigid body.	(2.229)
$\delta v_\xi(\xi, -\eta) = \delta v_\xi(\xi, \eta)$	Even function due to symmetry considerations.	(2.230)

$$\frac{\partial \delta v_\xi}{\partial \eta} + \frac{\partial v_\eta}{\partial \xi} + \left(\frac{\sinh \eta}{\cosh \eta - \cos \xi}\right)\delta v_\xi + \left(\frac{\sin \xi}{\cosh \eta - \cos \xi}\right)v_\eta = \frac{2\sinh \eta}{(\cosh \eta + 1)}\Delta\Psi \tag{2.231}$$

$$\frac{\partial \delta v_\xi}{\partial \xi} + \frac{\partial v_\eta}{\partial \eta} - \left(\frac{\sin \xi}{\cosh \eta - \cos \xi}\right)\delta v_\xi - \left(\frac{\sinh \eta}{\cosh \eta - \cos \xi}\right)v_\eta = 0 \tag{2.232}$$

2.8.2 Approximate Solution in Splitting of Variables

It is assumed that an approximate solution in splitting of variables exist, and therefore, the following changes are performed in Eq. (2.225):

$$\delta v_\xi(\xi, \eta) = \frac{f(\xi, \eta)}{(\cosh \eta - \cos \xi)} \tag{2.233}$$

$$v_\eta(\xi, \eta) = \frac{g(\xi, \eta)}{(\cosh \eta - \cos \xi)} \tag{2.234}$$

$$\frac{\partial g}{\partial \xi} + \frac{\partial f}{\partial \eta} = 2\sinh \eta\left(\frac{\cosh \eta - \cos \xi}{\cosh \eta + 1}\right)\Delta\Psi \tag{2.235}$$

$$\frac{\partial g}{\partial \xi} + \frac{\partial f}{\partial \eta} \cong \sinh \eta\left(1 - \cos \xi + \frac{\eta^2}{2}\right)\left(1 - \frac{\eta^2}{4}\right)\Delta\Psi \tag{2.236}$$

$$\frac{\partial g}{\partial \xi}+\frac{\partial f}{\partial \eta} \cong \sinh\eta\left\{(1-\cos\xi)+(1+\cos\xi)\left(\frac{\eta^2}{4}\right)\right\}\Delta\Psi \tag{2.237}$$

Consider the following splitting of variables:

$$f(\xi,\eta)=C(\xi)\left[\cosh\eta+F(\xi)\right]+D'(\xi)\eta_0^2\cosh\eta \tag{2.238}$$

$$g(\xi,\eta)=D(\xi)\left(\eta^2-\eta_0^2\right)\sinh\eta \tag{2.239}$$

Using Eqs. (2.238) and (2.239) in Eq. (2.237), the following functions are obtained:

$$C(\xi)+D'(\xi)\eta^2=\left\{(1-\cos\xi)+(1+\cos\xi)\left(\frac{\eta^2}{4}\right)\right\}\Delta\Psi \tag{2.240}$$

$$C(\xi)=(1-\cos\xi)\Delta\Psi \tag{2.241}$$

$$D'(\xi)=\frac{1}{4}(1+\cos\xi)\Delta\Psi \tag{2.242}$$

To calculate the remaining function $F(\xi)$ in Eq. (2.238), the condition of incompressibility in integral form across any section at constant ξ is applied:

$$\int_{-\eta_0}^{\eta_0}\delta v_\xi(\xi,\eta)\frac{\sqrt{e_2R}}{(\cosh\eta-\cos\xi)}d\eta=0 \tag{2.243}$$

$$\delta v_\xi(\xi,\eta)=\frac{f(\xi,\eta)}{(\cosh\eta-\cos\xi)}\cong\frac{f(\xi,\eta)}{(1-\cos\xi)}\left\{1-\frac{1}{(1-\cos\xi)}\left(\frac{\eta^2}{2}\right)\right\} \tag{2.244}$$

$$\int_{-\eta_0}^{\eta_0}\left\{C(\xi)\left[\cosh\eta+F(\xi)\right]+D'(\xi)\eta_0^2\cosh\eta\right\}\left\{1-\frac{1}{(1-\cos\xi)}\eta^2\right\}d\eta=0 \tag{2.245}$$

$$C(\xi)\left[\sinh\eta_0+F(\xi)\eta_0\right]+D'(\xi)\eta_0^2\sinh\eta_0-\frac{C(\xi)}{(1-\cos\xi)}\left[1+F(\xi)\right]\left(\frac{\eta_0^3}{3}\right)=0 \tag{2.246}$$

$$F(\xi)=-\frac{1}{C(\xi)}\left[C(\xi)+D'(\xi)\eta_0^2\right]\frac{\sinh\eta_0}{\eta_0} \tag{2.247}$$

The final expression for the longitudinal perturbation of the plastic velocity field is obtained, using Eqs. (2.241) and (2.242) along with Eqs. (2.247) and (2.238) in Eq. (2.244):

$$\delta v_{\xi}(\xi,\eta)=\Delta\Psi\left[1+\frac{1}{4}\left(\frac{1+\cos\xi}{1-\cos\xi}\right)\eta_0^2\right]\left\{\cosh\eta-\frac{\sinh\eta_0}{\eta_0}\right\}\left\{1-\frac{1}{(1-\cos\xi)}\left(\frac{\eta^2}{2}\right)\right\} \tag{2.248}$$

$$\delta v_{\xi}(\xi,\eta)\cong\Delta\Psi\left\{\cosh\eta-\frac{\sinh\eta_0}{\eta_0}\right\} \tag{2.249}$$

In the same manner, the following expression is computed for the transversal perturbation:

$$v_{\eta}(\xi,\eta)=\frac{1}{4}\frac{(\eta^2-\eta_0^2)}{(1-\cos\xi)}\sinh\eta\int_{\xi_0}^{\xi}(1+\cos\xi)\Delta\Psi(\xi)d\xi \tag{2.250}$$

This approximate solution fulfils all the boundary conditions, except the one defined by Eq. (2.229), and it will not be enforced upon. Instead, an elastoplastic transition is supposed to occur at the entrance and at the exit of the roll gap. In order to solve Eq. (2.232) for the incompressibility condition in differential form, the following change is made:

$$\delta v_{\xi}(\xi,\eta)=(\cosh\eta-\cos\xi)h(\xi,\eta) \tag{2.251}$$

$$v_{\eta}(\xi,\eta)=(\cosh\eta-\cos\xi)j(\xi,\eta) \tag{2.252}$$

and therefore, Eq. (2.232) takes the following form:

$$\frac{\partial h(\xi,\eta)}{\partial\xi}+\frac{\partial j(\xi,\eta)}{\partial\eta}=0 \tag{2.253}$$

Performing the second order Taylor expansion in Eq. (2.249), the function h is computed:

$$h(\xi,\eta)\cong\frac{\Delta\Psi}{(1-\cos\xi)}\frac{1}{2}\left\{\eta^2-\frac{\eta_0^2}{3}\right\} \tag{2.254}$$

Following the functional form in η of Eq. (2.250) to define the function j in Eq. (2.252) and using Eq. (2.253), the value of its amplitude is obtained:

$$j(\xi,\eta)=G(\xi)(\eta^2-\eta_0^2)\eta \tag{2.255}$$

$$G(\xi)=-\frac{1}{6}\frac{d}{d\xi}\frac{\Delta\Psi}{(1-\cos\xi)} \tag{2.256}$$

The transversal perturbation of the plastic velocity field is finally obtained, using Eqs. (2.255) and (2.256) in Eq. (2.252):

$$v_\eta(\xi,\eta) = -\frac{1}{6}(1-\cos\xi)\frac{d}{d\xi}\frac{\Delta\Psi}{(1-\cos\xi)}(\eta^2 - \eta_0^2)\eta \tag{2.257}$$

Equation (2.257) is considered instead of Eq. (2.250) because the local incompressibility is much more restrictive, and therefore, an inconsistency of order η^2 depending on Eq. (2.256) and its derivative appears in Eq. (2.240) when using this solution for the transversal perturbation. This inconsistency can be resolved in a much more laborious treatment of the solution that is left as an exercise (see Appendix). Because the shear structure is discontinuous at the neutral surface, the correction or the regularization of the plastic stress field is compelling in order to obtain a smooth velocity field along the full length of the roll gap.

2.8.3 The Vortical Nature of the Disturbed Flow in the Roll Gap

The most important feature of the calculations above is the prediction of the existence of a transversal perturbation for the plastic velocity field. When the derivative of the corrected shear structure around the neutral surface is very large and negative, this transversal perturbation is also large and negative (positive) in the upper (lower) part of the roll gap, in order to turn over the longitudinal perturbation and give rise to the rounding and the fall in pressure around the neutral point. This kinematic result is in close agreement with the prediction of the first order solution for the plastic stress field. The closer the conditions of rolling to the zero order solution, bigger is the gap in the velocity field at the neutral surface, taking in the limit of the form of a percussive motion: a sudden redistribution of the velocity field at a fixed position along with a singular stress field (Eqs. 2.166 and 2.167).

The longitudinal perturbation of the plastic velocity field changes sign at the neutral surface while the transversal perturbation takes maximum values around this surface. According to Eq. (2.243) there is no net flow of matter, and therefore, the disturbing motion predicted by the theory can be summarized in a vortex field. The material at the entrance of the roll gap is pushed forward by friction drag with the rolls and is delayed in centre, creating the parabolic shape of the longitudinal perturbation. Upon arrival at the neutral surface, the material flows from the outer regions towards the core of the strip, to compensate the delay in the central plane and nullify the longitudinal perturbation. Something similar but opposite in sign, happens from the neutral point to the exit of the roll gap. The material of the main flow is delayed in the outer regions of the strip by friction with the rolls, and flows towards the core of the strip at the neutral surface, creating a parabolic shape but opposite in sign to the former. In this description, an elastoplastic transition is supposed to occur at the entrance and at the exit of the roll gap.

Using again the scaling factor given by Eq. (2.221) in Eqs. (2.249) and (2.257), the velocity perturbations due to friction drag with the rolls (Figs. 2.8, 2.9 and 2.10) are finally obtained:

$$\delta v_{\xi}(\xi,\eta) = -\frac{v_2}{2}\Delta\Psi\left\{\eta^2 - \frac{\eta_0^2}{3}\right\} = v_2 \mathrm{A}(\xi)\left\{\eta^2 - \frac{\eta_0^2}{3}\right\} \tag{2.258}$$

$$v_{\eta}(\xi,\eta) = \frac{v_2}{6}(1-\cos\xi)\frac{d}{d\xi}\frac{\Delta\Psi}{(1-\cos\xi)}\left(\eta^2 - \eta_0^2\right)\eta = v_2 \mathrm{B}(\xi)\left(\eta^2 - \eta_0^2\right)\eta \tag{2.259}$$

The non-null component of the vortex field $\vec{\omega} = \nabla \times \vec{v}$ in bipolar cylindrical coordinates is:

$$\omega_z = \left(\nabla \times \vec{v}\right)_z = \frac{1}{g_{\xi} g_{\eta}}\left[\frac{\partial}{\partial \xi}\left(g_{\eta} v_{\eta}\right) - \frac{\partial}{\partial \eta}\left(g_{\xi} v_{\xi}\right)\right] \tag{2.260}$$

The expression for this vortex field, using Eqs. (2.258) and (2.259) and ignoring the contribution of the main velocity field given by Eq. (2.15) to the vorticity, is as follows:

$$\omega_z = \frac{v_2}{R}(1-\cos\xi)^2\left(\frac{\eta}{\eta_0}\right)\left[1 + \frac{1}{6}\left(\eta^2 - \eta_0^2\right)\frac{d^2}{d\xi^2}\right]\frac{\Delta\Psi}{(1-\cos\xi)} \tag{2.261}$$

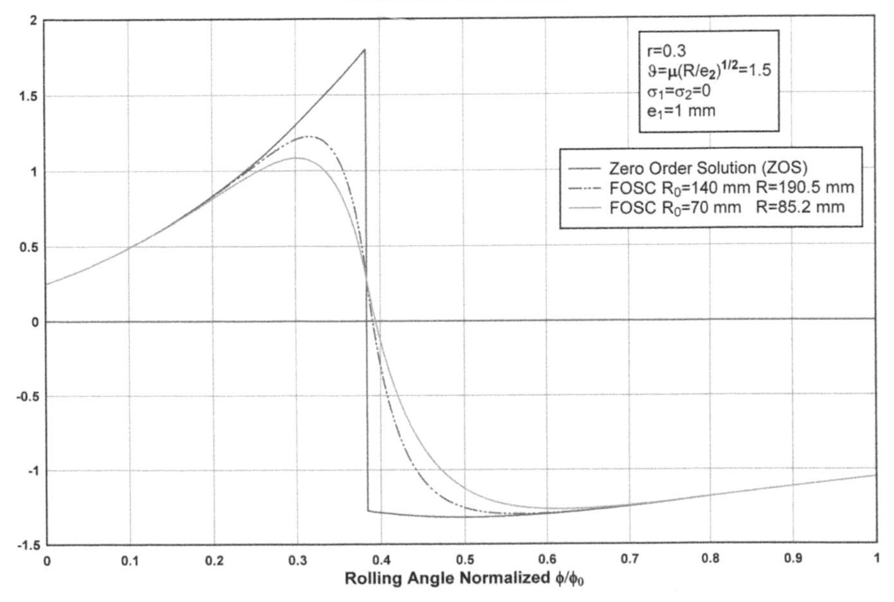

Figure 2.8 Amplitude $\mathrm{A}(\xi)$ for the longitudinal perturbation δv_{ξ}.

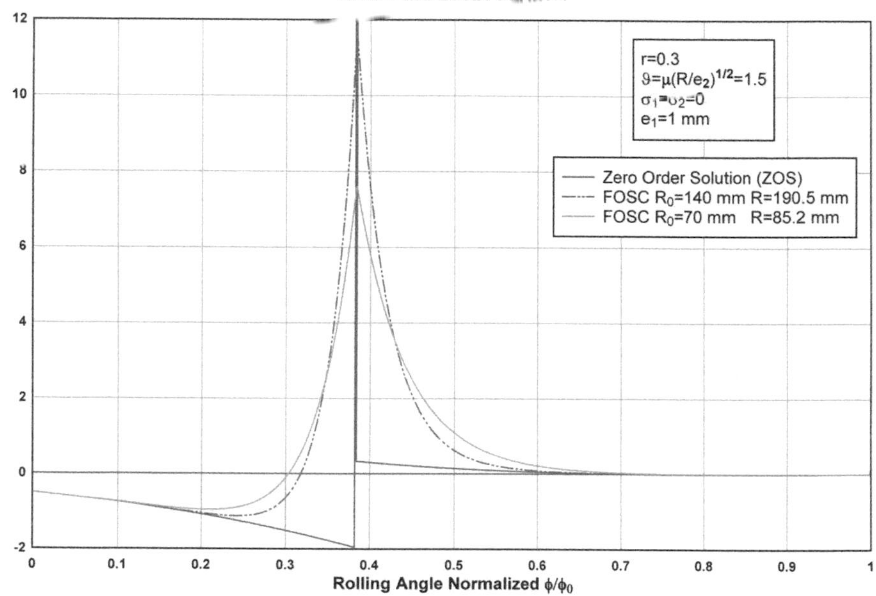

Figure 2.9 Amplitude B(ξ) for the transversal perturbation v_η.

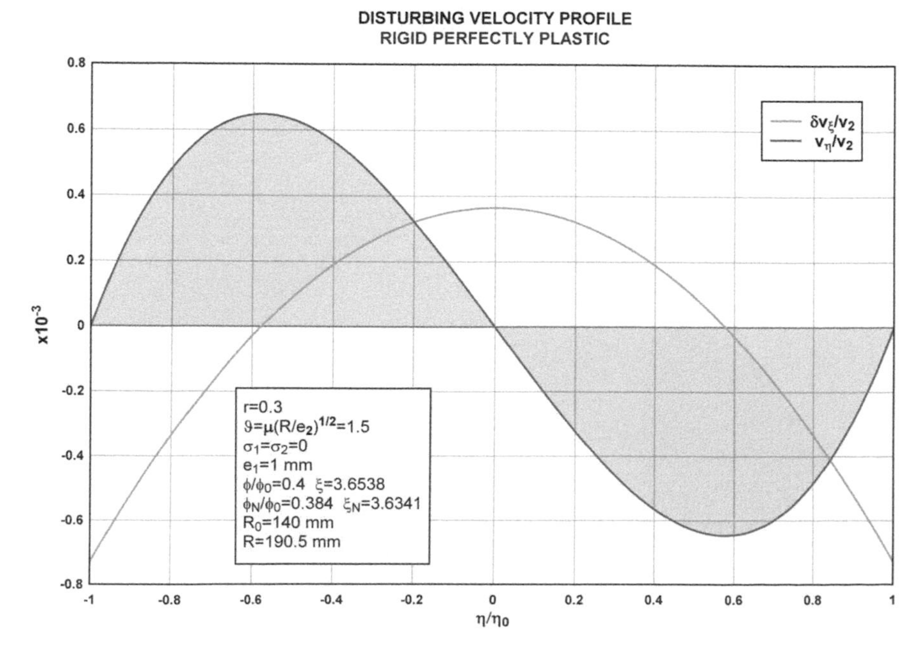

Figure 2.10 Normalized profile of the velocity perturbations.

Therefore, friction with the rolls is the source not only of the main plastic flow given by the velocity field defined by Eq. (2.15), but also of a disturbing motion confined within the roll gap and whose net effect is a vortex field. The field changes sign at the neutral surface and from the lower to the upper part of the roll gap, being null in the central plane of the strip. The vortex field defined by Eq. (2.261) is a proper motion of cold flat rolling.

Finally, when the frictional drag with the rolls becomes very high, the shear structure in Eq. (2.51) for the equivalent plastic strain rate and the influence of the vortex field on the structure of the plastic flow cannot be ignored, and therefore, new measures of strain and strain rate have to be considered and based on them, criteria to characterise the plastic flow in the roll gap (Echave 2011b).

2.9 Energetic Considerations

2.9.1 The Power of Deformation—Bulk Term with no Strain Hardening

In the rigid perfectly plastic case and when the shear structure is disturbing for the plastic flow, the following expression for the power of deformation applies:

$$P^D = \int_{\Omega_2} \bar{\sigma}\bar{D}dV = \sqrt{\frac{2}{3}}\sigma_C \int_{\Omega_2} \sqrt{D_{ij}D_{ij}}\,dV = \frac{2\sigma_C}{\sqrt{3}} \int_{\Omega_2} D_{\xi\xi}dV \equiv \int_{\Omega_2} \sigma_{ij}D_{ij}dV \quad (2.262)$$

where the strain rate is:

$$D_{\xi\xi} = -v_1\left(\frac{e_1}{e_2}\right)\left(\frac{\cosh\eta - \cos\xi}{a}\right)\frac{\sin\xi}{(\cosh\eta + 1)} = -D_{\eta\eta} \quad (2.263)$$

The elementary volume of the roll gap considering one unit in the z direction is:

$$dV = g_\xi g_\eta d\xi d\eta = \frac{a^2}{(\cosh\eta - \cos\xi)^2}d\xi d\eta \quad (2.264)$$

Therefore, the equation for the power of deformation is:

$$P^D = -\frac{2\sigma_C}{\sqrt{3}}v_1 a\left(\frac{e_1}{e_2}\right)\int_{-\eta_0}^{\eta_0}\int_{\pi}^{\xi_0}\left(\frac{\sin\xi}{\cosh\eta + 1}\right)\frac{d\xi d\eta}{(\cosh\eta - \cos\xi)} \quad (2.265)$$

and in the cold flat rolling approximation the following expressions are obtained:

$$P^{D} = -\frac{\sigma_{C}}{\sqrt{3}} v_{1} \sqrt{e_{2} R} \left(\frac{e_{1}}{e_{2}} \right) \int_{-\eta_0}^{\eta_0} \int_{\pi}^{\xi_0} \frac{\sin \xi}{(1-\cos \xi)} d\xi d\eta \tag{2.266}$$

$$P^{D} = -\frac{2\sigma_{C}}{\sqrt{3}} v_{1} \eta_{0} \sqrt{e_{2} R} \left(\frac{e_{1}}{e_{2}} \right) \ln \left(\frac{1-\cos \xi_{0}}{2} \right) \tag{2.267}$$

$$P^{D} = -\frac{2\sigma_{C}}{\sqrt{3}} v_{1} e_{1} \ln \left(\frac{1-\cos \xi_{0}}{2} \right) \tag{2.268}$$

$$P^{D} = -\frac{2\sigma_{C}}{\sqrt{3}} v_{1} e_{1} \ln (1-r) \tag{2.269}$$

The power of deformation in the cold flat rolling approximation, ignoring the influence of the shear strain rate on the flow structure, does not depend on the radius of the rolls.

2.9.2 Shear Power at the Entrance and at the Exit of the Roll Gap

The power to redistribute the velocity field at the entrance of the roll gap Σ_1^+ is:

$$P_{\Sigma_1^+}^{\delta v_1} = \int_{\Sigma_1^+} \sigma_{\xi\eta}(\xi_0,\eta) \Delta v_{\eta}(\xi_0,\eta) dS_{\xi} = -\int_{-\eta_0}^{\eta_0} \sigma_{\xi\eta}(\xi_0,\eta) v_1 \sin\alpha \frac{a d\eta}{(\cosh\eta - \cos\xi_0)} \tag{2.270}$$

$$P_{\Sigma_1^+}^{\delta v_1} = -\mu v_1 \left[\sigma_1 - \frac{2\sigma_C}{\sqrt{3}} \right] \sqrt{e_2 R} \int_{-\eta_0}^{\eta_0} \left(\frac{\eta}{\eta_0} \right) \frac{\sin\alpha}{(\cosh\eta - \cos\xi_0)} d\eta \tag{2.271}$$

$$P_{\Sigma_1^+}^{\delta v_1} \cong \mu v_1 \left[\sigma_1 - \frac{2\sigma_C}{\sqrt{3}} \right] \sqrt{e_2 R} \frac{\sin\xi_0}{(1-\cos\xi_0)^2} \int_{-\eta_0}^{\eta_0} \left(\frac{\eta^2}{\eta_0} \right) d\eta \tag{2.272}$$

$$P_{\Sigma_1^+}^{\delta v_1} = \frac{2}{3} \mu v_1 \eta_0^2 \left[\sigma_1 - \frac{2\sigma_C}{\sqrt{3}} \right] \sqrt{e_2 R} \frac{\sin\xi_0}{(1-\cos\xi_0)^2} \tag{2.273}$$

$$P_{\Sigma_1^+}^{\delta v_1} = \frac{2}{3} k v_1 e_1 \vartheta \eta_0^2 \left[1 - \frac{\sigma_1}{2k} \right] \sqrt{\frac{r}{1-r}} \propto \eta_0^2 \tag{2.274}$$

Considering also the shear power coming from the transversal perturbation of the velocity field defined by Eq. (2.259) at the entrance of the roll gap Σ_1^+, the following is obtained:

$$P_{\Sigma_1^+}^{v_\eta} = \int_{-\eta_0}^{\eta_0} \sigma_{\xi\eta}(\xi_0,\eta) v_{\eta}(\xi_0,\eta) \frac{a d\eta}{(\cosh\eta - \cos\xi_0)} \tag{2.275}$$

$$P_{\Sigma_1^+}^{v_\eta} = \mu v_1 \left(\frac{e_1}{e_2}\right) \mathrm{B}(\xi_0) \left[\sigma_1 - \frac{2\sigma_C}{\sqrt{3}}\right] \frac{\sqrt{e_2 R}}{(1-\cos\xi_0)} \int_{-\eta_0}^{\eta_0} \left[\eta^2 - \eta_0^2\right] \left(\frac{\eta^2}{\eta_0}\right) d\eta \tag{2.276}$$

$$P_{\Sigma_1^+}^{v_\eta} = \frac{4}{15} \vartheta k v_1 \frac{e_1}{(1-r)} \left[1 - \frac{\sigma_1}{2k}\right] \mathrm{B}(\xi_0) \eta_0^4 \cong 0 \tag{2.277}$$

In the same manner at the exit of the roll gap Σ_2^+, the shear power is:

$$P_{\Sigma_2^+}^{v_\eta} = \int_{-\eta_0}^{\eta_0} \sigma_{\xi\eta}(\pi,\eta) v_\eta(\pi,\eta) \frac{a d\eta}{(\cosh\eta + 1)} \tag{2.278}$$

$$P_{\Sigma_2^+}^{v_\eta} = \mu v_1 \left(\frac{e_1}{e_2}\right) \mathrm{B}(\pi) \left[\frac{2\sigma_C}{\sqrt{3}} - \sigma_2\right] \frac{1}{2} \sqrt{e_2 R} \int_{-\eta_0}^{\eta_0} \left[\eta^2 - \eta_0^2\right] \left(\frac{\eta^2}{\eta_0}\right) d\eta \tag{2.279}$$

$$P_{\Sigma_2^+}^{v_\eta} = -\frac{4}{15} \vartheta k v_1 e_1 \left[1 - \frac{\sigma_2}{2k}\right] \mathrm{B}(\pi) \eta_0^4 \cong 0 \tag{2.280}$$

All the shear power terms are disturbing compared with deformation and friction power.

2.9.3 Power Transmitted by Friction with the Rolls

This power delivered to the strip by both rolls is the only source of the cold rolling flow in the absence of tensions. Its expression as function of the shear structure is:

$$P^{fric} = 2 \int_{\pi}^{\xi_0} \sigma_{\xi\eta}(\xi,\eta_0) v_\xi(\xi,\eta_0) \frac{a d\xi}{(\cosh\eta_0 - \cos\xi)} \tag{2.281}$$

$$P^{fric} = -v_1 \left(\frac{e_1}{e_2}\right) \sqrt{e_2 R} \int_{\pi}^{\xi_0} \sigma_{\xi\eta}(\xi,\eta_0) d\xi \tag{2.282}$$

$$P^{fric} = -\left(\frac{v_1 e_1}{\eta_0}\right) \int_{\pi}^{\xi_0} \sigma_{\xi\eta}(\xi,\eta_0) d\xi = -\frac{\sigma_C}{\sqrt{3}} v_1 e_1 \int_{\pi}^{\xi_0} H'(\xi) d\xi \tag{2.283}$$

2.9.4 Power Dissipated due to Friction with the Rolls

Some of the power delivered by the rolls does not contribute to the structure of the plastic flow. Instead, it is dissipated into heat due to friction at the roll-strip interface. The expression for this power considering both rolls and in the cold flat rolling approximation is:

$$P^{heat} = -2 \int_{\pi}^{\xi_0} \sigma_{\xi\eta}(\xi,\eta_0) \left[v_\xi(\xi,\eta_0) - v_\xi(\xi_N,\eta_0)\right] \frac{a d\xi}{(\cosh\eta_0 - \cos\xi)} \tag{2.284}$$

$$P^{heat} = \frac{\sigma_C}{\sqrt{3}} v_1 e_1 \int_{\pi}^{\xi_0} H'(\xi) \left(\frac{\cos \xi_N - \cos \xi}{1 - \cos \xi} \right) d\xi \tag{2.285}$$

The angular velocity of the rolls according to Eq. (2.22) is:

$$v_{\xi}(\xi, \eta) = -\frac{(\cosh \eta - \cos \xi)(\cosh \eta_0 + 1)}{(\cosh \eta_0 - \cos \xi_N)(\cosh \eta + 1)} \omega R_0 \cong -\frac{(1 - \cos \xi)}{(1 - \cos \xi_N)} \omega R_0 \tag{2.286}$$

$$v_{\xi}(\pi, \eta) = -v_2 = -\frac{2\omega R_0}{(1 - \cos \xi_N)} \rightarrow \omega = \frac{v_1}{2R_0} \left(\frac{e_1}{e_2} \right) (1 - \cos \xi_N) \tag{2.287}$$

2.9.5 Total Power and Torque in the Drive

The total power delivered by the rolls is, therefore:

$$P^{rolls} = P^{fric} + P^{heat} = -\frac{\sigma_C}{\sqrt{3}} v_1 e_1 \int_{\pi}^{\xi_0} H'(\xi) \left(\frac{1 - \cos \xi_N}{1 - \cos \xi} \right) d\xi \tag{2.288}$$

In order to calculate the total power in the drive, the power to move the idle backup rolls along with the performance of the whole system has to be taken into account. To calculate the torque, only the power of Eq. (2.288) which corresponds to the specific torque of the work rolls is considered. The corresponding torque per roll and per unit width of the strip is:

$$T^{spec} = -\frac{\sigma_C}{2\sqrt{3}} \frac{v_1 e_1}{\omega} \int_{\pi}^{\xi_0} H'(\xi) \left(\frac{1 - \cos \xi_N}{1 - \cos \xi} \right) d\xi \tag{2.289}$$

Using Eq. (2.287) for the ω of the rolls, the specific torque (Figure 2.11) is obtained:

$$T^{spec} = -\frac{\sigma_C}{\sqrt{3}} e_2 R_0 \int_{\pi}^{\xi_0} \frac{H'(\xi)}{(1 - \cos \xi)} d\xi \equiv -R_0 \int_{\pi}^{\xi_0} \sigma_{\xi\eta}(\xi, \eta_0) \frac{\sqrt{e_2 R}}{(1 - \cos \xi)} d\xi \tag{2.290}$$

When the rolls are idle like in Steckel the specific torque is null, and the following important result to be used later is deduced:

$$\int_{\pi}^{\xi_0} H'(\xi) \left(\frac{1 - \cos \xi_N}{1 - \cos \xi} \right) d\xi = 0 \rightarrow \int_{\pi}^{\xi_0} H'(\xi) d\xi = \int_{\pi}^{\xi_0} H'(\xi) \left(\frac{\cos \xi_N - \cos \xi}{1 - \cos \xi} \right) d\xi \tag{2.291}$$

2.9.6 Contribution to the Torque due to Roll Flattening

Due to the offset between the centre of curvature of the flattened radius of the roll and its axis of rotation, there is a small contribution to the torque in the drive coming from the roll pressure profile. It is considered for the

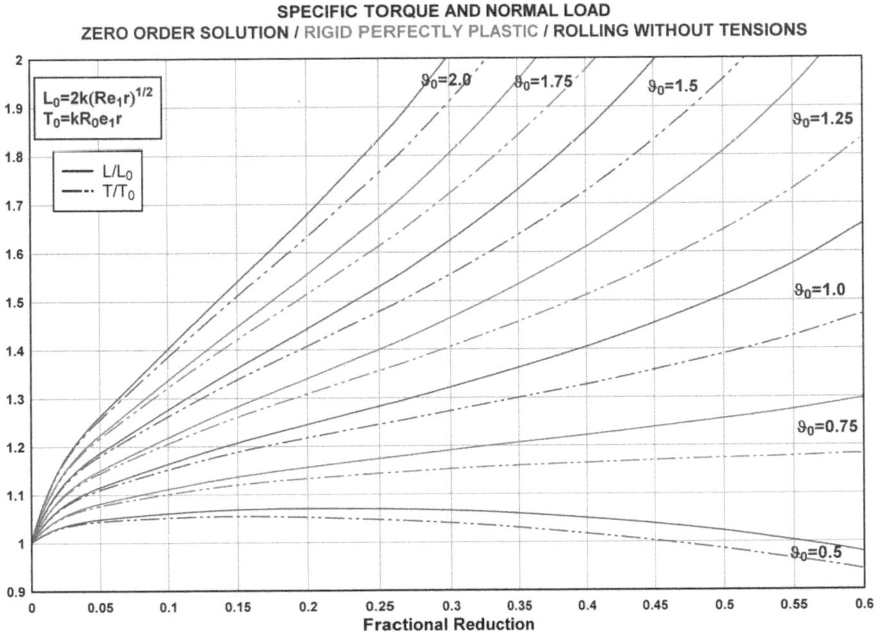

Figure 2.11 Specific torque and normal load.

calculations that the axis of the roll lies on the line between the middle point of the arc of contact with the strip and its centre of curvature, which lies on the vertical line through the exit of the roll gap or x-axis (Hill 1950). The lever arm with sign of the roll pressure profile at a rolling angle ϕ for a given ξ is:

$$\delta \cong (R - R_0)\left[\phi - \frac{1}{2}\phi_0\right] \cong (R - R_0)\left[\sin\phi - \frac{1}{2}\sin\phi_0\right] \tag{2.292}$$

$$\sin\phi = \sqrt{\frac{e_2}{R}}\frac{-\sin\xi}{(\cosh\eta_0 - \cos\xi)} \cong \sqrt{\frac{e_2}{R}}\frac{-\sin\xi}{(1-\cos\xi)} \tag{2.293}$$

The rolling angle corresponding to the entrance of the roll gap is:

$$\sin\phi_0 = \sqrt{\frac{e_2}{R}}\frac{-\sin\xi_0}{(1-\cos\xi_0)} = \eta_0\sqrt{\frac{r}{1-r}} \tag{2.294}$$

$$\delta(\xi) = \sqrt{\frac{e_2}{R}}(R - R_0)\left[\frac{1}{2}\frac{\sin\xi_0}{(1-\cos\xi_0)} - \frac{\sin\xi}{(1-\cos\xi)}\right] \tag{2.295}$$

$$T^{flatt} = -\sqrt{e_2 R}\int_{\pi}^{\xi_0}\sigma_{\eta\eta}(\xi,\eta_0)\,\delta(\xi)\frac{d\xi}{(1-\cos\xi)} \tag{2.296}$$

Therefore, the torque per roll and per unit width of the strip due to roll flattening is:

$$T^{flatt} = -e_2\left(R - R_0\right)\int_{\pi}^{\xi_0} \sigma_{\eta\eta}\left(\xi,\eta_0\right)\left[\frac{1}{2}\frac{\sin\xi_0}{\left(1-\cos\xi_0\right)} - \frac{\sin\xi}{\left(1-\cos\xi\right)}\right]\frac{d\xi}{\left(1-\cos\xi\right)} \tag{2.297}$$

2.9.7 Rolling Loads

The resultant loads on each roll, due to the pressure profile P and the shear stress distribution S, and projected on each Cartesian axis are as follows:

$$L_x^P = -\int_{\pi}^{\xi_0} \sigma_{\eta\eta}\left(\xi,\eta_0\right)\cos\phi\frac{\sqrt{e_2 R}}{\left(\cosh\eta_0 - \cos\xi\right)}d\xi \tag{2.298}$$

$$L_y^P = -\int_{\pi}^{\xi_0} \sigma_{\eta\eta}\left(\xi,\eta_0\right)\sin\phi\frac{\sqrt{e_2 R}}{\left(\cosh\eta_0 - \cos\xi\right)}d\xi = -\frac{1}{2}\left(\sigma_1 e_1 - \sigma_2 e_2\right) - L_y^S \tag{2.299}$$

$$L_x^S = -\int_{\pi}^{\xi_0} \sigma_{\xi\eta}\left(\xi,\eta_0\right)\sin\phi\frac{\sqrt{e_2 R}}{\left(\cosh\eta_0 - \cos\xi\right)}d\xi \approx \frac{1}{2}\phi_0\frac{T^{spec}}{R_0} \ll L_x^P \tag{2.300}$$

$$L_y^S = \int_{\pi}^{\xi_0} \sigma_{\xi\eta}\left(\xi,\eta_0\right)\cos\phi\frac{\sqrt{e_2 R}}{\left(\cosh\eta_0 - \cos\xi\right)}d\xi \cong -\frac{T^{spec}}{R_0} \tag{2.301}$$

where the rolling angle ϕ at a given ξ is defined as before:

$$\sin\phi = -\frac{y}{R} = \sqrt{\frac{e_2}{R}}\frac{-\sin\xi}{\left(\cosh\eta_0 - \cos\xi\right)} \cong \phi \tag{2.302}$$

In cold flat rolling $\cos\phi = 1$ with an error of less than 1%, except in cluster and laboratory mills where in general the cosine of the rolling angle must be taken into account using Eq. (2.302) for the sine, due to the fact that the angle of contact ϕ_0 could be much higher than 5°.

2.9.8 A Theorem on the Integral of the Shear Structure over the Roll Gap

Proposition: *In the cold flat rolling approximation and being friction perturbative for the plastic flow, shear power can be ignored, and the integral of the shear structure over the roll gap is independent of the rolling parameter ϑ, and depends only on the thickness reduction and tensions.*

Proof: Using the power balance defined by Eqs. (2.7) and (2.8), and the results for the powers given by Eqs. (2.269), (2.274) and (2.283), the following equation is obtained:

$$-\frac{\sigma_C}{\sqrt{3}} v_1 e_1 \int_{\pi}^{\xi_0} H'(\xi)\, d\xi + (\sigma_2 - \sigma_1) v_1 e_1 = -\frac{2\sigma_C}{\sqrt{3}} v_1 e_1 \ln(1-r) + \frac{2}{3} k v_1 e_1 \vartheta \eta_0^2 \left[1 - \frac{\sigma_1}{2k}\right] \sqrt{\frac{r}{1-r}} \tag{2.303}$$

where the power injected by front and back tensions has also been taken into account.

$$\int_{\pi}^{\xi_0} H'(\xi)\, d\xi = 2\ln(1-r) - \frac{2}{3} \vartheta \eta_0^2 \left[1 - \frac{\sigma_1}{2k}\right] \sqrt{\frac{r}{1-r}} + \left(\frac{\sigma_2 - \sigma_1}{k}\right) \tag{2.304}$$

$$\int_{\pi}^{\xi_0} H'(\xi)\, d\xi \cong 2\ln(1-r) + \left(\frac{\sigma_2 - \sigma_1}{k}\right) \tag{2.305}$$

This is a more convenient form of the power equation and the proposition is thus proved. The same result is obtained integrating over the roll gap Eq. (2.110) without the disturbing terms.

2.9.9 The Efficiency of the Cold Flat Rolling within the Roll Gap

The efficiency of the process ignoring the shear power term and considering only the forming process in the roll gap is:

$$Q(r, \vartheta) = \frac{P^D}{P^D + P^{heat}} = \frac{\ln(1-r)}{\ln(1-r) - \left(P^{heat}/2kv_1 e_1\right)} \tag{2.306}$$

The efficiency of the rolls in passing power by friction can be calculated in the same manner, using the result of Eq. (2.305):

$$R(r, \vartheta) = \frac{P^{fric}}{P^{rolls}} = \frac{P^{fric}}{P^{fric} + P^{heat}} = \frac{(\sigma_1 - \sigma_2) - 2k\ln(1-r)}{(\sigma_1 - \sigma_2) - 2k\ln(1-r) + \left(P^{heat}/v_1 e_1\right)} \tag{2.307}$$

In the absence of front and back tensions the rolls deliver all the power, and therefore, both efficiencies have the same expression. When front and back tensions are equal, and because the roll pressure is reduced, the power dissipated by friction decreases, and therefore, the efficiency thus defined increases for a given reduction and rolling parameter.

Although the power injected into the continuum as deformation is according to the right-hand side of Eq. (2.303) almost independent of the rolling parameter, the power dissipated into heat depends strongly on its value, and therefore, the cold flat rolling efficiency decreases steeply as this parameter increases. This is one of the main reasons to keep the friction coefficient as low as possible in cold rolling mill operation. Figure 2.12 is the

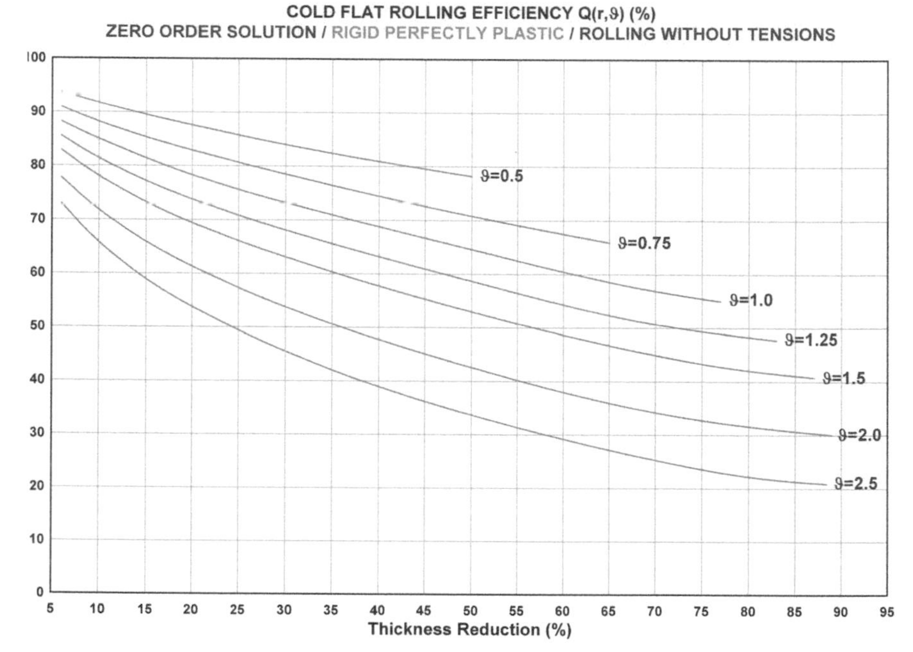

Figure 2.12 Cold flat rolling efficiency.

plot of the cold flat rolling efficiency versus thickness reduction using the zero order solution for the plastic stress field. Values in excess of 70% can be achieved in industrial mill operation, proving that the cold flat rolling is very efficient in passing power to the strip in a continuous and fast manner.

2.9.10 Influence of Front and Back Tensions

According to Eqs. (2.306) and (2.307), it may seem advantageous to roll with front and back tension, which is not the case from a perspective of global performance in the mill.

All that can be deduced from Eqs. (2.283) and (2.305) is that for equal amounts of front and back tensions, the power used in deformation remains unaltered while the specific torque in the drive diminishes, mainly due to a reduction in the heat dissipation. Front tension reduces the torque by displacing the neutral point towards the entrance of the roll gap, and therefore, decreasing the total amount of shear necessary to cold roll the strip. Quite the opposite occurs when back tension is applied, increasing in general the specific torque in the driving system. In both cases the roll pressure is reduced, and the shear stress due to friction is also reduced. Therefore, when power is applied to the roll gap by front tension, there is a direct saving in the drive but quite the opposite occurs when back tension is used. Consequently, all the possible savings in this case are due to a reduction in heat dissipation along the arc of contact with the rolls and in the driving system. The main advantage in rolling with tensions is a direct reduction in roll pressure, which means that the loads in the mill are also reduced. In general, it is better to roll with just the right amount of front tension to coil the strip.

2.9.11 Deformation Power—Bulk Term with Strain Hardening

The general expression for the power of deformation with strain hardening in the cold flat rolling approximation, and considering the shear structure as disturbing for the plastic flow is:

$$P^D = \int_{\Omega_2} \bar{\sigma}\bar{D}dV = \sqrt{\frac{2}{3}}\int_{\Omega_2} \sigma_C(\xi)\sqrt{D_{ij}D_{ij}}dV \equiv \int_{\Omega_2} \sigma_{ij}D_{ij}dV \tag{2.308}$$

$$P^D \cong \frac{2}{\sqrt{3}}v_1e_1\int_{\pi}^{\xi_0}\sigma_C(\xi)\frac{-\sin\xi}{(1-\cos\xi)}d\xi = \frac{2\sigma_C}{\sqrt{3}}v_1e_1\int_{\pi}^{\xi_0}\{1+f(\xi)\}\frac{-\sin\xi}{(1-\cos\xi)}d\xi \tag{2.309}$$

Splitting the rigid perfectly plastic term and integrating by parts, the following is obtained:

$$P^D = -\frac{2\sigma_C}{\sqrt{3}}v_1e_1\ln(1-r)+\frac{2\sigma_C}{\sqrt{3}}v_1e_1\int_{\pi}^{\xi_0}f(\xi)\frac{-\sin\xi}{(1-\cos\xi)}d\xi = P_0^D + P_1^D \tag{2.310}$$

$$P_1^D = -\frac{2\sigma_C}{\sqrt{3}}v_1e_1\left\{f(\xi_0)\ln(1-\cos\xi_0)-f(\pi)\ln 2-\int_{\pi}^{\xi_0}\ln(1-\cos\xi)f'(\xi)d\xi\right\} \tag{2.311}$$

The hardening function at the entrance of the roll gap is $f(\xi_0) = 0$, and therefore:

$$P_1^D = \frac{2\sigma_C}{\sqrt{3}}v_1e_1\left\{f(\pi)\ln 2+\int_{\pi}^{\xi_0}\ln(1-\cos\xi)f'(\xi)d\xi\right\} \tag{2.312}$$

$$P^D = -\frac{2\sigma_C}{\sqrt{3}}v_1e_1\left[\ln(1-r)-f(\pi)\ln 2-\int_{\pi}^{\xi_0}\ln(1-\cos\xi)f'(\xi)d\xi\right] \tag{2.313}$$

The power balance defined by Eq. (2.305) considering the strain hardening becomes:

$$\int_{\pi}^{\xi_0}F'(\xi)d\xi = 2\ln(1-r)-2f(\pi)\ln 2-2\int_{\pi}^{\xi_0}\ln(1-\cos\xi)f'(\xi)d\xi+\left(\frac{\sigma_2-\sigma_1}{k}\right) \tag{2.314}$$

The strain hardening can be incorporated in all the expressions for the powers and the torques obtained previously, replacing the shear structure

$H'(\xi)$ by the effective shear structure with strain hardening $F'(\xi) = H'(\xi)\{1 + f(\xi)\}$, and given by Eqs. (2.175) and (2.181).

2.9.12 Steckel Rolling or Drawing through Idling Rolls

In this special case, the torque induced by the strip passing through the idling rolls must be zero. Therefore, from Eq. (2.290) the following condition is obtained:

$$\int_{\pi}^{\xi_0} \frac{H'(\xi)}{(1-\cos\xi)} d\xi = 0 \rightarrow \int_{\pi}^{\xi_N} \frac{H'(\xi,\pi)}{(1-\cos\xi)} d\xi = -\int_{\xi_N}^{\xi_0} \frac{H'(\xi,\xi_0)}{(1-\cos\xi)} d\xi \tag{2.315}$$

and from Eq. (2.291) for the integral of the shear structure and the power balance given by Eq. (2.304), a direct expression for the relation between front and back tensions to draw the strip appears:

$$\left(\frac{\sigma_2-\sigma_1}{2k}\right) = -\ln(1-r) + \frac{1}{3}\vartheta\eta_0^2\left[1-\frac{\sigma_1}{2k}\right]\sqrt{\frac{r}{1-r}} + \frac{1}{2}\int_{\pi}^{\xi_0} H'(\xi)\left(\frac{\cos\xi_N - \cos\xi}{1-\cos\xi}\right) d\xi \tag{2.316}$$

The front tension in this case must account for the power dissipated by friction with the rolls. When this power is ignored, the following approximation applies:

$$\left(\frac{\sigma_2-\sigma_1}{2k}\right) \cong -\ln(1-r) + \frac{1}{3}\vartheta\eta_0^2\left[1-\frac{\sigma_1}{2k}\right]\sqrt{\frac{r}{1-r}} \tag{2.317}$$

Equation (2.316) has to be solved by iteration, neglecting the last term and giving a fix value to the back tension σ_1, a first estimation of the front tension σ_2 is obtained using Eq. (2.317), and therefore, the corresponding position of the neutral point with Eq. (2.142).

Now, the term of heat dissipation in Eq. (2.316) is estimated using Eqs. (2.140) and (2.141) for the corresponding branches of the shear structure, to obtain the new value for the front tension σ_2 and so on until convergence.

Usually, if the friction coefficient is small enough, Eq. (2.317) can be considered for engineering calculations. When the strain hardening cannot be ignored, the shear structure in the term of heat dissipation must be substituted by its effective counterpart $H'(\xi) \rightarrow F'(\xi)$, and instead of Eq. (2.317), the following expression applies:

$$\left(\frac{\sigma_2-\sigma_1}{2k}\right) \cong -\ln(1-r) + f(\pi)\ln 2 + \int_{\pi}^{\xi_0} \ln(1-\cos\xi) f'(\xi) d\xi + \frac{1}{3}\vartheta\eta_0^2\left[1-\frac{\sigma_1}{2k}\right]\sqrt{\frac{r}{1-r}} \tag{2.318}$$

A non-energetic approach to Steckel must rely on Eqs. (2.315), (2.140) and (2.141) for the corresponding branches of the shear structure, and on Eq. (2.142) for the position of the neutral point. The solution is again iterative. The calculation begins with the assumption that initially the neutral point is situated in the middle of the arc of contact with the rolls and then, to use Eq. (2.315) to estimate σ_2 for a fix value of σ_1, to finally recalculate the new position of the neutral point with Eq. (2.142) and so on until convergence.

2.10 Parametric Correction of the Rigid Perfectly Plastic Stress Field

So far, the first order plastic stress field has been obtained disturbing the zero order solution in the cold flat rolling approximation:

$$\sigma_{\xi\xi}^{(1)} = \sigma_{\xi\xi}^{(0)} - \frac{\sigma_C}{\sqrt{3}}\left[\Theta(\xi) - \Theta\left(\frac{\xi_0}{\pi}\right)\right]\left(\frac{\eta^2}{2}\right) \tag{2.319}$$

$$\sigma_{\eta\eta}^{(1)} = \sigma_{\eta\eta}^{(0)} - \frac{\sigma_C}{\sqrt{3}}\left[\Theta(\xi) - \Theta\left(\frac{\xi_0}{\pi}\right) - 2H'^2(\xi)\right]\left(\frac{\eta^2}{2}\right) \tag{2.320}$$

$$\Theta(\xi) = 2H'^2(\xi) + H''(\xi) - \left[I(\xi) - 2J(\xi) + 3K(\xi)\right] \tag{2.321}$$

where due to the disturbing procedure, $H'(\xi)$ in Eqs. (2.319), (2.320) and (2.321) corresponds to the shear structure to order zero. From the entrance of the roll gap to the neutral point is:

$$\begin{aligned} H'(\xi,\xi_0) = 2\vartheta\left\{\left[\frac{\sigma_1}{2k} - 1 - \ln(1-\cos\xi_0)\right]\exp\left[-\vartheta(\xi-\xi_0)\right] + \ln(1-\cos\xi)\right\} \\ -2\vartheta^2\exp[-\vartheta\xi]\int_{\xi_0}^{\xi}\exp[\vartheta\xi]\ln(1-\cos\xi)\,d\xi \end{aligned} \tag{2.322}$$

and from the neutral point to the exit of the roll gap its equation is:

$$\begin{aligned} H'(\xi,\pi) = -2\vartheta\left\{\left[\left(\frac{\sigma_2}{2k}\right) - 1 - \ln 2\right]\exp\left[\vartheta(\xi-\pi)\right] + \ln(1-\cos\xi)\right\} \\ -2\vartheta^2\exp[\vartheta\xi]\int_{\pi}^{\xi}\exp[-\vartheta\xi]\ln(1-\cos\xi)\,d\xi \end{aligned} \tag{2.323}$$

This function is discontinuous at the neutral surface, and therefore, the corresponding solution for the stress field to first order remains ill defined.

A possible way out is to consider the function in Eq. (2.321) without the term of derivative of the shear structure. From the entrance of the roll gap to the neutral point, the following roll pressure distribution is obtained:

$$\sigma_{\eta\eta}^{(1)} = \sigma_{\eta\eta}^{(0)} + \frac{\sigma_C}{\sqrt{3}} H'^2(\xi_0)\eta_0^2 + \frac{\sigma_C}{2\sqrt{3}}\left[I(\xi,\xi_0) - 2J(\xi,\xi_0) + 3K(\xi,\xi_0)\right]\eta_0^2 \tag{2.324}$$

and from the neutral point to the exit of the roll gap its equation is:

$$\sigma_{\eta\eta}^{(1)} = \sigma_{\eta\eta}^{(0)} + \frac{\sigma_C}{\sqrt{3}} H'^2(\pi)\eta_0^2 + \frac{\sigma_C}{2\sqrt{3}}\left[I(\xi,\pi) - 2J(\xi,\pi) + 3K(\xi,\pi)\right]\eta_0^2 \tag{2.325}$$

where the new position $\xi_N^{(1)}$ of the neutral point is obtained making Eqs. (2.324) and (2.325) equal. The field thus obtained has a "peak" of pressure (Fig. 2.6). It is evident, that in order to obtain a smooth pressure profile and with the right amount of fall in pressure around the neutral point without regularizing the slipping friction condition with the rolls (see Appendix), a parametric correction needs to be performed. In order to simplify the calculations, the following discontinuous pseudo[5] shear structure to first order corresponding to the pressure profile defined by Eqs. (2.324) and (2.325) is introduced. The differential equation of equilibrium of order zero is used, to define this "shear structure" as a transform of the first order pressure:

$$H'^{(1)}(\xi) = 2\left[\left(\frac{\sin\xi}{1-\cos\xi}\right) - \frac{d}{d\xi}\left(\frac{\sigma_{\eta\eta}^{(1)}}{2k}\right)\right] \qquad \forall\xi\in[\pi,\xi_0] \tag{2.326}$$

To smooth the shear structure, the following mono-parametric filter function is used as long as Eq. (2.305) for the power is fulfilled, where ξ_N is the corresponding neutral point:

$$\Gamma(\xi,\tau) = 1 - \exp\left[-\tau\left|\xi - \xi_N\right|\right] \qquad \tau > 0 \qquad \forall\xi\in[\pi,\xi_0] \tag{2.327}$$

The smoothened shear structure and its derivative at the neutral point are:

$$H_S'(\xi) = \Gamma(\xi,\tau)H'(\xi) \tag{2.328}$$

$$H_S''(\xi) = \Gamma'(\xi,\tau)H'(\xi) + \Gamma(\xi,\tau)H''(\xi) \tag{2.329}$$

[5] The function defined by Eq. (2.326) as a transform of the roll pressure profile to first order cannot be considered a proper shear structure, because it has no direct relation with the shear stress. To calculate the shear structure to first order, the friction condition with the rolls using the pressure profile to first order has to be applied.

$$H''_S(\xi_N,\xi_0)=\Gamma'(\xi_N,\tau)H'(\xi_N,\xi_0)=\tau H'(\xi_N,\xi_0) \tag{2.330}$$

$$H''_S(\xi_N,\pi)=\Gamma'(\xi_N,\tau)H'(\xi_N,\pi)=-\tau H'(\xi_N,\pi) \tag{2.331}$$

Therefore, the derivative at the neutral point is continuous because $H'(\xi_N, \pi) = -H'(\xi_N, \xi_0)$. To determine the parameter τ, Eq. (2.164) for the local fall in pressure is used:

$$\sigma_{\eta\eta}^{(1,3)}(\xi_N)=-\frac{\sigma_C}{2\sqrt{3}}H''(\xi_N)\eta_0^2=\frac{\sigma_C}{2\sqrt{3}}\tau H'(\xi_N,\pi)\eta_0^2 \tag{2.332}$$

The equation of equilibrium and the roll pressure profile using the smoothened shear structure are:

$$\frac{d\sigma_{\eta\eta}}{d\xi}=\frac{2\sigma_C}{\sqrt{3}}\left(\frac{\sin\xi}{1-\cos\xi}\right)-\frac{\sigma_C}{\sqrt{3}}H'_S(\xi) \tag{2.333}$$

$$\sigma_{\eta\eta}(\xi)=\sigma_{\eta\eta}^{(0)}(\pi)+\frac{2\sigma_C}{\sqrt{3}}\ln\left(\frac{1-\cos\xi}{2}\right)-\frac{\sigma_C}{\sqrt{3}}\int_\pi^\xi H'_S(\xi)d\xi \tag{2.334}$$

and therefore, the equation to determine the parameter τ is:

$$\sigma_{\eta\eta}^{(0)}(\xi_N)+\frac{\sigma_C}{2\sqrt{3}}\tau H'(\xi_N,\pi)\eta_0^2=\sigma_{\eta\eta}^{(0)}(\pi)+\frac{2\sigma_C}{\sqrt{3}}\ln\left(\frac{1-\cos\xi_N}{2}\right)-\frac{\sigma_C}{\sqrt{3}}\int_\pi^{\xi_N} H'_S(\xi)d\xi \tag{2.335}$$

where H'_s according to Eq. (2.328) depends on τ. Once the local fall in pressure given by Eq. (2.332) has been obtained, the following equation is used to calculate the parameter $\tau_\pi^{(1)}$ for the exit branch of the pseudo shear structure, translating to the new position of the neutral point $\xi_N^{(1)}$ the fall in pressure given by Eq. (2.332):

$$\sigma_{\eta\eta}^{(1)}(\xi_N^{(1)})+\sigma_{\eta\eta}^{(1,3)}(\xi_N)=\sigma_{\eta\eta}^{(1)}(\pi)+\frac{2\sigma_C}{\sqrt{3}}\ln\left(\frac{1-\cos\xi_N^{(1)}}{2}\right)-\frac{\sigma_C}{\sqrt{3}}\int_\pi^{\xi_N^{(1)}} H_S'^{(1)}(\xi)d\xi \tag{2.336}$$

Integrating over the roll gap the pseudo shear structure defined by Eq. (2.326) the following condition is obtained:

$$\int_\pi^{\xi_0} H'^{(1)}(\xi)d\xi=2\ln(1-r)+\left(\frac{\sigma_2-\sigma_1}{k}\right)+\left[H'^2(\pi)-H'^2(\xi_0)\right]\eta_0^2 \tag{2.337}$$

Therefore, the parameter $\tau_{\xi_0}^{(1)}$ corresponding to the entry branch of the pseudo shear structure must be calculated so as to fulfill Eq. (2.337). This condition guarantees that the correct value of the pressure at the entrance of the roll gap will be obtained using the corrected roll pressure profile, that is:

$$\sigma_{\eta\eta}(\xi_0) = \sigma_1 - \frac{2\sigma_C}{\sqrt{3}} + \frac{\sigma_C}{\sqrt{3}} H'^2(\xi_0)\eta_0^2 \tag{2.338}$$

The shear stress is calculated using the smoothened shear structure as follows:

$$\sigma_{\xi\eta}(\xi,\eta) = \frac{\sigma_C}{\sqrt{3}} H'_S(\xi)\eta \tag{2.339}$$

Finally, applying the inverse transform of equation (2.326), the corrected roll pressure profile to first order is obtained:

$$\sigma_{\eta\eta}(\xi) = \sigma_{\eta\eta}^{(0)}(\pi) + \frac{\sigma_C}{\sqrt{3}} H'^2(\pi)\eta_0^2 + \frac{2\sigma_C}{\sqrt{3}} \ln\left(\frac{1-\cos\xi}{2}\right) - \frac{\sigma_C}{\sqrt{3}} \int_{\pi}^{\xi} H_S'^{(1)}(\xi)d\xi \tag{2.340}$$

Therefore, although the pseudo shear structure in Eq. (2.340) is a function of class C^0 due to the discontinuity of its derivative at the first order neutral point, the corrected roll pressure profile depending on the integral of this function is of class C^1.

Figure 2.13 is a comparison between the zero order and the corrected first order solution for the roll pressure profile in rolling with and without tensions. The correction has been performed using the procedure described in this section. Two values for the un-deformed radius of the rolls have been considered and the corresponding flattening calculated with Hitchcock's equation. Both cases are within the cold flat rolling approximation, and the disturbing procedure can be considered as valid for the case with tensions and within the limit for the case without tensions. The most important feature of this plot is the fact, that the fall in pressure increases when the radius of the rolls decreases for a given final thickness.

Figure 2.14 gives the shear structures corresponding to the roll pressure distributions of Figure 2.13 for the case with tensions. Because of Eq. (2.305) for the power must be fulfilled, the area with sign under the corresponding shear structure must remain unaltered in the correction. Table 2.1 is the record of values of the different parameters used.

Based on this correction and in order to obtain negative pressure at the neutral surface, the conditions of rolling must be close to the zero order solution. Deriving the equation of equilibrium to order zero (Eq. 2.110), the following relation between the derivative of the shear structure and the curvature of the corresponding roll pressure profile is obtained:

$$H''(\xi) = -2\left[\frac{1}{(1-\cos\xi)} + \frac{d^2}{d\xi^2}\left(\frac{\sigma_{\eta\eta}}{2k}\right)\right] \tag{2.341}$$

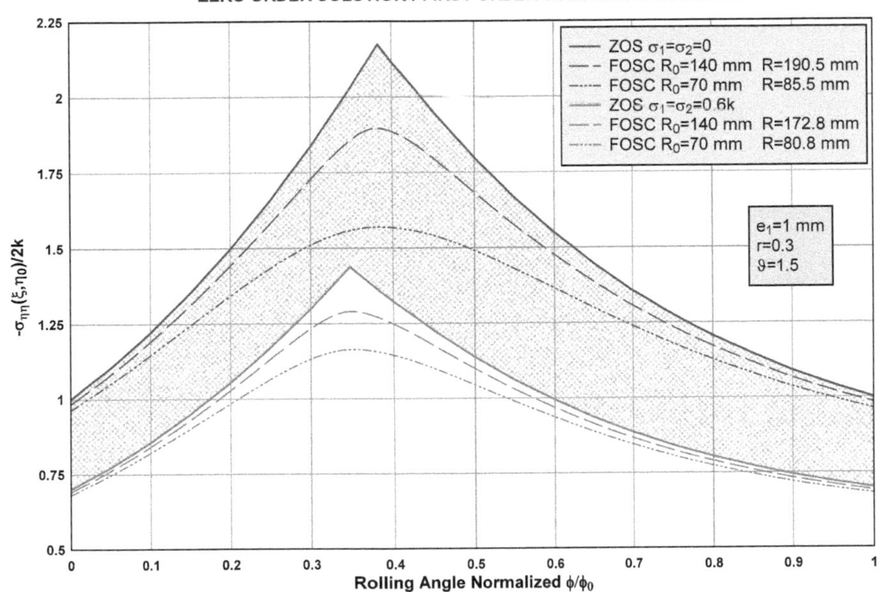

Figure 2.13 Corrected roll pressure distribution.

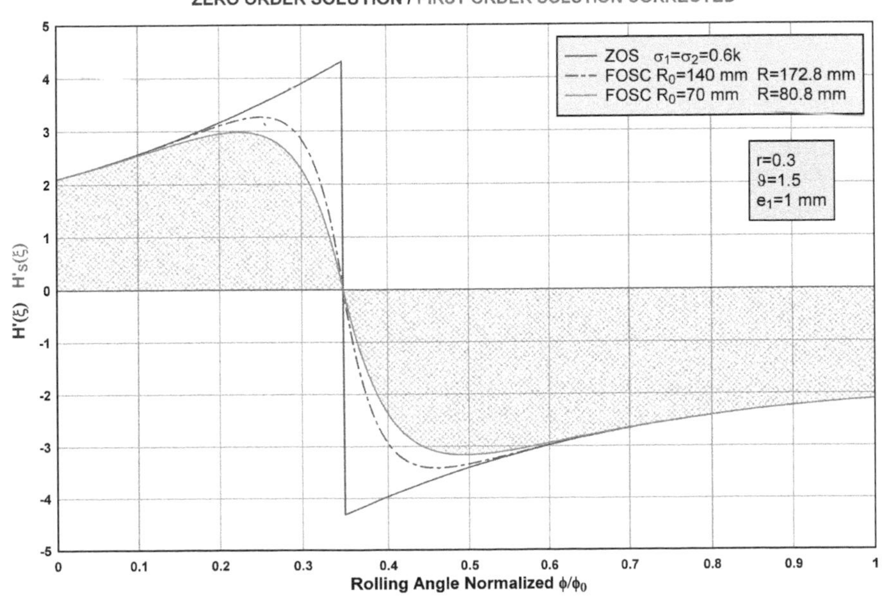

Figure 2.14 Corrected shear structure with tensions.

Table 2.1 Record of values used in the correction.

$\sigma_1=\sigma_2$	$R_0(mm)$	τ	$\tau_{\pi}^{(1)}$	$\tau_{\xi 0}^{(1)}$	I_{Integ}	ε (%)	ξ_N	$\xi_N^{(1)}$	$\sigma_{\eta\eta}^{(1,3)}/2k$
0	70	15.00	8.03	8.35	−0.7064	0.97	3.6341	3.6144	0.1997
0	140	22.28	24.90	25.35	−0.7101	0.45	3.6341	3.6280	0.1362
0.6	70	14.38	10.85	11.71	−0.7070	0.88	3.5908	3.5821	0.1343
0.6	140	21.37	21.25	22.12	−0.7106	0.37	3.5908	3.5871	0.0933

The value of the integrated shear structure over the roll gap using Eq. (2.305) is $I_c = -0.7133$ for all the cases.
I_{Integ} is the integral over the roll gap of H_s' and ε is the relative error compared with I_c.

Therefore, the smaller the radius of curvature of the roll pressure distribution at a given point, the larger and more negative the corresponding derivative of the shear structure. To approach the zero order solution in Figure 2.13, and in order to obtain the corresponding fall in pressure at the neutral surface, the following limit is calculated:

$$\lim_{\substack{\tau\to\infty \\ \eta_0\to 0}} \sigma_{\eta\eta}^{(1,3)}(\xi_N) = -\frac{\sigma_C}{2\sqrt{3}} H_S''(\xi_N)\eta_0^2 \cong \frac{\sigma_C}{\sqrt{3}} \frac{d^2}{d\xi^2}\left(\frac{\sigma_{\eta\eta}}{2k}\right)_{\xi=\xi_N} \qquad \eta_0^2 = \infty \cdot 0 \to +\infty \tag{2.342}$$

Therefore, the fall in pressure tends to zero as the peak of the zero order solution is approached, until eventually the indeterminacy in Eq. (2.342) tends to increase with no limit, and negative pressure and a non-plastic transition begin to appear. Due to the huge and localized tension, the corresponding de-cohesion in the material of the strip when passing through the neutral surface is established at this stage. In real practice to determine the right combinations of the rolling parameters for this meta-stable state to manifest in a mill, is a different matter.

2.11 Hitchcock's Equation

In order to calculate the elastic distortion of the rolls, the equation developed by Hitchcock (1935) is used. He replaced the actual roll pressure profile by an elliptical distribution giving the same total load. The change in shape due to the shear stress is ignored. Using the theory of the elastic contact due to Hertz, the resultant deformed surface is a cylinder of larger radius. This is a great and indeed useful mathematical simplification. More details can be found in Hill (1950). The result for the flattened radius of the rolls is:

$$\frac{1}{R_0} - \frac{1}{R} = \frac{P}{cL^2} \qquad c = \frac{\pi E}{16\left(1-\nu^2\right)} \tag{2.343}$$

The length of the arc of contact with one roll in the cold flat rolling approximation is:

$$L = \int_{\pi}^{\xi_0} \frac{\sqrt{e_2 R}}{(\cosh \eta_0 - \cos \xi)} d\xi \cong \sqrt{e_2 R} \int_{\pi}^{\xi_0} \frac{d\xi}{(1 - \cos \xi)} = -\sqrt{e_2 R} \cot(\xi_0/2) \quad (2.344)$$

$$L^2 = e_2 R \cot^2(\xi_0/2) = e_2 R \left(\frac{1 + \cos \xi_0}{1 - \cos \xi_0} \right) = e_2 R \left(\frac{r}{1 - r} \right) = R(e_1 - e_2) \quad (2.345)$$

and the total load along the arc of contact with the roll is:

$$P = -\int_{\pi}^{\xi_0} \sigma_{\eta\eta}(\xi, \eta_0) \frac{\sqrt{e_2 R}}{(\cosh \eta_0 - \cos \xi)} d\xi \quad (2.346)$$

Using Eq. (2.345) in Eq. (2.343), the usual form of Hitchcock's equation is obtained:

$$R = R_0 \left[1 + \frac{P}{c(e_1 - e_2)} \right] \quad (2.347)$$

Because the roll pressure distribution in Eq. (2.346) depends on the flattened radius of the rolls through the rolling parameter, Eq. (2.347) has to be solved by iteration.

In order to calculate the roll flattening, a yield stress of 400 MPa has been considered, unless another specific value is given. The rolls are of tool steel, and therefore, the elastic modulus is E = 210,000 MPa and the ratio of Poisson $v = 0.29$.

2.12 Conclusions

A new theory of cold flat rolling without considerations of spread is presented in the previous sections. Unlike in the classical theories and the theory by Orowan, the approach is kinematic and within the framework of the mechanics of the continuum media.

The choice of the bipolar cylindrical coordinates based on obvious symmetry considerations is a key factor of the theory. In this coordinate system the Levy-Mises flow rule, the Von Mises criterion, and the equations of internal equilibrium, lead to a successful coupling of the structure of the plastic flow and the stress field.

The shear structure is identified in the new theory as the function that controls the whole process of cold flat rolling.

The theory by Von Karman corresponds to the zero order solution.

The coupling of the structure of the plastic flow and the stress field is necessary in explaining some important features of the roll pressure distributions: the fall in pressure from the ends of the roll gap to the neutral point, the local fall around the neutral surface, and the corresponding rounding of the peak of the roll pressure profile. The new theory also explains the following reported phenomena: the existence of regions of negative pressure along the middle plane of the strip, the roll pressure profiles with more than a single maximum, and the separation between the neutral point and the maximum of pressure (see Appendix).

Because the shear structure is null at the neutral surface and the rest of the shear stresses are null under plane strain, the stress state predicted by the theory under certain rolling conditions tends to be hydrostatic, and therefore, non-plastic along this surface. Although this fact may be in contradiction to the assumption of the material to be in plastic flow in the roll gap, the theory is able to predict this singular state without any problem. In real practice, the amount of lateral spread, and therefore, the goodness of the plain strain hypothesis, is a limiting factor on the level of hydrostaticity that can be achieved by the stress state along the neutral surface.

Along the middle transversal plane of the strip and due to symmetry, the lateral spread is null and the flow bi-dimensional. Therefore, the closer the conditions of rolling to the zero order solution, the more hydrostatic is the stress state at the intersection of this plane and the neutral surface, and the corresponding pressure according to the theory must be more negative.

Based on these facts, two regimes of cold flat rolling exist: a stable regime with a smooth flow structure along the full length of the roll gap, and a meta-stable regime with a non-plastic transition and negative pressure at the neutral surface. Due to the lateral spread both regimes can coexist at the same time and in the same strip under the right rolling conditions. The meta-stable regime around the middle plane of the strip where the curvature at the maximum of the zero order roll pressure distribution and the hydrostaticity of the stress state are highest, to give way towards the edges of the strip to the stable regime as the lateral spread increases, and therefore, the curvature at the maximum of the roll pressure profile to order zero and the hydrostaticity decrease.

The classical theories fail in explaining these phenomena because they are, in general, not a good representative of the solution of the boundary value problem for the stress field in the roll gap. They ignore the dependence of the coordinate stresses with the transversal dimension, and therefore, the falls in pressure are filtered out from the outset. The theory by Orowan considers the stresses of a similar but different problem, and not much is gained in this sense. Therefore, although the cold flat rolling in the rigid perfectly plastic case is intrinsically isostatic under plane strain and with slipping friction, it has to be solved taking into account the flow structure, in order to obtain an analytical solution in perturbation theory that accounts for the right level of pressure and the phenomena behind the corresponding boundary value problem.

Boundary value problems for the stress field in the roll gap with boundary conditions of friction with the rolls containing the velocity field that make the problem statically undetermined due to the tribology, can also be solved with this kinematic approach (see Appendix).

The structure of the plastic flow in cold flat rolling can be decomposed into two independent motions: the main velocity field which is one-dimensional and is responsible for the net flow of matter, and a disturbing motion confined within the roll gap, and whose net effect is a vortex field. This vortex field is a characteristic motion of cold flat rolling.

From an energetic perspective the cold flat rolling is very efficient in passing power by friction to the strip, and although the pressure profile is highly dependent on the friction coefficient, the latter is just disturbing for the plastic flow.

In the cold flat rolling approximation and being friction perturbative for the plastic flow, shear power can be ignored, and the integral of the shear structure over the roll gap is independent of the rolling parameter, and depends only on the thickness reduction and on the tensions. This integral result is a special form of the law of conservation of the energy and must be fulfilled, when any correction or regularization of the plastic stress field is performed (see Appendix).

The new theory does not apply either to thin foil nor to temper rolling (skin pass). For the present case of AHSS, however it should be helpful for a more accurate calculation of the rolling loads and torque upon a correct characterization along the roll gap of the tribology (friction with the rolls) and the rheology (flow stress). It is clear, that a computer solution of the general boundary value problem is compelling in order to achieve this goal. The outcome of all this will be highly beneficial for the manufacturers of the corresponding rolling mills, that due to high strength of these steels, have to be more capable and the control of the rolling parameters more accurate, in order to obtain the final high quality products with required shape, profile and flatness.

Finally, quoting Egon Orowan from his 1943 paper on flat rolling:

> "In cases of disagreement between theory and experiment it is not clear which of the numerous simplifying assumptions and approximations are responsible; on the other hand, agreement can be the result of several errors cancelling each other. In fact, Siebel's theory with its crude mathematical simplifications was often found to agree with roll pressure measurements much better than

the theory put forward by Karman who, with the same physical assumptions, had used far better mathematical approximations".

The new theory should be at least a partial answer to the above quotation.

2.13 Further Development

Formulation of the general boundary value problem for the velocity and stress fields under plane strain considering the strain hardening, in order to obtain numerical solutions for the problem of the "roll gap" in full, using bipolar cylindrical coordinates. Extension with an adiabatic thermal field and its coupling with the dynamic flow stress.

Analytical and numerical investigation with alternative friction models. Development of new hardening models considering the thermal field in an adiabatic roll gap, with focus on high speed industrial rolling. Further investigation of the inertial range (see Appendix).

Keywords: Plastic, Cold, Roll, Self-consistent, Neutral, Non-plastic transition, Negative, Shear, Hydrostatic, Natural, Vortex, Adiabatic, High, Steckel, Inertial

Key of Abbreviations

R	Radius of the rolls (flattened).
R_0	Radius of the rolls (un-deformed).
e_1	Thickness of the sheet before rolling.
e_2	Thickness of the sheet after rolling.
v_1	Velocity of the sheet before rolling.
v_2	Velocity of the sheet after rolling.
ω	Angular velocity of the rolls.
Σ_1^+	Entrance of the roll gap $\xi = \xi_0$.
Σ_2^+	Exit of the roll gap $\xi = \pi$.
Σ_3^+	Arcs of contact between the rolls and the strip.
Ω_1	Metallic sheet before rolling.
Ω_2	Forming region or roll gap.
Ω_3	Metallic sheet after rolling.
$\partial\Omega_2$	Contour of the roll gap.
(ξ, η, z)	Bipolar cylindrical coordinates.
(x, y, z)	Cartesian coordinates.

ξ_N	Bipolar coordinate of the neutral surface.[6]
$\pm\eta_0$	Bipolar coordinates of the arcs of contact with the rolls.
$\pm x_0$	x-coordinates of the centres of curvature of the arcs of contact with the rolls.
dS_ξ	Surface element along a ξ-isosurface.[7]
dS_η	Surface element along a η-isosurface.[7]
$2a$	Distance between focal points in bipolar cylindrical coordinates.
σ_1	Back tension.
σ_2	Front tension.
σ_Y	Yield stress at the reference conditions in the quasi-static regime.
$\sigma_C = k\sqrt{3}$	Yield stress.
k	Yield stress in shear.
μ	Friction coefficient.
ϕ	Rolling angle of the classical theories.
ϕ_0	Rolling angle of the entrance of the roll gap or angle of contact.
p	Roll pressure distribution of the classical theories.
h	Height of a plane vertical section of the strip within the roll gap.
g_{ij}	Metric tensor.
v_i	Plastic velocity field.
t_i	Tension or projection of the stress tensor in a given direction $t_i = \sigma_{ij} n_j$.
$\sigma_{ij} \equiv \tilde{\sigma}$	Stress tensor.
$\bar{\sigma}$	Equivalent plastic stress.
D_{ij}	Symmetric component of the velocity gradient tensor or "strain rate".
$\bar{D}$	Equivalent plastic strain rate.
ε_{ij}	Plastic strain tensor.
$\bar{\varepsilon}$	Equivalent plastic strain.

[6] In bipolar cylindrical coordinates, the "neutral plane" is in fact a cylinder of a huge radius, and therefore, it will be referred to in the present chapter as the "neutral surface" instead. Properly speaking, it is a plane in the limit.

[7] The extensive magnitudes in this chapter are per unit of width of the strip: volume and surface elements, energies, powers, loads and torques. Therefore, volume and surface integrals are referred to a unit of width.

$\tilde{D}$	Average value of the equivalent plastic strain rate over the roll gap.
$\tilde{\varepsilon}$	Average value of the equivalent plastic strain over the roll gap.
A	Transversal area of the roll gap/Strain hardening coefficient.
$\dot{\lambda}$	Hardening parameter of the Levy-Mises flow rule.
$H'(\xi)$	Shear structure.
$\Psi(\xi)$	Second shear structure $\Psi(\xi) = H'(\xi)\sin\xi$.
$\Psi^{(0)}(\xi)$	Second shear structure of the undisturbed velocity field.
$C(\dot{\varepsilon}, \dot{\varepsilon}_t, T)$	Temperature-dependent strain rate sensitivity function.
ρ	Density of the metallic strip.
r	Thickness reduction.
ϑ	Rolling parameter.
$\vartheta_0 = \vartheta\sqrt{1-r}$	Second rolling parameter.
α	Angle between the input velocity v_1 and a streamline at Σ_1^+.
$\sigma_{ij}^{(0)}$	Zero order plastic stress field.
$\sigma_{ij}^{(1)}$	First order plastic stress field.
μ_{LB}	Minimum friction coefficient for the cold flat rolling to be feasible.
ε_{VM}^{max}	Maximum error in fulfilling the Von Mises criterion to order zero.
$\tilde{\varepsilon}_{VM}$	Average error in fulfilling the Von Mises criterion to order zero.
$f(\xi)$	Hardening function.
$F'(\xi)$	Effective shear structure considering the strain hardening.
$u_T = c_V T$	Thermal energy per unit of mass.
u_D	Energy per unit of mass due to micro-structural phenomena.
$\bar{\psi}$	Heat flux.
$\dot{q}$	Internal sources of heat per unit of mass.
c_V	Specific heat at constant volume.
t_{RG}	Time for a material particle to cross the roll gap.
$T(\xi)$	Adiabatic thermal field in the roll gap.
T_0	Temperature of the strip at the entrance of the roll gap.
T_M	Melting temperature of the strip.
T_R	Reference temperature below which the thermal softening cancels.

v_0	Threshold velocity of rolling for the roll gap to become adiabatic.
$Z_\beta(T)$	Thermal softening function.
$\Phi(\overline{\varepsilon})$	Strain hardening function.
$h(v_1/v_0)$	Function that controls the transition isothermal-adiabatic in the roll gap.
ω_z	Vortex field in the roll gap due to the friction drag with the rolls.
P^D	Bulk power of deformation.
$P_{\Sigma_1^+}$	Shear power at the entrance of the roll gap.
$P_{\Sigma_2^+}$	Shear power at the exit of the roll gap.
P^{fric}	Power used in deformation due to friction with the rolls.
P^{heat}	Power dissipated into heat due to friction with the rolls.
P^{rolls}	Total power delivered by the rolls.
T^{spec}	Specific torque per roll. Specific means per unit of width of the strip.
T^{flatt}	Specific torque per roll due to roll flattening.
δ	Lever arm of the pressure profile with respect to the axis of the roll.
L_x^P	x-component of the load on one roll due to the pressure profile.
L_y^P	y-component of the load on one roll due to the pressure profile.
L_x^S	x-component of the load on one roll due to the shear stress distribution.
L_y^S	y-component of the load on one roll due to the shear stress distribution.
$Q(r, \vartheta)$	Efficiency of the cold flat rolling process considering only the roll gap.
$R(r, \vartheta)$	Efficiency of the rolls in passing power by friction to the strip.
$\Gamma(\xi, \tau)$	Filter function to smooth the shear structure.
τ	Parameter of the filter function.
L	Length of the arc of contact with one roll.
E	Elastic modulus of the rolls.
ν	Poisson's ratio of the rolls.
$P \cong L_x^P$	Load along the arc of contact with the roll.
$T(r, \vartheta_0)$	Integral to calculate the fall in specific torque due to the fall in pressure.

$L(r, \vartheta_0)$	Integral to calculate the fall in normal load due to the fall in pressure.
$I^{flatt}(r, \vartheta_U)$	Integral to calculate the specific torque due to roll flattening.
E_K	Kinetic energy of the strip in the roll gap.

References

Al-Salehi, F.A.R., T.C. Firbank and P.R. Lancaster. 1973. An experimental determination of the roll pressure distributions in cold rolling. Int. J. Mech. Sci. 15: 693–710.

Bland, D.R. and H. Ford. 1948. The calculation of roll force and torque in cold strip rolling with tensions. Proc. Inst. Mech. Eng. 159(1): 144–163.

Ch'ien, W.Z. and C.T. Ch'en. 1952. Theory of rolling. Acta Sci. Sin. 1(2): 193–229.

Echave, G. 2011a. A new analytical model for cold rolling of sheet and strip. Int. J. Appl. Math. Mech. 7(10): 1–50.

Echave, G. 2011b. Plastic flow structure in cold flat rolling. Int. J. Appl. Math. Mech. 7(10): 51–70.

Echave, G. 2011c. Stress correction in cold flat rolling. Int. J. Appl. Math. Mech. 7(10): 71–96.

Hill, R. 1950. The mathematical theory of plasticity. Oxford University Press. First Edition.

Hitchcock, J.H. 1935. Roll neck bearings, App. I, ASME Research Publication.

Johnson, R. and W. Cook. 1983. A constitutive model and data for metals subjected to large strain, high strain rates, and high temperature. Proc. 7th Int. Symp. Ballistics, The Hague, the Netherlands 541–547.

Le, H.R. and M.P.F. Sutcliffe. 2006. A multi-scale model for friction in cold rolling of aluminium alloy. Tribol. Lett. 22(1): 95–104.

Nadai, A. 1939. The forces required for rolling steel strip under tension. J. Appl. Mech. ASME Transactions 6: 54–62.

Orowan, E. 1943. The calculation of roll pressure in hot and cold flat rolling. Proc. Inst. Mech. Eng. 150(4): 140–167.

Prandt, L. 1923. Anwendungsbeispiele zu einem Henckyschen satz über das plastische gleichgewicht (Examples of the application of Hencky's theory to the equilibrium of plastic solids), Z. ang. Math u. Mech. (ZAMM) 3: 401–406.

Siebel, E. 1925. Kraft und materialflub bei der bildsamen formanderung (Forces and material flow in plastic deformation). Stahl und Eisen. 45: 1563–1566.

Siebel, E. and W. Lueg. 1933. Untersuchungen über die spannungsverteilung im walzspalt (Studies on stress distribution in the roll gap). Mitt. K. W. Inst. Eisen. 15: 1–14.

Tabary, P.E., M.P.F. Sutcliffe, F. Porral and P. Deneuville. 1996. Measurements of friction in cold metal rolling. ASME J. Tribol. 118(3): 629–636.

Tselikov, A.I. 1939. Effect of external friction and tension on the pressure of the metal on the rolls in rolling. Metallurgy (USSR) 6: 61–76.

Von Karman, Th. 1925. Beitrag zur theorie des walzvorgangs (Contribution to the theory of rolling). Z. ang. Math u. Mech. (ZAMM) 5: 139–141.

Yamamoto, H., T. Uchimura and K. Yamada. 2002. Numerical simulation of friction coefficient and surface roughness in cold rolling of steel sheets. Trans. Jap. Soc. Mech. Eng. Series C 68(670): 1877–1882.

APPENDIX

Specific Torque to Roll the Strip and the Loads on the Rolling Mill

In order to perform the detailed calculations of the different magnitudes directly involved in cold rolling mill design, the following quantities corresponding to the torque and the load of a uniform pressure profile of value $2k$ are introduced:

$$T_0 = \frac{\sigma_C}{\sqrt{3}} R_0 e_1 r = R_0 \int_{\pi}^{\xi_0} \frac{2\sigma_C}{\sqrt{3}} \sin\phi \frac{\sqrt{e_2 R}}{(1-\cos\xi)} d\xi \tag{A.1}$$

$$L_0 = \frac{2\sigma_C}{\sqrt{3}} \sqrt{R e_1 r} = \int_{\pi}^{\xi_0} \frac{2\sigma_C}{\sqrt{3}} \frac{\sqrt{e_2 R}}{(1-\cos\xi)} d\xi \tag{A.2}$$

The non-dimensional expressions for the specific torque given in Eq. (2.290), and the disturbing torque due to roll flattening defined in Eq. (2.297) considering strain hardening are:

$$\frac{T^{spec}}{T_0} = -\left(\frac{1-r}{r}\right) \int_{\pi}^{\xi_0} \frac{F'(\xi)}{(1-\cos\xi)} d\xi \tag{A.3}$$

$$\frac{T^{flatt}}{T_0} = -2\left(\frac{1-r}{r}\right)\left(\frac{R}{R_0}-1\right) \int_{\pi}^{\xi_0} \frac{\sigma_{\eta\eta}}{2k}(\xi,\eta_0) \left[\frac{1}{2}\frac{\sin\xi_0}{(1-\cos\xi_0)} - \frac{\sin\xi}{(1-\cos\xi)}\right] \frac{d\xi}{(1-\cos\xi)} \tag{A.4}$$

where according to Hitchcock's equation (Eq. 2.347), the roll flattening is obtained:

$$\frac{R}{R_0} - 1 = \frac{P}{c(e_1 - e_2)} = -\frac{2k}{c}\left(\frac{1-r}{r}\right)\frac{1}{\eta_0} \int_{\pi}^{\xi_0} \frac{\sigma_{\eta\eta}}{2k}(\xi,\eta_0) \frac{d\xi}{(1-\cos\xi)} \tag{A.5}$$

The torque due to roll flattening is negative, and therefore, favours rolling. Its magnitude is in the range of a few percent of the specific torque.

In order to gain accuracy and stability in the calculation of the specific torque using Eq. (A.3), due to the high sensibility to the errors and to the position of the neutral point of the numerical quadrature involving the difference of two quantities of the same order of magnitude, the following change is made:

$$\int_{\pi}^{\xi_0} \frac{F'(\xi)}{(1-\cos\xi)} d\xi = \frac{1}{(1-\cos\xi_N)} \left\{ \int_{\pi}^{\xi_0} F'(\xi) d\xi - \int_{\pi}^{\xi_0} F'(\xi) \left(\frac{\cos\xi_N - \cos\xi}{1-\cos\xi} \right) d\xi \right\} \tag{A.6}$$

The integral of the effective shear structure over the roll gap is calculated using Eq. (2.314):

$$\int_{\pi}^{\xi_0} F''(\xi)d\xi = 2\ln(1-r) - 2f(\pi)\ln 2 - 2\int_{\pi}^{\xi_0}\ln(1-\cos\xi)f'(\xi)d\xi + \left(\frac{\sigma_2-\sigma_1}{k}\right) \tag{A.7}$$

Using Eq. (A.7) in Eq. (A.6) the sensibility to errors is largely reduced, because the integrand of the last term in Eq. (A.6) is null at the neutral point, does not change sign in all the interval, and its value is much smaller than the integrand of Eq. (A.3). The numerical procedure, thus, has been conditioned.

The non-dimensional expression for the load in the direction of the Cartesian x-axis defined by Eq. (2.298), and due to the roll pressure profile is:

$$\frac{L_x^P}{L_0} \cong -\sqrt{\frac{1-r}{r}}\int_{\pi}^{\xi_0}\frac{\sigma_{\eta\eta}}{2k}(\xi,\eta_0)\frac{d\xi}{(1-\cos\xi)} \qquad \cos\phi \cong 1 \tag{A.8}$$

Once the specific torque defined by Eq. (A.3) is known, the rest of the loads acting on the mill (section 2.9.7) can be obtained:

$$L_y^S = -\frac{T_0}{R_0}\left(\frac{T^{spec}}{T_0}\right) = -kre_1\left(\frac{T^{spec}}{T_0}\right) \tag{A.9}$$

$$L_y^P = \frac{ke_1}{2}\left[\frac{\sigma_2}{k}(1-r) - \frac{\sigma_1}{k}\right] - L_y^S \tag{A.10}$$

$$L_x^S \le \phi_0\frac{T_0}{R_0}\left(\frac{T^{spec}}{T_0}\right) = kre_1\sqrt{\frac{re_1}{R}}\left(\frac{T^{spec}}{T_0}\right) = kre_1\eta_0\sqrt{\frac{r}{1-r}}\left(\frac{T^{spec}}{T_0}\right) = -\eta_0\sqrt{\frac{r}{1-r}}L_y^S \tag{A.11}$$

When rolling without tensions, the following upper bound for this last expression applies:

$$L_x^S(\sigma_1=\sigma_2=0) \le kre_1\sqrt{\frac{re_1}{R}}\left(\frac{L_x^P}{L_0}\right) = \frac{re_1}{2R}L_x^P = \frac{1}{2}\left(\frac{r}{1-r}\right)\eta_0^2 L_x^P \le 0.005\left(\frac{r}{1-r}\right)L_x^P \tag{A.12}$$

In order to compare with the plots of specific torque and normal load in the literature, the calculations are performed in the less advantageous case for the normal load of rolling without tensions. Instead of the rolling parameter ϑ, the parameter ϑ_0 is used, defined as function of the former using the thickness reduction r as follows:

$$\vartheta_0 = \mu\sqrt{R/e_1} = \vartheta\sqrt{e_2/e_1} = \vartheta\sqrt{1-r} \to \vartheta = \frac{\vartheta_0}{\sqrt{1-r}} \tag{A.13}$$

Alternative Calculation of the Specific Torque

Using Eq. (2.171) and Eq. (2.172) of internal equilibrium to order zero with strain hardening in Eq. (A.3), an alternative equation for the specific torque as function of the corresponding roll pressure profile is obtained:

$$\frac{T^{spec}}{T_0}=-\left(\frac{1-r}{r}\right)\int_{\pi}^{\xi_0}\frac{F'(\xi)}{(1-\cos\xi)}d\xi \tag{A.14}$$

$$\frac{d}{d\xi}\frac{\sigma_{\xi\xi}}{2k}=\frac{d}{d\xi}\frac{\sigma_{\eta\eta}}{2k}+f'(\xi)=\left(\frac{\sin\xi}{1-\cos\xi}\right)\{1+f(\xi)\}-\frac{1}{2}F'(\xi) \tag{A.15}$$

$$\frac{T^{spec}}{T_0}=2\left(\frac{1-r}{r}\right)\int_{\pi}^{\xi_0}\frac{1}{(1-\cos\xi)}\left[\frac{d}{d\xi}\frac{\sigma_{\eta\eta}}{2k}+f'(\xi)-\left(\frac{\sin\xi}{1-\cos\xi}\right)\{1+f(\xi)\}\right]d\xi \tag{A.16}$$

$$\frac{T^{spec}}{T_0}=2\left(\frac{1-r}{r}\right)\int_{\pi}^{\xi_0}\left\{\frac{1}{(1-\cos\xi)}\frac{d}{d\xi}\frac{\sigma_{\eta\eta}}{2k}+\frac{d}{d\xi}\left[\frac{1+f(\xi)}{1-\cos\xi}\right]\right\}d\xi \tag{A.17}$$

$$\frac{T^{spec}}{T_0}=2\left(\frac{1-r}{r}\right)\int_{\pi}^{\xi_0}\frac{1}{(1-\cos\xi)}d\frac{\sigma_{\eta\eta}}{2k}+\frac{1}{r}-\left(\frac{1-r}{r}\right)\{1+f(\pi)\} \tag{A.18}$$

The following integral is now performed by parts:

$$\int_{\pi}^{\xi_0}\frac{\sigma_{\eta\eta}}{2k}\frac{\sin\xi}{(1-\cos\xi)^2}d\xi=-\frac{1}{(1-\cos\xi)}\frac{\sigma_{\eta\eta}}{2k}\Bigg]_{\pi}^{\xi_0}+\int_{\pi}^{\xi_0}\frac{1}{(1-\cos\xi)}d\frac{\sigma_{\eta\eta}}{2k} \tag{A.19}$$

$$\frac{1}{(1-\cos\quad)}\frac{\quad_{\eta\eta}}{2k}\Bigg]\quad=\frac{1}{4(1-r)}\left[\frac{\sigma_1}{k}-2\right]-\frac{1}{4}\left[\frac{\sigma_2}{k}-2\{1+\quad(\quad)\}\right] \tag{A.20}$$

and using Eqs. (A.19) and (A.20) in Eq. (A.18), the specific torque is obtained:

$$\frac{T^{spec}}{T_0}=2\left(\frac{1-r}{r}\right)\int_{\pi}^{\xi_0}\frac{\sigma_{\eta\eta}}{2k}\frac{\sin\xi}{(1-\cos\xi)^2}d\xi+\frac{1}{2r}\left[\frac{\sigma_1}{k}-\frac{\sigma_2}{k}(1-r)\right] \tag{A.21}$$

Using again the fact that in cold flat rolling $\cos\phi = 1$ with an error of less than 1%, Eq. (2.301) gives the following condition for the specific torque:

$$T^{spec}=-R_0\int_{\pi}^{\xi_0}\sigma_{\xi\eta}(\xi,\eta_0)\frac{\sqrt{e_2R}}{(1-\cos\xi)}d\xi=-R_0L_y^S \tag{A.22}$$

and from the equilibrium of forces acting on the boundaries of the strip in the roll gap along the Cartesian y-axis defined by Eq. (2.299), it is also possible to obtain Eq. (A.21):

$$T^{spec} = -R_0 \sqrt{e_2 R} \int_{\pi}^{\xi_0} \sigma_{\eta\eta} \frac{\sin \phi}{(1-\cos \xi)} d\xi + \frac{1}{2} R_0 \left(\sigma_1 e_1 - \sigma_2 e_2 \right) \tag{A.23}$$

$$\frac{T^{spec}}{T_0} = 2\left(\frac{1-r}{r}\right) \int_{\pi}^{\xi_0} \frac{\sigma_{\eta\eta}}{2k} \frac{\sin \xi}{(1-\cos \xi)^2} d\xi + \frac{1}{2r}\left[\frac{\sigma_1}{k} - \frac{\sigma_2}{k}(1-r)\right] \tag{A.24}$$

and in this way, the equivalence of Eqs. (A.24) and (A.14) is proved.

The values for the specific torque in the rigid perfectly plastic case are given in Table A.1, and for the normal load (Cartesian x-axis) in Table A.2. Figure 2.11 is a plot of these two tables.

Table A.1 Specific Torque T^{spec}/T_0.

Reduction	$\vartheta_0 = 0.5$	$\vartheta_0 = 0.75$	$\vartheta_0 = 1.0$	$\vartheta_0 = 1.25$	$\vartheta_0 = 1.50$	$\vartheta_0 = 1.75$	$\vartheta_0 = 2.0$
0	1	1	1	1	1	1	1
0.025	1.0336	1.0553	1.0773	1.0999	1.1231	1.1469	1.1714
0.05	1.0435	1.0757	1.1087	1.1428	1.1782	1.2150	1.2532
0.10	1.0525	1.1019	1.1530	1.2066	1.2632	1.3231	1.3864
0.15	1.0546	1.1197	1.1875	1.2597	1.3368	1.4197	1.5090
0.20	1.0527	1.1330	1.2176	1.3086	1.4074	1.5151	1.6328
0.25	1.0476	1.1435	1.2454	1.3565	1.4788	1.6142	1.7646
0.30	1.0400	1.1519	1.2723	1.4052	1.5539	1.7211	1.9100
0.35	1.0299	1.1588	1.2991	1.4565	1.6353	1.8400	2.0754
0.40	1.0175	1.1645	1.3269	1.5121	1.7264	1.9763	2.2695
0.45	1.0028	1.1693	1.3565	1.5741	1.8310	2.1371	2.5040
0.50	0.9855	1.1735	1.3893	1.6455	1.9552	2.3330	2.7971
0.55	0.9655	1.1775	1.4268	1.7305	2.1074	2.5804	3.1781
0.60	0.9422	1.1818	1.4717	1.8356	2.3019	2.9066	3.6972

Note that $\sigma_1 = \sigma_2 = 0$ when using the values of Tables A.1 to A.5.

Specific Torque and Normal Load considering the Roll Pressure Profile to First Order

Because the disturbing procedure fixes the shear structure to order zero, and therefore, the corresponding roll pressure profile must be used in Eq. (A.24), the specific torque is already fixed at this order. Nevertheless, a lower bound for the torque can be obtained using the pressure profile to first order given by Eqs. (2.324) and (2.325) in Eq. (A.24), ignoring the effect of the local fall in pressure around the neutral point as disturbing.

$$\frac{\sigma_{\eta\eta}^{(1)}}{2k}(\xi,\xi_0) = \frac{\sigma_{\eta\eta}^{(0)}}{2k}(\xi,\xi_0) + \frac{1}{2} H'^2(\xi_0)\eta_0^2 + \frac{1}{4}\Lambda(\xi,\xi_0)\eta_0^2 \tag{A.25}$$

Table A.2 Normal Load L_x^P/L_0.

Reduction	$\vartheta_0 = 0.5$	$\vartheta_0 = 0.75$	$\vartheta_0 = 1.0$	$\vartheta_0 = 1.25$	$\vartheta_0 = 1.50$	$\vartheta_0 = 1.75$	$\vartheta_0 = 2.0$
0	1	1	1	1	1	1	1
0.025	1.0359	1.0576	1.0798	1.1025	1.1258	1.1497	1.1742
0.05	1.0480	1.0806	1.1139	1.1483	1.1841	1.2212	1.2598
0.10	1.0615	1.1123	1.1646	1.2193	1.2770	1.3380	1.4026
0.15	1.0682	1.1363	1.2065	1.2810	1.3606	1.4461	1.5381
0.20	1.0706	1.1564	1.2452	1.3403	1.4434	1.5559	1.6788
0.25	1.0698	1.1742	1.2829	1.4006	1.5300	1.6734	1.8325
0.30	1.0660	1.1907	1.3212	1.4642	1.6239	1.8037	2.0066
0.35	1.0595	1.2065	1.3614	1.5337	1.7289	1.9527	2.2099
0.40	1.0502	1.2221	1.4049	1.6113	1.8498	2.1282	2.4546
0.45	1.0380	1.2382	1.4535	1.7011	1.9929	2.3411	2.7586
0.50	1.0224	1.2553	1.5094	1.8078	2.1679	2.6081	3.1495
0.55	1.0029	1.2745	1.5759	1.9391	2.3894	2.9557	3.6734
0.60	0.9787	1.2970	1.6583	2.1070	2.6821	3.4300	4.4114

$$\frac{\sigma_{\eta\eta}^{(1)}}{2k}(\xi,\pi)=\frac{\sigma_{\eta\eta}^{(0)}}{2k}(\xi,\pi)+\frac{1}{2}H'^2(\pi)\eta_0^2+\frac{1}{4}\Lambda(\xi,\pi)\eta_0^2 \tag{A.26}$$

Using as intermediate point the zero order instead of the first order neutral point $\xi_N \cong \xi_N^{(1)}$ in order to perform the integration in Eq. (A.24), the expression for the torque is:

$$\frac{T^{spec}}{T_0}\cong\left(\frac{T^{spec}}{T_0}\right)^{(0)} \tag{A.27}$$

$$+\eta_0^2\left(\frac{1-r}{r}\right)\left\{H'^2(\pi)\int_{\pi}^{\xi_N}\frac{\sin\xi}{(1-\cos\xi)^2}d\xi+H'^2(\xi_0)\int_{\xi_N}^{\xi_0}\frac{\sin\xi}{(1-\cos\xi)^2}d\xi\right\}$$

$$+\eta_0^2\left(\frac{1-r}{r}\right)\frac{1}{2}\left\{\int_{\pi}^{\xi_N}\Lambda(\xi,\pi)\frac{\sin\xi}{(1-\cos\xi)^2}d\xi+\int_{\xi_N}^{\xi_0}\Lambda(\xi,\xi_0)\frac{\sin\xi}{(1-\cos\xi)^2}d\xi\right\}$$

Subtracting Eq. (A.26) from Eq. (A.25) at the zero order neutral point the following is obtained:

$$\Lambda(\xi_N,\xi_0)-\Lambda(\xi_N,\pi)=2H'^2(\pi)-2H'^2(\xi_0)+\frac{4}{\eta_0^2}\left\{\frac{\sigma_{\eta\eta}^{(1)}}{2k}(\xi_N,\xi_0)-\frac{\sigma_{\eta\eta}^{(1)}}{2k}(\xi_N,\pi)\right\} \tag{A.28}$$

Integrating by parts the last term in Eq. (A.27), using Eq. (A.28) and the fact that the following differentials are equal $d\Lambda(\xi, \pi) = d\Lambda(\xi, \xi_0)$, the specific torque takes the following form:

$$\frac{T^{spec}}{T_0} = \left(\frac{T^{spec}}{T_0}\right)^{(0)}$$
$$+\frac{1}{2}\eta_0^2\left(\frac{1-r}{r}\right)\left\{H'^2(\pi) - H'^2(\xi_0)\left[\frac{1}{1-r}\right]\right\}$$
$$+2\left(\frac{1-r}{r}\right)\left\{\frac{\sigma_{\eta\eta}^{(1)}}{2k}(\xi_N,\xi_0) - \frac{\sigma_{\eta\eta}^{(1)}}{2k}(\xi_N,\pi)\right\}\left[\frac{1}{1-\cos\xi_N}\right]$$
$$+\frac{1}{2}\eta_0^2\left(\frac{1-r}{r}\right)\int_{\pi}^{\xi_0}\frac{1}{(1-\cos\xi)}\left\{H'(\xi)\left(\frac{3+\cos\xi}{1-\cos\xi}\right) - 2H'^2(\xi)\left(\frac{\sin\xi}{1-\cos\xi}\right)+\right.$$
$$\left.+3H'^3(\xi)\right\}d\xi \qquad \text{(A.29)}$$

Ignoring the term involving the first order pressure which happens to be positive while the rest are negative, the lower bound for the torque is obtained, where the torque to order zero can be considered as the upper bound. In order to perform the calculations, the following integral is defined and the fall in torque is considered apart:

Table A.3 Integral for the Specific Torque due to Roll Flattening $I^{flatt}(r, \vartheta_0)$.

Reduction	$\vartheta_0 = 0.5$	$\vartheta_0 = 0.75$	$\vartheta_0 = 1.0$	$\vartheta_0 = 1.25$	$\vartheta_0 = 1.50$	$\vartheta_0 = 1.75$	$\vartheta_0 = 2.0$
0	0	0	0	0	0	0	0
0.025	0.0022	0.0023	0.0024	0.0025	0.0027	0.0028	0.0029
0.05	0.0045	0.0049	0.0052	0.0056	0.0059	0.0062	0.0066
0.10	0.0091	0.0105	0.0117	0.0127	0.0139	0.0150	0.0163
0.15	0.0137	0.0167	0.0191	0.0215	0.0239	0.0265	0.0293
0.20	0.0181	0.0235	0.0277	0.0319	0.0363	0.0410	0.0462
0.25	0.0224	0.0309	0.0377	0.0444	0.0515	0.0594	0.0683
0.30	0.0263	0.0391	0.0492	0.0593	0.0704	0.0829	0.0970
0.35	0.0300	0.0481	0.0626	0.0775	0.0940	0.1130	0.1350
0.40	0.0331	0.0581	0.0784	0.0997	0.1239	0.1522	0.1858
0.45	0.0356	0.0694	0.0974	0.1275	0.1625	0.2044	0.2553
0.50	0.0374	0.0824	0.1207	0.1629	0.2134	0.2756	0.3531
0.55	0.0380	0.0976	0.1498	0.2093	0.2828	0.3761	0.4961
0.60	0.0371	0.1158	0.1874	0.2723	0.3811	0.5244	0.7153

$$T(r,\vartheta_0)=\int_{\pi}^{\xi_0}\frac{-1}{(1-\cos\xi)}\left\{H'(\xi)\left(\frac{3+\cos\xi}{1-\cos\xi}\right)-2H'^2(\xi)\left(\frac{\sin\xi}{1-\cos\xi}\right)+3H'^3(\xi)\right\}d\xi \tag{A.30}$$

$$\Delta\left(\frac{T^{spec}}{T_0}\right)=\frac{1}{2}\eta_0^2\left(\frac{1-r}{r}\right)\left\{\frac{\vartheta_0^2}{(1-r)^2}\left[2-\frac{\sigma_1}{k}\right]^2-\frac{\vartheta_0^2}{(1-r)}\left[2-\frac{\sigma_2}{k}\right]^2+T(r,\vartheta_0)\right\} \tag{A.31}$$

$$\left(\frac{T^{spec}}{T_0}\right)^{(0)}-\Delta\left(\frac{T^{spec}}{T_0}\right)<\frac{T^{spec}}{T_0}<\left(\frac{T^{spec}}{T_0}\right)^{(0)} \tag{A.32}$$

It is found that a good estimator for the torque to first order is in the middle of the interval:

$$\frac{T^{spec}}{T_0}\approx\left(\frac{T^{spec}}{T_0}\right)^{(0)}-\frac{1}{2}\Delta\left(\frac{T^{spec}}{T_0}\right) \tag{A.33}$$

The normal load defined by Eq. (A.8) can be calculated in the same manner. The final result is:

$$\begin{aligned}\frac{L_x^P}{L_0}=&\left(\frac{L_x^P}{L_0}\right)^{(0)}-\frac{1}{2}\eta_0^2H'^2(\xi_0)\\&-\sqrt{\frac{1-r}{r}}\left\{\frac{\sigma_{\eta\eta}^{(1)}}{2k}(\xi_N,\xi_0)-\frac{\sigma_{\eta\eta}^{(1)}}{2k}(\xi_N,\pi)\right\}\cot\left(\frac{\xi_N}{2}\right)\\&-\frac{1}{4}\eta_0^2\sqrt{\frac{1-r}{r}}\int_{\pi}^{\xi_0}\cot\left(\frac{\xi}{2}\right)\left\{H'(\xi)\left(\frac{3+\cos\xi}{1-\cos\xi}\right)-2H'^2(\xi)\left(\frac{\sin\xi}{1-\cos\xi}\right)+3H'^3(\xi)\right\}d\xi\end{aligned} \tag{A.34}$$

$$L(r,\vartheta_0)=\int_{\pi}^{\xi_0}\cot\left(\frac{\xi}{2}\right)\left\{H'(\xi)\left(\frac{3+\cos\xi}{1-\cos\xi}\right)-2H'^2(\xi)\left(\frac{\sin\xi}{1-\cos\xi}\right)+3H'^3(\xi)\right\}d\xi \tag{A.35}$$

$$\Delta\left(\frac{L_x^P}{L_0}\right)=\frac{1}{2}\eta_0^2\frac{\vartheta_0^2}{(1-r)}\left[2-\frac{\sigma_1}{k}\right]^2+\frac{1}{4}\eta_0^2\sqrt{\frac{1-r}{r}}L(r,\vartheta_0) \tag{A.36}$$

$$\left(\frac{L_x^P}{L_0}\right)^{(0)}-\Delta\left(\frac{L_x^P}{L_0}\right)<\frac{L_x^P}{L_0}<\left(\frac{L_x^P}{L_0}\right)^{(0)} \tag{A.37}$$

For the load the best estimator is found to be precisely the lower bound, because the terms of pressure to first order in Eq. (A.34) are smaller than in the case of the torque, and the influence of the rounding of the peak of pressure on the load cannot be completely ignored. Tables A.4 and A.5 are the integrals defined by Eqs. (A.30) and (A.35) for different values of

Table A.4 Integral for the Fall in Specific Torque $T(r, \vartheta_0)$.

Reduction	$\vartheta_0 = 0.5$	$\vartheta_0 = 0.75$	$\vartheta_0 = 1.0$	$\vartheta_0 = 1.25$	$\vartheta_0 = 1.50$	$\vartheta_0 = 1.75$	$\vartheta_0 = 2.0$
0	0	0	0	0	0	0	0
0.025	0.0869	0.1662	0.3018	0.4807	0.7350	1.0296	1.3837
0.05	0.1898	0.3856	0.6852	1.1304	1.7809	2.6754	3.7818
0.10	0.4328	0.9254	1.7549	3.1029	4.8662	7.6200	11.7339
0.15	0.7349	1.6366	3.2932	5.9462	10.3053	16.4437	26.7262
0.20	1.0989	2.6218	5.2678	10.0601	17.8035	29.5189	-
0.25	1.5450	3.7821	8.2473	16.1200	30.4117	52.5145	-
0.30	2.0664	5.3090	12.2057	24.7848	49.1214	-	-
0.35	2.7124	7.3198	17.6608	38.4409	80.1201	-	-
0.40	3.5152	9.9762	25.8487	60.7259	132.8764	-	-
0.45	4.4892	13.7313	37.3701	93.1108	-	-	-
0.50	5.7062	18.8912	55.2883	145.3975	-	-	-
0.55	7.2881	26.0857	84.4361	-	-	-	-
0.60	9.3187	37.1018	134.3284	-	-	-	-

The positions without number in Tables A.4 and A.5 are beyond the disturbing treatment of the plastic stress field.

Table A.5 Integral for the Fall in Normal Load $L(r, \vartheta_0)$.

Reduction	$\vartheta_0 = 0.5$	$\vartheta_0 = 0.75$	$\vartheta_0 = 1.0$	$\vartheta_0 = 1.25$	$\vartheta_0 = 1.50$	$\vartheta_0 = 1.75$	$\vartheta_0 = 2.0$
0	0	0	0	0	0	0	0
0.025	0.0662	0.1891	0.4322	0.8443	1.4892	2.4456	3.7514
0.05	0.1498	0.4330	0.9970	1.9831	3.5696	5.9708	9.4353
0.10	0.3600	1.0666	2.5332	5.2206	9.6601	16.7839	27.7477
0.15	0.6300	1.9212	4.7214	10.0107	19.3636	34.7725	59.9812
0.20	0.9657	3.0693	7.7075	16.9793	33.9469	63.3715	-
0.25	1.3805	4.5348	11.9365	27.1595	56.7958	110.4864	-
0.30	1.8834	6.4623	17.7589	42.0371	92.0312	-	-
0.35	2.5029	9.0114	25.9234	64.7008	148.9516	-	-
0.40	3.2677	12.4141	37.8617	100.4211	244.1866	-	-
0.45	4.2080	17.1217	55.2730	156.0000	-	-	-
0.50	5.3800	23.6747	82.2182	248.3689	-	-	-
0.55	6.8675	33.0299	126.0424	-	-	-	-
0.60	8.7625	47.2633	201.8743	-	-	-	-

the reduction and the rolling parameter, where the disturbing procedure leading to Eqs. (A.25) and (A.26) remains valid. Using these estimators, it is possible to obtain a more accurate value for the specific torque and the normal load, without the need to perform any calculation to first order.

When the strain hardening cannot be ignored, the disturbing terms coming from the second order Taylor expansion of Eqs. (2.205) and (2.206) have to be considered.

Parametric Expression for the Roll Flattening and General Calculation Procedure

Using Hitchcock's equation and Eq. (A.8) for the normal load, the roll flattening is obtained:

$$\frac{R}{R_0}-1=\frac{P}{c\left(e_1-e_2\right)}=-\frac{2k}{c}\left(\frac{1-r}{r}\right)\frac{1}{\eta_0}\int_{\pi}^{\xi_0}\frac{\sigma_{\eta\eta}}{2k}\left(\xi,\eta_0\right)\frac{d\xi}{\left(1-\cos\xi\right)} \tag{A.38}$$

$$\frac{R}{R_0}-1=\frac{2k}{c}\sqrt{\frac{1-r}{r}}\frac{1}{\eta_0}\left(\frac{L_x^P}{L_0}\right) \qquad \cos\phi\cong 1 \tag{A.39}$$

Equation (A.39) can be solved for the flattened radius of the rolls as function of the normal load making the following change: $x=\sqrt{R/R_0}$.

$$x^2-1=\frac{2k}{c}\sqrt{\frac{1-r}{r}}\sqrt{\frac{R_0}{e_2}}\left(\frac{L_x^P}{L_0}\right)x \qquad \rightarrow \qquad x^2-\Omega x-1=0 \tag{A.40}$$

$$\Omega=\frac{2k}{c}\frac{1}{\sqrt{r}}\left(\frac{L_x^P}{L_0}\right)\sqrt{\frac{R_0}{e_1}} \tag{A.41}$$

where Ω depends on the flattened radius of the rolls through the rolling parameter ϑ_0, which is necessary to calculate the normal load. Finding x in Eq. (A.40) the following is obtained:

$$x=\left(\frac{\Omega}{2}\right)+\sqrt{1+\left(\frac{\Omega}{2}\right)^2}\cong 1+\left(\frac{\Omega}{2}\right)+\frac{1}{2}\left(\frac{\Omega}{2}\right)^2=\frac{1}{2}\left\{1+\left(1+\frac{\Omega}{2}\right)^2\right\} \tag{A.42}$$

$$\frac{R}{R_0}-1=x^2-1=\Omega x=\Omega\left\{\left(\frac{\Omega}{2}\right)+\sqrt{1+\left(\frac{\Omega}{2}\right)^2}\right\}\cong\frac{\Omega}{2}\left\{1+\left(1+\frac{\Omega}{2}\right)^2\right\} \tag{A.43}$$

where the relative error in the approximation of Eq. (A.42) is less than 1% if $k < 0.01c$ for all the thickness reductions and rolling parameters of Table A.2. To calculate the estimators in Eqs. (A.31) and (A.36), it is necessary to obtain the bipolar coordinate of the flattened rolls:

$$\eta_0^2 = \frac{e_2}{R} = (1-r)\frac{e_1}{R_0}\frac{1}{x^2} = 4(1-r)\frac{e_1}{R_0}\left\{1+\left(1+\frac{\Omega}{2}\right)^2\right\}^{-2} \quad (A.44)$$

Therefore, knowing the thickness reduction r and the rolling parameter ϑ_0 the normal load is obtained using Table A.2, and the roll flattening and the coordinates of the arcs of contact with the rolls are calculated using Eqs. (A.43) and (A.44). Now in Eq. (A.4), the roll flattening is split up to obtain:

$$I^{flatt}(r,\vartheta_0) = 2\left(\frac{1-r}{r}\right)\int_{\pi}^{\xi_0}\frac{\sigma_{\eta\eta}}{2k}(\xi,\eta_0)\left[\frac{1}{2}\frac{\sin\xi_0}{(1-\cos\xi_0)} - \frac{\sin\xi}{(1-\cos\xi)}\right]\frac{d\xi}{(1-\cos\xi)} \quad (A.45)$$

$$\frac{T^{flatt}}{T_0} = -\left(\frac{R}{R_0}-1\right)I^{flatt}(r,\vartheta_0) = -\frac{\Omega}{2}\left\{1+\left(1+\frac{\Omega}{2}\right)^2\right\}I^{flatt}(r,\vartheta_0) \quad (A.46)$$

In this way, the specific torque due to roll flattening is also made completely parametric. The values for the integral defined by Eq. (A.45) are given in Table A.3 for the rigid perfectly plastic case in rolling without tensions.

The same procedure can be applied to the normal load to eliminate the flattened radius of the rolls, in order to obtain an explicit parametric expression that can be used in the computations.

$$L_x^P = \left(\frac{L_x^P}{L_0}\right)L_0 = 2k\sqrt{R_0 e_1 r}\left(\frac{L_x^P}{L_0}\right)\sqrt{\frac{R}{R_0}} = k\sqrt{R_0 e_1 r}\left(\frac{L_x^P}{L_0}\right)\left\{1+\left(1+\frac{\Omega}{2}\right)^2\right\} \quad (A.47)$$

$$L_x^P = k\sqrt{R_0 e_1 r}\left[1+\left\{1+\frac{k}{c}\frac{1}{\sqrt{r}}\left(\frac{L_x^P}{L_0}\right)\sqrt{\frac{R_0}{e_1}}\right\}^2\right]\left(\frac{L_x^P}{L_0}\right) \quad (A.48)$$

The final expressions for all the magnitudes derived from the stress field in the roll gap are:

$$T^{spec} = kR_0 e_1 r\left(\frac{T^{spec}}{T_0}\right) \quad (A.49)$$

$$L_x^P = k\sqrt{R_0 e_1 r}\left[1+\left\{1+\frac{k}{c}\frac{1}{\sqrt{r}}\left(\frac{L_x^P}{L_0}\right)\sqrt{\frac{R_0}{e_1}}\right\}^2\right]\left(\frac{L_x^P}{L_0}\right) \quad (A.50)$$

$$L_y^P = ke_1\left[\frac{\sigma_2}{2k}(1-r) - \frac{\sigma_1}{2k} + r\left(\frac{T^{spec}}{T_0}\right)\right] \quad (A.51)$$

$$L_x^S \leq 2kre_1\sqrt{\frac{re_1}{R_0}}\left[1+\left\{1+\frac{k}{c}\frac{1}{\sqrt{r}}\left(\frac{L_x^P}{L_0}\right)\sqrt{\frac{R_0}{e_1}}\right\}^2\right]^{-1}\left(\frac{T^{spec}}{T_0}\right) \tag{A.52}$$

$$L_y^S = -kre_1\left(\frac{T^{spec}}{T_0}\right) \tag{A.53}$$

$$T^{flatt} = -\frac{k^2}{c}e_1R_0\sqrt{r}\left(\frac{L_x^P}{L_0}\right)\sqrt{\frac{R_0}{e_1}}\left[1+\left\{1+\frac{k}{c}\frac{1}{\sqrt{r}}\left(\frac{L_x^P}{L_0}\right)\sqrt{\frac{R_0}{e_1}}\right\}^2\right]I^{flatt}(r,\vartheta_0) \tag{A.54}$$

$$\eta_0^2 = 4(1-r)\frac{e_1}{R_0}\left[1+\left\{1+\frac{k}{c}\frac{1}{\sqrt{r}}\left(\frac{L_x^P}{L_0}\right)\sqrt{\frac{R_0}{e_1}}\right\}^2\right]^{-2} \tag{A.55}$$

The intervals for the specific torque and the normal load to first order are:

$$\Delta\left(\frac{T^{spec}}{T_0}\right) = \frac{1}{2}\eta_0^2\left(\frac{1-r}{r}\right)\left\{\frac{\vartheta_0^2}{(1-r)^2}\left[2-\frac{\sigma_1}{k}\right]^2 - \frac{\vartheta_0^2}{(1-r)}\left[2-\frac{\sigma_2}{k}\right]^2 + T(r,\vartheta_0)\right\} \tag{A.56}$$

$$\Delta\left(\frac{L_x^P}{L_0}\right) = \frac{1}{2}\eta_0^2\frac{\vartheta_0^2}{(1-r)}\left[2-\frac{\sigma_1}{k}\right]^2 + \frac{1}{4}\eta_0^2\sqrt{\frac{1-r}{r}}L(r,\vartheta_0) \tag{A.57}$$

Finally, the relation between the friction coefficient μ and the rolling parameter ϑ_0 eliminating the flattened radius of the rolls is:

$$\vartheta_0 = \mu\sqrt{\frac{R}{e_1}} = \frac{\mu}{2}\sqrt{\frac{R_0}{e_1}}\left[1+\left\{1+\frac{k}{c}\frac{1}{\sqrt{r}}\left(\frac{L_x^P}{L_0}\right)\sqrt{\frac{R_0}{e_1}}\right\}^2\right] \tag{A.58}$$

In order to perform the calculations, the rolling parameter ϑ_0 must be obtained as function of the thickness reduction and the friction coefficient using Eq. (A.58). To start the iteration, the following initial value is considered:

$$\vartheta_0^{(0)} = \mu\sqrt{\frac{R_0}{e_1}} \rightarrow \vartheta_0^{(n+1)} = \frac{1}{2}\vartheta_0^{(0)}\left[1+\left\{1+\frac{k}{c}\frac{1}{\sqrt{r}}\left(\frac{L_x^P}{L_0}\right)^{(n)}\sqrt{\frac{R_0}{e_1}}\right\}^2\right] \tag{A.59}$$

Equation (A.59) is indeed an indirect way of solving Hitchcock's equation (Eq. 2.347).

Once the rolling parameter is obtained, the specific torque defined by Eq. (A.49) and the rest of the magnitudes to start the design of the cold rolling mill can be calculated. When the strain hardening cannot be ignored, the calculations of the specific torque, the normal load, the torque due to roll flattening, and the corresponding intervals coming from the fall in pressure to first order, have to be performed for every specific material taking into account the corresponding hardening function.

Inertial Fall in Pressure—The Inertial Range in High Speed Rolling

In our treatment of the problem the effect of the inertia has been ignored from the outset, due to the huge difference in numerical value between the yield stress and the density of the strip. Nevertheless, in high speed rolling and due to the quadratic dependence of the inertial term with the rolling speed, its effect can disturb the roll pressure profile. The equations of internal equilibrium considering the inertial convective term are:

$$\operatorname{div}\tilde{\sigma} = \rho \frac{D\vec{v}}{Dt} = \rho\left[\frac{\partial \vec{v}}{\partial t} + (\vec{v}\nabla)\vec{v}\right] = \rho(\vec{v}\nabla)\vec{v} \tag{A.60}$$

In bipolar cylindrical coordinates the components of Eq. (A.60) are:

$$(\operatorname{div}\tilde{\sigma})_\xi = \rho \frac{v_\xi}{g_\xi}\frac{\partial v_\xi}{\partial \xi} = \rho v_\xi D_{\xi\xi} = \rho \frac{v_1^2}{a}\left(\frac{e_1}{e_2}\right)^2 \left(\frac{\cosh\eta - \cos\xi}{\cosh\eta + 1}\right)^2 \sin\xi \tag{A.61}$$

$$(\operatorname{div}\tilde{\sigma})_\eta = -\rho \frac{v_\xi^2}{g_\xi g_\eta}\frac{\partial g_\xi}{\partial \eta} = \rho \frac{v_1^2}{a}\left(\frac{e_1}{e_2}\right)^2 \left(\frac{\cosh\eta - \cos\xi}{\cosh\eta + 1}\right)^2 \sinh\eta \tag{A.62}$$

In order to add these inertial terms to Eqs. (2.99) and (2.100) of internal equilibrium and according to Eq. (2.87) for the divergence of the stress tensor, it is necessary to multiply Eqs. (A.61) and (A.62) by the corresponding metric coefficient g_i. The following equations are obtained:

$$g_\xi (\operatorname{div}\tilde{\sigma})_\xi = \rho v_2^2 \frac{(\cosh\eta - \cos\xi)}{(\cosh\eta + 1)^2}\sin\xi \cong \rho \frac{v_2^2}{4}\sin\xi\left[(1-\cos\xi) + \cos\xi\left(\frac{\eta^2}{2}\right)\right] \tag{A.63}$$

$$g_\eta (\operatorname{div}\tilde{\sigma})_\eta = \rho v_2^2 \frac{(\cosh\eta - \cos\xi)}{(\cosh\eta + 1)^2}\sinh\eta \cong \rho \frac{v_2^2}{4}(1-\cos\xi)\eta \tag{A.64}$$

$$\frac{\partial \sigma_{\xi\xi}}{\partial \xi} - \frac{2\sigma_C}{\sqrt{3}}\left(\frac{\sin\xi}{1-\cos\xi}\right) + \frac{\sigma_C}{\sqrt{3}}H'(\xi) + \frac{\partial \sigma_{\xi\xi}^{(2,1)}}{\partial \xi} = \rho \frac{v_2^2}{4}\sin\xi\left[(1-\cos\xi) + \cos\xi\left(\frac{\eta^2}{2}\right)\right] \tag{A.65}$$

$$\frac{\partial \sigma_{\eta\eta}}{\partial \eta} + \frac{2\sigma_C}{\sqrt{3}}\frac{[1 - H'(\xi)\sin\xi]}{(1-\cos\xi)}\eta + \frac{\sigma_C}{\sqrt{3}}H''(\xi)\eta = \rho \frac{v_2^2}{4}(1-\cos\xi)\eta \tag{A.66}$$

Integrating Eq. (A.65) to order zero, the influence of the inertia is obtained:

$$\frac{\sigma_{\xi\xi}^{(0)}}{2k}(\xi)=q(\xi)-\frac{1}{2}H(\xi)+C \tag{A.67}$$

$$q(\xi)=\ln(1-\cos\xi)+\rho\frac{v_2^2}{8k}\cos\xi\left(\frac{1}{2}\cos\xi-1\right) \tag{A.68}$$

The solution for the roll pressure distribution using Eqs. (2.137) and (2.139) is straightforward:

$$\begin{aligned}\frac{\sigma_{\eta\eta}^{(0)}}{2k}(\xi,\xi_0)=&\left\{\frac{\sigma_1}{2k}-1-q(\xi_0)\right\}\exp\left[-\vartheta(\xi-\xi_0)\right]+q(\xi)\\&-\vartheta\exp\left[-\vartheta\xi\right]\int_{\xi_0}^{\xi}q(\xi)\exp\left[\vartheta\xi\right]d\xi\end{aligned} \tag{A.69}$$

$$\begin{aligned}\frac{\sigma_{\eta\eta}^{(0)}}{2k}(\xi,\pi)=&\left\{\frac{\sigma_2}{2k}-1-q(\pi)\right\}\exp\left[\vartheta(\xi-\pi)\right]+q(\xi)\\&+\vartheta\exp\left[\vartheta\xi\right]\int_{\pi}^{\xi}q(\xi)\exp\left[-\vartheta\xi\right]d\xi\end{aligned} \tag{A.70}$$

Therefore, the roll pressure profile due only to the inertia and without strain hardening is:

$$\begin{aligned}\Delta\frac{\sigma_{\eta\eta}^{(0)}}{2k}(\xi,\xi_0)=&\rho\frac{v_2^2}{8k}\left\{\cos\xi\left(\frac{1}{2}\cos\xi-1\right)-\cos\xi_0\left(\frac{1}{2}\cos\xi_0-1\right)\exp\left[-\vartheta(\xi-\xi_0)\right]\right\}\\&-\rho\frac{v_2^2}{8k}\vartheta\exp\left[-\vartheta\xi\right]\int_{\xi_0}^{\xi}\exp\left[\vartheta\xi\right]\cos\xi\left(\frac{1}{2}\cos\xi-1\right)d\xi=\rho\frac{v_2^2}{8k}G(\xi,\xi_0)\geq 0\end{aligned} \tag{A.71}$$

$$\Delta\frac{\sigma_{\eta\eta}^{(0)}}{2k}(\xi,\pi)=\rho\frac{v_2^2}{8k}\left\{\cos\xi\left(\frac{1}{2}\cos\xi-1\right)-\frac{3}{2}\exp\left[\vartheta(\xi-\pi)\right]\right\} \tag{A.72}$$

$$+\rho\frac{v_2^2}{8k}\vartheta\exp\left[\vartheta\xi\right]\int_{\pi}^{\xi}\exp\left[-\vartheta\xi\right]\cos\xi\left(\frac{1}{2}\cos\xi-1\right)d\xi=\rho\frac{v_2^2}{8k}G(\xi,\pi)\leq 0$$

The inertial disturbing terms in Eqs. (A.65) and (A.66) that add to the first order solution can be ignored in a zero order calculation and become relevant only for very high inertial speeds. The main effect of the inertia of the strip on the roll pressure profile is a progressive fall in pressure on the entry branch, and a progressive disturbing rise on the exit branch. Therefore, the neutral point moves progressively towards the exit of the roll gap as the speed of rolling increases. The inertial range begins, when this fall in pressure at the neutral point becomes of the same order of magnitude that the minimum fall in pressure to first order of section (2.6.1). Considering the upper bound for the constant fall in pressure when rolling without tensions, the following is obtained:

$$\Delta\frac{\sigma_{\eta\eta}^{(0)}}{2k}(\xi_N,\xi_0)=\rho\frac{v_2^2}{8k}G(\xi_N,\xi_0)\geq H'^2(\xi_0)\left(\frac{\eta_0^2}{2}\right)=2\vartheta^2\eta_0^2=2\mu^2 \qquad \text{(A.73)}$$

$$v_2\geq\frac{4\mu}{\sqrt{G(\xi_N,\xi_0)}}\left(\frac{k}{\rho}\right)^{1/2}\approx 25\div 75\ \text{m/s} \qquad \text{(for steel in the usual range of rolling)} \qquad \text{(A.74)}$$

Figure A.1 shows that the inertial range is at hand to present cold rolling technology and is more than a mere academic oddity. The upper curves correspond to the rolling without tensions of a steel of a yield stress of 400 MPa. Below an exit velocity of 50 m/s the zero order solution remains undisturbed. The curves corresponding to rolling with tensions are of copper with a yield stress of 200 MPa. The threshold velocity in this case is 25 m/s. The existence of a critical upper velocity $v_2^{\max}$ for the neutral point to reach the exit of the roll gap and the cold flat rolling to become unfeasible, manifests.

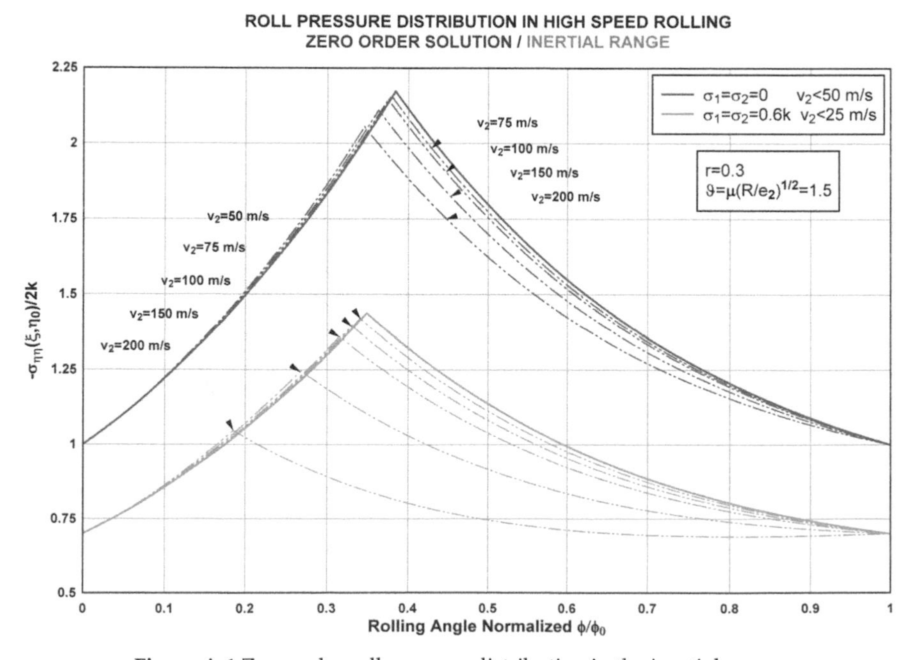

Figure A.1 Zero order roll pressure distribution in the inertial range.

The Equation of the Power in the Inertial Range

Equation (2.8) for the power has to be modified in the inertial range, so as to include the variation of kinetic energy within the roll gap in the following manner:

$$\oint_{\partial\Omega_2} t_i v_i dS = \sqrt{\frac{2}{3}} \int_{\Omega_2} \sigma_C \sqrt{D_{ij} D_{ij}} dV + \int_{\Sigma_1^+} \sigma_{\xi\eta}(\xi_0,\eta)\Delta v_\eta(\xi_0,\eta)\, dS_\xi + \frac{DE_K}{Dt} \quad \text{(A.75)}$$

where the last term in Eq. (A.75) is the material derivative of the kinetic energy within the roll gap. Making tacit use of the Euler identity for the material derivative of the Jacobian and the equation of continuity, this term is calculated as follows:

$$\frac{DE_K}{Dt} = \frac{D}{Dt}\int_{\Omega_2} \frac{\rho}{2} v_i^2 dV = \int_{\Omega_2} \frac{\rho}{2}\frac{Dv_i^2}{Dt} dV = \int_{\Omega_2} \frac{\rho}{2}\frac{Dv_\xi^2}{Dt} dV = \int_{\Omega_2} \rho v_\xi \frac{Dv_\xi}{Dt} dV \quad \text{(A.76)}$$

$$\frac{Dv_\xi}{Dt} = \frac{v_\xi}{g_\xi}\frac{\partial v_\xi}{\partial \xi} = \frac{v_2^2}{a}\left(\frac{\cosh\eta - \cos\xi}{\cosh\eta + 1}\right)^2 \sin\xi \quad \text{(A.77)}$$

$$\frac{DE_K}{Dt} = -\int_{\Omega_2} \rho \frac{v_2^3}{a}\left(\frac{\cosh\eta - \cos\xi}{\cosh\eta + 1}\right)^3 \sin\xi\, dV \quad \text{(A.78)}$$

$$\frac{DE_K}{Dt} = -\int_{-\eta_0}^{\eta_0}\int_{\pi}^{\xi_0} \rho \frac{v_2^3}{a}\left(\frac{\cosh\eta - \cos\xi}{\cosh\eta + 1}\right)^3 \sin\xi \frac{a^2}{(\cosh\eta - \cos\xi)^2} d\xi\, d\eta \quad \text{(A.79)}$$

Performing the integral of Eq. (A.79) to order zero the following is obtained:

$$\frac{DE_K}{Dt} = -\int_{-\eta_0}^{\eta_0}\int_{\pi}^{\xi_0} \rho \frac{v_2^3}{8} a(1-\cos\xi)\sin\xi\, d\xi\, d\eta \quad \text{(A.80)}$$

$$\frac{DE_K}{Dt} = -\frac{a}{4}\rho v_2^3 \eta_0 \int_{\pi}^{\xi_0} (1-\cos\xi)\sin\xi\, d\xi = \frac{a}{4}\rho v_2^3 \eta_0 \left[\frac{3}{2} - \cos\xi_0\left(\frac{1}{2}\cos\xi_0 - 1\right)\right] \quad \text{(A.81)}$$

$$\frac{DE_K}{Dt} = \frac{1}{2}\rho v_2^3 e_2 r(2-r) \quad \text{(A.82)}$$

Therefore, Eq. (2.304) for the power in the inertial range becomes:

$$\int_{\pi}^{\xi_0} H'(\xi)\, d\xi = 2\ln(1-r) - \frac{2}{3}\vartheta\eta_0^2\left[1 - \frac{\sigma_1}{2k}\right]\sqrt{\frac{r}{1-r}} + \left(\frac{\sigma_2 - \sigma_1}{k}\right) - \rho\frac{v_2^2}{2k} r(2-r) \quad \text{(A.83)}$$

and ignoring the shear power term at the entrance of the roll gap as disturbing:

$$\int_{\pi}^{\xi_0} H'(\xi)d\xi \cong 2\ln(1-r)+\left(\frac{\sigma_2-\sigma_1}{k}\right)-\rho\frac{v_2^2}{2k}r(2-r) \tag{A.84}$$

Similar considerations apply in the case with strain hardening, and therefore, the last term in Eq. (A.84) has to be added to the right-hand side of Eq. (A.7).

Specific Torque and Normal Load in the Inertial Range

In the same manner, Eq. (A.24) for the specific torque has to be modified in order to include the effect of the inertia. Using Eq. (A.65) of internal equilibrium to order zero and considering the strain hardening, the derivative of the pressure is obtained:

$$\frac{d}{d\xi}\frac{\sigma_{\eta\eta}}{2k}+f'(\xi)=\left(\frac{\sin\xi}{1-\cos\xi}\right)\{1+f(\xi)\}-\frac{1}{2}F'(\xi)+\rho\frac{v_2^2}{8k}(1-\cos\xi)\sin\xi \tag{A.85}$$

Therefore, the additional term in Eq. (A.24) due to the inertia is:

$$\Delta\frac{T^{spec}}{T_0}=-2\left(\frac{1-r}{r}\right)\int_{\pi}^{\xi_0}\rho\frac{v_2^2}{8k}\sin\xi d\xi=\rho\frac{v_2^2}{4k}\left(\frac{1-r}{r}\right)(\cos\xi_0+1)=\rho\frac{v_2^2}{2k}(1-r) \tag{A.86}$$

and the final expression for the specific torque in the inertial range is obtained as function of the corresponding roll pressure profile:

$$\frac{T^{spec}}{T_0}=2\left(\frac{1-r}{r}\right)\int_{\pi}^{\xi_0}\frac{\sigma_{\eta\eta}}{2k}\frac{\sin\xi}{(1-\cos\xi)^2}d\xi+\frac{1}{2r}\left[\frac{\sigma_1}{k}-\frac{\sigma_2}{k}(1-r)\right]+\rho\frac{v_2^2}{2k}(1-r) \tag{A.87}$$

Finally, Eq. (2.299) of equilibrium of forces acting on the strip at the boundaries of the roll gap has to be also modified in order to include the force due to the inertia. The expression for this force using Eq. (A.77) to order zero is:

$$F_y=\int_{\Omega_2}\rho\frac{Dv_\xi}{Dt}dV=\int_{-\eta_0}^{\eta_0}\int_{\pi}^{\xi_0}\rho\frac{v_2^2}{4}a\sin\xi d\xi d\eta=-\frac{\rho}{2}v_2^2e_2(\cos\xi_0+1)=-\rho v_2^2e_2r \tag{A.88}$$

and applying the second law of Newton in the form of D'Alembert to the material of the strip in the roll gap the following equation is obtained:

$$\sigma_2e_2-\sigma_1e_1-2L_y^S-2L_y^P+F_y=0 \tag{A.89}$$

$$L_y^P+\frac{1}{2}(\sigma_1e_1-\sigma_2e_2)+\frac{\rho}{2}v_2^2e_2r=-L_y^S=\frac{T^{spec}}{R_0} \tag{A.90}$$

Obviously, the torque calculated with Eq. (A.90) is given by Eq. (A.87). It must be pointed out, that Eq. (A.3) for the specific torque as function of the shear structure corresponding to the pressure profile given by Eqs. (A.69) and (A.70), maintains its full validity.

Using Eqs. (A.69) and (A.70) in Eq. (A.8) for the normal load and in Eq. (A.87) for the specific torque, both magnitudes are obtained in the inertial range.

Figures A.2 and A.3 are, respectively, the plots of normal load and specific torque for a material rigid perfectly plastic in rolling without tensions, for a value of the rolling parameter $\vartheta_0 = 1.5$ and several reductions and rolling speeds within the inertial range. The material of the strip is steel with a yield stress of 400 MPa which has been considered as constant during the whole range. The threshold velocity, for the inertial effects to manifest, and the critical velocity for the neutral point to reach the exit of the roll gap, are given in Table A.6. Beyond this maximum velocity the process under the same conditions is not feasible.

In order to obtain a properly normalized specific torque, Eq. (A.87) has been multiplied by the thickness reduction. It has to be mentioned, that although the normal load decreases with the rolling speed due to the inertial fall in pressure, the specific torque increases for all the range of reductions.

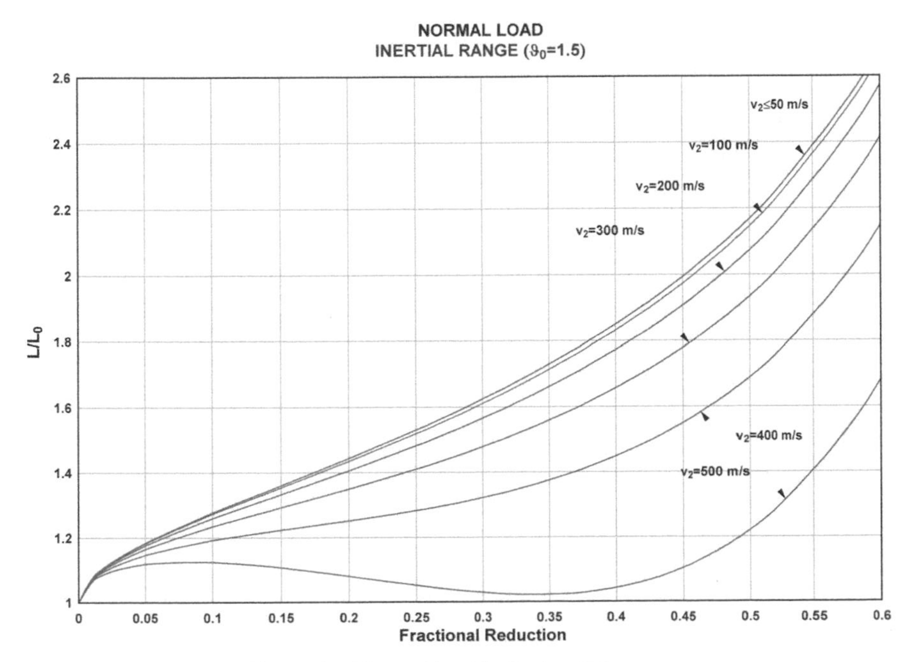

Figure A.2 Normal load in the inertial range.

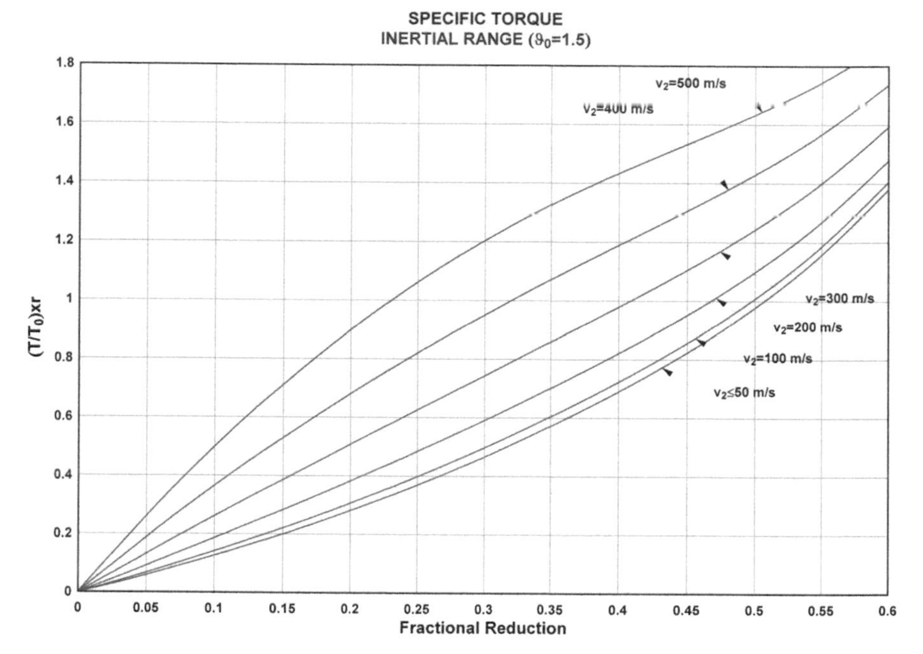

Figure A.3 Specific torque in the inertial range.

Table A.6 Threshold and Maximum Rolling Speeds ($\vartheta_0 = 1.5$).

Reduction	v_2^{min} (m/s)	v_2^{max} (m/s)
0	∞	∞
0.025	191	993
0.05	146	825
0.10	106	688
0.15	87	623
0.20	74	583
0.25	63	559
0.30	58	543
0.35	53	535
0.40	49	533
0.45	45	534
0.50	41	543
0.55	38	557
0.60	35	579

Consider the specific energy E necessary to cold roll a strip of final length L:

$$E = 2T^{spec}\omega t_L = \frac{\sigma_C}{\sqrt{3}} r e_1 L \left(\frac{T^{spec}}{T_0} \right) (1 - \cos \xi_N) \tag{A.91}$$

where the angular velocity of the rolls and the time to roll the strip of final length L are:

$$\omega = \frac{v_2}{2R_0} (1 - \cos \xi_N) \qquad t_L = L/v_2 \tag{A.92}$$

Therefore, although there is a general fall in pressure in the roll gap, the energetic expenditure is greater in the inertial range, due to the increment of kinetic energy in the continuum and the fact that the neutral point moves towards the exit of the roll gap increasing both, the specific torque and the last factor in Eq. (A.91). Consequently, from an energetic perspective, it is not efficient to cold roll in the inertial range.

It may be weird that inertial effects could play a role in future cold rolling mill design, but it must be taken into account that due to the incompressibility, in high speed rolling there is an unusual increase in momentum within the roll gap, that must translate into the corresponding fall in pressure, if the laws of Newton are to be fulfilled.

A Boundary Condition Due to Nadai

As an example of the application of the new theory to statically undetermined problems due to the boundary condition of friction with the rolls depending on the velocity field, the following hydrodynamic model due to Nadai (1939) is considered:

$$\sigma_{\xi\eta}(\xi, \eta_0) = \frac{\kappa}{\delta} \left[v_\xi(\xi_N, \eta_0) - v_\xi(\xi, \eta_0) \right] \tag{A.93}$$

where κ is the mean dynamic viscosity of the lubricating film and δ is its mean thickness. The shear stress at the interface is proportional to the relative velocity between the rolls and the strip. Using Eq. (2.102) for the shear stress and Eq. (2.15) for the velocity field, the following equation for the shear structure is obtained:

$$\frac{\sigma_C}{\sqrt{3}} H'(\xi)\eta_0 = v_2 \frac{\kappa}{\delta} \left(\frac{\cos \xi_N - \cos \xi}{\cosh \eta_0 + 1} \right) \cong v_2 \frac{\kappa}{2\delta} (\cos \xi_N - \cos \xi) \tag{A.94}$$

$$H'(\xi) = \frac{\kappa v_2}{2k\delta\eta_0} (\cos \xi_N - \cos \xi) \tag{A.95}$$

Therefore, the derivative and the integral of the shear structure are:

$$H''(\xi)=\frac{\kappa v_2}{2k\delta\eta_0}\sin\xi \tag{A.96}$$

$$H(\xi)-H(\pi)=\frac{\kappa v_2}{2k\delta\eta_0}\left[(\xi-\pi)\cos\xi_N-\sin\xi\right] \tag{A.97}$$

and using Eq. (2.125), the roll pressure profile to order zero is obtained:

$$\frac{\sigma_{\eta\eta}^{(0)}}{2k}(\xi,\pi)=\frac{\sigma_2}{2k}-1+\ln\left(\frac{1-\cos\xi}{2}\right)-\frac{\kappa v_2}{4k\delta\eta_0}\left[(\xi-\pi)\cos\xi_N-\sin\xi\right] \tag{A.98}$$

In order to obtain the equilibrium of forces in the roll gap, the position of the neutral point is calculated using the boundary condition of back tension as follows:

$$\frac{\sigma_{\xi\xi}^{(0)}}{2k}(\xi_0,\pi)=\frac{\sigma_1}{2k}=\frac{\sigma_2}{2k}+\ln\left(\frac{1-\cos\xi_0}{2}\right)-\frac{\kappa v_2}{4k\delta\eta_0}\left[(\xi_0-\pi)\cos\xi_N-\sin\xi_0\right] \tag{A.99}$$

$$\cos\xi_N^{(0)}=\frac{1}{(\xi_0-\pi)}\left\{\frac{4k\delta\eta_0}{\kappa v_2}\left[\left(\frac{\sigma_2-\sigma_1}{2k}\right)+\ln\left(\frac{1-\cos\xi_0}{2}\right)\right]+\sin\xi_0\right\} \tag{A.100}$$

The pressure profile is smooth across the roll gap because the shear structure is also smooth, and therefore, it has to be an explicit function of the position of the neutral point to be physically acceptable. Consequently, the entry branch of the pressure profile is redundant.

The position of the neutral point can also be calculated using Eq. (2.305) for the integral of the shear structure over the roll gap as follows:

$$\int_{\pi}^{\xi_0}H'(\xi)d\xi=\frac{\kappa v_2}{2k\delta\eta_0}\int_{\pi}^{\xi_0}(\cos\xi_N-\cos\xi)d\xi=2\ln(1-r)+\left(\frac{\sigma_2-\sigma_1}{k}\right) \tag{A.101}$$

$$\cos\xi_N^{(0)}=\frac{1}{(\xi_0-\pi)}\left\{\frac{4k\delta\eta_0}{\kappa v_2}\left[\left(\frac{\sigma_2-\sigma_1}{2k}\right)+\ln(1-r)\right]+\sin\xi_0\right\} \tag{A.102}$$

The longitudinal and transversal perturbations of the plastic velocity field are obtained using the following function in Eqs. (2.258) and (2.259):

$$\Delta\Psi=\frac{\kappa v_2}{2k\delta\eta_0}(\cos\xi_N-\cos\xi)\sin\xi-\left(\frac{3+\cos\xi}{4}\right) \tag{A.103}$$

Using the pressure profile given by Eq. (A.98) in Eq. (A.8), the normal load is obtained. The specific torque using Eq. (A.3) can be easily integrated:

$$\frac{T^{spec}}{T_0} = -\left(\frac{1-r}{r}\right)\int_{\pi}^{\xi_0} \frac{\kappa v_2}{2k\delta\eta_0}\left(\frac{\cos\xi_N - \cos\xi}{1-\cos\xi}\right)d\xi \tag{A.104}$$

$$\frac{T^{spec}}{T_0} = -\left(\frac{1-r}{r}\right)\frac{\kappa v_2}{2k\delta\eta_0}\left[(\xi_0 - \pi) + (1-\cos\xi_N)\cot\left(\frac{\xi_0}{2}\right)\right] \tag{A.105}$$

Deriving Eq. (A.98), the position of the maximum of pressure is:

$$\frac{d}{d\xi}\frac{\sigma_{\eta\eta}^{(0)}}{2k}(\xi = \xi_m, \pi) = \left(\frac{\sin\xi_m}{1-\cos\xi_m}\right) - \frac{\kappa v_2}{4k\delta\eta_0}\left[\cos\xi_N - \cos\xi_m\right] = 0 \tag{A.106}$$

and operating the following condition is obtained:

$$\cos\xi_N = \cos\xi_m + \frac{4k\delta\eta_0}{\kappa v_2}\left(\frac{\sin\xi_m}{1-\cos\xi_m}\right) \rightarrow \xi_N < \xi_m \tag{A.107}$$

Therefore, the neutral point is closer to the exit of the roll gap than the maximum of pressure.

Now the question arises as to how to calculate the pressure profile to first order for this model. To answer this question, Eq. (2.159) is subtracted from Eq. (2.157) to obtain the following:

$$\frac{\sigma_{\eta\eta}^{(1)}}{2k}(\xi,\pi) - \frac{\sigma_{\eta\eta}^{(1)}}{2k}(\xi,\xi_0) = -\frac{1}{2}\left[\Theta(\xi_0) - \Theta(\pi)\right]\left(\frac{\eta^2}{2}\right) \tag{A.108}$$

There is an offset of pressure between the two branches to first order and they never meet, not allowing in this manner the equilibrium of forces in the roll gap. Therefore, it is also compelling to use a single branch to calculate the pressure profile to first order. Consider the exit branch:

$$\frac{\sigma_{\eta\eta}^{(1)}}{2k}(\xi,\pi) = \frac{\sigma_{\eta\eta}^{(0)}}{2k}(\xi,\pi) - \frac{1}{2}\left[\Theta(\xi) - \Theta(\pi) - 2H'^2(\xi)\right]\left(\frac{\eta^2}{2}\right) \tag{A.109}$$

Applying the condition of back tension at the entrance of the roll gap maintaining the position of the neutral point $\xi_N^{(0)}$ according to the disturbing procedure, the following is obtained:

$$\frac{\sigma_{\xi\xi}^{(1)}}{2k}(\xi_0,\pi) = \frac{\sigma_1}{2k} - \frac{1}{2}\left[\Theta(\xi_0) - \Theta(\pi)\right]\left(\frac{\eta^2}{2}\right) \tag{A.110}$$

which is the expected value plus the offset of pressure given by Eq. (A.108). Therefore, in order to fulfill again the equilibrium of forces in the roll gap, the position of the neutral point must change in order to annul the offset of pressure according to the following equation:

$$\frac{\sigma_{\xi\xi}^{(1)}}{2k}(\xi_0,\pi)=\frac{\sigma_1}{2k}=\frac{\sigma_{\xi\xi}^{(0)}}{2k}(\xi_0,\pi)-\frac{1}{2}\left[\Theta(\xi_0)-\Theta(\pi)\right]\left(\frac{\eta^2}{2}\right) \tag{A.111}$$

Where the first term on the right-hand side of Eq. (A.111) is given by Eq. (A.99) and the function in the second term is given by Eq. (2.152) with the shear structure of Eq. (A.95). This is indeed an implicit equation for the new position of the neutral point, that according to Eq. (A.111) will be a function of the bipolar coordinate η. At the central plane of the strip $\eta = 0$, where the solution to order zero is valid $\xi_N = \xi_N^{(0)}$, the neutral point proper $\xi_N^{(1)}$ is obtained for $\eta = \eta_0$. Therefore, the larger the offset of pressure in Eq. (A.108), the larger is the shift in the position of the neutral point and larger the deviation of the neutral surface from a ξ-isosurface. It is interesting to note that this frictional model according to the theory does not provide room for negative pressure.

The same result for the new position of the neutral point can again be obtained from energetic considerations. Deriving and then integrating over the roll gap Eq. (2.156), the following is obtained:

$$\frac{\partial\sigma_{\xi\xi}^{(1)}}{\partial\xi}=\frac{d\sigma_{\xi\xi}^{(0)}}{d\xi}-\frac{\sigma_C}{\sqrt{3}}\Theta'(\xi)\left(\frac{\eta^2}{2}\right) \tag{A.112}$$

$$\frac{\partial}{\partial\xi}\frac{\sigma_{\xi\xi}^{(1)}}{k}=2\left(\frac{\sin\xi}{1-\cos\xi}\right)-H'(\xi)-\Theta'(\xi)\left(\frac{\eta^2}{2}\right) \tag{A.113}$$

$$\left.\frac{\sigma_{\xi\xi}^{(1)}}{k}\right|_{\pi}^{\xi_0}=\left(\frac{\sigma_1-\sigma_2}{k}\right)=2\ln(1-r)-\int_{\pi}^{\xi_0}H'(\xi)d\xi-\left[\Theta(\xi_0)-\Theta(\pi)\right]\left(\frac{\eta^2}{2}\right) \tag{A.114}$$

$$\int_{\pi}^{\xi_0}H'(\xi)d\xi=2\ln(1-r)+\left(\frac{\sigma_2-\sigma_1}{k}\right)-\left[\Theta(\xi_0)-\Theta(\pi)\right]\left(\frac{\eta^2}{2}\right) \tag{A.115}$$

$$\int_{\pi}^{\xi_0}H'\left(\xi,\xi_N^{(1)}\right)d\xi=\int_{\pi}^{\xi_0}H'\left(\xi,\xi_N^{(0)}\right)d\xi-\left[\Theta\left(\xi_0,\xi_N^{(1)}\right)-\Theta\left(\pi,\xi_N^{(1)}\right)\right]\left(\frac{\eta_0^2}{2}\right) \tag{A.116}$$

The position of the neutral point must be adjusted so as to fulfill Eq. (A.116).

The Roll Pressure Profiles with More than a Single Maximum or with Wavy-like Structure

As a further step in the new theory, the slipping friction condition can also be made smooth along the roll gap using for instance the following mono-parametric filter function:

$$\sigma_{\xi\eta}(\xi,\eta_0)=\mu\tanh\left[\tau(\xi-\xi_N)\right]\sigma_{\eta\eta}(\xi,\eta_0)\quad \tau>0\quad \forall\xi\in[\pi,\xi_0] \tag{A.117}$$

and the following differential equation instead of Eq. (2.134) appears:

$$F'(\xi)+\vartheta\tanh\left[\tau(\xi-\xi_N)\right]\left[F(\xi)-2\left\{\frac{\sigma_2}{2k}-1+\ln\left(\frac{1-\cos\xi}{2}\right)\right\}\right]=0 \tag{A.118}$$

whose solution $F(\xi)=H(\xi)-H(\pi)$ is:

$$F(\xi)=2\vartheta\left\{\cosh\left[\tau(\xi-\xi_N)\right]\right\}^{-\frac{\vartheta}{\tau}} \int_{\pi}^{\xi}\left\{\cosh\left[\tau(\xi-\xi_N)\right]\right\}^{\frac{\vartheta}{\tau}}\tanh\left[\tau(\xi-\xi_N)\right]\left\{\frac{\sigma_2}{2k}-1+\ln\left(\frac{1-\cos\xi}{2}\right)\right\}d\xi \tag{A.119}$$

and therefore, the corresponding pressure profile to order zero is:

$$\frac{\sigma_{\eta\eta}^{(0)}}{2k}(\xi,\pi)=\frac{\sigma_2}{2k}-1+\ln\left(\frac{1-\cos\xi}{2}\right)-\frac{1}{2}F(\xi) \tag{A.120}$$

The neutral point is obtained applying the boundary condition of back tension:

$$\frac{\sigma_{\xi\xi}^{(0)}}{2k}(\xi_0,\pi)=\frac{\sigma_1}{2k}=\frac{\sigma_2}{2k}+\ln\left(\frac{1-\cos\xi_0}{2}\right)-\frac{1}{2}F(\xi_0) \tag{A.121}$$

This is an implicit equation for the position of the neutral point to be solved by iteration. The calculation of the pressure profile to first order and the corresponding neutral point is similar to Nadai's model. The discontinuous version of the theory is recovered in the limit $\tau\to\infty$. This parameter is a measure of the width of the stick-slip transition around the neutral point and its relation with the level of pressure at the "peak", the relative velocity of slip, the surface roughness of rolls and strip, the properties of the lubricating film and other such factors is a matter of tribology. Therefore, as a last resort the tribology will point to the right combinations of rolling parameters for the negative pressure and the non-plastic transition to appear.

In this manner, the roll pressure profiles are reduced in the new theory to a single and smooth branch, and the "peak" of pressure disappears from the outset by simply regularizing the slipping friction condition with the rolls using a suitable filter function.

Figure A.4 is the plot of the roll pressure distributions in rolling with and without tensions using Eqs. (A.119) and (A.120) to order zero with the fall in pressure given by Eq. (A.109). Figure A.5 are the corresponding shear structures for the case without tensions and Table A.7 is the record of values used in the regularization. Unlike in the correction of the pressure distributions of Fig. 2.13 that simply fixes the level of pressure around the neutral point, an interesting feature appears when the fall due to the regularization of the friction condition with the rolls and the

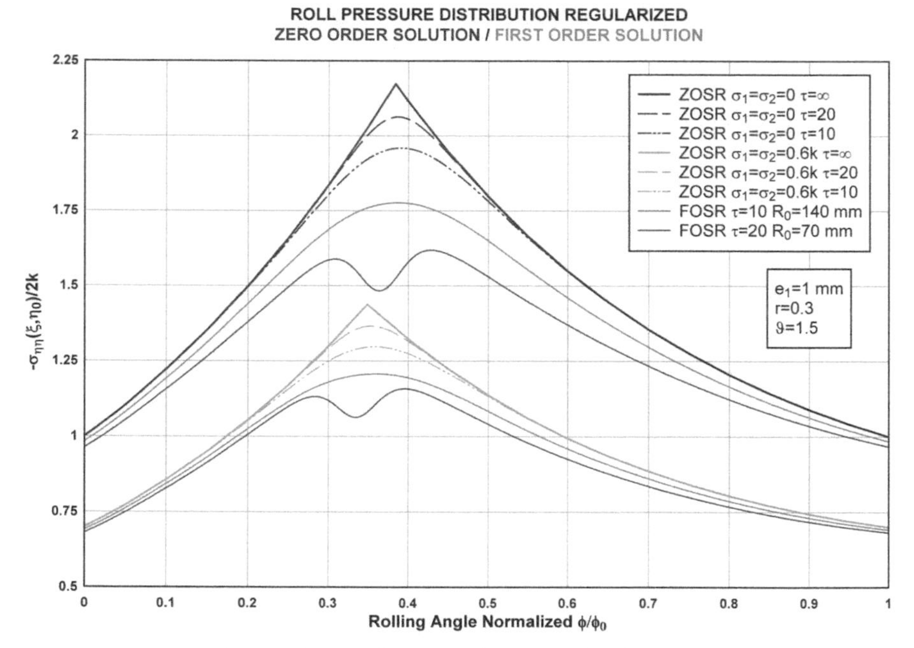

Figure A.4 Regularized roll pressure distribution.

Table A.7 Record of values for the regularized pressure profiles.

$\sigma_1 = \sigma_2$	$R_0(mm)$	$R(mm)$	τ	I_{Integ}	ε (%)	ξ_N	$\xi_N^{(1)}$	$\sigma_{\eta\eta}^{(1,3)}/2k$
0	70	86	20	–0.7135	0.03	3.6340	3.6105	0.2525
0	140	191	10	–0.7134	0.02	3.6336	3.6230	0.0573
0.6	70	81	20	–0.7133	0.00	3.5907	3.5741	0.1796
0.6	140	173	10	–0.7134	0.02	3.5901	3.5823	0.0420

The value of the integrated shear structure over the roll gap using Eq. (2.305) is $I_c = -0.7133$ for all the cases.

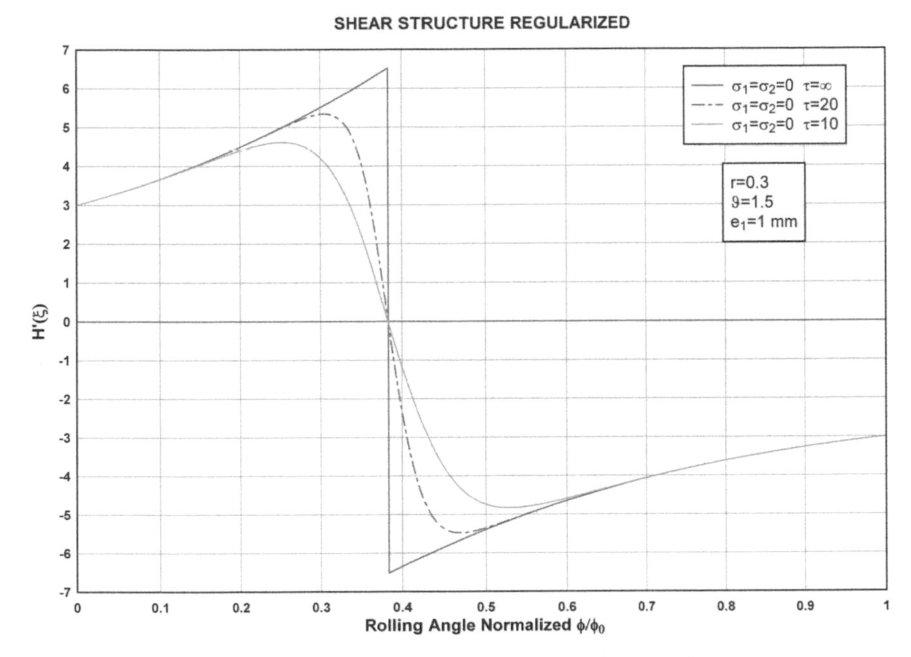

Figure A.5 Regularized shear structure without tensions.

fall due to the derivative of the shear structure given by Eq. (2.165) add up: a pressure profile with more than a single maximum under the right rolling conditions. Pressure distributions of this kind have been reported on previous experimental work on cold flat rolling (Al-Salehi et al. 1973).

According to Eq. (2.341), the derivative of the shear structure is proportional to the local curvature of the pressure profile to order zero. Therefore, wherever the curvature is sufficiently high, dips in the roll pressure distribution to first order could appear. It is obvious that due to this fact, the neutral point and the maximum of pressure could also be different. Apart from the possible influence of the lateral spread on the formation of pressure profiles with wavy-like structure, this could be under plane strain a good explanation for this curious phenomenon.

Exercise

In order to involve the reader with the new theory a little more, a further improvement in the calculation of the velocity perturbations due to friction drag with the rolls is proposed as an exercise:

Use the following expressions instead of Eqs. (2.238) and (2.239) to resolve the inconsistency in equation (2.240) when using the solution given by Eq. (2.257) for the transversal perturbation.

$$f(\xi,\eta) = C(\xi)\left[\cosh\eta + F(\xi)\right] + D'(\xi)\eta_0^2\cosh\eta \\ -E'(\xi)\left[\left(\eta^2-\eta_0^2\right)\cosh\eta + 2\left(\cosh\eta - \eta\sinh\eta\right)\right] \tag{A.122}$$

$$g(\xi,\eta) = \left[D(\xi)+E(\xi)\right]\left(\eta^2-\eta_0^2\right)\sinh\eta \tag{A.123}$$

Use Eq. (2.237) and the incompressibility condition across any section at constant ξ to prove that the functions $C(\xi)$ and $D'(\xi)$ are given by Eqs. (2.241) and (2.242) and to obtain the following expression for the function $F(\xi)$ that calibrates the longitudinal perturbation:

$$F(\xi) = -\frac{1}{C(\xi)}\left[C(\xi)+D'(\xi)\eta_0^2\right]\frac{\sinh\eta_0}{\eta_0} + \frac{E'(\xi)}{C(\xi)}\left[6\frac{\sinh\eta_0}{\eta_0} - 4\cosh\eta_0\right] \tag{A.124}$$

Compute the Taylor expansion of Eq. (A.122) to order η^4 to obtain the following result:

$$f(\xi,\eta) = \frac{1}{2}\left[C(\xi)+D'(\xi)\eta_0^2\right]\left[\eta^2 - \frac{\eta_0^2}{3}\right] - E'(\xi)\left[\frac{1}{4}\eta^4 - \frac{1}{2}\eta^2\eta_0^2 + \frac{7}{60}\eta_0^4\right] \tag{A.125}$$

where the quartic term of $C(\xi)$ has been ignored as irrelevant. Solve the differential equation of local incompressibility in the form given by Eq. (2.253) ignoring all the terms in η^4 that appear in the process to obtain the amplitude of Eq. (A.123):

$$D(\xi)+E(\xi) = -\frac{1}{6}(1-\cos\xi)^2\frac{d}{d\xi}\frac{\Delta\Psi}{(1-\cos\xi)} \tag{A.126}$$

Use Eq. (A.126) to demonstrate that the transversal perturbation (Eq. 2.259) remains unaltered while the longitudinal perturbation, ignoring the term in $D'(\xi)$ as irrelevant in Eq. (A.125) and with the appropriate rescaling factor of Eq. (2.221) takes the following form:

$$\delta v_\xi(\xi,\eta) = -\frac{v_2}{2}\Delta\Psi\left[\eta^2 - \frac{\eta_0^2}{3}\right] + v_2\frac{E'(\xi)}{(1-\cos\xi)}\left[\frac{1}{4}\eta^4 - \frac{1}{2}\eta^2\eta_0^2 + \frac{7}{60}\eta_0^4\right] \tag{A.127}$$

$$E'(\xi) = -\frac{1}{6}\frac{d}{d\xi}\left\{(1-\cos\xi)^2\frac{d}{d\xi}\frac{\Delta\Psi}{(1-\cos\xi)}\right\} - \frac{1}{4}(1+\cos\xi)\Delta\Psi \tag{A.128}$$

Finally, check again that Eq. (2.237) is thoroughly fulfilled using Eqs. (A.123) and (A.125).

The inconsistency appears now in the differential equation of local incompressibility and is also of order η^2 with respect to the corresponding principal term. Equation (A.127) is asymptotic with Eq. (2.258) and behaves better around the neutral surface where the transversal perturbation takes maximum values.

And thus the matter is closed, with these interesting examples of application of the new theory.

3

Estimation of Temperature Distribution in Flat Rolling

U.S. Dixit,[1,*] *V. Yadav*[1] and *A.K. Singh*[2]

ABSTRACT

Determination of temperature distribution in the roll and strip in the rolling process forms an important part of the rolling process modeling. Temperature affects the material properties, lubricant characteristics and wear during the rolling of advanced high strength steels (AHSSs). In this chapter, the procedures for determining the temperature distribution in rolling are reviewed. An integrated thermo-mechanical model for the determination of temperature distribution in flat rolling process is described in detail. The model carries out deformation analysis using finite element method (FEM) and thermal analysis by integral transform method. Some typical results are presented and compared with the experimental data. The integrated model described in this chapter is appropriate for the quick estimate of the temperature during rolling of metals including AHSSs. Some directions for further research are provided in the concluding section.

[1] Department of Mechanical Engineering, Indian Institute of Technology Guwahati, Assam, 781 039, India.

[2] Department of Civil Engineering, Indian Institute of Technology Guwahati, Assam, 781 039, India.

* Corresponding author: uday@iitg.ernet.in, usd1008@yahoo.com

3.1 Introduction

Flat rolling process is an important industrial bulk metal forming process. In this process, the thickness of a sheet or strip is reduced by passing it between two counter-rotating rolls. The difference between the sheet and strip is that the sheet has a greater width to thickness ratio than the strip. Sometimes the objective of the flat rolling is not to reduce the thickness, but to improve the mechanical, metallurgical and surface properties of the sheet metal. For example, temper rolling is employed to provide a degree of surface hardening, restore temper, prevent Lüders bands, impart a desired finish to the sheet and impart a degree of flatness (Roberts 1978). Similarly, the accumulative roll bonding (ARB) process is used for refining the grain microstructure (Saito et al. 1998). In ARB process, the rolled sheet is cut into two equal halves, stacked after surface cleaning and rolled again to 50% reduction. The rolled sheet is again cut, stacked and rolled to 50% reduction. This procedure is repeated several times for accumulating a large amount of reduction (strain), which helps in producing ultrafine grains of the rolled sheet.

Based on the working temperature in the rolling process, the process can be classified into three categories—hot, warm and cold. Cold rolling is performed at a temperature less than 0.3 T_m, warm rolling in the temperature range of 0.3–0.6 T_m, and hot rolling at a temperature greater than 0.6 T_m, where T_m is the melting point of the metal in Kelvin (Hawkins 1985). Due to plastic deformation and friction, the temperature of the sheet increases during the process. The proper estimation of temperature is important for all kinds of rolling processes, because it greatly influences the lubricant, contact behavior, roll wear and the properties of the product (Louaisil et al. 2009). Nowadays, rolling of advanced high strength steels (AHSSs) is gaining importance due to their extensive use in the automotive industry (Kuziak et al. 2008). AHSSs have high strength without sacrificing the ductility. It is recognized that the temperature is the most important parameter that must be controlled during and after hot rolling of AHSSs sheets. Sometimes, the mechanical properties and microstructure of the strip may be affected due to the temperature even in the cold rolling (Khan et al. 2004). In the hot rolling, consideration of temperature is a must as the mechanical and metallurgical properties are greatly influenced by the temperature. The knowledge of the strip and rolls temperature can provide insight into the metallurgical structure of the strip and the lubricant behavior (Tseng 1984). The cold rolling produces the sheet of better mechanical, metallurgical and surface qualities. The main advantage of the hot rolling is that it reduces the flow stress and strain hardening effect and thus reduces the roll torque and roll force. The warm rolling combines the advantages of hot and cold rolling (Subramanian and Bourell 1984). It requires lower loads and energies than the cold rolling and achieves better surface finish and dimensional accuracy than the hot rolling. The thermo-mechanical behavior of the strip and the rolls in warm rolling under steady-state condition has been studied by several researchers (Subramanian and Bourell 1984, Khalili et al. 2012, Koohbor 2015). The warm rolling is becoming an efficient means to produce flat steel products due to its ability to obtain the desired product properties at reduced cost. However, the improper cooling of the roll and the rolled sheet is of great anxiety to mill designers and operators. For proper design of roll cooling schedule, estimation of temperature is necessary and essential.

The objective of this chapter is to discuss the mathematical and numerical techniques for the estimation of the temperature distributions in the roll and sheet. As around 90% of the work of deformation gets converted into heat, it is necessary to carry out proper deformation modeling in order to estimate the temperature distribution. Hence, deformation modeling will also be briefly discussed. Section 3.2 describes the basic heat transfer equations required for thermal modeling of the rolling process. Section 3.3 reviews the various methods for solving the governing heat transfer equations to find out the temperature distributions in the roll and the strip. An integrated thermal model of flat rolling considering plane strain deformation is briefly described in Section 3.4. Section 3.5 concludes the chapter.

3.2 Governing Heat Transfer Equations of the Rolling Process

Modeling of the rolling process can be carried out by two approaches—Eulerian and updated Lagrangian. In the Eulerian formulation, a region fixed in space (called the control volume) is chosen as the domain for the analysis. In the updated Lagrangian formulation, the final deformed configuration is analyzed in several increments using the governing equations in the incremental form. The domain, the deformation, the stresses and the temperatures are updated at the end of the each increment. Heat transfer takes place by conduction in the strip as well as in the roll and by convection as well as radiation at the boundaries. The heat generation due to plastic deformation during rolling significantly affects the thermal behavior of strip (Kim et al. 2009).

3.2.1 Governing Heat Transfer Equation for Strip Temperature

For the updated Lagrangian formulation, the following governing differential equation of heat transfer can be used for finding out the temperature distribution in the strip:

$$\frac{\partial}{\partial x}\left(k_{sx}\frac{\partial T}{\partial x}\right)+\frac{\partial}{\partial y}\left(k_{sy}\frac{\partial T}{\partial y}\right)+\frac{\partial}{\partial z}\left(k_{sz}\frac{\partial T}{\partial z}\right)+\dot{Q}=\rho c_{ps}\frac{\partial T}{\partial t}, \tag{3.1}$$

where k_{si} is the thermal conductivity of the strip-material along i-axis, ρ is the density of the material, c_{ps} is the specific heat capacity of the material at constant pressure, $\dot{Q}$ is the rate of heat generation per unit volume, T is the temperature and t is the time. Equation (3.1) is derived from the principle of conservation of energy. The boundary condition at the free surface is given by

$$-k_{sn}\frac{\partial T}{\partial n}=h_c\left(T-T_\infty\right)+\sigma_{st}\varepsilon_r\left(T^4-T_\infty^4\right), \tag{3.2}$$

where h_c is the coefficient of convective heat transfer, σ_{st} is the Stefan-Boltzmann constant equal to 5.67×10^{-8} W/m^{2}K^4, ε_r is the total emissivity of the surface of the material from which the heat transfer takes place and T_∞ is the ambient temperature. In Eq. (3.2), the left hand term is the conductive heat flux given by

$$-k_{sn}\frac{\partial T}{\partial n}=-k_{sn}\left(\frac{\partial T}{\partial x}n_x+\frac{\partial T}{\partial y}n_y+\frac{\partial T}{\partial z}n_z\right), \tag{3.3}$$

where n_x, n_y, and n_z are the direction cosines of the normal to the surface. On the right hand side of Eq. (3.2), the first term represents the heat transfer due to convection and the second term represents the heat transfer due to radiation.

It is difficult to find out the heat flux at the roll-work interface due to unknown heat transfer coefficient at the interface. An exact analysis will require the modeling of strip and roll together with the lubricant. As a first order simplification, the heat generation due to friction is shared between the roll and the strip. Further, simplification may be achieved by assuming that the heat generated due to friction is shared between the roll and strip in a fixed proportion. For example, considering the value of heat conductivity of the roll to be twice that of the rolled material, Hatta et al. (1980) assumed that 60 to 70% friction heat goes in the rolls and the remaining heat goes in the rolled material. Wilson et al. (1989) analyzed the temperature distribution in the roll gap in rolling. In their model, the local value of the heat partition coefficient (λ, as per the notation of this chapter) is calculated by matching the temperature at the interface. If the local frictional heat flux generated is $\dot{q}_f$, the model assumes that the heat flow into the rolls is $\lambda\dot{q}_f$, whilst the heat flow into the strip is $(1-\lambda)\dot{q}_f$. If the entire frictional heat is transferred to the roll, the value of λ will be 1. A value of λ greater than 1 indicates that apart from the entire frictional heat, a part of the heat contained in the strip due to plastic deformation and initial heating also goes into the roll. Tseng (1984) and Tseng et al. (1990) also obtained λ by equating the temperature of both the roll and the strip at the interface. A similar procedure was followed by Chang (1999).

Another approach is to use the following boundary condition at the roll-strip interface:

$$-k_{sn}\frac{\partial T}{\partial n}=h_c\left(T_s-T_r\right) \tag{3.4}$$

where h_c is the coefficient of heat transfer that includes the effect of lubricant as well as contact resistance. T_s and T_r are the strip and roll temperatures at the interface, respectively.

A significant amount of heat generation takes place in the strip due to plastic deformation. The rate of heat generation per unit volume in the strip may be estimated as:

$$\dot{Q}=\eta\sigma_{ij}\dot{\varepsilon}_{ij} \tag{3.5}$$

where η lies in the range 0.85–0.95 (Jiang et al. 2004). When the strip comes out of the bite zone and is in the inter-stand zone or before the coiler, $\dot{Q}$ is zero.

In the Eulerian reference frame, the governing differential equation for heat transfer in the strip can be written as:

$$\frac{\partial}{\partial x}\left(k_x\frac{\partial T}{\partial x}\right)+\frac{\partial}{\partial y}\left(k_y\frac{\partial T}{\partial y}\right)+\frac{\partial}{\partial z}\left(k_z\frac{\partial T}{\partial z}\right)+\dot{Q}=\rho c_{ps}\left(\frac{\partial T}{\partial t}+v_x\frac{\partial T}{\partial x}+v_y\frac{\partial T}{\partial y}+v_z\frac{\partial T}{\partial z}\right). \tag{3.6}$$

Here, v_x, v_y and v_z are the x, y and z components of the velocity of the strip, respectively. Except in the bite zone, the components v_y and v_z are zero. Even in the bite zone, these components are negligible.

3.2.2 Governing Heat Transfer Equation for Roll Temperature

In the updated Lagrangian formulation, the governing differential equation for heat transfer in the orthotropic roll in the cylindrical system is expressed as:

$$\frac{\partial}{\partial r}\left(k_{Rr}\frac{\partial T}{\partial r}\right)+\frac{1}{r}\left(k_{Rr}\frac{\partial T}{\partial r}\right)+\frac{1}{r^2}\frac{\partial}{\partial\theta}\left(k_{R\theta}\frac{\partial T}{\partial\theta}\right)+\frac{\partial}{\partial z}\left(k_{Rz}\frac{\partial T}{\partial z}\right)=\rho c_{pR}\frac{\partial T}{\partial t}, \tag{3.7}$$

where k_{Ri} is the thermal conductivity of the roll in the ith direction, ρ is the density of the material, c_{pR} is the specific heat capacity of the material at constant pressure, $T\equiv T(r,\theta,z,t)$ is the temperature and t is the time. Usually, the material of the roll is considered isotropic. In the Eulerian reference frame, for the isotropic roll, the governing differential equation is:

$$\frac{\partial}{\partial r}\left(k_r\frac{\partial T}{\partial r}\right)+\frac{1}{r}\left(k_r\frac{\partial T}{\partial r}\right)+\frac{1}{r^2}\frac{\partial}{\partial\theta}\left(k_r\frac{\partial T}{\partial\theta}\right)+\frac{\partial}{\partial z}\left(k_r\frac{\partial T}{\partial z}\right)=\rho c_{pr}\left(\frac{\partial T}{\partial t}+\omega\frac{\partial T}{\partial\theta}\right), \tag{3.8}$$

where ω is the angular velocity of the roll and from now onwards notation k_r will be used to denote the thermal conductivity of an isotropic roll. The thermal properties and heat transfer coefficients are dependent on the temperature. Thus, in general, heat transfer analysis of the rolling processes is non-linear in nature. The solution must satisfy the initial condition $T = T_0(r, \theta, z)$ and the boundary conditions

$$k_r \frac{\partial T}{\partial r} + h_e\left(T - T_e\right) = \dot{q}\left(\theta, z\right) \text{ at outer surface,} \tag{3.9}$$

$$-k_r \frac{\partial T}{\partial r} + h_i\left(T - T_i\right) = 0 \text{ at inner surface.} \tag{3.10}$$

Here, $\dot{q}(\theta, z)$ is the heat flux entering in the roll, whilst h_e and h_i are the convective heat transfer coefficients at the outer and inner periphery of the work-roll, respectively. T_e and T_i are the ambient temperatures at the outer and inner periphery of the work-roll, respectively. The $\dot{q}(\theta, z)$ will be non-zero only at the bite zone. The effect of external cooling by water or air jet is taken into account through h_e.

3.3 Methods for Solving Heat Transfer Equations

In the literature, various methods have been developed for solving the governing heat transfer differential equation. The analytical and numerical methods are used to obtain the temperature field in the roll and the strip. Mostly, the analytical methods are based on separation of variables or integral transform techniques and the numerical methods include the finite difference method (FDM) or finite element method (FEM). Several researchers obtained the steady-state as well as the transient temperature distribution in the rolls and the strip, whilst many concentrated on the steady-state solutions. In the following subsections, the different methods for computing the temperature distribution in the rolls and the strip are summarized.

3.3.1 Method of Separation of Variables

In the method of separation of variables, the underlying assumption is that the solution function can be broken down to a product of two or more functions, which individually contain the lesser number of the variables than the original solution function. For example, in a one-dimensional heat conduction if the temperature is a function of the position x and time t, it is assumed that it can be decomposed in the following manner:

$$T(x, t) = F(x)G(t). \tag{3.11}$$

Substitution of the decomposed form in the governing partial differential equation and separating the terms on the basis of variables, different differential equations are obtained for different groups of variables. If all of them happen to be ordinary differential equations, they are solved by suitable methods for solving ordinary differential equations. If one or more are partial differential equations, further decomposition can be employed to get the ordinary differential equations or any other method may be employed for solving them. The constants present in the general solutions are determined by applying initial and boundary conditions. Invariably, the implementation of the boundary conditions provides an eigenvalue problem and the solution is obtained as the weighted summation of eigenfunctions. The coefficients associated with eigenfunctions are found out by applying initial conditions and by exploiting the orthogonality property of the eigenfunctions.

The mathematical model of Patula (1981) uses a classical approach of separation of variables with complex functions to obtain the steady-state temperature distribution in the roll. An Eulerian reference frame has been used. The axial heat conduction in the roll is neglected considering the length of the roll to be large. The thermal properties are considered to be temperature-independent. However, analysis of the strip has not been carried out.

Tseng et al. (1990) developed an analytical model to determine the temperature distribution of the roll and strip using an Eulerian approach. The strip temperature is obtained by solving the governing heat transfer equation by using the method of separation of variables. The roll temperature is estimated by using the Fourier integral technique. At the roll-strip interface, the average temperatures of the roll and the strip are matched. Thus, a perfect contact between the roll and the strip is assumed. Yuan (1985) has considered a scale layer between the roll and the strip. The heat generation due to the plastic deformation and friction is calculated numerically by Maslen and Tseng (1981) model, which is essentially a slab method model.

Arif et al. (2004) studied the temperature distribution in the roll and the strip based on the work of Patula (1981) and Tseng et al. (1990). However, the difference is that they considered a non-uniform heat input into the roll from the strip for the analysis. They carried out the deformation as well as steady-state thermal analysis of the strip. The flow stress model in the strip takes into account the effect of the strain, strain-rates and the temperature. However, the roll deformation is not considered. An error of not more than 5% is obtained by employing the assumption of uniform heat input to the roll.

3.3.2 Integral Transform Technique

The integral transform of a function $f(t)$ is defined by the following definite integral:

$$F(s) = \int_a^b K(s,t) f(t) \, dt. \tag{3.12}$$

This function $F(s)$ is called the integral transform and $K(s,t)$ is called the integral kernel of the transform. The integral transform method solves the partial differential equations along with initial and boundary conditions. The process consists of three main steps: (i) the given differential equation gets transformed in an algebraic form as a result of integral transformation, (ii) the algebraic equation is solved and (iii) the solution of the algebraic equation is transformed back to obtain the solution of the given problem. A special case of the integral transform is the Laplace transform, in which the kernel is $\exp(-st)$ and the limits a and b are 0 and ∞ respectively.

Fischer et al. (2001) used a Laplace transform technique to find the temperature distribution in a half-plane heated due to rolling/sliding contact. They neglected the heat conduction along the direction of relative motion. Fischer et al. (2004) proposed an approximate solution for plane strain hot rolling using Laplace transform technique and compared it with the FEM results of Sun et al. (1998) and Hwang et al. (2002). The maximum deviation is lesser than 10%.

A transient analysis was carried out by DesRuisseaux and Zerkle (1970) using the Laplace transform technique. They presented an analytical approach to compute transient temperature distributions and circumferential stresses in a work-roll surface. Guo (1998) proposed a Laplace transform method to obtain transient solution for a particular boundary condition. The Duhamel's theorem was used to obtain the complete solution for various boundary conditions. However, the thermal analysis of the strip was not carried out.

3.3.3 Finite Difference Method

The finite difference method (FDM) is numerical method for solving differential equations by approximating them to difference equations, in which finite differences approximate the derivatives. The domain is partitioned in space as well as in time and solution is calculated at the discrete points in the space and time. In essence, the differential equation gets converted into a system of simultaneous algebraic equations. To obtain the temperature between two adjacent grid points of the finite difference grid, the interpolation can be carried out.

Lahoti et al. (1978) presented an FDM model to find out the temperature distribution in the roll and the strip. Tseng (1984) developed an FDM model to estimate the temperature distribution in the roll and the strip. The input data for the heat generation was taken by the direct measurement of power. Further, 6.5% of the total 90% power was assumed to dissipate as friction heat. An Eulerian formulation is employed for minimizing the number of grid points used in the formulation. Considering high speed rolling, an upwind differencing scheme is chosen to overcome the numerical instability. It is assumed that the surface temperatures of the roll and the strip are equal due to negligible thermal resistance of the film at the interface, i.e.,

$$(T_r)_b = (T_s)_b, \tag{3.13}$$

where subscript b represents boundary. As per Eq. (3.13), the heat flux into the roll and the strip at the interface must equal the heat generated by friction, i.e.,

$$k_r \left(\frac{\partial T}{\partial r} \right)_{br} + k_s \left(\frac{\partial T}{\partial n} \right)_{bs} = \dot{q}_f, \tag{3.14}$$

where $\partial/\partial n$ represents differentiation along the normal of the strip boundary (positive outward), k_r and k_s are the thermal conductivity of the roll and the strip respectively, $\dot{q}_f$ is the rate of heat generation due to friction at the roll-strip interface. The boundary condition for remaining region of the roll circumference is:

$$k_r \frac{\partial T(R,\theta)}{\partial r} = h_c(\theta)\{T_\infty - T(R,\theta)\}, \tag{3.15}$$

where T_∞ is the ambient or coolant temperature and $h_c(\theta)$ is the convective heat transfer coefficient as a function of angular location θ. However, the method required too much computational time. Wilson et al. (1989) estimated temperature distributions at the roll-strip interface in cold rolling using an FDM model. Only advection parallel to the roll-strip interface and conduction along the thickness direction is considered. The heat distribution between the roll and the strip is obtained by matching the surface temperatures of the roll and the strip.

Chang (1999) used an FDM model coupled with the assumptions of a steady state rolling condition and a non-uniform heat flux in the deformation zone to predict the work-roll temperature field and the resulting stresses. Luo and Keife (1998) used an FDM model to investigate the temperature distribution in the roll and the foil, considering the effect of lubricant. Khan et al. (2004) studied the temperature distribution in the roll and the strip using FDM. FDM is a well-established technique for determining the temperature distribution in the roll and the strip. However, it requires a lot of computational time.

3.3.4 Finite Element Method

The finite element method (FEM) has attracted a number of researchers for modelling of the rolling process. It gives detailed information about the required roll torque, roll separating force, stress, strain, strain-rate as well as the roll and strip temperatures distribution of the process. The models can be two dimensional (plane strain) or three dimensional. FEM has been widely used for the modeling of deformation in the rolling. However, like FDM, FEM also requires too much computational time. It is to be noted that transient analysis in FEM follows the FDM procedure for time discretisation. Shifting of the preference of the researchers from FDM to FEM can also be attributed to the availability of a number of commercial FEM packages. A brief literature review on the estimation of temperature in rolling is presented below.

Hwang et al. (1993) carried out both deformation and temperature analyses using penalty rigid-viscoplastic FEM. An iterative scheme is adopted for metal flow and temperature in hot strip rolling. It is assumed that the friction heat is equally divided into the roll and the strip. Sheikh (2009) presented an FEM model coupled with an upper bound method to predict the temperature distribution during hot strip rolling. Khalili et al. (2012) used combined FEM and slab method to predict the thermo-mechanical behavior of work rolls and strip during warm strip rolling.

Apart from the steady-state analyses, transient analyses of roll and the strip are also discussed in the literature. A number of mathematical models are developed to estimate the transient temperature distribution in rolling. Lee et al. (2000) presented an FEM based three-dimensional thermal analysis of the roll. The deformation analysis was not carried out and a constant value of heat flux input into the roll was assumed. The authors have not discussed the issue of computational time in their paper. The predicted results match well with the model of Guo (1998) for two-dimensional case.

Pietrzyk and Lenard (1990) studied the effect of the temperature rise of the roll during flat rolling by FEM and experiments. They conducted the rolling experiments at different working temperatures and compared the results of FEM model with experimental data. Galantucci and Tricarico (1999) carried out the thermo-mechanical simulation of rolling process. They studied the sensitivity of rolling parameters, *viz.*, rolling speed, roll radius and reduction in strip thickness on the temperature. However, detailed description of the FEM formulation has not been provided.

Using FEM, Serajzadeh et al. (2002) obtained the transient temperature distribution in the roll and the strip in a multi-stand mill. They also conducted the experiments to validate the FEM model. The hybrid model to analyze the hot rolling was proposed by Serajzadeh (2006). The thermal and deformation analyses are carried out by FEM. During analysis, the flow stress data as a function of strain, strain-rate and temperature is supplied with the help of an artificial neural network (ANN). The model was examined on a low carbon steel strip. The computational results agree well with experimental data.

Kiuchi et al. (2000) developed a three-dimensional FEM model to investigate the temperature distribution in the roll and the strip. The heat generations due to friction and plastic deformation are estimated by FEM. Tudball and Brown (2006) developed a transient three-dimensional FEM model to estimate the temperature distribution in the hot rolling of steel. The model is validated with plant recorded data. Yu et al. (2012) studied the temperature distribution in the rolling of cold magnesium alloy strips with heated roll.

FEM and FDM based models provide a reasonably accurate temperature distribution in the roll and the strip, but these methods require a lot of computational time. The analytical solutions are usually in the form of summation of some infinite series, in which a judicious decision is required regarding the finite number of terms needed for convergence. In view of it, Yadav et al. (2013) proposed a methodology to enhance the computational efficiency of an FEM model. The methodology was tested on a FEM package. It can work well with other similar packages. More than 10 times reduction in computational time could be obtained without sacrificing the accuracy significantly.

3.4 A Plane Strain Thermo-mechanical Analysis of Flat Rolling

In this section, a two-dimensional thermo-mechanical analysis of the flat rolling process is described. The analysis comprises two modules–deformation module and thermal module. Deformation module is based on the Eulerian flow formulation of FEM. Thermal module uses analytical methods. Initially the deformation module takes the mechanical properties of the strip corresponding to the entry temperature. The deformation and friction powers provided by the deformation module are used as input in thermal module. From the output of the thermal module, the average temperature of the deformation zone can be obtained. Now, deformation module is activated again to carry out the analysis with mechanical properties corresponding to the average temperature of the deformation zone. This iterative procedure continues till the convergence is achieved, i.e., till the average temperature of deformation zone does not deviate by 1% in two consecutive iterations. A brief description of the deformation and thermal modules is provided in the following subsections.

3.4.1 Deformation Module

An FEM based Eulerian flow formulation is employed to analyze stresses, roll torque, roll-force, strain, strain-rate and forward slip considering plane strain condition. This formulation is based on steady-state deformation of the strip. Due to symmetry, only upper half domain is considered in the analysis. The FEM meshed domain of strip along with the boundary conditions is shown in Figure 3.1. A fine mesh is used in the deformation zone and coarse meshes are used in the inlet and exit zones. The element height is gradually reduced from the center to the surface of the strip. This is because the stress gradient is expected to be higher in the vicinity of the roll-strip interface. As the goal is to find out the powers (plastic deformation as well as friction), the elasticity is neglected and the plastic strain can be considered as the total strain. Roll deformation is taken into account using well-known Hitchcock's formula (Dixit and Dixit 2008).

The rigid-plastic material behavior is assumed. As per the Levy-Mises flow rule, the deviatoric part S_{ij} of the component of stress tensor σ_{ij} is related to the strain-rate $\dot{\varepsilon}_{ij}$ as

$$S_{ij} = 2\eta_1 \dot{\varepsilon}_{ij}, \tag{3.16}$$

where η_1 is the proportionality factor. It is to be noted that the plastic incompressibility condition is implicit in Eq. (3.16). The use of von Mises yield criterion provides

$$\eta_1 = \frac{\sigma_y}{3\dot{\varepsilon}_{eq}}, \tag{3.17}$$

where $\dot{\varepsilon}_{eq}$ is the equivalent strain-rate defined by

$$\dot{\varepsilon}_{eq} = \sqrt{\frac{2}{3}\dot{\varepsilon}_{ij}\dot{\varepsilon}_{ij}}. \tag{3.18}$$

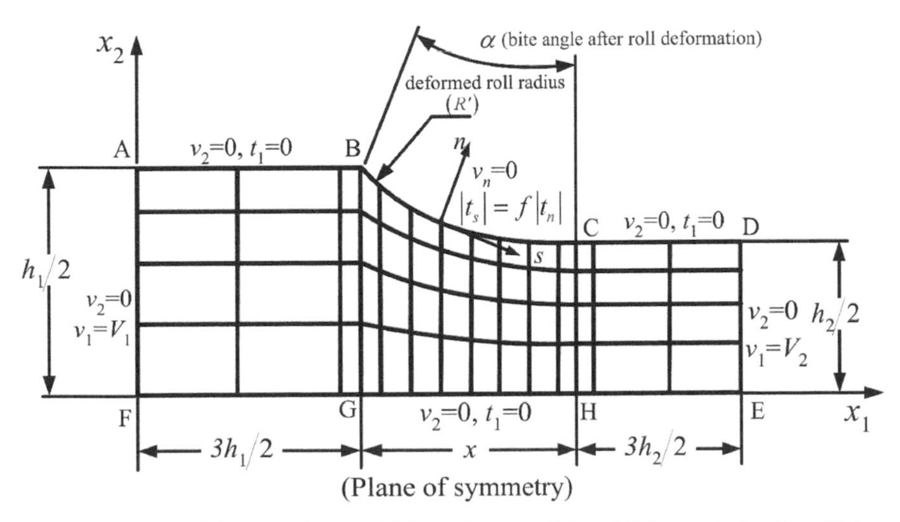

Figure 3.1 A meshed domain of strip with boundary conditions. With permission from Yadav et al. (2015). Copyright 2015 Elsevier.

In the above expression, the Einstein summation convention is employed. Considering plane strain, the indices i and j vary from 1 to 2. In unabridged form,

$$\dot{\varepsilon}_{eq} = \sqrt{\frac{2}{3}\left(\dot{\varepsilon}_{xx}^2 + \dot{\varepsilon}_{yy}^2 + 2\dot{\varepsilon}_{xy}^2\right)}. \tag{3.19}$$

The flow stresses of the strip may be described by Johnson-Cook (J-C) model (Johnson and Cook 1983):

$$\sigma_y = \left(A + B\varepsilon_{eq}^{n_1}\right)\left[1 + C\ln\frac{\dot{\varepsilon}_{eq}}{\dot{\varepsilon}_0}\right]\left[1 - \left(\frac{T - T_{amb}}{T_m - T_{amb}}\right)^{m_1}\right], \tag{3.20}$$

where A, B, C, m_1 and n_1 are the material constant, $\dot{\varepsilon}_0$ is the reference strain-rate, T_{amb} is the ambient temperature of the strip, and T_m is the melting temperature of the strip. The J-C model has a fairly wide range of applicability. The other models can also be employed without affecting the FEM procedure.

Two-dimensional continuity and momentum equation for the steady state process are solved, which are given in index notation as:

$$\frac{\partial v_i}{\partial x_i} = 0, \tag{3.21}$$

$$\frac{\partial \sigma_{ij}}{\partial x_j} \equiv -\frac{\partial p}{\partial x_i} + \frac{\partial S_{ij}}{\partial x_j} = 0, \tag{3.22}$$

where p is the hydrostatic pressure. The FEM procedure is similar to the cold rolling model of Dixit and Dixit (1996) except that the material models contain the effect of strain-rate and temperature. Wanheim and Bay (1978) friction model is used instead of Coulomb's model. At low normal pressure, the Wanheim and Bay (1978) model matches with the Coulomb's model and at high normal pressure, it approaches constant friction factor model. In between there is a smooth transition zone. In the notations of this chapter, f is the ratio of the magnitudes of tangential and normal tractions at the roll-work interface and μ is the equivalent Coulomb coefficient of friction. For further details about FEM procedure using Wanheim and Bay's friction model, the reader may refer the book by Dixit and Dixit (2008).

3.4.2 Thermal Module

In this section, thermal modeling of the roll and the strip is described using an analytical method. The solution of the governing heat transfer equation of the roll and the strip is presented separately. The temperature distribution in the roll is obtained by two different methods. The first method is based on the work of Fischer et al. (2004). The steady-state temperature distribution is obtained within the roll along the radial as well as circumferential directions. The second method presents the solution to obtain the transient temperature distribution within the roll using integral transform technique. In the following sub-sections, a brief description of both the approaches is presented.

3.4.2.1 An Approximate Method for Determining Steady-state Temperature Distribution in the Roll

The temperature distribution within the roll is obtained by solving two-dimensional heat transfer equation considering Eulerian reference frame. The following assumptions are made:

1) The heat transfer along the axial direction to the roll is negligible. Thus, the problem is essentially two-dimensional.
2) Thermal properties of the roll material are independent of the temperature. These are taken as the properties at average temperature.
3) Heat flux input into the roll is uniform along the roll-work interface.
4) The rotational speed of the roll is constant.

Figure 3.2 shows a rotating roll with heat flux at a fixed location. The governing heat conduction equation in the steady-state is:

$$\frac{\partial^2 T}{\partial x^2}+\frac{\partial^2 T}{\partial y^2}=\frac{\omega R}{\alpha_r}\frac{\partial T}{\partial x}, \tag{3.23}$$

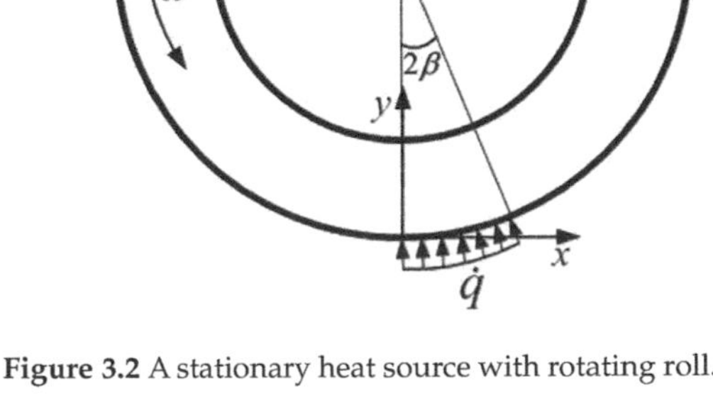

Figure 3.2 A stationary heat source with rotating roll.

where ω is the angular velocity of roll, R is the radius of roll and α_r is the thermal diffusivity of roll. The Peclet number (P_e), the ratio of advective transport rate to diffusive transport rate, is given by

$$P_e = \frac{\omega R^2 \beta}{2\alpha_r}. \tag{3.24}$$

For large Peclet number, the heat conduction only in the y (or radial) direction is relevant (Fischer et al. 2004). Thus, Eq. (3.23) becomes

$$\frac{\partial^2 T}{\partial y^2} = \frac{\omega R}{\alpha_r}\frac{\partial T}{\partial x}, \tag{3.25}$$

with initial condition:

$$T = T_0 \quad \text{at} \quad t = 0 \tag{3.26}$$

and boundary condition:

$$k_r \frac{\partial T}{\partial y} = \begin{cases} -\dot{q}(x) & \text{for } \; 0 \le x \le 2b\beta, \; y = 0 \\ 0 & \text{for } \; x > 2b\beta, \; y = 0 \end{cases}. \tag{3.27}$$

Considering non-dimensional coordinates with $x = 2\beta b\xi$, $y = \delta\zeta$, where δ is heat penetration depth given by (Fischer et al. 2004)

$$\delta = \sqrt{\frac{2b\beta\alpha_r}{\omega R}}, \tag{3.28}$$

the temperature distribution for zero initial temperature in the roll is obtained as (Fischer et al. 2004)

$$T(\xi,\zeta) = T_{\max}\left\{\sqrt{\xi}\exp\left(-\frac{\zeta^2}{4\xi}\right) - \frac{\sqrt{\pi}}{2}\zeta\left(1-\operatorname{erf}\left(\frac{\zeta}{2\sqrt{\xi}}\right)\right)\right\} \text{ for } 0 \le \xi \le 1, \tag{3.29}$$

$$T(\xi,\zeta) = T_{\max}\left\{\begin{array}{l} \sqrt{\xi}\exp\left(-\frac{\zeta^2}{4\xi}\right) - \frac{\sqrt{\pi}}{2}\zeta\left(1-\operatorname{erf}\left(\frac{\zeta}{2\sqrt{\xi}}\right)\right) \\ -\left(\sqrt{\xi-1}\exp\left(-\frac{\zeta^2}{4(\xi-1)}\right) - \frac{\sqrt{\pi}}{2}\zeta\left(1-\operatorname{erf}\left(\frac{\zeta}{2\sqrt{\xi-1}}\right)\right)\right) \end{array}\right\} \text{ for } \xi > 1, \tag{3.30}$$

where $T_{\max}$ is given as:

$$T_{\max} = \sqrt{\frac{8}{\pi}\frac{b\beta\alpha_r}{\omega R}}\frac{\dot{q}}{k_r}, \tag{3.31}$$

and $\dot{q}$ is the constant heat flux in the interval [0, 2β]. Equations (3.29) and (3.30) provide the temperature distribution for initial condition of zero temperature distribution. For the steady-state case, a constant temperature $T_o(\tau)$ is imposed at radial location corresponding to τ such that

$$T_o(\tau) + \frac{\int_0^{(\pi/\beta)} T(\xi,\tau)\,d\xi}{(\pi/\beta)} = T_{avg}(\tau), \tag{3.32}$$

where $T_{avg}(\tau)$ is the average temperature of the surface at the radial location corresponding to τ, which is obtained as follows.

At steady state, the heat balance provides (Yadav et al. 2014)

$$2\pi a h_i (T_a - T_i) + (2\pi - 2\beta) b h_e (T_b - T_e) = 2\beta b \dot{q}, \tag{3.33}$$

where $\dot{q}$, a, b, T_a, T_b, T_i, T_e, h_i and h_e are the heat input, inner radius of roll, outer radius of roll, average temperature of inner radius, average temperature of outer radius, ambient temperature at inner roll periphery, ambient temperature at outer roll periphery, convective heat transfer coefficient at the inner radius and convective heat transfer coefficient at the outer radius of the roll, respectively. As there are two unknowns *viz.*, T_a and T_b, one additional equation is needed, which is provided by the boundary condition at the inner periphery of the roll:

$$\left\{ -k_r \frac{\partial T}{\partial r} + h_i (T_a - T_i) \right\}_{r=a} = 0. \tag{3.34}$$

Substituting a finite difference approximation of temperature gradient in Eq. (3.34), the following equation is obtained:

$$h_i (T_a - T_i) = 2bk_r \frac{T_b - T_a}{b^2 - a^2}. \tag{3.35}$$

Solving Eqs. (3.33) and (3.35), the average temperatures T_a and T_b along the roll periphery are obtained. It can be assumed that average temperatures are linear function of radial distance and thus knowing the outer and inner average temperatures, the average temperature $T_{avg}(\tau)$ at any radial distance can be obtained. Equation (3.32) can now provide temperature $T_o(\tau)$ to be superimposed on the temperature distribution for different radial locations.

3.4.2.2 Series Solution using Integral Transform Technique for the Transient Temperature Distribution in the Roll

In this subsection, the solution of two-dimensional heat conduction equation with time-dependent, non-homogeneous boundary condition is presented using Lagrangian approach. Let the heat input be given at outer radius b in the zone 2β angular displacement as shown in Figure 3.3. Remaining outer surface of the roll is subjected to convective heat loss. The inner surface at radius a is subjected to convective loss only. The work roll is assumed to be rigid, fixed in space and heat source rotates with constant angular velocity ω.

Consider the roll as a hollow cylinder, the governing differential equation for heat conduction is given for the domain $a < r < b$ as:

$$\frac{\partial^2 T}{\partial r^2} + \frac{1}{r}\frac{\partial T}{\partial r} + \frac{1}{r^2}\frac{\partial^2 T}{\partial \theta^2} = \frac{1}{\alpha_r}\frac{\partial T}{\partial t} \tag{3.36}$$

where α_r is the thermal diffusivity of work-roll roll given by:

$$\alpha_r = \frac{k_r}{\rho_r c_{pr}}. \tag{3.37}$$

In Eq. (3.37), k_r is the thermal conductivity of roll material, ρ_r is the mass density and c_{pr} is the specific heat. The boundary conditions are expressed as:

$$k_r \frac{\partial T}{\partial r} + h_e\left(T - T_e\right) = \dot{q}\left(\theta\right) \text{ at } r = b, \tag{3.38}$$

$$-k_r \frac{\partial}{\partial} + h_i\left(T - T_i\right) = \text{ at } r = a, \tag{3.39}$$

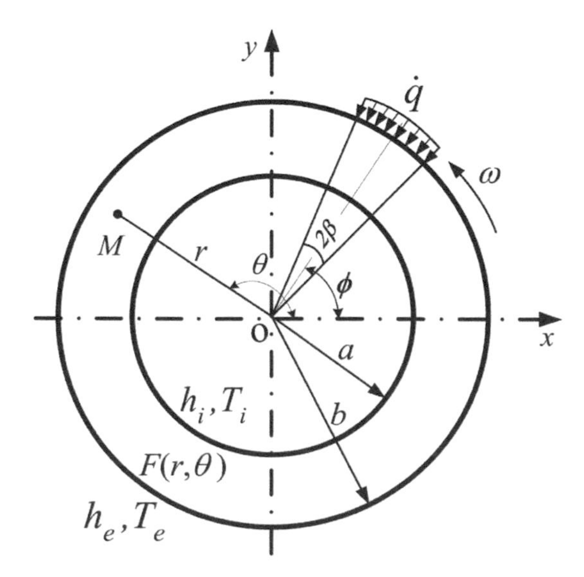

Figure 3.3 A moving heat sources with stationary roll.

where $\dot{q}(\theta)$ is the heat flux at the outer surface, h_e is the convective heat transfer coefficient at the outer surface, h_i is the convective heat transfer coefficient at the inner surface, T_i is the ambient temperature at the inner roll periphery and T_e is the ambient temperature at the outer roll periphery. The heat flux at the outer surface at any time is expressed as:

$$\dot{q}(\theta) = \begin{cases} \dot{q} & \text{for } \phi - \beta \leq \theta \leq \phi + \beta \\ 0 & \text{for } \phi + \beta \leq \theta \leq 2\pi + \phi - \beta \end{cases}, \tag{3.40}$$

where $\phi = \omega t$. The initial condition is expressed as:

$$T(r,\theta,t) = F(r,\theta) \quad \text{for } t = 0 \text{ in region } a < r < b, \tag{3.41}$$

where $F(r, \theta)$ is a specified function of r and θ.

The heat conduction equation is solved using the integral transform method consisting of the following steps (Özişik 1993):

i) Appropriate formula for the integral transform and corresponding inversion formula are developed.
ii) By the application of integral transformation, the partial derivatives with respect to the space variables are removed from the heat conduction equation, thus reducing it to an ordinary differential equation for the transform of temperature.
iii) The resulting ordinary differential equation is solved subject to the transformed initial condition. The desired solution is obtained by inverting the solution, available as the transform of the temperature.

The first step provides the inversion formula by employing the method of separation of variables in Eq. (3.36). The integral transform and inversion formula are given as (Yiannopoulos et al. 1997)

Integral transform:

$$\bar{T}(\beta_m, n, \phi, t) = \int_a^b \int_0^{2\pi} r R_n(\beta_m, r) \cos n(\theta - \phi) T(r,\theta,t) \, d\theta dr, \tag{3.42}$$

Inversion formula:

$$T(r,\theta,\phi,t) = \frac{\varepsilon}{\pi} \sum_{n=0}^{\infty} \sum_{m=1}^{\infty} \frac{R_n(\beta_m, r)}{N_n(\beta_m)} \bar{T}(\beta_m, n, \phi, t), \tag{3.43}$$

where,

$$\varepsilon = \begin{cases} 1/2 & \text{for } n = 0 \\ 1 & \text{for } n \geq 1 \end{cases}. \tag{3.44}$$

In Eqs. (3.42) and (3.43), n is positive integer or zero, and β_m, $m = 1,2,3,\ldots$, are the positive roots of the transcendental characteristics equations:

$$K_n L_n - V_n W_n = 0, \tag{3.45}$$

and $\bar{T}(\beta_m, n, \phi, t)$ is the integral transform of the function $T(r, \theta, t)$ with respect to the space variable r and θ. Appendix provides the detailed expression of the eigenfunctions $R_n(\beta_m, r)$, normality $N_n(\beta_m)$ and transcendental equation.

Solution of Eqs. (3.36)–(3.39) provides

$$T(r,\theta,t) = \frac{\varepsilon}{\pi}\sum_{n=0}^{\infty}\sum_{m=1}^{\infty}\frac{R_n(\beta_m,r)}{N_n(\beta_m)}\left[\tilde{\bar{F}}(\beta_m,n)\exp\left(-\alpha_r\beta_m^2 t\right) - \left\{\frac{A(\beta_m,n,t)}{\alpha_r\beta_m^2}\right\}\left\{1-\exp\left(-\alpha_r\beta_m^2 t\right)\right\}\right], \tag{3.46}$$

where,

$$A(\beta_m,n,t) = \alpha_r\left(-n\bar{T} - H_e\bar{T}_e b + \frac{\bar{q}(n)}{k_r}\right)R_n(\beta_m,b) + \alpha_r\left(n\bar{T} - H_e\bar{T}_e a\right)R_n(\beta_m,a), \tag{3.47}$$

$$\bar{q}(n) = \int_0^{2\pi} q(\theta)\cos n(\theta-\phi)\,\mathrm{d}\theta, \tag{3.48}$$

$$\tilde{\bar{F}}(\beta_m,n) = \int_a^b\int_0^{2\pi} rR_n(\beta_m,r)\cos n(\theta-\phi)F(r,\theta)\,\mathrm{d}\theta\,\mathrm{d}r \tag{3.49}$$

Substituting the expression in Eqs. (3.47)–(3.49) into Eq. (3.46), using the inversion formula given in Eqs. (3.42, 3.43) and after simplifying Eq. (3.46) using the Duhamel's theorem for time dependent boundary conditions problem, the temperature distribution is obtained as (Yadav et al. 2015)

$$\begin{aligned}
T(r,\theta,t) = {} & \frac{1}{2\pi}\sum_{m=1}^{\infty}\frac{R_0(\beta_m,r)}{N_0(\beta_m)}\exp\left(-\alpha_r\beta_m^2 t\right)\tilde{\bar{F}}(\beta_m,0) \\
& + \left(\pi H_e T_e + \frac{\dot{q}}{k_r}\beta\right)\sum_{m=1}^{\infty}\frac{R_0(\beta_m,r)}{F_0}\left(1-\exp\left(-\alpha_r\beta_m^2 t\right)\right) \\
& + \frac{\pi^2}{2}H_i a T_i\sum_{m=1}^{\infty}\frac{R_0(\beta_m,r)}{F_0}R_0(\beta_m,a)\left(1-\exp\left(-\alpha_r\beta_m^2 t\right)\right) \\
& + 2\frac{\dot{q}}{k_r}\sum_{n=1}^{\infty}\sum_{m=1}^{\infty}\frac{R_n(\beta_m,r)}{F_n}\frac{\sin n\beta}{n} \\
& \frac{\cos n(\theta-\omega t) - \lambda_n\sin n(\theta-\omega t) + \left(\lambda_n\sin n\theta - \cos n\theta\right)\exp\left(-\alpha_r\beta_m^2 t\right)}{1+\lambda_n^2},
\end{aligned} \tag{3.50}$$

where $\lambda_n = \omega n / \alpha_r \beta_m$ and the function F_n is defined in Appendix. Equation (3.50) was developed by Yadav et al. (2015) and is an extension of the work of Yiannopoulos et al. (1997) to incorporate non-uniform initial temperature.

3.4.2.3 A method for Determining Temperature Distribution in the Strip

The temperature distribution in the strip can be obtained by solving the heat conduction equation in the bite zone (BCHG) and the outer zone (CDEH) as shown in Figure 3.4. It is assumed that the plastic deformation occurs in the strip in the domain BCHG, although the actual plastic deformation zone depends on the process parameters. It is observed that about 90% of the work of plastic deformation gets converted into heat (Jiang et al. 2004, Khalili et al. 2012). The heat generated due to frictional work is distributed in the roll and the strip. Assume that the total heat Q_T comprises the heat due to plastic deformation, frictional heat and heat due to moving of hot strip. A portion of this heat $\lambda\, Q_T$ goes into the roll and $(1-\lambda)\, Q_T$ remains in the strip. The value of heat partition factor λ is obtained by matching the average surface temperature of the roll and the strip at roll-strip interface. During transient state, λ may vary with time. In that case, $\dot{q}$ in Eq. (3.50) will vary with time. One easy way is to carry out analysis in sequential intervals of time. In each interval $\dot{q}$ is considered as constant and $F(r, \theta)$ is taken as the temperature distribution at the end of the last time interval.

Considering the heat transfer in the longitudinal direction of the strip insignificant, Kim et al. (2009) solved one dimensional heat conduction equation by eigenfunction method. The following differential equation is solved for the bite zone:

$$\frac{\partial}{\partial y}\left(k_s \frac{\partial T(y,t)}{\partial y} \right) + \dot{Q}(y) = \rho_s c_{ps} \frac{\partial T(y,t)}{\partial t}, \tag{3.51}$$

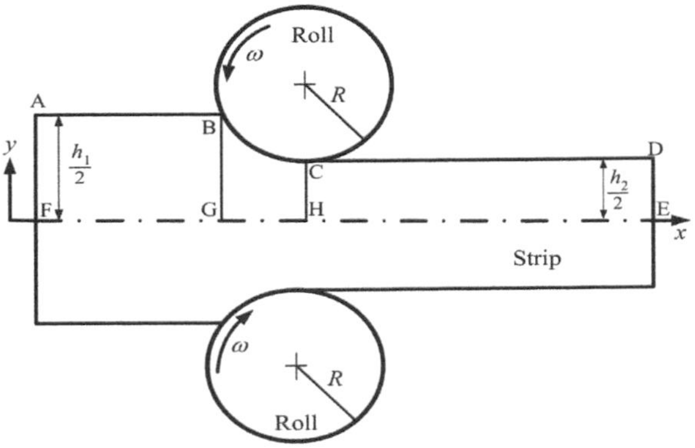

Figure 3.4 A schematic diagram of flat rolling.

with the boundary conditions:

$$\frac{\partial T(y,t)}{\partial y} = 0 \quad \text{at} \quad y = 0, \tag{3.52}$$

$$k_s \frac{\partial T(y,t)}{\partial y} = \dot{q}_s \quad \text{at} \quad y = h, \tag{3.53}$$

where ρ_s is the density, c_{ps} is the specific heat, k_s is the thermal conductivity and h is the average semi-thickness of the strip. The initial condition is:

$$T(y,0) = T_0(y). \tag{3.54}$$

Temperature distribution as per the coordinate system of Figure 3.4 is given by:

$$T_e(y,t) = \frac{1}{h}\int_0^h T_0(y)\,\mathrm{d}y + \frac{1}{h\rho_s c_{ps}}\left(\dot{q}_s + \int_0^h \dot{Q}\,\mathrm{d}y\right)t + \sum_{n=1}^{\infty}$$

$$\left[\left\{\exp\left(-\lambda_n^2 \frac{k_s}{\rho_s c_{ps}} t\right)\left\{a_n(0) + \frac{2(-1)^n}{h\rho_s c_{ps}}\int_0^t \left(\dot{q}_s \exp\left(\lambda_n^2 \frac{k_s}{\rho_s c_{ps}} t\right)\right)\mathrm{d}t\right\}\right\}\right.$$

$$\left. + \frac{2}{\lambda_n^2 k_s h}\left(1 - \exp\left(-\lambda_n^2 \frac{k_s}{\rho_s c_{ps}} t\right)\right)\int_0^h \dot{Q}\cos(\lambda_n y)\,\mathrm{d}y\right]\cos(\lambda_n y), \tag{3.55}$$

$$\lambda_n = \frac{n\pi}{h}, \tag{3.56}$$

$$a_n(0) = \frac{2}{h}\int_0^h T_0(y)\cos(\lambda_n y)\,\mathrm{d}y, \tag{3.57}$$

where $T_0(y)$ is the initial strip temperature, h is the average thickness of strip at deformation zone, k_s is the thermal conductivity of strip, ρ_s is the density of strip, c_{ps} is the specific heat of strip, $\dot{Q}$ is the heat generated per unit of volume and $\dot{q}_s$ is the average heat flow rate to the strip at the roll and strip interface. It is assumed that there is no variation of temperature along longitudinal direction in the deformation zone. The approximate time t_a spent by the strip in the roll bite can be calculated by dividing L with the average velocity of the strip. The time t_a is substituted in Eq. (3.55) to obtain the temperature in the roll bite. The rate of heat generated per unit volume $\dot{Q}$ and the average heat flux due to friction $\dot{q}_f$ are given by:

$$\dot{Q} = \frac{0.9P_P}{0.5(h_1 + h_2)Lw}, \tag{3.58}$$

$$\dot{q}_f = \frac{P_f}{wl_d}, \tag{3.59}$$

where P_p, P_f, h_1, h_2, L, w and l_d are the plastic deformation power, friction power, sheet thickness at inlet, sheet thickness at outlet, length of plastic deformation zone (=GH in Figure 3.4), width of the sheet and contact length of roll and strip respectively. The plastic deformation power P_p and friction power P_f are obtained from the deformation module. The P_p and P_f are estimated by an FEM based deformation module. It is assumed that there is no significant variation of temperature along longitudinal direction in the deformation zone. The approximate time t_a spent by the strip in the roll bite can be calculated by dividing L with the average velocity of the strip. This time t_a is put in Eq. (3.55) to obtain the temperature in the roll bite. Equation (3.55) provides approximate temperature in the bite zone, assuming that there is no significant change in the temperature in the longitudinal direction.

From the exit of roll bite to the temperature sensor (location where measurement of temperature is needed), the heat transfer is governed by:

$$\frac{\partial}{\partial y}\left(k_s \frac{\partial T(y,t)}{\partial y}\right) = \rho_s c_{ps} \frac{\partial T(y,t)}{\partial t}, \tag{3.60}$$

with boundary conditions:

$$\frac{\partial T(y,t)}{\partial y} = 0 \qquad \text{at } y = 0, \tag{3.61}$$

$$-k_s \frac{\partial}{\partial y}\left(\frac{\partial T(y,t)}{\partial y}\right) = h_a \{T(y,t) - T_a\} \quad \text{at } y = \frac{h_2}{2}, \tag{3.62}$$

where h_a is the heat transfer coefficient and h_2 is the exit thickness of the strip. The initial condition is:

$$T(y,0) = T_e(y,t_a). \tag{3.63}$$

The solution of the one-dimensional heat conduction problem is obtained by the method of separation of variables. The temperature distribution of strip at the exit side is given by (Kim et al. 2009):

$$T(y,t) = T_0 + \sum_{n=1}^{\infty} \exp\left(-\frac{k_s}{\rho_s C_{ps}} \lambda_n^2 t\right) \left\{ \frac{4\lambda_n \int_0^{\frac{h_2}{2}} T_e(y,0)\cos(\lambda_n y)\,dy - 4T_0 \sin\left(\lambda_n \frac{h_2}{2}\right)}{\lambda_n h_2 + \sin(\lambda_n h_2)} \right\} \cos(\lambda_n y), \tag{3.64}$$

where $T_e(y, 0)$ is the temperature at the exit of deformation zone that corresponds to time $t = 0$, T_0 is the temperature of the coolant, h_a is the convective heat transfer coefficient at the strip surface, and λ_n are obtained by solving the following transcendental equation:

$$k_s \lambda_n \sin\left(\lambda_n \frac{h_2}{2}\right) - h_a \cos\left(\lambda_n \frac{h_2}{2}\right) = 0. \tag{3.65}$$

Figure 3.5 shows a flowchart of the procedure for determining the temperature in rolling. The deformation analysis is carried out using FEM taking the mechanical properties at the initial strip temperature and the thermal analysis is carried out using analytical method. After the strip temperatures are obtained by thermal analysis, deformation analysis is again carried out at the mechanical properties evaluated at the updated temperature. This procedure is repeated till the convergence is obtained.

3.4.4 Typical Results and Discussion

The results of two-dimensional thermo-mechanical analyses to obtain the steady-state as well as transient temperature distribution of the roll and the strip are described. First, the temperature distribution in the roll is obtained by three different method *viz.*, approximate method based on Laplace transform, series solution using integral transform technique and FEM. The fast FEM model of Yadav et al. (2013) is used for comparison with analytical models. The thermal properties of the roll and strip are taken as independent of temperature. The material properties and the dimensions of the roll are tabulated in Table 3.1. Table 3.2 shows the comparison of peak temperature at different angular velocities of roll. It is observed that at high angular speed of the rolls, the differences of peak temperatures predicted by all three methods are very small. In all the cases, deviation between predictions is less than 10%.

Figure 3.6 shows the temperature distributions on the inner and outer surface of the roll at the angular velocity of 0.314 rad/s. Here, the inner roll radius a is 0.23 m and the outer roll radius b is 0.25 m. The other parameters are given in Table 3.1. It is observed that the approximate method predicts the larger error than the series solution using integral transform technique

Table 3.1 Material constants and geometric parameters for thermal analyses for roll steel.

Parameter	Value
Inner roll radius (a)	0.23 m
Outer roll radius (b)	0.25 m
Convective heat transfer coefficient at outer periphery (h_e)	260 W/m^2-°C
Convective heat transfer coefficient at inner periphery (h_i)	2.6 W/m^2-°C
Thermal conductivity (k_r)	52 W/m-°C
Thermal diffusivity (α_r)	0.144×10^{-4} m^2/s
Initial temperature at the inner and outer periphery (T_i,T_e)	30°C
Semi-bite angle (β)	8°
Heat flux ($\dot{q}$)	5×10^6 W/m^2

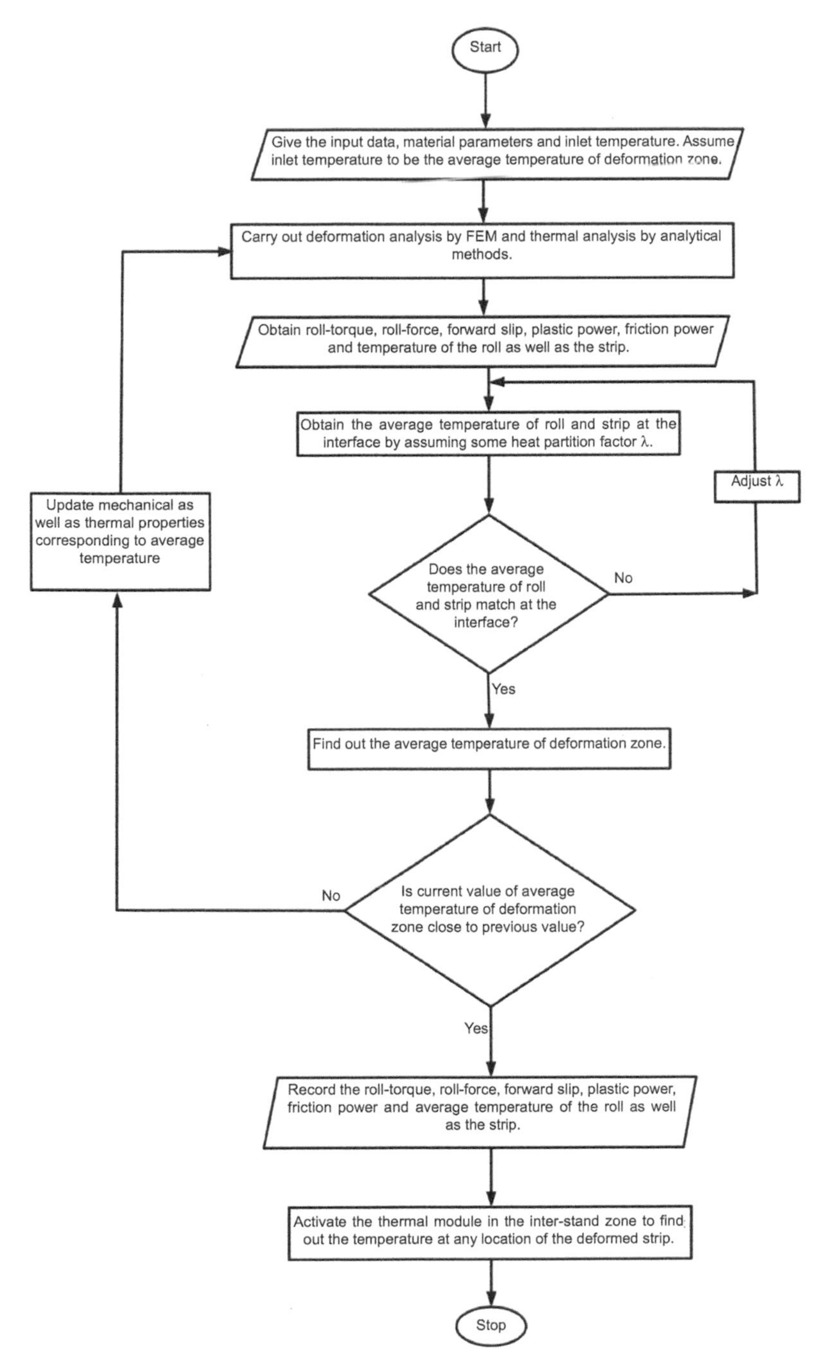

Figure 3.5 A flow chart illustrating the methodology of thermo-mechanical modeling of flat rolling.

Table 3.2 Comparison of maximum steady-state temperature of roll at different angular velocities.

S. No.	Angular velocity (ω rad/s)	Peak surface temperature in roll (°C)		
		Approximate method	Series solution using integral transform technique	FEM
1.	0.1	1420	1348	1360
2.	1	1055	1028	975
3.	10	940	931	896
4.	100	903	900	878

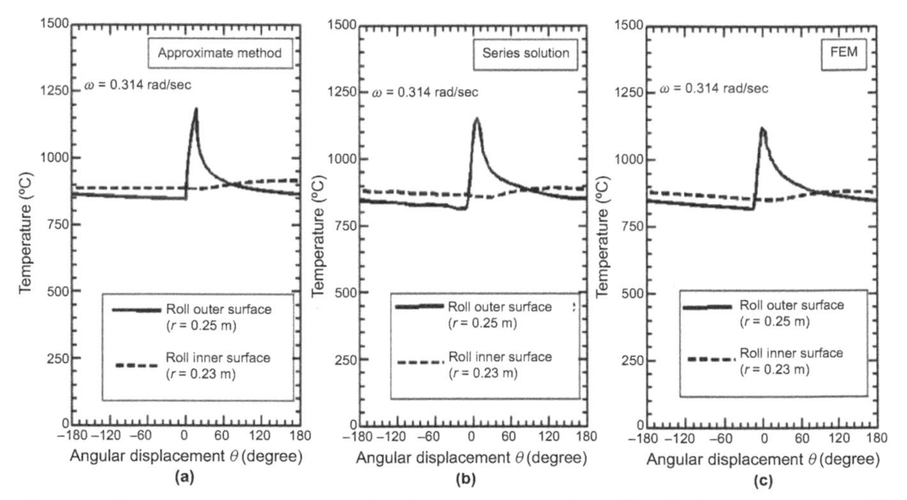

Figure 3.6 Steady-state temperature distributions in a rotating roll at the inner and outer radii for heat input of 5 MW/m², h_i = 2.6 W/m²-°C and h_e = 260 W/m²-°C: (a) Approximate method (b) Series solution using integral transform technique and (c) FEM.

in comparison to the FEM predictions. However, the maximum error is less than 6%.

The approximate method takes the minimum computational time compared with the other two methods. The series solution using integral transform technique involves the roots from the transcendental equation. These roots are function of the roll dimensions and convective heat transfer coefficient of the roll and need to be evaluated for change in these parameters. Table 3.3 shows the comparison of results with three methods.

For the transient analysis, the results of only series solution using integral transform technique are presented. The exit strip temperature is experimentally measured by a non contact sensor–IR Camera (Infra Tec hr Head Vario CAM 480SL) that captures the thermo-graphic image. IRBIS® 3plus analysis software is used for processing the thermo-graphic image.

Table 3.3 Comparison of three methods for carrying out the thermal analysis of the roll.

S. No.	Method	Computational screen time	Remarks
1.	Approximate	less than 10 second	• Steady-state temperature is obtained. • Constant initial temperature of roll (T_0) is used to obtain a closed form solution. • No provision for taking the temperature-dependent material property data. • The roll is assumed to be made of homogeneous material.
2.	Series solution using integral transform technique (transient as well as steady)	about 25–30 minutes	• Transient as well as steady-state temperature is obtained. • Initial temperature of roll can be considered as function of radial and circumferential coordinates. • Incremental analysis can be carried out (time varying heat input into the roll may be considered). • No provision for taking temperature dependent material property data. • Homogeneous material of the roll is assumed.
3.	FEM	about 10 minutes for steady-state analysis by the fast FEM proposed by Yadav et al. (2013); about 2 hours for transient analysis.	• Initial temperature of roll can be considered as function of radial and circumferential coordinates. • Temperature dependent material property data can be considered. • Composite or functionally graded material (FGM) can be considered.

The coefficient of friction is calculated on the basis of slip measurement. The details of experimental setup are described in (Yadav et al. 2015). The thermal properties of the strip and the roll are given in Table 3.4.
The flow stress of the strip is governed by Lenard and Malinowski (1993):

$$\sigma_f = \sigma_0 + n_2\varepsilon \quad \text{for } \varepsilon \geq 0.1, \tag{3.66}$$

$$\sigma_f = \sigma_1\left(\frac{\varepsilon}{0.1}\right)^{n_3} \quad \text{for } \varepsilon < 0.1, \tag{3.67}$$

where σ_0, σ_1, n_2 and n_3 are temperature-dependent parameters as per Table 3.5.

Table 3.6 shows the comparison of theoretical results with the experimental temperature measured at two places, 50 and 150 mm away from the roll bite on the surface of the exit strip. It is seen from Table 3.6 that the exit strip temperatures predicted by the model deviate by less than 6% from experimental results.

Table 3.4 Thermal properties for the roll and strip.

Material	**Thermal conductivity k (W/m-°C)**	**Specific heat c_p (J/kg-°C)**	**Density ρ (kg/m³)**	**Thermal diffusivity α (m²/s)**	**Reference**
Strip (Al 1100 H14)	218	904	2710	0.890×10^{-6}	Davis (1992)
Roll (Steel)	52	460	9850	0.144×10^{-4}	Davis (1990)

Table 3.5 Temperature dependent parameters for calculating flow stress of the strip.*

Temperature (°C)	**σ_1 (MPa)**	**σ_0 (MPa)**	**n_2**	**n_3**
22	150.38	153.78	34.065	0.256
100	146.23	149.64	34.090	0.195
200	135.90	138.56	26.637	0.175
300	111.29	113.85	25.623	0.363

* With permission from Lenard and Malinowski (1993), Copyright 2015 Elsevier.

Table 3.6 Comparison of exit strip temperature of model with experimental results for $h_1 = 5$ mm, $R = 100$ mm, $T_0 = 200$°C.

S. No.	**Percentage reduction (%r_d)**	**Exit strip temperature (°C)**					
		At 50 mm after the bite zone			**At 150 mm after the bite zone**		
		Experimental	**Series solution**	**% Error**	**Experimental**	**Series solution**	**% Error**
1.	12.4	172	175.2	−1.86	110	104.25	5.23
2.	19.2	183	188.4	−2.95	139	132.24	4.86
3.	25.4	198	199.8	−0.91	158	162.4	−2.78
4.	32.4	228	219.3	3.82	175	178.3	−1.88

One typical transient analysis of rolling process is carried out with focus on the variation of temperature with time. During transient state, the heat flux into the roll, $\dot{q}$, keeps on varying with time till the steady-state is achieved. The temperature of roll as well as $\dot{q}$ is updated after every 5 revolutions of roll using Eqs. (3.50). The flow stress of strip (steel) is governed by J-C model (Eq. (3.20)), where the constants of J-C model are taken from Meslin and Hamann (2003) and are as follows: $A = 598$ MPa, $B = 768$ MPa, $n_1 = 0.2092$, $C = 0.0137$, $m_1 = 0.807$. The reference strain rate ($\dot{\varepsilon}_0$) is 0.001 and melting temperature (T_m) is 1768 K. The initial temperature of roll and strip are taken as 15°C and 250°C, respectively. The outer radius of the roll (R) is 65 mm and the thickness of the roll is considered to be 20 mm. The thermal properties of the strip and the roll are given in Table 3.7

Table 3.7 Thermal properties of roll and strip made of steel.

Parameters	Roll	Strip
Thermal conductivity (W/m-°C)	52	40
Specific heat (J/kg-°C)	460	470
Density (kg/m³)	7850	7800
Thermal diffusivity (m²/s)	0.144×10^{-4}	0.11×10^{-4}

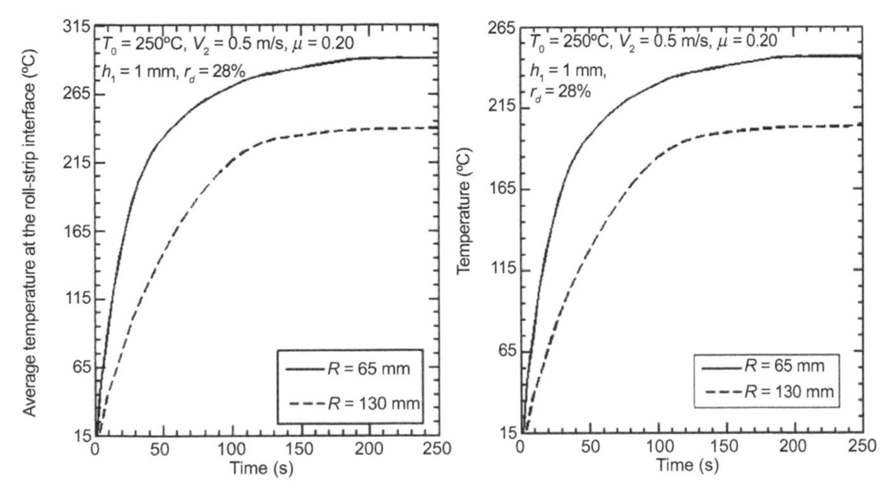

Figure 3.7 Effect of roll radius (a) average temperature at the roll-strip interface and (b) exit strip temperature 50 mm away from the roll bite for h_e = 260 W/m²-°C and h_a = 10 W/m²-°C.

on the basis of property data provided by Davis (1990). The computations are carried out for 28% reduction of 1 mm of strip (h_1 = 1 mm) taking the equivalent Coulomb coefficient of friction as 0.20. Figure 3.7 shows the variation of average temperature at the roll-strip interface and exit strip temperature with time.

In this work, the heat input into the roll is calculated on the basis of temperature matching at the roll-strip interface. It is based on the assumption that the heat transfer coefficient at the roll-work interface, h_c, is infinite, i.e., there is a perfect contact. Based on experimental finding, Hlady et al. (1995) proposed the following empirical expression for evaluating h_c:

$$\frac{h_c C}{k} = \left(\frac{P}{\sigma}\right)^m, \tag{3.68}$$

where k is the mean of the thermal conductivities of the roll and the strip, P is the mean roll pressure in kg/mm², σ is the flow stresses of the strip in MPa, m is the constant of the strip material and C is a surface roughness parameter of the roll material. Here, one example is presented to give an estimate of h_c.

The material constant m = 1.7 and surface roughness parameter C = 35 μm given by Hlady et al. (1995) is used to calculate the value of h_c with the input parameter given in Table 3.7. The value of h_c is obtained as 19.6×10^6 W/m^2-°C. The differences in the roll and the strip temperature at the interface is calculated using the following heat flux balance equation $q = h_c(T_s - T_r)$, where q denotes the heat input into roll from the strip at the steady-state condition. The value of q is 6.5×10^6 W/m^2. The temperature differences in the roll and the strip temperature ($T_s - T_r$) is equal to 0.33°C. This is very small. Hence, the assumption of same strip and roll temperature is justified.

3.5 Conclusion

In this chapter, an integrated model for flat rolling for obtaining the temperature of the exit strip using the FEM deformation module and separate thermal module is described. This model can be used for the estimation of temperature during the rolling of AHSSs sheets and strips. Three methods for determining the temperature distribution in the roll have been presented. These three methods are approximate method, series solution using integral transform method and an FEM model. A method to find out the temperature distribution in the strip by analytical method is described. The determination of temperature distribution through analytical methods for the roll as well as strip has the advantage in terms of computational time. The optimization and control of the process requires a computationally efficient model. It is to be mentioned that instead of an FEM model for deformation analysis, an analytical model, with necessary fine-tuning would have been used for further reducing the computational time. Nevertheless, the flow formulation of Dixit and Dixit (1996) is reasonably fast. In literature, a three-dimensional analytical model for the temperature distribution in a rolling process is not found. The development of such a model will highly benefit the rolling industry. There is a scope for extensive experimental study on the rolling of AHSSs sheets for fine-tuning the models for the prediction of temperature distributions in the roll and the sheet.

Keywords: Coefficient of friction, Duhamel's theorem, Eigenfunctions method, Eulerian flow formulation, Flat rolling, Forward slip, Frictional power, Johnson–Cook model, Integral transform technique, Method of separation of variables, Plastic deformation power, Roll temperature, Strip temperature, Steady-state temperature, Transient temperature

Key of Abbreviations

Roman letters

a,b	inner and outer roll radii of roll
A,B,C	constants of J-C material model
f	ratio of magnitudes of tangential and normal tractions
h_a	convective heat transfer coefficient at the surface of strip
h_e, h_i	convective heat transfer coefficient at the outer and inner periphery of roll, respectively
h_1, h_2	inlet and exit thickness of strip, respectively
k_r, k_s	thermal conductivity of roll and strip, respectively
c_{pr}	specific heat of roll
c_{ps}	specific heat of strip
J_n, Y_n	Bessel function of first and second kind, respectively
l_d	contact length of roll and strip at the interface
L	length of bite zone
m	positive integer in Eqs. (3.42)–(3.50)
n	positive integer or zero in Eqs. (3.42)–(3.50)
m_1, n_1	constants of J-C material model
n_2, n_3	temperature dependent parameters in Eqs. (3.66) and (3.67)
p	hydrostatic pressure
$\mathrm{P_e}$	Peclet number
P_f	power due to friction
P_p	power due to plastic deformation
$\dot{q}$	heat flux input into the roll
$\dot{q}_f$	heat flux due to friction
$\dot{q}_s$	heat flux entering into the strip
Q_T	total heat
$\dot{Q}$	rate of heat generation per unit volume
r-θ	polar coordinates in Eq. (3.50)
r_d	percentage reduction
R	undeformed roll radius
R'	deformed roll radius

S_{ij}	Deviatoric part of the stress tensor
t	time
t_a	time spent by strip at roll bite
t_n, t_s	interfacial normal and shear stress components
T	temperature
T_{amb}	ambient temperature
T_e, T_i	medium temperature at outer and inner periphery of roll, respectively
T_m	melting temperature
T_{max}	maximum temperature of the roll
T_r	temperature of the roll
T_s	temperature of the strip
$T_0(\tau)$	imposed temperature field on roll, used in Eq. (3.32)
T_0	initial temperature of the inlet strip
$T_0(y)$	initial temperature of the strip, used in Eqs. (3.55) and (3.57)
V_R	roll velocity
v_1, v_2	components of velocity vector
V_1, V_2	inlet and exit velocities of the strip, respectively
w	width of the strip
x-y	Cartesian coordinates of strip in Figure 3.4
x_1, x_2	Cartesian coordinates of strip in Figure 3.1

Greek letters

α	bite angle after roll deformation
α_r, α_s	thermal diffusivity of roll and strip, respectively
β	half arc of the heat source on the roll periphery
β_m	roots of the characteristics used in Eq. (3.50)
ε_{eq}	equivalent (plastic) strain
ε_r	total emissivity of the material
$\dot{\varepsilon}_{eq}$	equivalent strain-rate
$\dot{\varepsilon}_0$	reference strain-rate
$\dot{\varepsilon}_{ij}$	strain rate tensor
ϕ	the location of the heat source
η	fraction of the plastic deformation work dissipating as heat
η_1	proportionality factor in Levy-Mises flow rule

σ_{st}	Stefan-Boltzmann constant
λ	heat partition factor
λ_n	roots of the characteristics Eq. (3.65)
μ	equivalent Coulomb coefficient of friction
θ	angular displacement of roll
ρ_r, ρ_s	density of the roll and the strip, respectively
σ_f	flow stress of strip
ω	angular velocity of roll
ξ, ζ	Non-dimensional coordinates

References

Arif, A.F.M., O. Khan and S.M. Zubair. 2004. Prediction of roll temperature with a non-uniform heat flux at tool and workpiece interface. Heat Mass Transfer. 41: 75–94.

Chang, D.F. 1999. Thermal stresses in work rolls during the rolling of metal strip. J. Mater. Process. Tech. 94: 45–51.

Davis, J.R. 1990. ASM Handbook: Properties and Selection: Irons, Steels, and High-Performance Alloys. ASM Int. Handbook. Vol. 1: The Materials International Society, Material Park, Ohio, USA.

Davis, J.R. 1992. Properties and Selection: Nonferrous Alloys and Special-Purpose Materials. ASM Int. Handbook. Vol. 2: The Materials International Society, Material Park, Ohio, USA.

Dixit, U.S. and P.M. Dixit. 1996. A finite element analysis of flat rolling and application of fuzzy set theory. Int. J. Mach. Tools. Mf. 36: 947–969.

Dixit, P.M. and U.S. Dixit. 2008. Modeling of Metal Forming and Machining Processes: by Finite Element and Soft Computing Methods. Springer, London.

DesRuisseaux, N.R. and R.D. Zerkle. 1970. Temperature in semi-infinite and cylindrical bodies to moving heat sources and surface cooling. J. Heat Transfer. Trans. ASME 92: 456–464.

Fischer, F.D., E. Werner and K. Knothe. 2001. The surface temperature of a half plane heated by friction and cooled by Convection. Z. Angew. Math. Mech. 81: 78–81.

Fischer, F.D., W.E. Schreiner, E.A. Werner and C.G. Sun. 2004. The temperature and stress fields developing in rolls during hot rolling. J. Mater. Process. Tech. 150: 263–269.

Galantucci, L.M. and L. Tricarico. 1999. Thermo-mechanical simulation of a rolling process with an FEM approach. J. Mater. Process. Tech. 92/93: 494–501.

Guo, R. 1998. Two-dimensional transient thermal behavior of work rolls in rolling process. J. Manuf. Sci. Eng. Trans. ASME 120: 28–33.

Hawkins, D.N. 1985. Warm working of steels. J. Mech. Work. Technol. 11: 5–21.

Hatta, N., J. Kokado and H. Nishimura. 1980. Analysis of slab temperature change and rolling mill line length in quasi continuous hot strip mill equipped with two roughing mills and six finishing mills. Trans. ISIJ 21: 270–277.

Hlady, C.O., J.K. Brimacombe, I.V. Samarasekera and E.B. Hawbolt. 1995. Heat transfer in the hot rolling of metals. Metall. Mater. Trans. B 26B: 1019–1027.

Hwang, S.M., M.S. Joun and Y.H. Kang. 1993. Finite element analysis of temperatures, metal flow, and roll pressure in hot strip rolling. J. Manuf. Sci. Eng. Trans. ASME 115: 290–298.

Hwang, S.M., C.G. Sun, S.R. Ryoo and W.J. Kwak. 2002. An integrated FE process model for precision analysis of thermo-mechanical behaviors of rolls and strip in hot strip rolling. Comput. Methods Appl. Mech. Eng. 191: 4015–4033.

Jiang, Z.Y., A.K. Tieu, C. Lu, W.H. Sun, X.M. Zhang and X.H. Liu. 2004. Three dimensional thermomechanical finite element simulation of ribbed strip rolling with friction variation. Finite Elem. Anal. Des. 40: 1139–1155.

Johnson, G.R. and W.H. Cook. 1983. A constitutive model and data for metals subjected to large strains, high strain rates and high temperatures. In. Proc. of the Seventh Int. Symposium on Ballistics. Hagu. Netherlands 21: 541–547.
Khan, O.U., A. Jamal, G.M. Arshed, A.F.M. Arif and S.M. Zubair. 2004. Thermal analysis of a cold rolling process—A numerical approach. Numer. Heat Transfer. 46: 613–632.
Khalili, L., S. Serajzadeh and B. Koohbor. 2012. Thermomechanical behavior of work rolls during warm strip rolling. Metall. Mater. Trans. B. 43B: 1638–1648.
Kim, J., K. Lee and S.M. Hawang. 2009. An analytical model for the prediction of strip temperatures in hot strip rolling. J. Heat Transf. 52: 1864–1874.
Kiuchi, M., J. Yanagimoto and E. Wakamatsu. 2000. Overall thermal analysis of hot plate/ sheet rolling. CIRP Ann. 49: 209–213.
Koohbor, B. 2015. Finite element modeling of thermal and mechanical stresses in work-rolls of warm strip rolling process. Proc. IMechE. Part B: J. Engineering Manufacture. doi: 10.1177/0954405414564807.
Kuziak, R., R. Kawalla and S. Waengler. 2008. Advanced high strength steels for automotive industry. Arch. Civil. Mech. Eng. 8: 103–117.
Lahoti, G.D., S.N. Shah and T. Altan. 1978. Computer-aided analysis of the deformations and temperatures in strip rolling. J. Eng. for Industry. Trans. ASME 100: 159–166.
Lee, J.D., M.T. Manzari, Y.L. Shen and W. Zeng. 2000. A finite element approach to transient thermal analysis of work rolls in rolling process. J. Manuf. Sci. Eng. Trans. ASME 122: 706–715.
Lenard, J.G. and Z. Malinowski. 1993. Measurements of friction during the warm rolling of aluminum. J. Mater. Process. Tech. 39: 357–371.
Louaisil, K., M. Dubar, R. Deltombe, A. Dubois and L. Dubar. 2009. Analysis of interface temperature, forward slip and lubricant influence on friction and wear in cold rolling. Wear. 266: 119–128.
Luo, C. and H. Keife. 1998. A thermal model for the foil rolling process. J. Mater. Process. Tech. 74: 158–173.
Maslen, S.H. and A.A. Tseng. 1981. Program Rolling User's Manual, Report No. MMLTR818, Martin Marietta Laboratories, Baltimore, MD.
Meslin, F. and J.C. Hamann. 2003. The problem of constitutive equations for the modelling of chip formation: towards inverse methods. pp. 123–142. *In*: P. Boisse, T. Altan and K. Luttervelt (eds.). Friction and Flow Stress in Forming and Cutting. First South Asian Edition, London, UK.
Özişik, M.N. 1993. Heat Conduction. John Wiley & Sons. New York.
Patula, E.J. 1981. Steady-state temperature distribution in a rotating roll subject to surface heat fluxes and convective cooling. J. Heat Transfer Trans. ASME 103: 36–41.
Pietrzyk, M. and J.G. Lenard. 1990. The effect of the temperature rise of the roll on the simulation of the flat rolling process. J. Mater. Process. Tech. 22: 177–190.
Roberts, W.L. 1978. Cold rolling of steel. Marcel Dekker. New York.
Saito, Y., N. Tsuji, H. Utsunomiya, T. Sakai and R.G. Hong. 1998. Ultra-fine grained bulk aluminum produced by accumulative roll-bonding (ARB) process. Scripta Mater. 39: 1221–1227.
Serajzadeh, S., A.K. Tahiri and F. Mucciardi. 2002. Unsteady state work-roll temperature distribution during continuous hot slab rolling. Int. J. Mech. Sci. 44: 2447–2462.
Serajzadeh, S. 2006. A model for prediction of flow behavior and temperature distribution during warm rolling of a low carbon steel. Mater. Design. 27: 529–534.
Sheikh, H. 2009. Thermal analysis of hot strip rolling using finite element and upper bound methods. Appl. Math. Model. 33: 2187–2195.
Subramanian, E.V. and D.L. Bourell. 1984. Roll pressure modeling of multipass warm rolling of carbon steel. J. Appl. Metal Work 3: 267–71.
Sun, C.G., C.S. Yun, J.S. Chung and S.M. Hwang. 1998. Investigation of thermo mechanical behavior of a work roll and a roll life in hot strip rolling. Metall. Mater. Trans. A 29A: 2407–2424.

Tseng, A.A. 1984. A numerical heat transfer analysis of strip rolling. J. Heat Transfer Trans. ASME 106: 512–517.

Tseng, A.A., S.X. Tong, S.H. Maslen and J.J. Mills. 1990. Thermal behavior of aluminum rolling. J. Heat Transfer Trans. ASME 112: 301–308.

Tudball, A. and S.G.R. Brown. 2006. Practical finite element heat transfer modeling for hot rolling of steels. Iron Making and Steel Making. 33: 61–66.

Wanhiem, T. and N. Bay. 1978. A model for friction in metal forming processes. CIRP Ann. 27: 189–194.

Wilson, W.R.D., C.T. Chang and C.Y. Sa. 1989. Interface temperatures in cold rolling. J. Mater. Shaping Technol. 6: 229–240.

Yadav, V., J. Thakuria, A.K. Singh and U.S. Dixit. 2013. An approximate fast finite element analysis of temperature distribution in rolling. Int. J. Mech. Manuf. Sys. 6: 381–396.

Yadav, V., A.K. Singh and U.S. Dixit. 2014. An approximate method for computing the temperature distributions in roll and strip during rolling process. Proc. IMechE. Part B: J. Engineering Manufacture 228: 1118–1130.

Yadav, V., A.K. Singh and U.S. Dixit. 2015. Inverse estimation of thermal parameters and friction coefficient during warm flat rolling process. Int. J. Mech. Sci. 96-97: 182–198.

Yiannopoulos, A.C., N.K. Anifantis and A.D. Dimarogonas. 1997. Thermal stress optimization in metal rolling. J. Thermal Stresses 20: 569–590.

Yu, H., Q. Yu, J. Kang and X. Liu. 2012. Investigation on temperature change of cold magnesium alloy strips rolling process with heated roll. J. Mater. Eng. Perform. 21: 1841–1848.

Yuan, W.Y.D. 1985. On the heat transfer of a moving composite strip compressed by two rotating cylinders. J. Heat Transfer Trans. ASME 107: 541–548.

APPENDIX

Expression used in Eqs. (3.60)

Following expressions are used in the solution:

$$H_e = \frac{h_e}{k_r} \text{ and } H_i = \frac{h_i}{k_r}, \tag{A.1}$$

$$N_n(\beta_m) = \frac{2}{\pi^2} \frac{K_n^2 B_e - V_n^2 B_i}{\beta_m^2 K_n^2}, \tag{A.2}$$

$$F_n = \frac{K_n^2 B_e - V_n^2 B_i}{K_n^2}, \tag{A.3}$$

$$B_e = H_e^2 + \beta_m^2 \left\{1 - \left(\frac{n}{\beta_m b}\right)^2\right\} \text{ and } B_i = H_i^2 + \beta_m^2 \left\{1 - \left(\frac{n}{\beta_m a}\right)^2\right\}, \tag{A.4}$$

$$R_n(\beta_m, r) = L_n J_n(\beta_m r) - V_n Y_n(\beta_m r), \tag{A.5}$$

where K_n, L_n, V_n and W_n are defined as

$$K_n = \left(\frac{n}{a} - H_i\right) J_n(\beta_m a) - \beta_m J_{n+1}(\beta_m a), \tag{A.6}$$

$$L_n = \left(\frac{n}{b} + H_e\right) Y_n(\beta_m b) - \beta_m Y_{n+1}(\beta_m b), \tag{A.7}$$

$$V_n = \left(\frac{n}{b} + H_e\right) J_n(\beta_m b) - \beta_m J_{n+1}(\beta_m b), \tag{A.8}$$

$$W_n = \left(\frac{n}{a} - H_i\right) Y_n(\beta_m a) - \beta_m Y_{n+1}(\beta_m a). \tag{A.9}$$

Here J_n and Y_n are Bessel functions of the first and second kinds, respectively. Here n is zero or positive integer.

4

Cellular Automaton Modeling of the Microstructural Evolution in AHSS Hot Rolling

*Chengwu Zheng** and *Dianzhong Li*

ABSTRACT

Cellular automaton (CA) model is a promising tool for predicting and understanding the microstructure evolution at the grain scale. This chapter provides a brief overview on the status of its implementation in modeling the microstructural evolution in advanced high strength steel (AHSS) hot rolling. A general introduction is given first to the CA methodology itself. Then, an emphasis is placed on its application to the complex microstructural transformations closely associated with the processing route of AHSS hot rolling, including austenite recrystallization and austenite decomposition. Finally, a short commentary on future trends in CA modeling in AHSS hot rolling is made.

4.1 Introduction

In the past decades, the steel industry has undergone a revolutionary process by developing advanced high strength steels (AHSS) including dual phase (DP), complex phase (CP) and transformation induced plasticity (TRIP)

Shenyang National Laboratory for Materials Science, Institute of Metal Research, Chinese Academy of Sciences, 72 Wenhua Road, Shenyang 110016, China.
E-mail: dzli@imr.ac.cn
* Corresponding author: cwzheng@imr.ac.cn

steels, etc. with superior mechanical properties, such as higher tensile strength with improved formability, as compared to conventional high-strength low-alloy (HSLA) steels. This makes AHSS an attractive choice for modern vehicle designs given that the increase of strength enables weight reduction of vehicles leading to better fuel efficiency. The property improvements of AHSS result primarily from multiphase microstructures that combine the ferrite with other microstructural constituents, e.g., bainite, martensite and retained austenite (Sarkar and Militzer 2009). Consequently, better knowledge of the underlying microstructural phenomena associated with the processing routes appear to be particular significant for these advanced steels.

In steel industries, the demand of producing steels with consistent quality and high productivity make it crucial to predict and control the microstructure since the *microstructure engineering* could quantitatively link the operational parameters in the mills with the properties of the final product (Militzer 2007). Modeling of the microstructural evolution during steel strip hot rolling has received tremendous attentions. Starting with the pioneering work of Sellars and Whiteman (1979), mathematical modeling of hot strip rolling has attained a mature level with a wealth of models being available for conventional plain and HSLA steels (Militzer et al. 2000). Recently, such models also have been proposed for DP (Liu et al. 2007a) and TRIP steels (Liu et al. 2007b). However, currently available microstructure models are usually formulated at the macro-scale, i.e., the microstructure is described with a number of so-called internal state variables such as grain size, fraction recrystallized, fraction transformed, etc. They are mainly concerned with the quantitative but empirical relationships between microstructure and parameters of hot-strip mill using many experimentally derived, non-physical parameters to fit the simulation results to the measured values. Consequently, these models are inherently bound to typical experimental conditions as well as typical steel grades, and are thus of limited use when extended to a more advanced steel grades.

With the advancement of computer performance, it is now possible to model microstructure on mesoscale, i.e., on the length scale of the microstructure features using various mesoscopic approaches, such as cellular automaton (CA), Monte Carlo (Potts) method and phase field method, etc. (Raabe 1998). Modeling of the microstructure in such physically based manner, in which the model is described by thermo-physical properties of the material rather than parameters dependent on a specific experimental setup, can offer better insight into how the transformations occur under arbitrary conditions. These modeling strategies have the potential to not only provide the general average microstructure properties (e.g., volume fraction, grain sizes and overall transformation kinetics) but also describe the actual microstructure evolution, e.g., the morphology and spatial distributions of microstructure constituents. Among the various mesoscopic models, CA is more feasible for hot strip mill due to its intrinsic computational efficiency, allowing large volumes to be examined as well as its ready calibration to time and length scale. In particular, CA seems to be on the verge of providing a predictive tool that can be employed for various microstructural transformations in AHSS processing. Actually, the development of such microstructure-based CA models for AHSSs is particularly significant since these steels are more sensitive to process conditions than conventional steels. They will significantly facilitate the establishment of robust processing routes for these advanced steels.

In this chapter, the status of the CA modeling of the microstructural evolution in AHSS hot strip rolling will be briefly reviewed. Special attention is given to the microstructure-based CA models on the microstructural behavior of austenite in the hot rolling process. First, a brief general introduction to the CA methodology will be given. Subsequently, CA practices of austenite recrystallization and austenite decomposition that are closely associated with the processing route of AHSS hot rolling will be described. Finally, the chapter is concluded with short commentary on future trends in CA modeling and its application in steel hot rolling.

4.2 Cellular Automaton Methodology

4.2.1 Basics of Cellular Automaton

Before examining the application of CA to the microstructure phenomena in AHSS hot rolling, it is useful to provide a brief overview of the methodology itself and its microstructure-based implementation. For an in-depth description of the CA methodology and its underlying physics, the reader is referred to a number of recent reviews. The excellent review paper of Raabe (2002) provides fundamental introductions to CA and its applications in the field of microstructure study, with special attention given to the recrystallization phenomena. Reuther and Rettenmayr (2014) in their review of the CA models, emphasize and describe dendritic solidification of alloys. Janssens (2010) gives an introductory review to the CA modeling of moving grain boundaries in polycrystalline materials. Yang et al. (2011) summarizes the applications of CA in modeling many microstructural transformations in metal forming. The attention is placed on the underlying physical models of the transformations.

CA is an algorithm that describes the discrete spatial and temporal evolution of complex systems by applying local transformation rules to the cells of a regular (or non-regular) lattice (Raabe 2002, von Neumann 1963). The space of the modeling system is defined on a regular array of lattice points which typically represent volume portions. The property of each lattice site is defined in terms of a set of generalized state variables. For microstructure-based models these can be dislocation density, crystal orientation, concentration of the alloying elements, phase state, precipitation density, or any other quantity that the model requires. Each point assumes one out of a finite set of possible discrete states. The opening state of the CA can be defined by mapping the initial distribution of the chosen state variables onto the lattice.

The kinetics of the automaton results from changes in the state of the cells. They occur in accord with a local switching rule which determines the individual transforming probability of each cell as a function of its previous state and the state of the neighboring sites. The switching rule is designed to map the phenomenology of the underlying microstructural phenomena. The number, arrangement, and range of the neighbor sites used by the transformation rule for calculating a state switch determine the range of the interaction and the local shape of the areas that evolve. CA works in discrete time steps. After each time interval, the values of the state variables are updated simultaneously and synchronously for entire lattice of the automaton in mapping the new (or unchanged) values assigned to them through the transformation rule. Owing to these features, CA provides a discrete method to simulate the dynamical evolution of a complex system, which contains large numbers of similar components on the basis of their local interactions (Raabe 2002).

Generally speaking, a complete CA is typically characterized by following constituents:

- The geometry of the cells,
- The number and the kind of states that a cell can possess,
- The definition of the neighborhood of a cell,
- The transformation rules for a cell that determines the state of the cell in the next time step.

It should be noted that CA does not have restrictions in the type of elementary entities or transformation rules employed. As far as the abstract rules and properties of a general CA are transformed into a microstructure-related concept, such a model becomes phenomenologically sound to simulate the microstructure evolution phenomena in materials science. Due to its flexibility in considering a large variety of state variables and transformation laws, the CA approach is suited for variety of microstructural phenomena, particularly in the fields of recrystallization, grain growth, and phase transformation, etc. Moreover, the possible incorporation of the realistic or experimental input data, particularly of the thermodynamic and the kinetics coefficients, enables the CA model to quantitatively predict the overall kinetics, textures, microstructures on a real space and time scale.

4.2.2 Design of a Cellular Automaton Model

In this section, a two-dimensional (2D) CA model for the isothermal austenite-to-ferrite transformation in steels (Lan et al. 2004a) is introduced to illustrate how a CA is designed to describe the phase transformation process.

In a CA, the independent variables are time and space. The latter is discretized into equally shaped hexagonal cells, as shown in Figure 4.1a. Each cell represents a volume of real material characterized by certain attributes with a distinct phase, solute content and an orientation indicator. The neighbors of a cell are defined as the six nearest cells, i.e., the von Neumann's rule, as see in Figure 4.1b. In order to depict the decomposition of austenite into ferrite, six state variables are used on each lattice cell. They are (1) the orientation variable that represents the crystallographic orientations of the ferrite nuclei and the austenite grains, (2) the phase state variable that denotes whether the lattice site is ferritic, austenitic or contains α/γ interface, (3) the carbon concentration variable in the ferrite, (4) the carbon concentration in the austenite, (5) the average carbon concentration, and (6) the phase fraction variable quantifying the ferrite fraction transformed from austenite. The opening state of the automaton is defined in terms of unique distribution of austenitic phases with identical carbon concentration. Different austenitic grains are mapped as regions of identical crystal orientation. The most intricate point in CA modeling is in identifying appropriate phenomenological rules for the nucleation and growth events of the phase transition.

The nucleation event in the CA is usually treated by changing the phase state variable directly on selected cells. When a cell becomes nucleated, its phase state variable is immediately changed to α phase, and the carbon

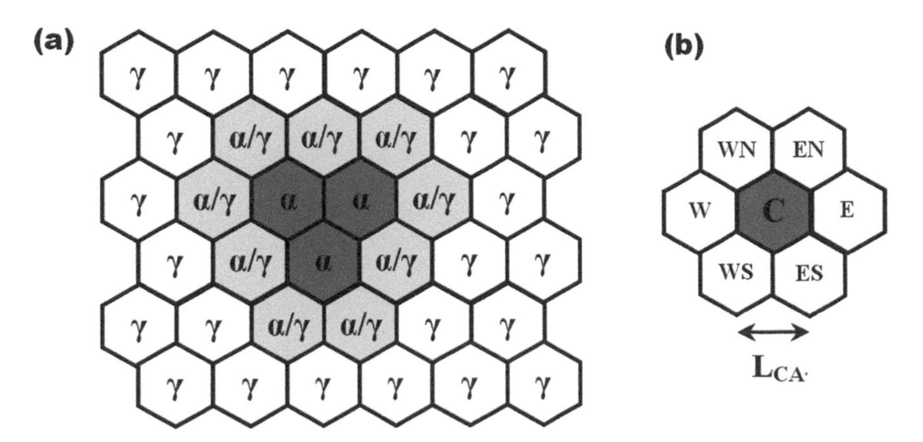

Figure 4.1 Schematic illustration of the hexagonal grid in the CA model (a) and the neighborhood of a cell (b) (Lan et al. 2004a).

concentration in the ferrite in this nucleated cell becomes the equilibrium value $c^{\alpha,eq}$ for ferrite phase at the underlying temperature. This cell will precipitate a certain amount of solute ($c^{\gamma}-c^{\alpha,eq}$) into its neighboring austenite cells and all its neighboring austenite cells change their phase state from austenite to γ/α interface.

After a ferrite grain nucleates at austenite grain boundaries, it immediately begins to grow into its neighboring interface cells with a velocity, v. At time t, the growth length $l_{i,j}^{t}$ of a ferrite cell (i, j) towards (k, l), one of its neighboring interface cells can be described as follows:

$$l_{i,j}^{t} = \int_{t_0}^{t} v_{i,j} dt \tag{4.1}$$

where t_0 is the time when cell (i, j) was nucleated. Then the fraction $f_{k,l}^{t}$ of ferrite phase at time t in an interface cell (k, l) caused by the growth of its neighboring ferrite cell (i, j) can be calculated by:

$$f_{k,l}^{t} = \frac{l_{i,j}^{t}}{L_{CA}}, \tag{4.2}$$

where L_{CA} is the length of a cell in the CA as seen in Figure 4.1b. The total fraction $F_{k,l}^{t}$ of ferrite phase in cell (k, l) at time t is the sum of the transforming ferrite fractions due to the growth of all its neighboring ferrite cells. In the interface cell (k, l), the carbon concentration $c_{k,l}^{\alpha}$ in ferrite is taken as the equilibrium concentration from phase diagram. The rejected carbon atoms from the growth of the ferrite phase transfer into the austenite phase, which increase the carbon concentration $c_{k,l}^{\gamma,t}$ in austenite in the interface cell (k, l). The average carbon concentration $c_{k,l}^{t}$ in the interface cell (k, l) is given by:

$$c_{k,l}^{t} = F_{k,l}^{t} \cdot c_{k,l}^{\alpha,t} + (1 - F_{k,l}^{t}) \cdot c_{k,l}^{\gamma,t}. \tag{4.3}$$

If the ferrite fraction $F_{k,l}^{t}$ in interface cell (k, l) becomes greater than unity, the phase state variable of cell (k, l) changes into α phase and its crystallographic orientation is selected randomly from those of its neighboring ferrite cells. It means that the growing ferrite grain successfully sweeps a transforming austenitic cell. Meanwhile this interface cell (k, l) will precipitate a certain amount of solute ($c^{\gamma} - c^{\alpha,eq}$) into its neighboring austenite cells and the received solute of its each neighboring interface cell is inversely proportional to average concentration in this cell.

4.3 Benchmarking of Cellular Automaton Modeling

Before discussing the CA applications in more detail, it is useful to further benchmark the CA model with the classic Johnson-Mehl-Avrami-Kolmogorov (JMAK) model (Kolmogorov 1937, Johnson and Mehl 1939, Avrami 1939). The JMAK theory is the fundamental for most macroscopic microstructure models, which is usually used in a semi-empirical manner to describe the kinetics of the individual microstructural reaction.

According to the JMAK theory, the transformation fraction f can be calculated by:

$$f = 1 - \exp(-b \cdot t^n) \quad (4.4)$$

where t is the transformation time, b is a temperature-dependent rate parameter and n is the Avrami-exponent which depends on the dimensionality of the system. Considering the JMAK scenario with nucleation site saturation, the Avrami-exponent n is two in 2D and three in 3D, respectively. Figure 4.2 shows the evolution of the grains growth as simulated using a 2D CA approach. As depicted in Figure 4.3, the CA simulation replicates the idealized Avrami-exponent $n = 2$. This benchmarking exercise provides a fundamental justification for the application of the CA method to the transformation kinetics and hence for its comparability with the macroscopic empirical microstructural models.

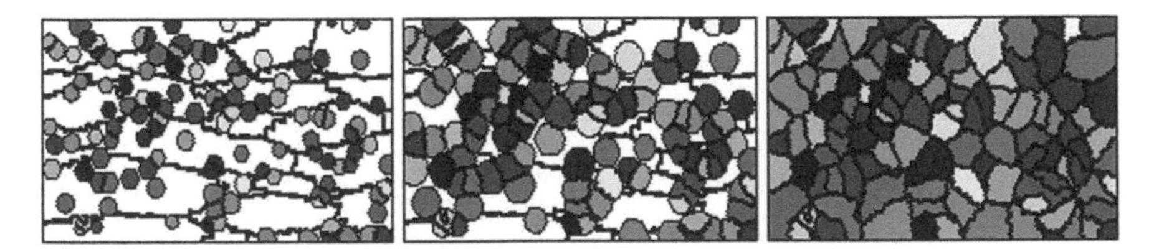

Figure 4.2 Time sequence of a 2D CA simulation of the JMAK transformation scenario with nucleation site saturation.

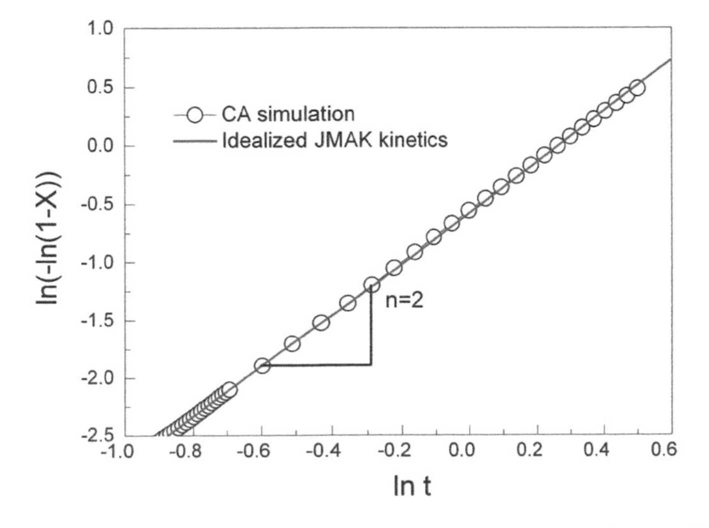

Figure 4.3 Comparison of idealized JMAK kinetics with CA simulation in 2D.

However, it should be noted that the JMAK theory has assumed an "ideal" scenario with homogeneous nucleation and isotropic growth of spherical grains, which is rather an exception in real transformations. Contrary to the JMAK theory, the CA model can give time and space dependent descriptions of the transformation processes. It is not restricted to assumption of statistical mean values and to homogeneous processes. Therefore, the CA method can be easily used to simulate more realistic transformation behaviors than the JMAK theory does.

4.4 Application of Cellular Automaton Modeling to the Microstructure Phenomena in AHSS Hot Rolling

4.4.1 Introduction

The proceedings in an AHSS hot-strip rolling mill can be subdivided into three principle stages: (1) reheating, (2) rolling (both rough and finish rolling), and (3) cooling (water cooling on the run-out table and coiling). The major metallurgical phenomenons which occur during these processing steps are: (1) recrystallization and grain growth of austenite, and (2) the austenite to ferrite transformation. Figure 4.4a shows a schematic diagram of the continuous cooling transformation (CCT) and the associated microstructural transitions by the conventional thermomechanical controlled processing in steels. When the austenite is deformed at a relative higher temperature, recrystallization might occur to refine the austenite grains. Once austenite is deformed within the non-recrystallization temperature region, a large amount of lattice defects which can be potential nucleation sites for ferrite transformation are introduced into the *pancaked* austenite. In this case, the ferrite transformation can be accelerated in subsequent CCT process.

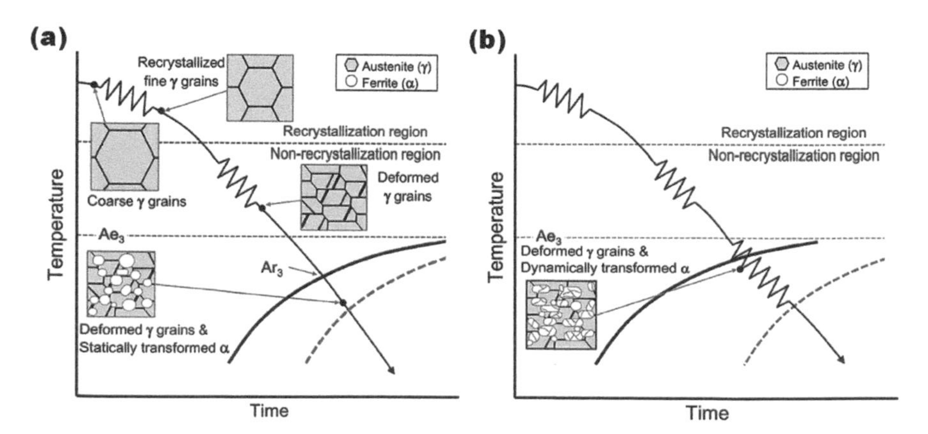

Figure 4.4 Schematic illustrations showing CCT diagrams and change in microstructures for static ferrite transformation (a) and dynamic ferrite transformation during deformation of austenite (b) (Park et al. 2014).

When a large deformation is applied at an even lower temperature as shown in Figure 4.4b, ferrite transformation could occur during the plastic deformation of austenite because the kinetics of ferrite transformation is greatly promoted by the deformation. This ferrite transformation has been termed as dynamic strain-induced transformation (DSIT). It has been proved that the DSIT mechanism can be realized in the industrial rolling processing to produce ultrafine ferrite (UFF) microstructures (Weng 2009, Beladi et al. 2007).

Obviously, the microstructural evolution of austenite during hot deformation and the run-out table cooling plays important roles in AHSS hot rolling processing. Design and optimization of the thermo-mechanical processes require the knowledge of the underlying metallurgical phenomena since they can be exploited to produce desired microstructures in various grades of AHSSs and hence to attain desired property combinations. Therefore, development of the comprehensive microstructure models according to these principal metallurgical phenomena in AHSS hot rolling appears to be important. In this context, physically-based CA simulations of the individual microstructural phenomena in austenite which take place during hot rolling processing, i.e., austenite recrystallization, austenite decomposition during continuous cooling, and the dynamic strain-induced phase transformation, are briefly reviewed in the following section.

4.4.2 Simulation of Austenite Recrystallization During Hot Rolling

The recrystallization of austenite is an important physical metallurgy phenomenon during controlled rolling of steels because the refinement in the recrystallized austenite grains remarkably influences the microstructural evolution during austenite decomposition. Finer austenite grains will decompose into finer microstructures and lead to favorable mechanical properties of the final products. Therefore, it is very important to develop quantitative descriptions of the processing of steels and their microstructure. During the hot strip rolling, three distinctly different recrystallization processes can potentially be operative, *viz* dynamic recrystallization (DRX) during deformation, static recrystallization (SRX) during the interpass time and meta-dynamic recrystallization (Meta-DRX) following DRX in the interpass time.

CA can readily be used to simulate the recrystallization in metallic materials. Hesselbarth and Göbel (1991) were the first to employ CA to describe the primary recrystallization. Goetz and Seetharaman (1998) presented the first CA model for DRX. Starting with their pioneering work, a number of sophisticated CA models have been proposed to simulate the SRX (Raabe 2002, Raabe and Hantcherli 2005) and DRX (Ding and Guo 2001) as well as the Meta-DRX (Kugler and Turk 2004). However, majority of these models appear to be depictive, which mainly present abilities of the developed tools to model the complex microstructural phenomena taking place during the recrystallization process. They have not yet been brought to a stage that reliable quantitative predictions can be made for actual steel processing. Here, only a few CA attempts dedicated to the recrystallization behavior of austenite which take place during the hot rolling processing are reviewed.

Dynamic recrystallization of austenite DRX is the main microstructural evolution mechanism in the rough rolling processing. Recently, a CA model was developed to depict the DRX behavior of austenite in a 0.13 wt.% C-0.19 wt.% Si-0.49 wt.% Mn steel (Ma et al. 2016). The hot deformation process with strain rate of $\dot{\varepsilon} = 0.1\ s^{-1}$ at the temperature of 900°C was considered. An empirical formula concerning the flow stress (σ) of plain steels was used to calculate the deformation stored energy, H, as follows:

$$\sigma = K(A\varepsilon^n - B\varepsilon)\dot{\varepsilon}^m \exp(\frac{Q_{def}}{RT}) \tag{4.5}$$

$$H = \frac{c\sigma^2 V_r}{a\mu} \tag{4.6}$$

where, ε and $\dot{\varepsilon}$ are the strain and strain rate, respectively, K, A and B are constants determined by material composition, N and m are the functions of material composition, R is the gas constant, T is the absolute temperature, Q_{def} is the deformation activation energy, V_r is the molar volume of austenite, and μ is the shear modulus of the material.

The nucleation rate for a DRX grain can be expressed as:

$$\dot{n} = C\dot{\varepsilon}^{\eta} \exp(-\frac{Q_{act}}{RT}), \tag{4.7}$$

where C is a constant, Q_{act} is the activation energy for nucleation, and the exponent η was set to 1 in the simulation model under consideration.

The velocity of the recrystallization front, V can be expressed as:

$$V = MP, \tag{4.8}$$

where P is the driving pressure for the recrystallization front movement derived from the difference of stored energy between the recrystallized grains and the unrecrystallized matrix. M is the grain boundary mobility, which can be calculated by:

$$M = M_0 \exp(-\frac{Q_b}{RT}), \tag{4.9}$$

where, M_0 is a pre-exponential factor, and Q_b is the activation energy for grain-boundary motion.

The simulated microstructural evolution of DRX is shown in Figure 4.5. From the CA simulation, the nucleation and grain growth process during DRX can be tracked. When the deformation proceeds and reaches the critical strain for triggering DRX, a new cycle of DRX will take place.

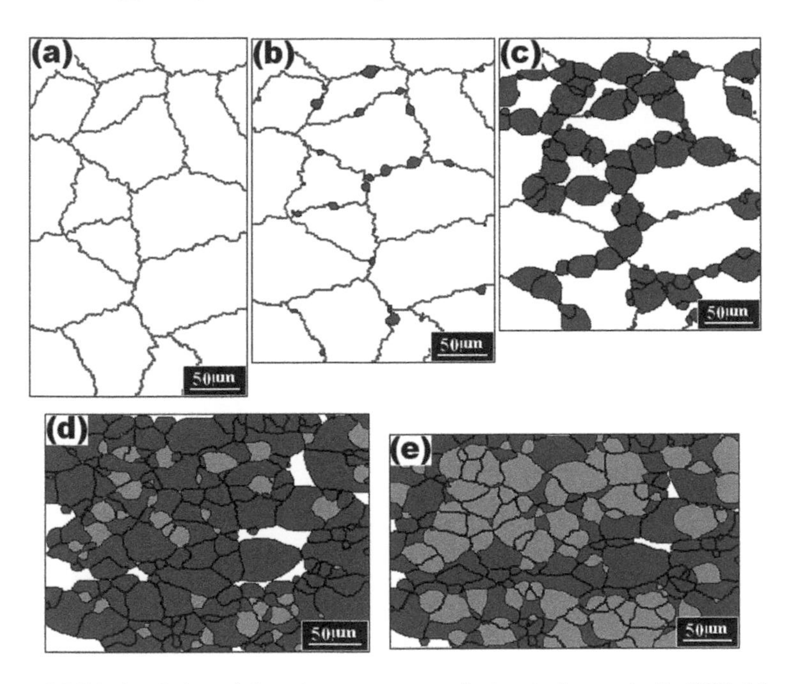

Figure 4.5 CA simulation of the microstructure evolution during austenite DRX: (a) ε = 0.1, (b) ε = 0.2, (c) ε = 0.3, (d) ε = 0.5, (e) ε = 0.6 (un-DRX: white, 1st cycle DRX: dark gray, 2nd cycle DRX: light gray) (Ma et al. 2016).

In order to verify the capability of the CA model, the simulation results are compared with the predictions by the empirical microstructure models, as shown in Figures 4.6–4.7. For this steel grade, the evolution of DRX fraction X_d, with the progress of strain, ε, can be predicted by (Weng 2009):

$$X_d = 1 - \exp\{-0.693[(\varepsilon - \varepsilon_c)/(\varepsilon_{0.5} - \varepsilon_c)]^2\} \tag{4.10}$$

$$\varepsilon_{0.5} = 1.144 \times 10^{-3} d_0^{\,0.25} \dot{\varepsilon}^{0.05} \exp(6420/T) \tag{4.11}$$

Where, ε_c is the critical strain for onset of DRX, $\varepsilon_{0.5}$ is the strain for 50 pct recrystallization, d_0 is the initial austenite grain size, and $\dot{\varepsilon}$ is the strain rate.

The austenite grain size of DRX D_{RX} can be calculated by:

$$\mathrm{D}_{RX} = 5.072 \times 10^3 \times Z^{-0.1555} \tag{4.12}$$

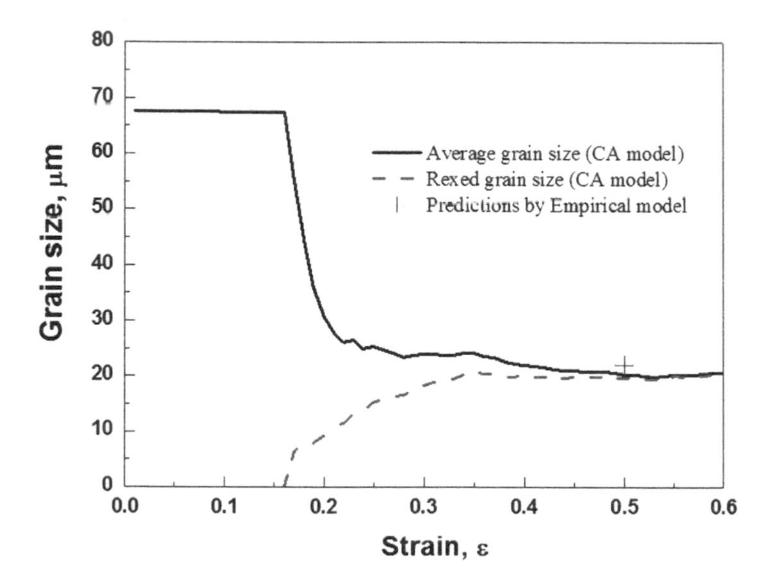

Figure 4.6 CA simulation of the grain size evolution during austenite DRX (Ma et al. 2016).

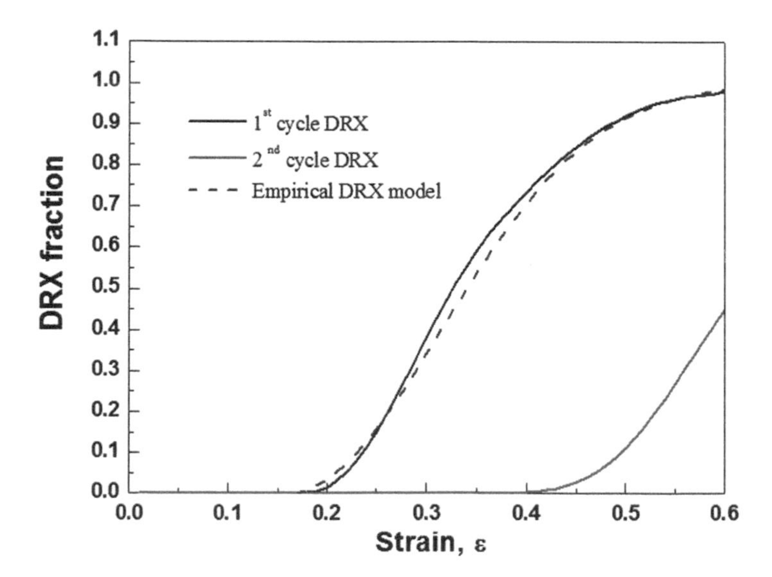

Figure 4.7 Comparison of the CA simulated DRX kinetics of austenite with the results predicted by the empirical microstructural model (Ma et al. 2016).

$$Z = \dot{\varepsilon} \cdot \exp(\frac{363155}{RT}) \tag{4.13}$$

Where, Z is the Zerner-Hollomon parameter.

The results indicate that the CA simulations agree very well with the predictions made with the macroscopic empirically-based models. Nevertheless, the macroscopic microstructure model can not provide a clear description about the evolution of the grain size. These results confirm the great possibility and effectiveness of the CA model for modeling the DRX of austenite during the hot rolling processing.

Static recrystallization of hot-deformed austenite

In most industrial productions, SRX is a dominant softening mechanism. SRX kinetics, grain size and grain growth during the period of interpass are important issues. Zheng et al. (2009a) used a 2D CA simulation to analyze the microstructure behavior of the austenite SRX following hot deformation by coupling with a crystal plasticity finite element modeling (CPFEM) in a 0.1 wt.% C-1.37 wt.% Mn-0.2 wt.% Si steel. The CPFEM simulation is capable of providing decent pictures of the deformation behavior of the austenite in the course of a thermo-mechanical process. The details and inhomogeneities of the deformed structure, including the stored energy distribution, grain topology and the crystallographic orientation (see Figure 4.8) strongly affect the subsequent SRX reactions.

The microstructure behavior of the austenite SRX is simulated by CA at various deformation stages. The nucleation rate of SRX is calculated using a phenomenological relation as follows:

$$\dot{n} = C_1\left(H - H_{min}\right)V_{\Omega}(t)\exp\left(\frac{-Q_A}{RT}\right) \tag{4.14}$$

where H is the stored energy of deformation, H_{min} is the minimum stored energy to cause nucleation which is determined by the critical strain for SRX, C_1 is a constant, and $V_{\Omega}(t)$ is the fraction of volume at which nucleation is still possible at time t.

The movement of the recrystallization front is treated as a strain-induced boundary migration process, for which the difference of the deformation energy between the deformed area and the recrystallized area provides the driving pressure.

By this coupling simulation, the SRX behavior of austenite can be clearly demonstrated. The results indicate that the spatial distribution of the SRX grains is inhomogeneous due to the heterogeneities of the stored deformation energy. Both the grain boundaries and the grain interiors with high stored energy could be successful sites for SRX nucleation. The increasing deformation provides high levels of the stored energy for both the SRX nucleation density and SRX grain growth, which refine the final grain size and accelerate the SRX kinetics markedly as shown in Figure 4.9. The result also indicates that the inhomogeneous distribution of SRX nuclei

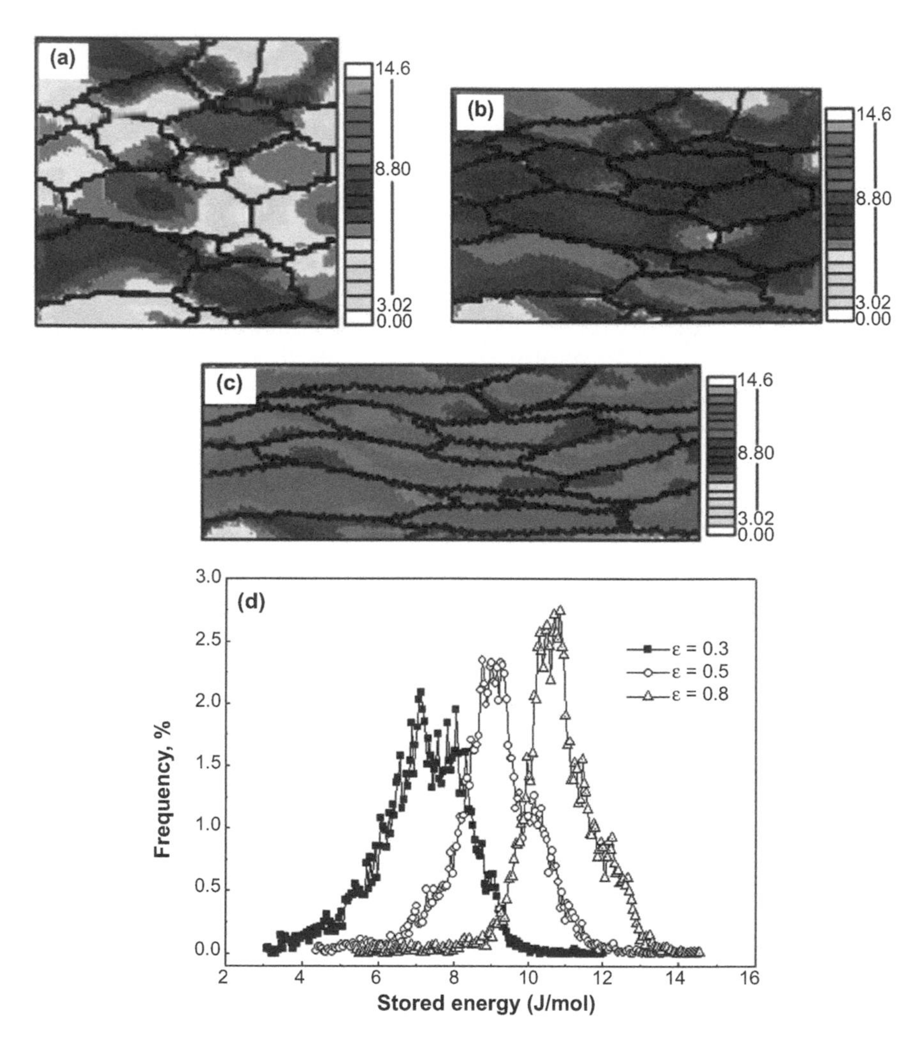

Figure 4.8 (a–c) Stored energy field (J/mol) due to hot deformation in austenite simulated by CPFEM at different logarithmic strains of ε = 0.3, 0.5, 0.8. (d) The statistical distribution of the stored energy at different logarithmic strains (Zheng et al. 2009a).

leads to a clear derivation of the recrystallization kinetics from the classic JMAK theory and results in smaller Avrami exponent than the ideal one.

Austenite recrystallization during multi-stand strip hot rolling

During hot strip rolling of steels, grain refinement from stand to stand takes place as long as the recrystallization is possible. The recrystallized austenite grain size has to be revealed on account of the large influence of the state of the austenite microstructure at the finish mill exit on the kinetics and microstructures of the austenite decomposition. In this context, a physically-based CA modeling was performed to simulate the microstructure evolution of the austenite recrystallization during multi-pass finishing hot rolling in a 0.143 wt.% C-0.21 wt.% Si-0.43 wt.% Mn steel (Zheng et al. 2008a). By coupling fundamental metallurgical principles, effects of the individual metallurgical phenomena related with the steel strip hot rolling, including SRX, Meta-DRX, DRX and grain growth has been integrated. All the processing parameters, including the pass time, strain, strain rate and temperature history, were derived from the actual finishing hot rolling mill line (as see in Table 4.1).

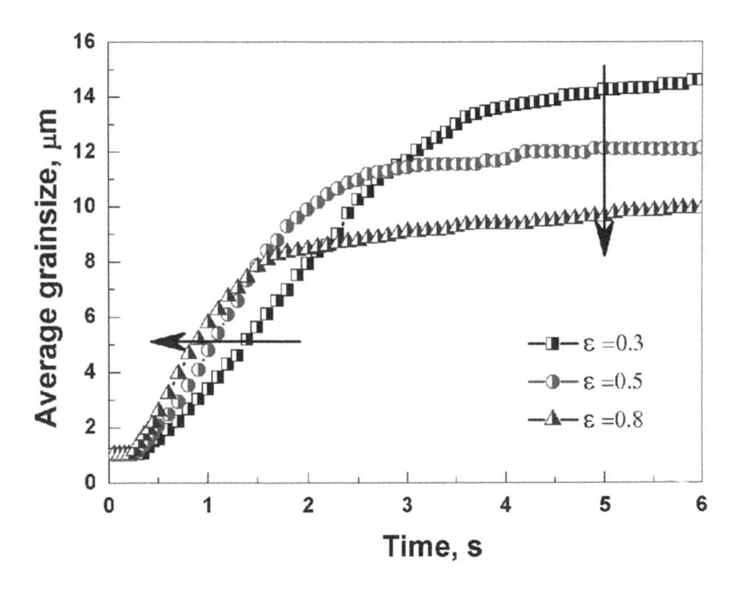

Figure 4.9 Variations of the average recrystallized grain size of austenite with time at different strain levels (Zheng et al. 2009a).

Table 4.1 Technology parameters used in the CA modeling of the multi-pass steel strip hot rolling.

Stand No.	Pass time(s)	Temperature(°C)	Strain	Strain-rate(s^{-1})
1	6.154	903	0.339	3.287
2	4.554	888	0.296	4.817
3	3.503	880	0.258	6.559
4	2.770	873	0.226	9.382
5	2.294	863	0.185	10.914
6	1.968	853	0.148	12.437
7	1.774	840	0.099	11.152

In this modeling, no attempt is made to distinguish between SRX and Meta-DRX or even DRX processes. For instance, the Meta-DRX is regarded as a process of the continuous growth of the recrystallized nuclei generated during deformation (see Figure 4.10). The microstructural process of the recrystallization is governed by the routine of the principal processing parameters, i.e., strain rate, strain, and temperature. Figure 4.11 shows the

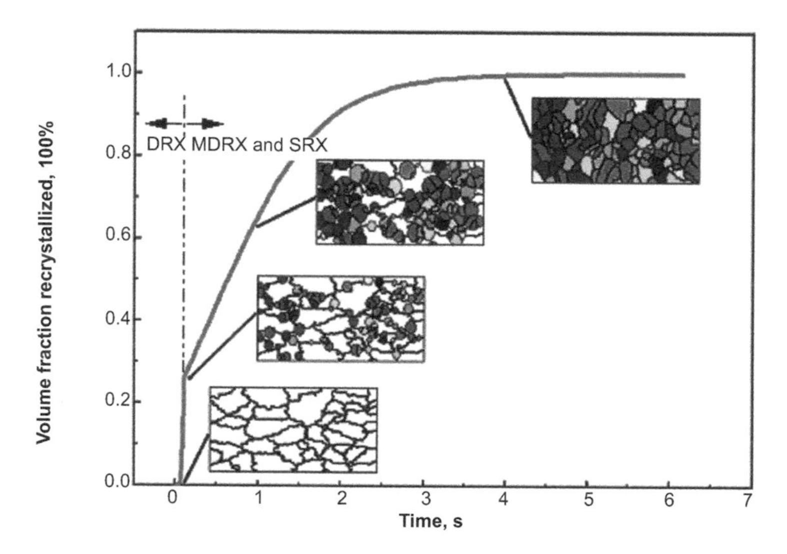

Figure 4.10 Simulated transformation kinetics concomitant with the microstructure changing during the 1st rolling pass (Zheng et al. 2008a).

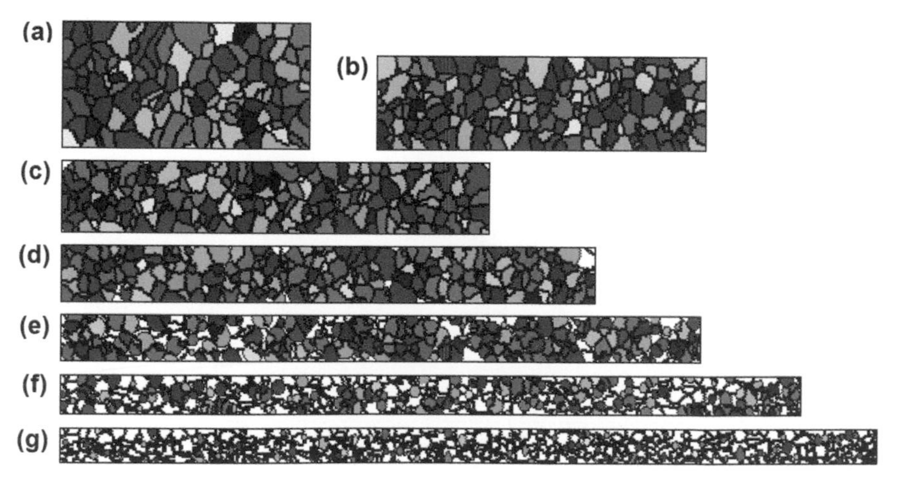

Figure 4.11 CA simulation of the microstructure changing at the end of each rolling pass: (a) 1st pass, (b) 2nd pass, (c) 3rd pass, (d) 4th pass, (e) 5th pass, (f) 6th pass, (g) 7th pass (Zheng et al. 2008a).

simulated microstructure at the end of each pass during the hot rolling process. By analyzing the microstructures, variation of the mean grain size and the recrystallized fraction with time can be quantitatively predicted (see Figure 4.12a–b). The results indicate that the average grain size is refined from their initial grain size of about 48 μm to about 11 μm after the seven-pass rolling. The simulation results are then compared with the predictions by the macroscopic empirically-based models and are found in good agreement. As expected, this physically-based CA modeling enables both quantitative and topographic predictions of the microstructural characteristics during the industrial hot-rolling processing.

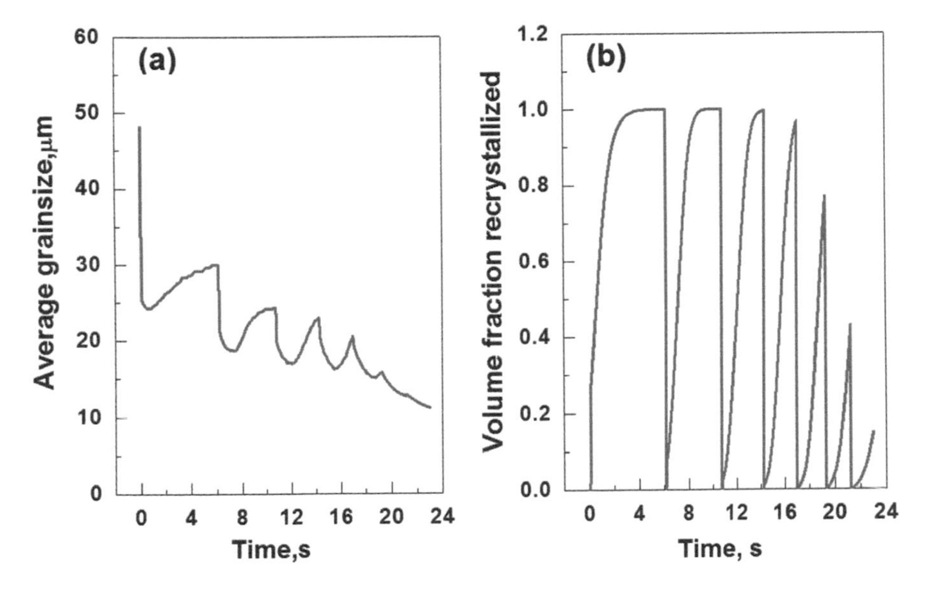

Figure 4.12 CA simulations of variations of the average grain size (a) and recrystallization fraction (b) with time during the multi-pass hot rolling (Zheng et al. 2008a).

4.4.3 Simulation of Austenite Decomposition During the Continuous Cooling Transformation

The austenite decomposition on the run-out table of a hot mill plays a dominant role regarding the final microstructure of the product. It is indeed the key metallurgical tool for AHSSs as it can be exploited to produce microstructures that are associated with significant improvement of the properties. An extensive body of CA simulations is available for the decomposition of austenite.

Kumar et al. (1998) were the first to use CA simulations to describe the formation of ferrite from austenite. In this model, the nucleation rate is defined by the Gaussian function:

$$I = \frac{N_{max}}{\sqrt{2\pi \Delta T_\sigma}} \int_0^{\Delta T} \exp[-\frac{1}{2}\frac{(\Delta T - \Delta T_{max})^2}{\Delta T_\sigma}] d(\Delta T) \tag{4.15}$$

Where, ΔT is the undercooling, N_{max} is the total density of ferrite nuclei, ΔT_σ is the standard deviation of gaussian distribution, and ΔT_{max} is the mean undercooling.

The growth rate of ferrite is determined following the solubility balance rule:

$$v = D_\gamma \frac{dC_\gamma}{dn} \frac{1}{C_\gamma^*(1-\kappa)}, \tag{4.16}$$

Where, dC_γ/dn is the concentration gradient of carbon in austenite in the direction perpendicular to the growing boundary, C_γ^* is the equilibrium carbon concentration, κ is the distribution coefficient of carbon in austenite and ferrite and D_γ is the carbon diffusion coefficient in austenite. This model was able to describe the competition between nucleation and early growth of ferrite.

Zhang et al. (2003) then developed another CA model using stochastic transition rules. In this model, the nucleation rate is defined with the classical nucleation theory:

$$I = K_1 D_\gamma (kT)^{-\frac{1}{2}} \exp(-\frac{K_2}{kT(\Delta G)^2}), \tag{4.17}$$

Where, K_1 is a constant related to the nucleation site density, K_2 is a constant related to austenite-ferrite interface energy, k is the Boltzmann constant, T is the temperature in Kelvin, and ΔG is the driving force for ferrite nucleation. The growth rate is calculated by assuming local equilibrium at the interface:

$$D_\gamma \frac{\partial C_\gamma}{\partial n} - D_\alpha \frac{\partial C_\alpha}{\partial n} = v(C_\alpha^{\alpha/\gamma} - C_\gamma^{\alpha/\gamma}) \tag{4.18}$$

Where, C_γ, C_α are the carbon concentrations in austenite and ferrite respectively, $C_\gamma^{\alpha/\gamma}$, $C_\alpha^{\alpha/\gamma}$ are the carbon concentrations of austenite and ferrite at the interface, and D_γ, D_α are the carbon diffusion coefficient in austenite and ferrite. The significant feature of this model is the implementation of incorporating local concentration changes into a nucleation or growth function.

Lan et al. (2004b) were the first who introduced deterministic transition rules into CA model for austenite to ferrite transformation. In this model, a mixed-mode growth kinetics model was adopted. The austenite-ferrite interface is hence at a non-equilibrium state which is mixed diffusion/interface controlled. The growth rate is described by:

$$v = MF_{\text{chem}}, \tag{4.19}$$

Where, M is the effective interface mobility and F_{chem} is the chemical driving force. The predicted CCT kinetics by this model agreed well with the experiments in C-Mn steels as shown in Figure 4.13.

Subsequently, Lan et al. (2005) investigated the decomposition behavior of the deformed, i.e., non-recrystallized, austenite in C-Mn steel by coupling the CA model with a CPFEM simulation of the hot compression. The CPFEM simulation provides the quantitative microstructural information of the deformed austenite as the input for the CA model, including crystal orientations, heterogeneous distributions of the stored deformation energy and the pancaked grain morphology. In CA simulation, the stored energy of deformation (F_{def}) is then incorporated as an additional driving force for phase transformation. The growth rate is hence described as:

$$v = M(F_{\text{chem}} + F_{\text{def}}). \tag{4.20}$$

This simulation demonstrates the grain refinement and acceleration of overall kinetics due to the austenite deformation from the aspect of microstructural behaviors (see Figure 4.14).

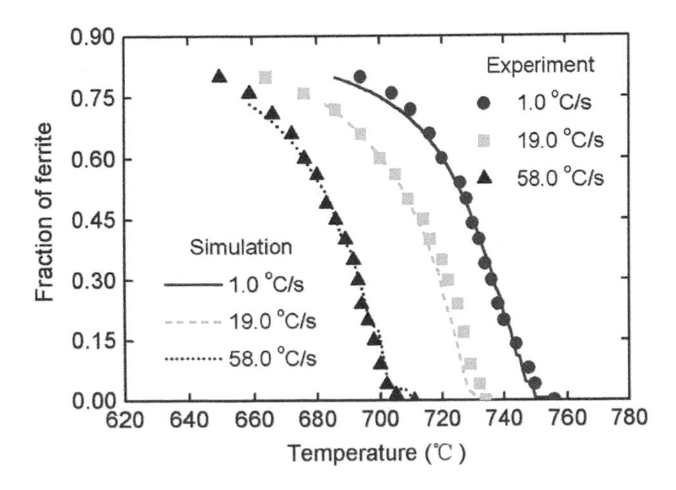

Figure 4.13 CA simulated kinetics of ferrite formation in a 0.17 wt.% C-0.012 wt.% Si-0.74 wt.% Mn steel and the experimental results during continuous cooling transformation with different cooling rates (Lan et al. 2004b).

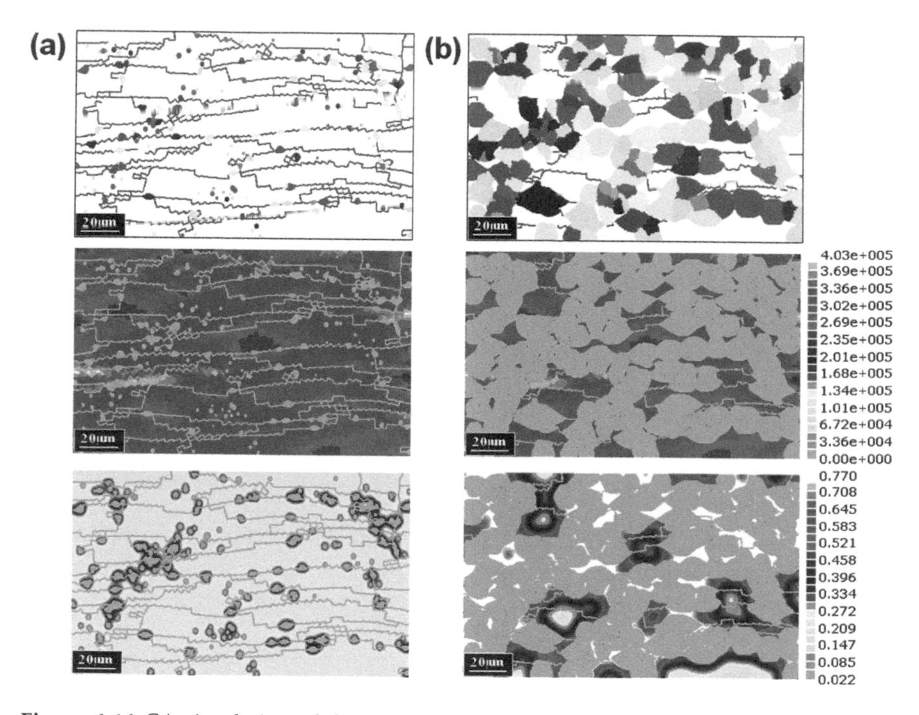

Figure 4.14 CA simulation of the microstructure evolution during deformed austenite decomposition into ferrite at different temperatures in a 0.19 wt.% C-1.46 wt.% Mn-0.445 wt.% Si steel: (a) 714°C and (b) 609°C. The material is compressed at 850°C up to a logarithmic strain of 0.6 with a strain rate of 10 s^{-1}, then cooled at a rate of 15°C s^{-1} until the ferrite transformation completed. The figures show the changes in the transformed microstructure (the upper images), the stored energy of deformation (the middle images, Jm^{-3}) and the carbon concentration distribution (the lower images, wt.%) (Lan et al. 2005).

It should be noted that available CA models for austenite decomposition mainly work on the formation of the polygonal ferrite upon cooling from the austenitic temperature range. Nevertheless, a few CA models attempted to simulate the complex phase transformations that occur during the run out table cooling of AHSS hot rolling, i.e., martensite, bainite transformation, etc.

4.4.4 Simulation of Dynamic Strain-induced Austenite-ferrite Transformation and its Post-dynamic Kinetics

In the thermo-mechanical practice of steel design, deformation of the undercooled metastable austenite just above the transformation temperature (Ar3) is found to provoke strain-induced transformation and to produce ferrite grain sizes of 1–3 μm or less (Beladi et al. 2007). Under this condition, the transformation is believed to be dynamic and not static, i.e., the concurrent deformation and phase transformation are necessary. This kind of ferrite transformation occurring during deformation of austenite has been termed as dynamic strain-induced austenite-ferrite transformation (Weng 2009). Due to its conceptual simplicity and efficiency for grain refinement, and the associated improvement in strength and toughness at low costs, the DSIT mechanism has attracted much attention (Dong and Sun 2005). However, it is a complex mechanism involving a number of physical processes including diffusion of solute atoms, evolution of dislocation, propagation of grain boundaries, phase transition, DRX and their interactions. This typical transformation phenomena are not included in most of the macroscopic microstructure models.

Recently, Zheng et al. (2009b, 2012) developed a mesoscopic CA model to investigate the DSIT and the post-dynamic transformation (post-DT) by prudently interweaving the metallurgical effects of hot deformation, phase transformation, and recrystallization phenomena. In order to clarify the distinct characteristics of DSIT, two transformation scenarios have been considered for comparison: the DSIT, in which the sample is subjected to a uni-axial engineering thickness reduction of 80 pct at a strain rate of 10 s^{-1} at 780°C and the *isothermal transformation*, during which the sample is isothermally held for a long time at 780°C till the ferrite transformation is completed.

Figure 4.15 shows the microstructure of the DSIT, derived, both from the simulation (Figures 4.15a–c) and from the optical microscopic (Figures 4.15d–f). It can be seen that DSIT ferrite grains nucleate preferably at the prior austenite grain boundaries at the early stage of transformation and then originate close to the γ/α interface with the deformation continuing. This typical nucleation mode leads to a higher nucleation density compared to the isothermal transformation. Figures 4.16a–b show the simulated carbon-concentration fields in the DSIT and the isothermal transformation, respectively. In the DSIT, a faster kinetics is obtained on the moving γ/α interface due to the role of the deformation and thus more carbon atoms are rejected due to the moving γ/α interface. These carbon atoms easily pile up near the γ/α interface to form a narrow carbon-rich layer, which in turn decreases the driving pressure for the interface migration and suppress the growth of the ferrite grains. It implies that the growth of ferrite will be limited during the DSIT. Therefore, the ferrite refinement derived from DSIT can be reasonably considered as the result of increasing ferrite nuclei density and the limited ferrite growth. As a result, the ultimate DSIT microstructure is formed with the duplex phases of finer ferrite grains and the retained austenite as seen in Figure 4.16b. Fine DSIT ferrite grains occupy the majority of the specimen with some small-sized islands/flakes of the retained austenite dispersing in the matrix, which are mainly located at the grain boundaries and triple points between the DSIT ferrite grains. The predicted characteristic of the DSIT microstructure agrees well with the experimental optical micrograph (Zheng et al. 2008b). The results also show that the processing variables (e.g., deformation strain, strain rate, and initial

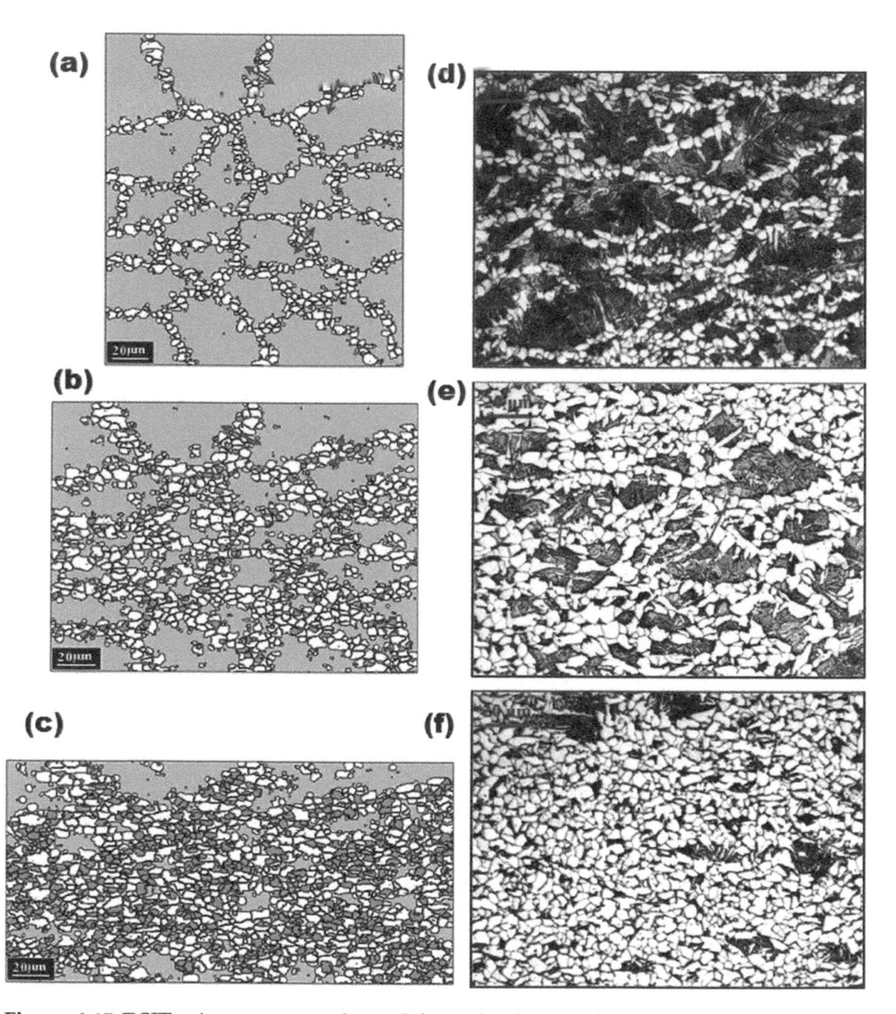

Figure 4.15 DSIT microstructures derived from the CA simulation at different strains ((a) $\varepsilon = 0.3$, (b) $\varepsilon = 0.5$, (c) $\varepsilon = 0.7$) and from the optical micrographs ((d) $\varepsilon = 0.3$, (e) $\varepsilon = 0.5$, (f) $\varepsilon = 0.7$) during the deformation of $\dot{\varepsilon} = 10\ s^{-1}$, T = 780°C. The material is a 0.13 wt.% C-0.19 wt.% Si-0.49 wt.% Mn steel (Zheng et al. 2009b).

austenite grain size in the present modeling) influence the evolution of the DSIT ferrite and the characteristics of the resultant microstructure markedly.

Subsequently, the modified CA model was used to investigate the metallurgical mechanisms of the post-DT (Zheng et al. 2012). Specifically, this model takes into account the effect of local solute redistribution on the migration of the γ/α interface, which allows both the continuing transformation from austenite to ferrite and the reverse transformation to be studied. Figure 4.17a shows the simulated kinetics of ferrite formation

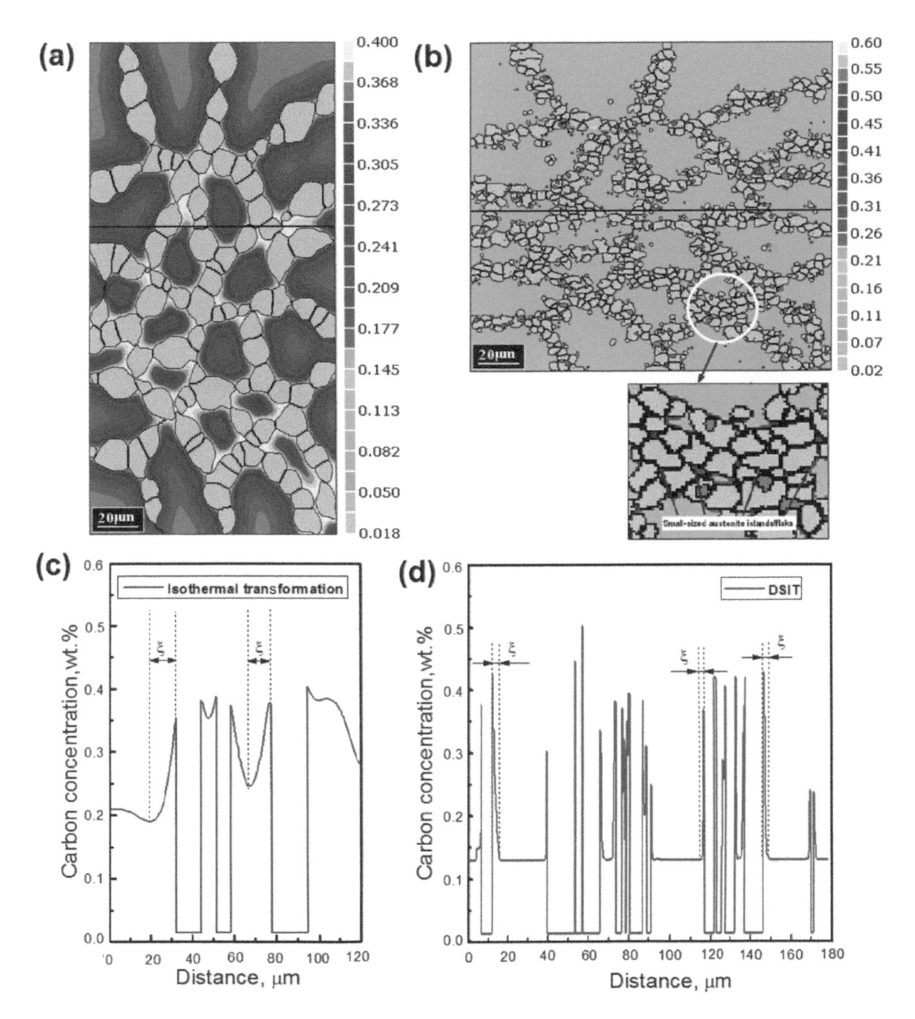

Figure 4.16 CA simulations of the carbon concentration fields during the isothermal decomposition of the un-deformed austenite at the time of 10 s under the isothermal temperature T = 780°C (a), the DSIT when ε = 0.4 under a deformation of $\dot{\varepsilon} = 10\ s^{-1}$, T = 780°C (b) and the carbon concentration distributions of isothermal γ-α transformation (c) and DSIT (d) along the black straight lines shown in (a) and (b). The color scale denotes the carbon concentration (wt.%) (Zheng et al. 2009b).

during post-DT when the DSIT final strain is 0.8. The volume fraction of ferrite decreases rapidly at the early stage of the post-DT and then rises again during subsequent isothermal soaking. In effect, the transformation depends strongly on the local distribution of the carbon concentration within the simulation domain as the drive towards local chemical equilibrium determines the driving force for it. By analyzing the locally acting chemical conditions as seen in Figure 4.17b–c, it is seen that continuing transformation

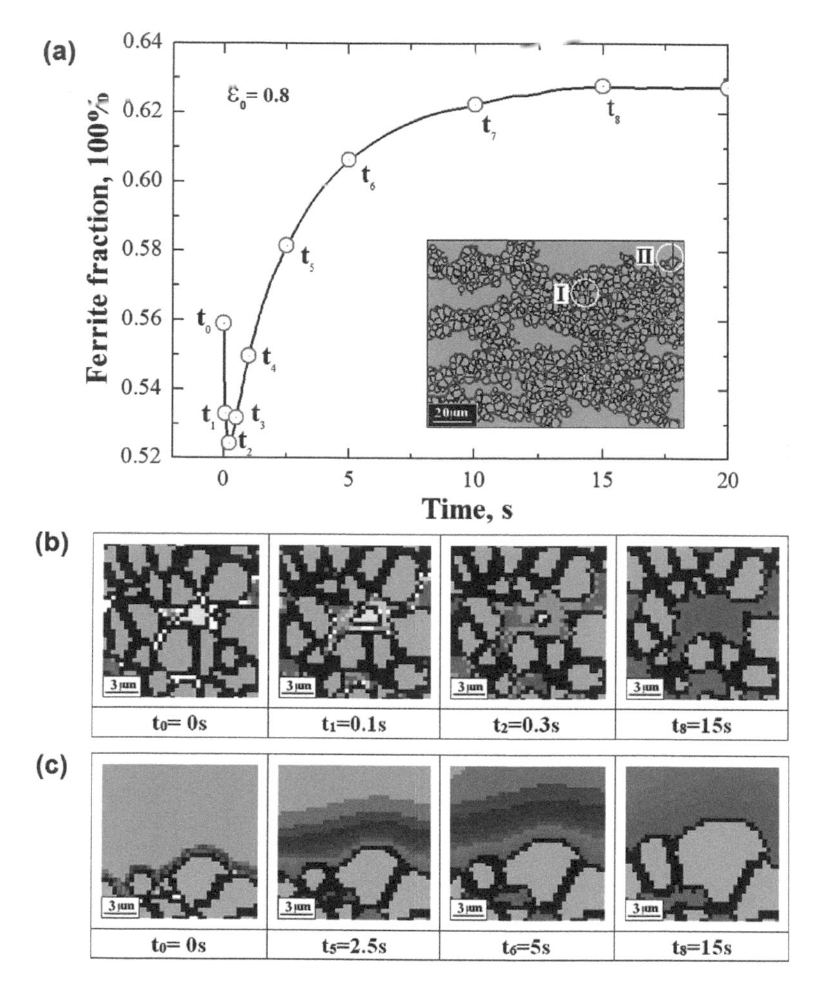

Figure 4.17 CA simulated transformation kinetics of ferrite during the post-DT at the DSIT final strain of 0.8 (a); Snapshots at different times of the carbon concentration field in reverse transformation (b) and continuing austenite-to-ferrite transformation (c) in the selected regions shown in (a). The material is a 0.17 wt.% C-0.27 wt.% Si-0.71 wt.% Mn steel (Zheng et al. 2012).

from retained austenite to ferrite and the reverse transformation can occur simultaneously in the same microstructure. Competition does exist between the austenite-ferrite transformation and the reverse transformation during post-DT after DSIT. And the overall transformation kinetics can be understood as a dynamic equilibrium between these two mechanisms. The simulation also suggests that these two competing transformation processes are strongly affected by the DSIT final strain, i.e., the ferrite fraction formed during DSIT: it determines which occurs preferentially for the continuing transformation from retained austenite to ferrite and the reverse transformation until the final thermodynamic equilibrium is reached. The influence of the DSIT final strain on the grain size of ferrite and the characteristics of the resultant microstructure is also discussed. Figure 4.18 shows the initial and resultant microstructures during post-DT at

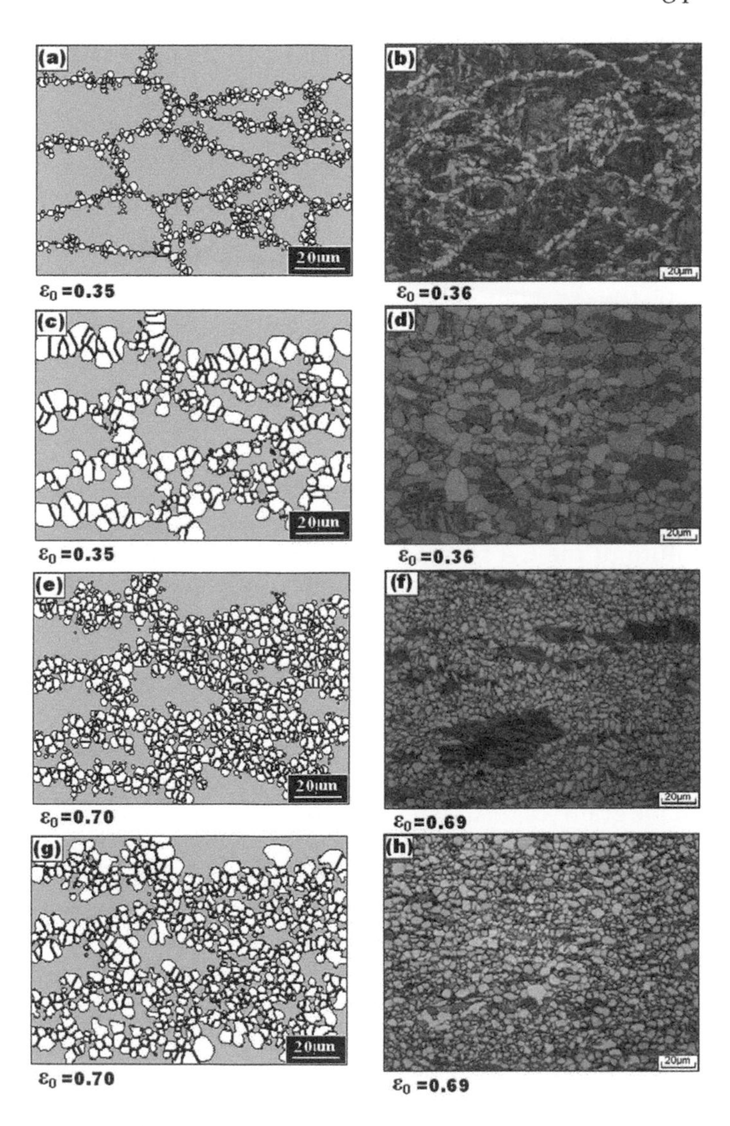

Figure 4.18 Initial DSIT microstructures (a–b, e–f) and the resultant post-DT microstructures (c–d, g–h) derived from the CA simulation (the left images) and from the optical micrographs (the right images) during the post-dynamic isothermal holdings of T = 780°C at two DSIT final deformation levels (ε_0). In the simulated micrographs, ferrite is white, austenite is gray. In the optical micrographs, ferrite is light, martensite which has been austenite before quenching, is dark (Zheng et al. 2012).

the two DSIT final deformation strain levels, as obtained from the simulation and metallographic observations of water-quenched specimens. It is seen that the simulated phase arrangement and the grain structure resemble the metallographic results both in the initial and final microstructures at the two DSIT final strains. Thus, the CA modeling is able to reproduce the post-DT after DSIT satisfactorily and also match the experimental results very well.

4.5 Summary and Outlook

A concise review is given on the application of CA in modeling the microstructure evolution during AHSS hot rolling, with particular attention placed on the CA depictions about the individual transformation behaviors of the austenite during/after hot deformations. The results presented indicate that the CA is a very promising tool for the modeling of the microstructural evolution for AHSS hot rolling. From a fundamental stand point, better understanding of the dynamics and the morphology of microstructure that arise from the complex microstructure physics can be made. From a practical point of view, the microstructure parameters such as grain size, fraction transformed, and fraction recrystallized that determine the mechanical properties of the materials can be predicted.

The future of the CA method in AHSS hot rolling is most likely to develop models to predicate the evolution of microstructure through the entire process, from hot rolling to run-out table cooling and coiling, of industrial relevance. However, the real industrial processes can only be encountered in complex physical transformations. And many of the influencing factors are either not exactly known or sufficiently well defined to apply CA models. For example, owing to the ongoing debate about the reaction mechanisms, the bainite transformation still remains exclusive to be modeled at the grain scale. Another challenge arises from the fact that some individual microstructure processes or interacting sub-processes can range from different length scales. The austenite-to-pearlite transformation is an important example in this regard. The individual pearlitic lamellae structure is usually much smaller than the austenite grains. Thus, it is conceivable that, for this type of mesoscale problem, a suitable hybrid approach is required to bridge the length scales. And the development of such hybrid modeling approaches is bound to remain ahead of the application of CA to simulate the complex microstructure phenomena in AHSS productions.

Overall, the CA approach provides a promising tool for quantitative and explicit simulation of microstructure evolution in AHSS hot rolling. However, many challenging tasks still need to be undertaken to bring CA modeling of the microstructure evolution during industrial processing onto a satisfactory level.

Acknowledgements

Chengwu Zheng gratefully acknowledges the financial support from National Natural Science Foundation of China (NSFC) under Grant No. 51371169 and No. 51401214.

Keywords: Cellular automaton, Austenite recrystallization, Austenite decomposition, Mesoscopic modeling, Hot rolling

References

Avrami, M. 1939. Kinetics of phase change I - general theory. J. Chem. Phys. 7: 1103–1112.

Beladi, H., G.L. Kelly and P.D. Hodgson. 2007. Ultrafine grained structure formation in steels using dynamic strain induced transformation processing. Inter. Mater. Rev. 52: 14–28.

Dong, H. and X.J. Sun. 2005. Deformation induced ferrite transformation in low carbon steels. Curr. Opin. Solid State Mat. Sci. 9: 269–276.

Ding, R. and Z.X. Guo. 2001. Coupled quantitative simulation of microstructural evolution and plastic flow during dynamic recrystallization. Acta Mater. 49: 3163–3175.

Goetz, R.L. and V. Seetharaman. 1998. Modeling dynamic recrystallization using cellular automata. Scripta Mater. 38: 405–413.

Hesselbarth, H.W. and I.R. Göbel. 1991. Simulation of recrystallization by cellular automata. Acta Metall. Mater. 39: 2135–2143.

Janssens, K.G.F. 2010. An introductory review of cellular automata modeling of moving grain boundaries in polycrystalline materials. Math. Comput. Simul. 80: 1361–1381.

Johnson, W.A. and R.F. Mehl. 1939. Reaction kinetics in processes of nucleation and growth. Trans. AIME 135: 416–442.

Kolmogorov, A.N. 1937. Statistical theory of crystallization of metals. Izv Akad Nauk USSR-Ser Matemat, 1: 355–359 (in Russian).

Kugler, G. and R. Turk. 2004. Modeling the dynamic recrystallization under multi-stage hot deformation. Acta Mater. 52: 4659–4668.

Kumar, M., R. Sasikumar and P.K. Nair. 1998. Competition between nucleation and early growth of ferrite from austenite-Studies using cellular automaton simulations. Acta Mater. 46: 6291–6303.

Lan, Y.J., D.Z. Li and Y.Y. Li. 2004a. Modeling austenite-ferrite transformation in low carbon steel using cellular automaton method. J. Mater. Res. 19: 2886–2877.

Lan, Y.J., D.Z. Li and Y.Y. Li. 2004b. Modeling austenite decomposition into ferrite at different cooling rate in low-carbon steel with cellular automaton method. Acta Mater. 52: 1721–1729.

Lan, Y.J., N.M. Xiao, D.Z. Li and Y.Y. Li. 2005. Mesoscale simulation of deformed austenite decomposition into ferrite by coupling a cellular automaton method with a crystal plasticity finite element model. Acta Mater. 53: 991–1003.

Liu, D.S., F. Fazeli and M. Militzer. 2007a. Modeling of microstructure evolution during hot strip rolling of dual phase steels. ISIJ Int. 47: 1789–1798.

Liu, D.S., F. Fazeli, M. Militzer and W.J. Poole. 2007b. A microstructure evolution model for hot rolling of a Mo-TRIP steel. Metall. Mater. Trans. A. 38A: 894–908.

Ma, X., C.W. Zheng, X.G. Zhang and D.Z. Li. 2016. Microstructural depictions of austenite dynamic recrystallization in a low carbon steel: A cellular automaton model. Acta Metall. Sin. -Engl. Lett. 29: 1127–1135.

Militzer, M. 2007. Computer simulation of microstructure evolution in low carbon sheet steels. ISIJ Int. 47: 1–15.

Militzer, M., E.B. Hawbolt and T.R. Meadowcroft. 2000. Microstructural model for hot strip rolling of high-strength low-alloy steels. Metall. Mater. Trans. A. 31A: 1247–1259.

Park, N., L. Zhao, A. Shibata and N. Tsuji. 2014. Dynamic ferrite transformation behaviors in 6Ni-0.1C steel. JOM. 66: 765–773.

Raabe, D. 1998. Computional Materials Science: The Simulation of Materials, Microstructural and Properties. Wiley-VCH, Weinheim.
Raabe, D. 2002. Cellular automata in materials science with particular reference to recrystallization simulation. Annu. Rev. Mater. Res. 32: 53–76.
Raabe, D. and L. Hantcherli. 2005. 2D cellular automaton simulation of the recrystallization texture of an IF sheet steel under consideration of Zener pinning. Comput. Mater. Sci. 34: 299–313.
Reuther, K. and M. Rettenmayr. 2014. Perspective for cellular automata for the simulation of dendritic solidification-A review. Comput. Mater. Sci. 95: 213–220.
Sarkar, S. and M. Militzer. 2009. Microstructure evolution model for hot strip rolling of Nb-Mo microalloyed complex phase steel. Mater. Sci. Technol. 25: 1134–1146.
Sellars, C.M. and J.A. Whiteman. 1979. Recrystallization and grain growth in hot rolling. Met. Sci. 13: 187–194.
von Neumann, J. 1963. Papers of John von Neumann on Computing and Computer Theory. *In*: W. Aspray and A. Burks (eds.). Reprint Series on the History of Computing. Vol. 12. MIT Press, Cambridge, MA.
Weng, Y. 2009. Ultra-Fine Grained Steels. Springer. NewYork.
Yang, H., C. Wu, H.W. Li and X.G. Fan. 2011. Review on cellular automata simulations of microstructure evolution during metal forming process: Grain coarsening, recrystallization and phase transformation. Sci. China-Technol. Sci. 54: 2107–2118.
Zhang, L., C.B. Zhang, Y.M. Wang and H.Q. Ye. 2003. A cellular automaton investigation of the transformation from austenite to ferrite during continuous cooling. Acta Mater. 51: 5519–5527.
Zheng, C.W., N.M. Xiao, D.Z. Li and Y.Y. Li. 2008a. Microstructure prediction of the austenite recrystallization during multi-pass steel strip hot rolling: A cellular automaton modeling. Comput. Mater. Sci. 44: 507–514.
Zheng, C.W., D.Z. Li, S.P. Lu and Y.Y. Li. 2008b. On the ferrite refinement during the dynamic strain-induced transformation: A cellular automaton modeling. Scripta Mater. 58: 838–841.
Zheng, C.W., N.M. Xiao, D.Z. Li and Y.Y. Li. 2009a. Mesoscopic modeling of austenite static recrystallization in a low carbon steel using a coupled simulation method. Comput. Mater. Sci. 45: 568–575.
Zheng, C.W., L.H. Hao, N.M. Xiao, D.Z. Li and Y.Y. Li. 2009b. Numerical simulation of dynamic strain-induced austenite-ferrite transformation in a low carbon steel. Acta Mater. 57: 2956–2968.
Zheng, C.W., D. Raabe and D.Z. Li. 2012. Prediction of post-dynamic austenite-to-ferrite transformation and reverse transformation in a low-carbon steel by cellular automaton modeling. Acta Mater. 60: 4768–4779.

5

Thermomechanical Processing of TRIP Steels

Adam Grajcar

ABSTRACT

This chapter addresses the most important issues associated with thermo-mechanical processing of transformation induced plasticity (TRIP) steels. TRIP steels are one of the most promising groups of advanced high strength steels (AHSSs) for automotive industry in terms of light weight opportunities and crash worthiness. In this paper, main types of low-carbon and medium-carbon TRIP-assisted steels are analyzed. The review concentrates on the design of the thermo-mechanical processing schedules considering flow stress values, softening kinetics of austenite and multi-step physical simulation of hot strip rolling. The effect of chemical composition on the hot deformation resistance is addressed. The design of the semi-industrial hot rolling lines is also characterized. Some of the problems occurring in the industrial practice are identified. The austenite decomposition of various types of TRIP steels is monitored as a function of hot deformation conditions and multi-step cooling paths. Finally, the challenges concerning an industrial production of hot rolled TRIP steel strips are described.

Silesian University of Technology, Institute of Engineering Materials and Biomaterials, 18a Konarskiego Street, 44-100 Gliwice, Poland.
E-mail: adam.grajcar@polsl.pl

5.1 Introduction

A family of advanced high strength steels covers multiphase steels consisting of different mixtures of soft and hard phases. A soft phase constitutes a matrix whereas a hard structural constituent is the reinforcement. The best combination of strength and plasticity are shown by AHSSs which contain ferrite, bainite and retained austenite. In these multiphase steels ferrite is a soft matrix, bainite is reinforcement and retained austenite transforms into martensite during sheet's forming. The interactions between hard and soft phases and the strain-induced martensitic transformations ensure a high work hardening rate resulting in an additional increase of strength and plasticity. However, a condition to retain austenite at room temperature for low-C and medium-C steels is to control precisely the temperature of sheets during cooling followed by hot-working. It is possible only under controlled conditions of thermomechanical rolling.

Thermomechanical processing of transformation induced plasticity steels is one of the present determinants of modern metallurgical approaches for manufacturing high-quality value-added steel products. This technology is essential particularly for steel sheets used in the automotive industry. However, there are also some attempts to adopt the thermomechanical processing for production of TRIP steel rods. The TRIP-aided steels combine successfully high strength, plasticity and crashworthiness; the latter one is especially important for the passive safety of cars and other vehicles (Gronostajski et al. 2010). Since the thermomechanical processing combines hot rolling and direct cooling from a finishing rolling temperature, a precise monitoring of both hot deformation conditions and cooling profiles on run-out tables is required to obtain a desired sheet profile and multiphase microstructure.

TRIP steels include a variety of different grades. These multiphase steels can be produced as the steels with ferritic (Adamczyk and Grajcar 2005), bainitic (Sugimoto et al. 2000) and martensitic (Fonstein 2015) microstructures. Their most characteristic feature is a presence of retained austenite, which transforms into martensite upon cold straining. The strain-induced martensitic transformation prevents a strain localization leading to delaying necking in a tensile-tested specimen or thinning of a drawn sheet part (Mesquita et al. 2013). It is possible due to high work strengthening accompanying this martensitic transformation (De Cooman 2004). It is obvious that most of TRIP steels are produced as cold rolled sheets subjected subsequently to intercritical annealing (Demeri 2013). They are characterized by small thickness and a good surface quality.

One of the disadvantages of cold-rolled TRIP steels is their relatively low stretch-flangeability, which is assessed by means of the hole-expanding test (De Cooman and Speer 2012). Sugimoto et al. (2004) found that sheet edges have especially poor flanging behaviour and stretch-flangeability.

Takahashi (2003) reported that hole-expanding behaviour of sheet edges can be improved by the homogeneity of the microstructure resulting in a better local ductility. In cold-rolled ferrite-based steels there is a big difference in hardness between recrystallized ferrite and formed strain-induced martensite. In industrial practice it often leads to cracking of formed sheet edges (Demeri 2013). One of the sulutions to improve the stretch-flangeability is to use bainite-based steels because it decreases the difference in hardness between bainite and strain-induced martensite (Sugimoto et al. 2000). Another idea is to replace cold-rolled sheets by hot-rolled substitutes (Takahashi 2003). In such hot-rolled sheet steels the ferritic matrix is hardened by precipitation strengthening, grain refinement and strain accumulation caused by finishing hot rolling below the non-recrystallization temperature (T_{nr}) of austenite (Basuki and Aernoudt 1999). Hence, the homogeneity of the microstructure increases and the local ductility and resulting hole-expanding behaviour are improved. The hot-rolled sheets are required for underbody elements of cars (rocker arms, wheels, etc.) (Takahashi et al. 2003).

5.2 Types of TRIP Steels

Typical TRIP steels consist of ferrite (as a matrix), bainite and retained austenite. The requirement for the presence of retained austenite is to produce carbide-free bainite and a gradual enrichment of the austenite (γ) phase in carbon (Bojarski and Bold 1974). The first one is ensured by a proper selection of chemical composition and the latter one can be achieved by applying a multi-step controlled cooling following hot-working (Eberle et al. 1999). Representative microstructures of the TRIP steels with a different matrix are shown in Figure 5.1. Figure 5.1a presents the most popular TRIP steels containing a ferrite phase as a matrix. In such a type of the microstructure, retained austenite is present both as blocky grains and layers located in bainitic islands (Mohamadizadeh et al. 2016). In medium-C TRIP steels a ferrite matrix is replaced by carbide-free bainite. In this case a majority of the γ phase is interlath retained austenite of various thicknesses, as shown in Figure 5.1b. The bainite matrix can also be formed in low-C steels using direct cooling from a finishing rolling temperature or by adding alloying elements, which increase steel's hardenability. Figure 5.1c shows a bainitic-austenitic microstructure of medium-Mn steel. In this case retained austenite occurs as small blocky granules and its rest is located between bainitic ferrite laths.

The stabilization of retained austenite to room temperature requires increased amounts of Si, Al or Si-Al (Petrov et al. 2007). These elements prevent carbide precipitation during an isothermal holding step, which is applied at a bainitic transformation range (Grajcar 2007). Aluminium, silicon and manganese contents affect the progress of phase transformations on a

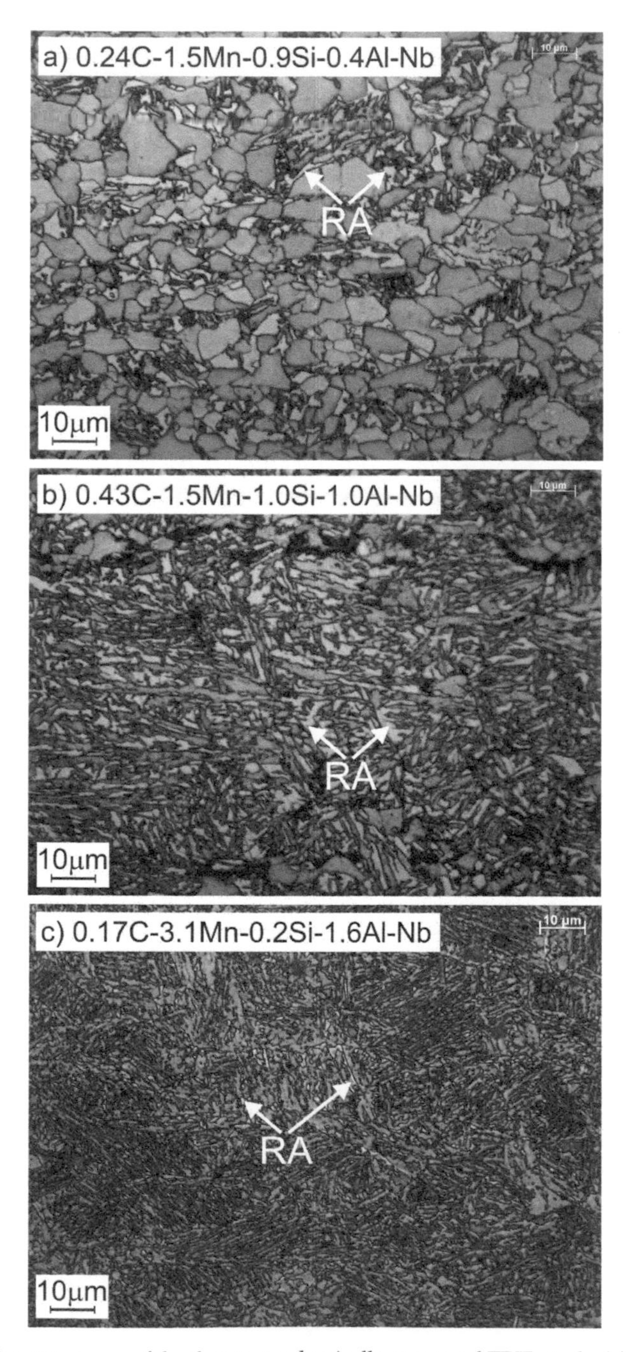

Figure 5.1 Microstructures of the thermomechanically processed TRIP steels: (a) ferrite-based steel, (b) medium-C bainite-based steel, (c) low-C medium-Mn bainite-based steel; etched in sodium metabisulfite.

run-out table. The alloying elements also determine a level of deformation resistance during hot rolling and kinetics of softening processes in plastically deformed austenite.

A total content of alloying elements in different types of TRIP steels ranges from 2.6 to app. 6.5 wt.% as can be seen in Table 5.1. In general, the rolling forces increase with increasing content of Mn, Si and Cr in conventional low-alloyed steels (Opiela 2014). The hot deformation resistance rises further with the addition of microalloyed elements, i.e., Ti, Nb, V and Mo (Opiela and Grajcar 2012). This behaviour is more complex in TRIP steels due to a large amount of Si and Al affecting the hot deformation process strongly. Numerous works (Pereda et al. 2015) show that the effects exerted by these elements do not follow a linear (or other) relationship and very often change rapidly with small modification of chemistry.

5.3 Physical Simulation of Hot Rolling

Industrial rolling has to be always preceded by physical simulation tests reflecting complex industrial temperature-time-strain profiles using compression or torsion simulators. Basic tests include the identification of flow stress levels and softening kinetics of austenite. The fundamental significance has multi-step hot strip rolling schedules reflecting practical industrial passes with particular strains, strain rates, temperatures, inter-pass times, etc. Additionally, the force-energetic data must be accompanied by simulations of microstructural evolution of deformed austenite with subsequent prediction of the decomposition of pancaked austenite on run-out table (Garbarz et al. 2012).

Flow stress. Basic data on hot deformation behaviour of steels are gathered during continuous compression or torsion tests determined for a few deformation temperatures and strain rates. These data are useful for a proper selection of strain values, strain rates and pass temperatures of thermomechanical hot strip rolling. A few representative stress-strain curves for two model grades of TRIP steels registered for different deformation conditions are shown in Figure 5.2. An effect of the compression temperature and strain rate on a shape of σ-ε curves is typical. Stress values increase with decreasing deformation temperature and increasing strain rate (Grajcar 2008). A shape of the curves indicates that dynamic recrystallization takes place only for the lowest Zener-Holomon parameter values. A clear peak of ε_p (corresponding to a maximum value of flow stress) followed by decreasing flow stress can only be observed for both steels deformed at a relatively high temperature of 1150°C at a strain rate of 0.1 s^{-1}. The medium-Mn steel has higher flow stress and critical strain for the initiation of dynamic recrystallization in comparison to the 0.43C-1.5Mn-1Si-1Al-Nb-Ti steel. A level of flow stress rapidly increases with decreasing deformation temperature to 850°C (Figure 5.2). The difference between two analyzed

Table 5.1 Different types of TRIP steels and total amount of alloying elements.

Matrix of TRIP steel	Examples	Total amount of Mn+Si+Al	Other elements	Type of retained austenite	Source
ferrite	0.23C-1.5Mn-0.25Si-0.8Al	2.6	0.18Cr, 0.21Cu	blocky, interlath	(Adamczyk et al. 2008)
	0.19C-1.5Mn-1.6Si	3.1	0.2Mo		(Liu et al. 2007)
micro-alloyed ferrite	0.2C-1.6Mn-1.3Si-0.5Al-Nb	3.4	0.25Mo, 0.04Nb	blocky, interlath	(Ranjan et al. 2015)
	0.24C-1.5Mn-0.9Si-0.4Al	2.8	0.03Nb, 0.02Ti		(Grajcar 2010)
bainite	0.2C-1.5Mn-1.5Si	3.0	-	interlath	(Sugimoto et al. 2004)
	0.43-1.5Mn-1Si-1Al	3.5	0.03Nb, 0.01Ti		(Grajcar 2015)
medium-Mn bainite or martensite	0.16C-4.7Mn-1.6Al-0.2Si	6.5	0.2Mo	interlath	(Grajcar et al. 2012a)
	0.2C-3Mn-1.5Si	4.5	-		(Sugimoto et al. 2015)
martensite	0.2C-1.5Mn-1.5Si	3.0	-	interlath	(Sugimoto et al. 2015)

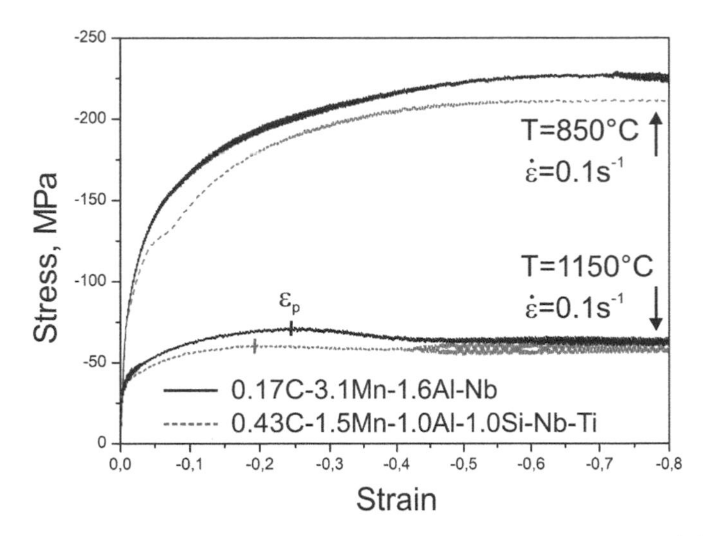

Figure 5.2 Effect of chemical composition on stress-strain curves registered at a deformation temperature of 1150°C and 850°C and at a strain rate of 0.1 s^{-1}.

steels is maintained, almost at a constant in a whole range of compression. It is clear that there is no peak initiating dynamic recrystallization. Instead there is a dynamic equilibrium state between work hardening and dynamic recovery, which removes the excess of dislocations.

Numerous flow curves registered in a wide range of temperature and strain rate for different TRIP steel grades can be found elsewhere for compression tests (Poliak and Siciliano 2004) and torsion tests (Zubialde et al. 2009). It is clear that the increase in strain rate leads to the increase of flow stress (Figure 5.3a). Moreover, Figure 5.3b indicates that the increase in Mn content results in the growth of flow stress. However, an effect of chemical composition is more complex than it is suggested by Figure 5.3b. The reason for that are opposite and masked effects of various alloying additions in TRIP steels (Grajcar and Kuziak 2011a). The hot deformation behaviour is relatively simple for binary Fe-Mn alloys containing less than 10% Mn (Cabanas et al. 2006). The increase of the Mn content increases flow stress and retards dynamic recrystallization (peak of ε_p is shifted right). However, the hot working behaviour is more complicated in the presence of higher Al additions (typical for TRIP steels). Grajcar et al. (2012c) reported that the strengthening effect of Mn in such steels is further limited to about 5%. They showed that the increase of Mn content in a range from 3% to 5% in the presence of app. 1.5% Al does not significantly change the shape of σ-ε curves and thus the hot deformation behaviour. These results are in good agreement with earlier data reported by Poliak and Siciliano

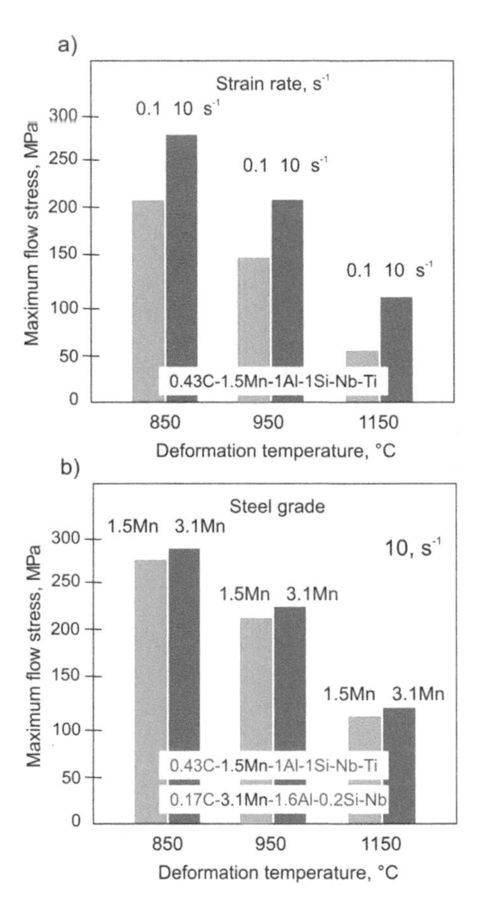

Figure 5.3 Effect of the strain rate (a) and chemical composition (b) on maximum values of flow stress measured in the continuous compression tests.

(2004), who noticed that the addition of 1.7% Al to 0.2 C-1.8 Mn-0.25 Si steel influences the restoration processes in the austenite and favours dynamic recrystallization.

The situation complicates even more for Nb-microalloyed austenite with increased amounts of Mn and Al. The effect of Nb on stress-strain curves in high-strength low-alloy (HSLA) steels is well known. Niobium dissolved in the austenite retards static and dynamic recrystallization during hot-working. Small particles of Nb (C,N) further increase flow stress and delay recrystallization, being obstacles to migration of grain boundaries (Bleck and Phiu-On 2005). However, the effect of Nb on TRIP steels containing increased contents of Mn, Al and/or Si can be completely different than that on HSLA steels. It is well known (Siciliano and Poliak 2005) that the increase of Mn content in Nb-microalloyed steels delays the precipitation of

Nb (C,N). By contrast, silicon shows an accelerating effect. Koh et al. 1998, Zubialde et al. 2009, reported an acceleration of dynamic recrystallization in high-Al Nb-microalloyed steels. The increased Al addition, loweres the stacking fault energy of the austenite, that leads to an increase in the net driving force for boundary migration. Niobium raises flow stress similarly as an effect of Al but its interaction with austenite is opposite. Poliak and Siciliano (2004) suggested that Nb and Al interfere with each other when added simultaneously, so that the dominating strengthening mechanism and resultant acceleration or delaying of dynamic recrystallization strongly depends on both Al and Nb concentrations. Moreover, the delaying of the softening by Nb depends on the state of this element under conditions of hot working (dissolved or precipitated). Grajcar et al. (2014b) showed that higher is the Al and Mn content, stronger is the solute drag effect of niobium whereas an effect from precipitates disappears.

Concluding, the flow stress levels are higher when compared to HSLA steels but dynamic recrystallization behaviour depends on above-mentioned interactions between the chemistry and deformation conditions. Hence, the hot deformation behaviour of TRIP steels is very complex and changes rapidly with small modifications of chemical composition. It refers especially to microalloyed steels, in which the state of microalloyed additions depends on opposite and concurrent interactions of Mn and Si as well as Mn and Al.

Softening kinetics. Softening kinetics curves are crucial to assess the progress of recrystallization of austenite between successive deformation steps. At the same time the potential residual strain can be evaluated. Representative σ-ε curves for different TRIP steel grades have been published by Grajcar and Kuziak (2011b) and Pereda et al. (2015). Results from double-hit compression tests are utilized to obtain curves showing relationships between a recrystallization extent and time. Such curves for some TRIP steel grades are shown in Figure 5.4. The data can be well fitted with the sigmoidal curves, corresponding to the Avrami-type relationship. It is interesting to note that the softening kinetics of the steel containing the highest Mn content is faster compared to other two steels containing a lower total content of alloying elements. Reason for this is same as given above while analyzing the flow stress behaviour and interactions between Nb microaddition and other alloying elements. While considering the effect of Nb it is required to distinguish between a solute drag effect of Nb in solid solution and a pinning effect caused by precipitates. It is well known (Poliak and Siciliano 2004) that increasing the Mn content delays the precipitation of Nb(C,N). Manganese enhances the solubility of NbC and NbN in austenite and decreases Nb diffusivity. Finally, it leads to delaying of a rate of Nb(C,N) precipitation. Moreover, Al addition enhances this phenomenon because this element suppresses precipitation of Nb(C,N) due to its strong chemical

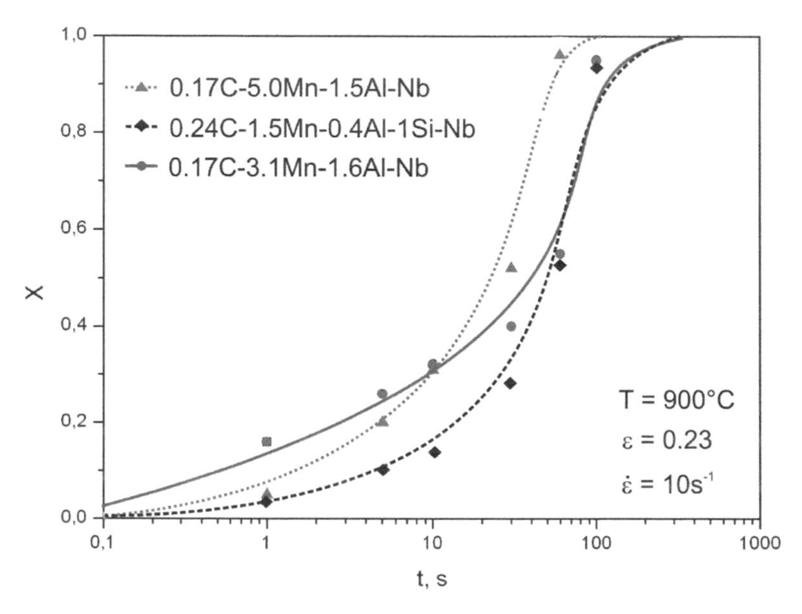

Figure 5.4 Effect of chemical composition on a softening fraction of austenite (X).

affinity to nitrogen (Skolly and Poliak 2005). The smaller content of Mn in the 0.17C-3.1Mn-1.6Al-Nb and thus the smaller chemical limitation of potential precipitation of particles containing Nb results in delaying the progress of static recrystallization of this steel.

Suikkanen et al. (2012) noticed that Al addition to 0.2C-2Mn-0.6Cr steels retards static recrystallization more strongly than silicon alloying. Pereda et al. (2015) reported that in 0.2C-2Mn-(1-2)% Al Nb-microalloyed steels at temperatures higher than 1000°C the retardation of the softening kinetics is caused by Nb in solid solution whereas at lower temperatures NbC precipitates lead to incomplete softening for the 1% Al steel. The increase of Al content to 2% results in a complex interaction between softening, strain-induced NbC precipitation and $\gamma \rightarrow \alpha$ phase transformation. The number of identified NbC particles were relatively small, confirming the suppressing effect of Al addition on the precipitation of carbonitrides. The complex carbonitrides containing Nb and Ti were also revealed by Grajcar (2014) in 0.24C-1.5Mn-0.4Al-1Si-Nb-Ti steel. However, their size was relatively coarse and thus a softening curve does not show a plateau that often occurrs in micro alloyed steels (Rodriguez-Ibabe 2007). Anyway, the time for 50% static recrystallization of this steel is equal to app. 60 s. The results measured by the double-hit compression show good agreement with metallographic measurements shown in Figure 5.5. Fine recrystallized grains are formed both at grain boundaries of austenite and within deformation bands (Grajcar and Opiela 2008b).

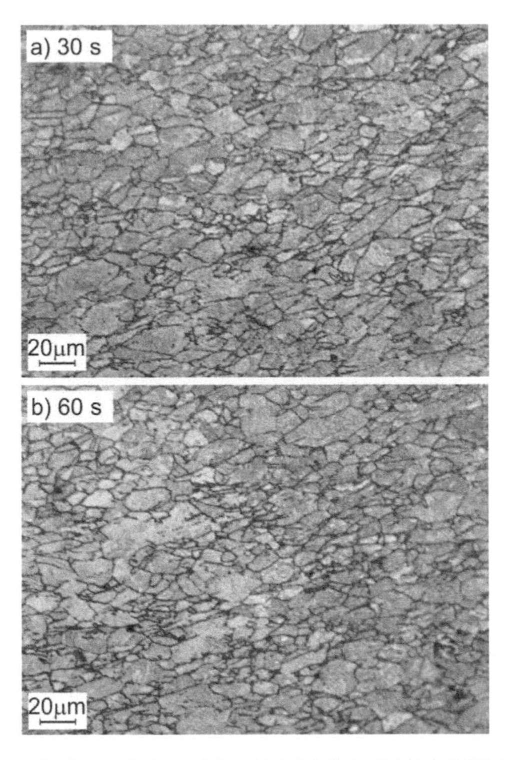

Figure 5.5 Prior austenite boundaries of the 0.24C-1.5Mn-0.4Al-1Si-Nb-Ti steel deformed at 900°C to true strain of 0.23 at a strain rate of 10 s^{-1} and isothermally held at this temperature for 30 s (a) and 60 s (b).

Multi-step simulation of hot strip rolling. A proper industrial implementation of new thermomechanical schedules in the industry requires earlier physical simulations of hot strip rolling. The real conditions can be satisfactorily reflected by multi-step compression or torsion tests, which simulate complex industrial temperature-time-strain cycles using small samples (Hadasik et al. 2006). Physical simulation is usually integrated with mathematical and numerical models describing the microstructure evolution during hot-working and cooling (Pietrzyk et al. 2014). One of the major tasks of the physical simulation is to obtain strongly deformed fine-grained austenite prior to multi-step cooling from a finishing rolling temperature. The presence of large total grain boundary area of austenite and deformation bands of high dislocation density guarantees significant refinement of ferrite and bainite, formed during run-out table cooling (Timokhina et al. 2003).

A major parameter influencing a final deformation of γ phase is the non-recrystallization temperature of austenite (Grajcar et al. 2015a). This temperature can be calculated approximately for the HSLA steel chemistry

(Suwanpinij et al. 2012). Unfortunately, the calculation of the T_{nr} for AHSSs (including TRIP steels) is useless because of richer alloying additions and complex interactions between Mn, Si and Al. The calculation gets more complicated with the addition of microalloying elements. In general, it is well known that adding Nb to the steel increases the non-recrystallization temperature, below which, no complete static recrystallization occurs between successive rolling passes. Below T_{nr} the strain is accumulated with increasing passes leading to austenite pancaking (Bandyopadhyay et al. 2011). Table 5.2 contains a few representative T_{nr} temperatures for different types of HSLA steels and AHSSs. For example, these temperatures are in a range from 930 to 1000°C for different grades of HSLA microalloyed with Nb and Ti (Rodriguez-Ibabe 2007). There is a relatively simple rule in HSLA steels. The higher addition of microalloyed elements higher is the T_{nr}. In some cases there are exceptions, especially when the effects of strain, strain rates and inter-pass times are considered.

Aretxabaleta et al. (2014) reported that the non-recrystallization austenite temperature for 0.21-2Mn-1Al TRIP steel is equal from 950 to 890°C. In general, the T_{nr} decreases with increasing inter-pass time and pass reduction. The non-recrystallization temperature of austenite rises rapidly to 1050–1070°C with an addition of 2% Al (Table 5.2). It shows that a range of the T_{nr} is larger for AHSSs and thus the hot rolling is more difficult to predict

Table 5.2 Non-recrystallization austenite temperatures (T_{nr}) for different types of HSLA steels and AHSSs.

Steel type	T_{nr}, °C	Method of determination	Source
Nb and Ti microalloyed HSLA	930–1000	Torsion	(Rodriguez-Ibabe 2007)
0.19C-1.5Mn-0.15Si	910	Compression	(Zarei-Hanzaki et al. 1995)
0.22C-1.5Mn-1.5Si-0.035Nb	930	Compression	(Zarei-Hanzaki et al. 1995)
0.20C-1.6Mn-1.3Si-0.5Al-0.25Mo-0.04Nb	968	Torsion	(Ranjan et al. 2015)
0.20C-1.6Mn-0.5Si-1.2Al-0.25Mo-0.04Nb	966	Torsion	(Ranjan et al. 2015)
0.20C-1.6Mn-0.5Si-1.2Al-0.25Mo-0.04Nb-0.07P	970	Torsion	(Ranjan et al. 2015)
0.21-2Mn-1Al	950–890	Torsion	(Aretxabaleta et al. 2014)
0.20-2Mn-1Al-0.028Nb	1055–1010	Torsion	(Aretxabaleta et al. 2014)
0.20-2Mn-2Al	1056–1067	Torsion	(Aretxabaleta et al. 2014)

in terms of rolling resistance and microstructure evolution. A problem is more complex in microalloyed austenite. For example, Zarei-Hanzaki et al. (1995) reported T_{nr} was app. 930°C for 0.22C-1.5Mn-1.5Si-0.035Nb steel. Ranjan et al. (2015) showed that this crucial temperature increases to app. 970°C for various combinations of total additions of Al, Si, P and Mo in Nb-microalloyed 0.20C-1.6Mn-Si-Al-0.25Mo-0.04Nb steels. The simple deduction of the T_{nr} temperature can be confusing because in Nb-microalloyed austenite an increase of this temperature can be caused by a solute drag effect and/or precipitation of Nb(C,N), as described above while analyzing the softening kinetics behaviour.

Physical simulation of hot strip rolling has been done by Liu et al. (2007) for 0.19C-1.5Mn-1.6Si-0.2Mo steel under torsion conditions. The idea was that the first high deformation step at the highest temperature should simulate rough rolling. After a long time needed for a complete static recrystallization there are 6 or 7 finishing rolling steps realized under conditions of decreasing strain and inter-pass time and increasing strain rate (Thomas et al. 2011). Liu et al. (2007) reported that significant softening occurs between rough rolling stand and the first two finish rolling stands followed by partial recrystallization. Finally, austenite is continuously work hardened in the last three passes with a negligible amount of inter-pass softening. Equivalent stresses reach 190 MPa at a final deformation temperature of 890°C and at a strain rate of 1 s^{-1}.

It can be seen from Figure 5.6, the flow stress is a little bit higher for 0.24C-1.5Mn-1Si-0.4Al-Nb steel at a comparable deformation temperature

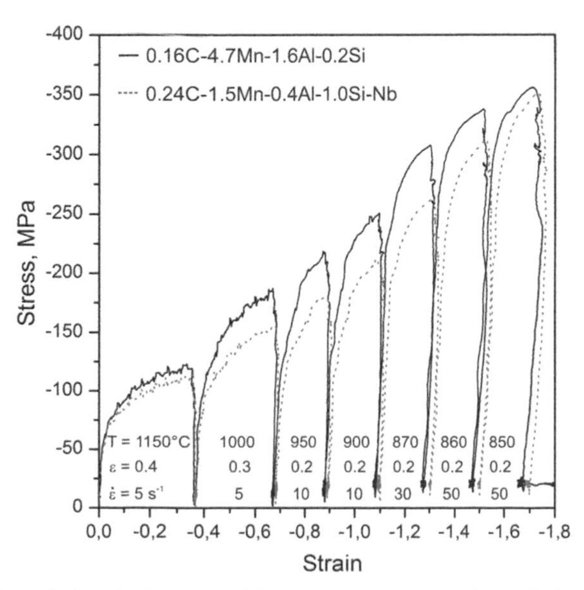

Figure 5.6 Effect of chemical composition on σ-ε curves registered during multi-step compression of specimens under conditions of plane state of strain.

(app. 900°C). The increase of the flow stress is due to Nb microaddition and a higher strain rate of 10 s^{-1}. The flow stress strongly rises with a further decrease in the deformation temperature to 850°C and an increase of strain rate to 50 s^{-1}. The richer alloying in the 0.16C-4.7Mn-1.6Al-0.2Si steel results in a further increase of deformation resistance expressed as a rise in flow stress in Figure 5.6. It is interesting that the highest difference in deformation resistance occurs from the second to sixth finishing rolling stand. It is due to rich Mn and Al alloying. However, the flow stress values for a final stand are almost the same (app. 350 MPa). It is caused by strain-induced precipitation of (Ti,Nb) (C,N) in the steel containing microalloyed elements (Grajcar 2014). The increase of the finishing deformation temperature from 850 to 950°C results in a significant reduction of deformation resistance (Figure 5.7) due to a higher extent of recrystallization between final passes (Grajcar et al. 2014b).

A final result of plastic deformation below the non-recrystallization temperature is a pancaked austenite showing high density of dislocations and other structural defects (Ryu et al. 2002). Two examples of the microstructures formed after 4-step compression (4 × 0.25 strain) of the 0.43C-1.5Mn-1Al-1Si-Nb-Ti steel are presented in Figure 5.8. The micrographs show ferrite films located along prior austenite boundaries after deformation at finishing rolling temperature (T_{fr}) equal to 850°C (Figure 5.8a) and 750°C (Figure 5.8b). A final deformation temperature

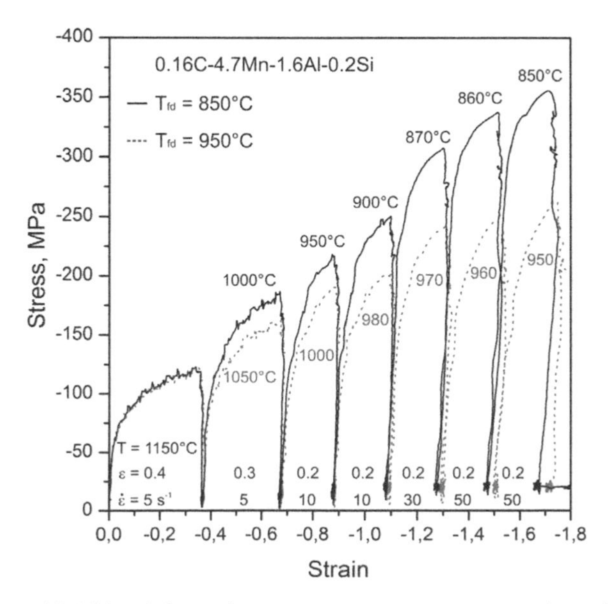

Figure 5.7 Effect of finishing deformation temperature on σ-ε curves registered during multi-step compression of specimens under conditions of plane state of strain.

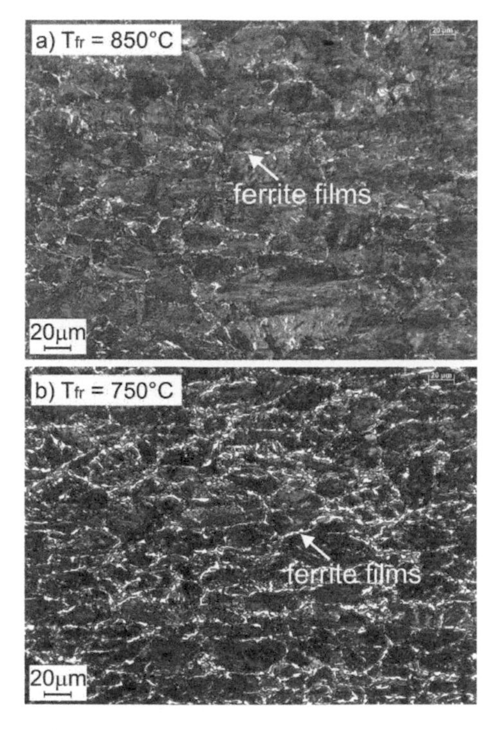

Figure 5.8 Ferrite films along prior austenite boundaries of the 0.43C-1.5Mn-1Al-1Si-Nb-Ti steel formed after 4-step compression at a finishing deformation temperature (T_{fr}) of 850°C (a) and 750°C (b).

affects pancaking of the austenite, which is much stronger after completing compression at 750°C. At lower temperature a fraction of strain-induced ferrite decorating prior austenite boundaries increases too. A type of the formed microstructure influences the decomposition of austenite during a multi-step cooling required for TRIP steel sheets (Skalova et al. 2006).

5.4 Austenite Decomposition

Cooling strategies. Output parameters of hot-working are, the final austenite grain size and residual strain describing the degree of accumulated strain in the austenite (Pietrzyk et al. 2010). The crucial problem for modeling of microstructure evolution of TRIP steel sheets is decomposition of austenite during run-out table cooling and coiling. The austenite decomposition models follow the concept of a sequential transformation model with sub-models for transformation start, ferrite growth, bainite start, bainite growth and martensite formation. The most advanced phenomenological model describing the bainite transformation and retained austenite amount in TRIP steels was developed by Liu et al. (2007). The first model was satisfactory except the prediction of carbide precipitation influencing the amount and stability of retained austenite, which are crucial for mechanical properties of TRIP steels (Krizan et al. 2015). Difficulties are due to a complex character of bainitic transformation, which can result in formation of different morphologies of bainite: granular bainite, degenerate upper bainite, degenerate lower bainite (Zajac et al. 2005). The most advanced model for the formation of bainitic ferrite and carbides in TRIP steels was developed by Fazeli and Militzer (2012). A nucleation-growth based model was used to describe the simultaneous formation of bainitic ferrite and cementite precipitation.

Figure 5.9 shows thermomechanical processing strategies used for production of hot-rolled: ferrite-matrix TRIP steel (Figure 5.9a), strain-induced ferrite-matrix TRIP steel (Figure 5.9b) and bainitic TRIP steel (Figure 5.9c). The formation of multiphase microstructure strongly depends on a strict control of the sheet temperature leaving a final rolling stand. In general, once final rolling is complete, sheets are cooled in a laminar way up to app. 700°C. After that the sheets are held for a few seconds (or longer) to form a required volume fraction of polygonal ferrite. Since this transformation is controlled by the diffusion it depends significantly on a chemical composition of steels (Zaefferer et al. 2004). Moreover, it is well known that higher is the strain accumulation in austenite faster is the transformation kinetics (Grajcar et al. 2014a). The next step is to cool the steel to a range of bainitic transformation. Rest of the austenite, enriched in carbon due to the $\gamma \rightarrow \alpha$ transformation transforms into bainitic ferrite. It is a critical step of cooling because this influences a furher enrichment of the γ phase in carbon. That is why the martensite's start temperature of the γ phase ($M_{s\gamma}$) continuously decreases as a function of time (Figure 5.9a). If the $M_{s\gamma}$ is lower than room temperature it means the presence of retained austenite in a final multiphase microstructure (Wang and Van der Zwaag 2001). In typical TRIP steels it is possible to stabilize app. 10–15% of retained austenite (DeArdo et al. 2005).

An acceleration of ferrite formation can be achieved by decreasing a final rolling temperature (Hosseini et al. 2015). It leads to enhanced strain-induced ferrite formation and enables first step of laminar cooling to be omitted. The strain accumulation in the strongly pancaked austenite (Figure 5.8b) shifts a ferritic bay both to the shorter time and higher temperature (Figure 5.9b). Production of TRIP-assisted bainitic steels requires more simple cooling strategy, shown in Figure 5.9c. These steels are usually characterized by richer alloying (Grajcar and Lesz 2012). Hence, the ferrite and pearlite areas are shifted to longer times. Bainitic-austenitic mixtures possess better homogeneity of the microstructure compared to ferrite-based steels and thus they have improved local ductility, which is required during formation of stretch-flanging operations (Hashimoto et al. 2004). Recently,

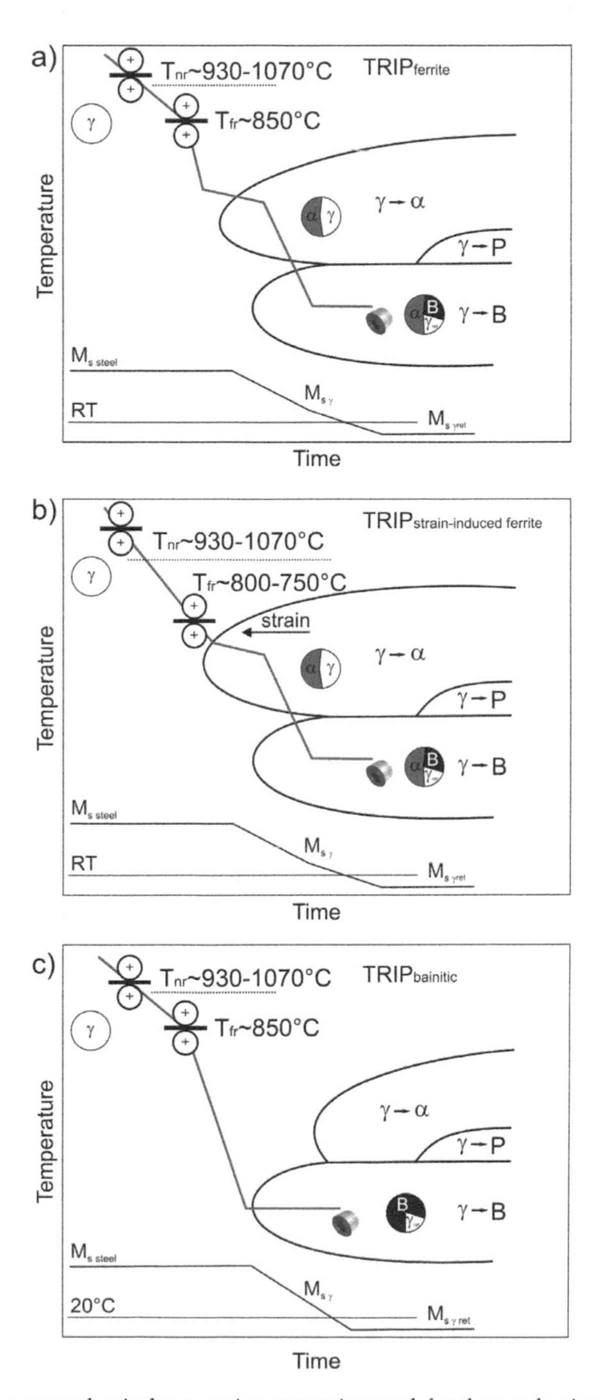

Figure 5.9 Thermomechanical processing strategies used for the production of hot-rolled: ferrite-matrix TRIP steel (a), strain-induced ferrite-matrix TRIP steel (b), bainitic TRIP steel (c).

ultra-high strength steel sheet grades are also produced by quenching and subsequent partitioning of carbon from martensite into retained austenite (Futch et al. 2012).

Effect of finishing deformation temperature. One of the most important parameters of thermomechanical processing of TRIP steels is the final deformation temperature. As shown in Figure 5.8, it influences a final prior austenite microstructure. Figure 5.10 presents an effect of finishing rolling temperature on the microstructure of 0.17C-3.3Mn-1.7Al-0.2Si-0.2Mo steel formed after 7-step compression and isothermal holding of specimens at 450°C for 300 s. It is clear that all the microstructures are characterized by a significant grain refinement but the grain size decreases with decreasing final deformation temperature. The high hardenability of steel (high Mn content) results in ferrite-free microstructures consisting of bainite and retained austenite. The reduction of final deformation temperature leads to banding of microstructural constituents (Figure 5.10c). In the given example these martensite bands deteriorate ductility of steel products (Grajcar and Radwanski 2014). On the other hand, the increase in final deformation temperature results in a rise of grain size (Figure 5.10a). However, it has a minor effect on the volume fraction of retained austenite (Grajcar et al. 2014b). Timokhina et al. (2008) reported that the bainite morphology in TRIP steels has a more pronounced effect on their mechanical behaviour than the refinement of the microstructure. Many works show that the most critical effect on final mechanical properties is the amount of retained austenite and its stability (Fonstein et al. 2005). Therefore, the contribution of grain refinement is smaller than in conventional thermomechanically processed HSLA steels. The major role play the bainite morphology and the stability of retained austenite (Timokhina et al. 2004).

Effect of accumulated strain. A hot deformation history influences a prior austenite microstructure (Figure 5.8) as well as the amount of accumulated strain and resulting phase transitions upon cooling. Figure 5.11 shows two micrographs of 0.17C-3.3Mn-1.7Al-0.2Si-0.2Mo bainite-based TRIP steels formed after finishing rolling at 850°C and subsequent isothermal holding of specimens at 450°C for 300 s. One can notice that the grain refinement is stronger for 7-step compression (total true strain equal to 1.7) (Figure 5.11a) compared to the one obtained after 4-step compression with a total true strain of 1 (Figure 5.11b). As suggested above the grain refinement is not the most important direct parameter influencing the mechanical properties of TRIP steels. However, as it can be seen higher is the grain refinement better is the distribution of retained austenite. Moreover, for some chemical compositions the large blocky grains of austenite are not sufficiently refined and are transformed into martensite or martensite-austenite (M-A) constituents. It indicates that the refinement of the microstructure affects

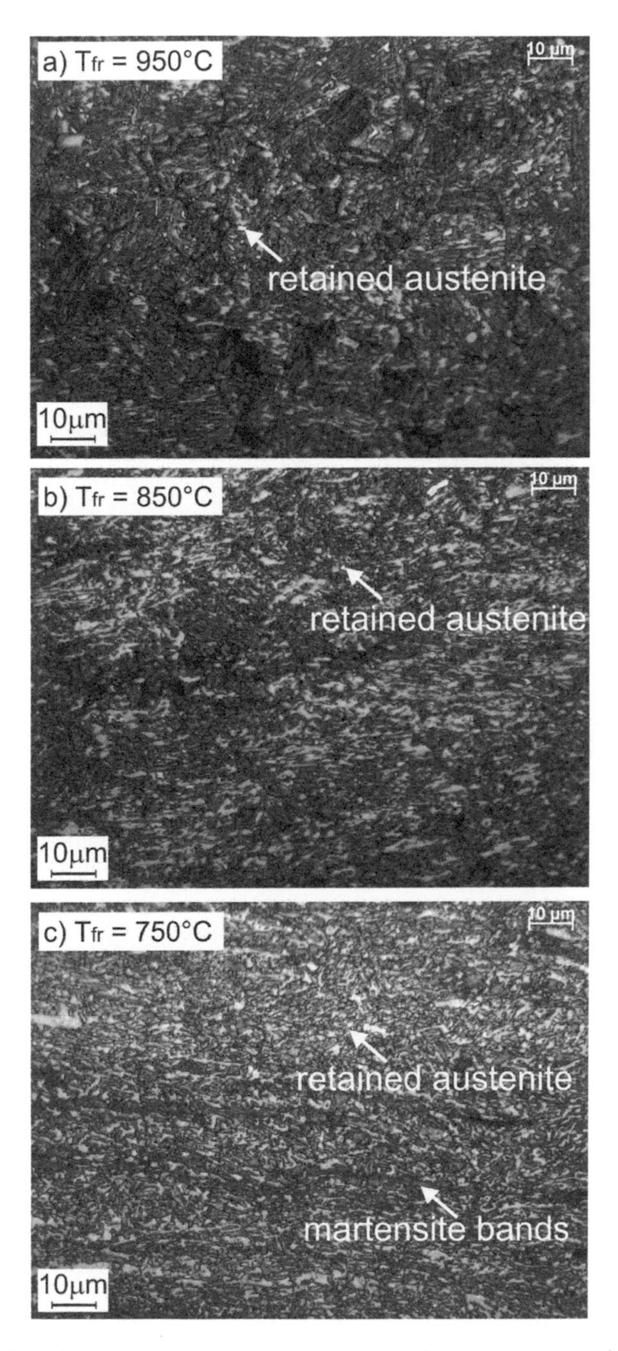

Figure 5.10 Effect of finishing rolling temperature (T_{fr}) on the microstructure of 0.17C-3.3Mn-1.7Al-0.2Si-0.2Mo steel formed after 7-step compression and isothermal holding of specimens at 450°C for 300 s.

Figure 5.11 Effect of deformation schedule on the microstructure of 0.17C-3.3Mn-1.7Al-0.2Si-0.2Mo steel formed after finishing rolling at 850°C and isothermal holding of specimens at 450°C for 300 s.

the stability of retained austenite and in turn this factor is important for a proper gradual progress of strain-induced martensite transformation upon cold forming of sheets (Grajcar et al. 2015b).

Effect of slow cooling. For conventional TRIP steels a key parameter is to produce some fraction of polygonal ferrite as a matrix. Figure 5.12 shows an effect of slow cooling in the $\gamma \rightarrow \alpha$ transformation range on the microstructure of 0.24C-1.5Mn-0.4Al-1Si-Nb-Ti steel formed after 4-step compression (4 x 0.25 strain) at a strain rate of 10 s^{-1} and a finishing rolling at 750°C. It can be seen that direct cooling from 750°C does not lead to formation of ferrite (Figure 5.12a). Ferrite appears after applying 25 s of slow cooling between 750°C and 650°C (Figure 5.12b). This promotes, in turn, a higher volume fraction of retained austenite (white grains). The presence of the

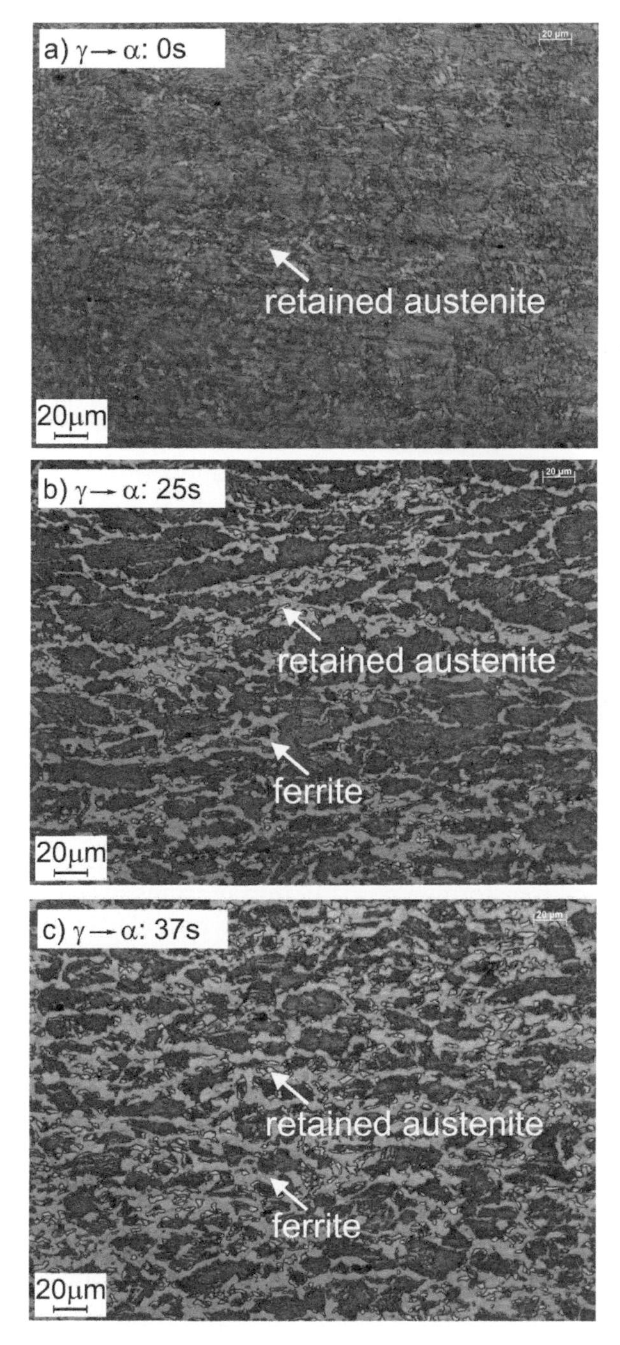

Figure 5.12 Effect of slow cooling in the $\gamma \rightarrow \alpha$ transformation range on the microstructure of 0.24C-1.5Mn-0.4Al-1Si-Nb-Ti steel formed after finishing rolling at 750°C and isothermal holding of specimens at 350°C for 600 s.

50% volume fraction of ferrite (typical for TRIP steels) requires more time (app. 37 s – Figure 5.12c). A cumulative positive result is the stabilization of the highest amount of retained austenite (Grajcar et al. 2011). Considering industrial opportunities of cooling facilities it might be difficult to ensure such cooling conditions (Sprock et al. 2010). However, the solution of the acceleration of strain-induced ferrite formation can be higher strain rates and higher total strains applied under industrial conditions (Poliak and Bhattacharya 2014).

Effect of isothermal holding temperature. The most important parameter of multi-step cooling is an isothermal holding temperature at a bainitic transformation range (Grajcar and Krzton 2009). This temperature means coiling temperature in industrial practice and thus it determines coiling forces. From a metallurgical point of view the austempering temperature determines a refinement of microstructure and diffusion rate of carbon (Garcia-Mateo et al. 2003).

The lower austempering temperature is the stronger grain refinement should be expected (Bhadeshia 2001). However, it requires longer time to stabilize retained austenite due to the decrease in the diffusion rate of carbon. The kinetics of bainite formation can be accelerated by the overall refinement of the microstructure and Al addition (Timokhina et al. 2011).

Figure 5.13 presents an effect of isothermal holding temperature (T_B) on the microstructure of 0.24C-1.5Mn-0.4Al-1Si-Nb-Ti steel formed after finishing rolling at 850°C and isothermal holding of specimens at T_B for 600 s. The bainitic islands and grains of retained austenite are uniformly distributed within the ferrite matrix but a volume fraction of γ phase strongly depends on an isothermal holding temperature. One can see that at 250°C austenite grains transformed partially into martensite (Figure 5.13a) in a final stage of cooling from T_B (Grajcar and Krzton 2011). It means that a slow diffusion rate of carbon was not sufficient to enrich the austenite and keep this phase stable at room temperature. The conditions for stabilization of retained austenite are better with increasing T_B to a range from 300 to 350°C but an amount of γ phase is still relatively small in conventional TRIP steels (Grajcar and Opiela 2008a). This range of temperature is often used for nanocrystalline bainitic-austenitic mixtures but in steels with a carbon content of 0.8 wt.% (Timokhina et al. 2011).

A high volume fraction of retained austenite occurs at 400°C, where a diffusion rate of C is relatively fast to enrich the austenite in carbon within 600 s (Figure 5.13b). However, the best conditions for stabilization of retained austenite occur at 450°C. Under these conditions retained austenite is located both among ferrite grains and inside bainitic islands (Figure 5.13c). The identification of interlath retained austenite requires microscopic techniques of better resolution (Hausmann et al. 2013). The optimum austempering temperature for most TRIP steels ranges from 400

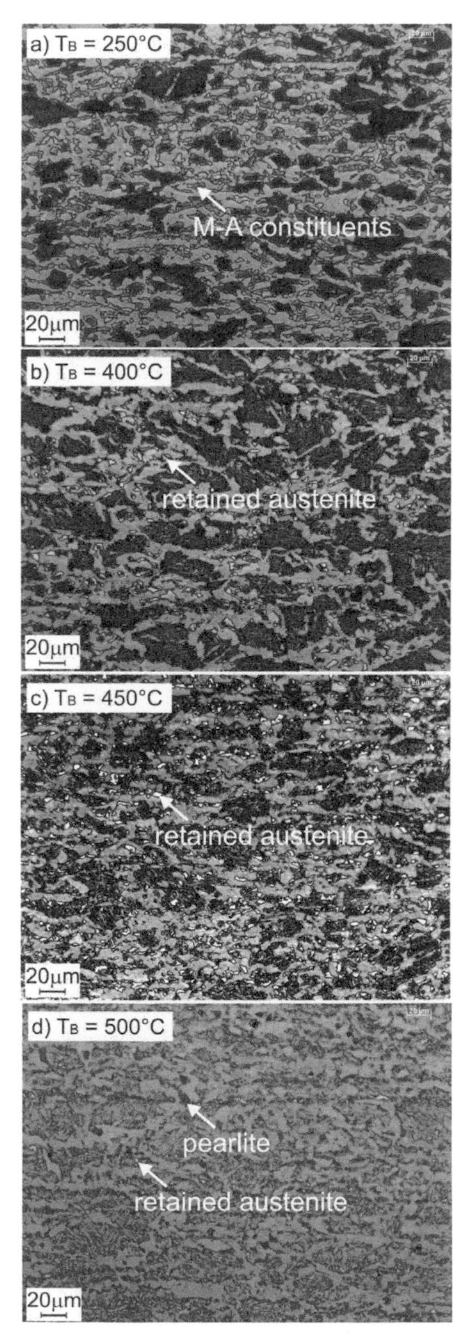

Figure 5.13 Effect of isothermal holding temperature on the microstructure of 0.24C-1.5Mn-0.4Al-1Si-Nb-Ti steel formed after finishing rolling at 850°C and isothermal holding of specimens at T_B for 600 s.

to 450°C (Jung et al. 2011). A further increase of the T_B temperature leads to the transformation of almost all austenite grains located in a ferritic matrix into degenerated pearlite (Figure 5.13d). The stable austenite remains only in bainitic areas. This is due to higher concentration of carbon of this morphological type of retained austenite (Jirkova et al. 2014). The steel sheets held between 450 and 500°C show usually worse mechanical properties. The stability of retained austenite can be slightly improved after addition of Mo, which is the only element effectively suppressing carbide precipitation in this temperature region (Samek et al. 2006).

In cold-rolled and intercritically annealed steel sheets the additional stabilization of retained austenite is a result of the enrichment of the austenite in Mn during an intercritical annealing step (De Moor et al. 2010). Under these conditions the isothermal holding step at a bainitic transformation range can be avoided and the sheets are continuously cooled to room temperature (Lee et al. 2011). However, the relatively long time at the intercritical temperature can not be accepted in industrial concepts of thermomechanical processing. Hence, manganese increases rather hardenability of thermomechanically rolled sheets and it is not partitioned between ferrite and austenite. An austempering step is thus required. Therefore, thermomechanically processed medium-Mn steels can be applied for bainitic-austenitic products (Grajcar et al. 2012b) or in martensitic-austenitic concepts (Futch et al. 2012).

An effect of isothermal holding temperature on the microstructure of 0.17C-3.3Mn-1.7Al-0.2Si-0.2Mo medium-Mn steel formed after finishing rolling at 850°C and isothermal holding of specimens at T_B for 600 s is shown in Figure 5.14. The overall refinement of the microstructure can be visible by decreasing an isothermal holding temperature. The morphology of retained austenite changes from blocky type at 450°C (Figure 5.14a), through mixed type at 400°C (Figure 5.14b) to more interlath at 350°C (Figure 5.14c). At the highest isothermal holding temperature M-A constituents also occur due to a hampering effect of Mn addition on the enrichment of the austenite in carbon (Takahashi and Bhadeshia 1991).

Effect of isothermal holding time. There is a simple relationship between an isothermal holding temperature and time. Lower the temperature longer is the time required to sufficiently enrich the austenite in carbon (Krizan and De Cooman 2014). Figure 5.15 presents an effect of isothermal holding time (t_B) on the microstructure of 0.24C-1.5Mn-0.4Al-1Si-Nb-Ti steel formed after finishing rolling at 850°C and isothermal holding of specimens at 350°C for t_B ranging from 60 to 1800 s. The 60 s time is not enough to stabilize the austenite in carbon. Thus, almost all austenite grains got transformed into martensite forming M-A constituents (Figure 5.15a). The best conditions to stabilize retained austenite occur for the time of app. 600 s (Figure 5.15b). A further increase of isothermal holding time to

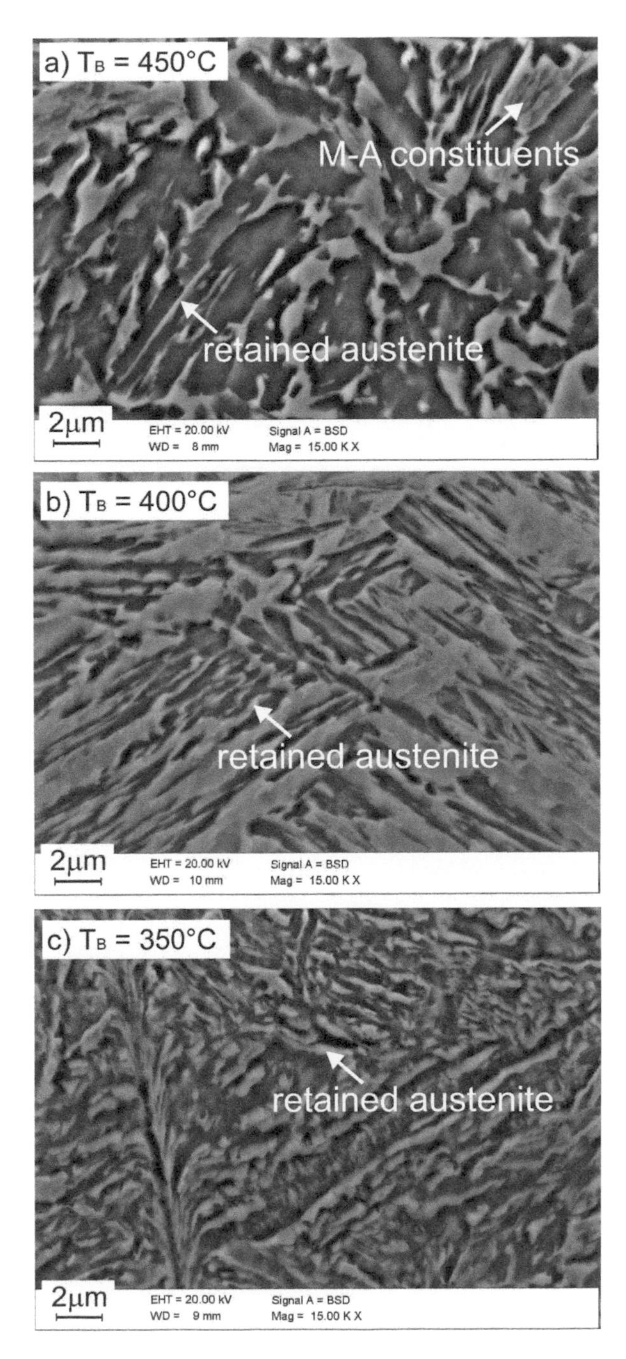

Figure 5.14 Effect of isothermal holding temperature on the microstructure of 0.17C-3.3Mn-1.7Al-0.2Si-0.2Mo steel formed after finishing rolling at 850°C and isothermal holding of specimens at T_B for 600 s.

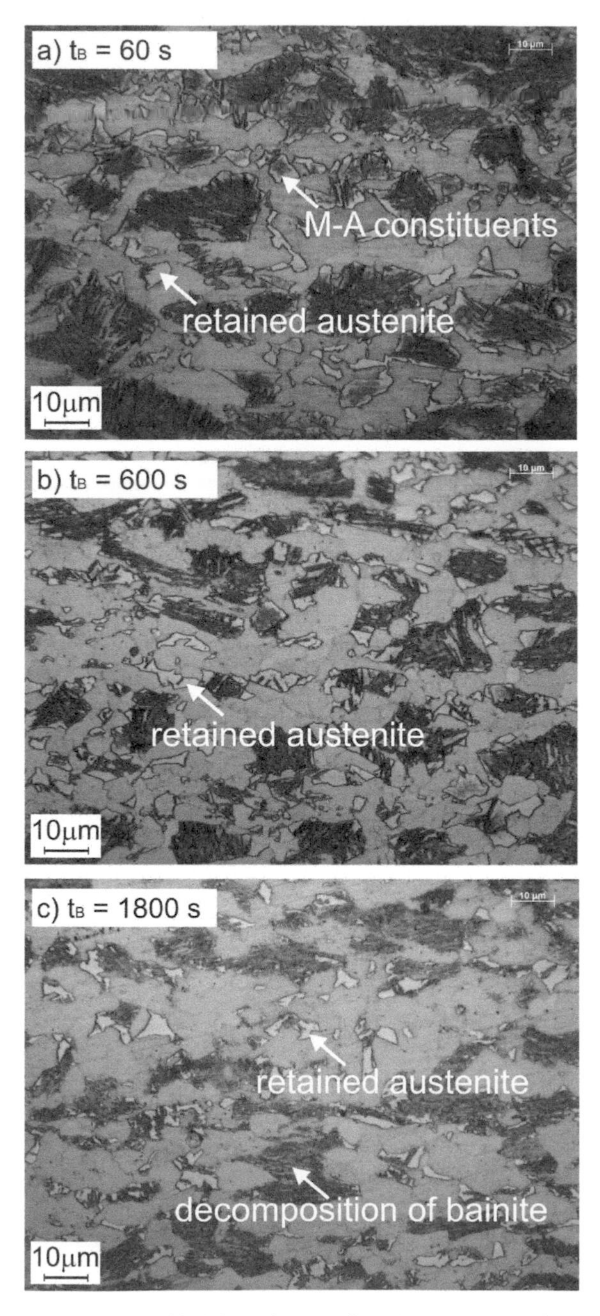

Figure 5.15 Effect of isothermal holding time on the microstructure of 0.24C-1.5Mn-0.4Al-1Si-Nb-Ti steel formed after finishing rolling at 850°C and isothermal holding of specimens at 350°C for t_B.

1800 s initiates the decomposition of bainitic regions and retained austenite (Figure 5.15c). Carbides are formed under these conditions (Li and Wu 2006). The sensitivity of the microstructure on changes in the amount of retained austenite rapidly increases at higher temperatures because of much higher diffusivity of carbon (Tsukatani et al. 1991).

5.5 Semi-industrial Simulation of Hot Rolling

Physical simulation of thermomechanical processing can be further reflected in semi-industrial hot strip rolling lines (Kuc et al. 2012). This approach allows avoiding a large scale of experimentation and disturbances in production schedules when industrial trials are performed (Garbarz et al. 2012). Semi-industrial conditions differ from the industrial ones in number of passes, rolling reductions, strain rates, etc. (Grajcar et al. 2013). The major difference exists in sizes of sheets and hence the heat transfer conditions (Suwanpinij et al. 2012). However, these differences can be extrapolated with a relatively good accordance to industrial conditions.

Recently, Grajcar et al. (2013) proved that the semi-industrial hot rolling line consisting of two-high reversing mill, roller tables with isothermal heating panels, cooling devices and controlling-recording systems (Institute for Ferrous Metallurgy, Gliwice, Poland) enables the efficient semi-industrial simulation of hot strip rolling of medium-Mn TRIP-assisted steel sheets. Continuous registration of force-energetic parameters of hot rolling makes it possible to effectively measure pressure forces and to estimate deformation resistance. Figure 5.16 shows the registered rolling forces during the semi-

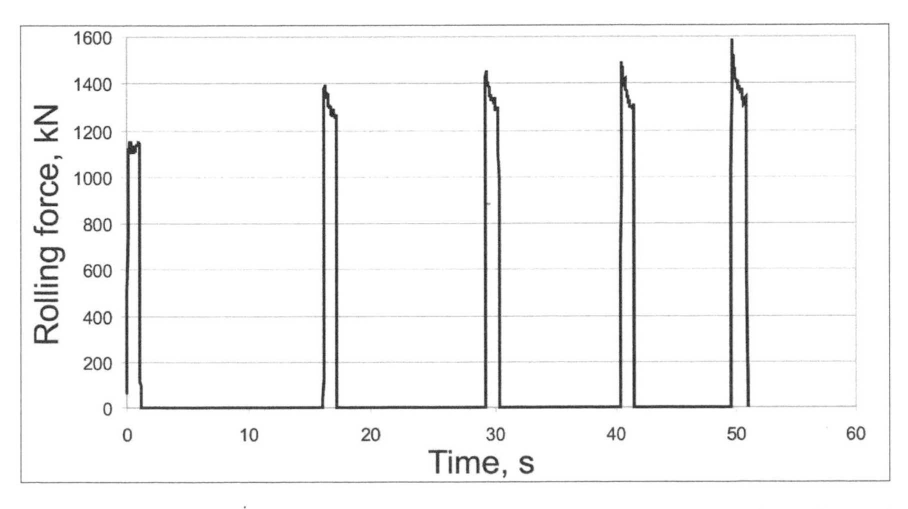

Figure 5.16 Rolling forces registered during the semi-industrial 5-pass hot rolling of 0.17C-3.3Mn-1.7Al-0.2Si-0.2Mo steel sheets at a finishing rolling temperature of about 750°C and at a strain rate of app. 10 s^{-1}.

industrial 5-pass hot rolling of 0.17C-3.3Mn-1.7Al-0.2Si-0.2Mo steel sheets at a finishing rolling temperature of about 750°C and at a strain rate of app. 10 s^{-1}. The pass temperatures, rolling reductions (0.21, 0.21, 0.20, 0.19 and 0.14) and time parameters applied reflect industrial conditions satisfactorily. It is obvious that the strain values and strain rates are higher whereas final inter-pass times are shorter in the industrial practice. However, the produced sheet samples are characterized by comparable microstructures when compared to physical simulation results obtained in the laboratory scale using the Gleeble system (Grajcar et al. 2014a).

The temperatures of the successive passes range from 1050 to 750°C (Figure 5.17). Isothermal panels located at both sides of a rolling stand decrease effectively the cooling rate of hot-rolled strips and enable obtaining a high surface quality sheet samples with a thickness of 4.5 mm (Grajcar et al. 2013).

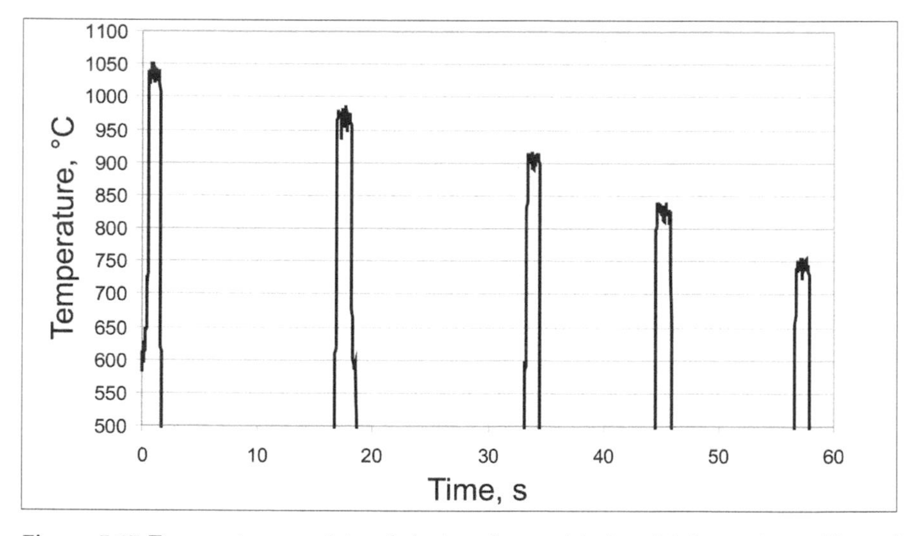

Figure 5.17 Temperatures registered during the semi-industrial 5-pass hot rolling of 0.17C-3.3Mn-1.7Al-0.2Si-0.2Mo steel sheets at a finishing rolling temperature of about 750°C and at a strain rate of app. 10 s^{-1}.

5.6 Industrial Trials on TRIP Steels

Poliak and Bhattacharya (2014) reported some deformation resistance data for several types of AHSS steels at processing temperatures typical for industrial hot strip rolling mills. The strain rate corrected maximum stress is between 150 and 340 MPa, what is in relatively good agreement with the data provided in Figures 5.6 and 5.7. It should be noticed that the data in Figures 5.6 and 5.7 should also be corrected for the strain rate, which reach up to 100–150 s^{-1} in final passes of hot strip rolling (Siciliano and Leduc 2005).

Conventional thermomechanical processing strategies (recrystallization controlled rolling and no-recrystallization controlled rolling) can not be fully realized in industrial finishing rolling of AHSS due to delayed initiation of static recrystallization (Poliak and Bhattacharya 2014). A special attention should be directed to combined effects of increased Mn, Al and Si additions. The most complex effect is exerted by aluminium, which is added up to 2% in TRIP steels (Zhang et al. 2006). Skolly and Poliak (2005) compared changes of mean flow stress between HSLA steel and Al-alloyed Nb-microalloyed 0.2C AHSS steel, in a six-stand finishing hot strip mill. They identified a pronounced softening in pass 5 in both steels (as a result of dynamic recrystallization caused by strain accumulation) followed by intensive hardening in the last pass. Al increases also considerably the A_1 and A_3 temperatures (Kuziak et al. 2008). It can lead in some cases to uncontrollable variation in rolling loads due to ferrite formation during finishing rolling (Skolly and Poliak 2005).

The crucial requirement for the production of hot rolled TRIP steel sheets is to control precisely cooling profiles on a run-out table. The major constraints of existing hot strip rolling lines include a limited run-out table length and the heat extraction capabilities (Sprock et al. 2010). A multi-step cooling profile needed for the production of ferrite-based TRIP steel sheets is shown in Figure 5.18. Taking into account that a finishing rolling temperature ranges usually from 900 to 850°C, the time of the first

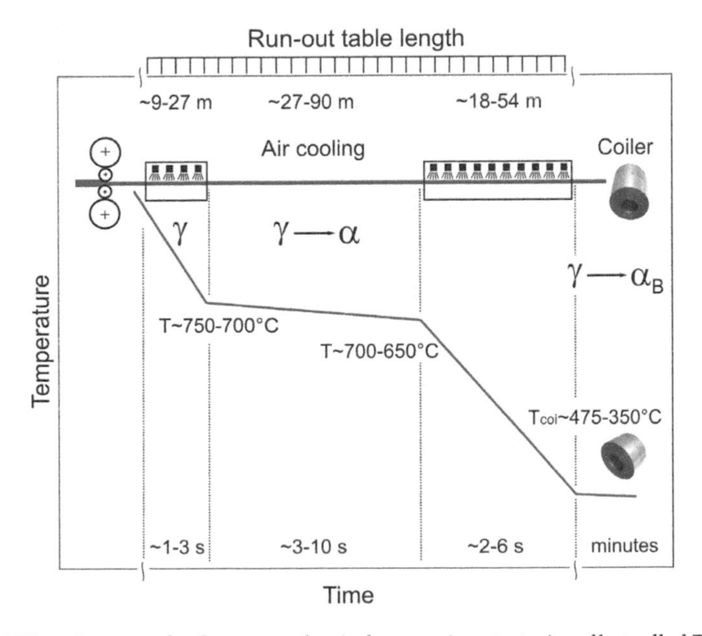

Figure 5.18 Requirements for thermomechanical processing strategies of hot rolled TRIP steel sheets on run-out table.

laminar cooling is ~1–3 s. A relatively long time is required to produce a sufficient fraction of polygonal ferrite. This time strongly depends on a steel chemistry (Wang et al. 2011). Typical laminar cooling systems make it possible to apply a cooling speed of 50–100°C/s. It means that 2 to 6 s are required to cool the steel sheet from 700–650°C to coiling temperature at a bainitic temperature range. The longer time means the longer run-out table is required, which could be difficult in existing industrial lines. Shortening of the run-out table length requirements can be achieved when compact cooling facilities are employed (Sprock et al. 2010). A lack of the $\gamma \rightarrow \alpha$ phase transformation enables the production of bainitic-austenitic steels using shorter run-out tables.

5.7 Summary

An efficient production of thermomechanically processed TRIP steel sheets requires both strict hot-working strategies and demanding cooling schedules. A complex chemistry and industrial limitations decide that technological windows are relatively narrow and should be precisely adjusted for a specific steel grade. One of the major problems of the thermomechanical processing of TRIP steels is a variety of chemistry and hence the different procedures which have to be applied in production of Si-alloyed, Al-alloyed or/and Mn-alloyed steels. The modification of parameters of hot-working and controlled cooling must also take into account effects of microalloyed elements. A final response of the specific composition can be very different since the alloying elements and microadditions show very often opposite effects. The production of TRIP steels with stable mechanical and technological properties requires modern technological lines or upgraded existing lines which enable the effective utilization of various effects of accumulated strain, finishing rolling conditions, cooling strategies and coiling conditions for tailoring sheets' properties.

Keywords: Thermomechanical processing, TRIP steel, Hot rolling, Flow stress, Multistep cooling, Retained austenite, Carbide-free bainite, Physical simulation, Austenite decomposition

References

Adamczyk, J. and A. Grajcar. 2005. Structure and mechanical properties of DP-type and TRIP-type sheets obtained after the thermomechanical processing. J. Mater. Proc. Technol. 162-163: 267–274.

Adamczyk, M., D. Kuc and E. Hadasik. 2008. Modelling of structure changes in TRIP type steel during hot deformation. Arch. Civ. Mech. Eng. 8: 5–13.

Aretxabaleta, Z., B. Pereda and B. Lopez. 2014. Multipass hot deformation behaviour of high Al and Al-Nb steels. Mater. Sci. Eng. A 600: 37–46.

Bandyopadhyay, P.S., S.K. Ghosh, S. Kundu and S. Chatterjee. 2011. Evolution of microstructure and mechanical properties of thermomechanically processed ultrahigh-strength steel. Metall. Mater. Trans. A 42: 2742–2752.

Basuki, A. and E. Aernoudt. 1999. Influence of rolling of TRIP steel in the intercritical region on the stability of retained austenite. J. Mater. Proc. Tech. 89-90: 37–43.
Bhadeshia, H.K.D.H. 2001. Bainite in Steels, Transformations, Microstructure and Properties. Second Edition, The University Press, Cambridge.
Bleck, W. and K. Phiu-On. 2005. Microalloying of cold-formable multi phase steel grades. Mater. Sci. Forum 500-501: 97–112.
Bojarski, Z. and T. Bold. 1974. Structure and properties of carbide-free bainite. Acta Metall. 22: 1223–1234.
Cabanas, N., N. Akdut, J. Penning and B.C. De Cooman. 2006. High-temperature deformation properties of austenitic Fe-Mn alloys. Metall. Mater. Trans. A 37A: 3305–3315.
De Ardo, A.J., J.E. Garcia, M. Hua and C.I. Garcia. 2005. A new frontier in microalloying: advanced high strength, coated sheet steels. Mater. Sci. Forum 500-501: 27–38.
De Cooman, B.C. 2004. Structure-properties relationship in TRIP steels containing carbide-free bainite. Curr. Opin. Solid St. Mater. Sci. 8: 285–303.
De Cooman, B.C. and J.G. Speer. 2012. Fundamentals of Steel Product Physical Metallurgy. Association for Iron and Steel Technology, Pittsburgh.
Demeri, M.Y. 2013. Advanced High-Strength Steels. Science, Technology, and Applications. ASM International, The Materials Information Society, Ohio.
De Moor, E., P.J. Gibbs, J.G. Speer and D.K. Matlock. 2010. Strategies for third-generation advanced high-strength steel development. Iron Steel Tech. 7: 133–144.
Eberle, K., P. Cantinieaux and P. Harlet. 1999. New thermomechanical strategies for the production of high strength low alloyed multiphase steel showing a transformation induced plasticity (TRIP) effect. Steel Res. 70: 233–238.
Fazeli, F. and M. Militzer. 2012. Modelling simultaneous formation of bainitic ferrite and carbide in TRIP steels. ISIJ Int. 52: 650–658.
Fonstein, N. 2015. Advanced High Strength Sheet Steels. Springer International Publishing, Switzerland.
Fonstein, N., O. Yakubovsky, D. Bhattacharya and F. Siciliano. 2005. Effect of niobium on the phase transformation behaviour of aluminum containing steels for TRIP products. Mater. Sci. Forum 500-501: 453–460.
Futch, D.B., G.A. Thomas, J.G. Speer and K.O. Findley. 2012. Thermomechanical simulation of hot rolled Q&P sheet steels. Iron Steel Technol. 9: 101–106.
Garbarz, B., W. Burian and D. Wozniak. 2012. Semi-industrial simulation of in-line thermomechanical processing and heat treatment of nano-duplex bainite-austenite steel. Steel Res. Int. Special Edition: Metal Forming 2012: 1251–1254.
Garcia-Mateo, C., F.G. Caballero and H.K.D.H. Bhadeshia. 2003. Acceleration of low-temperature bainite. ISIJ Int. 43: 1821–1825.
Grajcar, A. 2007. Hot-working in the $\gamma+\alpha$ region of TRIP-aided microalloyed steel. Arch. Mater. Sci. Eng. 28: 743–750.
Grajcar, A. 2008. Structural and mechanical behaviour of TRIP-type microalloyed steel in hot-working conditions. J. Achiev. Mater. Manuf. Eng. 30: 27–34.
Grajcar, A. 2010. Microstructural features of thermomechanically processed C-Mn-Si-Al-Nb-Ti multiphase steel. Proc. 2nd Int. Conf. on Super-High Strength Steels. Peschiera del Garda, Italy: 10 pages.
Grajcar, A. 2014. Thermodynamic analysis of precipitation processes in Nb-Ti-microalloyed Si-Al TRIP steel. J. Therm. Anal. Calorim. 118: 1011–1020.
Grajcar, A. 2015. Microstructure evolution of advanced high-strength TRIP-aided bainitic steel. Mater. Tehnol. 49: 715–720.
Grajcar, A. and H. Krzton. 2009. Effect of isothermal bainitic transformation temperature on retained austenite fraction in C-Mn-Si-Al-Nb-Ti TRIP-type steel. J. Achiev. Mater. Manuf. Eng. 35: 169–176.
Grajcar, A. and H. Krzton. 2011. Effect of isothermal holding temperature on retained austenite fraction in medium-carbon Nb/Ti-microalloyed TRIP steel. J. Achiev. Mater. Manuf. Eng. 49: 391–399.

Grajcar, A. and K. Radwanski. 2014. Microstructural comparison of the thermomechanically treated and cold deformed Nb-microalloyed TRIP steel. Mater. Tehnol. 48: 679–683.
Grajcar, A. and M. Opiela. 2008a. Influence of plastic deformation on CCT-diagrams of low-carbon and medium-carbon TRIP steels. J. Achiev. Mater. Manuf. Eng. 29: 71–78.
Grajcar, A. and M. Opiela. 2008b. Microstructure evolution in thermomechanically processed TRIP-type microalloyed steel. Proc. 3rd Int. Conf. on Thermomechanical Processing of Steels. Padua, Italy: 11 pages.
Grajcar, A. and R. Kuziak. 2011a. Dynamic recrystallization behavior and softening kinetics in 3Mn-1.5Al TRIP steels. Adv. Mater. Res. 287-290: 330–333.
Grajcar, A. and R. Kuziak. 2011b. Softening kinetics in Nb-microalloyed TRIP steels with increased Mn content. Adv. Mater. Res. 314-316: 119–122.
Grajcar, A. and S. Lesz. 2012. Influence of Nb microaddition on a microstructure of low-alloyed steels with increased manganese content. Mater. Sci. Forum 706-709: 2124–2129.
Grajcar, A., R. Kuziak and W. Zalecki. 2011. Designing of cooling conditions for Si-Al microalloyed TRIP steel on the basis of DCCT diagrams. J. Achiev. Mater. Manuf. Eng. 45: 115–124.
Grajcar, A., M. Kaminska, M. Opiela. P. Skrzypczyk, B. Grzegorczyk and E. Kalinowska-Ozgowicz. 2012a. Segregation of alloying elements in thermomechanically rolled medium-Mn multiphase steels. J. Achiev. Mater. Manuf. Eng. 55: 256–264.
Grajcar, A., R. Kuziak and W. Zalecki. 2012b. Third generation of AHSS with increased fraction of retained austenite for the automotive industry. Arch. Civ. Mech. Eng. 12: 334–341.
Grajcar, A., R. Kuziak. W. Ozgowicz and K. Golombek. 2012c. Physical simulation of thermomechanical processing of new generation advanced high strength steels. Comp. Methods Mater. Sci. 12: 115–129.
Grajcar, A., P. Skrzypczyk, D. Wozniak and S. Kolodziej. 2013. Semi-industrial simulation of hot rolling and controlled cooling of Mn-Al TRIP steel sheets. J. Achiev. Mater. Manuf. Eng. 57: 38–47.
Grajcar, A., P. Skrzypczyk and D. Wozniak. 2014a. Thermomechanically rolled medium-Mn steels containing retained austenite. Arch. Metall. Mater. 59: 1691–1697.
Grajcar, A., P. Skrzypczyk. R. Kuziak and K. Golombek. 2014b. Effect of finishing hot-working temperature on microstructure of thermomechanically processed Mn-Al multiphase steels. Steel Res. Int. 85: 1058–1069.
Grajcar, A., P. Skrzypczyk and M. Morawiec. 2015a. Physical simulation of hot strip rolling of Nb-microalloyed medium-Mn steels. Proc. 6th Int. Conf. on Modelling and Simulation of Metallurgical Processes in Steelmaking. Bardolino, Italy: 8 pages.
Grajcar, A., W. Kwasny and W. Zalecki. 2015b. Microstructure-property relationships in TRIP aided medium-C bainitic steel with lamellar retained austenite. Mater. Sci. Technol. 31: 781–794.
Gronostajski, Z., A. Niechajowicz and S. Polak. 2010. Prospects for the use of new-generation steels of the AHSS type for collision energy absorbing components. Arch. Metall. Mater. 55: 221–230.
Hadasik, E., R. Kuziak, R. Kawalla, M. Adamczyk and M. Pietrzyk. 2006. Rheological model for simulation of hot rolling of new generation steel strip for automotive industry. Steel Res. Int. 77: 927–933.
Hashimoto, S., S. Ikeda, K. Sugimoto and S. Miyake. 2004. Effects of Nb and Mo addition to 0.2%C-1.5%Si-1.5%Mn steel on mechanical properties of hot rolled TRIP-aided steel sheets. ISIJ Int. 44: 1590–1598.
Hausmann, K., D. Krizan, K. Spiradek-Hahn, A. Pichler and E. Werner. 2013. The influence of Nb on transformation behavior and mechanical properties of TRIP-assisted bainitic-ferritic sheet steels. Mater. Sci. Eng. A 588: 142–150.
Hosseini, S.M.K., A. Zarei-Hanzaki and S. Yue. 2015. Effects of ferrite phase characteristics on microstructure and mechanical properties of thermomechanically-processed low-silicon content TRIP-assisted steels. Mater. Sci. Eng. A 626: 229–236.
Jirkova, H., L. Kucerova, D. Aisman and B. Masek. 2014. Optimization of the Q-P process parameters for low alloyed steels with 0.2% C. Arch. Metall. Mater. 59: 1205–1210.

Jung, J., S.J. Lee, S. Kim and B.C. De Cooman. 2011. Effect of Ti additions on micro-alloyed Nb TRIP steel. Steel Res. Int. 82: 857–865.

Koh, H.J., S.K. Lee, S.H. Park, S.J. Choi, S.J. Kwon and N.J. Kim. 1998. Effect of hot rolling conditions on the microstructure and mechanical properties of Fe-C-Mn-Si multiphase steels. Scripta Mater. 38: 763–768.

Krizan, D. and B.C. De Cooman. 2014. Mechanical properties of TRIP steels microalloyed with Ti. Metall. Mater. Trans. A 45A: 3481–3492.

Krizan, D., K. Spiradek-Hahn and A. Pichler. 2015. Relationship between microstructure and mechanical properties in Nb-V microalloyed TRIP steel. Mater. Sci. Technol. 31: 661–668.

Kuc, D., E. Hadasik, G. Niewielski, I. Schindler, E. Mazancova, S. Rusz and P. Kawulok. 2012. Structural and mechanical properties of laboratory rolled steels high-alloyed with manganese and aluminium. Arch. Civ. Mech. Eng. 12: 312–317.

Kuziak, R., R. Kawalla and S. Waengler. 2008. Advanced high strength steels for automotive industry. Arch. Civ. Mech. Eng. 8: 103–117.

Lee, S., S.J. Lee, S. Santhosh Kumar, K. Lee and B.C. De Cooman. 2011. Localized deformation in multiphase, ultra-fine-grained 6 pct Mn transformation-induced plasticity steel. Metall. Mater. Trans. A 42: 3638–3651.

Li, Z. and D. Wu. 2006. Effects of hot deformation and subsequent austempering on the mechanical properties of Si-Mn TRIP steels. ISIJ Int. 46: 121–128.

Liu, D., F. Fazeli, M. Militzer and W.J. Poole. 2007. A microstructure evolution for hot rolling of a Mo-TRIP steel. Metall. Mater. Trans. A 38A: 894–909.

Mesquita, R.A., R. Schneider, K. Steineder, L. Samek and E. Arenholz. 2013. On the austenite stability of a new quality of twinning induced plasticity steel. Exploring new ranges of Mn and C. Metall. Mater. Trans. A 44: 4015–4019.

Mohamadizadeh, A., A. Zarei-Hanzaki, S. Mehtonen, D. Porter and M. Moallemi. 2016. Effect of intercritical thermomechanical processing on austenite retention and mechanical properties in a multiphase TRIP-assisted steel. Metall. Mater. Trans. A 47A: 436–449.

Opiela, M. 2014. Effect of thermomechanical processing on the microstructure and mechanical properties of Nb-Ti-V microalloyed steel. J. Mater. Eng. Perform. 23: 3379–3388.

Opiela, M. and A. Grajcar. 2012. Hot deformation behavior and softening kinetics of Ti-V-B microalloyed steels. Arch. Civ. Mech. Eng. 12: 327–333.

Pereda, B., Z. Aretxabaleta and B. Lopez. 2015. Softening kinetics in high Al and high Al-Nb-microalloyed steels. J. Mater. Eng. Perform. 24: 1279–1293.

Petrov, R., L. Kestens, A. Wasilkowska and Y. Houbaert. 2007. Microstructure and texture of a lightly deformed TRIP-assisted steel characterized means of the EBSD technique. Mater. Sci. Eng. A 447: 285–297.

Pietrzyk, M., J. Kusiak, R. Kuziak, L. Madej, D. Szeliga and R. Golob. 2014. Conventional and multiscale modeling of microstructure evolution during laminar cooling of DP steel strips. Metall. Mater. Trans. A 45A: 5835–5851.

Pietrzyk, M., L. Madej, L. Rauch and R. Golob. 2010. Multiscale modelling of microstructure evolution during laminar cooling of hot rolled DP steel. Arch. Civ. Mech. Eng. 10: 57–67.

Poliak, E.I. and D. Bhattacharya. 2014. Aspects of thermomechanical processing of 3rd generation advanced high strength steels. Mater. Sci. Forum 783-786: 3–8.

Poliak, E.I. and F. Siciliano. 2004. Hot deformation behavior of Mn-Al and Mn-Al-Nb steels. Proc. MS&T'2004. Pittsburgh, USA: 39–45.

Ranjan, R., H. Beladi, S.B. Singh and P.D. Hodgson. 2015. Thermo-mechanical processing of TRIP-aided steels. Metall. Mater. Trans. A 46A: 3232–3247.

Rodriguez-Ibabe, J.M. 2007. Thin Slab Direct Rolling of Microalloyed Steels. Trans Tech Publications Ltd., Stafa-Zuerich.

Ryu, H.B., J.G. Speer and J.P. Wise. 2002. Effect of thermomechanical processing on the retained austenite content in a Si-Mn transformation-induced-plasticity steel. Metall. Mater. Trans. A 33A: 2811–2816.

Samek, L., E. De Moor, J. Penning and B.C. De Cooman. 2006. Influence of alloying elements on the kinetics of strain-induced martensitic nucleation in low-alloy multiphase high-strength steels. Metall. Mater. Trans. A 37: 109–121.

Siciliano, F. and E.I. Poliak. 2005. Modeling of the resistance to hot deformation and the effects of microalloying in high-Al steels under industrial conditions. Mater. Sci. Forum 500-501: 195–202.

Siciliano, F. and L.L. Leduc. 2005. Modeling of the microstructural evolution and mean flow stress during thin slab casting/direct rolling of niobium microalloyed steels. Mater. Sci. Forum 500-501: 221–228.

Skalova, L., R. Divizova and D. Jandova. 2006. Thermo-mechanical processing of low-alloy TRIP-steel. J. Mater. Proc. Technol. 175: 387–392.

Skolly, R.M. and E.I. Poliak. 2005. Aspects of production hot rolling of Nb microalloyed high Al high strength steels. Mater. Sci. Forum 500-501: 187–194.

Sprock, A., M.J. Peretic and J.G. Speer. 2010. Compact cooling as an alternative to alloying for production of DP/TRIP steel grades. Iron Steel Technol. 7: 170–177.

Sugimoto, K., H. Tanino and J. Kobayashi. 2015. Impact toughness of medium-Mn transformation-induced plasticity-aided steels. Steel Res. Int. 86: 1151–1160.

Sugimoto, K., J. Sakaguchi, T. Iida and T. Kashima. 2000. Stretch-flangeability of a high-strength TRIP type bainitic sheet steel. ISIJ Int. 40: 920–926.

Sugimoto, K., M. Tsunezawa, T. Hojo and S. Ikeda. 2004. Ductility of 0.1-0.6C-1.5Si-1.5Mn ultra high-strength TRIP-aided sheet steels with bainitic ferrite matrix. ISIJ Int. 44: 1608–1614.

Suikkanen, P.P., V.T.E. Lang, M.C. Somani, D.A. Porter and L.P. Karjalainen. 2012. Effect of silicon and aluminium on austenite static recrystallization kinetics in high-strength TRIP-aided steels. ISIJ Int. 52: 471–476.

Suwanpinij, P., U. Prahl, W. Bleck and R. Kawalla. 2012. Fast algorithms for phase transformations in dual phase steels on the hot strip mill run out table (ROT). Arch. Civ. Mech. Eng. 12: 305–311.

Takahashi, M. 2003. Development of high strength steels for automobiles. Nippon Steel Tech. Rep. 88: 2–7.

Takahashi, M. and H.K.D.H. Bhadeshia. 1991. A model for the microstructure of some advanced bainitic steels. Mater. Trans. Jap. Inst. Metals 32: 689–696.

Takahashi, M., T. Hayashida, H. Taniguchi, O. Kawano and R. Okamoto. 2003. High strength hot-rolled steel sheets for automobiles. Nippon Steel Tech. Rep. 88: 8–12.

Thomas, G.A., J.G. Speer and D.K. Matlock. 2011. Quenched and partitioned microstructures produced via Gleeble simulations of hot-strip mill cooling practices. Metall. Mater. Trans. A 42: 3652–3659.

Timokhina, I.B., E.V. Pereloma, H. Beladi and P.D. Hodgson. 2008. A study of the strengthening mechanism in the thermo-mechanically processed TRIP/TWIP steel. Proc. 3rd Int. Conf. on Thermomechanical Processing of Steels. Padua, Italy: 10 pages.

Timokhina, I.B., H. Beladi, X.Y. Xiong, Y. Adachi and P.D. Hodgson. 2011. Nanoscale microstructural characterization of a nanobainitic steel. Acta Mater. 59: 5511–5522.

Timokhina, I.B., P.D. Hodgson and E.V. Pereloma. 2003. Effect of deformation schedule on the microstructure and mechanical properties of a thermomechanically processed C-Mn-Si transformation-induced-plasticity steel. Metall. Mater. Trans. A 34A: 1599–1609.

Timokhina, I.B., P.D. Hodgson and E.V. Pereloma. 2004. Effect of microstructure on the stability of retained austenite in transformation-induced-plasticity steels. Metall. Mater. Trans. A 35: 2331–2341.

Tsukatani, I., S. Hashimoto and T. Inoue. 1991. Effects of silicon and manganese addition on mechanical properties of high-strength hot-rolled sheet steel containing retained austenite. ISIJ Int. 31: 992–1000.

Wang, J. and S. Van der Zwaag. 2001. Stabilization mechanisms of retained austenite in transformation-induced plasticity steel. Metall. Mater. Trans. A 32: 1527–1538.

Wang, X., L. Du, H. Xie, H. Di and D. Gu. 2011. Effect of deformation on continuous cooling phase transformation behaviors of 780 MPa Nb-Ti ultra-high strength steel. Steel Res. Int. 82: 1417–1424.

Zaefferer, S., J. Ohlert and W. Bleck. 2004. A study of microstructure, transformation mechanisms and correlation between microstructure and mechanical properties of a low alloyed TRIP steel. Acta Mater. 52: 2765–2778.

Zajac, S., V. Schwinn and K.H. Tacke. 2005. Characterisation and quantification of complex bainitic microstructures in high and ultra-high strength linepipe steels. Mater. Sci. Forum 500-501: 387–94.

Zarei-Hanzaki, A., P.D. Hodgson and S. Yue. 1995. Hot deformation characteristics of Si-Mn TRIP steels with and without microalloy additions. ISIJ Int. 35: 324–331.

Zhang, M., L. Li, R.Y. Fu, D. Krizan and B.C. De Cooman. 2006. Continuous cooling transformation diagrams and properties of micro-alloyed TRIP steels. Mater. Sci. Eng. A 438-440: 296–299.

Zubialde, R., P. Uranga, B. Lopez and J.M. Rodriguez-Ibabe. 2009. Dynamic recrystallization of a Nb bearing Al-Si TRIP steel. Proc. MS&T'2009. Pittsburgh, USA: 605–616.

6

Newly-Developed High-Manganese Fe–Mn–(Al, Si) Austenitic TWIP and TRIP Steels

Leszek A. Dobrzański, Janusz Mazurkiewicz, Wojciech Borek* and *Małgorzata Czaja*

ABSTRACT

The first part of this chapter consists of the description of a general development in conditions of materials used in vehicle manufacturing industries and the general characteristics of steels used for car parts ensuring enhanced passive passenger safety. Next, the general concept of our research on high manganese steels exhibiting TRIP and TWIP effects is presented. This chapter investigates high manganese austenitic steels containing 18–25% of Mn, 1–3% of Si, approx. of 3% Al and some microadditions of Nb and Ti, and with a diverse concentrations of carbon from 0.08 to 0.73% C, in order to determine and describe structural mechanisms decisive for increasing the store of cold plastic deformation energy of such steels. The general discussion on high manganese steels exhibiting TRIP and TWIP effects are presented in the end of this chapter.

Silesian University of Technology, Institute of Engineering Materials and Biomaterials, Konarskiego 18A, 44-100 Gliwice, Poland.

* Corresponding author: leszek.dobrzanski@polsl.pl

6.1 General Development Conditions of Materials Used in Vehicle Manufacturing Industries

Top priorities worldwide are to improve living standards, ensure the best possible healthcare, the appropriate standards and quality of daily food, the reasonable use of all products consumed to meet everyday needs. There is also a need to protect the environment we live in and to make available the necessary energy, educational opportunities and unconstrained acquisition and transmission of information via the Internet, which is not only important for education but also benefits cultural and knowledge resources. It is equally important to provide safe means of transport for transporting people and for carriage of various goods. The above tasks are underlying multiannual and multinational programmes and strategies aimed at creating knowledge- and innovation-based economy. Innovations understood as precious and novel ideas are the basis of economic development for production, distribution and implementation of knowledge, which are the product and the main driver of sustainable development bringing the highest added value (Dobrzański and Dobrzańska-Danikiewicz 2013). Multifaceted aspects are required to achieve the above goals, including, in particular, extensive research programmes in which the field of materials science plays an important role. For all products satisfying human needs, raw materials have to be processed into products, the process of which is called manufacturing. Manufacturing consists of making products using raw materials in various processes, with various machines and in operations managed in line with a well elaborated plan. The process of manufacturing thus consists of the proper utilisation of resources: materials, energy, capital and labour. These days, manufacturing is a comprehensive activity involving people performing different jobs and professions using different machines, facilities and tools, differently automated, including computers and robots. The purpose of manufacturing is always to satisfy clients' market needs in consistency with an established strategy of a manufacturing enterprise or organisation utilising its existing capacities and equipment. The technical aspects of introducing a given product to the market by a manufacturing organisation relate to industrial design, engineering design, production preparation, manufacturing and maintenance service. The following three fields can therefore be distinguished in the process of introducing products to the market:

- marketing and sales;
- product development;
- production and manufacture.

Adequate decisions have to be made in each of the fields depending on the implementation phase of tasks associated with preparation of products to be introduced to the market. The first product design phase is industrial design in which product functions are generally described and a product concept is developed, including only its external form, colour and possibly general ideas as to how to assemble the key components. The subsequent phases include engineering design and production preparation. Engineering design, where the design of a manufacturing system and of products can be differentiated, is not a stand-alone activity. The reason for this is that engineering design influences all other phases of introducing a given product to the market, and also these phases precondition the product design. Engineering design encompasses the three equally vital and inseparable elements:

- structural design whose purpose is to develop a shape and geometrical features of a product satisfying human needs;
- material design to guarantee the required durability of a product or product components made of engineering materials with the required physiochemical and technological properties;
- technological design of a process to achieve the required geometrical features and properties for particular product elements and also their correct interaction after assembly considering the volume of production, level of automation and computer aid, while ensuring also the lowest possible product costs.

All the contemporary tasks relating to the design of new avant-garde products require the knowledge of materials science aspects, and most often such products have to be designed according to the defined needs and very often completely original and new materials have to be designed.

Aspects related to transport, manufacturing and operation of means of transport are also very important for the issues discussed, also in quantitative terms. It is a great technical and R&D challenge to develop and investigate new, avant-garde materials which, after application in vehicles, especially in passenger cars, could contribute to improved passive safety of passengers, and consequently, their wide spread industrial use. Travel safety is especially important in the era of mass transportation. People commute to work every day and travel long distances and also travels are connected with active leisure, including mass tourism, holidays and recreation. Logistics and the mass carriage of raw materials and ready products is also very significant as the consumption-based pattern of life has become common due to improved general economic conditions. Fluctuations in economic circumstances at the global market have a relatively small effect here. This is driving a constant demand for vehicles, notably passenger cars. The figures presented by Euler Hermes, a transaction hedging and debt recovery company, show that the global production of cars in 2017 will be more than 100 million units. The OICA, i.e., an international organisation of motor vehicle manufacturers, reports that the total of 89.5 million cars was produced in 2014 worldwide, which is a 34% increase since 2005. It is projected that production in China has reached 20 million vehicles.

Considering very high global production rates of cars and vehicles and a constantly growing group of vehicles on roads as well as a current average global index of cars 174 per 1,000 people, as reported by the OICA, which is an increase of +21% since 2005, all this is unfortunately leading to a rising number of road collisions and accidents. The number of car and other transport accidents is at a worryingly high level, despite multiple initiatives pursued globally to reduce the scale of such adverse effects. This forces many countries in the world to establish and enforce relevant programmes to counteract this very adverse situation. Road safety depends on multiple factors, including legislative solutions, transport infrastructure, road users' qualifications and behaviour, road and weather conditions and road safety improvement factors inherent to design solutions of vehicles and safety systems incorporated in such vehicles, which comes down to the two interlinked factors:

- active safety and
- passive safety.

Active safety depends on a myriad of factors aimed at reducing the probability of collision or accident by actively assisting the driver while driving a vehicle and by improving certain dynamic properties of a car. The factors also include the technical condition of a vehicle and vehicle equipment and an engine power reserve allowing to accelerate when in danger. Trucks also should be fitted with electronic stability control systems and good rear visibility and other active safety components. Passive safety systems are employed to ensure the highest possible driver and passenger safety, and, also the safety of other traffic users, e.g., pedestrians, cyclists and motorcyclists against injuries sustained in accidents or collisions, which are dangerous for life and health. Passive safety can be intrinsic or extrinsic and encompasses multiple elements and aspects. Active and passive vehicle safety system solutions are commonly regarded as structural solutions. In fact, the problem is much deeper and pertains to engineering design, including not only structural design but also technological and material design of all the mentioned active and passive safety components in vehicles, and generally in any means of transport.

6.2 General Characteristic of Steels Used for Car Parts Ensuring Enhanced Passive Passenger Safety

All vehicle users (also the biggest players of the global car markets) are, by the nature of things, interested in aspects related to the means of transport, and in particular in the associated production aspects. Apart from vehicle manufacturers, producers of trailers and semi-trailers and other transport equipment, producers of railway, aviation and water means of transport, are also interested in passenger safety. An important aspect of car production is a systematic effort to reduce car mass, mainly to lower fuel consumption and an adverse environmental effect of fumes, which is closely linked to introducing new, more durable engineering materials for cars' structural and body parts. Certain international projects have been successful, starting with the ULSAB project launched in 1994, through ULSAC and ULSAS, including ULSAB–AVC (Table 6.1).

Table 6.1 The scope of international projects for improvement of car safety and performance.

No.	Abbreviation	Title	Scope	Ref.
1	ULSAB	Ultra Light Steel Auto Body	reduction of body parts mass by approx. 25% and improvement of car operational safety and comfort	(WorldAutoSteel 2015a)
2	ULSAC	Ultra Light Steel Auto Closure	body closure parts, e.g., doors, hatches	(WorldAutoSteel 2015b)
3	ULSAS	Ultra Light Steel Auto Suspension	suspension	(WorldAutoSteel 2015c)
4	ULSAB–AVC	Ultra Light Steel Auto Body - Advanced Vehicle Concepts	development of light car consuming 3 L of fuel per 100 km and emitting limited CO_2 concentration to atmosphere	(WorldAutoSteel 2002, 2015d)

Due to considerable consumption of metallic materials for car production, mainly steel, many leading steel companies have participated in such initiatives. The motor industry has been developing in several directions: on one hand, vehicle power train systems are being designed and improved, on the other hand, materials are being developed for constructing not only mechanical systems but also car bodies. The listed programmes have incontestably contributed to implementation of many high strength materials, application of modern material formation methods and state-of-the-art technologies of joining body parts and other car components. Earlier, corrosion resistance and low cost had been the key criterion of body material selection. Now manufacturers have started to put more and more emphasis on passive vehicle safety over the last two decades (Anderson 2008, Borek 2010, Kafka 2012, Price et al. 2006, Van Hecke 2008), and development trends over the last four decades in this field are shown in Table 6.2.

Table 6.2 Variations in material selection criteria for car body steel in 1980–2020 (Anderson 2008, Van Hecke 2008).

Year	Key vehicle aspect	Limitations	Body structure materials emerging
1980	cost, style, corrosion resistance	rigidity	IF steel for assembled parts, zinc coatings
1990	cost, accident (door pressure for side impact, reduced repair costs for insurance)	rigidity	HSLA, C-Mn, BH (martensitic steels, DP for bumpers and door bars)
2000	cost, fuel consumption (mass), collision behaviour (side impact, compatibility for passenger car and SUV, rollover)	body behaviour in collision (absorption of impact energy, door indent inside the car)	multi-phase steels, DP, martensitic, TRIP I
2020	fuel consumption: 35 mpg CAFE (USA) fume emission: 0.9 g/km (Europe)	body behaviour in collision (absorption of impact energy, door indent inside the car)	3rd generation of AHSS, TWIP, L-IP

Symbols: SUV – Sport Utility Vehicle, mpg; CAFE – miles per gallon Corporate Average Fuel Economy; IF – Interstitial-Free steel; HSLA – High Strength Low Alloy steel; C-Mn – carbon-manganese steels; BH – Bake Hardening steels; DP – Dual Phase steels; TRIP I – Transformation Induced Plasticity - low manganese steels, where plastic deformation occurs by martensitic transformation; AHSS – Advanced High Strength Steel; TWIP – Twinning Induced Plasticity - high manganese austenitic steels where plastic deformation occurs by mechanical twinning; L-IP – light steels with induced plasticity with Al additive

New steels for the car sector have been developed in projects stated in Table 6.1 and the existing projects are systemised and split into three groups (Figure 6.1):

- Low Strength Steels (LSS);
- High Strength Steels (HSS); and
- Advanced High Strength Steel (AHSS) (Keeler and Kimchi 2014, WorldAutoSteel 2015a, 2015b).

Table 6.3 compares elongation and tensile strength of particular steel types.

Much higher quality requirements for automotive vehicles and other means of transport are imposed these days as compared to the situation seen just a few years ago. This holds true not only for comfort and aesthetics, but mainly travel safety and operation of all automotive vehicles and means of transport in general. Safety is approached seriously by car manufacturers as signified, among others, by the importance manufacturers and consumers attach to positions in safety ratings, e.g., Euro NCAP (European New Car Assessment Programme) (EuroNCAP 2016).

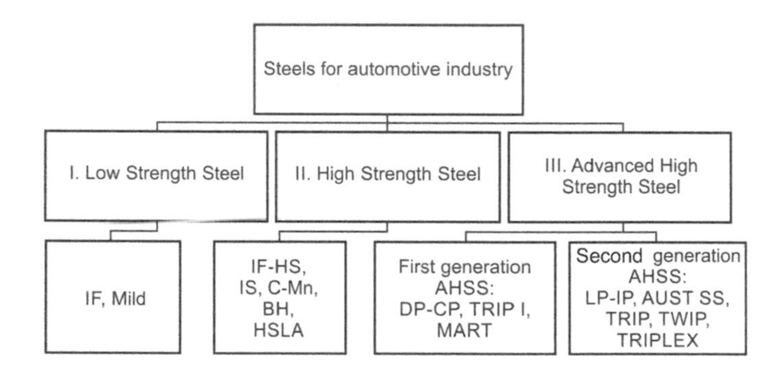

Figure 6.1 Classification of steels for motor industry (Anderson 2008, Dobrzański 2006, Keeler and Kimchi 2014, WorldAutoSteel 2015a,b); IF: Interstitial-Free steel, MILD: medium alloy steels, IF-HS: Interstitial-Free with microadditives, IS: Isotropic Steel, C-Mn: carbon-manganese steels, BH: Bake Hardening, HSLA: High Strength Low Alloy steel, DP-CP: Dual Phase Complex Phase, TRIP: Transformation Induced Plasticity, MART: martensitic steels, LP-IP: Light steels with Induced Plasticity, AUST SS: Austenitic Strength Steel, TWIP: TWinning Induced Plasticity, TRIPLEX: three-component Fe-C-Mn-Al steels.

Table 6.3 Comparison of elongation and tensile strength of steels for car industry (Anderson 2008, Frommeyer and Brüx 2003, Keeler and Kimchi 2014) (symbols as in Figure 6.1).

Type of steel	**UTS, MPa**		**ε, %**		**Evaluation**
	min.	**max.**	**min.**	**max.**	
IF	150	250	40	55	↓
MILD	180	300	27	37	↓
BH	200	400	25	35	↓
IF-HS	220	400	30	43	↓
IS	280	400	32	38	↓
C-Mn	250	600	15	30	↓
HSLA	300	800	10	25	→
DP-CP	500	100	7	25	↓
MART	700	1400	5	15	↓
AUST SS	850	1000	40	50	→
TRIP I	500	1000	15	32	→
TRIP II	1000	1250	40	60	↑
TWIP	750	1150	50	70	↑
TRIPLEX	1000	1200	50	65	↑
Lowest value					**Highest value**

The modern automotive industry imposes very stringent requirements for manufacturing technologies and for steels used for bodies, structural parts and other road vehicles components. All the materials applied for car bodies have to meet requirements for specific tensile strength and rigidity and good plastic properties responsible for good press-formability ensuring the required shape. An appropriate store of cold plastic deformation energy, i.e., impact energy absorption, is also important but only then material density reduces fuel consumption. These requirements result from the constant rise in safety standards to develop a controlled plastic deformation zone damaged according to a defined scenario of the maximum absorption of high impact energy, and also from the necessity to decrease car mass, thus fuel consumption and CO_2 emission (Anderson 2008, Keeler and Kimchi 2014). By choosing adequately the chemical composition and manufacturing technology ensuring a structure that achieves a beneficial combination of strength and plastic properties of steel, new groups of steels for use in the car industry have been created in the last forty years. A formability index, corresponding to the tensile strength times the maximum steel elongation, was established to compare the favourable strength and plastic properties of such steels (Keeler and Kimchi 2014, WorldAutoSteel 2015a, 2015b, 2015c, 2015d). A special group of passive safety solutions are additional car body skeleton reinforcements with controlled plastic deformation zones during car collision and side car reinforcements enabling to absorb as much energy during a side car impact as possible (Figure 6.2), as well as a special truck cabin construction and its mounting ensuring the absorption of maximum energy during accident or collision in controlled plastic deformation zones.

Figure 6.2 Schematic examples of side door and car body skeleton reinforcements with controlled plastic deformation zones during car collision (corporate promotional materials).

A concept of steel development has been established to satisfy such sophisticated requirements for mainly passive car safety (Dobrzański 2011). An unconventional approach to steel cracking was presented. As a result of the acting strain, the material cracks into parts, which is associated with, respectively, the initiation and propagation of cracking (Dobrzański 2006, 2011, 2015, Dobrzański and Borek 2010a, 2010b, 2011a, 2011b, 2012a, 2012b, 2012c, 2012d, Dobrzański et al. 2008a, 2008b, 2008c, 2009, 2010, 2011, 2012, 2013a, 2013b, 2014a, 2014b, 2014c, 2015a, 2015b, 2015c, 2016a, 2016b). Brittle cracking is characterised by a very small portion of plastic deformation and small energy absorbed before complete fracture. Ductile cracking is however characterised by a very high portion of plastic deformation and energy absorbed before complete fracture. Ductile cracking is a property that describes the ability of the material featuring such cracks to withstand cracking. The following are the factors having an effect on cracking ductility: temperature, strain rate, dependency between strength and material ductility and the presence of notches (strain concentrators) on the material surface. A field under the curve (integral) of stress variations in the function of deformation is a measure of cracking ductility. Plastic materials have much higher cracking ductility than brittle materials. Fracture mechanics is modeling the possible cases of brittle cracking (linear elastic fracture mechanics), brittle cracking with small share of plastic and ductile cracking (Dobrzański 2006). An opposite approach poses a question whether it is possible to use the plastic deformation energy of engineering materials to prevent their damage before their decohesion process takes place (Dobrzański 2011). This question became the basis for a series of several studies (Dobrzański et al. 2011, 2016c). The idea is that no bodily injuries are sustained in serious car accidents, despite a very severe car damage. For this reason, the purpose of the investigations mentioned was to determine the ability of high manganese steels, having an austenitic structure with the addition of Al and Si, to prevent cracking by, respectively, inducing twinning, a martensite transformation in micro areas and precipitation processes in cold plastic deformation. For this, the considerable absorption of energy and cracking prevention is required, after completion, in prior, of structural changes and phase transitions, which are accompanying the processing processes by hot or cold plastic deformation and heat treatment (Borek 2010, Czaja 2014, Dobrzański et al. 2011, 2016c, Jagiełło 2016). This is accompanied also by other works (Grajcar 2009), connected in particular with adding Ti, Nb and B micro additives to such steels (Dobrzański et al. 2011, 2016c, Grajcar 2009), allowing to additionally refine the structure of such steels by applying thermomechanical treatment in a manufacturing process, thus hardening them additionally. A synergic interaction of the mentioned structural effects in the said group of high manganese steels, having an austenitic structure, makes it possible to produce steels with greatest energy absorption by triggering, respectively, plastic deformation mechanisms and phase transitions and precipitation processes during dynamic cold plastic deformation in a road accident. This undoubtedly has an effect on the imitation of fatalities and permanent body injuries, thus significantly enhancing passive safety of car passengers and drivers which can always suffer from a car accident. For this reason, the research efforts undertaken concentrate on selecting the suitable chemical composition and the manufacturing and processing process—including adequate plastic working, thermomechanical treatment and heat treatment of this special group of steels—to enable the maximum possible activation of the mentioned mechanisms, phase transitions and structural changes preventing cracking by absorbing the relevant, high energy directly in a road accident, if one occurs. Overall, the idea is to find a possibility of using the energy, which is necessarily released in a car accident, to activate, plastic deformation, phase transitions and precipitation processes of steels with the adequately selected chemical and phase composition to limit a portion of energy consumed directly for cracking processes, hence to ensure biggest feasible cracking prevention. The knowledge of the basis of plastic deformation and structural phenomena accompanying plastic and thermomechanical treatment, discussed in numerous references, is useful for the initial analysis of the issues taken up (Dobrzański 2006, 2012, Lin et al. 2012, WorldAutoSteel 2015c).

The steels and alloys used these days and in the future for fabrication of car sheets are shown in Figure 6.1 by comparing dependencies between strength and elongation of the mentioned groups of steels (Table 6.3, Figure 6.3).

Over the decades, the way to increase steel strength and plasticity at the same time was to create new steels having an A2 lattice structure—ferritic and martensitic steels (Dobrzański 2006, 2012, 2013), meanwhile, an idea is now emerging in the related references (Dobrzański 2011, 2013, Dobrzański et al. 2011) stating that opportunities have already been practically exploited to optimise the chemical composition and technological conditions of properties formation for steels having the A2 lattice structure.

It is thought, however, that major progress in achieving the products combining possibly high strength and plastic properties can be ensured by new steels having the austenitic A1 structure containing Mn in a concentration of over 25% as well as Si and Al. Advanced high manganese steels can respond to increasingly higher challenges currently imposed on the car industry. It is only in few cases that high manganese steels may be an alternative for the well-developed steels with the A2 lattice structure having established uses, also in the automobile branch. Such steels, as complementary engineering materials, which are still being developed and intensively studied, may ensure significant progress, especially in

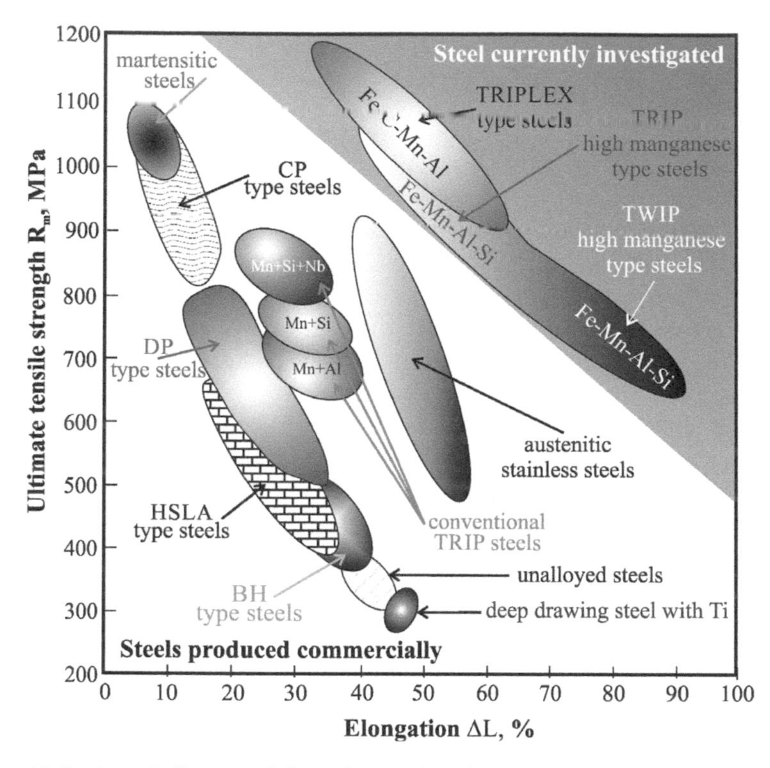

Figure 6.3 Steels and alloys used these days and in the future for fabrication of car sheets (provided in (Dobrzański 2013) acc. to S. Keeler's data); grey field for steels with A1 lattice structure.

automotive applications, also by fulfilling the assumptions of the mentioned concept of preventing fractures in structural parts and bodywork of a car during an accident or road collision by absorbing a large store of the then produced energy, to induce structural changes and phase transitions, occurring in the conditions of dynamic plastic deformation of critical parts of the car made of such steels. Steels with a significant concentration of Mn having the A1 structure and with a special tendency of hardening as a result of plastic deformation, indeed belong to classical materials science as manganese austenitic steels containing 1.1–1.3% of C and 12–13% of Mn, discovered by an English steelmaker, R.A. Hadfield, and have been known since 1882 (Dobrzański 2006, Karaman et al. 2000, Petrov 2006). The benefits of using high manganese steels having an austenitic structure were again brought to attention in the last two decades of the 20th century as intensive mechanical twinning mechanisms or a martensitic transformation that could be induced in such steels. This ensured, most often, the simultaneous improvement of strength and plastic properties, and thus achieving an increased store of energy unprecedented for other steels, allowing to mould parts with a complicated shape or to release energy during plastic deformation at high rates, in particular during a road accident (Bracke et al. 2009, Brüx et al. 2002, Cabañas et al. 2006). Numerous researches were carried out (Bracke et al. 2009, Dobrzański and Borek 2010a, 2011b, 2012d, Dobrzański et al. 2008a, 2008c, Frommeyer and Brüx 2006, Grässel et al. 2000, Hamada 2007, Kliber et al. 2008, 2009, Mazancová et al. 2009, Timokhina et al. 2002, 2008, Vercammen et al. 2004) for the new generation of high manganese steels with an austenitic structure. At present, high manganese austenitic steels fall into the following groups, according to their chemical compositions and basic structural mechanisms decisive for strengthening (Adamczyk and Grajcar 2005, Adamczyk and Opiela 2002, Allain et al. 2004a, 2004c, Berrahmoune et al. 2004, Bouaziz and Guelton 2001, Bracke et al. 2006, 2009, Brüx et al. 2002, Cugy et al. 2005, De Cooman 2004, De Meyer et al. 1999, Frommeyer and Brüx 2006, Frommeyer et al. 2000, 2003, Grässel et al. 2000, Hamada 2007, Kliber et al. 2008, 2009, Kwon et al. 2010, Lis and Gajda 2006, Mazancová et al. 2007, 2008, 2009, Sabet et al. 2008, Scott et al. 2006, Takechi 2000, Timokhina et al. 2008, Vercammen et al. 2004):

- Hadfield steel containing between 1.1 and 1.3% of C and between 12 and 13% of Mn with a high tendency of strain hardening connected with the formation of microtwins discussed earlier;
- steels containing between 0.05 and 0.15% of C and 15–22% of Mn and Si and Al with a diverse concentration, in which a multiple TRIP (TRansformation Induced Plasticity) effect occurs, where a martensitic transformation is induced in austenite during cold plastic deformation, ensuring high strength properties of such steels (Figure 6.3); (it is worth noting that the TRIP effect also exists in Cr-Ni austenitic steels (Lagneborg 1964, Talonen et al. 2005), but they are too costly to use them in the motor sector);
- steels containing between 0.02 and 0.65% of C and 20–30% of Mn and Al and Si with a diverse concentration, characterised by the TWIP (TWinning Induced Plasticity) effect consisting of intensive mechanical twinning in austenite during plastic deformation securing such steels' high plastic properties (Figure 6.3);
- steel containing between 0.5 and 2% of C, 18–35% of Mn, 8–12% of Al and 3–6% of Si, whereas the total concentration of Al and Si must not be below 12%, known as TRIPLEX, due to a three-phase structure represented by austenite grains γ-Fe(Mn,Al,C), dispersion carbon precipitates κ-$(Fe,Mn)_3AlC_{1-x}$ and ferrite α-Fe(Mn,Al), in which a dislocation slip, twinning, transformation into martensite ε and martensite α occurs in austenite during plastic deformation, depending on the chemical composition of steels and deformation conditions, which ensures indirect properties for such steels as compared to TRIP and TWIP steels (Figure 6.3), and also the density of 6.5–7 g/cm^3, lower than steels without Al added.

The following factors significantly influence the mechanical properties of high manganese steels with an austenitic structure:

- chemical composition of steel;
- plastic deformation rate;
- plastic deformation temperature.

The chemical composition of steel is the main factor decisive for the set of properties of the discussed steels which respond so well to the car industry's requirements, by having an effect on stacking-fault energy (SFE) of austenite γ A1, and thus also on the progress of phase transitions and plastic deformation mechanisms in such steels. Because of the dependency between SFE and dislocation ability to climb and cross slip, the dependency is crucial for behaviour of the plastically deformed material, for the value of the stored plastic deformation energy and for the progress of polygonisation and recrystallisation. Plastic deformation in steels with the A1 crystallographic structure and high SFE is, therefore, easier and the stored plastic deformation energy is smaller, and peripheral dislocation climb and cross slip occurs more easily. It is opposite in steels with low SFE due to a tendency of dislocation pile-ups, e.g., by the grain boundaries or particles of alien phases. As plastic deformation is rising, shear and transition bands are formed, and in steels with high SFE, cellular dislocation tangle systems are created separating the cells nearly free of dislocations, which decrease as dislocation tangles densify as the degree of plastic deformation steel is growing (Christian and Mahajan 1995, Dobrzański 2006, WorldAutoSteel 2015c). Small SFE energy (<20 mJ/m^2) encourages a martensitic transformation suppressed by the growth of SFE (to 25 mJ/m^2). If SFE energy of austenite in ambient temperature is smaller than 20 mJ/m^2, martensite ε A3 (Basuki and Aernoudt 1999) is produced during plastic deformation. Higher SFE energy is accompanied by a growing tendency of mechanical twinning having an intensity which is decreasing from high—with being arranged uniformly as a result of a nearly homogenous plastic deformation—to small, due to impeded dissociation of dislocations into particle dislocations and with the resulting domination of hardening by slip of total dislocations, along with SFE energy increased from 25 to 60 mJ/m^2. In case of an indirect value of SFE energy, mechanical twinning and dislocation slip occurs simultaneously in steels (Christian and Mahajan 1995, Diani and Parks 1998, Hamada 2007, Shin et al. 2001). The carbon present in all steels reaches, for the discussed group of steels, a concentration of 0.02–1.2% and is a chemical element stabilising austenite γ A1 in the presence of Mn (Hamada 2007, Hamada et al. 2007a, 2007b, Kannan et al. 2008, Kliber et al. 2008, 2009). Two groups of the discussed steels can be classified according to the concentration of C:

- steels with a concentration of 0.01–0.1% C, with the strength of R_m = 600–700 MPa, yield point of $R_{p0.2}$ = 250–450 MPa, and elongation A_r

= 50–60% (Brüx et al. 2002, Frommeyer and Brüx 2006, Frommeyer et al. 2003, Hamada 2007, Hamada et al. 2007a, Huang et al. 2006, Sabet et al. 2008, Vercammen et al. 2004);

- steels with a concentration of 0.5–1.2% C with the strength of R_m = 700–900 MPa, yield point of $R_{p0.2}$ = 200–300 MPa, exhibiting larger uniform elongation reaching even up to 80% (Allain et al. 2004a, 2004b, 2004c, Barbier et al. 2009, Bracke et al. 2006, 2009, Bouaziz et al. 2008, Grässel et al. 2000, Kliber et al. 2008, 2009, Scott et al. 2006).

The main alloy element in the analysed steels is Mn (Allain et al. 2004a, Brüx et al. 2002, Frommeyer and Brüx 2006, Frommeyer et al. 2000, 2003, Grässel et al. 2000, Hamada 2007, Vercammen et al. 2004) and its concentration is conditioning the SFE energy value in austenite γ A1 in such steels (Bracke et al. 2006, Idrissi et al. 2009, Lee and Choi 2000, Mazancová et al. 2009, Schumann 1972, Volosevich et al. 1976), and also mutual proportions between the concentration of C and Mn (Allain et al. 2004b), which is directly determining phase transition taking place in such steels during plastic deformation. At the Mn concentration of 5–10%, the steels discussed exhibit the structure of ferrite α A2 and austenite γ A1. The structure of austenite γ A1 and martensite α′A2 is ensured by increasing its concentration to 22%. Only when steel contains 22–35% of Mn, full stabilisation of γ A1 austenite structure is seen.

If the concentration is not higher than 25% of Mn, the TRIP effect of steel hardening may occur in steel during cold plastic deformation as a consequence of a martensitic transformation. As plastic deformation is growing, the volumetric fraction of individual phases in the steel structure is changing: the fraction of austenite γ A1 is dropping and the fraction of martensite α′A2 is rising, with relatively small changes in the volumetric fraction of martensite ε A3 (Brüx et al. 2002, Frommeyer et al. 2000, 2003, Fujita and Ueda 1972, Grässel et al. 2000, Han et al. 2009, Hilditch and Speer 2004, Hyun and Park 2004, Jacques et al. 2001, Jee et al. 2004). Steel in such a state exhibits the R_m strength of about 900 MPa and elongation A of up to 40% (Frommeyer et al. 2003).

If the concentration of Mn is increased from 15 to 25%, it weakens R_m tensile strength from 900 to 600 MPa and increases elongation A from 40 to 80%. For steels with a concentration higher than 25% of Mn, austenite γ A1 is stable during cold plastic deformation and the TWIP effect may take place (Liu et al. 1994). Deformation twins in austenite γ A1 are increasing the rate of strain hardening because, together with grain boundaries, they form barriers for the movement of deformations. Steel in such a state exhibits the R_m strength of about 650 MPa and elongation A of up to about 80%.

If the concentration of Mn is further increased, the plastic properties of steel are not further influenced. 3% of Al and 3% of Si added at the same time have a beneficial effect on steel properties at a concentration of above

25% of Mn (Grässel et al. 2000). Both these elements have influence on hardening of steel. The growth of SFE energy and the stability of austenite γ A1 are however influenced by adding Al and thus suppress a martensitic transformation (De Meyer et al. 1999, Hamada 2007, Han and Hong 1997, Mazancová et al. 2009). The addition of Si lowers SFE energy, by intensifying growth in the density of stacking faults and hence encouraging the activation of the γ A1 > ε A3 type martensitic transformation (Bracke et al. 2009, Brüx et al. 2002, Diani and Parks 1998, Grässel et al. 2000, Qin and Bhadeshia 2008, Vercammen et al. 2004). The rate of plastic deformation in case of TRIP steels (e.g., X2MnAlSi15-3-3), in which a martensitic transformation is induced during cold plastic deformation, does not significantly influence the strength of R_m and yield point of $R_{p0.2}$, and reduces elongation by 10–15% though. This stems from the fact that stretching with high plastic deformation rate leads to the adiabatic heating of a specimen and higher SFE energy. In case of TWIP steels with plasticity induced by twinning (e.g., X3MnAlSi25-3-3), if the rate of plastic deformation rises to 10^{-1} s^{-1} at constant temperature, this will increase the yield point $R_{p0,2}$ from 250 to 450 MPa, the tensile strength R_m from 600 to 800 MPa, with declining elongation A. After reaching a minimum value, a further rise in the rate of plastic deformation does not cause major changes in elongation. Due to intensive mechanical twinning, the total elongation is A = 80% with the strength of R_m = 800 MPa (Brüx et al. 2002, Diani and Parks 1998, Frommeyer and Brüx 2006, Frommeyer et al. 2000, 2003, Grässel et al. 2000, Hamada 2007).

The temperature of plastic deformation has an effect on the mechanical properties of high manganese austenitic steels due to rise in SFE energy of austenite along with the rising plastic deformation temperature (Byun et al. 2004, Hamada 2007, Hamada et al. 2007a, 2007b). Three temperature intervals can be distinguished for TRIP steels (e.g., X2MnAlSi15-3-3) in which a martensitic transformation is induced during cold plastic deformation (Brüx et al. 2002, Edmonds et al. 2006, Frommeyer and Brüx 2006, Frommeyer et al. 2000, 2003, Grässel et al. 2000, Hamada 2007).

- 150–400°C in which strength and plastic properties do not change, and plastic deformation occurs as a result of dislocation slip;
- 80–150°C where strength and plastic properties are greatly ameliorated as a consequence of austenite being transformed into martensite (elongation increased to 60%) and the neck in the specimens being stretched is formed later;
- 80–100°C where strength properties are heightening with a significant reduction in steel plasticity (elongation A = 30%) associated with a higher rate of a martensitic transformation induced by plastic deformation and its completion in an early stage of plastic deformation.

Two temperature intervals can be distinguished for TWIP steels (e.g., X3MnAlSi25-3-3) in which mechanical twinning takes place in cold plastic deformation (Brüx et al. 2002, Frommeyer et al. 2000, 2003, Grässel et al. 2000, Hamada 2007, Vercammen et al. 2004):

- 20–400°C, where—as plastic deformation temperature is going down—tensile strength R_m and yield point $R_{p0.2}$ increase. Steel elongation in the plastic deformation temperature of 400°C A = 40–50% and as temperature is decreased by about 20°C, it increases to A = 90%, due to intensive mechanical twinning; the steel achieves its most advantageous properties at the plastic deformation temperature of 20°C;
- below 20°C where the fraction of deformation twins is rising significantly, thus finishing twinning in an early stage of plastic deformation and reduced steel plasticity.

Table 6.4 compares the properties of high manganese austenitic steels of the TRIP, TWIP and TRIPLEX type intended for the automotive industry.

By choosing adequately the chemical composition and manufacturing technology ensuring a structure allowing it to achieve a beneficial combination of strength and plastic properties of steel, new groups of steels for use in the car industry have been created over the last forty years. A formability index corresponding to the tensile strength times the

Table 6.4 Comparison of the properties of high manganese steels for applications in car industry (Anderson 2008, Mazancová et al. 2012).

Property	TRIP	TWIP	TRIPLEX
Carbon concentration, %	up to 0.2	about 0.1–0.6	0.7–1.2
Manganese concentration, %	12–20	15–35	18–28
Al concentration, %	2–4	2–4	8.5–12
Si concentration, %	2–4	2–4	2–4
Structure before plastic deformation	austenitic	austenitic	austenitic-ferrous
Structural mechanisms and transformations occurring during cold plastic deformation	austenite transformation into martensite	mechanical twinning	dislocation slip, mechanical twinning
Structure after plastic deformation	austenite + martensite	austenite	austenite + ferrite + carbide κ
Elongation, %	40–60	40–90	45–75
Tensile strength, MPa	1100	700–1000	1100

tensile strength and maximum elongation was established to compare the combination of favourable strength and plastic properties of such steels (Keeler and Kimchi 2014, WorldAutoSteel 2015a, 2015b, 2015c, 2015d). The newly established high manganese austenitic steels are characterised by the high index and at the same time very high tensile strength (reaching even 1100 MPa), as well as long elongation (to 60%) for all the steels used in the car industry. Conventional steels with a low concentration of carbon worked plastically by cold and hot rolling have been used mostly for car body structures until recently. The heat treatment of such steels, i.e., annealing after cold plastic working and cooling, is carried out in a way so as to obtain a ferritic structure. Thermo mechanical treatment is required for new generations of steels and an austenitic structure has to be achieved, in which mechanisms occur during cold plastic deformation, which prevents the cracking of parts made of such steel (Anderson 2008, Keeler and Kimchi 2014, WorldAutoSteel 2015a, 2015b, 2015c, 2015d). A special characteristic of the newly developed high manganese austenitic steels is the favourably high formability index, guaranteeing a store of plasticity in controlled energy absorption zones in operation, especially during a sudden action of unpredicted loads existing in cars during an accident or road collision (WorldAutoSteel 2015a). This special characteristic is very important in case of a collision or car accident when it is most needed that the body material exhibits two completely contrary characteristics (ASP 2010, Borek 2010, Opbroek 2009):

- high ductility due to which maximum impact energy can be absorbed upon material deformation;
- maximum part stability maintained.

High manganese steels, which meet the above factors to the fullest extent, may revolutionise this part of car industry which is involved in body production, as steels used till date for such parts showed only one of the mentioned characteristics. The aim of works conducted over newly developed high manganese austenitic steels is not to remove conventional steels from the market, but to create an alternative for structural parts and parts reinforcing a car body which are loaded most and which require the highest store of plasticity upon collision or accident. The first generation of steels from the third group AHSS is gradually put into production, and their share is gradually rising. Newly created high manganese austenitic TRIP, TWIP and TRIPLEX steels (Grässel et al. 2000, Keeler and Kimchi 2014, WorldAutoSteel 2015a, 2015b, 2015c, 2015d) are the second generation of advanced high strength steels (AHSS). The steels are being intensively investigated by several global research institutions and many researchers (Dobrzański et al. 2011, 2016c, Ferguson et al. 2009, Frommeyer et al. 2000, 2003, Santos et al. 2011) (Dobrzański 2006, 2011, 2015, Dobrzański and Borek

2010a, 2010b, 2011a, 2011b, 2012a, 2012b, 2012c, 2012d, Dobrzański et al. 2008a, 2008b, 2008c, 2009, 2010, 2011, 2012, 2013a, 2013b, 2014a, 2014b, 2014c, 2015a, 2015b, 2015c, 2016a, 2016b), including those presented in this chapter.

6.3 General Concept of Own Research into High Manganese Steels Exhibiting TRIP and TWIP Effects

The presented state of the art indicates that the newly established groups of high manganese steels having the austenitic structure are an interesting research material. The comparative and preliminary results of investigations of the store of energy used for cold plastic deformation (a field under the dynamic stretching curve) show that the high-manganese steels, in fact, demonstrate unrivalled these properties in comparing to other steels used to date. Therefore the research creates an opportunity to optimise properties selected for the steels, especially considering extreme conditions of dynamic plastic deformation corresponding to road accidents and collisions; and manufacturers, and especially car users, imposing high vehicle passive safety requirements, certainly must be interested in this. It is a challenge to study the interesting structural phenomenons occurring in such steels during technological processes of plastic deformation as well as in laboratory tests simulating extreme conditions existing in road accidents and collisions and explain issues related to materials sciences related to such processes.

In the framework of own works presented in this chapter, interactions between all the factors which are likely to influence structure formation and an advantageous relationship between high strength ($R_m \geq 1000$ MPa) and high plastic properties ($A \geq 60\%$) were identified, with the highest possible store of cold plastic deformation energy of at least 300 MJ/m^3. This is preconditioned by a surface integral under a cold plastic deformation curve and by cracking prevention of newly created austenitic TRIP (TRansformation Induced Plasticity), TWIP (TWinning Induced Plasticity) and TRIPLEX steels (with a diversified fraction of ferrite, austenite and finely dispersive precipitates of carbides). The basic research thesis proposed assumes that multiple factors have an effect on an increased store of cold plastic deformation energy and thus prevention of cracking for the analysed groups of steels and such factors act in opposite ways sometimes. When each factor is thoroughly recognised, and when mutual interactions are determined between them and the determined trends are captured in an adequate model, the proposed research problem can be solved fully (Figure 6.4). The factors analysed include austenite grain size, which is adjustable in a controlled manner by heat and thermo mechanical treatment, allowing the controlled progress of dynamic, meta dynamic or even static recrystallisation. The chemical composition of the studied steels is also important for the analysed factors, which is conditioning austenite stacking-

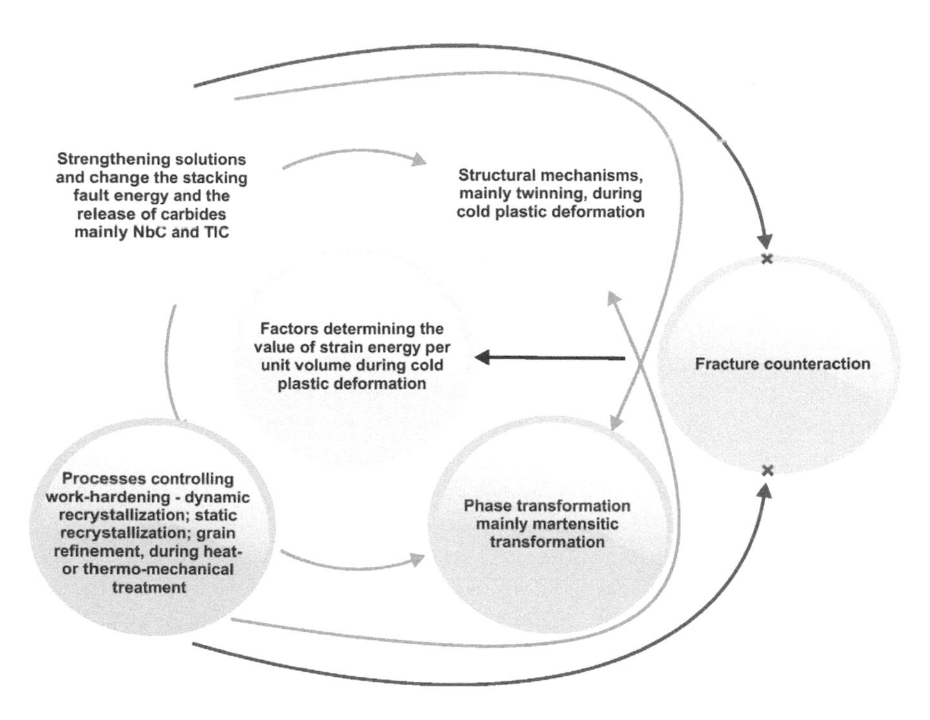

Figure 6.4 Scheme of interaction of factors influencing structure formation and advantageous relationship between high strength of the studied steels and their high plastic properties with the highest possible store of cold plastic deformation energy.

fault energy. The investigations of high manganese steels undertaken to date (Dobrzański 2006, 2011, 2015, Dobrzański and Borek 2010a, 2010b, 2011a, 2011b, 2012a, 2012b, 2012c, 2012d, Dobrzański et al. 2008a, 2008b, 2008c, 2009, 2010, 2011, 2012, 2013a, 2013b, 2014a, 2014b, 2014c, 2015a, 2015b, 2015c, 2016a, 2016b) and the literature data (Allain et al. 2004a, Barbier et al. 2009, Bouaziz et al. 2008, Bracke et al. 2009, Brüx et al. 2002, Cabañas et al. 2006, Dini et al. 2010a, 2010b, Frommeyer and Brüx 2006, Frommeyer et al. 2000, 2003, Hamada 2007, Idrissi et al. 2009, Kliber et al. 2008, Kwon et al. 2010, Lagneborg 1964, Li and Wu 2008, Li et al. 2008, Mazancová et al. 2009, Mazurkiewicz 2013, Niendorf et al. 2009, Vercammen et al. 2004) indicate that, due to a varied concentration of Mn and Si and Al additives, such steels are characterised by a different value of austenite stacking-fault energy, thus have a different tendency of a martensitic transformation or mechanical twinning during cold plastic deformation. On the other hand, the refining of the austenite structure depends on introducing Nb and Ti microadditives which, by the release of dispersion carbides, are inhibiting the growth of grains during heat and thermo mechanical treatment and are decisive for precipitation hardening during treatment and/or cold plastic deformation. The structural mechanisms and phase transitions activated during the cold plastic deformation of the studied steels, having a preformed structure, are the factors decisive for the value of the store of cold plastic deformation energy and thus for cracking prevention for the studied group of steels. Such factors include, respectively: the TRIP effect consisting of steel hardening as a result of a martensitic transformation induced with cold plastic deformation, the TWIP effect utilising the intensive progress of mechanical twinning as a result of cold plastic deformation and the TRIPLEX effect in steels with a multi-phase structure. As long as the steel does not crack during cold plastic deformation, its plastic deformation is possible along with the resulting increase in the store of cold plastic deformation energy. It is therefore important to establish in what conditions the studied steels undergo cracking to prevent such processes and to use the energy delivered to the material, e.g., during external impact loads, only for plastic deformation in the considered range of external loads. This is another factor analysed, however, as long as it does not occur during cold plastic deformation, the situation should be considered favourable. The principal research task presented in this chapter is, therefore, to examine interdependencies between structural mechanisms (twinning mainly) and phase transitions (martensitic transformation mainly or/and carbides release processes) during cold plastic deformation and the structural condition of such steels caused by the desired refining of austenite grains a result of controlled progress of recrystallisation (dynamic mainly, but also metadynamic or static) during prior thermomechanical and/or heat treatment, which is also dependent on the microadditives of strongly carbide-forming elements (such as, e.g., Nb and Ti). The principal research task is to determine also a synergic interaction of the mentioned factors, depending on the type of TRIP, TWIP or TRIPLEX steel, on an increased store of cold plastic deformation energy and cracking prevention and their importance for formulating such materials' other mechanical properties. The ultimate outcome of the said investigations is to acquire information which, in fact, creates the function of many variables—the analysis of which—despite a systematic programme of the planned investigations, is very difficult, as confirmed by the current but unsatisfactory state of the art in this field. It is currently impossible to indicate optimum values of the mentioned factors and their mutual combinations due to the maximisation criterion of the store of cold plastic deformation energy and cracking prevention. The presented results are the reasons for elaborating a general model of interdependencies of the mentioned mechanisms, phase transformations and structure versus to the chemical composition of the examined group of steels.

The essence of the research outcomes presented in this chapter is to recognise the mechanisms and sequence of structural transformations taking place in such steels during cold plastic deformation, the basis of structure development, a favourable dependency between high strength

($R_m \geq 1000$ MPa) and high plastic properties ($A \geq 60\%$) for a possibly highest store of cold plastic deformation energy which is determined by the surface integral under a cold plastic deformation curve. The essence is also the prevention of cracking, by applying heat and thermo mechanical treatment and by creating a model interlinking the chemical composition of steel, the conditions of thermo mechanical, heat treatment and plastic working to the structure and mechanical properties of steels.

The investigations were carried out for more than ten newly developed own species of TRIP, TWIP or TRIPLEX steels (Dobrzański 2006, 2011, 2015, Dobrzański and Borek 2010a, 2010b, 2011a, 2011b, 2012a, 2012b, 2012c, 2012d, Dobrzański et al. 2008a, 2008b, 2008c, 2009, 2010, 2011, 2012, 2013a, 2013b, 2014a, 2014b, 2014c, 2015a, 2015b, 2015c, 2016a, 2016b), of which chemical composition shown in Table 6.5 was selected. The table shows the selected results of investigations for TRIP and TWIP steels.

A thermo mechanical processing simulator DSI Gleeble 3800 by DSI was used to simulate heat and thermo mechanical treatment. The simulator has enabled to determine hardening curves of the examined steels, and the effect of deformation temperature and rate on mechanisms controlling the progress of strain hardening of the developed steels. A hot pressing process, consisting of several phases, simulating the actual plastic deformation conditions of the studied group of steels, was designed and carried out in the next stage of plastomeric tests. The experiment was also carried out using the thermo mechanical processing simulator (Figure 6.5).

Dependencies were established, based on the outcomes of preliminary tests, between plastic deformation rate, degree of deformation, and intervals between the subsequent plastic working operations, and the structure, the

Table 6.5 Chemical composition of high manganese Fe-Mn-Al steels.

Steel symbol	Type of steel	Chemical composition, % mas.					
		C	Mn	Si	Al	Nb	Ti
X11MnSiAl17-1-3	TRIP	0.11	17.55	1.17	3.37	–	–
X11MnSiAlNbTi18-1-3		0.11	18.25	1.20	3.29	0.027	0.025
X11MnSiAl25-1-3	TWIP	0.11	24.93	1.20	3.24	–	–
X8MnSiAlNbTi25-1-3		0.08	24.60	0.91	3.10	0.040	0.024
X13MnSiAlNbTi25-3-3		0.13	25.1	3.50	3.3	0.050	0.018
X73MnSiAlNbTi25-1-3		0.73	25.50	1.30	3.20	0.047	0.027
X105MnAl24-11	TRIPLEX	1.05	23.83	0.10	10.76	–	–
X98MnAlNbTi24-11		0.98	23.83	0.20	10.76	0.048	0.019

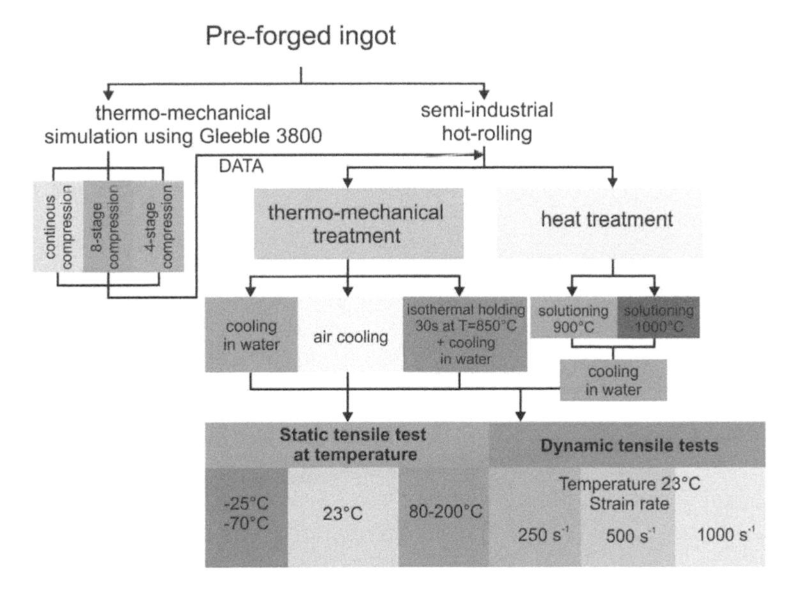

Figure 6.5 Division of research material.

measure of which is shape and size of grains, and mechanical properties of the newly developed steels. Thermo mechanical treatment was carried out in a Gleeble 3800 simulator and on a line for semi-industrial simulation of rolling. The degrees of deformation, plastic deformation rates and intervals between subsequent deformations were selected according to the conditions which exist during hot rolling. Following the last plastic working at 850°C, the material was cooled directly in water, air or was subjected to isothermal heating at the temperature of the last treatment of 850°C for 30 s and was then cooled in water.

The following specimens were prepared for plastometric examinations:

- cylindrical specimens with the dimensions of ∅ 10 × 12 mm for high-temperature continuous pressing and for simulation of thermo mechanical treatment consisting of four-phase pressing;
- perpendicular specimens dimensioned 20 × 35 × 15 mm for simulating thermo mechanical treatment consisting of eight stages of pressing.

Thermo mechanical treatment was carried out in the Gleeble 3800 simulator consisting of eight or four stages of hot pressing of perpendicular or cylindrical specimens. The routine involving eight stages of pressing in the Gleeble 3800 simulator was presented in Figure 6.6 (variants I, II, III), while Table 6.6 lists the conditions of eight stages of hot pressing of perpendicular prism-shaped specimens.

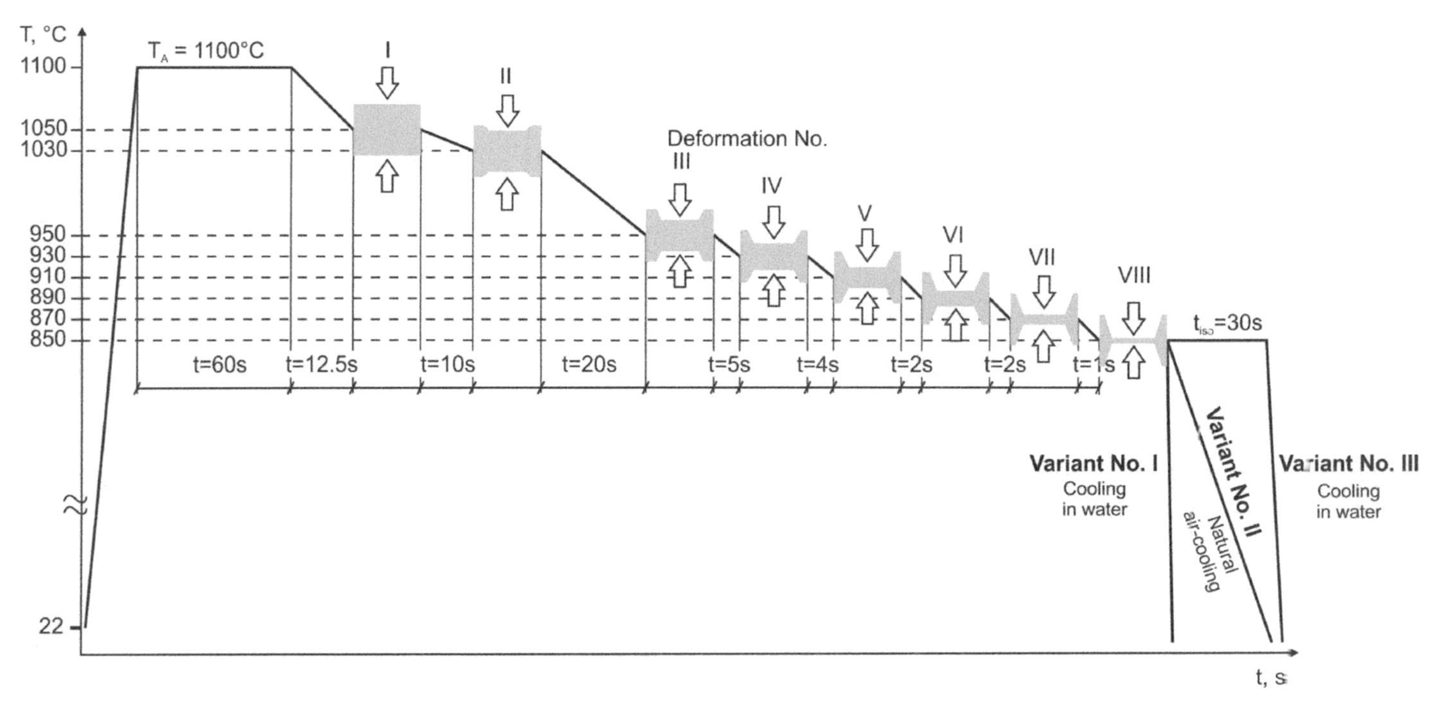

Figure 6.6 Pattern of thermoplastic treatment of perpendicular prism-shaped specimens subject to plastic working in the Gleeble 3800 simulator T_A: austenitisation temperature, t_{iso}: isothermal heating time at the temperature of the last working of 850°C.

Table 6.6 Hot thermo mechanical treatment conditions (Gleeble 3800 simulator) consisting of eight stages of hot pressing of perpendicular specimens dimensioned 15 × 20 × 35 mm.

<table>
<tr><th colspan="5">Heating</th></tr>
<tr><td>Heating temperature,°C</td><td>Heating rate, °C/s</td><td colspan="2">Heating time, s</td><td>Colling rate to deformation temperature, °C/s</td></tr>
<tr><td>1100</td><td>3</td><td colspan="2">30 (60)</td><td>4</td></tr>
<tr><th colspan="5">Deformation</th></tr>
<tr><td>No.</td><td>T_D, °C</td><td>ε</td><td>$\dot{\varepsilon}$, s^{-1}</td><td>Cooling rate, °C/s</td></tr>
<tr><td>I</td><td>1050</td><td>0.4</td><td>5</td><td>2</td></tr>
<tr><td>II</td><td>1030</td><td>0.3</td><td>5</td><td>4</td></tr>
<tr><td>III</td><td>950</td><td>0.25</td><td>10</td><td>4</td></tr>
<tr><td>IV</td><td>930</td><td>0.25</td><td>10</td><td>5</td></tr>
<tr><td>V</td><td>910</td><td>0.2</td><td>80</td><td>4</td></tr>
<tr><td>VI</td><td>890</td><td>0.2</td><td>100</td><td>5</td></tr>
<tr><td>VII</td><td>870</td><td>0.2</td><td>200</td><td>5</td></tr>
<tr><td>VIII</td><td>850</td><td>0.2</td><td>250</td><td>see below</td></tr>
<tr><th colspan="5">Cooling</th></tr>
<tr><td>No.</td><td>Temperature range,°C</td><td colspan="2">Cooling medium</td><td>Heating time prior to cooling, s</td></tr>
<tr><td>I</td><td rowspan="3">850 → room temperature</td><td colspan="2">water</td><td>0</td></tr>
<tr><td>II</td><td colspan="2">air</td><td>0</td></tr>
<tr><td>III</td><td colspan="2">water</td><td>30</td></tr>
</table>

The actual deformation in a crushing test is defined by the following dependency:

$$\varphi = \ln\left(\frac{h_1}{h_0}\right) \tag{6.1}$$

where:

h_0 – initial height of specimen;
h_1 – final height of specimen.

A simulation of thermo mechanical treatment encompassing four stages of pressing of cylindrical specimens was undertaken with the actual deformation of 4 × 0.23. The maximum actual deformation was about 1.8. A pattern and conditions of four-stage pressing are shown in Figure 6.6 (variants IV, V i VI) and in Table 6.7.

The dependencies established with the Gleeble 3800 simulator were used to design and perform thermo mechanical treatment in the selected

Table 6.7 Hot thermo mechanical treatment conditions (Gleeble 3800 simulator) consisting of four stages of hot pressing of cylindrical specimens dimensioned Ø 10 × 12 mm.

<table>
<tr><th colspan="7">Heating</th></tr>
<tr><td colspan="3">Heating temperature,°C</td><td>Heating rate, °C/s</td><td colspan="2">Heating time, s</td><td>Colling rate to deformation temperature,°C/s</td></tr>
<tr><td colspan="3">1100</td><td>3</td><td colspan="2">30</td><td>–</td></tr>
<tr><th colspan="7">Deformation</th></tr>
<tr><td colspan="2">No.</td><td>T_D,°C</td><td>ε</td><td>$\dot{\varepsilon}$, s^{-1}</td><td>Cooling rate,°C/s</td><td>Heating time, s</td></tr>
<tr><td colspan="2">I</td><td>1100</td><td>0.23</td><td>7</td><td>5</td><td>10</td></tr>
<tr><td colspan="2">II</td><td>1050</td><td>0.23</td><td>8</td><td>10</td><td>10</td></tr>
<tr><td colspan="2">III</td><td>950</td><td>0.23</td><td>9</td><td>14</td><td>7</td></tr>
<tr><td colspan="2">IV</td><td>850</td><td>0.23</td><td>10</td><td>–</td><td>–</td></tr>
<tr><th colspan="7">Cooling</th></tr>
<tr><td>No.</td><td colspan="3">Temperature range,°C</td><td colspan="2">Cooling medium</td><td>Heating time prior to cooling, s</td></tr>
<tr><td>IV</td><td colspan="3" rowspan="3">850 → room temperature</td><td colspan="2">water</td><td>0</td></tr>
<tr><td>V</td><td colspan="2">air</td><td>0</td></tr>
<tr><td>VI</td><td colspan="2">water</td><td>30</td></tr>
</table>

optimum conditions of hot plastic working in terms of the obtained strength properties. Hot rolling was carried out with a line for semi-industrial simulation of rolling processes.

Three sheets with the dimensions of 5 × 185 × 600 mm were subjected to hot rolling in four passes within the temperature range of 1150 to 850°C to the thickness of 2–3 mm. A diagram of hot rolling is presented in Figure 6.7 (variants VII, VIII and IX).

Hot rolling conditions are presented in Table 6.8. A flow chart of the LPS line equipment assembly is illustrated in Figure 6.8.

Plastic deformation rates for particular passes were calculated according to the dependency (2) given by Ekelund.

$$\dot{\varepsilon} = \frac{2V_w}{h_1 + h_2} \cdot \sqrt{\frac{\Delta h}{R}} \tag{6.2}$$

where:

V_w – peripheral speed of rollers equal to 0.74 m/s,
h_1, h_2 – thickness of specimens before and after the pass,
Δh – absolute draft,
R – radius of cylinders of 0.275 m.

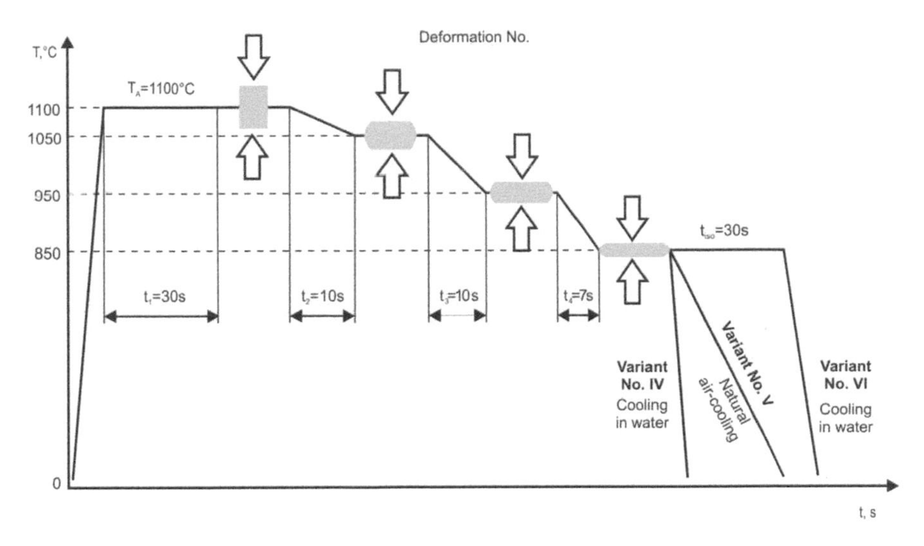

Figure 6.7 Pattern of four-stage thermomechanical treatment of cylindrical specimens of the investigated steels subject to plastic working in the Gleeble 3800 simulator and of hot rolling on semi-industrial simulation of rolling processes, T_A: austenitisation temperature, t_{iso}: isothermal heating time at the temperature of the last working of 850°C.

Table 6.8 Rolling programme of test sections of steel.

Pass no.	Plastic working temperature,°C	Thickness before pass, mm	Thickness after pass, mm	Absolute degree of deformation, mm	Actual deformation
1	1100	5	4	0.1	0.23
2	1050	4	3.2	0.8	0.23
3	950	3.2	2.55	0.65	0.23
4	850	2.55	2	0.55	0.23

It was agreed, based on the formula that plastic deformation rates for subsequent passes are, respectively: 9.5, 10, 10.3 and 10.1 s^{-1} and these are the similar values as those applied for plastically deformed specimens during tests in the Gleeble simulator.

Another important stage of tests are static tensile strength tests and dynamic rupture tests which are an important point for checking whether the research assumptions and actions described above are correct. Apart from the effect of thermo mechanical treatment and the effect of deformation rate on the strength properties, the impact of deformation temperature in the range of –80 to 200°C was also scrutinised. A static tensile test was carried out with the Zwick Z/100 tensile testing machine. A deformation rate of the

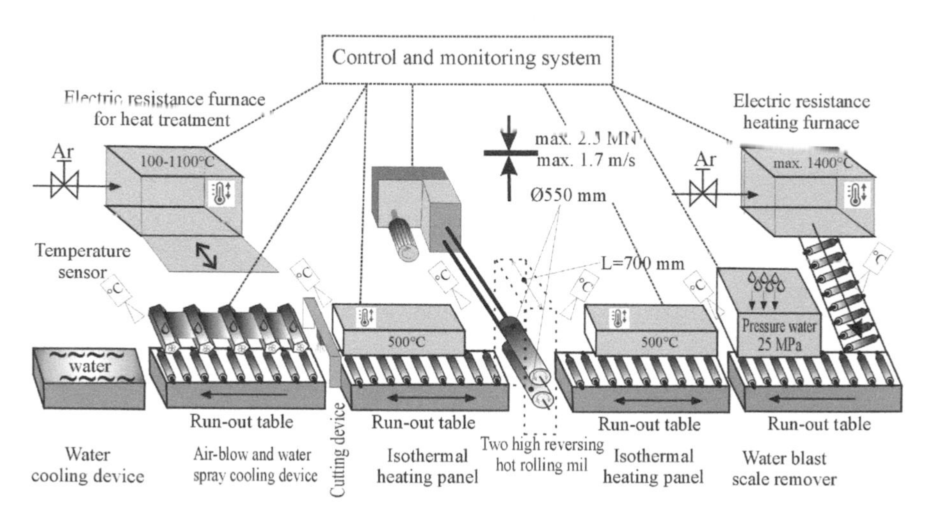

Figure 6.8 Arrangement diagram of hot rolling and thermomechanical treatment devices and auxiliary equipment.

tested specimen was 0.01 s^{-1}. The test was carried out at room temperature using an analogue extension meter fitted directly to a flat specimen. The specimens after thermo mechanical treatment were subjected to a tensile test at room temperature to a predefined elongation of 5, 10, 20, 30% and until rupture to determine the progress of a martensitic transformation or twinning connected with plastic deformation. The tests of mechanical properties at reduced and elevated temperature were carried out with a tensile strength machine MTS Insight 10 kN. The temperature of the stretched material is measured with an extension metre. The tests were carried out at an elevated temperature of 80, 100, 150 and 200°C, and at a reduced temperature of –25 and –70°C. The tests of mechanical properties in a dynamic tensile test were made with a rotary hammer RSO from WPM Lipsk. The specimens for the tests of dynamic properties were made from the investigated steels subjected to thermo mechanical treatment. The investigations of dynamic properties were performed at a deformation rate of 250, 500 and 1000 s^{-1}. The manner of performing a dynamic tensile test was described in (Dobrzański et al. 2016b, Niechajowicz and Tobota 2008). A general view of a stand for a dynamic tensile test with a single-pin dynamometer together with a fitted specimen is given in Figure 6.9.

The values of agreed stress and relative deformation in a specimen according to the one-dimensional theory of wave motion in a measuring pin were determined by means of the following formulas (6.3), (6.4):

$$\sigma_n(t) = \frac{F(t)}{A_0} = \frac{A_{bar}}{A_0} E_{bar} \varepsilon_g \left(t + \frac{a}{c} \right) \tag{6.3}$$

$$\varepsilon_n(t) = \frac{1}{L_0}\int_0^t \left[V_A(\tau) - c\varepsilon_g\left(\tau + \frac{a}{c}\right)\right] d\tau \tag{6.4}$$

where:

E_{bar} – Young's modulus of pin,
A_{bar} – cross section of pin,
c – velocity of elastic wave in measuring pin,
L_0 – gauge length of specimen,
A_0 – cross section of specimen,
a – distance of glued extensometers from the end of measuring pin,
ε_g – extensometer sector deformation,
t – time.

Specimens with the shape and dimensions as in Figure 6.10 were used in a dynamic tensile test.

Heat treatment was also performed for comparative purposes, which—as it appears—is not very relevant technically, hence it was not included in the test results presented in this chapter.

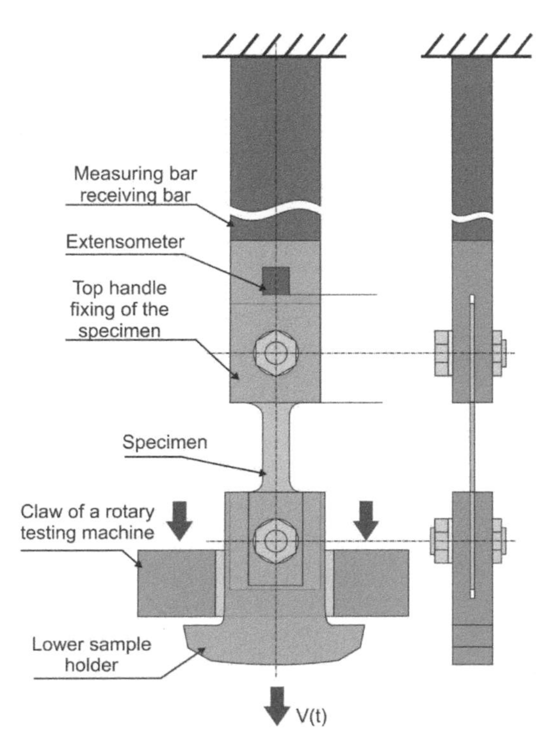

Figure 6.9 General view of stand for dynamic tensile test with single-pin dynamometer together with a fitted specimen for dynamic tensile test.

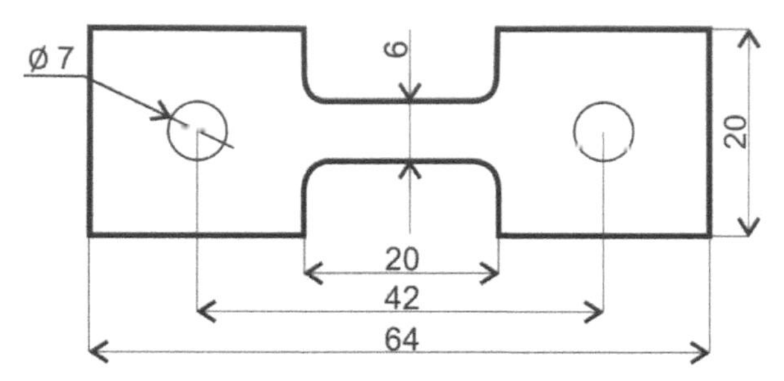

Figure 6.10 Shapes and dimensions of specimens used for dynamic stretching.

The material obtained in the final stage of the experimental technological process and at its particular stages was subject to an extensive structural assessment, using also a high resolution transmission electron microscope (HRTEM) together with visualisation and analysis of structural mechanisms decisive for the properties of the investigated materials directly during stretching in a microscope chamber using an adapter for micro scale deformations of thin foils in HRTEM. A unique body of cognitive information relating to steels and their hardening mechanisms is obtained by evaluating structural mechanisms at a nano scale. The investigations in this chapter were carried out using, in particular, a FEI TITAN 80–300 high resolution transmission electron microscope together with GIF, a ZEISS Supra 35 scanning microscope with EDS, WDS and EBSD, a LEICA metallographic microscope with an image analysis system, a Zeiss LSM 5 confocal microscope, an XE-100 atomic force microscope by Park System, a PANANALITICAL X-ray diffractometer with an adapter for texture examinations.

The FEI TITAN 80–300 high resolution transmission electron microscope is now fitted with a unique adapter for testing stresses and material deformations *in situ* to the nanometric and micrometric scale with the simultaneous testing of material response and the generated transformations connected with, most of all, mechanical deformation of material. A specimen preparation technique is however required for the described specialised holder for a specific material (in this case for high manganese steels of TRIP and TWIP type) which must feature a high precision of workmanship. The preparation of thin foils for tests using a specialised tensiometric holder by Gatan enables to set an *in situ* deformation of the investigated high manganese steels also ensures good quality areas for observation of structural changes in the studied material. The Gatan company, by offering the device, also proposes a technical solution for the specimen sample, which is one of many solutions used for such type of holders. Apart from the specimen's shape and dimensions, one of the most difficult tasks is to achieve the very shape of the specimen while maintaining its appropriate thickness in a central place allowing it to screen the tested material with a beam of electrons. The proposal presented in Figure 6.11 was one of the specimen shapes possible for use and execution.

As accurate dimensions of the ready specimen have to be obtained and as it needs to be secured against accidental damage, a special device was designed enabling to set the plate and ensuring the guiding of drills making small diameter holes in hard high manganese steels with the hardness of above 400 HV (Figure 6.12). Specimen preparation time was greatly shortened by using the presented device and the proper dimensional accuracy of specimens was also ensured. The device guides the drills and this protects against decalibration of opening dimensions and opening axis spacing. A technology of preparing thin foils for the mentioned tests of this type of steels was established and it was verified practically with several dozens of specimens for investigations in HRTEM. No preferences or differences in the degree of difficulty of preparation of TRIP or TWIP steels were found.

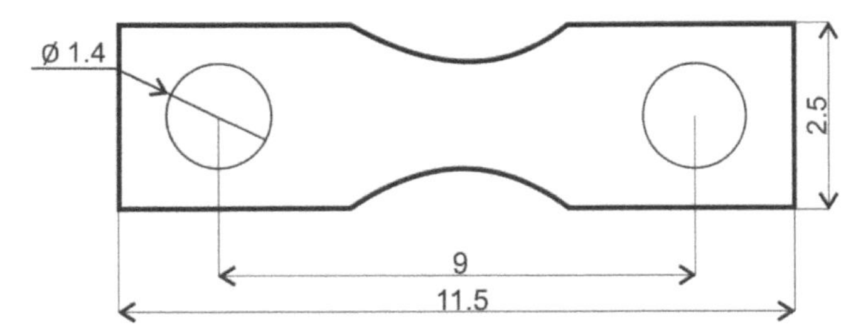

Figure 6.11 Sketch of shape and dimensions of alternative specimen for tensiometric holder.

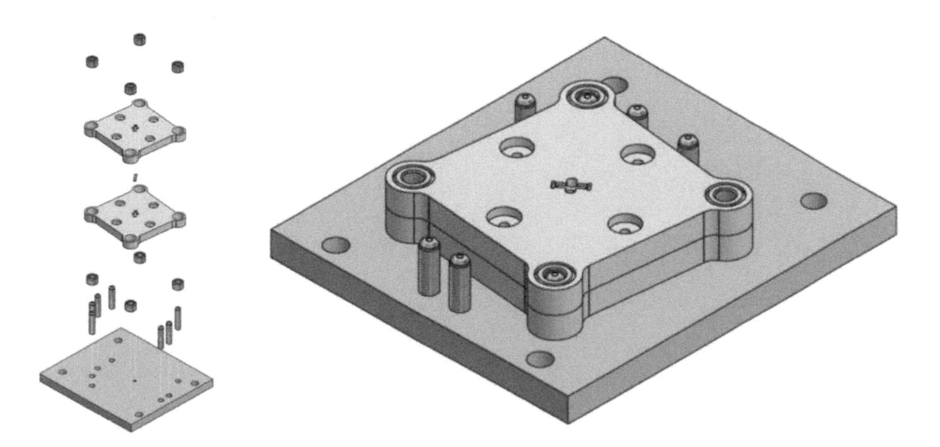

Figure 6.12 Drawing of the final version of device for making openings in the specimen.

6.4 Description of Outcomes of the Present Research of High Manganese Steels Exhibiting TWIP Effects

The newly developed high manganese X8MnSiAlNbTi25-1-3 and X73MnSiAlNbTi25-1-3 steels in the initial state after open die forging, an ingot on flat sections, with the thickness of 20 mm exhibit a structure of homogenous austenite with numerous twins (Figure 6.13) with the similar average grain size of, respectively, 62 μm and 58 μm. Precipitations of the (Ti,Nb)C_xN_y and (Ti,Nb)C-type were identified in both the steels, which were effectively blocking the growth of austenite grains, especially during thermo mechanical treatment of the studied steels.

Limit strain and yield stress are the basic processes characterising the susceptibility of material to forming. In order to determine the range of yield stress values for the investigated steels, continuous hot pressing tests were performed at the temperature of 850, 950 and 1050°C and the strain rates of 0.1, 1 and 10 s^{-1}, with the Gleeble 3800 simulator. The stress-strain curves of the investigated steels obtained as a result of plastometric tests are shown in Figure 6.14.

The range of yield stresses spans between 136 and 357 MPa for X8MnSiAlNbTi25-1-3 steels and between 130 and 458 MPa for X73MnSiAlNbTi25-1-3 steels. An increase in yield stress together either with a reduction in pressing temperature or with an increase in the rate of plastic strain is accompanied by a shift towards higher values of deformation corresponding to the maximum yield stress value. As the rate of plastic strain is increasing at a constant temperature of deformation, the yield stress rises significantly. And when the temperature of hot deformation rises, the maximum yield stress value moves on stress–strain diagrams towards smaller deformation values. For each test temperature and for all strain rates for X8MnSiAlNbTi25-1-3 steel, the maximum yield stress is not clearly visible. It is best seen for the strain rate of 1 s^{-1}, where a flow stress is

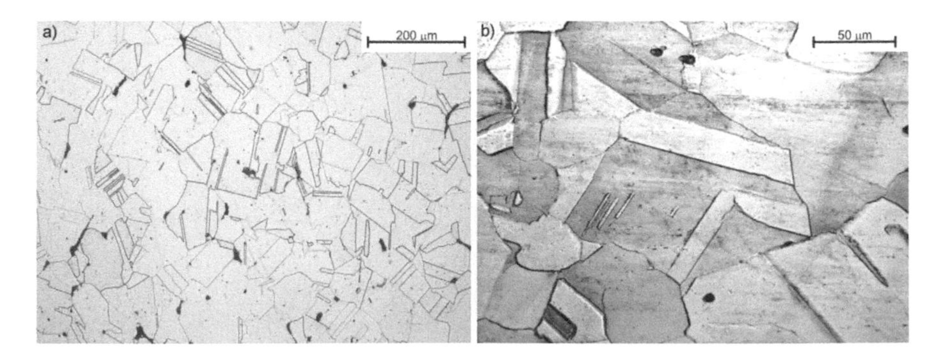

Figure 6.13 Structure in the initial state after open die forging (a) X8MnSiAlNbTi25-1-3 steel, (b) X73MnSiAlNbTi25-1-3 steel.

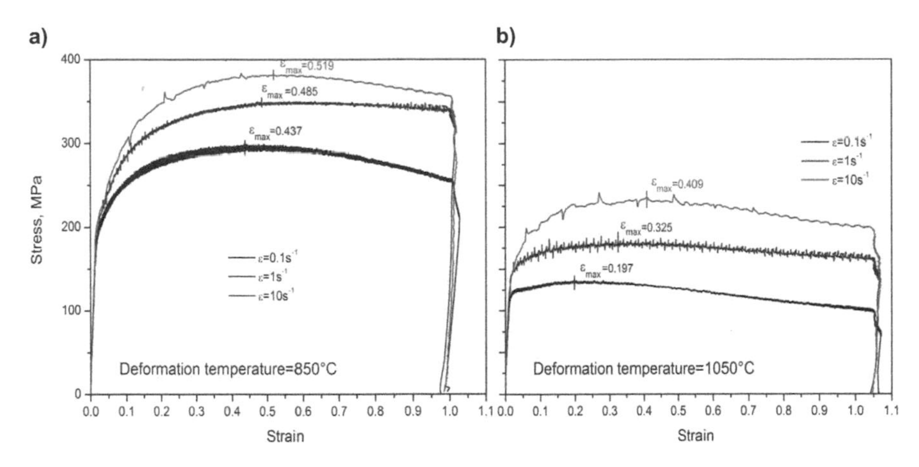

Figure 6.14 The strain-stress curves of continuous pressing tests of the high manganese austenitic X8MnSiAlNbTi25-1-3 TWIP type steel in the temperature of 850 (a) and 1050°C (b) with 0.1, 1 and 10 s^{-1} strain rate obtained using Gleeble 3800 thermo-mechanical simulator.

almost constant, which is a result of the state of equilibrium between strain hardening and dynamic processes that eliminates the effects of deformation, i.e., dynamic recovery or dynamic recrystallisation. The effect of dynamic recrystallisation occurs to the highest degree in X73MnSiAlNbTi25-1-3 steel, strongest at the cooling rate of 10 s^{-1}, and it intensifies as temperature rises. Table 6.9 shows the technological properties of material which are satisfactorily describing the material susceptibility to the planned process.

The stress-strain curves of the investigated plastically deformed steels according to eight stages of hot pressing of perpendicular prism-shaped specimens are shown in Figure 6.15, and the plastic strain conditions were selected to enable the hot rolling of the investigated steels in industrial production. An actual deformation of 0.4 in the first deformation at the temperature of 1050°C, and 0.3, 0.25 and 0.2 in the subsequent stages of deformation for the investigated steels was applied in eight stages of the simulated thermo mechanical treatment. Dynamic recrystallisation and meta dynamic recrystallisation then occurs in intervals between individual deformations, as indicated by the maximum stress values for the curves σ-ε—especially in the temperature of 1050°C. Stress values in the first stage of deformation for the both investigated steels are comparable to the stress values obtained in one-stage hot pressing at the temperature of 1050°C with a strain rate of 10 s^{-1}. The maximum stress value in the 1st stage of deformation is similar regardless of the steel grade and is 210 MPa for X8MnSiAlNbTi25-1-3 steel and 225 MPa for X73MnSiAlNbTi25-1-3 steel. In the last, 8th stage of hot plastic deformation for the both investigated steels, the maximum stress is almost twice higher than in the 1st stage and is, respectively, 425 and 475 MPa. As a result of lowering the deformation

Table 6.9 The effect of deformation temperature and rate on yield stress values σ of investigated steels.

Steel	Deformation temperature,°C	$\dot{\varepsilon}$, s^{-1}	ε_{max}	σ, MPa
X8MnSiAlNbTi25-1-3	850	0.1	0.488	285
		1	0.574	327
		10	0.612	357
	950	0.1	0.412	201
		1	0.449	246
		10	0.46	294
	1050	0.1	0.251	136
		1	0.326	180
		10	0.364	230
X73MnSiAlNbTi25-1-3	850	0.1	0.330	331
		1	0.488	395
		10	0.486	458
	950	0.1	0.222	220
		1	0.281	280
		10	0.310	360
	1050	0.1	0.200	130
		1	0.189	194
		10	0.264	263

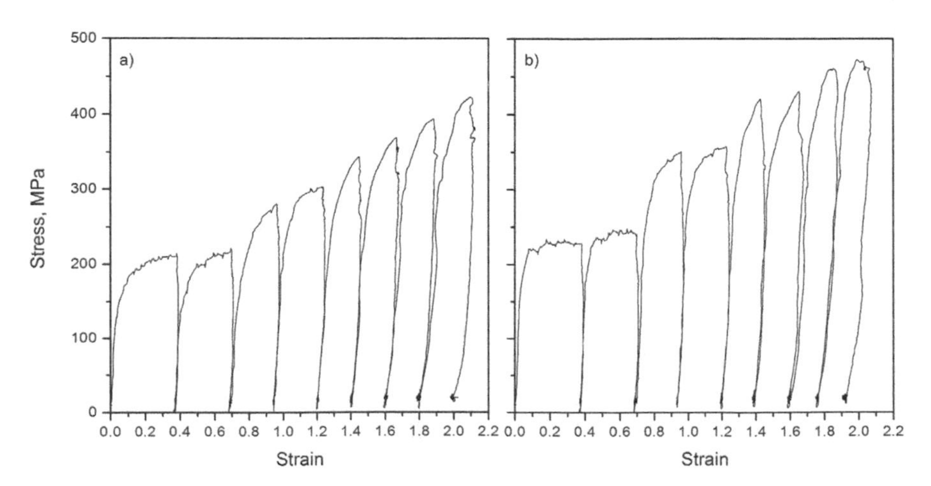

Figure 6.15 Stress-strain curves of eight stages of hot pressing in conditions similar to plane state of strain of (a) X8MnSiAlNbTi25-1-3 steel; (b) X73MnSiAlNbTi25-1-3 steel.

values in the successive stage of thermo mechanical treatment and as a result of lowering the deformation temperature, only the metadynamic or static recrystallisation taking place between particular deformations are the processes eliminating the consequences of strain hardening (Figure 6.15). The analogous results were obtained as a result of four stages of hot pressing of axially symmetric specimens, simulating the final hot rolling passes using the Gleeble simulator (Figure 6.16).

An X-ray qualitative phase analysis carried out has doubtlessly confirmed the presence of austenite in the structure of the investigated X8MnSiAlNbTi25-1-3 and X73MnSiAlNbTi25-1-3 steels, both, in the state after thermo mechanical treatment in semi-industrial conditions, and after the simulated treatment comprised of four stages (Figure 6.17).

Metallographic examinations reveal that dynamic recrystallisation occurring in the first and second stage of hot plastic deformation and static and meta dynamic recrystallisation occurring in intervals between successive deformations is the primary process which is eliminating the consequences of strain hardening (Figure 6.18). The application of free air cooling after hot plastic working (variant B) does not cause substantial changes in the average size of grains in the investigated steels as compared to the variant A. As a result of isothermal heating after the last plastic deformation in the temperature of 850°C for 30 s according to the variant C of working, a fine-crystalline structure of statically or metadynamically recrystallised grains is achieved (Figure 6.18).

The results of tests of properties in a static tensile test of X8MnSiAlNbTi25-1-3 and X73MnSiAlNbTi25-1-3 steel subjected to thermo mechanical treatment consisting of semi-industrial rolling with cooling

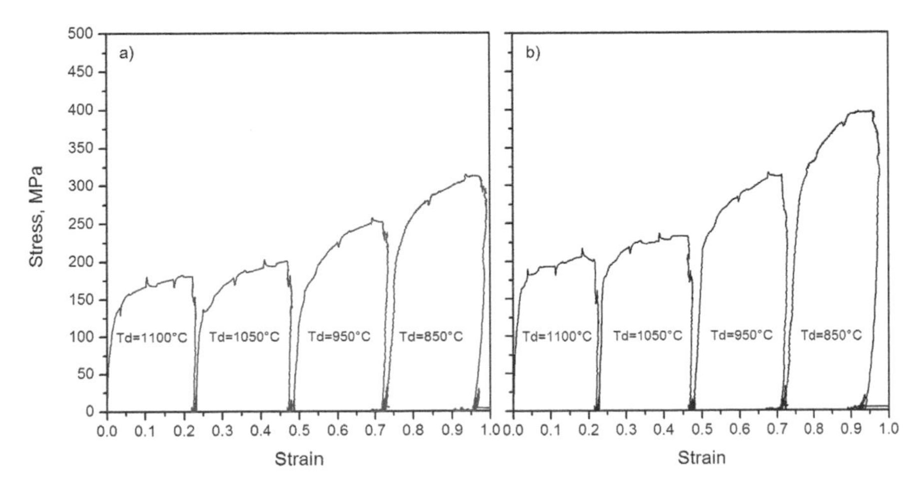

Figure 6.16 Stress-strain curves of four stages of hot pressing with actual deformation 4 x 0.23 of (a) X8MnSiAlNbTi25-1-3 steel; (b) X73MnSiAlNbTi25-1-3 steel.

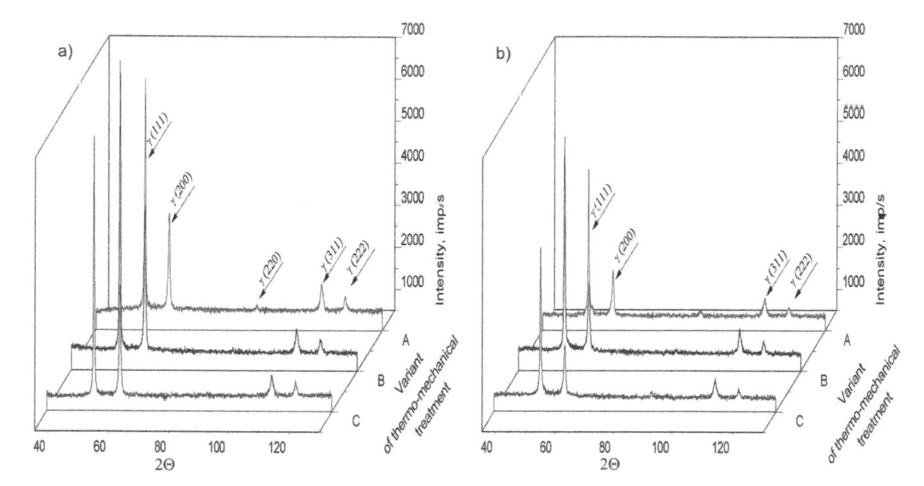

Figure 6.17 X-ray diffraction patterns of (a) X8MnSiAlNbTi25-1-3 steel after hot rolling in semi-industrial conditions and (b) X73MnSiAlNbTi25-1-3 steel subjected to four stages of hot pressing with deformation $\varepsilon = 0.23$ on the simulator; with water cooling (A), with natural air cooling (B), heated isothermally for 30 s at the temperature of final deformation and cooled in water (C).

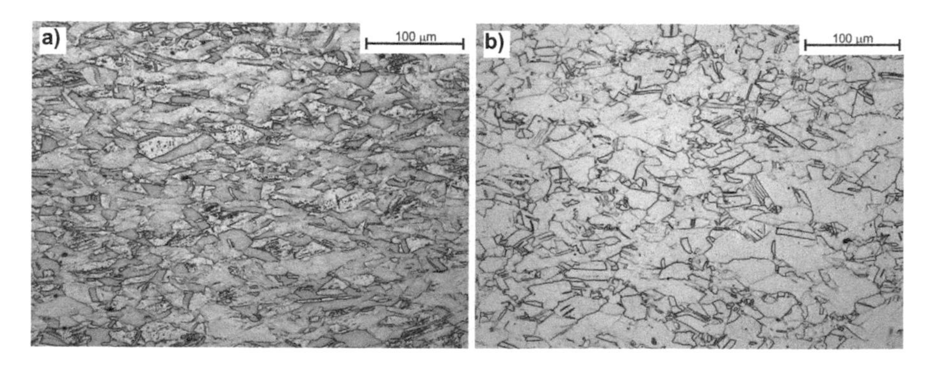

Figure 6.18 Structure of X8MnSiAlNbTi25-1-3 steel, (a) after four stages of hot plastic deformation with the Gleeble 3800 metallurgic simulator with cooling in water (cooling variant A), (b) after eight stages of hot plastic deformation with the Gleeble 3800 metallurgic simulator heated isothermally for 30 s at the temperature of completion of plastic deformation and cooling in water (C cooling variant).

according to 3 variants: A, B and C and 2 heat treatment variants consisting of super saturation at the temperature of 900 and 1000°C (respectively, variants E and F) for 60 minutes with cooling in water indicate that X73MnSiAlNbTi25-1-3 steel exhibits higher yield strength $R_{p0,2}$ by 9 to 25% depending on the variant analysed (Table 6.10). The investigated steels subjected to thermo mechanical treatment indicate highly differentiated $R_{p0,2}$ values (230–602 MPa). Both, the strain energy per unit volume as well as a

Table 6.10 Variations in values of mechanical properties of the investigated steels after hot rolling in semi-industrial conditions with cooling in water (A) in the function of test temperature.

Steel	Test temperature,°C	–70	–25	20	80	100	150	200
X8MnSiAlNbTi25-1-3	Tensile strength R_m, MPa	693	636	559	518	503	497	496
	Yield strength $R_{p0,2}$, MPa	557	509	475	456	443	435	419
	Relative elongation A, %	23	49	45	30	25	19	18
	Elongation at highest force A_g, %	19	45	28	25	16	13	13
	Strain energy per unit volume E_{ZP}, MJ/m³	148	293	239	153	124	88	81
	$R_{p0,2}/R_m$	0,8	0,8	0,85	0,88	0,88	0,88	0,84
X73MnSiAlNbTi25-1-3	Tensile strength R_m, MPa	726	702	688	608	558	612	561
	Yield strength $R_{p0,2}$, MPa	633	612	597	519	485	523	478
	Relative elongation A, %	23	38	35	28	25	16	12
	Elongation at highest force A_g, %	17	44	43	34	32	20	15
	Strain energy per unit volume E_{ZP}, MJ/m³	171	339	273	324	290	110	229
	$R_{p0,2}/R_m$	0,87	0,87	0,87	0,85	0,87	0,85	0,85

yield point of the investigated steels are 10–20% higher as a consequence of thermo mechanical treatment than after ordinary heat treatment where steel is heated again to the super saturation temperature and then cooled in water, in complete detachment from the applied plastic working (Table 6.11).

The obtained results of strength tests for both the steels cooled according to variant A, indicate a high effect of room temperature and a temperature of –70, –25, 80, 100, 150 and 200°C on their properties, especially on the yield strength $R_{p0,2}$. The highest values were obtained at the temperature of –70°C, and lowest at the temperature of 200°C for the both steels, respectively, 557 and 419 MPa for X8MnSiAlNbTi25-1-3 and 633 and 478 MPa for X73MnSiAlNbTi25-1-3. The maximum tensile strength R_m of the investigated steels at room temperature is 587 MPa for X8MnSiAlNbTi25-1-3 steel (variant B) and 698 MPa for X73MnSiAlNbTi25-1-3 steel (variant C). Analogously as for the yield point, tensile strength is also subjected to significant changes along with changes in the test temperature. The maximum values were obtained at the temperature of –70°C, of 693 MPa for X8MnSiAlNbTi25-1-3 steel and 726 MPa for X73MnSiAlNbTi25-1-3 steel and lowest at the temperature of 200°C, respectively, 496 and 561 MPa.

Table 6.11 Variations in values of mechanical properties of the investigated steels at room temperature according to thermo mechanical treatment and heat treatment variant.

Variant of treatment	Strain rate s^{-1}	Tensile strength Rm, MPa		Yield strength $R_{p0,2}$, MPa		Relative elongation A, %		Strain energy per unit volume E_{ZP}, MJ/m^3	
		steel Φ	steel Ω	steel Φ	steel Ω	steel Φ	steel Ω	steel Φ	steel Ω
A	0.01	559	688	475	597	45	43	239	273
	250	709	846	717	762	56	50	486	545
	500	790	898	768	784	44	37	425	528
	1000	909	1098	840	965	48	39	432	574
B	0.01	587	667	513	583	25	41	190	254
C		572	698	503	602	43	39	235	301
D		554	630	376	413	38	37	192	208
E		498	577	230	293	48	40	205	218

Steel Φ: X8MnSiAlNbTi25-1-3; Steel Ω: X73MnSiAlNbTi25-1-3
D – supersaturation from 900°C for 1 h
E – supersaturation from 1000°C for 1 h

A higher concentration of carbon has an advantageous effect on the strength properties of the investigated steels, however, it has an adverse effect on plastic properties. X8MnSiAlNbTi25-1-3 steel shows the highest relative elongation of A = 44.5% after thermo mechanical treatment with cooling according to variant A. A slightly lower elongation value of 42.5% corresponds to the C variant of cooling. The highest tensile strength value after cooling according to variant B corresponds to lower elongation of approx. 24%. Differences in the elongation value for X73MnSiAlNbTi25-1-3 steel are smaller and reach the value of 39.3% (variant C) to 42.5% (variant A). In case of the both examined steels, the highest elongation occurs at the temperature of –25°C and at room temperature (20°C) and, for X8MnSiAlNbTi25-1-3 steel, it is, respectively, approx. 49 and 44%, and for X73MnSiAlNbTi25-1-3 steel, approx. 43%. The highest elongation of about 18% takes place at 200°C for X8MnSiAlNbTi25-1-3 steel and about 16% at –70 and 200°C for X73MnSiAlNbTi25-1-3 steel.

If the rate of plastic strain at room temperature is increased by, respectively, 250, 500 and 1000 s^{-1}, which corresponds to rapture rates of, respectively, 4, 7 and 14 m/s, the total elongation is increased to the maximum value of 56% for X8MnSiAlNbTi25-1-3 steel and 50% for X73MnSiAlNbTi25-1-3 steel at the strain rate of 250 s^{-1} and decreased to

15% (X8MnSiAlNbTi25-1-3 steel) and to 25% (X73MnSiAlNbTi25-1-3 steel) with a further increase in the strain rate. An increase in the plastic strain rate increases the yield point $R_{p0,2}$ of the investigated X8MnSiAlNbTi25-1-3 steel from 475 to 840 MPa, and for X73MnSiAlNbTi25-1-3 steel, from 597 to 965 MPa, which increases by 60–77% as compared to static conditions. An increase in the rate of plastic strain to 1000 s^{-1} increases tensile strength approx. from 660 to 910 MPa for X8MnSiAlNbTi25-1-3 steel and from 690 to 1100 MPa for X73MnSiAlNbTi25-1-3 steel (Table 6.11).

The strain energy per unit volume, E_{zp}, is the energy which can be accumulated in dynamic loading, for, e.g., in a road collision of vehicles, calculated as a field area under the actual stress-actual strain curve for X8MnSiAlNbTi25-1-3 and X73MnSiAlNbTi25-1-3 steel, is presented in Table 6.11, and Figure 6.19. It also shows the examples of curves of static and dynamic elongation: actual stress-actual strain with the strain energy per unit volume determined, as a field area under the elongation curve for X8MnSiAlNbTi25-1-3 steel. Depending on the test temperature and heat treatment conditions, the strain energy per unit volume values for X8MnSiAlNbTi25-1-3 steel span over the range of 81–293 MJ/m³, and 110–339 MJ/m³ forX73MnSiAlNbTi25-1-3 steel. X8MnSiAlNbTi25-1-3 steel exhibits the highest average value of strain energy per unit volume of 239 MJ/m³ at room temperature in the state after four stages of hot rolling and cooling in water (variant A), and X73MnSiAlNbTi25-1-3 steel the value of 301 MJ/m³ after isothermal heating for 30 s and cooling in water (variant C). The both investigated steels demonstrate much lower E_{zp} values at –70°C of, respectively, 148 MJ/m³ for X8MnSiAlNbTi25-1-3 steel and

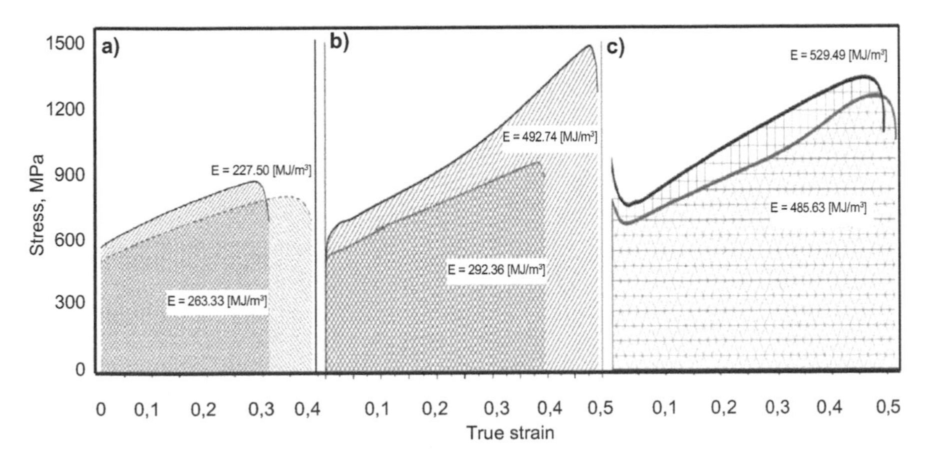

Figure 6.19 The results of the strain energy per unit volume of the high-manganese austenitic X8MnSiAlNbTi25-1-3 of the TWIP type steel (a) after static tensile tests at room temperature and after the successive cold deformation, (b) after static tensile tests in reduced temperature of deformation to –25°C and –70°C, (c) after dynamic tensile tests with strain rate 500 s^{-1} and 250 s^{-1}.

171 MJ/m^3 for X73MnSiAlNbTi25-1-3 steel, and the lowest of 81 MJ/m^3 for A steel and 150°C (110 MJ/m^3) for X73MnSiAlNbTi25-1-3 steel at an elevated temperature of 200°C. X73MnSiAlNbTi25-1-3 steel, containing more carbon, exhibits much higher resistance to drop in E_{zp} value at an elevated temperature then X8MnSiAlNbTi25-1-3 steel. The investigated steels reach the most advantageous plastic properties in the plastic deformation temperature between –25°C and room temperature. In case of plastic strain at room temperature at high rates, the strain energy per unit volume value is increased for X8MnSiAlNbTi25-1-3 steel by 80% to 431 MJ/m^3 at a strain rate of 1000 s^{-1} and by 110% to 574 MJ/m^3 for X73MnSiAlNbTi25-1-3 steel. The maximum value of strain energy per unit volume is exhibited by X73MnSiAlNbTi25-1-3 steel for the highest strain rate applied in the tests of 1000 s^{-1} (Table 6.11). X8MnSiAlNbTi25-1-3 steel shows the strain energy per unit volume lower by approx. 10% for each of the applied cold strain rates.

Figure 6.20 shows the structures of the investigated X8MnSiAlNbTi25-1-3 and X73MnSiAlNbTi25-1-3 steels in the state after hot rolling with the degree of deformation of 20%, cooling in water (A) or in air (B) and isothermal

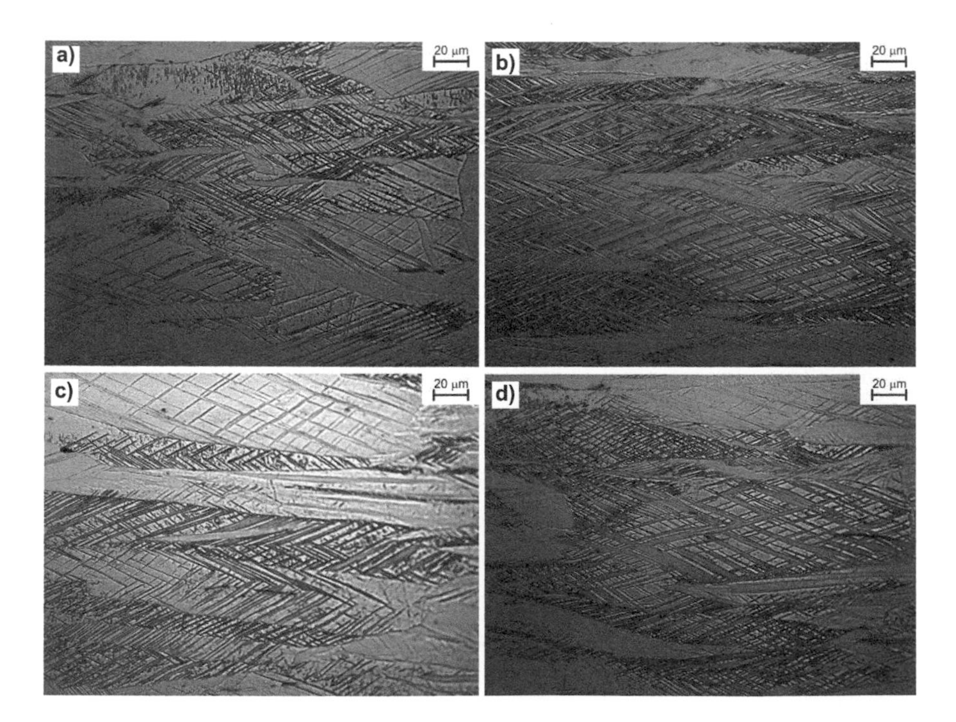

Figure 6.20 Structure of X8MnSiAlNbTi25-1-3 steel after hot rolling in semi-industrial conditions with the degree of plastic deformation of 0.2 with cooling in water (A), and then stretched statically until elongation (a) 10%; (b) 30%; heated isothermally for 30 s at the temperature of completion of deformation and cooled in water (C), and then stretched statically until elongation (c) 10% and (d) 30%; LM.

heating for 30 s and then cooling in water (C), after a static tensile test. The structure of both the steels subjected to a static tensile test to the set plastic deformation of 5, 10, 20, 30%, and until specimen rupture, is represented by austenite. In case of both the steels, even a small deformation of 5 or 10% leads to the elongation of grains in the direction of acting tensile forces. The presence of intersecting slip bands is found in austenite grains and in annealing twins. By increasing specimen elongation to 20–30%, the intersecting slip bands and deformation twins are densified in a static tensile test (Figure 6.20). The structures of the specimens deformed until rupture are characterised by austenite grains strongly elongated in the direction of stretching and by deformed austenite grains with a large density of slip bands and deformation twins.

It was found by examining thin foils in a transmission electron microscope that the structure of the newly developed high manganese austenitic X8MnSiAlNbTi25-1-3 and X73MnSiAlNbTi25-1-3 steels in the state after thermo mechanical treatment, and then after a static tensile test, consists mainly of strongly deformed austenite grains with a high density of dislocation with numerous twins with various intersecting slip systems (Figures 6.21 and 6.22).

The basic mechanism of TWinning Induced Plasticity TWIP closely linked to cold plastic deformation by the activation of twinning in the intersecting systems was confirmed with a high-resolution electron transmission microscope (Figure 6.23).

A qualitative analysis of the structure at a nanometric scale, EBSD (Electron Backscatter Diffraction Analysis), in a scanning electron microscope of X8MnSiAlNbTi25-1-3 and X73MnSiAlNbTi25-1-3 steel after

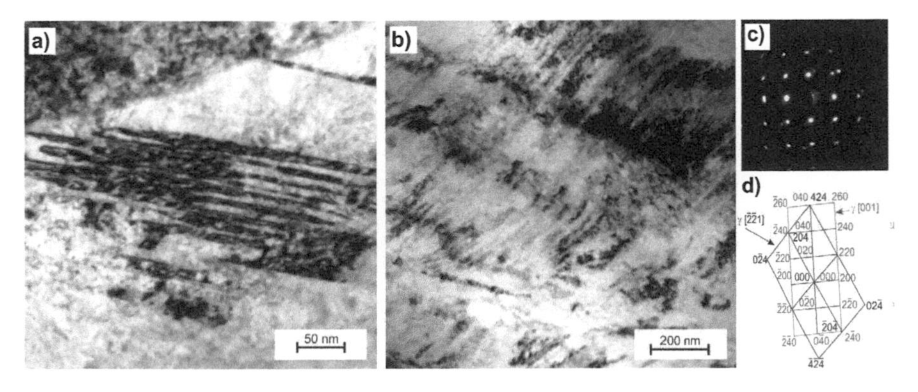

Figure 6.21 Deformation twins in the structure of thin foils after hot rolling in semi-industrial conditions with the degree of plastic deformation of 0.2 with cooling in water (A), and then stretched statically (a) from X8MnSiAlNbTi25-1-3 steel to elongation of 20%; (b) from X73MnSiAlNbTi25-1-3 steel to elongation of 5%; (c) diffraction patterns from the area as in fig. (b); (d) diffraction pattern solution; [001] Feγ: axis of crystallographic band of matrix and of its corresponding [221] Feγ: axis of crystallographic band of twins.

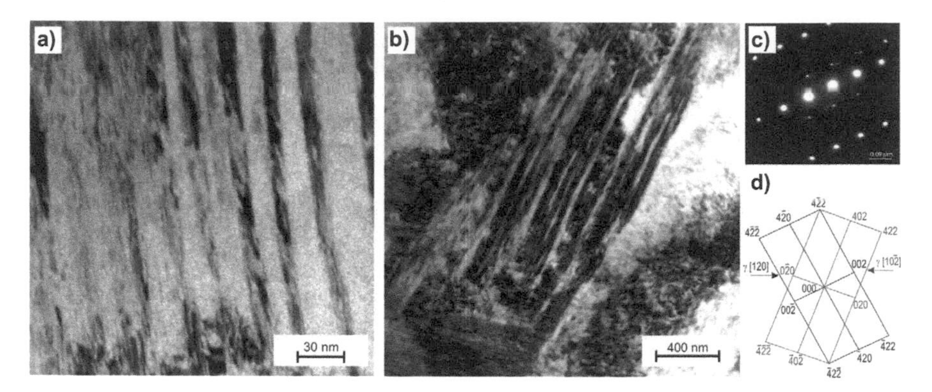

Figure 6.22 Structure of thin foils from X73MnSiAlNbTi25-1-3 steel after hot rolling in semi-industrial conditions with the degree of plastic deformation of 0.2 with cooling in water (A), and then stretched statically until rupture; (a) deformation twins; (b) deformation twins intersecting in two different systems; (c) diffraction pattern from the area as in fig. (b; d) diffraction pattern solution.

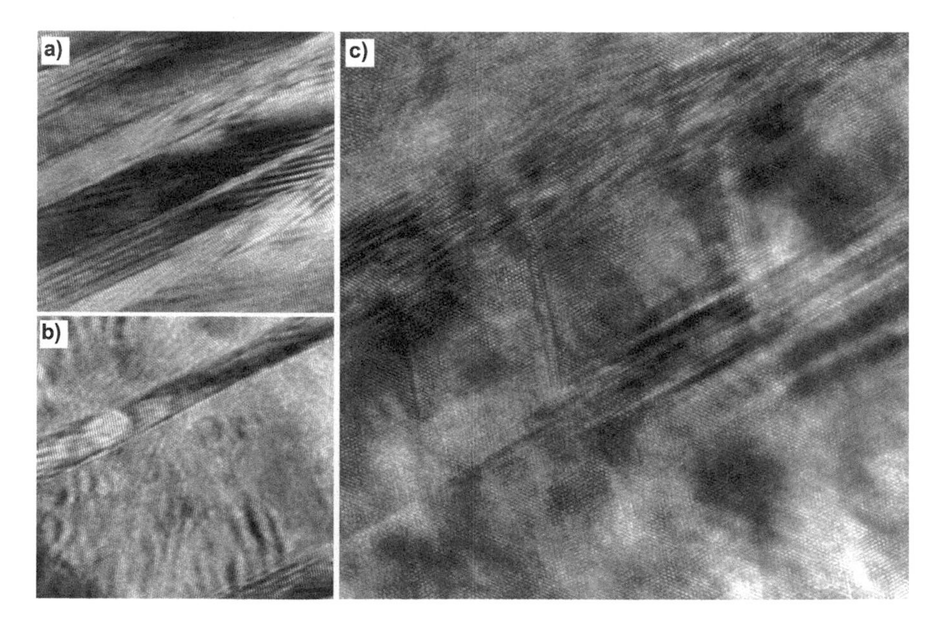

Figure 6.23 Deformation twins in structure of thin foils of X73MnSiAlNbTi25-1-3 steel after hot rolling in semi-industrial conditions with the degree of plastic deformation of 0.2 with cooling in water (A), and then stretched statically until rupture; HRTEM (a) (b) parallel systems; (c) systems of intersecting twins in two different systems.

hot rolling in semi-industrial conditions with cooling in water (variant A) stretched statically and dynamically at room temperature (20°C) was made. This analysis confirms the presence of twins in different intersecting systems and of mutually intersecting slip bands and deformation bands. It is shown in the produced crystallographic orientation maps and maps of crystallographic misorientation angles between grains, for instance in Figures 6.24 and 6.25.

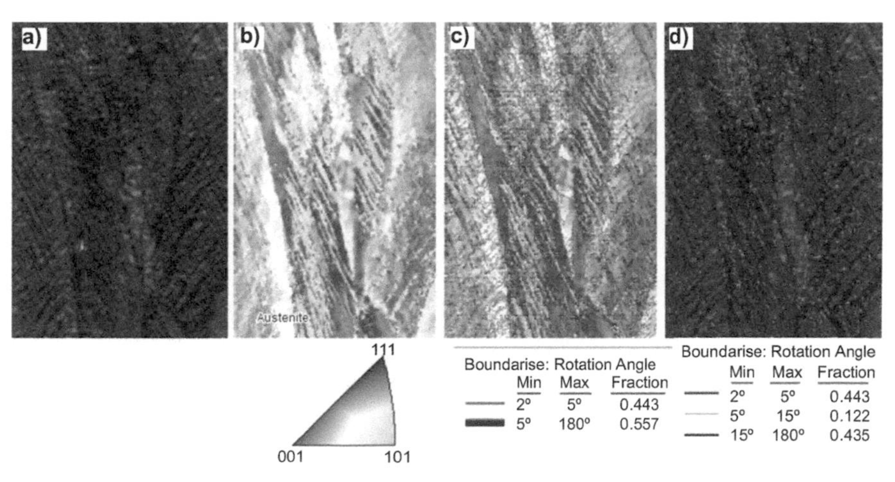

Figure 6.24 (a) Structure of X73MnSiAlNbTi25-1-3 steel after hot rolling in semi-industrial conditions with cooling in water (variant A) stretched dynamically at room temperature (20°C) with a rate of 500 s^{-1}; (b) crystallographic orientation map of structure from fig. (a; c) crystallographic orientation map with misorientation angles marked in the range of 2–5° (54%), 5–180° (46%); (d) map of crystallographic misorientation angles between grains, angles are marked in the range of 2–5° (53.6%), 5–15° (17.7%) and 15–180° (28.7%).

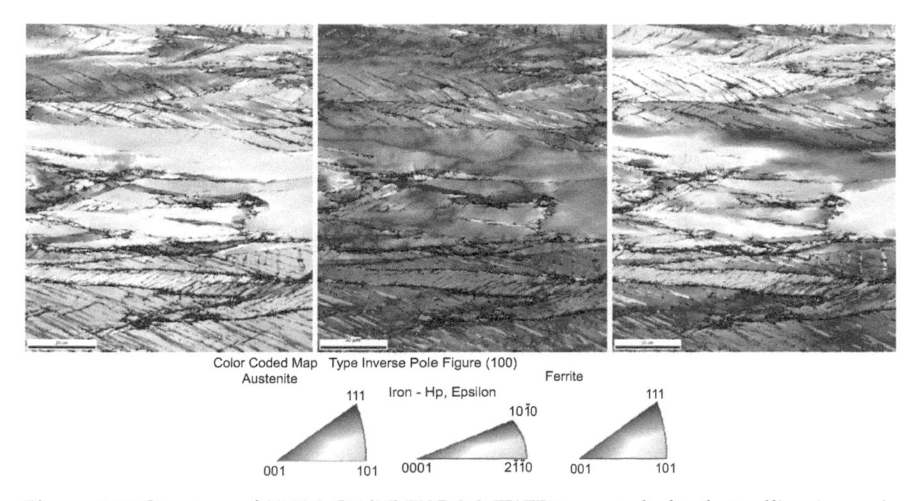

Figure 6.25 Structure of X13MnSiAlNbTi25-3-3 TWIP-type steel after hot rolling in semi-industrial conditions with cooling in water (variant B) stretched dynamically at room temperature (20°C) with a rate of 1000 s^{-1}, maps made by EBSD technique - IPF of maps in the specimen stretching direction.

Figures 6.26 and 6.27 present unique results of the effect of stretching in real time in a TEM chamber of thin foils made of X13MnSiAlNbTi25-3-3 steel in the state after forging. The next images in the dark field (Figure 6.26) and bright field (Figure 6.27) present the successive stages of the structure accompanying the growing plastic deformation, manifested by the appearance of a twin limit and the constantly rising dislocation density.

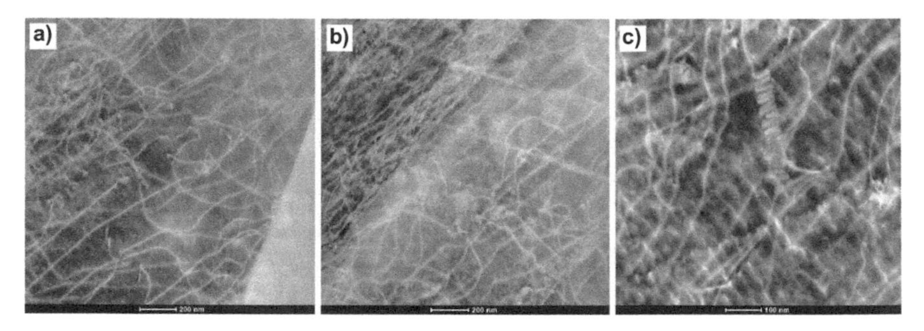

Figure 6.26 Structure of X13MnSiAlNbTi25-3-3 steel in the state after forging, an image produced as a result of cold plastic deformation in TEM chamber – dark field; (a–c) successive stages of plastic deformation in successive stages of specimen stretching, structures show a limit of twins and dislocations with higher and higher density.

6.5 Description of Results of Own Investigations into High Manganese Steels Exhibiting TRIP Effect

Newly developed high manganese X11MnSiAl17-1-3 steel exhibits after casting a homogenous austenite structure with numerous annealing twins (Figure 6.28). The average size of austenite grains is 150–200 μm. A homogenous austenitic structure was confirmed with an X-ray phase analysis.

The investigated steel was subjected to heat and thermo mechanical treatment and plastomeric tests were made with the Gleeble 3800 simulator. The continuous pressing of the investigated steel was performed for temperatures ranging between 850 and 1050°C for different strain rates $\dot{\varepsilon}$ of 0.1, 1 or 10 s^{-1} (Figure 6.29a). The range of yield stresses σ_p spans between 120 and 360 MPa. As the yield stress growes and test temperature falls, there is shift of deformation ε_{max} towards higher values. Increased plastic working velocity at a constant deformation temperature increases the yield stress (Figure 6.29b).

Figure 6.30 shows stress-strain curves recorded while deforming the investigated X11MnSiAl17-1-3 steel in treatment consisting of eight and four stages of hot pressing. No large differences exist between the particular

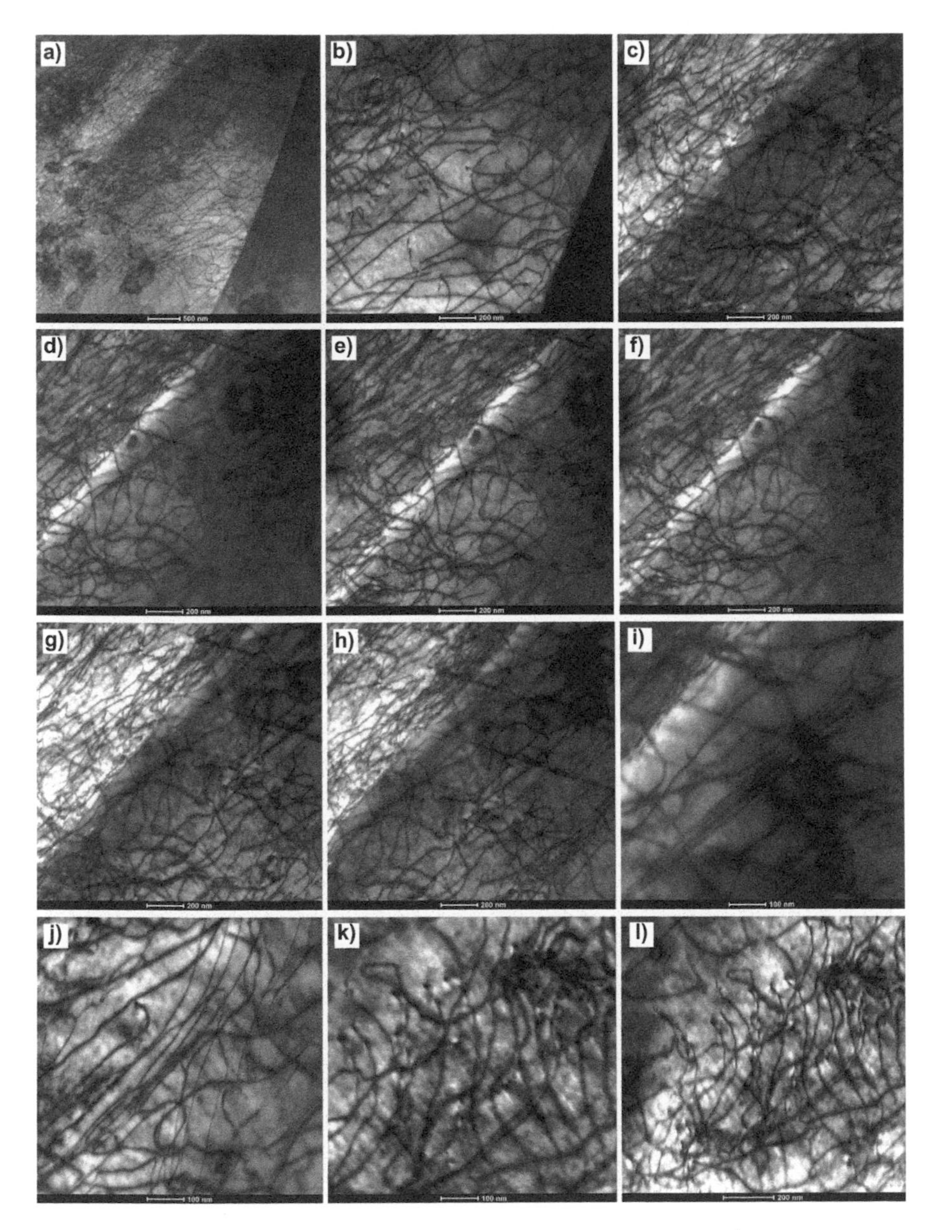

Figure 6.27 Structure of X13MnSiAlNbTi25-3-3 steel in the state after forging, an image produced as a result of cold plastic deformation in TEM chamber – bright field; (a) initial state, (b–l) successive stages of plastic deformation in successive stages of specimen stretching, structures show a limit of twins and dislocations with higher and higher density.

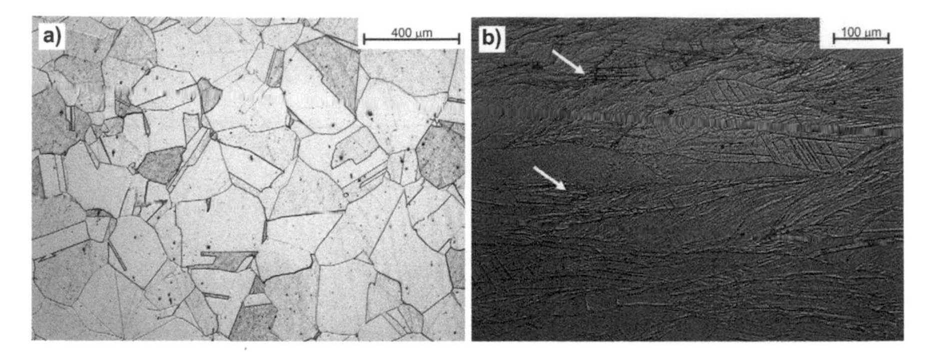

Figure 6.28 Austenitic structure of X11MnSiAl17-1-3 steel (a) in the state after casting (light microscope); (b) after hot rolling in four passes in semi-industrial conditions, with degree of deformation of 4 x 0.23 and cooling in water after isothermal heating for 30 s at the temperature of the last working, the arrows indicate recrystallised grains; LM.

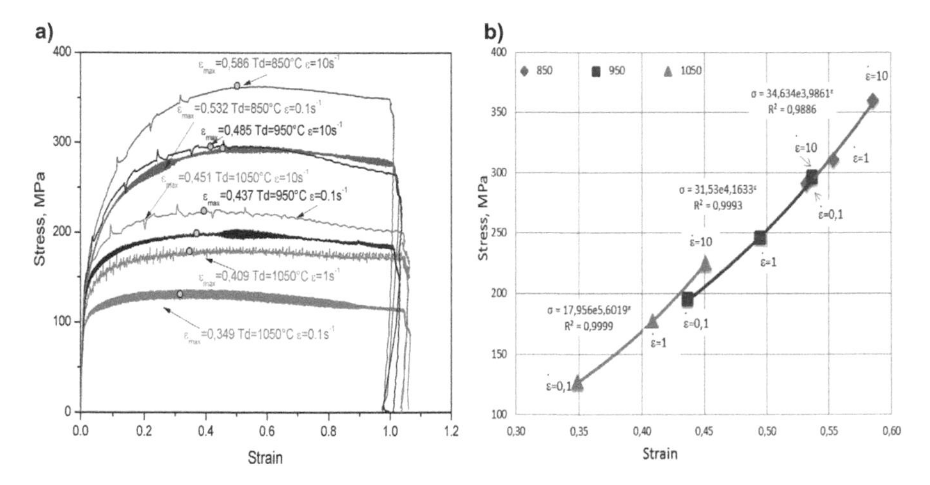

Figure 6.29 (a) Effect of strain rate on the shape of stress-strain curves of X11MnSiAl17-1-3 steel with strain rate of $\dot{\varepsilon}=0.1$, 1 and 10 s^{-1}, deformation temperature: 850, 950 and 1050°C; (b) progression of the function determining the dependency of the value from which dynamic recrystallisation begins (ε_{max}) on plastic strain rate (ε) and deformation temperature (correlation coefficient of $R^2=0.99$).

curves corresponding to three variants of thermo mechanical treatment. A maximum stress on σ–ε curves occurs at the maximum real deformation value of 0.2 during pressing in eight stages after reducing plastic deformation temperature. The yield stress value within the deformation temperature of 1100 to 950°C is comparable with the values obtained in a continuous pressing test at a rate of 50 s^{-1}. The yield stress value within

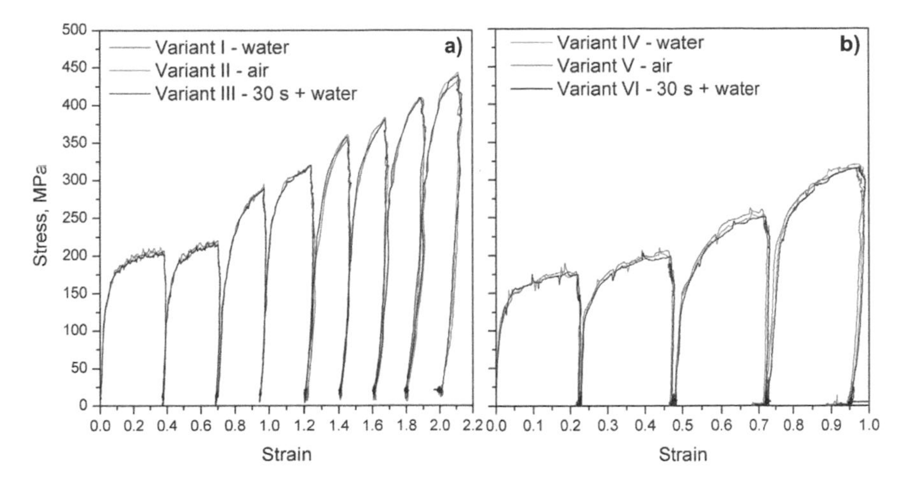

Figure 6.30 Stress-strain curves of hot pressing of perpendicular prism-shaped specimens made of X11MnSiAl17-1-3 steel in conditions similar to plane state of strain (a) eight stages; (b) four stages.

the deformation temperature of 1100 to 950°C during pressing in four stages is comparable with the values obtained in a continuous pressing test (Figure 6.30b).

The investigated steel subjected to thermo mechanical treatment in eight stages and cooled in water immediately after last deformation at the temperature of 850°C is characterised by an equally refined austenite grain with the size of 5–10 μm. If four stages of hot pressing are applied, the grain size for the investigated steel is bigger than in case of pressing in eight stages. After hot rolling, the steel structure is represented by austenite grains strongly elongated in the direction of rolling, containing mechanical twins. The tests performed with X-ray qualitative phase analysis methods confirm that after thermo mechanical treatment, consisting of hot pressing in several stages, both in the Gleeble 3800 laboratory simulator as well as in semi-industrial conditions, the steel possesses a homogenous structure of the phase γ. This was confirmed on the basis of results of tests made with a high-resolution electron transmission microscope (Figure 6.31).

Figure 6.32 shows the results of tests of mechanical and structural properties of the investigated steel after heat and thermo mechanical treatment in a static tensile test at room temperature and of strain energy per unit volume of cold plastic deformation, so-called strain energy per unit volume E_{ZP} calculated in a field area under the actual stress curve.

The structure of steel after cold deformation in static conditions is represented by austenite grains which, already at small deformation of 5 or 10%, undergo elongation in the direction of the stretching force with the occurring deformation twins intersected with numerous slip bands. As

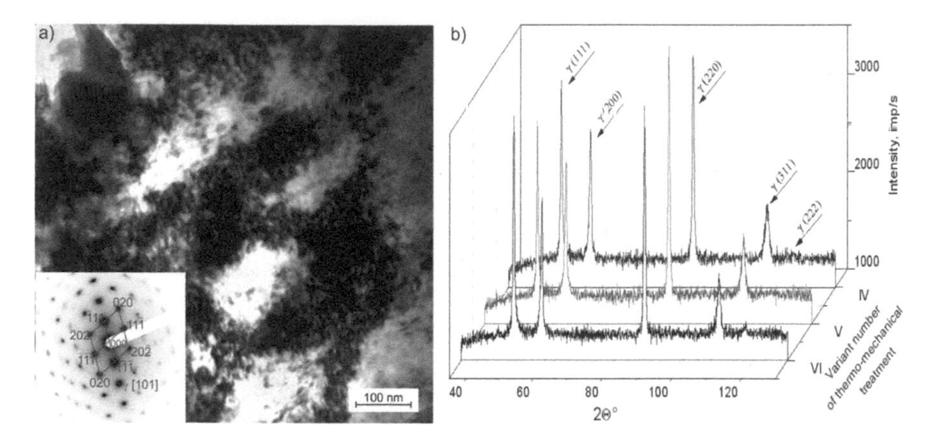

Figure 6.31 (a) Structure of X11MnSiAl17-1-3 steel in the state after hot rolling with the degree of deformation of 4 x 0.23 and after cooling in water; TEM with diffraction solution; (b) X-ray diffraction patterns of steels subjected to four stages of hot pressing in the Gleeble thermomechanical treatment simulator with the deformation of 4 x 0.23, and then cooled in water (variant IV); in air (variant V); or heated isothermally at the temperature of final deformation for 30 s and cooled in water (variant VI).

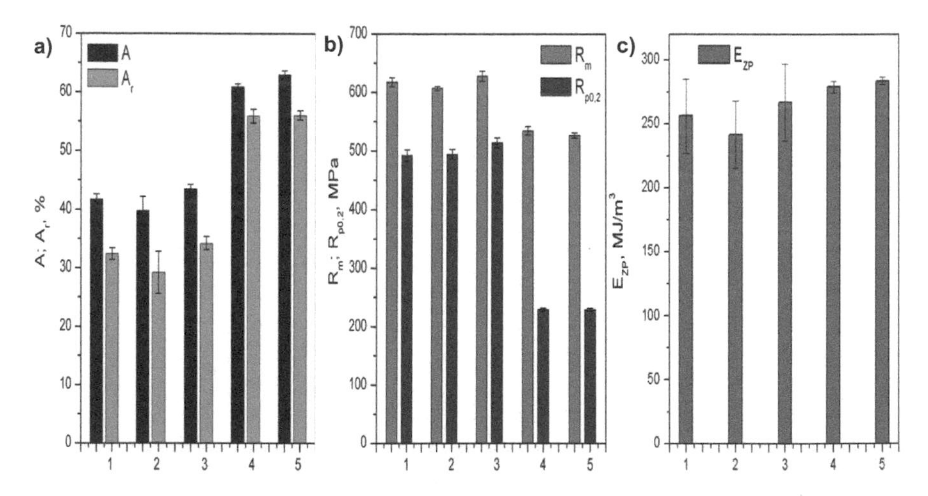

Figure 6.32 Tests results of mechanical properties with the marked deviation of X11MnSiAl17-1-3 steel after thermomechanical and heat treatment (a) tensile strength R_m and yield point $R_{p0,2}$, (b) total elongation A and uniform elongation A_r, (c) strain energy per unit volume E_{ZP} with standard deviations marked; 1,2,3 – thermomechanical treatment: 1 – cooling in water, 2 – cooling in air; 3 – isothermal 30 s and cooling in water; 4,5 – heat treatment: 4 – 900°C, 5 – 1000°C.

the elongations of specimens increase to 20% and 30%, highly densified deformation twins and slip bands exist in austenite grains (Figure 6.33). The structure of the steel deformed statistically until rupture is represented by austenite grains strongly deformed and elongated with a large density of deformation twins and slip bands. It was found through tests in a high resolution transmission electron microscope HRTEM that static cold plastic deformation causes the partial transformation of austenite (Figure 6.34) into martensite. This relates to martensite α′ (Figure 6.35), and martensitc ε (Figure 6.36). The presence of a martensitic phase in the structure was also confirmed with an X-ray qualitative phase analysis (Figure 6.37). No substantial effect of the type of cooling on the fraction of phases in steel was found.

Table 6.12 lists the test results of mechanical properties at a reduced and elevated temperature (–70 to 200°C) of X11MnSiAl17-1-3 steel after four stages of hot rolling with deformation degree of 4 × 0.23 and after cooling in water, whereas Figure 6.38 presents representative strain curves for the total

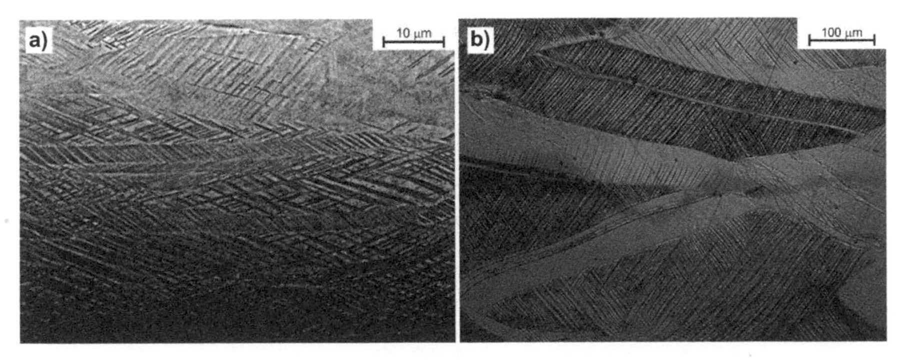

Figure 6.33 Structure of austenitic X11MnSiAl17-1-3 steel with numerous twinning deformations and intersecting deformation bands, in the state after hot rolling with the degree of deformation of 4 x 0.23, (a) subject to elongation to 30%, SEM; (b) after isothermal heating for 30 s after last deformation at 850°C and followed by cooling in water, and then subjected to static tensile test; specimen deformed to the elongation of 30%; LM.

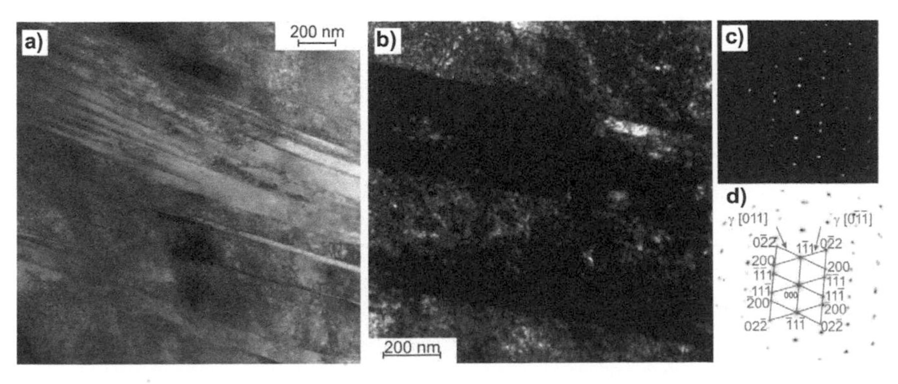

Figure 6.34 Deformation twins in the structure of X11MnSiAl17-1-3 steel in the state after hot rolling and subjected to static tensile test, TEM: (a) bright field; (b) another specimen, dark field from reflection 200 γ, (c) diffraction, (d) diffraction solution from fig. c.

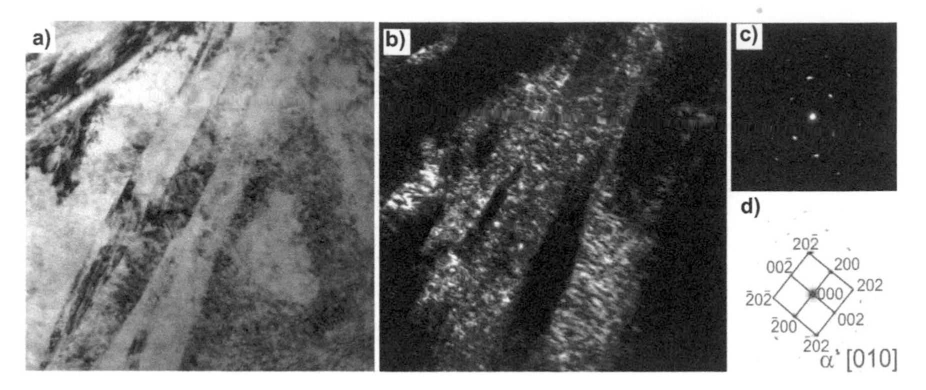

Figure 6.35 Structure of X11MnSiAl17-1-3 steel in the state after hot rolling and static tensile test, TEM: (a) bright field, (b) dark field from reflection 20$\bar{2}$ α'; (c) diffraction, (d) diffraction solution in fig. c.

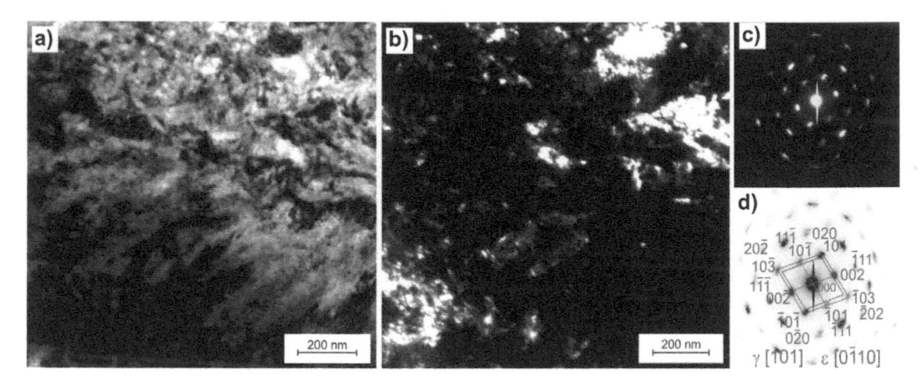

Figure 6.36 Structure of X11MnSiAl17-1-3 steel in the state after hot rolling and subjected to static tensile test, TEM: (a) bright field, (b) dark field from reflection 220 α', (c) diffraction, (d) diffraction solution in fig. c.

specimen elongation with the determined strain energy per unit volume of cold plastic deformation, as a field area under the strain curve σ_{rz}-ε_{rz}.

The strength and plastic properties within the test temperature range of 150 to 200°C change to a small extent only. Strength and plastic properties are clearly heightened as the test temperature is lowered from 100°C to 80°C. At room temperature, total and uniform elongations reach the value of 40% and 32%, respectively. Tensile strength is also rising explicitly (R_m = 620 MPa). If temperature is further reduced from 23 to –25 and then to –70°C, elongation is slightly decreased by 30% and then it suddenly grows to 53%. Tensile strength is also on the constant rise to 850 MPa. Plastic deformation temperature is also significantly influencing the strain energy per unit volume of cold plastic deformation energy E_{ZP} determined by a field area under the strain curve (Table 6.12). As the temperature increases from

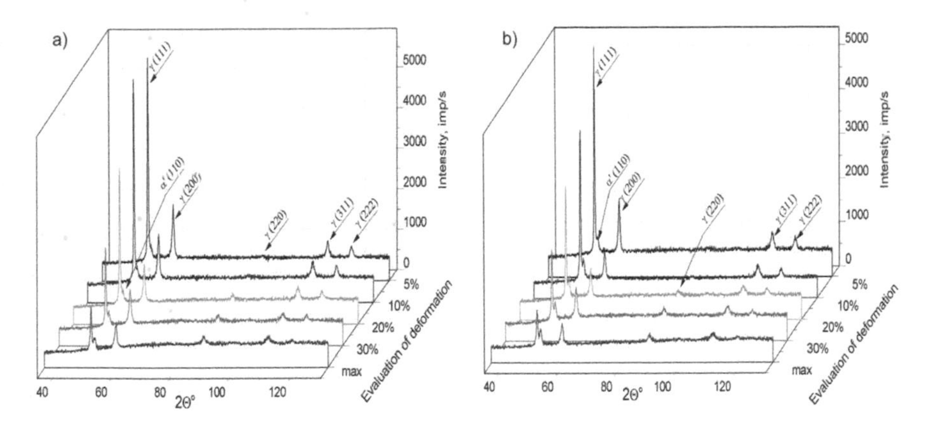

Figure 6.37 X-ray diffraction patterns of X11MnSiAl17-1-3 steel specimens after hot rolling with deformation degree of 4 x 0.23 and (a) cooling in water, (b) isothermal heating for 30 s, subjected to a static tensile test to the set elongation.

Table 6.12 The effect of deformation temperature and strain rate on mechanical properties: yield strength $R_{p0.2}$, tensile strength R_m, total elongation A and strain energy per unit volume of cold plastic deformation E_{ZP} of the investigated steel X11MnSiAl17-1-3 after thermo mechanical treatment.

Test temperature,°C	Strain rate, s^{-1}	$R_{p0.2}$, MPa	R_m, MPa	A, %	E_{ZP}, MJ/m³
–70		472	838	51	
–25	0.1	571	796	32	
23		486	612	42	256
	250	827	794	55	425
23	500	909	820	62	462
	1000	942	826	64	481
80		461	519	40	
100		481	528	30	
150	0.1	443	503	11	
200		401	436	2	

100 to 200°C, the E_{ZP} gets lower value of 8 to 150 MJ/m³. As the deformation temperature falls to 80°C, a value of the average strain energy per unit volume increases. E_{ZP} equals approx. 256 MJ/m³ at room temperature. When deformation temperature is decreased to –25°C, the E_{ZP} value drops to 232 MJ/m³, and E_{ZP} suddenly rises to 353 MJ/m³ at the temperature of –70°C.

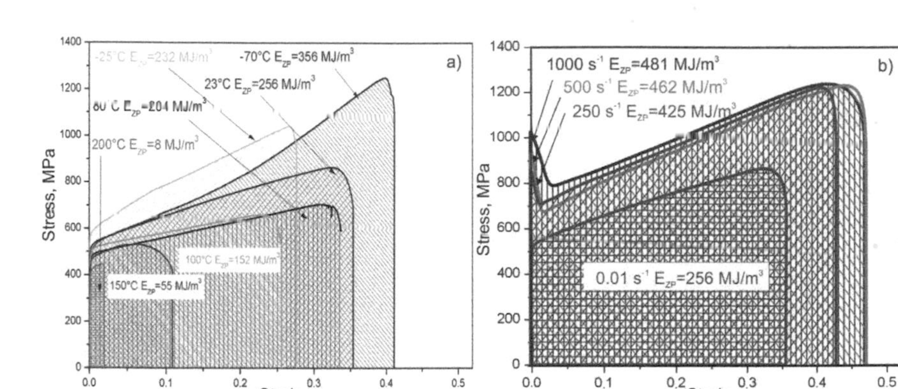

Figure 6.38 The results of the strain energy per unit volume of the high-manganese austenitic X11MnSiAl17-1-3 of the TRIP type steel (a) after static tensile tests at room, cryogenic and elevated temperature, (b) after dynamic tensile tests with strain rates 0.01, 250, 500 and 1000 s^{-1}.

The investigated high manganese steel subjected to plastic deformation at the temperature of –25°C possesses an austenitic structure with scarce slip bands. The slip bands were clearly densified after lowering the temperature to –70°C (Figure 6.39). Steel deformed plastically in a static tensile test at the temperature heightened to 23°C has numerous and densely intersecting slip bands and mechanical twins. Steel is deformed by twinning in two systems (Figure 6.39b) when temperature is increased to 80–100°C. Steel is deformed by mechanical twinning if the temperature is further elevated between 150 and 200°C.

An austenite and martensite structure was revealed for X11MnSiAl17-1-3 steel through tests results for steel thin foils after completion of mechanical tests at a lower and higher temperature. The structures of martensite α' and ε were revealed in diffraction tests of the steel subjected to a static tensile test at a temperature reduced to –25°C (Figure 6.40).

A static tensile test and a dynamic rupture test were performed with specimens from the investigated steel subjected to hot rolling, respectively with the plastic strain rate of 0.01, 250, 500 and 1000 s^{-1}. The outcomes of the tests are collated in Table 6.12. The actual representative stress-strain curves—actual strain σ_{rz}-ε_{rz} with the range of total specimen elongation with the determined strain energy per unit volume as a field area under the strain curve are shown in Figure 6.38. Those results were compared with the tests results for such steels in the conditions of static plastic deformation obtained during a static tensile test. The yield point $R_{p0,2}$ of X11MnSiAl17-1-3 steel reaches the highest value of 942 MPa during fast deformation of 1000 s^{-1}. The same steel deformed plastically at a lower strain rate has the $R_{p0,2}$ value of 827–909 MPa. This steel, when deformed in static conditions, has a nearly half lower yield point $R_{p0,2}$ of approx. 486 MPa. The elongation of the investigated high manganese austenitic steel X11MnSiAl17-1-3 in the

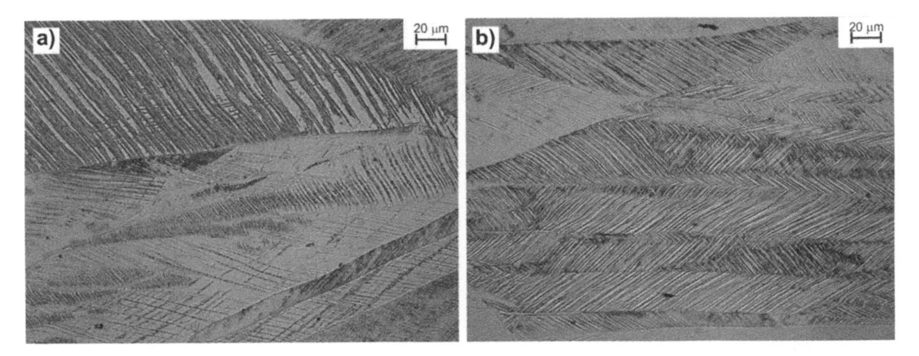

Figure 6.39 Structures of austenite X11MnSiAl17-1-3 steels after thermomechanical treatment in semi-industrial conditions, deformed plastically in a static tensile test at reduced temperature: (a) –70°C; (b) at temperature heightened to 100°C, LM.

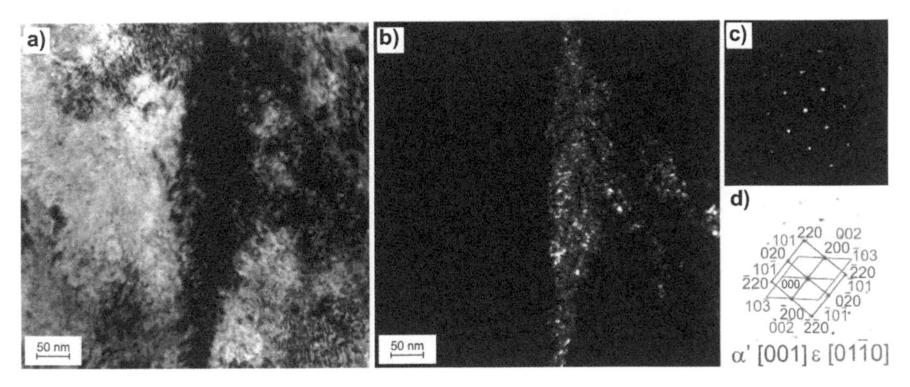

Figure 6.40 Structure of X11MnSiAl17-1-3 steel in the state after hot rolling and static tensile test at temperature reduced to –25°C, TEM: (a) bright field, (b) dark field from reflection 220 α′, (c) diffraction, (d) diffraction solution in fig. c.

conditions of static cold plastic deformation is about 42%, and may reach even 64% if the strain rate is increased to 1000 s^{-1}. An increased strain rate also leads to increased strain energy per unit volume of cold plastic deformation E_{ZP} (Figure 6.38 and Table 6.12). The highest value of strain energy per unit volume of cold plastic deformation energy for a strain rate of 1000 s^{-1} equals to 481 MJ/m^3 and is nearly two times higher than the value of E_{ZP} obtained in static conditions. This corresponds to the best strength (R_m = 942 MPa) and plastic ($R_{p0,2}$ = 826 MPa) properties obtained in such test conditions.

X11MnSi17-1-3 steel after a static and dynamic tensile test is characterised by an austenitic structure with numerous mechanical twins and intersecting slip bands. Intersecting slip bands dominate in the steel deformed from the velocity rate of 0.1 s^{-1} (Figure 6.41). Mechanical twins in the steel deformed with the rate of 250 s^{-1} are 40–60 μm wide and the distance between particular slip bands is approx. 7–10 μm. If a strain rate is increased to 500 s^{-1}, slip bands are densified and twinning intensity grows. When steel is deformed with the rate of 1000 s^{-1}, this influences the occurrence of numerous slip bands and mechanical twins in the material structure (Figure 6.41).

Twinned austenite is mainly found in the steel deformed in a dynamic tensile test (Figure 6.42), after deformation at a rate of 250 s^{-1} and also martensite ε and α′ (Figure 6.43), and austenite and martensite α′ occurs after deformation at a rate of 500 and 1000 s^{-1}. The Kurdjumov-Sachs crystallographic relationship was identified between austenite and martensite $(011)\gamma||(111)\alpha'$, with the directions of $[110]\gamma||[010]\alpha'$and $[\bar{1}01]\gamma||[\bar{1}11]\alpha'$ (Figure 6.44). Martensite α′ was identified on the diffractions attained from the marked areas with a high-resolution electron transmission

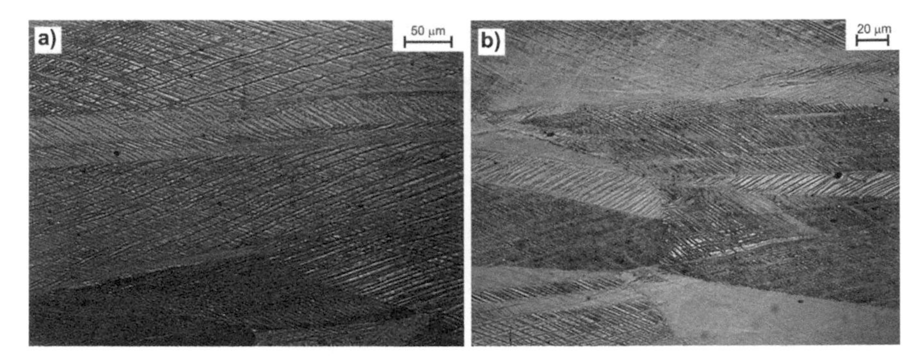

Figure 6.41 Structures of austenitic X11MnSiAl17-1-3 steel after thermomechanical treatment in semi-industrial conditions; light microscope; steel after cold plastic deformation at a rate of (a) 0,1 s^{-1}, (b) 1000 s^{-1}; LM.

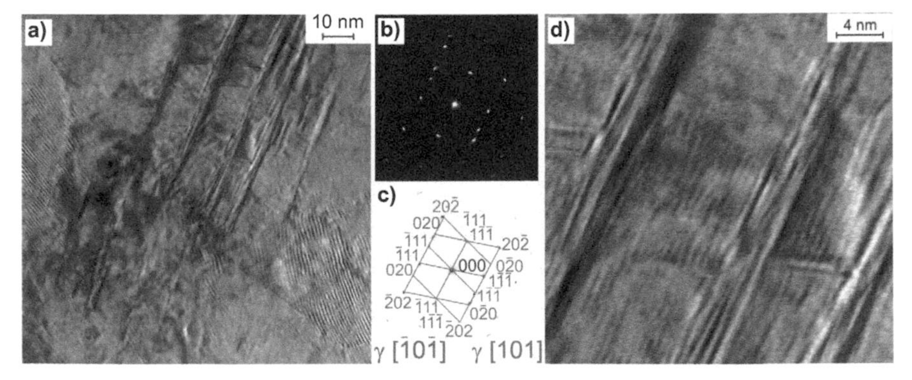

Figure 6.42 Deformation twins in the structure of X11MnSiAl17-1-3 steel in the state after hot rolling and dynamic rapture test, (a) HRTEM; (b) diffraction, (c) diffraction solution from fig. (b; d) HRTEM.

microscope HRTEM (Figures 6.44, 6.45 and 6.47). The presence of austenite and martensite α′ was confirmed by the X-ray phase analysis method (Figure 6.46).

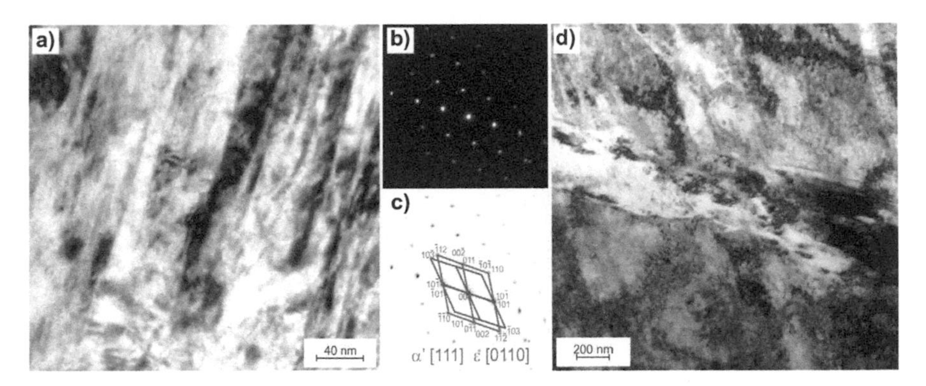

Figure 6.43 Structure of X11MnSiAl17-1-3 steel after rolling and dynamic rapture test; (a) TEM, bright field, (b) diffraction, (c) diffraction solution from fig. (b; d) TEM, bright field.

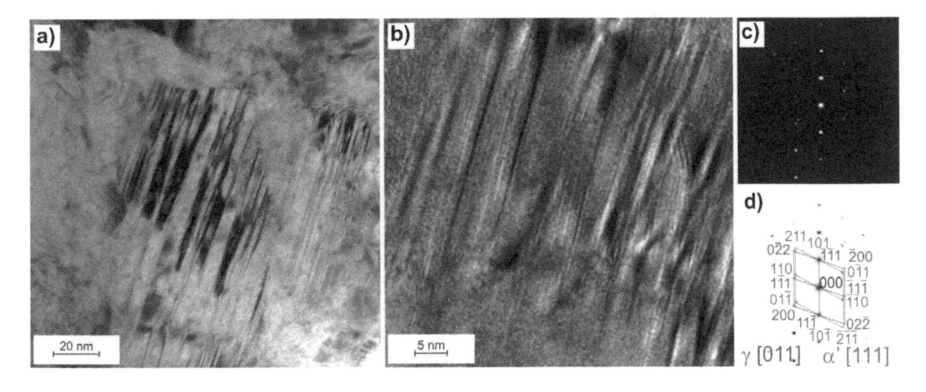

Figure 6.44 Structure of X11MnSiAl17-1-3 steel in the state after hot rolling and dynamic rapture test, an image produced in a transmission electron microscope; (a) bright field; (b) high-resolution image with HRTEM, (c) diffraction, (d) diffraction solution from fig. c.

6.6 General Discussion on High Manganese Steels Exhibiting TRIP and TWIP Effects

This chapter depicted a synergic effect of structural mechanisms, mainly twinning and precipitation of carbides as well as phase transformations, mainly martensitic transformation, during cold plastic deformation of newly developed custom high manganese TWIP and TRIP steels on an increase in the strain energy per unit volume of cold plastic deformation energy. Very extensive tests were carried out starting with a simulation of hot plastic

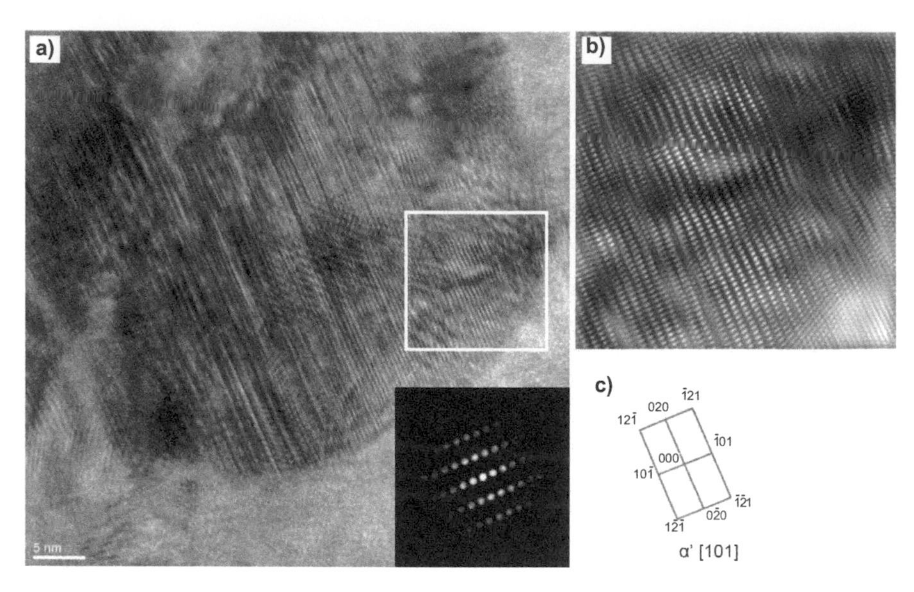

Figure 6.45 Structure of martensite α′ in X11MnSiAl17-1-3 steel with diffraction image, HRTEM; (b) image produced with the Fourier transform from the area marked with white square in fig. (a, c) diffraction solution from fig. a.

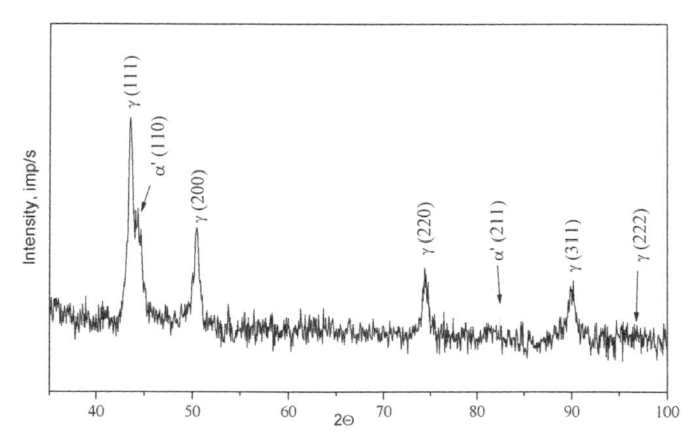

Figure 6.46 Representative X-ray diffraction pattern of specimen made of X11MnSiAl17-1-3 steel, after hot rolling with deformation degree of 4 x 0.23, cooling in water and dynamic deformation at a rate of 1000 s^{-1}.

deformation with the Gleeble 3800 simulator and heat treatment and thermo mechanical treatment of the investigated steels on an experimental Line for Semi-Industrial Simulation LPS. A static and dynamic deformation tests in the temperature range of –70 to 200°C, respectively with the plastic strain rate of 0.01, 250, 500 and 1000 s^{-1} were made also. Finally the

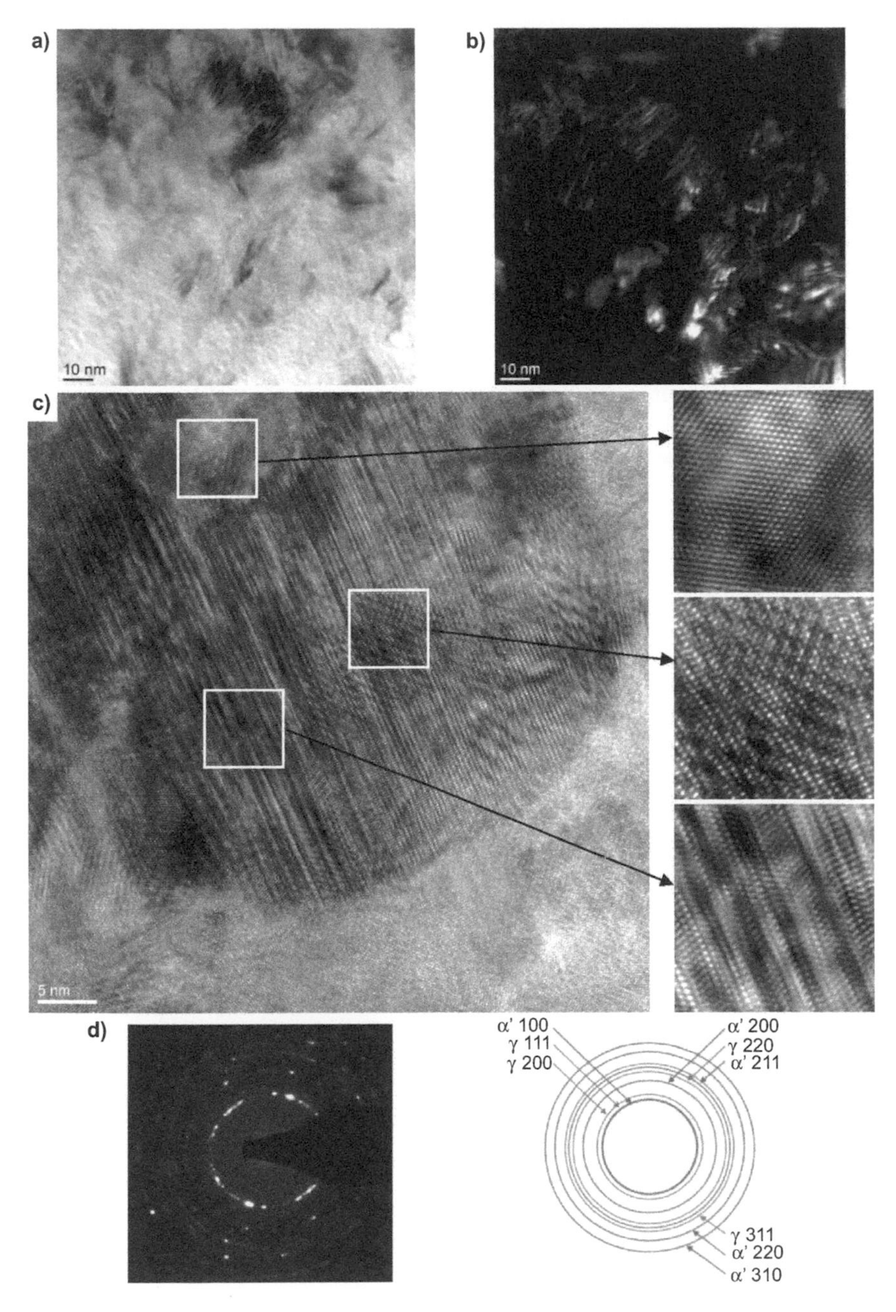

Figure 6.47 Structure of X11MnSiAl17-1-3 steel obtained by observation with high-resolution transmission electron microscope (HRTEM) (a) bright field, (b) dark field from reflection 100 γ, (c) HRTEM high-resolution image with magnification obtained using the Fourier transform, (d) diffraction from the whole area c with solution.

comprehensive structural examinations with diffraction and spectral tests and the high-resolution transmission electron microscopy technique inclusively were made. The chemical compositions of the investigated steels were purposefully selected based on theoretical premises in such a way so that their stacking-fault energy SFE is different, by selecting TWIP X8MnSiAlNbTi25-1-3, X73MnSiAlNbTi25-1-3 and TRIP X11MnSiAl17-1-3 steels for tests. Some results of tests of other steels were also presented for the whole body of newly created custom grades. The intended differentiation of dominant structural effects essential for the mechanical properties of the investigated steels was achieved by planning steel selection in such a way. By controlling the conditions of hot plastic treatment for the investigated steels, a structure of a dynamically recovered, dynamically recrystallised or partially statically and/or metadynamically recrystallised austenite can be created. The structure of the newly developed steels can be considerably refined, by employing thermo mechanical treatment consisting of hot plastic deformation and adjustable cooling, especially for a higher concentration of carbon participating in the processes of the release of dispersion carbides created with micro additives and to ensure the expected strength properties. Interactions were identified as a result of the described investigations between all the factors likely to have an effect on structure forming and an advantageous relationship was identified between high strength of at least $R_m \geq 1000$ MPa, and high plastic properties represented by elongation $A \geq 60\%$ with the possibly highest strain energy per unit volume of cold plastic deformation, determined by a surface integral under a cold plastic deformation curve. The ranges between 190 and 575 MJ/m^3 has been demonstrated, depending on the chemical composition of steel and steel structure hot formation conditions and subsequently simulating working conditions, including a high strain rate in a broad temperature range, especially below 0°C.

It was found that depending on the type of steel, the main causes of growth of the strain energy per unit volume of cold plastic deformation energy in dynamic conditions include:

- the basic mechanism of TWinning Induced Plasticity (TWIP) (Figure 6.48) connected with cold plastic deformation through the activation of mechanical twinning in the intersecting systems and through the mutual intersection of slip bands and deformation twins in austenite grains and annealing twins; its diagram was prepared by analogy to the diagram of structural changes in 50.8 alloy at.% NiTi of Nitinol type given in the work Karaman et al. (2005);
- the mechanism of TRansformation Induced Plasticity TRIP (Figure 6.48) (in case of the investigated steels aiding the interrelated mechanism of TWinning Induced Plasticity TWIP) connected with the progress of a martensitic transformation and directly into phase α′, and with

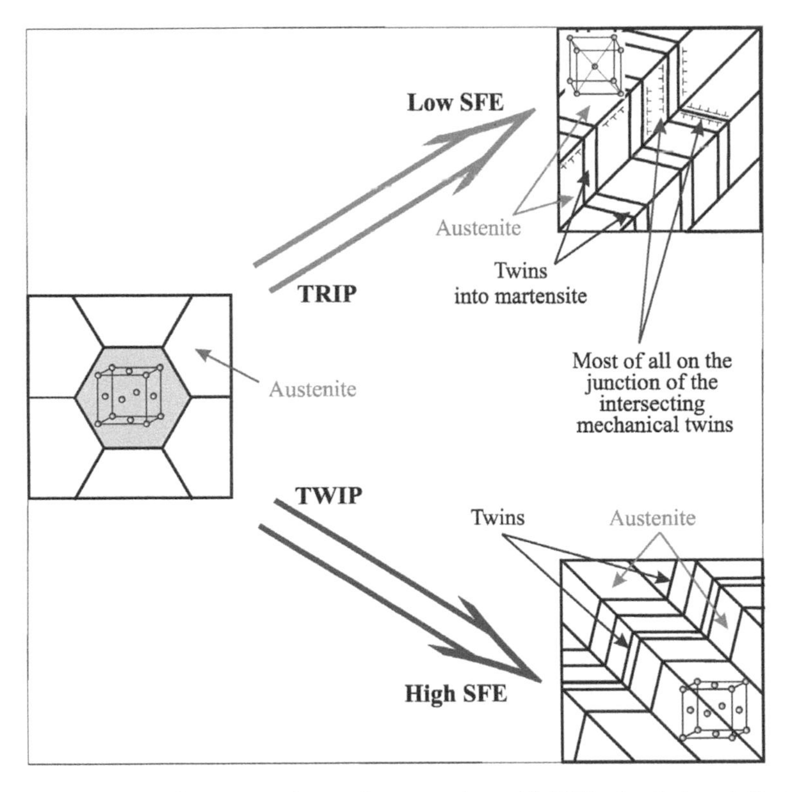

Figure 6.48 Scheme of structure change in connection with TWinning Induced Plasticity TWIP effect consisting of intensive mechanical twinning in austenite during cold plastic deformation and with TRansformation Induced Plasticity TRIP effect, connected with the progress of martensitic transformation directly into phase α′, and with intermediate phase ε, where martensite, mainly in the form of laths, is placed in austenite grains, and most of all on the junction of the intersecting mechanical twins.

an indirect phase ε; the TRIP mechanism is a result of cold plastic deformation, where martensite, mainly in the form of laths, is placed in austenite grains, and most of all on the junction of the intersecting mechanical twins created already at the stage of early deformation phase (ε = 5% or 10%); the process is progressing as it is increased (ε = 20% and 30%) until rupture, also in dynamic conditions; the diagram is prepared the same as the previous one.

The analyses, information and results of the tests concerning the newly developed custom steels indicate that:

- an appropriately selected chemical composition with a high concentration of manganese and carbon in the right mutual proportions with an additive of aluminium and silicon and micro additives of niobium and titanium,

- appropriately elaborated hot plastic deformation conditions and thermo mechanical working conditions,

have a synergic effect by ensuring:

- unit volume of cold plastic deformation energy,
- their high strength properties and
- substantial elongation,

as a result of refining an austenite structure and intensification of, accordingly, mechanical twinning in austenite as a result of TWinning Induced Plasticity TWIP and a local martensitic transformation as a result of TRansformation Induced Plasticity TRIP.

An increase in the strain energy per unit volume of cold plastic deformation energy of the investigated steels sets a basis for applying them in the automotive industry for sheets and structural components of cars as reinforcements and controlled plastic deformation zones, behaving in a controlled and pre-defined manner, during dynamic cold plastic deformation occurring, in particular, in a road accident. It is therefore possible to use the plastic deformation energy of the investigated steels in such conditions to prevent rupture until their plastic deformation lasts. Fractures are prevented in structural parts and bodywork of a car during an accident or road collision by absorbing a large store of the then produced energy. This verifies positively the concept of preventing the fractures of structural parts and bodywork of a car during an accident or road collision by absorbing a large store of the then produced energy, to induce structural changes and phase transitions, occurring in the conditions of dynamic plastic deformation of critical parts of the car made of such steels. High strain energy per unit volume ensures energy absorption and prevents premature fracture of the damaged car components. It has been demonstrated that the increased strain energy per unit volume of cold plastic deformation energy is essential for prevention of the cracking of the analysed groups of steels.

Acknowledgements

The investigations were made mainly in the framework of the Project "MATWITRIS—Structural foundation of counteracting cracking by increase of the cold plastic strain energy margin of the newly developed high-manganese steels of TRIP, TWIP and TRIPLEX types", funded by the DEC-2012/05/B/ST8/00149 of the Polish National Science Centre in the framework of the "OPUS" competitions, headed by Prof. Leszek A. Dobrzański.

The authors wish to extend special thanks to Dr. Eugeniusz Hajduczek, Ph.D. Eng. for help in the technical preparation of this chapter and Dr. Mirosława Pawlyta, Ph.D. Eng. for help in the performance of tests using a high-resolution transmission electron microscope.

Keywords: High manganese steels, Mechanical twinning, Transformation, Store of plasticity, Thermomechanical treatment, Impact tension, Metallurgical simulator

References

Adamczyk, J. and A. Grajcar. 2005. Structure and mechanical properties of DP-type and TRIP-type sheets obtained after the thermomechanical processing. Journal of Materials Processing Technology 162-163: 23–27.

Adamczyk, J. and M. Opiela. 2002. Obróbka cieplno-mechaniczna blach grubych ze stali konstrukcyjnej Cr-Mo z mikrododatkami Nb, Ti i B. Inżynieria Materiałowa 6: 717–723.

Allain, S., J.P. Chateau and O. Bouaziz. 2004a. A physical model of the twinning-induced plasticity effect in a high manganese austenitic steel. Materials Science and Engineering A 387-389: 143–147.

Allain, S., J.P. Chateau, O. Bouaziz, S. Migot and N. Guelton. 2004b. Correlations between the calculated stacking fault energy and the plasticity mechanisms in Fe-Mn-C alloys. Materials Science and Engineering A 387-389: 158–162.

Allain, S., J.P. Chateau, D. Dahmoun and O. Bouaziz. 2004c. Modeling of mechanical twinning in a high manganese content austenitic steel. Materials Science and Engineering A. 387-389: 272–276.

Anderson, D. 2008. Application and Repairability of Advanced High-Strength Steels. Use, growth and repair for AHSS. American Iron and Steel Institute.

ASP. 2010. Advanced High-Strength Steel Applications, Design and Stamping Process Guidelines. Auto/Steel Partnership. Southfield, Michigan.

Barbier, D., N. Gey, S. Allain, N. Bozzolo and M. Humbert. 2009. Analysis of the tensile behavior of a TWIP steel based on the texture and microstructure evolutions. Materials Science and Engineering A 500: 196–206.

Basuki, A. and E. Aernoudt. 1999. Influence of rolling of TRIP steel in the intercritical region on the stability of retained austenite. Journal of Materials Processing Technology 89-90: 37–43.

Berrahmoune, M.R., S. Berveiller, K. Inal, A. Moulin and E. Patoor. 2004. Analysis of the martensitic transformation at various scales in TRIP steel. Materials Science and Engineering A 378: 304–307.

Borek, W. 2010. Znaczenie bliźniakowania w kształtowaniu struktury i własności stali austenitycznych wysokomanganowych. Ph.D. Thesis, Silesian University of Technology, Gliwice, Poland.

Bouaziz, O. and N. Guelton. 2001. Modelling of TWIP effect on work-hardening. Materials Science and Engineering A 319-321: 246–249.

Bouaziz, O., S. Allain and C. Scott. 2008. Effect of grain and twin boundaries on the hardening mechanisms of twinning-induced plasticity steels. Scripta Materialia 58: 484–487.

Bracke, L., G. Mertens, J. Penning, B.C. De Cooman, M. Liebeherr and N. Akdut. 2006. Influence of phase transformations on the mechanical properties of high-strength austenitic Fe-Mn-Cr steel. Metallurgical and Materials Transactions A 37: 307–317.

Bracke, L., K. Verbeken, L. Kestens and J. Penning. 2009. Microstructure and texture evolution during cold rolling and annealing of a highMn TWIP steel. Acta Materialia 57: 1512–1524.

Brüx, U., G. Frommeyer, O. Grässel, L.W. Meyer and A. Weise. 2002. Development and characterization of high strength impact resistant Fe-Mn-(Al-Si) TRIP/TWIP steels. Steel Research 73: 294–298.

Byun, T.S., N. Hashimoto and K. Farrell. 2004. Temperature dependence of strain hardening and plastic instability behaviors in austenitic stainless steels. Acta Materialia 52: 3889–3899.

Cabañas, N., N. Akdut, J. Penning and B.C. DeCooman. 2006. High-temperature deformation properties of austenitic Fe-Mn alloys. Metallurgical and Materials Transactions A 37: 3305–3315.

Christian, J.W. and S. Mahajan. 1995. Deformation Twinning. Progress in Materials Science 39: 1–157.

Cugy, P., A. Hildenbrand, M. Bouzekri, D. Cornette, S. Göklü and H. Hofmann. 2005. A super-high strength Fe-Mn-C austenitic steel with excellent formability for automobile applications. CD-ROM. SAE Commercial Vehicle Engineering Congress and Exhibition, Roma, Italy.
Czaja, M. 2014. Mechanizmy przeciwdziałania pękaniu stali austenitycznej X11MnSiAl17-1-3 w warunkach dynamicznego odkształcenia plastycznego na zimno. Ph.D. Thesis, Silesian University of Technology, Gliwice, Poland.
De Cooman, B.C. 2004. Structure – properties relationship in TRIP steels containing carbide-free bainite. Current Opinion in Solid State & Materials Science 8: 285–303.
De Meyer, M., D. Vanderschueren and B.C. De Comman. 1999. The influence of substitution of Si by Al on the properties of cold rolled C-Mn-Si TRIP steels. ISIJ International 39: 813–822.
Diani, J.M. and D.M. Parks. 1998. Effects of strain state on the kinetics of strain-induced martensite in steels. Journal of the Mechanics and Physics of Solids 35: 1613–1635.
Dini, G., A. Najafizadeh, R. Ueji and S.M. Monir-Vaghefi. 2010a. Improved tensile properties of partially recrystallized submicron grained TWIP steel. Materials Letters 64: 15–18.
Dini, G., A. Najafizadeh, R. Ueji and S.M. Monir-Vaghefi. 2010b. Tensile deformation behavior of high manganese austenitic steel: The role of grain size. Materials and Design 31: 3395–3402.
Dobrzański, L.A. 2006. Materiały inżynierskie i projektowanie materiałowe. Podstawy nauki o materiałach i metaloznawstwo. Wydanie II zmienione i uzupełnione. WNT, Warszawa.
Dobrzański, L.A. 2011. TWIP & TRIP mechanisms importance for the fracture counteraction in high-manganese austenitic steels (invited lecture). 21st International Congress of Mechanical Engineering, COBEM 2011. Natal, Brazil.
Dobrzański, L.A. 2012. Podstawy nauki o materiałach. Wydawnictwo Politechniki Śląskiej, Gliwice.
Dobrzański, L.A. 2013. Metaloznawstwo opisowe. Wydawnictwo Politechniki Śląskiej, Gliwice.
Dobrzański, L.A. 2015. Applications of newly developed nanostructural and microporous materials in biomedical, tissue and mechanical engineering. Archives of Materials Science and Engineering 76: 53–114.
Dobrzański, L.A. and W. Borek. 2010a. Hot-working of advanced high-manganese austenitic steels. Journal of Achievements in Materials and Manufacturing Engineering 43: 507–526.
Dobrzański, L.A. and W. Borek. 2010b. Hot-Working Behaviour of Advanced High-Manganese C-Mn-Si-Al Steels. Materials Science Forum 654-656: 266–269.
Dobrzański, L.A. and W. Borek. 2011a. Hot deformation and recrystallization of advanced high-manganese austenitic TWIP steels. Journal of Achievements in Materials and Manufacturing Engineering 46: 71–78.
Dobrzański, L.A. and W. Borek. 2011b. Hot-rolling of high-manganese Fe-Mn-(Al,Si) TWIP steels. pp. 117–120. Conference Proceedings, 8th International Conference on Industrial Tools and Material Processing Technologies, ICIT&MPT 2011. Ljubljana, Slovenia.
Dobrzański, L.A. and W. Borek. 2012a. Hot-rolling of advanced high-manganese C-Mn-Si-Al steels. Materials Science Forum 706-709: 2053–2058.
Dobrzański, L.A. and W. Borek. 2012b. Thermo-mechanical treatment of Fe-Mn-(Al,Si) TRIP/ TWIP steels. Archives of Civil and Mechanical Engineering 12: 299–304.
Dobrzański, L.A. and W. Borek. 2012c. Mechanical properties and microstructure of high-manganese TWIP, TRIP and TRIPLEX type steels. Journal of Achievements in Materials and Manufacturing Engineering 55: 230–238.
Dobrzański, L.A. and W. Borek. 2012d. Structure and properties of high manganese TWIP, TRIP & TRIPLEX steels. pp. 1–10. Proceedings of the 11th Global Congress on Manufacturing and Management (GCMM2012), Auckland, New Zealand.
Dobrzański, L.A. and A.D. Dobrzańska-Danikiewicz. 2013. Kształtowanie struktury i własności powierzchni materiałów inżynierskich. Wydawnictwo Politechniki Śląskiej, Gliwice.
Dobrzański, L.A., A. Grajcar and W. Borek. 2008a. Influence of hot-working conditions on a structure of high-manganese austenitic steels. Journal of Achievements in Materials and Manufacturing Engineering 29: 139–142.
Dobrzański, L.A., A. Grajcar and W. Borek. 2008b. Hot-working behaviour of high-manganese austenitic steels. Journal of Achievements in Materials and Manufacturing Engineering 31: 7–14.

Dobrzański, L.A., A. Grajcar and W. Borek. 2008c. Microstructure evolution and phase composition of high-manganese austenitic steels. Journal of Achievements in Materials and Manufacturing Engineering 31: 218–225.

Dobrzański, L.A., A. Grajcar and W. Borek. 2009. Microstructure evolution of high-manganese steel during the thermomechanical processing. Archives of Materials Science and Engineering 37: 69–76.

Dobrzański, L.A., A. Grajcar and W. Borek. 2010. Microstructure evolution of C-Mn-Si-Al-Nb high-manganese steel during the thermomechanical processing. Materials Science Forum 638-642: 3224–3229.

Dobrzański, L.A. et al. 2011. Przemiany strukturalne w nowo opracowanych wysokomanganowych stalach austenitycznych o wysokiej wytrzymałości i zwiększonej zdolności pochłaniania energii z modelem komputerowym predykcji ich własności wykorzystującym metody sztucznej inteligencji. Project N N507 287936. Silesian University of Technology, Gliwice.

Dobrzański, L.A., W. Borek and M. Ondrula. 2012. Thermo-mechanical processing and microstructure evolution of high-manganese austenitic TRIP-type steels. Journal of Achievements in Materials and Manufacturing Engineering 53: 59–66.

Dobrzański, L.A., W. Borek, M. Czaja and J. Mazurkiewicz. 2013a. Structure of X11MnSiAl17-1-3 steel after hot-rolling and Gleeble simulations. Archives of Materials Science and Engineering 61: 13–21.

Dobrzański, L.A., M. Czaja, W. Borek and K. Labisz. 2013b. Influence of thermo-plastic deformation on grain size of high-manganese austenitic X11MnSiAl17-3-1 steel. Journal of Achievements in Materials and Manufacturing Engineering 61: 169–174.

Dobrzański, L.A., W. Borek and J. Mazurkiewicz. 2014a. Structure and mechanical properties of high-manganese steels. pp. 199–218. *In*: S. Hashmi (ed.). Comprehensive Materials Processing, Vol. 2; G.F. Batalha (ed.). Materials Modeling and Characterization. Elsevier Ltd.

Dobrzański, L.A., W. Borek and J. Mazurkiewicz. 2014b. Mechanical properties of high-manganese austenitic TWIP-type steel. Materials Science Forum 783-786: 27–32.

Dobrzański, L.A., M. Czaja, W. Borek, K. Labisz and T. Tański. 2014c. Influence of hot-working conditions on a structure of X11MnSiAl17-1-3 steel. Advanced Materials Research 1036: 122–127.

Dobrzański, L.A., W. Borek and J. Mazurkiewicz. 2015a. Influence of thermo-mechanical treatments on structure and mechanical properties of high-Mn steel. Advanced Materials Research 1127: 113–119.

Dobrzański, L.A., W. Borek and J. Mazurkiewicz. 2015b. TWIP mechanism in processing of high-manganese austenitic steel. Journal of Achievements in Materials and Manufacturing Engineering 71: 22–27.

Dobrzański, L.A., M. Czaja, W. Borek, K. Labisz and T. Tański. 2015c. Influence of hot-working conditions on a structure of X11MnSiAl17-1-3 steel for automotive industry. International Journal of Materials and Product Technology 51: 264–280.

Dobrzański, L.A., W. Borek and J. Mazurkiewicz. 2016a. Mechanical properties of high-Mn austenitic steel tested under static and dynamic conditions. Archives of Metallurgy and Materials 61 (in press).

Dobrzański, L.A., W. Borek and J. Mazurkiewicz. 2016b. Effect of strain deformation rates on forming the structure and mechanical properties of high-manganese austenitic TWIP steels. Advances in Materials and Processing Technologies 2 (in press).

Dobrzański, L.A. et al. 2016c. Podstawy strukturalne przeciwdziałania pękaniu przez zwiększenie zapasu energii odkształcenia plastycznego na zimno nowo opracowanych wysokomanganowych stali typu TRIP, TWIP i TRIPLEX. Project UMO-2012/05/B/ST8/00149.Silesian University of Technology, Gliwice.

Edmonds, D.V., K. He, F.C. Rizzo, B.C. De Cooman, D.K. Matlock and J.G. Speer. 2006. Quenching and partitioning martensite - A novel steel heat treatment. Materials Science and Engineering A 438-440: 25–34.

EuroNCAP. 2016. The Official Site of The European New Car Assessment Programme. http://www.euroncap.com/en (accessed May 4, 2016).

Ferguson, D., W. Chen, T. Bonesteel and J. Vosburgh. 2009. A look at physical simulation of metallurgical processes, past, present and future. Materials Science and Engineering A 499: 329–332.

Frommeyer, G. and U. Brüx. 2003. Max-Planck-Institute for iron research ultimate tensile strength. Dusseldorf.

Frommeyer, G. and U. Brüx. 2006. Microstructures and mechanical properties of high-strength Fe-Mn-Al-C light-weight TRIPLEX steels. Steel Research International 77: 627–633.

Frommeyer, G., E.J. Drewes and B. Engl. 2000. Physical and mechanical properties of iron-aluminium-(Mn, Si) lightweight steels. Revue de Métallurgie. 1245–1253.

Frommeyer, G., U. Brüx and P. Neumann. 2003. Supra-ductile and high-strength manganese-TRIP/TWIP steels for high energy absorption purposes. ISIJ International. 43: 438–446.

Fujita, H. and S. Ueda. 1972. Stacking faults and F.C.C. (γ) → H.C.P. (ε) transformation in 18/8-type stainless steel. Acta Metallurgica. 20: 759–767.

Grajcar, A. 2009. Struktura stali C-Mn-Si-Al kształtowana z udziałem przemiany martenzytycznej indukowanej odkształceniem plastycznym. Wydawnictwo Politechniki Śląskiej, Gliwice.

Grässel, O., L. Krüger, G. Frommeyer and L.W. Meyer. 2000. High strength Fe-Mn-(Al,Si) TRIP/TWIP steels development – properties – application. International Journal of Plasticity 16: 1391–1409.

Hamada, A.S. 2007. Manufacturing, mechanical properties and corrosion behaviour of high-Mn TWIP steels. Oulu University Press, Oulu.

Hamada, A.S., L.P. Karjalainen and M.C. Somani. 2007a. The influence of aluminium on hot deformation behaviour and tensile properties of high-Mn TWIP steels. Materials Science and Engineering A 467: 114–124.

Hamada, A.S., L.P. Karjalainen, M.C. Somani and R.M. Ramadan. 2007b. Deformation mechanisms in high-Al bearing high-Mn TWIP steels in hot compression and in tension at low temperatures. Materials Science Forum 550: 217–222.

Han, Y.S. and S.H. Hong. 1997. The effect of Al on mechanical properties and microstructures of Fe-32Mn-12Cr-xAl-0.4C cryogenic alloys. Materials Science & Engineering A 222: 76–83.

Han, H.N., C.-S. Oh, G. Kim and O. Kwon. 2009. Design method for TRIP-aided multiphase steel based on a microstructure-based modelling for transformation-induced plasticity and mechanically induced martensitic transformation. Materials Science and Engineering A 499: 462–468.

Hilditch, T. and J. Speer. 2004. Influence of volume change on plasticity relationship in TRIP sheet steel. Steel Grips 3: 209–211.

Huang, B.X., X.D. Wang, Y.H. Rong, L. Wang and L. Jin. 2006. Mechanical behavior and martensitic transformation of an Fe-Mn-Si-Al-Nb alloy. Materials Science and Engineering A 438-440: 306–311.

Hyun, J. and S.H. Park. 2004. Decomposition of retained austenite during coiling process of hot rolled TRIP-aided steels. Materials Science and Engineering A 379: 204–209.

Idrissi, H., L. Ryelandt, M. Veron, D. Schryvers and P. Jacques. 2009. Is there a relationship between the stacking fault character and the activated mode of plasticity of Fe–Mn-based austenitic steels? Scripta Materialia 60: 941–944.

Jacques, P.J., E. Girault, A. Martens, B. Verlinden, J. van Humbeeck and F. Delannay. 2001. The developments of cold–rolled TRIP–assisted multiphase steels. Al–alloyed TRIP-assisted multiphase steels. ISIJ International 41: 1068–1074.

Jagiełło, A. 2016. Struktura, własności i perspektywy rozwojowe stali X11MnSiAl25-1-3 do zastosowań motoryzacyjnych (in progress). Ph.D. Thesis, Silesian University of Technology, Gliwice, Poland.

Jee, K.K., J.H. Han and W.Y. Jang. 2004. Measurement of volume fraction of ε martensite in Fe-Mn based alloys. Materials Science and Engineering A 378: 319–322.

Kafka, P. 2012. The Automotive Standard ISO 26262, the innovative driver for enhanced safety assessment & technology for motor cars. Procedia Engineering 45: 2–10.

Kannan, M.B., R.K. Singh Raman and S. Khoddam. 2008. Comparative studies on the corrosion properties of a Fe–Mn–Al–Si steel and an interstitial-free steel. Corrosion Science 50: 2879–2884.

Karaman, I., H. Sehitoglu, A.J. Beaudoin, Y.I. Chumlyakov, H.J. Maier and C.N. Tomé. 2000. Modeling the deformation behavior of Hadfield steel single and polycrystals due to twinning and slip. Acta Materialia 48: 2031–2047.

Karaman, I., G.G. Yapici, Y.I. Chumlyakov and I.V. Kireeva. 2005. Deformation twinning in difficult-to-work alloys during severe plastic deformation. Materials Science and Engineering A 410-411: 243–247.

Keeler, S. and M. Kimchi (eds.). 2014. Advanced High-Strength Steels Application Guidelines Version 5.0. WorldAutoSteel. http://www.worldautosteel.org/projects/advanced-high-strength-steel-application-guidelines/(accessed May 3, 2016).

Kliber, J., T. Kursa and I. Schindler. 2008. The influence of hot rolling on mechanical properties of high-Mn TWIP steels. CD-ROM. Proceedings of the 3rd International Conference on Thermomechanical Processing of Steels – TMP'2008. Padua, Italy.

Kliber, J., K. Drozd and I. Mamuzic. 2009. Stress-strain behaviour and softening in manganese TWIP steel tested in thermal-mechanical simulator. HutnickeListy. 3: 31–36.

Kwon, O., K. Lee, G. Kim and K. Chin. 2010. New trends in advanced high strength steel developments for automotive application. Materials Science Forum 638-642: 136–141.

Lagneborg, R. 1964. The martensite transformation in 18%Cr-8%Ni steels. Acta Metallurgica 12: 823–843.

Lee, Y.K. and C.S. Choi. 2000. Driving force for $\gamma \rightarrow \varepsilon$ martensitic transformation and stacking fault energy of γ in Fe-Mn binary system. Metallurgical and Materials Transactions A 31: 355–360.

Li, Z. and D. Wu. 2008. Influence of hot rolling conditions on the mechanical properties of hot rolled TRIP steel. Journal of Wuhan University of Technology—Materials Science Edition 23: 74–79.

Li, Z., D. Wu and H. Lu. 2008. Effect of thermomechanical processing on mechanical properties of hot rolled multiphase steel. Journal of Iron and Steel Research, International 15: 55–60.

Lin, J., D. Balint and M. Pietrzyk (eds.). 2012. Microstructure evolution in metal forming processes: Modelling and applications. Woodhead Publishing.

Lis, A.K. and B. Gajda. 2006. Modelling of the DP and TRIP microstructure in the CMnAlSi automotive steel. Journal of Achievements in Materials and Manufacturing Engineering 15: 127–134.

Liu, S.K., L. Yang, D.G. Zhu and J. Zhang. 1994. The influence of the alloying elements upon the transformation kinetics and morphologies of ferrite plates in alloy steels. Metallurgical and Materials Transactions 25: 1991–1996.

Mazancová, E., Z. Jonšta and K. Mazanec. 2007. Properties of high manganese Fe-Mn-Al-C alloys. Archives of Materials Science 28: 90–94.

Mazancová, E., Z. Jonšta and K. Mazanec. 2008. Structural metallurgy properties of high manganese Fe-Mn-Al-C alloy. HutnickeListy. 61: 60–63.

Mazancová, E., I. Schindler and K. Mazanec. 2009. Stacking fault energy analysis from point of view of plastic deformation response of the TWIP and TRIPLEX alloys. HutnickeListy 3: 55–58.

Mazancová, E., I. Ruziakand I. Schindler. 2012. Influence of rolling conditions and aging process on mechanical properties of high manganese steels. Archives of Civil and Mechanical Engineering 12: 142–147.

Mazurkiewicz, J. 2013. Struktura i własności stali wysokomanganowych MnSiAlNbTi25-1-3 o zwiększonym zapasie energii odkształcenia plastycznego na zimno. Open Access Librar 7: 1–139.

Niechajowicz, A. and A. Tobota. 2008. Application of flywheel machine for sheet metal dynamic tensile test. Archives of Civil and Mechanical Engineering 8: 129–137.

Niendorf, T., C. Lotze, D. Canadinc, A. Frehn and H.J. Maier. 2009. The role of monotonic pre-deformation on the fatigue performance of a high-manganese austenitic TWIP steel. Materials Science and Engineering A 499: 518–524.

Opbroek, E.G. (ed.). 2009. Advanced High Strength Steel (AHSS) Application guidelines. World AutoSteel.
Petrov, Y.N., V.G. Gavriljuk, H. Berns and F. Schmalt. 2006. Surface structure of stainless and Hadfield steel after impact wear. Wear. 260: 687–691.
Price, C.J., N.A. Snooke and S.D. Lewis. 2006. A layered approach to automated electrical safety analysis in automotive environments. Computers in Industry 57: 451–461.
Qin, B. and H.K.D.H. Bhadeshia. 2008. Plastic strain due to twinning in austenitic TWIP steels. Materials Science and Technology 24: 969–973.
Sabet, M., A. Zarei-Hanzaki and S. Khoddam. 2008. An investigation to the hot deformation behavior of high-Mn TWIP steels. CD-ROM. Proceedings of the 3rd International Conference on Thermomechanical Processing of Steels – TMP'2008. Padua, Italy.
Santos, D.B., A.A. Saleh, A.A. Gazder, A. Carman, D.M. Duarte, É.A.S. Ribeiro, B.M. Gonzalez and E.V. Pereloma. 2011. Effect of annealing on the microstructure and mechanical properties of cold rolled Fe–24Mn–3Al–2Si–1Ni–0.06C TWIP steel. Materials Science and Engineering A 528: 3545–3555.
Schumann, V.H. 1972. Martensitische Umwandlung in austenitischen Mangan-Kohlenstoff-Stählen. Neue Hütte. 17: 605–609.
Scott, C., S. Allain, M. Faral and N. Guelton. 2006. The development of a new Fe-Mn-C austenitic steel for automotive applications. Revue de Metallurgie. 103: 293–302.
Shin, H.C., T.K. Ha and Y.W. Chang. 2001. Kinetics of deformation induced martensitic transformation in a 304 stainless steel. Scripta Materialia. 45: 823–829.
Takechi, H. 2000. Application of IF based sheet steels in Japan. pp. 1–12. Proceedings of the International Conference on the Processing, Microstructure and Properties of IF Steels. Pittsburgh, USA.
Talonen, J., P. Nenonen, G. Pape and H. Hanninen. 2005. Effect of strain rate on the strain-induced $\gamma \rightarrow \alpha'$ – martensite transformation and mechanical properties of austenitic stainless steels. Metallurgical and Materials Transactions A 36: 421–432.
Timokhina, I.B., P.D. Hodgson and E.V. Pereloma. 2002. Effect of alloying elements on the microstructure-property relationship in thermomechanically processed C-Mn-Si TRIP steels. Steel Research 73: 274–279.
Timokhina, I.B., E.V. Pereloma, H. Beladi and P.D. Hodgson. 2008. A study of the strengthening mechanism in the thermomechanically processed TRIP/TWIP steel. CD-ROM. Proceedings of the 3rd International Conference on Thermomechanical Processing of Steels – TMP'2008. Padua, Italy.
Van Hecke, B. 2008. The susceptibility of stainless steel to plastic working. Materials and Applications 8, Euro Inox.
Volosevich, P.Y., V.N. Grindnev and Y.N. Petrov. 1976. Manganese influence on stacking-fault energy in iron-manganese alloys. Physics of Metals and Metallography 42: 126–130.
Vercammen, S., B. Blanpain, B.C. De Cooman and P. Wollants. 2004. Cold rolling behaviour of austenitic Fe-30Mn-3Al-3Si TWIP-steel: the importance of deformation twinning. Acta Materialia. 52: 2005–2012.
WorldAutoSteel. 2002. ULSAB-AVC. Advanced Vehicle Concepts. Overview Report. Safe, affordable, fuel efficient vehicle concepts for the 21st century designed in steel. http://www.autosteel.org/~/media/Files/Autosteel/Programs/ULSAB-AVC/avc_overview_rpt_complete.pdf (accessed May 2, 2016).
WorldAutoSteel. 2015a. The UltraLight Steel Auto Body (ULSAB) Programme. http://www.worldautosteel.org/projects/ulsab/ (accessed May 2, 2016).
WorldAutoSteel. 2015b. The Ultralight Steel Auto Closure (ULSAC) Programme. http://www.worldautosteel.org/projects/ ulsac-2/ulsac/ (accessed May 2, 2016).
WorldAutoSteel. 2015c. The UltraLight Steel Auto Suspension (ULSAS) Programme. http://www.worldautosteel.org/projects/ ulsas/ (accessed May 2, 2016).
WorldAutoSteel.2015d. The UltraLight Steel Auto Body – Advanced Vehicle Technology (ULSAB-AVC) Programme. http://www.worldautosteel.org/projects/ulsab-avc-2/ (accessed May 2, 2016).

7

Transformation-Induced Plasticity Steel and Their Hot Rolling Technologies

Zhihui Cai,[1] *Jingwei Zhao*[2] and *Hua Ding*[1,*]

ABSTRACT

Transformation-induced plasticity (TRIP) steels are one type of advanced high strength steels (AHSSs) which have extensive applications in the automotive industry. TRIP steels offer an excellent combination of strength and ductility as a result of their unique microstructural characterizations. The purposes of this chapter are to review the current knowledge on the three generations of TRIP steels which are categorized by different Mn contents, and introduce their hot rolling technologies. The chemical compositions, heat treatment technology, microstructure and mechanical properties of each category of TRIP steels are firstly analyzed, followed by a discussion on the factors affecting the stability of austenite which plays a significant role in TRIP effect. Finally, hot rolling and the related technologies of TRIP steels being currently investigated worldwide are introduced with a purpose of producing high quality TRIP steel products with required microstructure and mechanical properties.

[1] School of Materials Science and Engineering, Northeastern University, Shenyang 110819, China.

[2] School of Mechanical, Materials and Mechatronic Engineering, University of Wollongong, Wollongong, NSW 2522, Australia.

* Corresponding author: dingneu@163.com

7.1 Introduction

It is one of the greatest trends of this century to conserve natural resources and prevent global warming. One of the most direct and effective way in the modern automotive industry of achieving this is by reducing the weight of automobiles, and therefore lowering the rate of fuel consumption and greenhouse gas emissions. In an attempt to respond to such objectives of the automotive sector, the steel industry has been developing new advanced high strength steel (AHSS) grades whose unique metallurgical properties and processing methods enable the automotive industry to meet requirements for safety, efficiency, emissions, manufacturability, durability and quality at a relatively low cost. In recent decades, energy conservation, environmental protection and security are the main factors considered by the automotive manufacturers. There are three principal reasons for the current trend to use more AHSSs in vehicles:

- The reduction of vehicle weight by increasing the use of AHSSs with thinner gauge, leading to reduced fuel consumption and emissions.
- The improvement of security of vehicles, which leads to a better passenger safety by an enhanced crash-worthiness.
- The prominent advantages over the light-weight materials, in particular Al and Mg alloys, and plastics.

The product of strength and elongation (PSE) indicates the balance of strength and ductility for AHSSs in engineering applications. There are several types of AHSSs, such as dual phase (DP) (Son et al. 2005) and transformation-induced plasticity (TRIP) steels with 15 ± 10 GPa% (Jacques et al. 2007, Lani et al. 2007, De Cooman 2004, Shi et al. 2010), are categorized as the first generation automotive steels, whereas the high alloyed steels with PSE of 60 ± 10 GPa%, such as twinning-induced plasticity (TWIP) steels, are classified as the second generation automotive steels. These two classes of steels have attracted limited use in the automotive industry because of inherent disadvantages, for instance, the PSE of the first generation steels is low, while the second generation steels are too expensive. Recently, a number of studies (Lee et al. 2011a, Gibbs et al. 2011, Merwin 2007, Suh et al. 2009, Lee et al. 2011b, Cai et al. 2013a, Cai et al. 2014a, Aydin et al. 2013) have turned towards medium Mn content (5–12%) TRIP steels, which are considered as the promising third generation automotive steels with PSE ≥ 30 GPa%.

The TRIP effect was observed by Zackay et al. (1967) and Tamura et al. (1970) in highly-alloyed homogeneous metastable austenitic steels and TRIP steels. As demonstrated in Figure 7.1, the TRIP effect derives from deformation induced transformation of retained austenite to martensite (De Cooman 2004). This results in work hardening and hence delays the

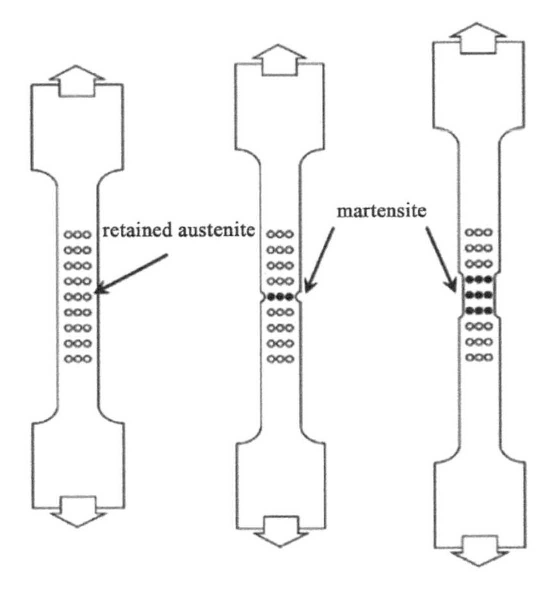

Figure 7.1 Schematic of the TRIP-aided plasticity mechanism in TRIP steels. During straining the retained austenite transforms to martensite. The austenite is replaced by a high strength martensite, and the transformation is associated with a volume expansion.

onset of necking, eventually leading to a higher total elongation. TRIP steel is characterized by enhanced ductility at a very high strength level (Mintz and Wright 1970). TRIP effect enhances the mechanical properties by two mechanisms of (Bellhouse and McDermid 2010, Jacques et al. 2001b) (i) composite strengthening via the formation of hard martensite particles, and (ii) formation of dislocations around newly formed martensite regions as a result of the volume expansion during the austenite-to-martensite transformation. The TRIP effect depends on the amount and degree of stability of retained austenite.

TRIP steels for auto-making are primarily produced by hot rolling, plus heat treatment if necessary. Conventional low-to-high-strength steels, including mild, interstitial-free, bake-harden able and high-strength low-alloy steels, can be produced by the existing mature rolling technologies because they have simpler structures. Differently, TRIP steels are a group of high strength steels with specified chemical compositions. A change in chemical compositions, such as Mn content, may induce variations of microstructural features and mechanical properties of TRIP steels. Due to the metallurgical complexity of TRIP steels, it is impossible to roll all of the TRIP steels in a rolling mill with the same rolling technology to achieve the required microstructure and mechanical properties. Each kind of TRIP steel has unique applications in vehicles, and specified rolling technologies should be developed to produce a high quality TRIP steel product where it might be best employed to meet performance demands for the automotive parts.

In this chapter, TRIP steels are classified into three categories based on Mn contents: conventional low Mn content TRIP steels (Mn≤2.5 wt.%) (Samek et al. 2006, Wang et al. 2014a, Wang et al. 2014b), medium Mn content (in the range of 5–12 wt.%) TRIP steels (Shi et al. 2010, Lee et al. 2011a, Gibbs et al. 2011, Merwin 2007, Suh et al. 2009, Lee et al. 2011b, Cai et al. 2013a, Cai et al. 2014a, Aydin et al. 2013) and high Mn content (in the range of 15–20 wt.%) TRIP steels (Hecker et al. 1982, Tomita and Iwamoto 1995, Tomota et al. 1986, Grassel et al. 2000). The chemical compositions, heat treatment technology, microstructure and mechanical properties of each category of TRIP steels are firstly analyzed, then, the factors affecting the stability of austenite are discussed. Finally, hot rolling and the related technologies of TRIP steels being currently investigated worldwide are introduced with a purpose of producing high quality TRIP steel products with required microstructure and mechanical properties.

7.2 Low Mn TRIP Steels

7.2.1 Chemical Composition

Conventional low Mn TRIP-aided steels are characterized by a very low content of alloying elements. For instance, in current 800 MPa TRIP steel, the total content of alloying elements is about 3.5 wt.% (De Cooman 2004). The chemical compositions of conventional low Mn TRIP steels are usually designed based on the original 0.12–0.55 wt.% C, 0.2–2.5 wt.% Mn and 0.4–1.8 wt.% Si concept proposed by Matsumura et al. (1987a, 1987b, 1991, 1992).

The C content plays a significant role in the chemical compositions and mechanical properties. It should be enriched as much as possible in the retained austenite in order to make it stable. The laboratory TRIP steels could have a C content as high as 0.4 wt.%, whereas current TRIP steels contain typically 0.20–0.25 wt.% C or less in consideration of weld ability (De Cooman 2004). Mn is an austenite stabilizer. It can lower the temperature at which cementite starts to precipitate. Mn also lowers the activity coefficient of C in ferrite and austenite and increases the C solubility in ferrite. Si significantly increases the activity coefficient of C in both ferrite and austenite and reduces the C solubility in ferrite. Si inhibits the formation of cementite during the isothermal bainitic transformation. The minimum level of Si needed to effectively suppress cementite formation is probably about 0.8 wt.%. Nevertheless, a high Si content is not only detrimental to surface quality, but also leads to poor galvanizing ability (Mintz 2001, Ferjutz and Davis 1993). To avoid this problem, Si is partially replaced by

Al. Similarly, Al greatly retards cementite formation. Al decreases the C activity coefficient in ferrite and increases the solubility of C inferrite. More importantly, Al accelerates bainite formation (Garcia-Mateo et al. 2003). However, reduced solid solution hardening is one of the disadvantages of Al alloying in TRIP steels (Jacques et al. 2001a).

The tensile strength of conventional low Mn TRIP steels is in the range of 500–800 MPa. In order to achieve a strength level of 1 GPa or higher, micro-alloying elements (Nb, Ti and V, etc.) are usually added individually or in specific combinations. The micro-alloying additions benefit the refinement of microstructure and formation of carbide precipitates.

7.2.2 Heat Treatment

The microstructure of TRIP-aided steels can be achieved by carrying out a two-stage heat treatment after cold rolling, as shown in Figure 7.2 (Zhang et al. 2012). The first stage of heat treatment is performed at a slightly higher temperature in the $\alpha + \gamma$ two-phase region to form a microstructure of about 50% austenite and 50% ferrite. A fast cooling rate after annealing is employed to avoid any major ferrite formation. The final transformation is carried out isothermally in the bainite region (second stage heat treatment). During bainite formation, carbon diffuses into austenite islands. The enrichment of carbon in the austenite increases its thermal stability and, consequently, the austenite can be retained upon cooling to the room temperature (Lani et al. 2007, De Cooman 2004, Shi et al. 2010, Lee et al. 2011a).

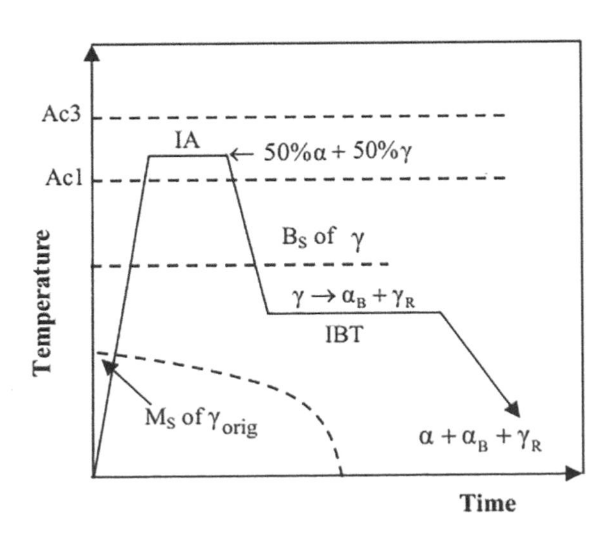

Figure 7.2 Schematic representation of the thermal cycle used to obtain the typical TRIP microstructure. (IA: intercritical annealing, IBT: isothermal bainite treatment, α_B: bainite, γ_R: retained austenite.)

The mechanical properties of low Mn TRIP steels are expected to be excellent by innovative quenching and partitioning (Q&P) process proposed by Speer et al. (2003). As shown in Figure 7.3, Q&P process involves: (i) a fast quenching to a temperature (T_q) which is much higher than room temperature and between the martensite-start (M_s) and martensite-finish (M_f) temperatures so as to form carbon-supersaturated martensite and untransformed austenite; and (ii) a subsequent partitioning at or above T_q to accomplish the diffusion of carbon from supersaturated martensite into retained austenite so as to keep carbon-enriched retained austenite stable during subsequent cooling to room temperature. During Q&P process, the formation of cementite may cause brittlement of TRIP steels. Q&P process makes use of C partitioning between quenched martensite and retained austenite to extend the strength range of current TRIP steels.

The formation of carbides is assumed to be avoided in Q&P process based on Speer's "Constrained Carbon Paraequilibrium" (CCE) theory (Speer et al. 2003), which excludes the potential of precipitation strengthening. Hsu et al. (2007) proposed a quenching-partitioning-tempering (Q-P-T) process based on Q&P theory to utilize the effect of precipitation strengthening. In Q-P-T steels, carbide forming elements (such as Nb, Mo and V) are added to form fine stable carbides at the tempering temperature to strengthen the steels (Wang et al. 2009, Zhong et al. 2009). The carbon content in Q-P-T steels increases to compensate the carbon consumption caused by carbide formation in martensite. Meanwhile, it should be mentioned that partitioning occurs during tempering in Q-P-T process.

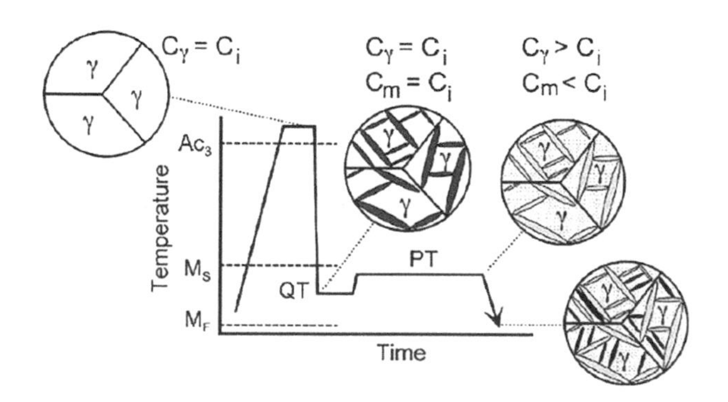

Figure 7.3 Schematic diagram of the Q&P process (Speer et al. 2003).

7.2.3 Microstructure and Mechanical Properties

The microstructure of conventional low Mn TRIP steels consists of ferrite as the dominant phase (55–65%), retained austenite (<20%), bainite (25–35%), and occasionally a small amount of martensite (De Cooman 2004, Zhong et al. 2009, Samek et al. 2006). Low Mn TRIP steels exhibit good combinations of strength (500–1000 MPa) and elongation (15–40%) together with relatively high PSE (~20 GPa%) (Zhou et al. 2011).

Q&P TRIP steels consisting of martensite laths and retained austenite exhibit much higher strength but lower ductility than conventional TRIP steels. For example, a Fe-0.19C-1.63Si-1.59Mn-0.036Al-0.03Cr (wt.%) TRIP steel subjected to Q&P process possesses a high strength of 1400 MPa and an adequate elongation of 10% with PSE of 14 GPa% (Streicher et al. 2004), and a Fe-0.43C-2.0Si-0.59Mn-1.33Cr-0.03Mo-0.03Nb (wt.%) cold-rolled steel treated by Q&P process exhibits a strength of 1907 MPa and an elongation of 17% with excellent PSE of 32 GPa% (Jirkova et al. 2014).

The micro structural constituents of Q-P-T samples consist of lath martensite, retained austenite and dispersedly distributed carbides. Recent researches (Wang et al. 2009, Zhong et al. 2009) show that Q-P-T low Mn TRIP steels exhibit excellent mechanical properties. For example, a Fe-0.2C-1.5Mn-1.5Si-0.05Nb-0.13Mo (wt.%) cold-rolled sheet treated by Q-P-T process shows tensile strength of 1500 MPa and total elongation of 16% with PSE of 24 GPa% (Zhong et al. 2009). And a Fe-0.25C-1.5Mn-1.2Si-1.5Ni-0.05Nb (wt.%) steel subjected to Q-P-T process covers a wide range of strength (1200~1500 MPa) and elongation (14–18%), and shows PSE of 21–22 GPa% (Zhou et al. 2011).

7.2.4 Work Hardening Behavior

The relationship between micro structure and mechanical properties can be reinforced through the study of work-hardening behavior. Work hardening ability could be reflected by the derivation of the relationships proposed by Hollomon (1945), Ludwik (1909), Swift (1952), etc. The parameters involved in these relationships, particularly the work hardening exponent n, have been correlated to microstructural constituents and deformation behavior.

The methods of analysis of the tensile stress-strain curves are summarized as follows:

(1) The Hollomon analysis

The work hardening exponent (n) was derived from Hollomon equation:

$$\sigma = K\varepsilon^n \tag{7.1}$$

The logarithmic form of Eq. (7.1) differentiated with respect to ε is

$$n = \frac{\varepsilon}{\sigma}\frac{d\sigma}{d\varepsilon} \tag{7.2}$$

where σ is the true stress, ε is the true strain, and K is a constant. When $d\sigma/d\varepsilon = \sigma$, then $n = \varepsilon_u$ (necking criterion), where ε_u is the maximum value of uniform elongation.

(2) The Crussard-Jaoul (C-J) analysis

The C-J analysis (Crussard 1953, Jaoul 1957) is based on the following Ludwik relation (Ludwik 1909):

$$\sigma = \sigma_0 + K\varepsilon^{n_E} \tag{7.3}$$

Where σ_0 is a material constant, and n_E is an experimental exponent obtained experimentally. The logarithmic form of Eq. (7.3) differentiated with respect to ε is:

$$\ln(d\sigma / d\varepsilon) = \ln(kn_E) + (n_E - 1)\ln\varepsilon \tag{7.4}$$

Thus, n_E could be determined by plugging the experimental results into Eqs. (7.3) and (7.4). The constant n_C can be calculated from the following Eq. (7.5) (Tomita and Okabayashi 1985):

$$n_c = [\sigma_u / (\sigma_u - \sigma_0)]\varepsilon_u \tag{7.5}$$

where σ_u is the maximum uniform true stress. When n_E is nearly consistent with n_C, the C-J analysis is appropriate for describing the work-hardening behavior.

(3) The modified C-J analysis

The modified C-J analysis (Reed-Hill et al. 1973) is based on the Swift formula (Swift 1952). In this case, the stress-strain relationship is:

$$\varepsilon = \varepsilon_0 + c\sigma^n \tag{7.6}$$

where ε_0 is the initial true strain, and c is the material constant. The logarithmic form of Eq. (7.6) differentiated with respect to ε is:

$$ln(d\sigma/d\varepsilon) = (1-n)\ ln\sigma - ln(cn) \tag{7.7}$$

Likewise, n_E could be determined by plugging the experimental results into Eqs. (7.6) and (7.7). The constant n_C can be calculated from the following Eq. (7.8) (Tomita and Okabayashi 1985):

$$n_c = 1/(\varepsilon_u - \varepsilon_0) \tag{7.8}$$

When n_E is nearly consistent with n_C, the modified C-J analysis is suitable for describing the work-hardening behavior.

The Hollomon analysis has been used extensively for revealing the work hardening behavior of metals and alloys (Shi et al. 2010, Cai et al. 2014a, Wang et al. 2009, Zhong et al. 2009). However, this relationship has limitations in analyzing those alloys which show more than one deformation process. It was proven in our previous work that the C-J analysis was applicable in an austenite-ferrite dual-phase steel (Cai et al. 2013b). It was reported that modified C-J analysis is appropriate for steels having mixed structure of ferrite, martensite and a small amount of retained austenite (Jha et al. 1987).

Take Fe-0.25C-1.5Mn-1.2Si-1.5Ni-0.05Nb (wt.%) Q-P-T TRIP steel for example (Zhou et al. 2011), Figure 7.4 illustrates the variation of instantaneous work-hardening exponent (n) and retained austenite fraction (f) with true strain (ε) during tensile tests. It can be seen that there are three stages in the work-hardening behavior of experimental steel: (1) $\varepsilon < 0.015$, n rapidly drops to the minimum value, while f only has a slight decrease from 12% to 10.5%; (2) $0.015 < \varepsilon < 0.07$, corresponding to the uniform tension stage, n almost remains constant, while f continuously decreases from 10.5% to 7%, which is attributed to TRIP effect induced by the transformation from retained austenite to martensite; and (3) $\varepsilon < 0.07$, both n and f decrease at the same time, showing a weak TRIP effect. This work-hardening behavior of Q-P-T steel is similar to that of TRIP steels (Timokhina et al. 2003, Timokhina et al. 2004, Santos et al. 2009), indicating that the TRIP effect of retained austenite in Q-P-T steels plays a crucial role on the combination of high strength and high ductility.

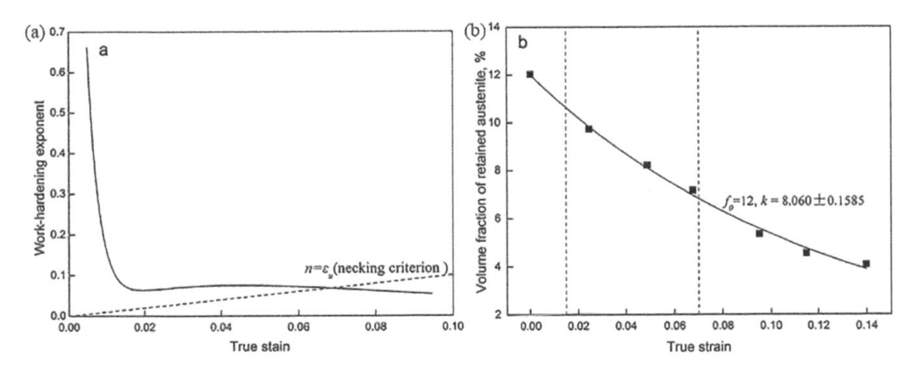

Figure 7.4 Instantaneous work-hardening exponent curve (a) and variation of retained austenite fraction (b) of Q-P-T sample (Zhou et al. 2011).

7.3 High Mn TRIP/TWIP Steels

7.3.1 Composition

High Mn steels, which contain 15–20 wt.% Mn, 2–4 wt.% Si, 2–4 wt.% Al and small amount of C, Ni, N, Cr and some other micro-alloying elements, exhibit TWIP effect during deformation (Grassel et al. 2000). This mechanism leads to excellent combinations of mechanical properties. However during deformation of steels with TWIP compositions, the transformation of austenite-to-martensite occurs as a competing deformation mechanism, leading to occurrence of a combined TWIP and TRIP effect.

The phase transformation is correlated to the stacking fault energy (SFE) in the austenitic matrix. Two different transformation paths of "austenite

(FCC) → α-ferrite (BCC)" and "austenite → ε-martensite (HCP) → α-ferrite" have been reported (Sato et al. 1982). Low SFE (≤20 mJ/m^2) favors the "austenite → ε-martensite" phase transformation, whereas in the SFE range of 20 ~ 40 mJ/m^2 such phase transformation is suppressed and TWIP effect is typically observed (Sato et al. 1989). Alloys with high SFE benefit mechanical twinning formation instead of phase transformation. SFE depends on temperature and chemical compositions, and it has a major influence on the mechanical properties of alloys. Additions of Mn and Al increase SFE and, therefore, strongly suppress "austenite → ε-martensite" transformation (Ishida and Nishizawa 1974). In contrast, Si decreases SFE and sustains "austenite → ε-martensite" transformation during cooling and deformation (Schramm and Reed 1975). A model of SFE calculation was proposed by Allain et al. (2004).

7.3.2 Microstructure and Mechanical Properties

The microstructure constituents of TRIP/TWIP steels depend on the Mn content, leading to different mechanical properties. For example, the metastable austenitic Fe-15Mn-3Al-3Si (wt.%) TRIP steel with certain volume fractions of ferrite and martensite exhibits high strength (1100 MPa) and enhanced tensile ductility (55%) via multiple martensitic "austenite → ε-martensite → α-ferrite" transformations at ambient temperature (Frommeyer and Brux 2003). The microstructure constituents of Fe-19Mn-3Al-3Si-0.04C TRIP/TWIP steel mainly comprise of austenite with some annealing twins, and martensite or ferrite islands distribute dispersedly at the grain boundaries of austenite. During the deformation process, more deformation twins will form, a tensile strength of about 800 MPa and ductility of about 65% could be achieved (Ding et al. 2011). The work of Grassel et al. (2000) indicated that high Mn TRIP steel (Fe-20Mn-3Si-3Al) could exhibit a total elongation of about 82% and an ultimate tensile strength of about 830 MPa at room temperature. The formation of α-martensite enhanced the tensile elongation due to retardation of local necking. The TWIP-steel (Fe-25Mn-3Si-3Al) exhibits a total elongation of about 92% and ultimate tensile strength of 650 MPa at room temperature (Grassel et al. 2000). The excellent mechanical properties are due to massive twinning in the austenitic matrix during deformation. Comparing with TRIP steels, TRIP/TWIP or TWIP steels exhibit lower tensile strength but much better ductility.

7.3.3 Work Hardening Behavior

Figure 7.5 shows true strain-stress plot and the responding work hardening rate in a Fe-19Mn-3Al-3Si-0.04C TRIP/TWIP steel (Ding et al. 2011). The initial work hardening rate in elastic stage (stage I) is high, and it decreases rapidly with increasing strain. The work hardening rate declines slightly in

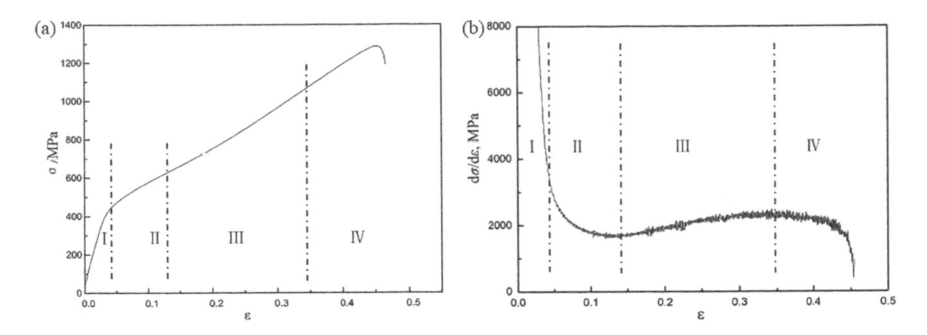

Figure 7.5 True σ-ε curve of Fe-19Mn-3Al-3Si-0.04C steel (a) and the responding work hardening rate (b) (Ding et al. 2011).

stage II. In the true strain range from 0.14 to 0.35, the stress-strain curve is characterized by an upward trend, and the corresponding work hardening rate in stage III is enhanced. Based on the XRD results, there was no phase transformation in this period. However, massive deformation twins were discovered through TEM. Therefore, the enhancement of work hardening rate was a result of twinning. In the final stage with strain of 0.35–0.45, the interaction between TRIP effect and TWIP effect makes the work hardening rate decrease slowly.

7.4 Medium Mn TRIP Steels

7.4.1 Composition and Mechanical Properties

Fe-Mn-C ternary alloy systems have been extensively investigated (Shi et al. 2010, Lee et al. 2011a, Gibbs et al. 2011, Merwin 2007, Suh et al. 2009, Lee et al. 2011b, Miller 1972, Luo et al. 2011). For instance, Miller (1972) established the possibility of TRIP steels containing medium Mn content. The microstructural constituents of a Fe-0.11C-5.7Mn TRIP steel consisting of ferrite and 29 vol.% austenite may achieve a tensile strength of 878 MPa and a total elongation of 34%. Motivated by their success of medium Mn alloy design in obtaining high strength-high ductility combination, recent research (Shi et al. 2010, Lee et al. 2011a, Gibbs et al. 2011, Merwin 2007, Suh et al. 2009, Lee et al. 2011b, Miller 1972, Luo et al. 2011) has been focusing on Fe-(5-7)Mn-(0.1-0.2)C TRIP steels with large volume fraction (20–40%) of austenite, and it is suggested that the mechanical properties can be improved with increase in Mn and C contents, which leads to an increase in the fraction of austenite. For example, Luo et al. (2011) reported that Fe-5Mn-0.2C (wt.%) steel exhibited combinations of tensile strength of 850–950 MPa and ductility of 20–30%. Merwin (2007) achieved a high tensile strength (1018 MPa) and a large total elongation (32%) on a Fe-7Mn-0.1C (wt.%) steel. Shi et al. (2010) reported 1420 MPa tensile strength with 31% total elongation on a Fe-7Mn-0.2C (wt.%) steel.

In recent studies, Al was added in medium Mn TRIP steels to optimize austenite stability by suppressing cementite formation (Bhattacharyya et al. 2011). Furthermore, Al in TRIP steels encourages the growth of intercritical ferrite (Suh et al. 2007) and facilitates the presence of δ-ferrite during solidification and contributes to excellent tensile properties (Yi et al. 2010, Chatterjee et al. 2007). Ferrite is a soft phase with good ductility and helps to stabilize the austenite phase (Cai et al. 2013b, 2014c). In this regard, Li et al. (2015) reported that Fe-6Mn-0.2C-1.6Al steel was characterized by tensile strength of 1040 MPa and ductility of 41%. Suh et al. (2010) reported that Fe-6Mn-0.1C-3Al (wt.%) steel demonstrated an excellent combination of high tensile strength (1000 MPa) and ductility (30%). Park et al. (2013) achieved a tensile strength of 949 MPa and a total elongation of 54% on a Fe-8Mn-0.2C-5Al (wt.%) steel. Cai et al. (2013b) achieved a tensile strength of 998 MPa and a total elongation of 67% on a Fe-11Mn-0.2C-4Al (wt.%) steel.

7.4.2 Heat Treatment

Generally in heat treatment, the medium Mn TRIP steels are first soaked in the austenitic domain and quenched to obtain a completely martensitic structure. Next, an intercritical annealing step followed by air cooling enables stabilization of a large fraction of retained austenite. During intercritical annealing, martensite reverts to austenite and is referred as "austenite reverted transformation" (ART) (Shi et al. 2010, Lee et al. 2011a, Gibbs et al. 2011, Merwin 2007, Suh et al. 2009, Lee et al. 2011b). Retained austenite is obtained by successive enrichment of Mn and C in the reversed austenite during the intercritical annealing process.

However, ART-annealing was not applicable to our experimental steels (Fe-8/11Mn-0.2C-4Al), and quenching and tempering (Q&T) was envisioned by us an alternative and effective heat treatment (Cai et al. 2013a, Cai et al. 2013b, Cai et al. 2014a, Cai et al. 2014c, Li et al. 2015). Tempering is often used to relieve the residual stress. However, it generally leads to the decomposition of retained austenite into ferrite and cementite, which is detrimental to ductility. It was reported that (Luo et al. 2010, Tong et al. 2014) the amount of retained austenite decreased with an increase in tempering temperature in steels containing 2 wt.% Mn. A large fraction of austenite continued to remain in Fe-0.2C-5Mn (wt.%) steel after tempering at 400°C (Xu et al. 2012). The decomposition temperature of austenite increased with increase in Mn content (Thomas 1978), which suggested that the thermal stability of austenite can be enhanced by the enrichment of Mn. Thus, it is possible for austenite to remain stable in medium Mn steels during the tempering process. This has been further confirmed in our research on Fe-11Mn-0.2C-4Al. After tempering at 200°C for 20 min, the ductility of the samples can be significantly improved, and there is no decrease in strength (Cai et al. 2014c).

7.4.3 Microstructure

Generally, the microstructure constituents of Fe-Mn-C ternary alloys with medium Mn consist of austenite and ferrite. The typical microstructure of Fe-0.2C-5Mn steel is illustrated in Figure 7.6a (Shi et al. 2010). The lamellar austenite locates between adjacent ferrite laths. For the Fe-(6-12)Mn-(3-6) Al-0.2C (wt.%) steels, the microstructure comprises of δ-ferrite, α-ferrite and austenite (Cai et al. 2014c, Cai et al. 2013b, Li et al. 2015), moreover, the amount of δ-ferrite increases with the increase of Al content. Figure 7.6b shows the microstructure of hot rolled Fe-11Mn-0.2C-4Al (wt.%) steel. As

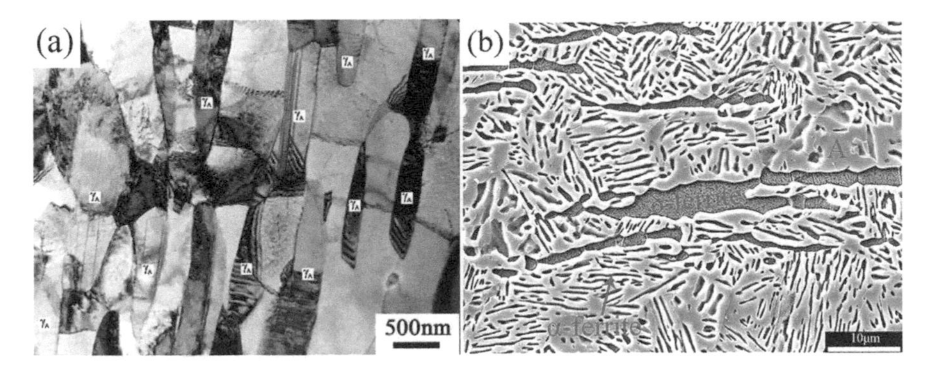

Figure 7.6 TEM micrograph of Fe-0.2C-5Mn steel (a) (Shi et al. 2010) and SEM micrograph of Fe-11Mn-0.2C-4Al steel (b).

labeled in Figure 7.6b, the streaks spreading out as layers during hot-rolling are δ-ferrite, and can be retained during the following heat treatment.

7.4.4 Discontinuous TRIP effect

According to the previous studies (Shi et al. 2010, Arlazarov et al. 2012), steels with 5–7% Mn exhibited three stages of work hardening rate (WH) evolution: in stage 1, WH rapidly decreases; in stage 2, WH increases; and in stage 3, there is a continuous drop of WH. The majority of studies reported in the literature (Shi et al. 2010, Jha et al. 1987, Arlazarov et al. 2012, Dan et al. 2008) primarily relate the first stage to deformation of ferrite, the second intermediate stage characterized by a higher work hardening rate to the occurrence of TRIP effect, and the final stage to possible straining of ferrite and martensite, because the martensitic transformation is inactive in this stage. WH obtained from the tensile test of a hot-rolled Fe-11Mn-4Al-0.2C (wt.%) steel quenched from 800°C followed by tempering at 200°C (hence forward labeled as 800T sample) is presented in Figure 7.7a. A similar three-stage WH described above is observed, but an obvious difference is that it decreases with marked fluctuations in stage III. These fluctuations are directly connected to the observed serrations in the stress-strain plot.

Stress-strain plot of the 800T sample is presented in Figure 7.7b. It is seen that the flow curve is characterized by serrations and are of interest in medium-manganese TRIP steels and high Mn TWIP steels (Dastur and Leslie 1981, Qian et al. 2013, Koyama et al. 2011, Lee et al. 2011c), where they were ascribed to dynamic strain aging, resulting from dynamic interactions of mobile dislocations with solute atoms, during tensile deformation. However, serrations in the flow curve of the 800 T sample are quite different. As shown in Figure 7.7c, the serrations are characterized by abrupt stress drops to below the general level of the flow curve (point A_1 drops to A_3), followed by a slow rise of the stress (point A_3 rises to A_2). The fluctuation between points A_1 and A_2 corresponds to a strain of about 0.016, and the fluctuations between B_1 and B_2, C_1 and C_2 correspond to a strain of about 0.023 and 0.030, respectively.

Study on stage III in Figure 7.7c enables us to have a good understanding of the discontinuous TRIP effect. The strain from peak points A_1 to B_1 can be divided into two stages. In the first stage (A_1-A_2), martensitic transformation is activated when a certain critical stress is attained, and then transformation continuously takes place in the remaining austenite of similar stability. In this stage, the WH increases rapidly (Figure 7.7a) because of TRIP effect and corresponds to the abrupt drop in stress followed by a gradual and slow increase (Figure 7.7c). The abrupt drop is attributed to the onset of TRIP effect which relaxes and transfers the local stress to the surrounding ferrite and austenite. On the other hand, the gradual and slow increase is a consequence of competition between the increase in tensile stress and stress relaxation and transfer induced by the TRIP effect. When the stress rebounds to the point A_2 which has the same stress value as A_1 (Figure 7.7c), the austenite in the deformation zone with similar degree of stability is transformed to martensite. The deformation of the soft ferrite phase induced by the transferred stress produces an additional strain range from A_1 to A_2. The second stage (A_2-B_1) concerns accumulation of stress. In this stage, the WH decreases (Figure 7.7a) and rebounds when the stress is adequately large (point B_1), initiating a second round of TRIP effect, corresponding to the stage of B_1-B_2. Considering that the stress at point B_1 is larger than A_1, it is highly likely that austenite with a higher degree of stability is activated

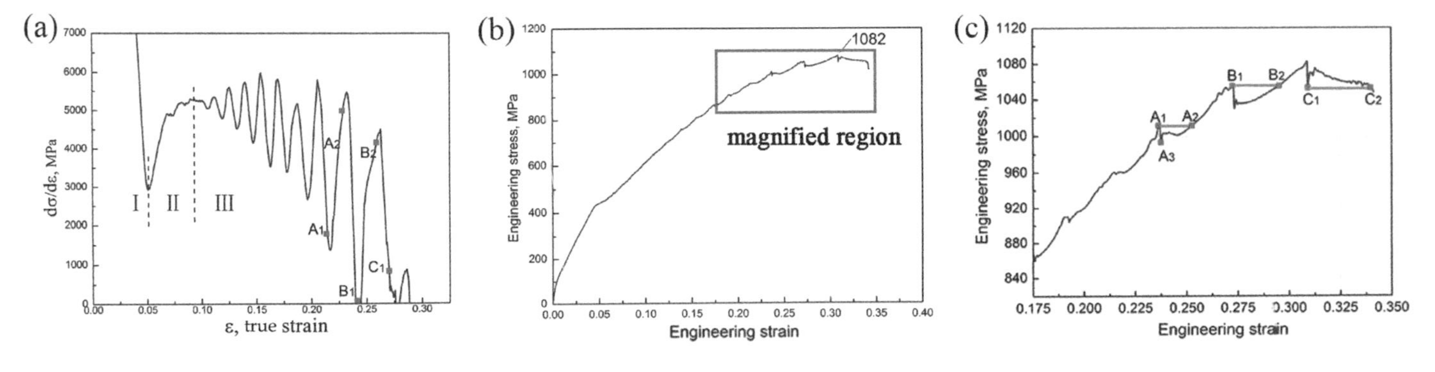

Figure 7.7 Work hardening rate and strain-stress plot of sample heat treated at 800°C + tempered at 200°C (800T sample). (a) Work hardening rate of the 800T sample, (b) strain-stress plot of 800T sample, (c) partial magnification of strain-stress plot (Cai et al. 2014b).

to transform such that a higher stress is transferred into the surrounding ferrite resulting in larger strain. In this case, the strain range of B_1 to B_2 (0.023), contributed from TRIP effect, is larger than that of A_1 to A_2 (0.016). Thus, transformation of austenite with higher stability into martensite produces higher strain.

Comparing the microstructures of other medium Mn TRIP steels (Lee et al. 2011a, Gibbs et al. 2011, Merwin 2007, Xu et al. 2012) with those in our study, there are two important factors for the discontinuous TRIP effect. First, the martensitic transformation leads to a volume expansion, which causes plastic deformation of the δ-ferrite and the ferrite laths and consequently introduces relaxation and transfer of concentrated local stress. Second, austenite with different degrees of stability is responsible for the TRIP effect that occurs discontinuously when a critical stress is attained.

7.5 Austenite Stability

7.5.1 Factors Affecting Austenite Stability

A number of investigations have suggested that there is an optimum stability for the retained austenite (Sugimoto et al. 1992a, Sugimoto et al. 1992b, Garcia-Mateo and Caballero 2005). Austenite stability against mechanically-induced transformation to martensite is known to depend on chemical composition (primarily the carbon content) (Syn et al. 1978, Takahashi and Bhadeshia 1991, Tomita and Okawa 1993), grain size (Jung et al. 2011) and morphology (Garcia-Mateo et al. 2003) of the austenite. The lamellar and granular retained austenite is much more stable than equiaxed one (Bhadeshia and Edmonds 1983). Chiang et al. (2011) indicated that the lamellar microstructure exhibits superior sustained work hardening behavior at higher strain, while the equiaxed microstructure has a relatively rapid rate of transformation.

M_S is the thermal martensite start temperature. The stability of retained austenite can be characterized by M_S^σ temperature, at which the stress required to trigger stress-assisted transformation attains the yield strength (Haidemenopoulos and Vasilakos 1997). In the M_S-M_S^σ temperature range, martensite nucleation on existing nucleation sites is enhanced by stress, i.e., stress-assisted transformation of austenite to martensite takes place. In the M_S^σ-M_d temperature range, martensite formation occurs on nucleation sites introduced by plastic deformation, such as slip band intersections, where strain induced transformation of austenite to martensite occurs. No transformation takes place at temperatures higher than the M_d temperature. Therefore, austenite stability is influenced by the deformation temperature, and accounts for the different martensite transformation mechanisms (stress-assisted or strain-induced).

Furthermore, the stability of austenite could be quantified by the following equation (Shi et al. 2010, Zhou et al. 2011):

$$f_\gamma = f_{\gamma 0}\exp(-k\varepsilon) \tag{7.9}$$

where $f_{\gamma 0}$, f_γ and k are the initial austenite fraction, the austenite fraction at strain ε, and the mechanical stability of austenite, respectively. A higher value of k corresponds to higher driving force for transformation and lower austenite stability.

7.5.2 Schmid Factor and Morphology

Schmid factor of austenite plays an important role in determining austenite stability during deformation. In order to understand well the influence of orientation on austenite stability, Schmid factors of austenite in 800T sample before and after tensile deformation were calculated by EBSD.

Figure 7.8 shows that the relative volume fractions of the austenite grains with the Schmid factors smaller than 0.40 are about 9% before

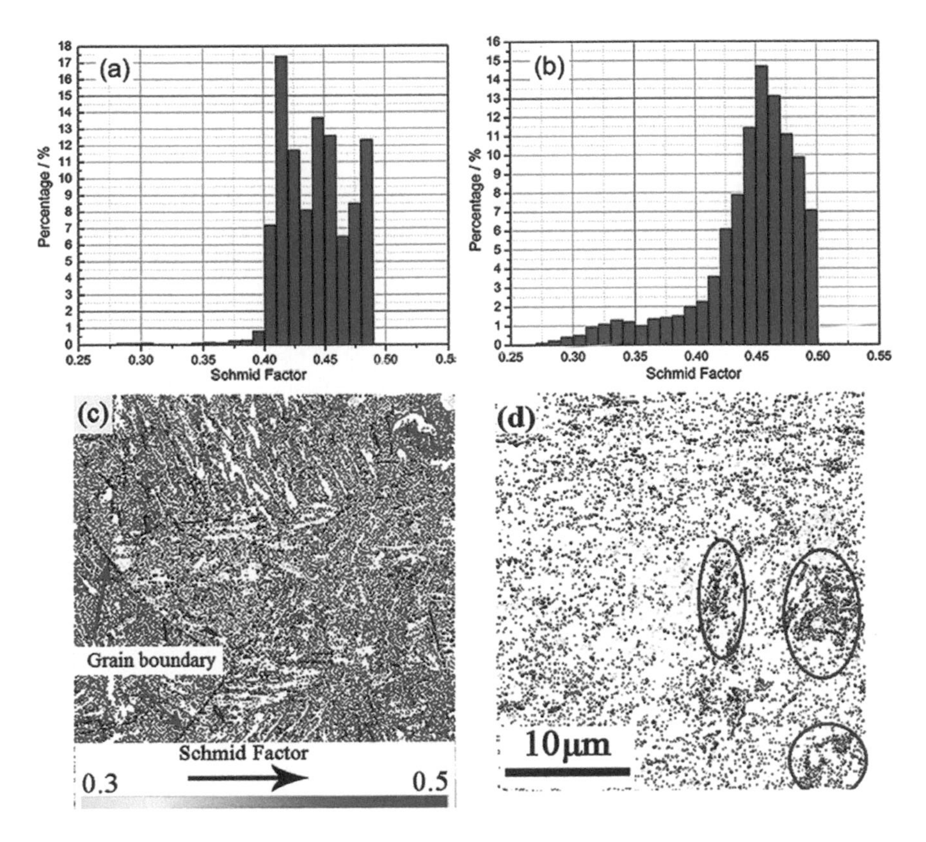

Figure 7.8 Schmid factors and Schmid factors distribution map of the austenite grains in 800T sample. (a) (c) before tensile deformation; (b) (d) after tensile deformation (Cai et al. 2013a).

deformation (Figure 7.8a) and 15% after tensile deformation (Figure 7.8b), indicating that most of the austenite grains having the Schmid factors smaller than 0.40 do not transform to martensite during deformation. Figure 7.8c reveals that the blocky-shape austenite grains with a greater Schmid factors have a high priority in martensitic transformation during deformation. The result is similar to that obtained by Seo et al. (2012) on a Fe-3.5Mn-5.9AL-0.4C steel. However, the austenite grains with greater Schmid factors are not necessarily transformed during deformation. Compared with the marked regions in Figures 7.8d, 7.9b and 7.10, it can be found that the <111> orientation austenite grains with greater Schmid factors are still retained after tensile failure, and the morphology turns out to be granular and film shapes. Therefore, it can be concluded that

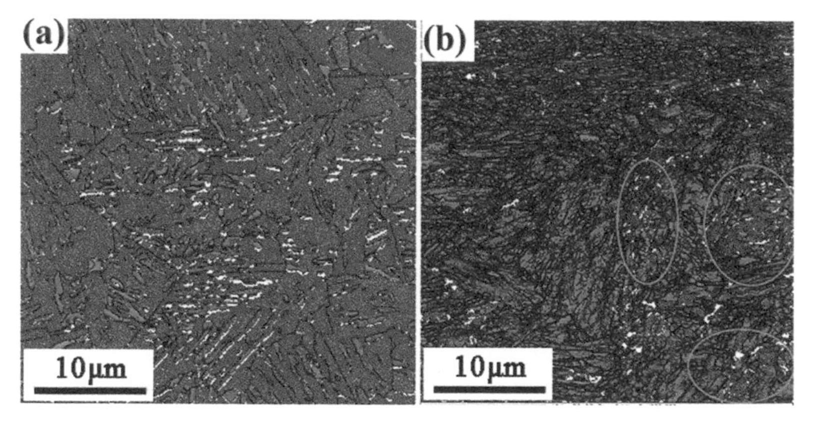

Figure 7.9 Phase mappings of 800T sample through EBSD (a) the undeformed one, (b) the fractured one. (Retained austenite appearing as red phase, and BCC phase appearing as gray phase.) (Cai et al. 2013a.)

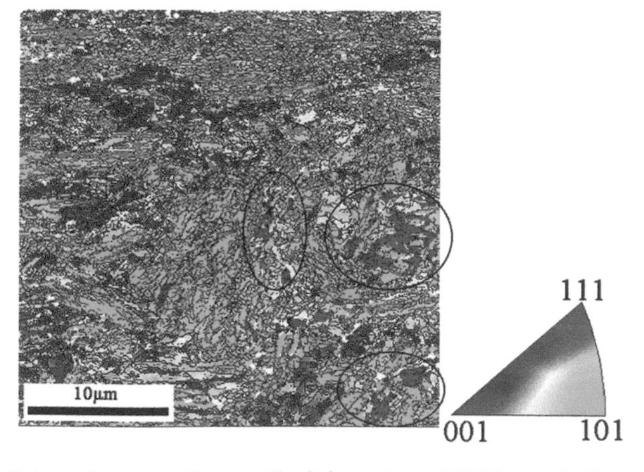

Figure 7.10 Orientation map after tensile deformation of 800T sample (Cai et al. 2013a).

the morphology of austenite grains plays a more important role in the mechanical stability.

7.5.3 Chemical Composition or Grain Size

The chemical composition has a strong influence on austenite stability. An increase in C and Mn contents, which are strong austenite stabilizers, increases the stability of austenite at room temperature (Wang and Zwaag 2001, Lee et al. 2011d). A reduction in austenite grain size can effectively increase austenite stability by suppressing martensite transformation (Sugimoto et al. 1993, Lee and Lee 2005). We have focused on defining the critical factor that governs austenite stability, chemical composition or grain size. To solve this question, the cold-rolled Fe-11Mn-4Al-0.2C steel was selected as the experimental steel.

It is clear that austenite grain size of the heat-treated samples increases with increase in the intercritical hardening temperature (Figure 7.11),

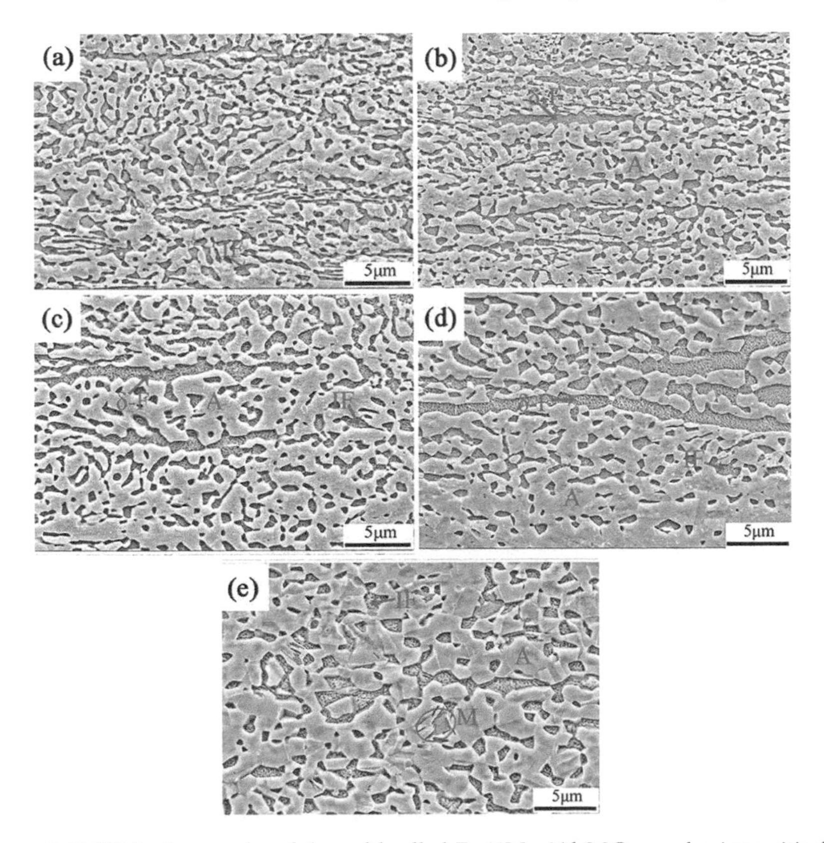

Figure 7.11 SEM micrographs of the cold-rolled Fe-11Mn-4Al-0.2C samples intercritically hardening at different temperatures. (a) 730°C, (b) 750°C, (c) 770°C, (d) 800°C and (e) 850°C (A: austenite, IF: intercritical ferrite, δ-F: δ ferrite, M: martensite) (Cai et al. 2015).

and is the primary reason for the decrease in austenite stability. C and Mn concentrations in austenite presumably decrease with the increase in intercritical hardening temperature, which also decrease austenite stability. In an attempt to clarify the dominant factor that governs austenite stability, the as-cold rolled Fe-11Mn-4Al-0.2C samples were intercritical annealed at 750°C for 3, 7 and 10 min, respectively (henceforth referred as 3-min sample, 7-min sample and 10-min sample), and then immediately quenched in water. The mechanical properties and austenite fraction prior to and after tensile tests of the three samples are listed in Table 7.1. It is clear that both UTS and TE decrease with annealing time. Comparing the austenite fraction prior to and after tensile tests, the 3-min sample exhibited superior mechanical properties and significantly contributed by the TRIP effect. In contrast, the two other samples exhibit insignificant TRIP effect. Thus, it is inferred that austenite stability increases with the increase of annealing time.

Figure 7.12 summarizes the WH behavior of the three samples. A similar WH behavior is observed for all the three samples in stages 1 and 2. However, the serrated behavior in stage 3 becomes smooth with the increase of annealing time. Based on our previous work (Cai et al. 2014b), the fluctuation in stage 3 has resulted from discontinuous TRIP effect because of the presence of austenite with different degrees of stability. The increase in grain size with annealing time (Figure 7.13) decreases austenite stability. Accordingly, we envisage that the chemical elements, Mn and C, are enriched and uniformly distributed in the austenite phase with the increase of annealing time.

The enrichment of carbon in austenite (X_C) is estimated using Eq. (7.10) (Van Dijk et al. 2005):

$$\alpha_\gamma = 0.3556 + 0.00453X_C + 0.000095X_{Mn} + 0.00056X_{Al} \tag{7.10}$$

Table 7.1 Mechanical properties and austenite fraction prior to and after tensile tests of the samples intercritical annealing at 750°C for 3, 7 and 10 min.

	3 min	**7 min**	**10 min**
YS (MPa)	699	696	683
UTS (MPa)	899	867	860
TE (%)	70.4	54	53.1
UTS × TE (GPa %)	63.3	46.8	45.7
f_0 (vol. %)	65.7	62.4	67.0
f_1 (vol. %)	27.1	52.4	51.4

YS: yield strength, UTS: ultimate tensile strength, TE: total elongation, f_0: fraction of austenite prior to tensile tests, f_1: fraction of austenite {after} tensile tests.

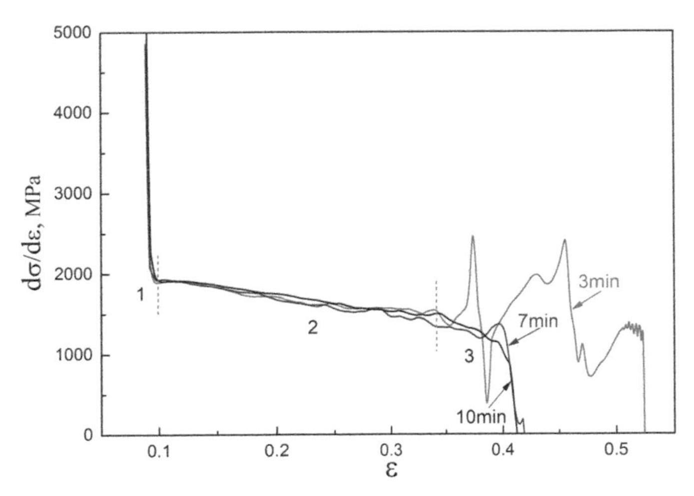

Figure 7.12 Work hardening rate of the cold-rolled Fe-11Mn-4Al-0.2C samples intercritical annealing at 750°C for 3 min, 7 min and 10 min, respectively, and then quenching (Cai et al. 2015).

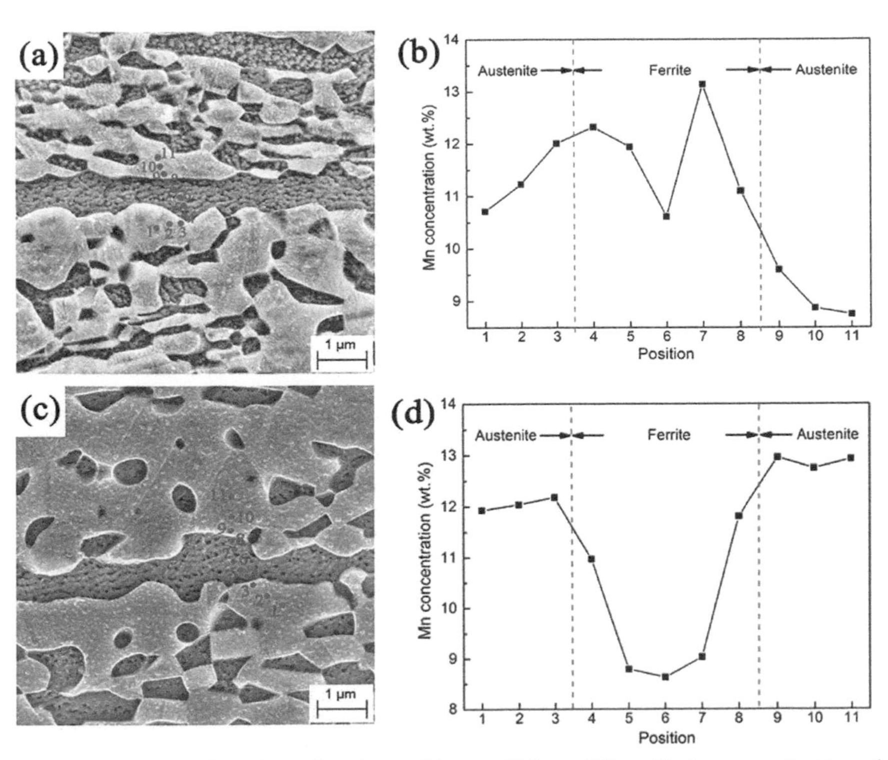

Figure 7.13 SEM-EDS analysis showing evidence of Mn partitioned between austenite and δ-ferrite. SEM micrographs of samples intercritical annealing at 750°C for 3 min (a) and 10 min (c), (b) and (d) are the corresponding Mn distribution by EDS (Cai et al. 2015).

where X_C, X_{Mn}, and X_{Al} are the concentrations of carbon, manganese, and aluminum in austenite, respectively, in wt.%. α_γ is the austenite lattice parameter, in nm, which can be calculated from Eqs. (7.11) and (7.12) (Xu et al. 2012).

$$\alpha_\gamma = (\lambda / 2 \sin \theta) \times \sqrt{h^2 + k^2 + l^2} \tag{7.11}$$

$$d_{hkl} = \lambda / 2 \sin \theta \tag{7.12}$$

where λ is the X-ray wave length, θ is the diffraction angle, d_{hkl} is the inter planar spacing, and h, k, l are the lattice parameters. The estimated carbon concentrations (X_C) of the three samples are 0.35, 0.29 and 0.23, respectively, which decrease with annealing time, and are contrary to our assumption that carbon is enriched in austenite with the increase of annealing time.

Figure 7.13 shows the distribution of Mn in and around austenite grains of the 3-min sample and 10-min sample. EDS has been used to estimate the approximate partitioning of Mn. Based on Figure 7.13b and 7.13d, Mn is partitioned to austenite from δ-Ferrite during intercritical annealing. In the case of the 3-min sample, the Mn concentration is non-uniform in a single austenite grain, and the average Mn concentration varies significantly in different austenite grains. The statistical results of ten randomly selected grains show that intercritical ferrite (IF) (12.1 wt.%) has a higher Mn content than austenite (11.2 wt.%). In the 10-min sample, the Mn concentration is uniform in a single austenite grain, and no significant difference in Mn concentration is observed in different austenite grains. The statistical results of ten randomly selected grains indicate that austenite (12.3 wt.%) has a higher Mn concentration than IF (9.8 wt.%). This implies that non-uniform Mn distribution in austenite of the 3-min sample results in different degrees of stability of austenite, because of which serrated work-hardening behavior is observed in the stage 3 of the 3-min sample. Moreover, the enhancement of austenite stability is attributed to Mn diffused from ferrite (both IF and δ-ferrite) to austenite more sufficiently with the increase of annealing time.

On the basis of the WH behavior of the samples annealed at 750°C, 770°C, and 800°C for 3 min in Figures 7.12 and 7.14, it can be found that the initiation of stage 3 corresponds to strains of 0.34, 0.25, and 0.10, respectively, and it occurs over a total strain range of 0.18, 0.25 and 0.26, respectively. Based on the numerical results and the above discussion, it is inferred that austenite becomes less stable and Mn atoms diffuse more unevenly with the increase of intercritical annealing temperature. Comparing the microstructure of the samples heat-treated at 770°C and 800°C prior to and after tensile tests (Figure 7.15), it is obvious that after tensile tests, the size of the majority of the retained austenite grains is fine. Moreover, as marked in Figure 7.15, some randomly selected austenite grains have lower Mn content than their adjacent martensite grains, which has been confirmed by using EDS. Thus, it is inferred that austenite grain size is the

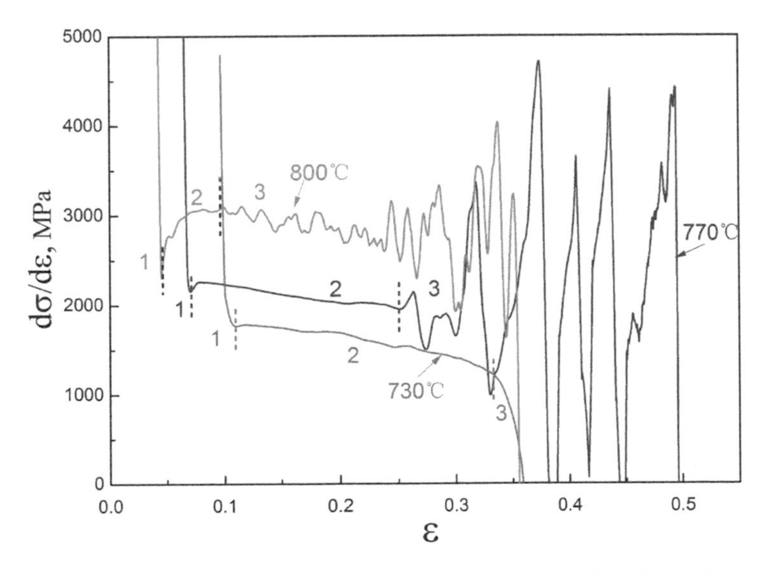

Figure 7.14 Work hardening rate of the cold-rolled Fe-11Mn-4Al-0.2C samples intercritical hardening at 730°C, 770°C and 800°C, respectively (Cai et al. 2015).

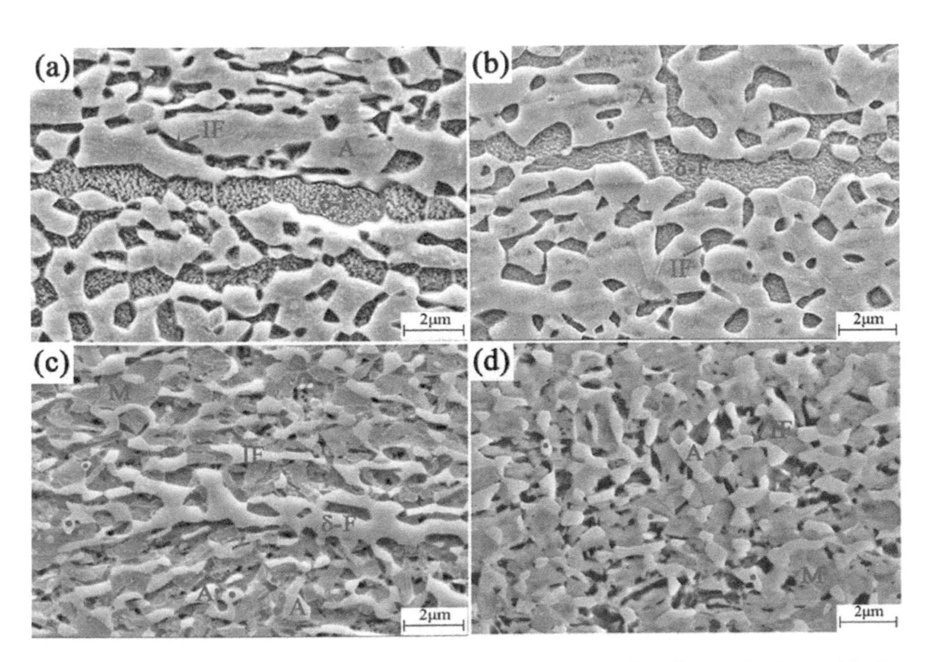

Figure 7.15 SEM micrographs of the cold-rolled Fe-11Mn-4Al-0.2C samples intercritically hardening at 770°C (a) and 800°C (b) prior to tensile tests, (c) and (d) are the corresponding sample after tensile test. The red dots represent austenite, and the green rectangles represent martensite (A: austenite, IF: intercritical ferrite, δ-F: δ ferrite, M: martensite) (Cai et al. 2015).

critical factor in influencing austenite stability, and the diffusion of Mn at different intercritical annealing temperatures results in austenite with different degrees of stability. On prolonged annealing time, Mn diffuses uniformly, while the increase in annealing temperature has an opposite effect. In conclusion, the critical factor that governs the stability of austenite in the investigated cold rolled TRIP steel is grain size.

7.6 Hot Rolling of TRIP Steels

7.6.1 Necessity of Hot Rolling

The excellent mechanical properties of TRIP steels come from the strain-induced transformation of retained austenite to martensite at room temperature, which may lead to redistribution of stresses and composite effect during deformation. An excellent balance between strength and ductility can be achieved if the TRIP effect can be triggered at an appropriate level by optimal stabilization of retained austenite. The amount of retained austenite is a key factor in controlling the final mechanical properties of TRIP steels. It has been shown that an increase in the volume fraction of retained austenite increases the strain-hardening coefficient. However, a large amount of retained austenite does not necessarily induce a higher uniform elongation because sufficient carbon content is required to stabilize the large amount of retained austenite (Lee et al. 2004). If the retained austenite undergoes strain-induced martensitic transformation at very low strains or is very resistant to such transformation, the TRIP effect will be minimal. Therefore, the volume fraction of retained austenite and its stability must be carefully controlled in order to obtain the required combination of strength and ductility of TRIP steels.

For a given TRIP steel, another way to improve the stability of retained austenite is to reduce its grain size to an appropriate level. The work of Wang and Zwaag (2001) indicated that the decrease in retained austenite grain size induced a significant decrease in the martensitic transformation starting temperature. A retained austenite with a grain size smaller than 0.01 μm is useless for TRIP steels since it will not transform to martensite, while that with a grain size larger than 1 μm may be equally useless, since it will immediately transform to martensite upon cooling or during application of small stress.

In order to achieve the best combination of mechanical strength and ductility of TRIP steels, other microstructural constituents including ferrite, bainite and/or martensite should also be properly controlled in addition to the volume fraction and stability of retained austenite. A TRIP steel with adequate retained austenite and optimal multiphase microstructural combination is essential for obtaining the best overall performance. For obtaining the multiphase microstructure of TRIP steels, a mixed microstructure of austenite and ferrite with nearly the same volume fraction of about 50% is usually achieved before the isothermal bainitic treatment (Li et al. 2013). For conventional hot-rolled process of TRIP steels, the control of such mixed microstructure is commonly heat-treated by adjusting cooling rate and time through multi-stage cooling after hot rolling of austenite (Hashimoto et al. 2004, Pereloma et al. 1999). In order to obtain an optimal multiphase microstructure, cold-rolled TRIP steels are usually processed with a two-step heat treatment, i.e., intercritical annealing and isothermal bainite treatment (Fu et al. 2014). Heat treatment technologies can normally yield the best combination of strength-ductility in TRIP steels. However, the key drawback of heat treatment is the difficulties in the precise realization in industry, limiting the mass production of TRIP steels. Under such background, thermo-mechanical controlled hot rolling technology has been proposed to generate an optimal multiphase microstructure and thus achieve the required mechanical properties of TRIP steels (Hanzaki et al. 1995, Li and Wu 2006a, Li and Wu 2006b, Li and Wu 2008, Yang et al. 2010).

7.6.2 Hot Rolling and the Related Technologies

Hot rolling is a necessary processing method in the production of TRIP steels in industry, and a selection of hot rolling procedures becomes extremely important as it may affect the microstructure and mechanical properties of the final products. In hot rolling, TRIP steels are generally suffered controlled rolling followed by controlled cooling to the bainitic transformation temperature, at which coiling of rolled TRIP steel strip is performed (Hanzaki et al. 1995, Pereloma et al. 1999). In the industrial production of TRIP steels, austenite may transform to bainite during the coiling process after hot rolling. Hot coils cool slowly, and they may maintain temperatures of about 350–450°C for a few hours or even days. During this process, the metastable retained austenite can be decomposed into other phases (Sandvik 1982). Coiling temperature is an important factor affecting the grain size and phase composition of hot rolled TRIP steels, and a change in coiling temperature may induce variations of microstructural constituents (Xiong et al. 2010). In order to identify the decomposition behavior of retained austenite during coiling process, Jun et al. (2004) conducted a study on hot rolling and coiling of a 0.2C-1.5Mn-2Si TRIP-aided steel by laboratory simulation on a Gleeble 3500 thermal-mechanical test simulator. Four-pass hot rolling was firstly performed at temperatures from 1000 to 800°C with reduction of 20% for each pass. Then, coiling was simulated by salt bath with holding temperatures of 350, 400 and 450°C, and holding time of 5, 20, 60, 120 and 480 min, respectively. Their results indicated that retained austenite was thermally stable up to 350°C, but it could be decomposed into cementite, ferrite and pearlite at temperatures higher than 370°C. Therefore, coiling temperature of hot rolled TRIP-aided steels should not be higher than 370°C in order to avoid deterioration of formability and mechanical properties due to decomposition of retained austenite.

Rolling in the intercritical region, i.e., two-phase ($\alpha + \gamma$) area, has a substantial effect on the microstructures and mechanical properties of TRIP-aided steels. It increases dislocation density, and arranges it in a sub grain substructure which in its turn affects the mechanisms occurring during the following ageing treatment conduced on the run-out table. The work of Basuki and Aernoudt (1999) indicated that rolling in the intercritical region stabilized retained austenite against strain induced transformation on room temperature straining. The increased dislocation density, grain refinement and carbon enrichment were the main factors that govern the retained austenite stability, and consequently the strength and work-hardening behavior of the TRIP steel. Zhang et al. (2013) found that intercritical rolling resulted in a micro duplex microstructure in TRIP steel with ferrite and austenite lath thickness continuously thinned and gradually rotated to be parallel to the rolling plane with rolling strain, and during which the austenite was most retained after intercritical rolling. The mechanical properties of TRIP steel could be greatly enhanced by intercritical rolling due to the refinement of microstructure and the TRIP effects contributed by the increased volume fraction of austenite phase.

The finish rolling temperature plays an important role in the formation of multiphase microstructure of TRIP steels. Mi et al. (2013) found that more austenite grains existed when TRIP steels were processed at low finish rolling temperature, while at high finish rolling temperature, more recrystallized retained austenite would be formed. However, there was no significant difference in mechanical properties of tensile strength and total elongation under different finish rolling temperatures. Hou et al. (2011) showed that the volume fractions of ferrite and retained austenite in a low Si TRIP steel increased with a decrease in finish rolling temperatures, and the enhanced strength could be obtained due to the combined effects of fine grain strengthening, dislocation strengthening and precipitation strengthening. Based on the work conducted by Hou et al. (2011), a lower finish rolling temperature benefited the acquirement of excellent mechanical properties of TRIP steel.

Cooling methods after hot rolling have great effect on the final microstructure of TRIP steels. Wu et al. (2008) investigated the effects of cooling patterns on the microstructural evolution and the stability of retained austenite of a low Mn TRIP steel by thermo-mechanical controlled hot rolling. Figure 7.16 shows the processing schedules. In their study, the hot rolled TRIP steel was suffered from a three-step cooling pattern with an order of "water cooling→air cooling→water cooling". A multiphase microstructure containing polygonal ferrite, granular bainite and retained austenite and excellent mechanical properties of TRIP steel can be obtained

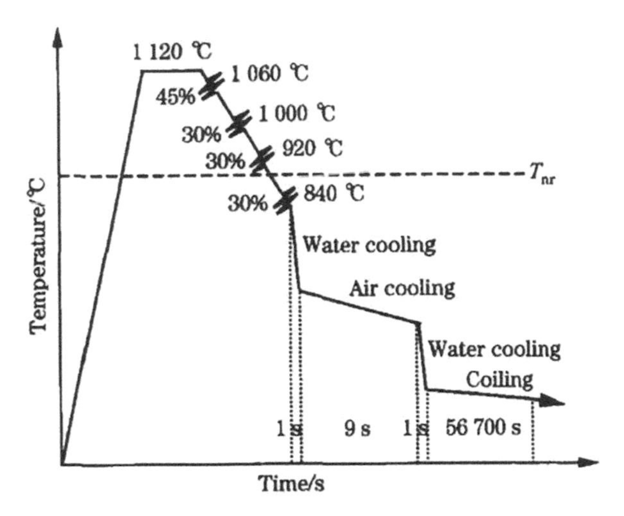

Figure 7.16 Schematic of Thermo-mechanical controlled processing (T_{nr}: non-recrystallization temperature) (Wu et al. 2008).

through the three-step cooling after hot rolling. A careful control of cooling schedules after hot rolling may result in an enhanced stability of the retained austenite and consequently promote the TRIP effect of TRIP steels (Wu et al. 2006, 2008).

Q&P aims to create mixtures of carbon-depleted martensite and carbon-enriched retained austenite through a controlled thermal treatment process, and is widely used to produce high-strength steels consisting of martensitic matrix and enhanced level of retained austenite (Speer et al. 2003, Edmonds et al. 2006). However, most investigations to date have been focusing on the hot-rolling off-line Q&P heat treatment or the heat treatment of cold-rolled sheet. In order to realize hot rolling on-line heat treatment and to save energy by replacing the reheating process with making use of the residual deformed temperature to achieve the Q&P heat treatment, Tan et al. (2014) applied hot-rolling direct quenching and partitioning (HDQP) technology to process a low carbon TRIP steel with required microstructure and mechanical properties. The HDQP technology combines thermo-mechanical control process (TMCP) and ultra-fast cooling technology, and can be easily controlled by the dynamic partitioning procedures and the isothermal partitioning procedures in the laboratory environment. The dynamic partitioning process has a potential in the improvement of the mechanical properties of a HDQP steel with higher UTS and excellent PSE, but significantly more work is still required to optimize the processing parameters with the purpose of producing a variety of TRIP steel grades and to realize transition from scientific research in laboratory to hot rolling production in industry.

7.7 Summary

A comprehensive overview has been given on the three generations of TRIP steels with different Mn contents and their hot rolling technologies. Focus has been made on the heat treatment technology on each category of TRIP steels in order to achieve the excellent combination of strength and ductility under optimized heat treatment schedules. As the stability of austenite plays significant role in the TRIP effect, factors including composition, morphology and grain size of austenite affecting the stability of austenite have been discussed. In order to implement production in industry, hot rolling and the related technologies on TRIP steels have been induced with a purpose of producing high quality TRIP steel products with required microstructure and mechanical properties.

Results presented indicate that the microstructure of conventional low Mn TRIP steels consists of ferrite as the dominant phase, retained austenite, bainite, and occasionally a small amount of martensite. Low Mn TRIP steels exhibit good combinations of strength (500–1000 MPa) and elongation (15–40%) together with relatively high PSE (near 20 GPa%). In contrast, Q&P TRIP steels consisting of martensite laths and retained austenite exhibit much higher strength (1200–2000 MPa) but lower ductility (10–20%) than conventional TRIP steels. The microstructure constituents of high Mn TRIP steels include austenite and martensite or ferrite islands. High Mn TRIP steels demonstrate lower tensile strength (650–900 MPa) but much better ductility (65–95%) than low Mn TRIP steels. The medium Mn TRIP steels with austenite-ferrite mixed microstructure are characterized by excellent combination of tensile strength of 850–1500 MPa and elongation of 30–70%. Austenite stability plays a significant role in TRIP effect. Morphology of austenite plays a more important role than orientation in the mechanical stability, and grain size is found to be the most critical factor in the austenite stability. A careful design of hot rolling schedules and an optimization of the related processing parameters benefit the transition of scientific research in laboratory to hot rolling production in industry of a variety of TRIP steels.

keywords: TRIP steel, heat treatment, austenite stability, microstructure, mechanical properties, work hardening behavior

References

Allain, S., J.-P. Chateau, O. Bouaziz, S. Migot and N. Guelton. 2004. Correlations between the calculated stacking fault energy and the plasticity mechanisms in Fe-Mn-C alloys. Mater. Sci. Eng. A 387-389: 158–162.

Arlazarov, A., M. Goune, O. Bouaziz, A. Hazotte, G. Petitgand and P. Barges. 2012. Evolution of microstructure and mechanical properties of medium Mn steels during double annealing. Mater. Sci. Eng. A 542: 31–39.

Aydin, H., E. Essadiqi, I.H. Jung and S. Yue. 2013. Development of 3rd generation AHSS with medium Mn content alloying compositions. Mater. Sci. Eng. A 564: 501–508.

Basuki, A. and E. Aernoudt. 1999. Influence of rolling of TRIP steel in the intercritical region on the stability of retained austenite. J. Mater. Process. Tech. 89-90: 37–43.
Bellhouse, E.M. and J.R. McDermid. 2010. Effect of continuous galvanizing heat treatments on the microstructure and mechanical properties of high Al-Low Si transformation induced plasticity steels. Metall. Mater. Trans. A 41A: 1460–1473.
Bhadeshia, H.K.D.H. and D.V. Edmonds. 1983. Bainite in silicon steels: new composition–property approach. Metal. Sci. 17: 411–419.
Bhattacharyya, T., S.B. Singh, S. Das, A. Haldar and D. Bhattacharjee. 2011. Development and characterisation of C-Mn-Al-Si-Nb TRIP aided steel. Mater. Sci. Eng. A 528: 2394.
Cai, Z.H., H. Ding, R.D.K. Misra, H. Kong and H.Y. Wu. 2014c. Unique impact of ferrite in influencing austenite stability and deformation behavior in a hot-rolled Fe-Mn-Al-C steel. Mater. Sci. Eng. A 595: 86–91.
Cai, Z.H., H. Ding, R.D.K. Misra and H. Kong. 2014b. Unique serrated flow dependence of critical stress in a hot-rolled Fe-Mn-Al-C steel. Scr. Mater. 71: 5–8.
Cai, Z.H., H. Ding, X. Xue, J. Jiang, Q.B. Xin and R.D.K. Misra. 2013b. Significance of control of austenite stability and three-stage work-hardening behavior of an ultra high strength-high ductility combination transformation-induced plasticity steel. Scr. Mater. 68: 865–868.
Cai, Z.H., H. Ding, X. Xue and Q.B. Xin. 2013a. Microstructural evolution and mechanical properties of hot-rolled 11% manganese TRIP steel. Mater. Sci. Eng. A 560: 388–395.
Cai, Z.H., H. Ding, Z.Y. Ying and R.D.K. Misra. 2014a. Microstructural Evolution and Deformation Behavior of a Hot-Rolled and Heat Treated Fe-8Mn-4Al-0.2C Steel. J. Mater. Eng. Perform. 23: 1131–1137.
Cai, Z.H., H. Ding, R.D.K. Misra and Z.Y. Ying. 2015. Austenite stability and deformation behavior in a cold-rolled transformation-induced plasticity steel with medium manganese content. Acta Mater. 84: 229–236.
Chatterjee, S., M. Murugananth and H.K.D.H. Bhadeshia. 2007. δ TRIP steel. Mater. Sci. Technol. 23: 819–827.
Chiang, J., B. Lawrence, J.D. Boyd and A.K. Pilkey. 2011. Effect of microstructure on retained austenite stability and work hardening of TRIP steels. Mater. Sci. Eng. A 528: 4516–4521.
Crussard, C. 1953. Rapport entre la formeexacte des courbes de traction des métauxet les modifications concomitantes de leur structure. Rev. Met. (Paris) 10: 697–710.
Dan, W.J., S.H. Li, W.G. Zhang and Z.Q. Lin. 2008. The effect of strain-induced martensitic transformation on mechanical properties of TRIP steel. Mater. Des. 29: 601–612.
Dastur, Y.N. and W.C. Leslie. 1981. Mechanism of Work Hardening in Hadfield Manganese Steel. Metall. Trans. A 12: 749–759.
De Cooman, B.C. 2004. Structure-properties relationship in TRIP steels containing carbide-free bainite. Curr. Opin. Solid State Mater. Sci. 8: 285–303.
Ding, H., H. Ding, D. Song, Z.Y. Tang and P. Yang. 2011. Strain hardening behavior of a TRIP/TWIP steel with 18.8% Mn. Mater. Sci. Eng. A 528: 868–873.
Edmonds, D.V., K. He, F.C. Rizzo, B.C. De Cooman, D.K. Matlock and J.G. Speer. 2006. Quenching and partitioning martensite-A novel steel heat treatment. Mater. Sci. Eng. A 438-440: 25–34.
Ferjutz, K. and J.R. Davis. 1993. ASM Handbook, Welding, Brazing and Soldering, ASM Int. 6: 405–407.
Frommeyer, G. and U. Brux. 2003. Supra-Ductile and High-Strength Manganese-TRIP/TWIP Steels for High Energy Absorption Purposes. P. Neumann, ISIJ Int. 43: 438–446.
Fu, B., W.Y. Yang, L.F. Li and Z.Q. Sun. 2014. Effect of bainitic transformation temperature on the mechanical behavior of cold-rolled TRIP steels studied with *in situ* high-energy X-ray diffraction. Mater. Sci. Eng. A 603: 134–140.
Garcia-Mateo, C. and F.G. Caballero. 2005. The role of retained austenite on tensile properties of steels with bainitic microstructures. Mater. Trans. 46: 1839–1846.
Garcia-Mateo, C., F.G. Caballero and H.K.D.H. Bhadesia. 2003. Acceleration of low-temperature bainite. ISIJ Int. 43: 1821–1825.

Gibbs, P.J., E. De Moor, M.J. Merwin, B. Clausen, J.G. Speer and D.K. Matlock. 2011. Austenite stability effects on tensile behavior of manganese-enriched-austenite transformation-induced plasticity steel. Metall. Mater. Trans. A 42A: 3691–3702.

Grassel, O., L. Kruger, G. Frommeyer and L.W. Meyer. 2000. High strength Fe-Mn-(Al, Si) TRIP/TWIP steels development—properties—application. Int. J. Plast. 16: 1391–1409.

Haidemenopoulos, G.N. and A.N. Vasilakos. 1997. On the thermodynamic stability of retained austenite in 4340 steel. J. Alloy. Compd. 247: 128–133.

Hanzaki, A.Z., P.D. Hodgson and S. Yue. 1995. Hot deformation characteristics of Si-Mn TRIP steels with and without Nbmicroalloy additions. ISIJ Int. 35: 324–331.

Hashimoto, S., S. Ikeda, K.I. Sugimoto and S. Miyake. 2004. Effects of Nb and Mo addition to 0.2%C–1.5%Si1.5%Mn steel on mechanical properties of hot rolled TRIP-aided steel sheets. ISIJ Int. 44: 1590–1598.

Hecker, S.S., M.G. Stout, K.P. Staudhammer and J.L. Smith. 1982. Effects of strain state and strain rate on deformation-induced transformation in 304 stainless steel: Part I. Magnetic measurements and mechanical behavior. Metall. Trans. A 13A: 619–626.

Hollomon, J.H. 1945. Tensile deformation. AIME Trans. 162: 268–290.

Hou, X.Y., Y.B. Xu, Y.F. Zhao and D. Wu. 2011. Microstructure and mechanical properties of hot rolled low silicon TRIP steel containing phosphorus and vanadium. J. Iron. Steel Res. Int. 18(11): 40–45.

Hsu, T.Y., Y.W. Chang, N.J. Kim and C.S. Lee (eds.). 2007. Proceedings of 6th Pacific Rim International Conference on Advanced Materials and Processing. Material Science Forum. Cheju Island. South Korea, pp. 2283–2286.

Ishida, K. and T. Nishizawa. 1974. Effect of alloying elements on stability of epsilon iron. Trans. Jpn. Inst. Met. 15: 225–231.

Jacques, P.J., E. Girault, A. Mertens, B. Verlinden, J. Van Humbeeck and F. Delannay. 2001a. The developments of cold-rolled TRIP-assisted multiphase steels. Al-alloyed TRIP-assisted multiphase steels. ISIJ Int. 41(9): 1068–1074.

Jacques, P.J., Q. Furnémont, F. Lani, T. Pardoen and F. Delannay. 2007. Multiscale mechanics of TRIP-assisted multiphase steels: I. Characterization and mechanical testing. Acta Mater. 55: 3681–3693.

Jacques, P., Q. Furnémont, A. Mertens and F. Delannay. 2001b. On the sources of work hardening in multiphase steels assisted by transformation-induced plasticity. Philos. Mag. A 81: 1789–1812.

Jaoul, B. 1957. Etude de la forme des courbes de deformation plastique. J. Mech. Phys. Solids. 5: 95–114.

Jha, B.K., R. Avtar, V.S. Dwivedi, V. Ramaswamy and J. Mater. 1987. Applicability of modified Crussard-Jaoul analysis on the deformation behaviour of dual-phase steels. Sci. Lett. 6: 891–893.

Jirkova, H., B. Masek, M.F.X. Wagner, D. Langmajerova, L. Kucerova, R. Treml and D. Kiener. 2014. Influence of metastable retained austenite on macro and micromechanical properties of steel processed by the Q&P process. J. Alloy. Compd. 615: S163–168.

Jun, H.J., S.H. Park, S.D. Choi and C.G. Park. 2004. Decomposition of retained austenite during coiling process of hot rolled TRIP-aided steels. Mater. Sci. Eng. A 379: 204–209.

Jung, Y.S., Y.K. Lee, D.K. Matlock and M.C. Mataya. 2011. Effect of grain size on strain-induced martensitic transformation start temperature in an ultrafine grained metastable austenitic steel. Metal. Mater. Int. 17: 553–556.

Koyama, M., T. Sawaguchi, T. Lee, C. Lee and K. Tsuzaki. 2011. Work hardening associated with ε-martensitic transformation, deformation twinning and dynamic strain aging in Fe-17Mn-0.6C and Fe-17Mn-0.8C TWIP steels. Mater. Sci. Eng. A 528: 7310–7316.

Lani, F., Q. Furnémont, T. Van Rompaey, F. Delannay, P.J. Jacques and T. Pardoen. 2007. Multiscale mechanics of TRIP-assisted multiphase steels: II. Micromechanical modelling. Acta Mater. 55: 3695–3705.

Lee, C.G., S.-J. Kim, T.-H. Lee and S. Lee. 2004. Effects of volume fraction and stability of retained austenite on formability in 0.1C-1.5Si-1.5Mn-0.5Cu TRIP-aided cold-rolled steel sheet. Mater. Sci. Eng. A 371: 16–23.

Lee, S., S.-J. Lee and B.C. De Cooman. 2011a. Austenite stability of ultrafine-grained transformation-induced plasticity steel with Mn partitioning. Scr. Mater. 65: 225–228.

Lee, S., S.-J. Lee, S.S. Kumar, K. Lee and B.C. De Cooman. 2011b. Localized deformation in multiphase, ultra-fine-grained 6 PctMn transformation-induced plasticity steel. Metall. Mater. Trans. A 42A: 3638–3651.

Lee, S.J., J. Kim, S.N. Kane and B.C. De Cooman. 2011c. On the origin of dynamic strain aging in twinning-induced plasticity steels. Acta Mater. 59: 6809–6819.

Lee, S.J., S. Lee and B.C. De Cooman. 2011d. Mn partitioning during the intercritical annealing of ultrafine-grained 6% Mn transformation-induced plasticity steel. Scri. Mater. 64: 649–652.

Lee, S.J. and Y.K. Lee. 2005. Effect of austenite grain size on martensitic transformation of a low alloy steel. Mater. Sci. Forum. 475-479: 3169–3172.

Li, L.F., X.J. Zhang, W.Y. Yang and Z.Q. Sun. 2013. Microstructure and mechanical properties of a low-carbon Mn-Si multiphase steel based on dynamic transformation of undercooled austenite. Metall. Mater. Trans. A 44: 4337–4345.

Li, Z. and D. Wu. 2006a. Effects of hot deformation and subsequent austempering on the mechanical properties of Si-Mn TRIP steels. ISIJ Int. 46: 121–128.

Li, Z. and D. Wu. 2006b. Effect of thermomechanical controlled processing on the microstructure and mechanical properties of Fe-C-Mn-Si multiphase steels. ISIJ Int. 46: 1059–1066.

Li, Z. and D. Wu. 2008. Influence of hot rolling conditions on the mechanical properties of hot rolled TRIP steel. J. Wuhan Uni. Technol.-Mater. Sci. Ed. 23: 74–79.

Li, Z.C., H. Ding and Z.H. Cai. 2015. Mechanical properties and austenite stability in hot-rolled 0.2C-1.6/3.2Al-6Mn-Fe TRIP steel. Mater. Sci. Eng. A 639: 559–566.

Luo, H.W., J. Shi, C. Wang, W.Q. Cao, X.J. Sun and H. Dong. 2011. Experimental and numerical analysis on formation of stable austenite during the intercritical annealing of 5 Mn steel. Acta Mater. 59: 4002.

Luo, Y., J.M. Peng, H.B. Wang and X.C. Wu. 2010. Effect of tempering on microstructure and mechanical properties of a non-quenched bainitic steel. Mater. Sci. Eng. A 527: 3433–3437.

Ludwik, P. 1909. Elemente der TechnogyschenMechanik, Springer, Berlin, Germany, 32–52.

Matsumura, O., Y. Sakuma and H. Takechi. 1987a. Retained Austenite in 0.4C-Si-1.3Mn steel sheet intercritically heated and austempered. ISIJ Int. 27: 570–579.

Matsumura, O., Y. Sakuma and H. Takechi. 1987b. TRIP and its kinetics aspects in austempered 0.4C-1.5Si-0.8Mn steel. Script Metall. 21: 1301–1306.

Matsumura, O., Y. Sakuma and H. Takechi. 1991. Mechanical properties and retained austenite in intercritically heat-treated bainite-transformed steel and their variation with Si and Mn additions. Metall. Trans. A 22A: 489–498.

Matsumura, O., Y. Sakuma and H. Takechi. 1992. Enhancement of elongation by retained austenite in intercritically annealed 0.4C-1.5Si-0.8Mn steel. Trans. ISIJ. 32: 1014–1020.

Merwin, M.J. 2007. Low-carbon manganese TRIP Steels. Mater. Sci. Forum. 539-543: 4327–4332.

Miller, R.L. 1972. Ultrafine-grained microstructures and mechanical properties of alloy steels. Metall. Trans. 3: 905–912.

Mintz, B. 2001. Hot dip galvanising of transformation induced plasticity and other intercritically annealed steels. Int. Mater. Rev. 46: 169–197.

Mintz, B. and J.C. Wright. 1970. Tensile and press-forming properties of 18 percent Ni maraging steel at temperatures up to 300°C. J. Iron Steel Inst. 208: 401.

Mi, Z.L., H.T. Jiang, Z.C. Li, M.F. Chen and Z.G. Wang. 2013. Effect of finishing rolling temperature on microstructure and mechanical properties of microalloyed TRIP steels. J. Iron Steel Res. Int. 20(10): 75–80.

Park, S.J., B. Hwang, K.H. Lee, T.H. Lee, D.W. Suh and H.N. Han. 2013. Microstructure and tensile behavior of duplex low-density steel containing 5 mass% aluminum. Scr. Mater. 68: 365–369.

Pereloma, E.V., I.B. Timokhina and P.D. Hodgson. 1999. Transformation behaviour in thermomechanically processed C-Mn-Si TRIP steels with and without Nb. Mater. Sci. Eng. A 273-275: 448–452.

Qian, L.H., P.C. Guo, F.C. Zhang, J.Y. Meng and M. Zhang. 2013. Abnormal room temperature serrated flow and strain rate dependence of critical strain of a Fe-Mn-C twin-induced plasticity steel. Mater. Sci. Eng. A 561: 266–269.

Reed-Hill, R.E., W.R. Cribb and S.N. Monteiro. 1973. Concerning the analysis of tensile stress-strain data using log $d\sigma/d\varepsilon_p$ versus log σ diagrams. Metall. Trans. 4: 2665–2667.

Samek, L., E. De Moor, J. Penning and B.C. De Cooman. 2006. Influence of alloying elements on the kinetics of strain-induced martensitic nucleation in low-alloy, multiphase high-strength steels. Metall. Mater. Trans. 37: 109–124.

Sandvik, B.P.J. 1982. The bainite reaction in Fe-Si-C alloys: The secondary stage. Metall. Trans. A 13A: 789–800.

Santos, D.B., R. atilde, P.P. Barbosa, D. Oliveira and E.V. Pereloma. 2009. Mechanical behavior and microstructure of high carbon si-mn-cr steel with trip effect. ISIJ Int. 49: 1592–1600.

Sato, A., K. Soma and T. Mori. 1982. Hardening due to pre-existing ε-Martensite in an Fe-30Mn-1Si alloy single crystal. Acta Metall. 30: 1901–1907.

Sato, K., M. Ichinose, Y. Hirotsu and Y. Inoue. 1989. Effects of deformation induced phase transformation and twinning on the mechanical properties of austenitic Fe-Mn-Al alloys. ISIJ Int. 29: 868–877.

Schramm, R.E. and R.P. Reed. 1975. Stacking fault energies of seven commercial austenitic stainless steels. Metall. Trans. A 6: 1345.

Seo, C.-H., K.H. Kwon, K. Choi, K.-H. Kim, J.H. Kwak, S. Lee and N.J. Kim. 2012. Deformation behavior of ferrite-austenite duplex lightweight Fe-Mn-Al-C steel. Scri. Mater. 66: 519–522.

Shi, J., X. Sun, M. Wang, W. Hui, H. Dong and W. Cao. 2010. Enhanced work-hardening behavior and mechanical properties in ultrafine-grained steels with large-fractioned metastable austenite. Scr. Mater. 63: 815–818.

Son, Y.I., Y.K. Lee, K.T. Park, C.S. Lee and D.H. Shin. 2005. Ultrafine grained ferrite–martensite dual phase steels fabricated via equal channel angular pressing: Microstructure and tensile properties. Acta Mater. 53: 3125–3134.

Speer, J.G., D.K. Matlock, B.C. De Cooman and J.G. Schroth. 2003. Carbon partitioning into austenite after martensite transformation. Acta Mater. 51: 2611–2622.

Streicher, A.M., J.G. Speer, D.K. Matlock, B.C. De Cooman and M.A. Baker (eds.). 2004. International Conference on Advanced High Strength Sheet Steels for Automotive Applications. Colorado. United States, pp. 51–62.

Sugimoto, K., M. Kobayashi and S. Hashimoto. 1992a. Ductility and strain-induced transformation in a high-strength transformation-induced plasticity-aided dual-phase steel. Metall. Trans. A 23A: 3085–3091.

Sugimoto, K., M. Misu, M. Kobayashi and H. Shirasawa. 1993. Effects of second phase morphology on retained austenite morphology and tensile properties in a TRIP-aided dual-phase steel sheet. ISIJ Int. 33: 775–782.

Sugimoto, K., N. Usui, M. Kobayashi and S. Hashimoto. 1992b. Effects of volume fraction and stability of retained austenite on ductility in TRIP-Aided dual-phase steel sheets. ISIJ Int. 32: 1311–1318.

Suh, D.W., S.J. Park, C.S. Oh and S.J. Kim. 2007. Influence of partial replacement of Si by Al on the change of phase fraction during heat treatment of TRIP steels. Scr. Mater. 57: 1097–1100.

Suh, D.W., S.J. Park, H.N. Han and S.J. Kim. 2009. Influence of Al on microstructure and mechanical behavior of Cr-containing transformation-induced plasticity steel. Metall. Mater. Trans. A 41: 3276–3281.

Suh, D.W., S.J. Park, T.H. Lee, C.S. Oh and S.J. Kim. 2010. Influence of Al on the microstructural evolution and mechanical behavior of low-carbon, manganese transformation-induced-plasticity steel metall. Mater. Trans. A 41: 397–408.

Swift, H.W. 1952. Plastic instability under plane stress. J. Mech. Phys. Solids 1: 1–18.

Syn, C.K., B. Fultz and J.W. Morris, Jr. 1978.Mechanical stability of retained austenite in tempered 9Ni steel. Metall. Trans. A 9: 1635–1642.

Takahashi, M. and H.K.D.H. Bhadeshia. 1991. A model for the microstructure of some advanced bainitic steels. Mater. Trans. JIM 32: 689–696.

Tamura, I., T. Maki and H. Hato. 1970. Morphology of strain-induced martensite and the transformation-induced plasticity in Fe-Ni and Fe-Cr-Ni alloys. Trans. Iron Steel Inst. Jpn. 10: 163–172.

Tan, X., Y. Xu, X. Yang, Z. Liu and D. Wu. 2014. Effect of partitioning procedure on microstructure and mechanical properties of a hot-rolled directly quenched and partitioned steel. Mater. Sci. Eng. A 594: 149–160.

Thomas, G. 1978. Retained austenite and tempered martensite embrittlement. Metal. Trans. A 9A: 439–450.

Timokhina, I.B., P.D. Hodgson and E.V. Pereloma. 2003. Effect of deformation schedule on the microstructure and mechanical properties of a thermomechanically processed C-Mn-Si transformation-induced plasticity steel. Metall. Mater. Trans. A 34 A: 1599–1609.

Timokhina, I.B., P.D. Hodgson and E.V. Pereloma. 2004. Effect of microstructure on the stability of retained austenite in transformation-induced-plasticity steels. Metall. Mater. Trans. A 35A: 2331–2341.

Tomita, Y. and K. Okabayashi. 1985. Tensile stress-strain analysis of cold worked metal and steels and dual-phase steels. Metal. Trans. A. 16: 865–872.

Tomita, Y. and T. Iwamoto. 1995. Constitutive modeling of trip steel and its application to the improvement of mechanical properties. Int. J. Mech. Sci. 12: 1295–1305.

Tomita, Y. and T. Okawa. 1993. Effect of microstructure on mechanical properties of isothermally bainite-transformed 300M steel. Mater. Sci. Eng. A 172: 145–151.

Tomota, Y., M. Strum and J.W. Morris. 1986. Microstructural dependence of Fe-high Mn tensile behavior. Metall. Trans. A 17A: 537–547.

Tong, M.W., P.K.C. Venkatsurya, W.H. Zhou, R.D.K. Misra, B. Guo, K.G. Zhang and W. Fan. 2014. Structure-mechanical property relationship in a high strength microalloyed steel with low yield ratio: The effect of tempering temperature. Mater. Sci. Eng. A 609: 209–216.

Van Dijk, N.H., A.M. Butt, L. Zhao, J. Sietsma, S.E. Offerman, J.P. Wright and S.V.D. Zwaag. 2005. Thermal stability of retained austenite in TRIP steels studied by synchrotron X-ray diffraction during cooling. Acta Mater. 53: 5439–5447.

Wang, C., H. Ding, M.H. Cai and B. Rolfe. 2014a. Characterization of microstructures and tensile properties of TRIP-aided steels with different matrix microstructure. Mater. Sci. Eng. A 610: 65–75.

Wang, C., H. Ding, M.H. Cai and B. Rolfe. 2014b. Multi-phase microstructure design of a novel high strength TRIP steel through experimental methodology. Mater. Sci. Eng. A 610: 436–444.

Wang, J. and S.V.D. Zwaag. 2001. Stabilization mechanisms of retained austenite in transformation-induced plasticity steel. Metall. Mater. Trans. A 32: 1527–1539.

Wang, X.D., N. Zhong, Y.H. Rong, T.Y. Hsu, L. Wang and J. Mater. 2009. Novel ultrahigh-strength nanolath martensitic steel by quenching-partitioning-tempering process. Res. 24: 260–267.

Wu, D. and Z. Li. 2006. Effect of thermomechanical controlled processing on the microstructure and mechanical properties of Fe-C-Mn-Si multiphase steels. ISIJ Int. 46: 1059–1066.

Wu, D., Z. Li and H.S. Lv. 2008. Effect of controlled cooling after hot rolling on mechanical properties of hot rolled TRIP steel. J. Iron Steel Res. Int. 15(2): 65–70.

Xiong, Z.L., Q.W. Cai and H.T. Jiang. 2010. Heredity characteristic from hot rolled microstructure to annealed microstructure in high strength TRIP steels. J. Iron. Steel Res. Int. 17(10): 38–44.

Xu, H.F., J. Zhao, W.Q. Cao, J. Shi, C.Y. Wang, J. Li and H. Dong. 2012. Heat treatment effects on the microstructure and mechanical properties of a medium manganese steel (0.2C–5Mn). ISIJ Int. 52: 868–873.

Yang, W., L. Li, Y. Yin, Z. Sun and X. Wang. 2010. Hot-rolled TRIP steels based on dynamic transformation of undercooled austenite. Mater. Sci. Forum. 654-656: 250–253.

Yi, H.L., S.K. Ghosh, W.J. Liu, K.Y. Lee and H.K.D.H. Bhadeshia. 2010. Non-equilibrium solidification and ferrite in δ-TRIP steel. Mater. Sci. Technol. 26: 817–823.

Zackay, V.F., E.R. Parker, D. Fahr and R. Busch. 1967. The enhancement of ductility in high-strength steels. Trans. ASM 60: 252–259.

Zhang, R., W.Q. Cao, Z.J. Peng, J. Shi, H. Dong and C.X. Huang. 2013. Intercritical rolling induced ultrafine microstructure and excellent mechanical properties of the medium-Mn steel. Mater. Sci. Eng. A 583: 84–88.

Zhang, Z.C., K.-I. Manabe, Y. Li and F.X. Zhu. 2012. Effect of isothermal bainite treatment on microstructure and mechanical properties of low-carbon TRIP seamless steel tube. Steel Res. Int. 83: 645–652.

Zhong, N., X.D. Wang, L. Wang and Y.H. Rong. 2009. Enhancement of the mechanical properties of a Nb-microalloyed advanced high-strength steel treated by quenching-partitioning-tempering process. Mater. Sci. Eng. A 506: 111–116.

Zhou, S., K. Zhang, Y. Wang, J.F. Gu and Y.H. Rong. 2011. High strength-elongation product of Nb-microalloyed low-carbon steel by a novel quenching-partitioning-tempering process. Mater. Sci. Eng. A 528: 8006–8012.

8

Hot Rolling Practice of Multiphase Steels

Mohamed Soliman and *Heinz Palkowski**

ABSTRACT

In hot strip rolling, thermo-mechanical processing (TMP) implies various types of controlled rolling where temperature, strain rate, rolling reductions and cooling rates are carefully selected to produce the target austenite microstructure and to prevail the phase transformation. In this respect, TMP is considered to be an essential tool for the development of multiphase steels. For the production of these steels, TMP should be designed to ensure a controlled type and extent of transformation on the run-out table by precise control of strip temperature and rolling speed. This chapter sheds light on the production of various types of advanced high strength steels (AHSSs) produced applying TMP. The AHSSs considered are dual phase (DP), transformation induced plasticity (TRIP), bainitic pipeline- and ultra-fine bainite steels. Various aspects of hot rolling of these steels are investigated aiming at contributing to understanding of the microstructural evolution during TMP and correlating them to the mechanical properties.

Institute of Metallurgy, Clausthal University of Technology, D38678 Clausthal-Zellerfeld, Germany.

E-mail: mohamed.soliman@tu-clausthal.de

* Corresponding author: heinz.palkowski@tu-clausthal.de

8.1 Introduction

In spite of the development of sophisticated composite materials in recent years, steel is still the essential industrial manufacturing material. Nevertheless, increasingly in many areas this position is being challenged by other materials such as plastics, ceramics and composites. The steel industry has been facing this challenge by reinventing steel over time and pioneering a whole new class of materials that are over 50% stronger than a decade ago (www.worldautosteel.com). These new generations of steels have been developed in order to respond to the market requirements for steels exhibiting a good combination of high strength and good impact toughness balanced with adequate weldability. At the core of this development lies the AHSSs. AHSSs are produced by controlling the processing methods to obtain steels with defined amounts and distributions of phases. While classical single phase steels either have good formability (ferritic steels) or strength (martensitic steels), multiphase steels combine good formability properties with high strength and have therefore become important construction materials, especially in the automotive industry. AHSSs like DP or TRIP steels have shown high potential especially for automotive applications.

It is a common practice to utilize TMP to achieve the desired final microstructures of the AHSSs and hence their final properties. TMP means the control of the reheating temperature, the deformation schedule and the cooling rate to achieve a defined phase distribution and structure fines. The purpose of applying a deformation schedule is to obtain an optimum grain refinement and it is therefore necessary to maximize the area of austenite grain boundary per unit volume at the onset of phase transformations.

The stored energy due to the accumulated dislocations during hot deformation is generally lowered by three processes: recovery, recrystallization and grain growth (Djaic and Jonas 1972, Bradley et al. 1977). Recovery and recrystallization can take place during and after deformation and to distinguish them they are called dynamic and static, respectively. During TMP, the material generally stops recrystallizing below a certain temperature. This temperature is often called the recrystallization stop temperature (RXST) or the non-recrystallization temperature (T_{nRX}). The T_{nRX} denotes the temperature above which static recrystallization (SRX) occurs between the passes (Yue et al. 1997). It depends on the deformation, the cooling rate and the inter-pass times in the rolling process. Below this temperature recrystallization is retarded due to strain-induced precipitation of second-phase particles.

Micro-alloying elements (Ti, Nb or V) can also be used in the production of DP (Davies 1979, Speich et al. 1983). Adding small amounts of micro-alloying elements retards the recrystallization process (De-Ardo 1984). The reheating of the slabs should be at a temperature that the microalloying elements become a solution. Subsequently, during the TMP, the temperature is reduced, so microalloying elements form nitrides, carbides and/or carbo-nitrides which effectively retards recovery and recrystallization leading to finer austenite grains.

The microstructural evolution during TMP involves accumulation, annihilation and rearrangement of dislocations, recrystallization and grain growth. The microstructural evolution that occurs in the material is dependent on the alloying elements, the amount of the reductions, the strain rate, the temperature and the length of the holding times between the reductions (Bäcke 2009). Varying processing parameters during TMP and the chemical composition has a major influence on the microstructure of steels and consequently on their properties.

The TMP typically consists of four stages, namely:

1. **Austenitization:** It is a common practice to heat steels upto the range of 1200°C to 1320°C prior to hot working. This aims at reducing the strength in order to limit the working force, thereby reducing the loads imposed on rolling mills. With the introduction of carbide/nitride forming micro-alloying elements, with their restricted solubility, it is pertinent to relate the dissolution of the carbide nitrides to soaking on hot working temperatures. Nevertheless, the grain size is drastically coarsened by soaking at such temperatures.
2. **Roughing stage:** This stage is characterized by refining of the austenite grain due to repeated cycles of work hardening and the recrystallization process.
3. **Finishing stage** (controlled hot working): The austenite is deformed in the non-recrystallization temperature regime, resulting in a significant refinement of the final microstructure.
4. **Controlled cooling and coiling:** After the last rolling step, accelerated cooling is applied to suppress the formation of polygonal ferrite and/or pearlite and to facilitate the formation of lower-temperature transformation products such as martensite and different types of bainite.

This chapter presents a packet including the results of research activities undertaken on TMP of four different types of AHSSs steels, namely the dual phase (DP), transformation induced plasticity (TRIP), pipeline and fine bainite steels. The packet provides an insight on how the TMP can be adjusted to develop the desired microstructure in each type of steel. The effect of TMP-parameters on the microstructure development and tensile properties is also discussed.

8.2 Experimental Methods and Details

8.2.1 Thermo-mechanical Simulator

The Simulation of the hot rolling process was performed on a TTS-820 thermo-mechanical simulator. The TTS-820 has three different experimental setups corresponding to three different modes of deformations namely, flat compression, torsion and cylindrical compression. For the current investigation, the flat compression mode is used. The experimental setup for the flat-compression mode at TTS-820 is shown in Figure 8.1. To carry out simulation process, sheathed type S Pt/Pt-10% Rh thermocouple wires with a nominal diameter of 0.2 mm are individually spot welded in the center of the sample's surface. The sample is then placed on two ceramic rollers and fixed from the upper side by two ceramic rods. During compression, the two ceramic rods move upwards allowing the sample to extend in its longitudinal direction. Two deformation stamps upset the specimen at its middle. Heating is carried out using an induction coil, whereas cooling is accomplished through four gas nozzles, symmetrically positioned to the left and right of the specimen. The specimen is cooled with the desired cooling rate using helium. The simulator is equipped with an optical measuring system to monitor the length change of the specimen during the process, so that the phase transformation of the specimen can be observed.

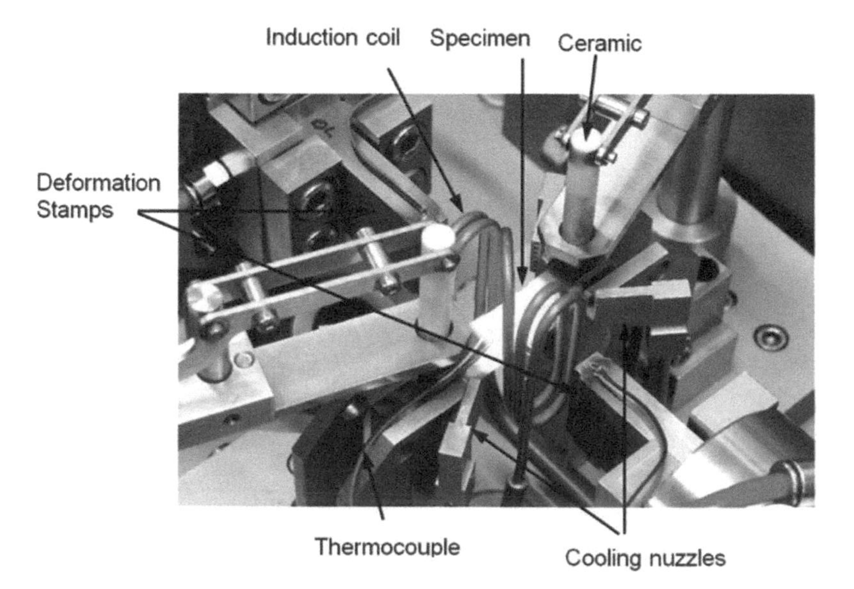

Figure 8.1 The experimental setup for the flat-compression mode at TTS820.

8.2.2 Light and Scanning Electron Microscopy

Light optical microscopy (LOM) and scanning electron microscopy (SEM) of the samples after applying various processing conditions were performed on sections cut at central position of the deformed zone and perpendicular to the width-direction of the sample.

After mounting, the samples were ground using standard abrasive grinding papers. Final polishing was done using 1.0 μm and 0.05 μm alumina, respectively. The microstructure was developed by etching with 2% Nital. After etching, the samples were rinsed with ethyl alcohol and dried under a warm air drier. To reveal the martensite and retained austenite constituents (M/A), samples were etched for 25–35 s using LePera tint etchant. The LePera's etchant composed of two parts:

1. 1 g of sodium metabisulfite + 100 ml distilled water.
2. 4 g of picric acid + 100 ml ethyl alcohol.

The two components are kept separately until use, being mixed together in equal parts by volume (LePera 1980, Girault et al. 1998).

Microstructure and precipitates study was also carried out using scanning electron microscopy (SEM) on 3% nital etched samples.

8.2.3 Tensile Testing

The tensile tests were conducted on a 250 kN computerized universal testing machine (UTS) equipped with a video extensometer applying a crosshead speed of 5 mm/min.

8.3 Dual Phase Steel: Influence of Martensite Content and Cooling Rate

Dual phase steels are characterized by a microstructure consisting of about 75–90 vol% ferrite (α-iron) with the remainder being a mixture of martensite, lower bainite, and retained austenite. They are essentially low carbon steels that are thermo-mechanically processed to have better formability than ferrite-pearlite steels of similar tensile strength (Rashid 1981). In general, the production of DP sheet steels is possible using different process routes. In the conventional route DP steels are produced by intercritical annealing (producing $\alpha + \gamma$ microstructure) followed by severe cooling/quenching, resulting in a soft ferrite matrix containing hard martensite particles (Huang et al. 2004). The intercritical anneal can be followed either by finishing rolling for hot strip product or cold rolling for cold rolled annealed (CRA) or hot dip galvanized (HDG) products. In thermomechanical process routes the strips are finished rolled in the two phase fields of austenite and ferrite and then cooled down towards low temperatures. Due to the lower amounts of alloying elements higher cooling rate during quenching in that scheme is needed (Heller and Nuss 2005).

Aim of the study: This investigation aimed at designing microstructural variations in DP steels and assessing the cooling rate influenced by such microstructural variations. The elements of microstructural control through TMP in DP steels along with a consideration of microstructural effects on mechanical properties, with emphasis on cooling rate during $\gamma \rightarrow \alpha$ transformation is discussed.

8.3.1 Investigated Material

DP steel used in this study was delivered by as rough rolled plate with a thickness of 50 mm. The chemical composition of the steel is listed in Table 8.1. The specimens were taken with their longitudinal axes parallel to the rolling direction and their thicknesses parallel to plate.

Table 8.1 Chemical composition of the steels (wt.%).

C	Si	Mn	Cr	P	N	Al
0.06	0.10	1.30	0.60	0.04	0.006	0.035

8.3.2 Thermo-mechanical Schedule

Figure 8.2 illustrates schematically the employed TMP schedule for simulating the final steps of the finishing rolling. The experiments were performed using a flat compression setup of the deformation simulator (Figure 8.1). The dimensions of the flat compression sample are shown in Figure 8.3. Heat transfer to the shoulders of the flat compression samples was reduced by two holes.

All flat compression specimens had been austenitized at a temperature of $T_A = 1000°C$ for austenitization time $t_A = 180$ s. The hot deformation parameters of the last three deformation steps had been selected according to industrial conditions and kept fixed. The strain rate of each deformation step was $\dot{\varphi} = 10\ s^{-1}$. The first and second deformation steps took place in the recrystallization region; above the recrystallization-stop temperature (T_{nRX}) followed by a further step in the non-recrystallization region. The intermediate time between the deformation steps was 5 s. After the last deformation step the specimens were cooled below Ms in two stages. First, they were cooled at different cooling rates to 'fast cooling' start temperature (T^{FC}) in $\gamma \rightarrow \alpha$ transformation region, until required fraction of ferrite and austenite was obtained. The specimens are then cooled at a cooling rate of ~ 100 K/s to the room temperature.

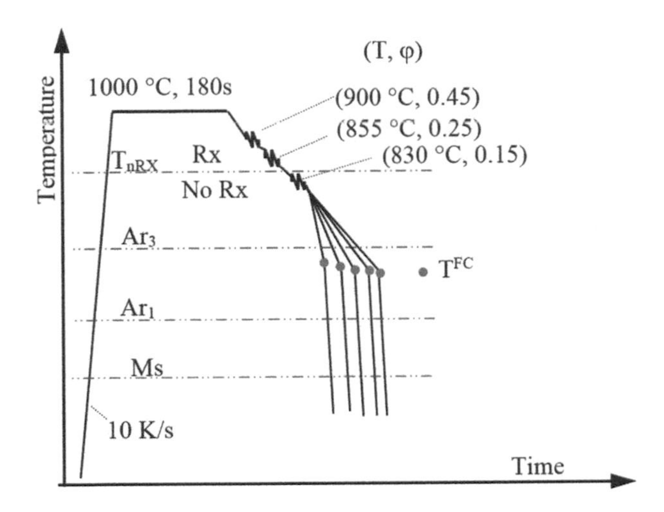

Figure 8.2 Schematic drawing of the applied thermo-mechanical schedules.

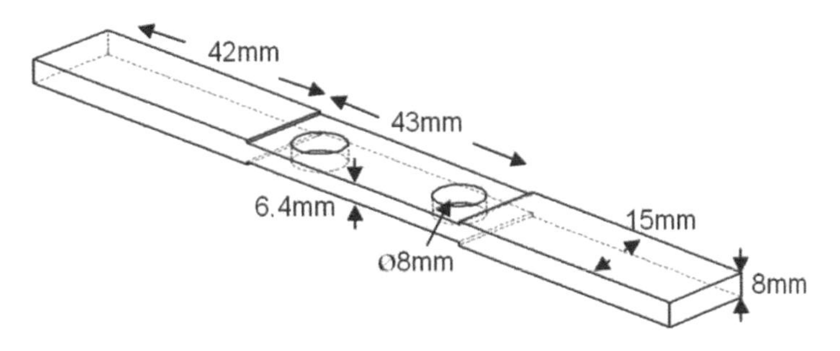

Figure 8.3 Sample dimensions for the flat compression test.

In order to obtain DP steels with prescribed amounts of ferrite and martensite the appropriate T^{FC} temperatures must be determined from which the specimens have to be fast cooled. Varying TMP parameters affects T^{FC} temperature. The accelerated cooling of the specimen from T^{FC} below M_S results in formation of martensite from the untransformed austenite. Therefore, adjustment of the martensite volume fraction (MVF) is possible by variation of T^{FC}. Dilatometric measurements had been performed to define T^{FC} depending on the cooling rate and MVF. The methodology adopted for determining T^{FC} by means of the dilatometric measurements is described elsewhere (Palkowski et al. 2007).

The estimate T^{FC} values corresponding to three different values of 90%, 80% and 70% of ferrite in the matrix are listed in Table 8.2. The effect of cooling rate on Ar_1, Ar_3 and Ms is also given in this table.

Table 8.2 Determined Ar_3, Ar_1, appropriate T^{FC} for $f_\alpha = 70, 80$ and 90%, $f_\gamma = 10, 20$ and 30% and Ms obtained during accelerated cooling for different schedules.

Cooling rate (Ks^{-1})	Ar_3 (°C)	Ar_1 (°C)	$T^{FC, 90\% \alpha}$ (°C)	$T^{FC, 80\% \alpha}$ (°C)	$T^{FC, 70\% \alpha}$ (°C)	M_S (°C)
1	785	651	687	704	714	427
10	775	643	679	697	705	420
20	769	635	672	688	696	414
30	756	626	663	681	690	408
40	742	623	656	671	682	403

The data from Table 8.2 indicate that the cooling rate reveals a strong influence on Ar_3 and Ar_1 during the $\gamma \rightarrow \alpha$ transformation as well as on Ms during the $\gamma \rightarrow \alpha'$ transformation. It can be concluded that higher cooling rate results in lower Ar_3, Ar_1, T^{FC} and Ms. With increasing cooling rate from 1 to 10 K/s the Ar_3 and Ar_1 decrease by about 10 and 8 K, respectively. The specimen having lowest cooling rate of 1 K/s shows the highest Ar_3, Ar_1 and Ms while the lowest Ar_3, Ar_1 and M_S are obtained for the specimen with highest cooling rate of 40 $K{\cdot}s^{-1}$. A comparison of the lowest cooling rate at 1 $K{\cdot}s^{-1}$ to the highest cooling rate at 40 $K{\cdot}s^{-1}$ shows a difference of about 43 K for Ar_3 and about 30 K for Ar_1.

8.3.3 Microstructure Evolution

The TMP was designed to produce DP microstructures with different martensite volume fractions MVFs. LOM investigations were performed to study the microstructure evolution after TMP. Figure 8.4 shows exemplarily LOM images corresponding to samples containing 10, 20 and 30% of MVF with a cooling rate of 10 Ks^{-1}. It is revealed that three DP microstructures are characterized by similar cooling rates (10 Ks^{-1}) but exhibit different MVFs. The estimated MVFs in the final microstructure based on dilatation curves are in good agreement with the quantitative analysis. All images show a classical DP microstructure with relatively globular martensite islands embedded in the ferrite matrix phase. Large martensite islands can clearly be observed, but they often show dark substructures either within or in their direct surroundings. In addition, such a dark phase can also be observed at the boundaries between two neighbors. Some bainitic phases could also be detected in the microstructure.

For samples, cooled during the $\gamma \rightarrow \alpha$ transformation with different cooling rates and the same MVFs, no significant differences with regard to the morphology, phase distribution and grain size could be found.

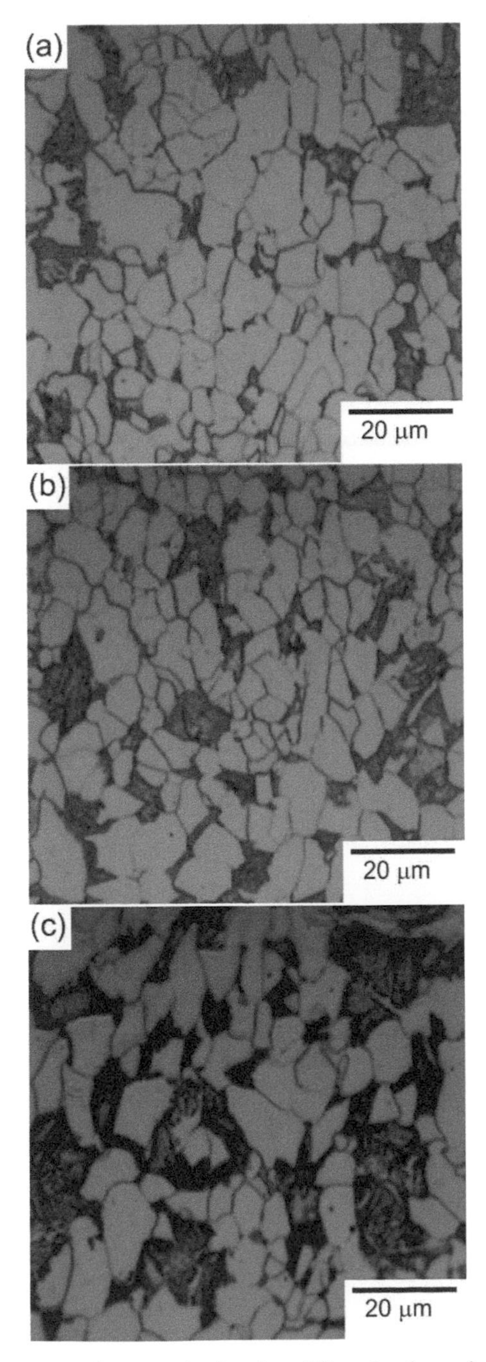

Figure 8.4 Microstructures of DP steels showing different volume fractions of martensite obtained after TMP (a) fα = 90%, fα′ = 10%; (b) fα = 80%, fα′ = 20% and (a) fα = 70%, fα′ = 30%; cooling rate of the samples in the first cooling stage was 10 K/s.

8.3.4 Tensile Properties

Tensile specimens with special geometry were machined out of flat compression specimens to determine the mechanical properties. Figure 8.5 shows flat compression specimens before and after hot deformation as well as after machining. In general, three specimens were tested for each condition and the results were averaged.

The mechanical behavior of thermo-mechanically processed material with respect to cooling rate and MVF is shown in Figure 8.6. As the MVF increases the strength values increase steeply to higher strength level as shown in Figure 8.6a. Total elongation decreases by increasing the MVF as shown in Figure 8.6b. This behavior is commonly interpreted in terms of local dislocation accumulation (Balliger and Gladman 1981) introduced by the martensitic transformation together with the increase in the quantity of the strong martensite phase. On the other hand, as the plastic strain of the martensite phase is negligible, the total elongation to fracture is reduced with increasing MVF.

Higher cooling rates after the last deformation step to fast cooling start temperature during the $\gamma \rightarrow \alpha$ transformation yield higher strength level. The increase of strength with the cooling rate indicates that there is an increase of the stress required for dislocation movement (Matlock et al. 2003). Wu et al. have pointed out that the recovery and recrystallization mechanisms are suppressed during the TMP when a higher cooling rate is applied after the last deformation (Wu et al. 2008). This results in a high dislocation density in the final microstructure, which in turn increases the strength of the DP steels.

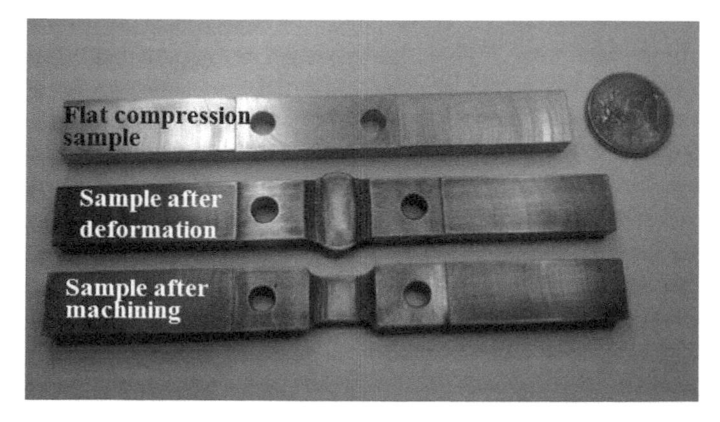

Figure 8.5 Flat compression specimens before and after hot rolling simulation and after machining for tension tests.

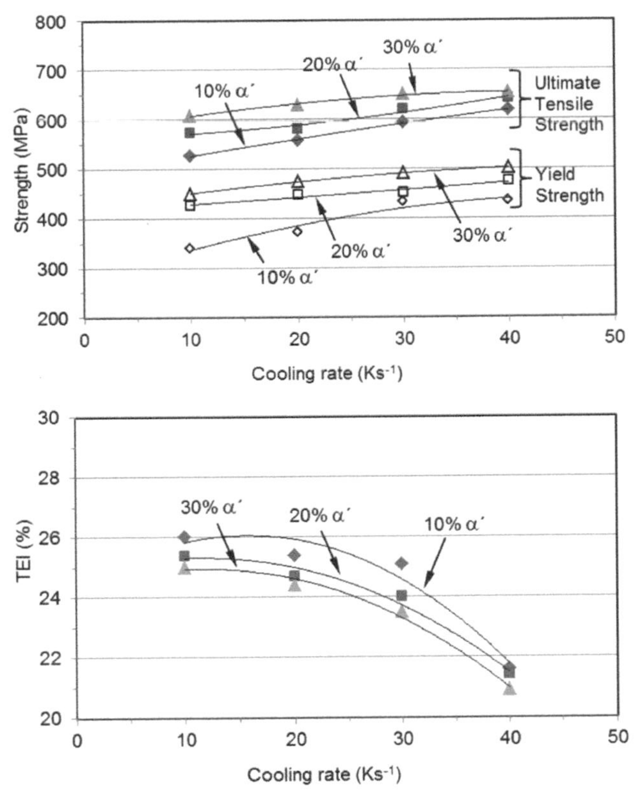

Figure 8.6 Influence of the MVF and the cooling rate after TMP on tensile and yield strength (top) and total elongation (bottem).

8.3.5 Conclusions

DP steels with different MVFs were obtained by TMCP, cooling from the last deformation step to TFC at different cooling rates and quenching the samples from TFC to RT. The influence of the cooling rate in the first cooling stage and the MVF on the mechanical properties and the BH behavior was studied. The major conclusions drawn from the present investigation are as follows:

1. The estimated MVF in the final microstructure is in good quantitative agreement with the experimental data. It is estimated that quenching of specimens from different TFC temperatures results in different DP microstructures containing different MVFs at a given cooling rate.
2. An assessment of the microstructures obtained in the thermo-mechanically processed samples with respect to MVF and cooling rate allows selection of processing parameters required to develop the specified DP microstructures.

3. Increasing the cooling rate after the last deformation step to T^{FC} during the $\gamma \rightarrow \alpha$ transformation results in higher strength and lower ductility after TMP.

8.4 TRIP Steel: Influence of Ferrite Content and Hot Rolling Parameters

The remarkable strength-ductility balance in transformation-induced-plasticity (TRIP) steels results from the occurrence of the TRIP phenomenon during deformation (Zackay et al. 1967). The coexistence of austenite with a certain microstructural stability is of vital importance in order for this phenomenon to occur and, hence, to achieve the desired properties. The austenite retention is usually obtained by combining effects of chemical composition and typical heat treatment. In this respect, adding large amounts of silicon to TRIP steel ensures that cementite precipitation is unlikely to occur in the microstructure during bainite formation (Matsumura et al. 1987). The absence of cementite ensures that the carbon will enrich the austenite rather than form cementite plates. Therefore, after the bainite transformation finishes by further cooling to room temperature (RT), the austenite is stabilized. Jeong and Chung concluded that austenite retention in these low-alloyed steels is almost impossible with Si concentrations much below 1 wt.% (Jeong and Chung 1992).

Aim: The present study aims at ascertaining how Al content variations, the prior hot-rolling conditions, and heat-treatment parameters affect the microstructure and mechanical properties of the cold-rolled Mn-Si-Al TRIP-aided steel alloyed with Mo-Nb.

8.4.1 Studied Materials

The Si-Al-Mo-Nb steel studied in this work was produced in the laboratory. The two studied alloys steels A1 and A2, differ in their Al content. The chemical composition is given in Table 8.3. The alloys contain Mo and are microalloyed with Nb. The Nb in solid solution has been found to improve the TRIP properties; Mo retards the precipitation of Nb(C, N), thus potentially improving the effectiveness of Nb as a TRIP enhancer (Bouet et al. 1998, Jiao et al. 2002). Furthermore, Mo retards austenite transformation to both ferrite and pearlite, affecting more manageable process control (Capdevila et al. 2005).

Table 8.3 Chemical compositions of the steels used (wt.%).

Alloy	C	Si	Mn	Al	Mo	Nb	P	S	N
A1	0.278	0.852	1.48	0.228	0.401	0.041	0.037	0.020	0.006
A2	0.254	0.869	1.37	0.643	0.383	0.040	0.030	0.021	0.008

8.4.2 Hot Rolling Schedules

The cast ingots were hot rolled in four passes as shown in Figure 8.7 from a thickness of 19 mm to 4 mm, and with a true strain value $\varphi = 0.38$ for each pass. The temperature was continually monitored with a pyrometer during the final cooling of the hot rolled strips in air. It took between 20 to 31 min to cool the strips from the finish rolling temperature to 350°C. The deformation was designed respecting the T_{nRX} such that all the three possibilities were covered, namely all deformations conducted above T_{nRX}, below T_{nRX} and mixtures of above and below T_{nRX}. The T_{nRX} was determined using the method described elsewhere (Bai et al. 1993). The estimated values of T_{nRX} for A1 and A2 are 865°C and 886°C, respectively.

The schedule resulted in microstructure formed from the recrystallized-austenite is denoted by "R", whereas that resulted in microstructure formed from pancaked-austenite is denoted by "P". The "RP" schedule is for the microstructure resulting from the recrystallized- and then pancaked-austenite. Figure 8.8 shows the microstructure obtained after the different hot-rolling schedules. Acicular ferrite (AF) dominates and polygonal ferrite (PF) is also present.

8.4.3 Heat Treatment and Microstructure Formation

The heat treatment had been conducted on the mechanical testing samples and on samples for microconstituents' investigations using salt baths. This had been done by intercritical annealing in Durferrit GS 540/R2 and austempering in Durferrit AS 140 salt baths. A phase content of 70, 50, and 30% PF was required at the end of intercritical annealing. In

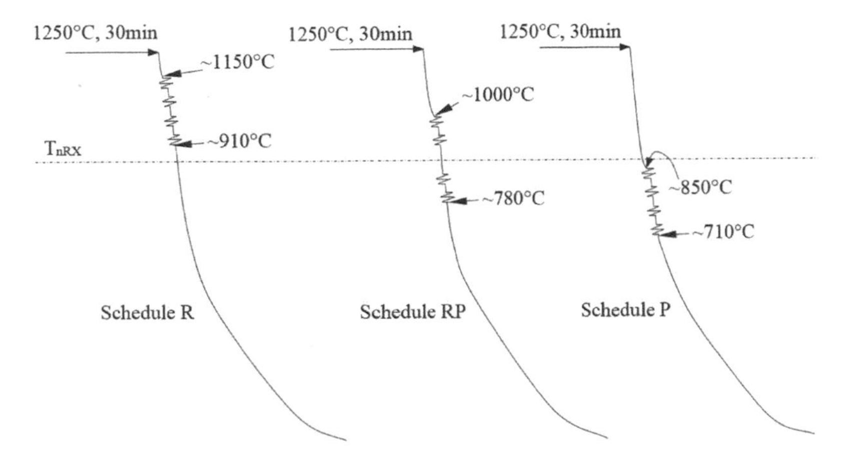

Figure 8.7 Schematic representation of the hot rolling schedules.

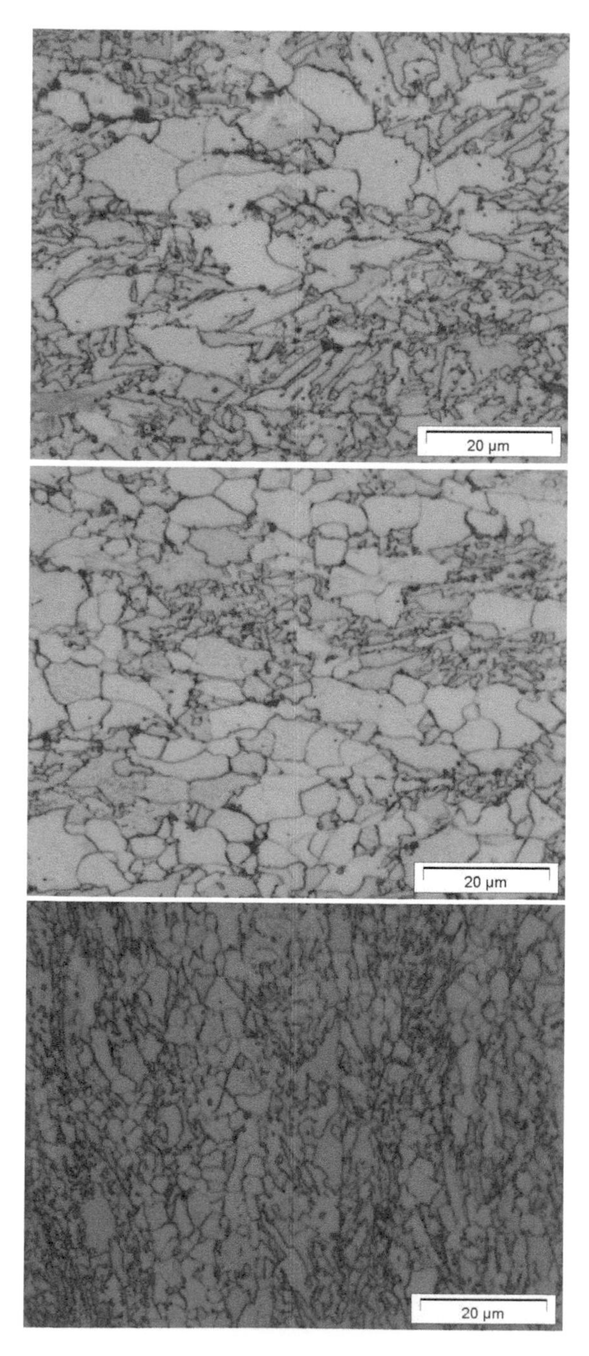

Figure 8.8 Structure refinement as a result of applying different rolling schedules on Steel A2; resulted from applying schedules R, RP and P, respectively (etchant: nital). Micrographs were obtained at the middle section of the hot-rolled material.

equilibrium conditions, this phase distribution is obtained in the intercritical annealing temperatures (T_A) shown in Table 8.4. The intercritical annealing temperature was chosen on the basis of dilatometric measurements as described in (Soliman et al. 2009). The intercritical annealing temperatures were implemented in combination with austempering temperatures, T_B, of 365°C, 400°C, and 435°C. The isothermal holding time for each of the two steps was 8 min. After austempering, the samples were quenched in water.

Figure 8.9 shows representative microstructural results using the same magnification for Steel A2, after applying the two-step annealing conditions. In all specimens, the minor microstructure is the martensite austenite (MA) structure (white areas). Figure 8.9b to d show the influence of the intercritical annealing temperature on the microstructure, for specimens austempered at 400°C. An increase in the intercritical annealing temperature resulted in a decrease in the PF amount to reach approximately the expected amount (30, 50, and 70%). A concurrent increase in the amount of the MA phase was observed. It can also be inferred from Figure 8.9 that making use of the beneficial effect of the rolling below T_{nRX}, a pronounced finer cold-rolled TRIP-aided steel structure was produced. Thus, controlling the hot-rolling schedule, prior to cold-rolling and heat-treatment of TRIP-aided steel, has a decisive effect on a structure refinement, and results in a finer structure.

Table 8.4 Intercritical annealing temperatures (T_A) for A1 and A2 (°C).

	70% PF	50% PF	30% PF
Steel A1	742	768	800
Steel A2	770	810	849

8.4.4 Mechanical Properties

Figure 8.10 shows the effect of the processing route on the ultimate tensile strength (UTS), yield strength (YS), and the total elongation percent (El) for steels A1 and A2, respectively. The YS values plotted in the histograms are the lower yield strengths or the 0.2% offset YS (0.2% YS) values, in the case of the absence of a yield point. It can be concluded that it is the PF percent (the intercritical annealing condition) that has the most pronounced effect on the mechanical properties of both steels. The UTS values increase with a decrease in the PF percent. This can be explained by the higher austenite content that formed during the intercritical annealing. A higher amount of γ_i results in a greater amount of high-strength bainite, after the isothermal bainitic transformation step.

Since TRIP-aided steels combine high strength and high ductility, their mechanical properties are often characterized by the product of the tensile strength and the total elongation (UTS × % El) which is known as the formability index (Barbe´ et al. 2002, Hashimoto et al. 2004).

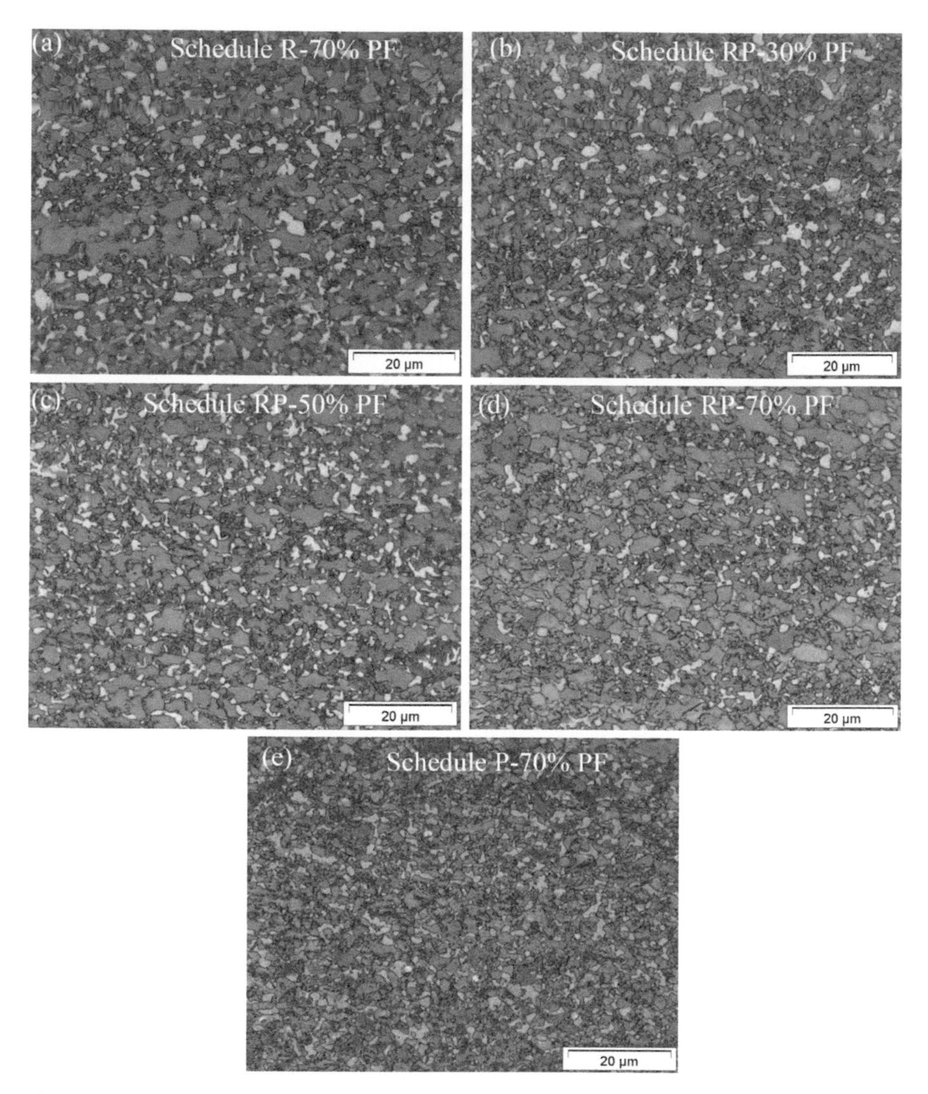

Figure 8.9 Different TRIP steel microstructures obtained as a result of using different hot rolling schedules (a) schedule R, (b), (c) and (d) schedule RP, (e) schedule P. Microstructures are for steel A2.

At each combination of PF content and hot rolling condition, the average formability index was calculated from the values recorded at the three corresponding T_B temperatures. Figure 8.11 shows a histogram comparing these averages. This figure shows that the highest formability index was obtained for the steels annealed to 50% PF.

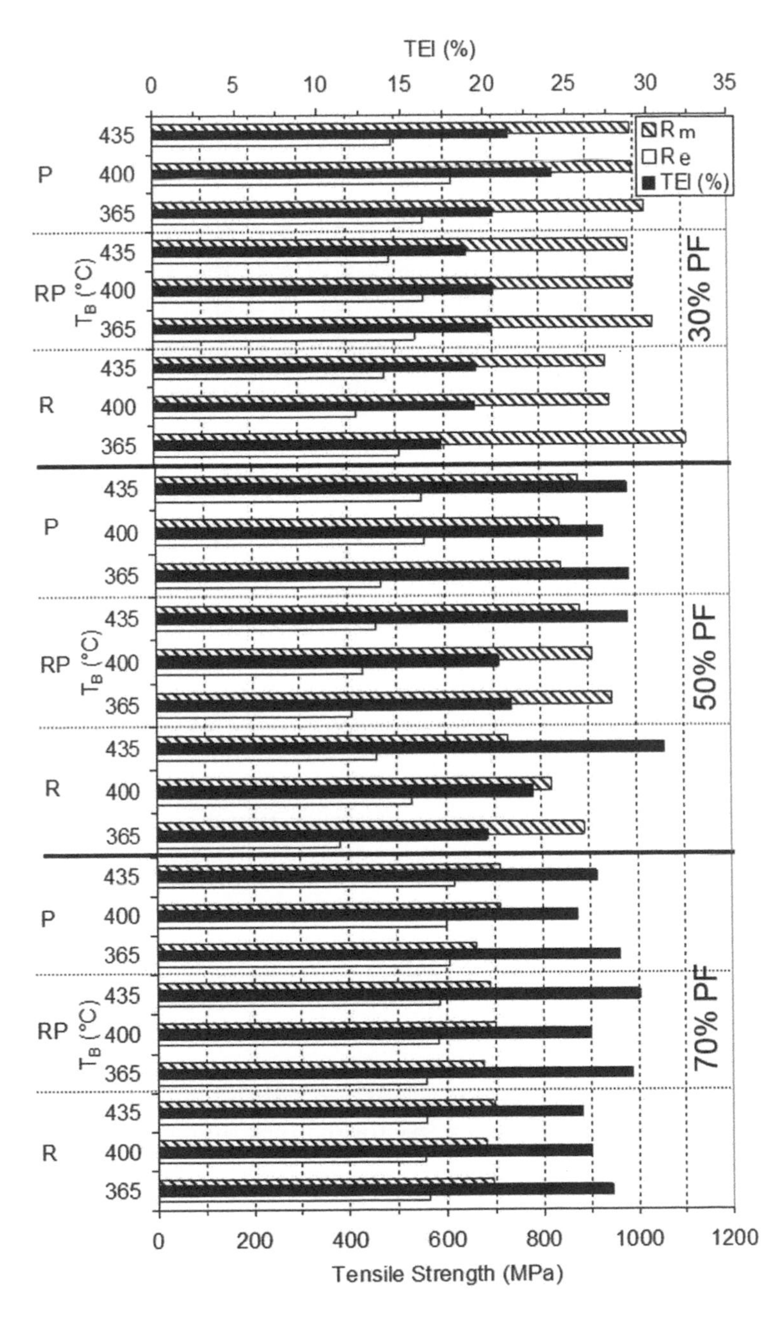

Figure 8.10 Map of UTS vs. pct El, for all combinations of investigated processing routes for steels A1 and A2.

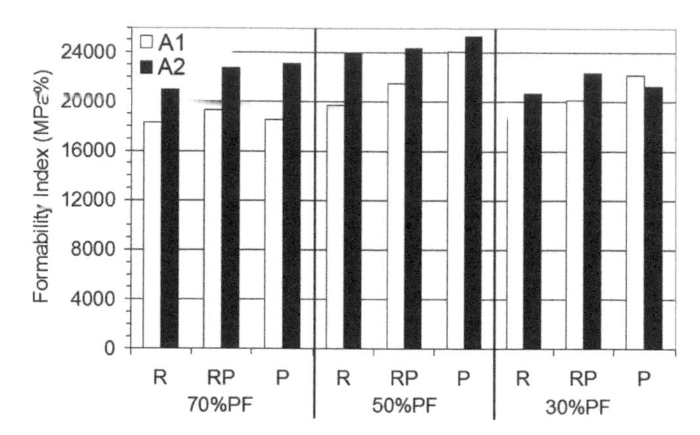

Figure 8.11 The average formability index at each combination of PF-content and hot-rolling condition; calculated from the values recorded at the three corresponding T_B temperatures.

This figure also shows that increasing the deformation below T_{nRx} results in increasing the formability index. This improvement is due to microstructural refinement (see Figure 8.9). In addition to the well-known effect of grain refinement on improving mechanical properties, the greater stability of the smaller retained austenite grains has a further improving effect. Thus, by controlling the deformation temperature and the degree of deformation below T_{nRX} during the hot-rolling process, it was possible to improve the strength-ductility balance of the cold-rolled TRIP-aided steel. On the other hand, the formability index increased by increasing the aluminum content.

8.4.5 Conclusions

1. By making use of the beneficial effect of the hot rolling below T_{nRX}, a pronounced finer cold-rolled TRIP-aided steel structure was produced. This resulted in an improvement in the strength-ductility balance of the steel.
2. Increasing the aluminum content not only reduces the cementite stability but also enhances the formability though increasing the ductility.
3. For the given alloying elements, the most promising microstructures with respect to the strength-ductility balance are those containing 50% polygonal ferrite.

8.5 Bainitic Pipeline Steel: Influence of Rough Rolling Parameters

In order to respond to the requirements for pipeline steels to exhibit a good combination of high strength and good low temperature impact toughness balanced with adequate weldability, new generations of low carbon micro-alloyed steels have been developed (Williams et al. 1995, Cizek 2001). In these steels, the carbon is reduced to below 0.09 wt.% to improve the weldability and weld toughness. The strength loss due to a low C content is compensated by an alloy design philosophy based on the advanced use of cost effective micro-alloying elements, such as Nb, Ti and B in conjunction with moderate levels of other alloying elements, such as Mn, Si, Cr, Mo and Cu (Suikkanen 2009). These alloying and micro-alloying elements and thermo-mechanical processing contribute to the increase in strength via microstructural refinement, precipitation hardening and solid solution strengthening as well as strengthening through microstructural modification.

Aim: In the presented work an investigation on bainitic pipe line steel is carried out to study the effect of parameters in the roughing stages on the microstructure evolution and final mechanical properties. The subsequent finishing and cooling stages were kept unchanged.

8.5.1 Material and Specimens Preparation

The study is carried out on samples machined out of a transfer bar of API X80 pipeline steel with a thickness of 52 mm. The chemical composition of the studied material is given in Table 8.5. All the flat compression samples for the thermo-mechanical simulation were machined from this slab. Taking their longitudinal axes parallel to the rolling direction of the transfer bar.

Table 8.5 Chemical composition of the studied material (wt.%).

C	Si	Mn	Cr	Mo	Ti	Nb	S	P	N
0.055	0.3	1.84	0.18	0.26	0.026	0.101	0.0008	0.014	0.006

8.5.2 Thermo-Mechanical Simulation

The simulation of the hot rolling process was performed on a TTS-820 thermo-mechanical simulator described in Section 2.1. The samples were subjected to the thermo-mechanical schedule designated in Figure 8.12. In this schedule, specimens were austenitized at T_A of 1285°C and subjected to one deformation step with a true strain value of φ_v at T_V. The austenite status at this stage—regarding the prior austenite grain size (PAGS) and precipitation—simulates the condition of austenite after the roughing process. The subsequent three deformation steps are to simulate the finishing rolling process, the time between roughing and finishing is designated in the figure by t_V. The studied parameters are varied according to the values listed in Table 8.6. The finishing rolling parameters and the

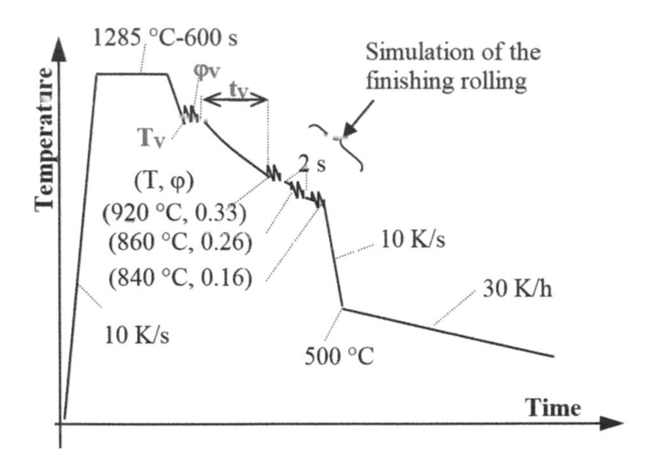

Figure 8.12 Schematic drawing of the applied thermo-mechanical schedule.

Table 8.6 Combination of parameters studied (Figure 8.2).

T_A (°C)	1285														
T_V (°C)	1000								1100						
φ_V (–)	0.3					0.5			0.3			0.5			
t_V (s)	5	30	60	120	180	5	60	180	5	60	180	5	60	180	

subsequent cooling strategy were kept unchanged throughout all the applied simulation processes. The parameters in Table 8.6 are considered for varying the austenite status before entering the finishing mill.

8.5.3 Microstructure Evolution

A combination of LOM and SEM integrating EDX were used to characterize the microstructures of the specimens.

φ_V = 0.3: The effects of T_V and t_V on the microstructure for the samples austenitised at T_A = 1285°C and deformed at T_V with $\varphi_V = 0.3$ are shown in Figures 8.13 and 8.14. The microstructure is predominantly a mixture of acicular ferrite (AF) and granular bainite (GB). The combined effect of the applied thermo-mechanical schedule with the alloying concept avoided the formation of the polygonal ferrite microstructure.

The AF is characterized by non-smooth outer boundaries (jaggy) with featureless white interior areas (Tafteh 2011). The term AF steel was firstly described by Smith et al. in the early 1970s (Smith et al. 1972) and has been widely accepted in pipeline engineering. Up to date there are still controversies and uncertainties on the metallographically identification and

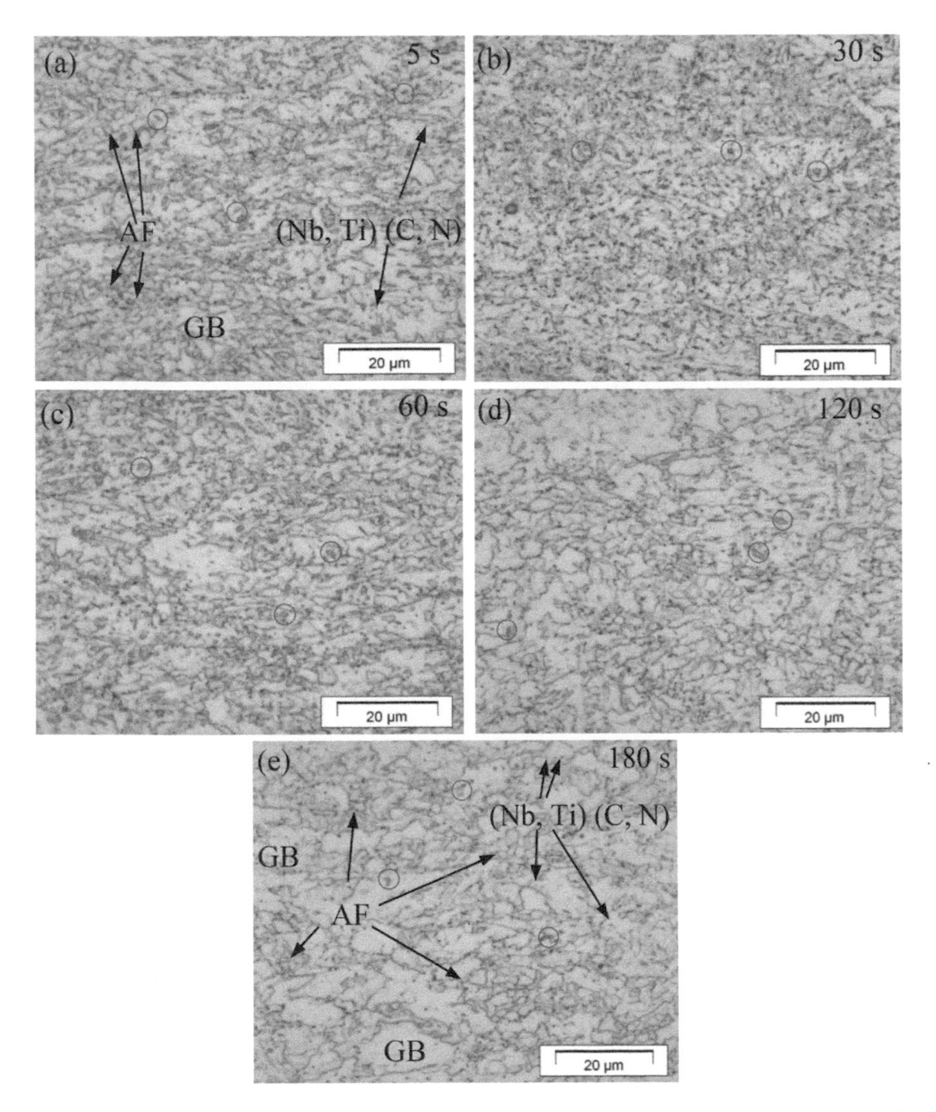

Figure 8.13 Effect of t_V on the microstructure for samples austenitized at T_A = 1250°C and deformed at T_V = 1000°C with φ_V = 0.3. Etchant: Nital.

classification of the phases. The structure of acicular ferrite is sometimes considered as bainite and sometimes as quasi-polygonal ferrite (Xiao et al. 2006, Wei et al. 2009). The acicular ferrite microstructure is confused from the optical micrographs, but it is well distinguishable in the scanning electron micrographs Figure 8.14. It has been well accepted that the AF microstructure comes from a mixed diffusion and shear transformation mode during continuous cooling. This transformation begins at a temperature slightly above the upper bainite temperature transformation

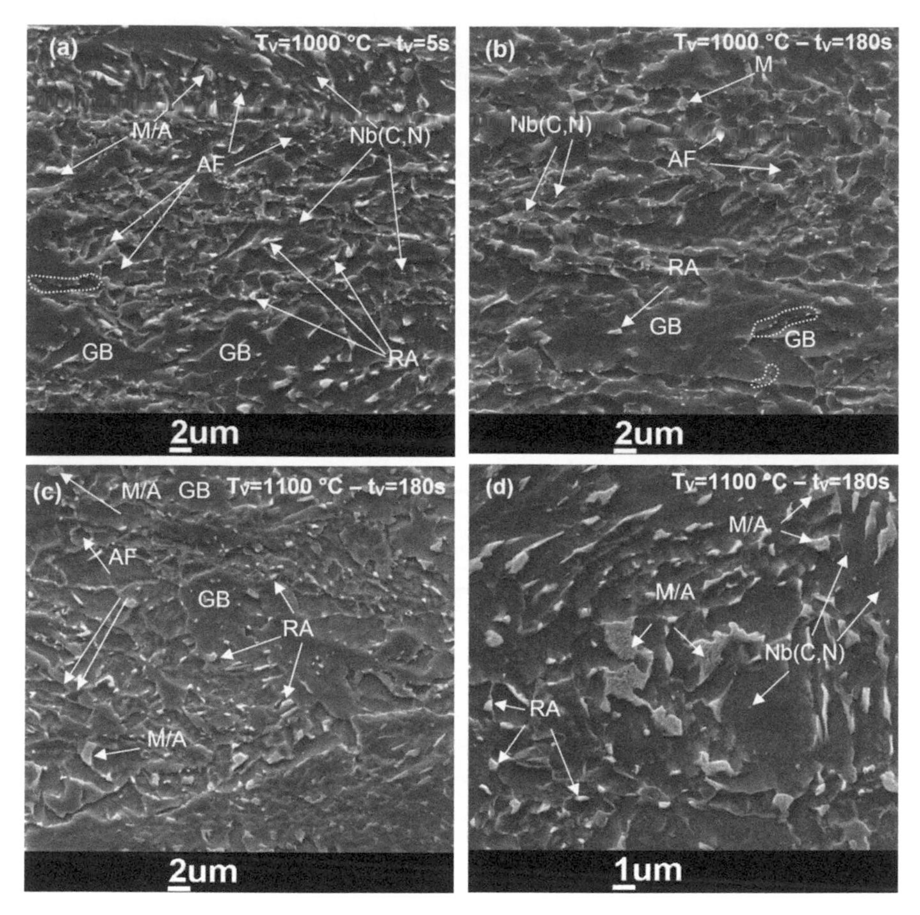

Figure 8.14 Effect of T_V and t_V on the microstructure for the samples austenitized at T_A = 1285°C and deformed at T_V with φ_V = 0.3. Etchant: Nital.

region. The AF presents as an assemblage of interwoven non-parallel ferrite laths with high density of tangled dislocations (Lee et al. 1987, Wei et al. 2009). These interwoven ferrite laths can be distinguished in the transmission electron micrographs TEM (Yakubtsov and Boyd 2001, Zhao et al. 2002). Achieving AF microstructure in pipeline steel results in a superior properties combination, such as high strength, excellent toughness, good H2S resistance (Zhao et al. 2003).

The GB, on the other hand, is characterized by non-smooth outer boundaries with interior areas featured with residuals and high carbon phases, dispersed within the grain (Park et al. 1992). GB is known to be transformed at lower temperatures than AF (Tamura et al. 1988).

The micrographs of Figure 8.13 show that the microstructure is predominantly a mixture of acicular ferrite (AF) and granular bainite (GB). The microstructures for the samples having t_V = 180 s (Figure 8.13e) are dominated by the GB structure. For t_V = 5 s (Figure 8.13a), the microstructures show more AF and finer GB than that obtained for t_V = 180 s. The domination of the GB structure is also observed in the microstructures of the samples with t_V = 60 s and 120 s. The sample with t_V = 30 s shows more or less similar microstructural features to that for t_V = 5 s. The very tiny phase, e.g., the encircled phase in Figure 8.13, is defined as a martensite/austenite (M/A) phase; this is confirmed by SEM investigations. The occurrence of tiny martensite/austenite (M/A) phase is more pronounced for t_V = 5 s and 30 s than for t_V = 60 s, 120 s and 180 s. The shorter cooling time between the roughing and finishing resulted in finer and/or pancaked prior austenite grains which motivated the formation of both, AF and fine M/A phases. The microstructural-features, previously mentioned, can clearly be observed (Figure 8.14).

Furthermore, a successful distinction of martensite (M) from the retained austenite (RA) is possible. Figure 8.14 shows that most of the RA (the phase that does not reveal substructure) has a size below ~1 µm, the larger grains, on the other hand, transformed partially to martensite during cooling. The M/A phase areas consist of central regions of martensite bordered by retained austenite. It is reported in (Brandt and Olson 1993, Soliman and Palkowski 2008) that sufficiently small particles might not contain an effective martensite nucleation site. The limited resolution of the LOM does not allow resolving the very small RA phase. Nevertheless, the larger martensite particles can be resolved.

Therefore, decreasing the delay time between the roughing- and finishing-stage yields more AF and a finer and more dispersive M/A phase. The amount of M/A on the other hand is not significantly affected.

Starting the finishing rolling stage a pancaked structure (t_V = 5 s) is expected to result in a higher density substructure and dislocations in the austenite after finishing rolling compared to the finer recrystallized austenite structure (t_V = 180 s). It is previously reported that the high density substructure and dislocation increase the nucleation site for the acicular ferrite and promote the acicular ferrite transformation (Xiao et al. 2006).

On the other hand, the precipitated carbides shown in Figure 8.14, according to their globular shape, are considered as (Nb, Ti) (C,N) (Pereloma et al. 2001). They are well distributed in all microstructures. There is a frequent observation of concentrations of the precipitates at the grain-boundaries of the GB (e.g., marked with dashed line shapes in Figure 8.14a and b). This can be attributed to the fact that the GB show traces of the prior austenite grain boundaries (Tafteh 2011). The metallographic investigations show that the highest AF volume fraction is observed for the samples processed with T_V = 1000°C and t_V = 5 s. For this condition AF has a volume fraction ~ 62%. On the other hand, a value of ~32% was the lowest volume fraction of AF and recorded by T_V = 1100°C and t_V = 180 s.

φ_V = 0.5: The combined effects of T_V and t_V on the microstructure for the samples austenitised at T_A = 1285°C and deformed at T_V with φ_V = 0.5 are shown in Figures 8.15 and 8.16. Similar microstructure constituents to that observed for φ_V = 0.3 (Figures 8.13 and 8.14) are observed here as well. A remarkable observation when comparing Figure 8.16 with Figure 8.14 is that increasing φ_V from 0.3 to 0.5 results in a significant refinement of M/A and a radical decrease in its quantity. Increasing the deformation during the roughing stage may result in stimulating the transformation kinetics during cooling and consequently a radical decrease in the untransformed austenite that forms the M/A microconstituent.

The scanning electron micrograph in Figure 8.16 shows the typical feature of the constituents relevant to the samples having t_V = 180 s, which is the domination of the granular bainite. The characteristic feature of GB is the lack of carbides in the microstructure. Instead, the carbon that is partitioned from the parent austenite stabilizes the residual austenite, so that the microstructure contains transformation products that transform from the carbon-enriched austenite. The SEM micrographs show two of these products which are debris of cementite and M/A.

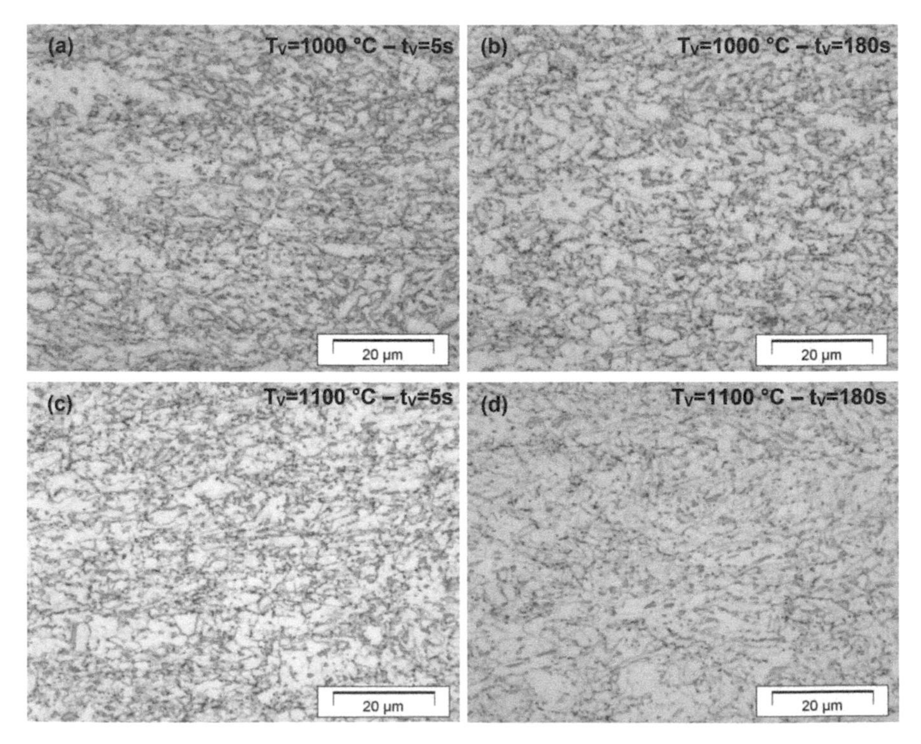

Figure 8.15 Effect of T_V and t_V on the microstructure for the samples austenitized at T_A = 1285°C and deformed at T_V with φ_V = 0.5. Etchant: Nital.

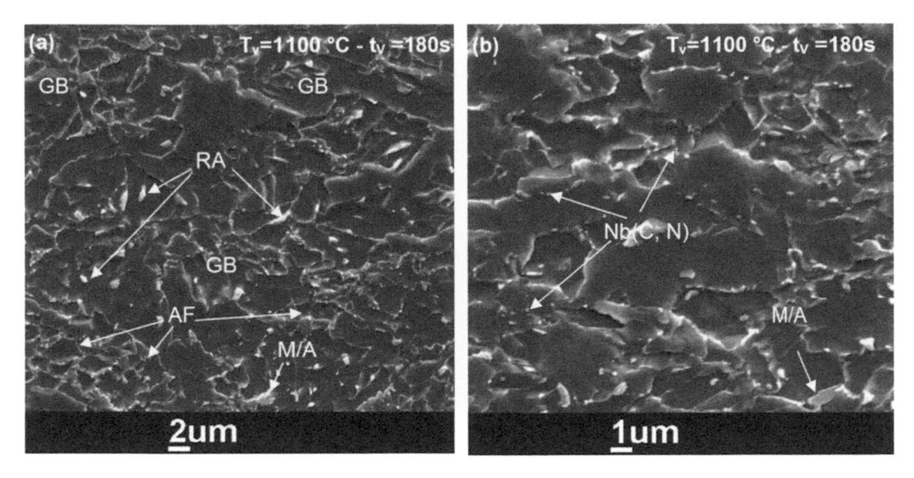

Figure 8.16 Scanning electron micrographs of the prescribed treatment conditions - $\varphi_V = 0.5$. Etchant: Nital.

8.5.4 Mechanical Properties

Figure 8.17 shows the stress-strain curves for the different treatment conditions. It becomes clear from Figure 8.17 that the most dominating factor affecting the tensile properties is the t_V. Generally, the incremental decrease in t_V results in an incremental increase in ultimate tensile strength (R_m) and yield strength (R_p). For the microstructural study it is demonstrated that t_V is the most dominating factor affecting the microstructure evolution. The tensile results fit close to the microstructure observations. The samples subjected to lower t_V have significantly higher quantity of AF and finer M/A. It is reported that if the ultra-fine AF microstructures are obtained in pipe line steel both their strength and toughness are increase (Han et al. 2010, Xiao et al. 2006). AF is radiated in many different directions from nucleation sites; therefore acicular ferrite tends to form interlocked microstructures. Propagating cracks are then deflected when they encounter a differently oriented acicular ferrite plate. This gives rise to superior mechanical properties (Cao 2010).

However, it seems that this behavior has a saturation point after which t_V has a limited/slight effect on R_m and R_p. The saturation point for material deformed at a low temperature (T_V = 1000°C) is not reached at t_V = 60 s (Figure 8.17a and c) but rather at t_V = 120 s (Figure 8.17a). A value of 60 s for t_V was enough to attain the saturation point for materials deformed at high temperature (Figure 8.17b and d).

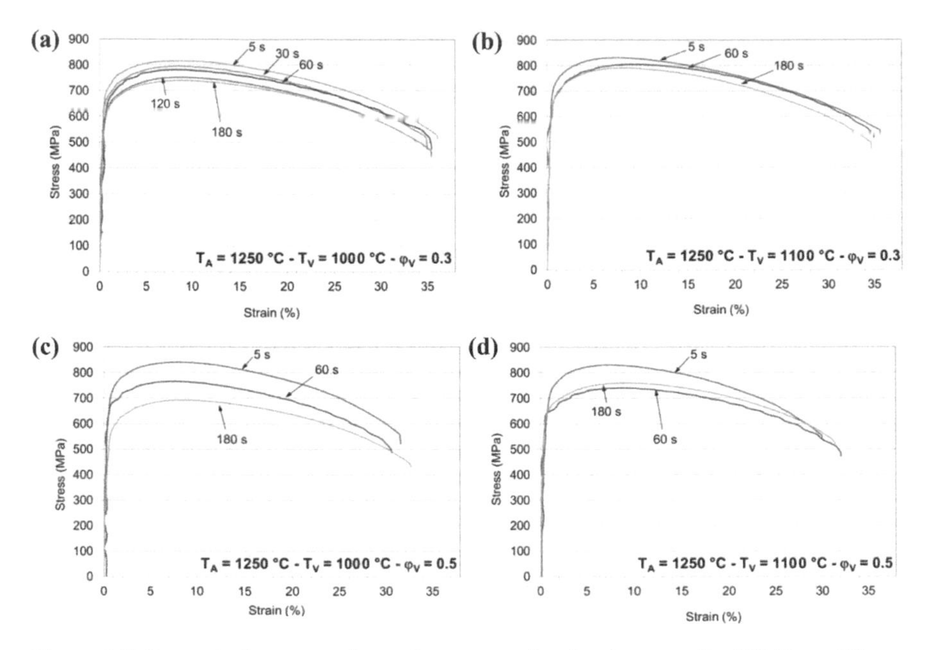

Figure 8.17 Stress-strain curves of samples processed under the prescribed TMP conditions.

8.5.5 Conclusions

Hot deformation rolling simulation tests were carried out on samples of a low-carbon CMnNbMoTi pipeline steel following a close-to-industrial schedule. In this physical simulation, specimens were heated up to austenitization temperature of 1285°C and subjected to one deformation step having a true strain value of φ_V at T_V. The cooling interval between roughing and finishing is designated by (t_V). The austenite status after this cooling interval, regarding the prior austenite grain size (PAGS) and precipitates, simulates the condition of austenite before entering the finishing mill. The finishing parameters and the subsequent cooling strategy were kept unchanged throughout all the applied simulation processes. The following conclusions can be drawn from the current study:

1. The gradual increase in t_V results in a gradual increase of the granular bainite phase on the expense of the aciculare ferrite. This results in an incremental decrease in R_m and R_p with increasing t_V. The effect of t_V on the total elongation (TEl) was limited.
2. However, the behavior mentioned under conclusion number 1 seems to have saturation level after which the t_V has a limited/insignificant effect on the Rm and Rp. This saturating value of t_V is process parameter dependent.

3. Increasing the deformation during roughing from $\varepsilon_V = 0.3$ to 0.5 resulted in stimulating the transformation kinetics during cooling and consequently a radical decrease in the untransformed austenite that forms the M/A microconstituent.

8.6 Ultra-fine Bainite Steel: Influence of Rolling Parameters

Carbide-free bainite in steels is composed of fine plates of bainitic ferrite separated by carbon-enriched austenite films with adequate resistance to cementite precipitation. Suppression of cementite precipitation during bainite transformation is possible through adequate addition of silicon which has very low solubility in cementite thus greatly retards its formation (Deliry 1965, Soliman and Palkowski 2008). The attractive mechanical performance of microstructures which contain carbide-free bainite is now well-established. One of these promising microstructures is of novel high-strength steel developed by isothermal transformation of austenite to produce very thin alternating plates of bainitic ferrite and austenite (Soliman and Palkowski 2007, Soliman et al. 2010). The high strength of these steels is due to their very fine structure. The observed refinement is a consequence mainly of the ability of high carbon content and low transformation temperature to enhance the strength of the austenite. The bainite-plates become thinner as the yield strength of the austenite, from which they are formed, increases (Bhadeshia 2001).

Aim: This work presents an investigation on the production of the ultra-fine bainite structure in flat products by applying thermo-mechanical processing on the material. Thermo-mechanically processed ultra-fine bainite steel is produced in a thermo-mechanical simulator through a defined combination of deformation-steps and temperature control. A dilatometry system is used to analyze the transformation kinetics during the thermo-mechanical processing.

8.6.1 Material Preparation

The steel presented in this work was obtained from sand casting process in the laboratory by induction melting in air and solidifying in Y-blocks with a size of 45 × 190 × 230 mm. The blocks were machined to 40 × 187 × 227 mm, to remove the cast surface structure, and then heated up to 1200°C for one hour in an argon atmosphere. The material was then hot-rolled to a final thickness of 12 mm then cooled in furnace to the room temperature (RT). The chemical composition of the investigate alloys are given in Table 8.7.

Table 8.7 Chemical composition of the investigated alloys.

	C	Si	Mn	Mo	Al	Cr	Co	P	S	N
F1	0.319	1.71	2.63	0.25	0.65	1.35	1.52	0.028	0.015	0.018
F2	0.563	1.66	2.53	0.25	0.63	1.31	1.48	0.027	0.019	0.017

8.6.2 Thermo-Mechanical Processing and Microstructure Evolution

The TM cycle shown in Figure 8.18 was applied on alloys F1 and F2 using thermo-mechanical simulator TTS820. The bainite transformation in terms of change in length for the two alloys for $T_2 = 1000°C$ is given in Figure 8.19. For alloy F1, it was not possible to perform the bainite transformation at temperatures ≤ 300°C because of its relatively high Ms, which is measured to be 306°C.

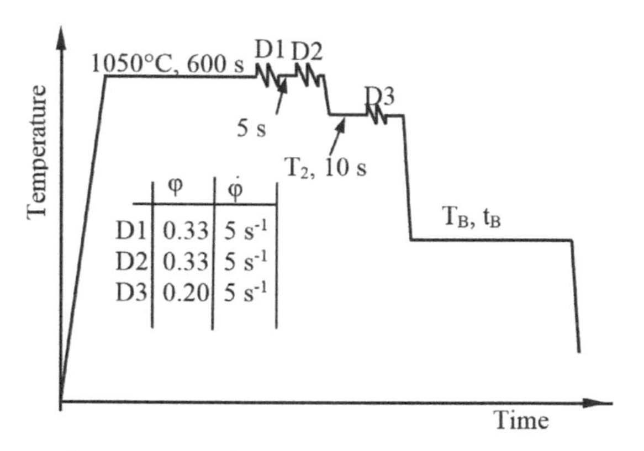

Figure 8.18 Applied TM cycle using Bähr TTS820.

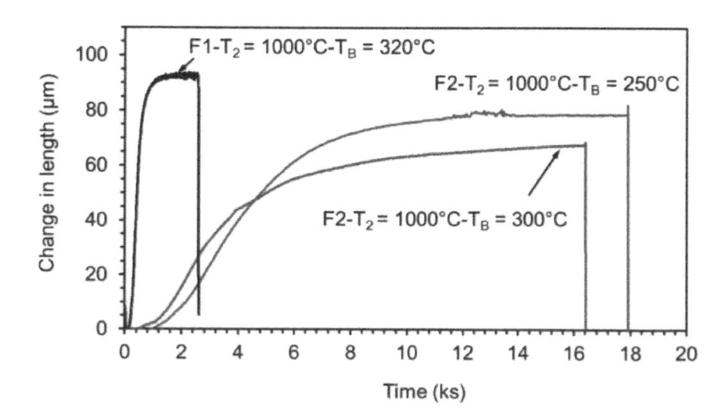

Figure 8.19 Bainite transformation kinetics in terms of change in length during holding at - $T_2 = 1000°C$.

The effect of applying the last deformation step, i.e., D3 below the non-recrystallization temperature is studied for alloy F1. Figure 8.20 compares the bainite-transformation kinetics observed for three different values of T_2. Deformation of alloy A1 at 850°C moved the Ms from 306°C to record 286°C. So it was possible to obtain bainite for this alloy at 300°C. Representative structures obtained from alloys F1 and F2 are given in Figure 8.21. As clearly seen in Figure 8.21, it is not possible to distinguish individual ferrite plates within the thin aggregates using LOM. Therefore, SEM investigations had been conducted. Figure 8.21e to h show that the ferrite aggregates build up from many ferrite plates in the same crystallographic orientation.

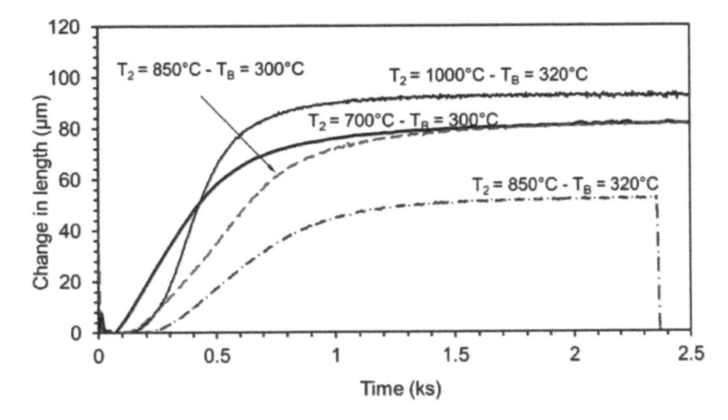

Figure 8.20 Effect of T2 on the bainite-transformation kinetics for alloy F1.

8.6.3 Tensile Properties

Figure 8.22 shows the true stress-true strain curves observed for material F1 and F2. F2 showed better strength due to its high carbon content compared to alloy F1. The higher carbon level strengthens the parent austenite, from which the bainite is formed. It is expected that the bainite-plates will become thinner as the yield strength of the austenite, from which they are formed, increases (Bhadeshia 2001). However, the high carbon content retarders the bainite transformation kinetics as shown in Figure 8.19.

It is clear from Figure 8.22 also that alloy F1 recorded higher values of both strength and ductility when performing the bainite transformation at 300°C rather than at 320°C. Allowing the bainite transformation at 300°C as a consequence of performing the last deformation-step below TnRx resulted in opening new limits of the mechanical properties of this alloy.

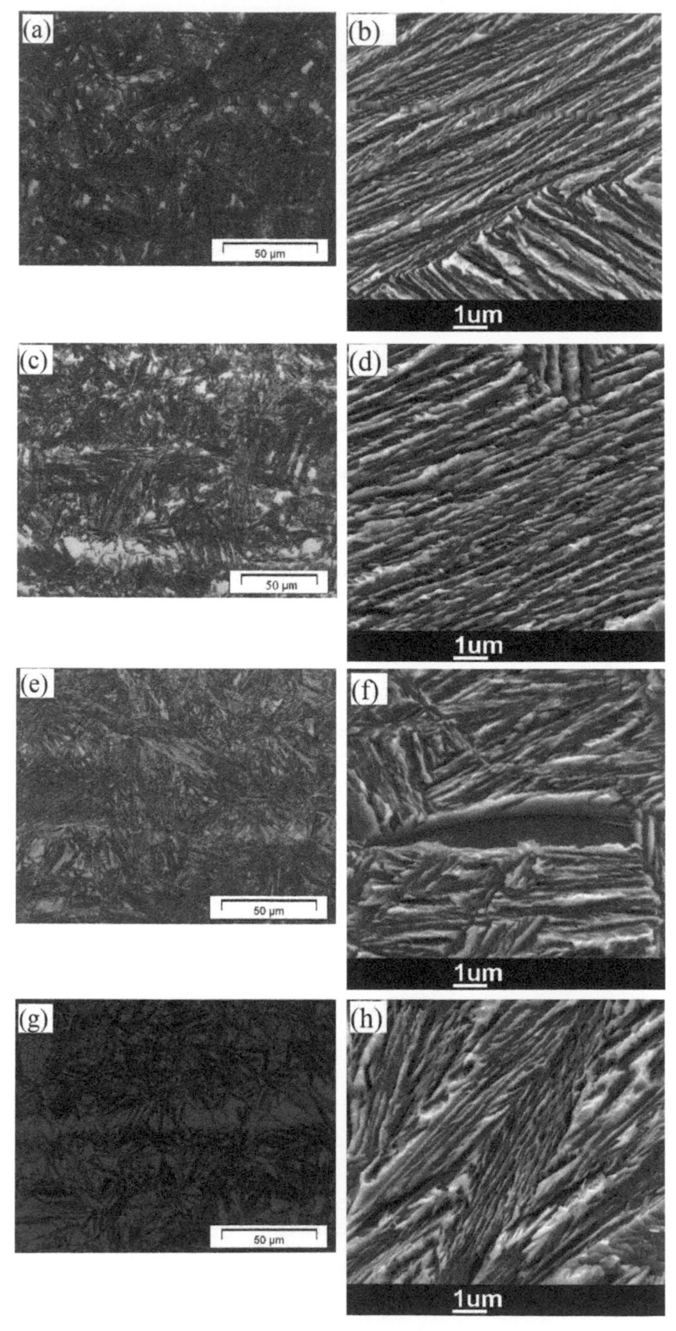

Figure 8.21 Light optical and scanning electron micrographs observed after austempering for alloy F1: (a) and (b) $T_2 = 1000°C - T_B = 320°C$, (c) and (d) $T_2 = 850°C - T_B = 320°C$ and (e) and (f) $T_2 = 700°C - T_B = 300°C$ and for alloy F2: (g) and (h) $T_2 = 1000°C–T_B = 300°C$.

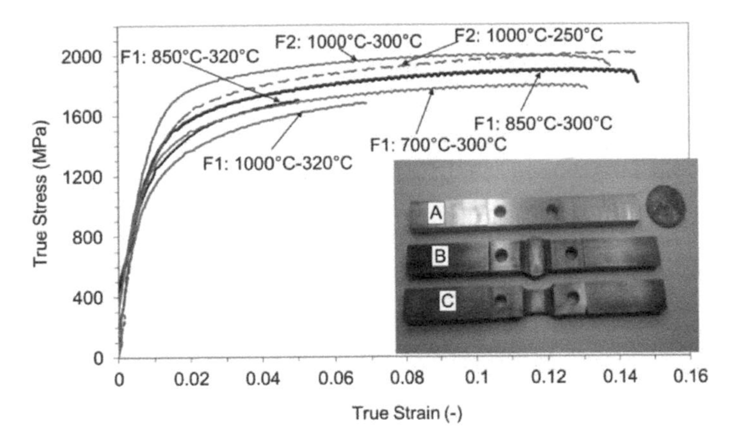

Figure 8.22 True stress-true strain curves alloys F1 and F2 and photo of the samples. A: before -, B: after TM processing and C: prepared for tensile test.

8.6.4 Conclusions

Ultra-fine bainite structure had been obtained in two steels by isothermal transformation at low temperatures between 250°C and 320°C. The conclusions drawn from the present investigation are as follows:

1. It is possible to produce the ultra-fine bainite in flat products like sheets and plates by thermo-mechanical processing.
2. Controlling the deformation process during TMP shifts the Ms to lower values and consequently allowed obtaining finer bainite by isothermal holding at temperatures lower than that of the undeformed materials
3. New limits of the mechanical properties can be obtained as a consequence of conclusion no. 2.

8.7 Summary

The processing methodologies and the alloying concepts to produce four types of AHSSs, namely dual phase, transformation induced plasticity, bainitic pipeline and ultra-fine bainite steels by TMP are overviewed. It is shown during this study that:

1. An assessment of the microstructure development during the thermo-mechanically processing using dilatometric measurements with respect to the extent of phase-transformation allows selection of processing parameters required to develop the specified microstructures. For instance, different specified quantities of martensite are produced in DP steel by defining the quenching temperatures during continuous cooling. For TRIP steel, isothermal holding temperatures corresponding to specified quantities of ferrite in the final structure are identified. In

both TRIP and fine bainite steels, the isothermal holding periods during bainite transformation are specified.

2. New boundaries of the mechanical properties were established through control of the TMP-parameters. For the bainitic pipeline steel, it is shown that the shorter cooling time between the roughing and finishing resulted in finer and/or pancaked prior austenite grains which motivated the formation of both, AF and fine M/A phases. This resulted in enhancing R_m and R_p without significant effect on (TEl). For the fine bainite steel, it is shown that decreasing the temperature of last deformation step shifted Ms to lower values and consequently allowed obtaining finer bainite by isothermal holding at lower temperatures. The finer structure obtained as a consequence of this showed better strength ductility balance.

Keywords: Multiphase steels, thermo-mechanical processing, physical simulation, dual phase steel, TRIP steel, pipeline steel, ultra-fine bainite steel, bake hardening, phase transformation, dilatometry, microstructure evolution, Tensile Properties

References

Bäcke, L. 2009. Modeling the microstructural evolution during hot deformation of microalloyed steels. Ph.D. Thesis. Royal Institute of Technology. Stockholm.

Bai, D.Q., S. Yue, W.P. Sun and J.J. Jonas. 1993. Effect of deformation parameters on the no-recrystallization temperature in Nb-bearing steels. Metall. Trans. A 24: 2151–2159.

Balliger, N.K. and T. Gladman. 1981. Work-hardening of dual-phase steels. Metal Science. 15: 95–108.

Barbé, L., L. Tosal-Martínez and B.C. De Cooman. 2002. Effect of phosphorus on the properties of a cold rolled and intercritically annealed TRIP-aided steel. Proc. Int. Conf. on TRIP-Aided High Strength Ferrous Alloys. Gent, pp. 147–151.

Bhadeshia, H.K.D.H. 2001. Bainite in Steel, IOM commercial Ltd., London, 2nd Edition.

Bouet, M., J. Root, E. Es-Sadiqi and S. Yue. 1998. Effect of Mo in Si-Mn Nb bearing TRIP steels. Materials Science Forum. 284-286: 319–326.

Bradley, J.R., J.M. Rigsbee and H.I. Aaronson. 1977. Growth kinetics of grain-boundary ferrite allotriomorphs in Fe-C alloys. Metall. Trans. A 8: 323–333.

Brandt, M.L. and G.B. Olson. 1993. Bainitic stabilization of austenite in low alloy steels. Iron Steelmaker. 5: 55–60.

Cao, Z.G., Y.P. Bao, Z.H. Xia, D. Luo, A.M. Guo and K.M. Wu. 2010. Toughening mechanisms of a high-strength acicular ferrite steel heavy plate. International Journal of Minerals Metallurgy and Materials. 17: 567–572.

Capdevila, C., F.G. Caballero and C. García De Andrés. 2005. Neural network model for isothermal pearlite transformation. Part II: Growth Rate: ISIJ Int. 45: 238–247.

Cizek, P. 2001. Transformation behaviour and microstructure of an API X80 line-pipe steel subjected to simulated thermomechanical processing. Metal 2001. Ostrava. Czech Republic.

Davies, R.G. 1979. Early Stages of yielding and strain aging of a vanadium-containing dual-phase steel. Metallurgical Transactions. A 10: 1549–1555.

De-Ardo, A.J. 1984. Fundamental aspects of the physical metallurgy of thermomechanical processing of steel. International Conference on Physical Metallurgy of Thermomechanical Processing of Steels and Other Metals. Thermec-88. Vol. 1, Tokyo, 6–10 June, pp. 20–29.

Deliry, J. 1965. Mem. Sci. Rev. Metall. 527: 7–8.
Djaic, R.A.P. and J.J. Jonas. 1972. Static recrystallization of austenite between intervals of hot working. Journal of the Iron and Steel Institute. 210: 256–261.
Girault, E., P. Jacques, Ph. Harlet, K. Mols, J. Van Humbeeck, E. Aernoudt and F. Delannay. 1998. Metallographic methods for revealing the multiphase microstructure of TRIP-assisted steels. Mater. Char. 40: 111–118.
Han, S.Y., S.Y. Shin, S. Lee, N.J. Kim, J.H. Bae and K. Kim. 2010. Effects of cooling conditions on tensile and Charpy impact properties of API X80 Line pipe Steels. Metall. and Mater. Trans. A 41: 329–340.
Hashimoto, S., S. Ikeda, K. Sugimoto and S. Miyake. 2004. Effects of Nb and Mo addition to 0.2%C-1.5%Si-1.5%Mn steel on mechanical properties of hot rolled TRIP-aided steel sheets. ISIJ Int. 44: 1590–1598.
Heller, T. and A. Nuss. 2005. Effect of alloying elements on microstructure and mechanical properties of hot rolled multiphase steels. Ironmaking & Steelmaking. 32: 303–308.
Huang, J., W.J. Poole and M. Militzer. 2004. Austenite formation during intercritical annealing. Metall. and Mater. Trans. A 35: 3363–3375.
Jeong, W.C. and J.H. Chung. 1992. HSLA steels processing, properties and applications, ed. by G. Tither and Z. Shoubua. Min., Met. and Mat. Soc., Warrendale, PA: 305–311.
Jiao, S., F. Hassani, R.L. Donaberger, E. Essadiqi and S. Yue. 2002. The effect of processing history on a cold rolled and annealed Mo–Nb microalloyed TRIP Steel. ISIJ Int. 42: 299–303.
Lee, J.L., M.H. Hon and G.H. Cheng. 1987. The intermediate transformation of Mn–Mo–Nb steel during continuous cooling. J. Mater. Sci. 22: 2767–2777.
LePera, F.S. 1980. Improved etching technique to emphasize martensite and bainite in high-strength dual phase steel. Journal of Metals. 32: 38–39.
Matlock, D.K., D.M. Bruce and J.G. Speer. 2003. Strengthening mechanisms and their applications in extremely low C steels. International Forum for the Properties and Applications of IF Steels, pp. 118–127.
Matsumura, O., Y. Sakuma and H. Takechi. 1987. Enhancement of elongation by retained austenite in intercritical annealed 0.4C-1.5Si-0.8Mn steel. Trans. ISIJ. 27: 570–579.
Palkowski, H., M. Soliman and G. Kugler. 2007. Effect of thermomechanical processing parameters on bake hardening ability of hot rolled dual phase steels. Proc. TMS Annual Meeting & Exhibition. Orlando, USA.
Park, S.H., S. Yue and J.J. Jonas. 1992. Continuous-cooling-precipitation kinetics of Nb(CN) in high-strength low-alloy steels. Metallurgical Transactions A. 23: 1641–1651.
Pereloma, E.V., B.R. Crawford and P.D. Hodgson. 2001. Strain-induced precipitation behaviour in hot rolled strip steel. Mater. Sci. Eng. A 299: 27–37.
Rashid, M.S. 1981. Dual phase steels. Ann. Rev. Mater. Sci. 11: 245–266.
Smith, Y.E., A.P. Coldren and R.L. Cryderman. 1972. Toward improved ductility and toughness. Climax Molybdenum Company, Japan Ltd., Tokyo, pp. 119–142.
Soliman, M. and H. Palkowski. 2008. On factors affecting the phase transformation and mechanical properties of cold-rolled transformation-induced-plasticity–aided steel. Metall. Mater. Trans. A 39: 2513–2527.
Soliman, M. and H. Palkowski. 2007. Ultra-fine bainite structure in hypo-eutectoid steels. ISIJ International. 47: 1703–1710.
Soliman, M., A. Asadi and H. Palkowski. 2010. Role of dilatometer in designing new bainitic steels. Advanced Materials Research. 89-91: 35–40.
Soliman, M., B. Weidenfeller and H. Palkowski. 2009. Metallurgical phenomena during processing of TRIP steel. Steel Research International. 80: 57–65.
Speich, G.R., A.J. Schwoeble and G.P. Huffman. 1983. Tempering of Mn and Mn-Si-V dual-phase steels. Metall. Trans. A 14: 1079–1087.
Suikkanen, P. 2009. Development and processing of low carbon bainitic steels. Academic dissertation. Acta Univ. Oul., OulunYliopisto, Oulu.
Tafteh, R. 2011. Austenite decomposition in a X80 line pipe steel. Master Thesis. The University of British Columbia.

Tamura, I., H. Sekine, T. Tanaka and C. Ouchi. 1988. Thermomechanical processing of high-strength low-alloy steels, Butterworth & Co. Ltd., London, pp. 80–100.
Wei, W., Y. Shan and K. Yang. 2009. Study of high strength pipeline steels with different microstructures. Materials Science and Engineering. A 502: 38–44.
Williams, J.G., C.R. Killmore, F.J. Barbaro, A. Meta and L. Fletcher 1995. Modern technology for ERW linepipe steel production (X60 to X80 and beyond). Proc. Int. Conf. Microalloying'95, Warrendale, USA, pp. 117–139.
Wu, D., Z. Li and H.S. Lue. 2008. Effect of controlled cooling after hot rolling on mechanical properties of hot rolled TRIP steel. Steel Research International. 15: 65–70.
www.worldautosteel.org/why-steel/safety/facing-the-challenge-for-crash-safety/
Xiao, F.R., B. Liao, Y.Y. Shan, G.Y. Qiao, Y. Zhongb, C. Zhang and K. Yang. 2006. Challenge of mechanical properties of an acicular ferrite pipeline steel. Materials Science and Engineering. A 431: 41–52.
Yakubtsov, I.A. and J.D. Boyd. 2001. Bainite transformation during continuous cooling of low carbon microalloyed steel. Materials Science and Technology. 17: 296–301.
Yue, S., A. Di-Chiro and A.Z. Hanzaki. 1997. Thermomechanical processing effects on C-Mn-Si TRIP steels. Jom-Journal of the Minerals Metals & Materials Society. 49: 59–61.
Zackay, V.F., E.R. Parker, D. Fahr and R. Bush. 1967. The enhancement of ductility in high-strength steels. T. Am. Soc. Metal. 60: 252–259.
Zhao, M.C., B. Tang, Y.Y. Shan and K. Yang. 2003. Role of microstructure on sulfide stress cracking of oil and gas pipeline steels. Metall. Mater. Trans. 34: 1089–1096.
Zhao, M.C., K. Yang and Y. Shan. 2002. The effects of thermo-mechanical control process on microstructures and mechanical properties of a commercial pipeline steel. Materials Science and Engineering. A 335: 14–20.

9

Hot Rolling System Design for Advanced High Strength Steels (AHSSs)

Priyadarshan Manohar

ABSTRACT

Advanced high strength steels (AHSSs) find significant applications in oil and gas pipelines, offshore platforms, ship building and of course the automotive industry. The envelop for property and performance of these steels is ever increasing due to increasing standards for vehicle safety, crash worthiness, fuel efficiency, emission control, manufacturability, formability, durability and quality at low cost. This is made possible through the design and optimization of chemical composition, thermomechanical processing and specialized heat treatments. This chapter describes different types of AHSSs, their typical compositions, applications and their significantly different processing routes. Fundamentals of strain softening mechanisms and applicable constitutive equations are presented. Industrially significant manufacturing processes and heat treatments are described citing specific examples of AHSSs using those manufacturing routes. Finally, two case studies are presented with complete descriptions of hot rolling system design and optimization of high strength low alloy plates and hot strip rolling for automotive AHSS.

Robert Morris University, Engineering Department, JJ 119B, 6001 University Boulevard, Moon Township, PA - 15108, USA.
E-mail: manohar@rmu.edu

9.1 Introduction

There are ever increasing pressures on steel industry to improve the properties, performance, and reliability while at the same time to decrease (or at least not to increase) the cost of commercially available steels. Newer light-weight materials such as carbon fiber reinforced plastics (CFRP), metallic foams and advanced aluminum grades are competing with steels in this regard. This materials competition is intense especially in the consumer markets such as in automotive industry because the consumer expectations and also the legal standards for car safety (crash resistance), fuel economy, vehicle performance and environmental impact (e.g., tailpipe emissions) continue to rise, which is driving the need for new and improved steels. The global steel industry has met this need through the development of new Advanced High Strength Steels (AHSSs), whose unique metallurgical properties and processing capabilities enable the automotive industry to meet these requirements, while keeping costs down (Tamarelli 2011). AHSSs have been defined as a class of multiphase steels which contain ferrite, martensite, bainite, and/or retained austenite in sufficient quantities to produce excellent combination of mechanical properties such as strength, toughness and formability (Horvath 2004). The unique combination of high strength and high formability of AHSS results primarily from their high strain hardening capabilities. Other than the mechanical properties, other factors such as manufacturability, weldability, corrosion resistance and fatigue strength have also been important in the search for improved steels not only in the automotive but also in heavy engineering industry such as gas and oil pipelines and off shore platforms. The alloy composition of these steels must be carefully designed in view of the thermomechanical processes (TMPs) that will be employed for their manufacture. Thus materials design coupled with precisely controlled heating, deformation and cooling processes enables the achievement of desired multiphase microstructures and unique properties of AHSS that find diverse industrial applications.

9.1.1 Major AHSS Products

The low carbon structural steel products can be classified based on their mechanical properties as shown in Figure 9.1:

The major classes of steels given in (Figure 9.1) can be described as follows (Keeler 2004, Keeler and Kimchi 2014):

- Low Strength Steels: Tensile Strength <270 MPa. Examples in this category include mild steels and interstitial free low strength (IF/IFLS) steels.
- Conventional High Strength Steels (Conventional HSS): Tensile Strength 270–700 MPa. Examples in this category include Interstitial Free High Strength (IFHS) steels, Carbon–Manganese (C-Mn) steels,

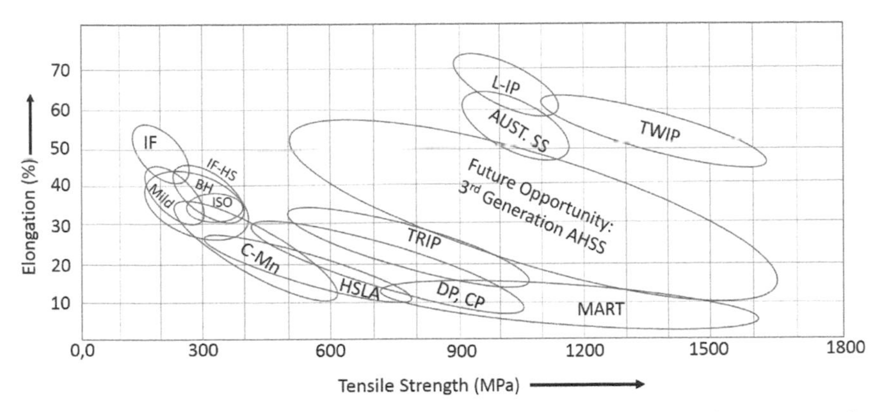

Figure 9.1 Strength and ductility property ranges for HSLA and AHSS steels as compared to other steel grades employed in automotive industry (after Krupitzer 2010).

HSLA steels, Bake Hardenable (BH) steels, and Isotropic (ISO or IQ —isotropic quality) steels. These high-strength Isotropic Steel (I-steel) sheets show direction-independent ("isotropic") material qualities and have a very good deep drawing behavior. They are used for the manufacture of interior components as well as outdoor panels for cars.

- Advanced High Strength Steels (***1st generation AHSS***): ferrite based (ferrite strengthening through grain refinement and precipitation) Yield Strength > 550 MPa, examples include Dual Phase (DP) steels, Complex Phase (CP) steels, ferritic–bainitic (FB) steels, Martensitic (MART or MS) steels, Transformation-Induced Plasticity (TRIP) steels, and hot formed (HF) steels.
- Ultra-High Strength Steels (UHSS, now called ***2nd Generation AHSS***): austenite based steels, high Tensile Strength >780 MPa and high ductility steels. Examples in this category are Aluminum-added Lightweight with Induced Plasticity (L-IP) steels, Shear band formation Induced Plasticity (SIP) steels, Austenitic Stainless Steels (Aust SS) and TWinning-Induced Plasticity (TWIP) steels. These steels have significant amounts of alloying additions making them much more expensive and hence economically unsuitable for automotive applications.
- High Strength, Ductile, Formable and Low Cost Steels: multi-phase based steels (now called ***3rd Generation AHSS***): Tensile Strength >750 MPa. Examples include Boron-Modified steels, Hot Formed (HF) steels and Q&P (Quench and Partitioning) steels. Various strengthening mechanisms such as grain refinement, precipitation hardening, dispersion strengthening, solid solution hardening and phase transformations with a well-designed amount of retained austenite, are

employed to achieve the desired combination of mechanical properties (Matlock and Speer 2008).

- Giga Pascal Steels (now suggested as ***4th Generation AHSS***): Tensile Strength > 2000 MPa, total elongation > 20%—these steels are currently under development (Nicholas 2014).

Each steel type has unique microstructural features, alloying additions, processing requirements, advantages and challenges associated with its use. Each steel type has unique applications where it might be best deployed to meet the performance demands of the automotive part. Some of the compositions and applications of AHSS and HSLA steels are shown in Table 9.1. The present book chapter deals with the hot deformation and rolling of AHSS and HSLA steels to achieve the mechanical properties depicted in Figure 9.1 along with the microstructures and other characteristics listed in Table 9.1.

Formability is one of the crucially important properties for AHSS. The formability is measured by many tests including the standard tension test, and more specialized test such as hole expansion test, bulge test, bend test and cupping test that assess stretching, bending, stretch-bending, deep drawing and flanging characteristics of AHSS (Billur and Altan 2012). Formability data as measured by % hole expansion (% HE) indicates that formability decreases as strength increases, highlighting one of the most famous conflicts in materials science: while both properties are necessary, ductility decreases as strength increases. This effect is shown in Figure 9.2, the curve is sometimes referred to as a "banana curve".

One of the reasons this strength vs. ductility conflict occurs is due to the fact that higher strength phases such as martensite and bainite are needed to increase the strength of the steels but these phases are inherently not as ductile as ferritic microstructures due to their higher dislocation content. To resolve this conflict and develop a better combination of strength and ductility in 3rd generation AHSS it is proposed to have multiple-phase steels such as high ductility austenite along with high strength martensite (M + A) combination. The strength and ductility of the three stable constituent phases in AHSSs are given in Table 9.2.

Depending on alloy composition, hot processing and cooling, the AHSS will have multi-phase microstructures and their properties could be estimated using the property data of the constituent phases shown in Table 9.2. In addition to the strength of these steels, there are many other important mechanical properties of AHSSs that need to be considered. Some of these properties are listed in Table 9.3.

Several strengthening and hardening mechanisms are often used in various combinations to develop desired mechanical properties, fatigue strength or dent resistance in AHSS (Tamarelli 2011). Strengthening

Table 9.1 Typical properties and applications of some key AHSS, HSLA and Conventional steel grades (after Tamarelli 2011).

Grade	Microstructure	Composition: Alloying additions	Mechanical properties YS/UTS (MPa)	Formability*	Weldability*	Automotive applications
DP (Dual Phase)	Ferrite + martensite islands	C, Si, P for strength; Mn, Cr, Mo, V, Ni for hardenability	Good toughness and %E, 300/500; 500/800; 1150/1270	5	4	Crash resistance, energy absorption, beams, cross members
FB (Ferrite–Bainite)	Ferrite + fine bainite	Al, B, Nb and/or Ti	Fatigue resistance, good hole expansion, 350/450 450/600	5	4	Complex parts, vibration loads, suspension/chassis parts
SF (Stretch Flangeable)	Ferrite + very fine bainite	Al, B, Nb and/or Ti	Fatigue resistance, good hole expansion, 570/640 600/780	5	4	As above
CP (Complex Phase)	Ferrite + bainite + martensite + some retained austenite	Similar to DP but with Nb, Ti and V for grain refinement and Rx control prior to phase transformation	High UTS 500/800 1050/1470	4	3+	Crash safety, wear resistance, body structure, chassis, suspension parts
MART/MS (Martensitic)	Martensite + some ferrite/bainite	Si, Cr, Mn, B, Ni, Mo, V to increase hardenability	Highest YS 950/1200 1250/1500	3	4	Strong, lightweight parts, bumper reinforcements, door intrusion
TRIP (Transformation Induced Plasticity)	Martensite + bainite + atleast 5% retained austenite	More C, Si and other alloying elements	Fatigue and Durability, 350/600 600/980	5	3	Load cycle, cross and long members, crash resistance

Table 9.1 contd. ...

... Table 9.1 contd.

Grade	Microstructure	Composition: Alloying additions	Mechanical properties YS/UTS (MPa)	Formability*	Weldability*	Automotive applications
HF (Hot Formed)	Martensitic	0.002–0.005% B, Mn, Al	1050/1500 YS > 1500	5	4	Complex parts, limited repairability
TWIP (Twinning Induced Plasticity)	Austenitic (twin boundaries)	17–24% Mn Stretchability, deformation resistance	instantaneous hardening, 500/980	5	4	Complex parts, high cost, front rail
HSLA (High Strength Low Alloy)	Ferrite + pearlite or complex	C, Mn and Ti, Nb, V	YS 350–700/UTS 450–780	5	5	Suspension parts, wheels, chassis
BH (Bake Hardenable)	Ferrite + precipitation hardening	C, Mn	210/340 260/370	5	5	Door outers, hood, deck lids
IF (Interstitial Free Low Strength)	Ferritic	No alloying elements	260/410 300/420	5	5	Body structures, closures
Mild (Mild Steels)	Ferrite + pearlite	C, Mn	140/270	5	5	Body structures

Notes: %E = % Elongation; *Likert Scale for processability: 5 = excellent; 4 = good; 3 = fair; 2 = below average; 1 = poor.

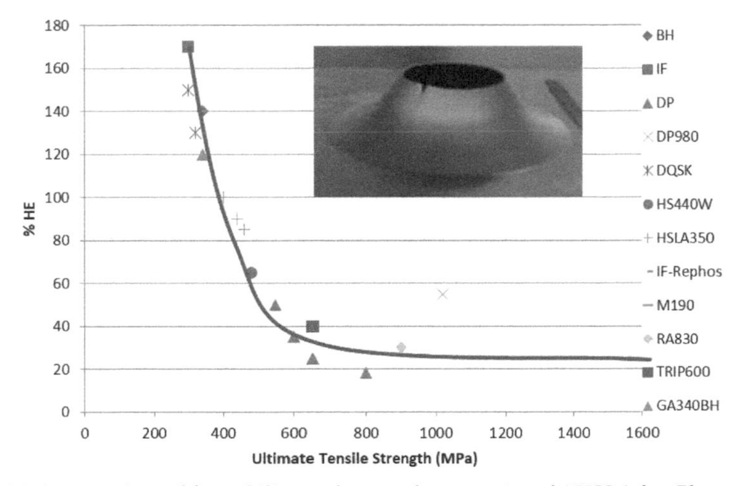

Figure 9.2 An overview of formability and strength properties of AHSS (after Bhattacharya 2006).

Table 9.2 Properties of stable constituent phases in AHSS (Matlock 2006).

Stable constituent	**Strength (UTS in MPa)**	**Ductility (Uniform True Strain)**
Ferrite (IF Steel)	300	0.30
Austenite (Mn Steel)	640	0.60
Martensite (Low C Steel)	2,000	0.08

mechanisms typically work by hindering or impeding the movement of dislocations through the steel, and include the following:

- *Solid solution strengthening*: When another species is added to form a solid solution, the interstitial or substitutional atoms form localized strain fields that can increase the strength and hardenability, although they may simultaneously decrease ductility (C-Mn steels).
- *Grain refinement*: As dislocations travel through a material, they tend to pile up at grain boundaries, preventing further plastic deformation. As grain size decreases, the effective area of grain boundaries increases, increasing the strength of the material (HSLA steels).
- *Work hardening* (*a.k.a. strain hardening*): As a result of cold working (rolling, drawing, bending, etc.), dislocations in steel become more entangled, preventing their relative movement. Work hardening typically increases YS, UTS, and hardness, but often has an adverse effect on ductility and toughness (cold rolled steels).

Table 9.3 Typical mechanical properties of some commercial AHSS grades (Schaeffler 2005).

Steel Grade	YS (MPa)	UTS (MPa)	Total EL (%)	*n* Value* (5–15%)	*r-Bar***	*K* Value* (MPa)
BH 210/340	210	340	34–39	0.18	1.8	582
BH 260/370	260	370	29–34	0.13	1.6	550
DP 280/600	280	600	30–34	0.21	1.0	1,082
IF 300/420	300	420	29–36	0.2	1.6	759
DP 300/500	300	500	30–34	0.16	1.0	762
HSLA 350/450	350	450	23–27	0.14	1.1	807
DP 350/600	350	600	24–30	0.14	1.0	976
DP 400/700	400	700	19–25	0.14	1.0	1,028
TRIP 450/800	450	800	26–32	0.24	0.9	1,690
DP 500/800	500	800	14–20	0.14	1.0	1,303
CP 700/800	700	800	10–15	0.13	1.0	1,380
DP 700/1000	700	1,000	12–17	0.09	0.9	1,521
Mart 950/1200	950	1,200	5–7	0.07	0.9	1,678
Mart 1250/1520	1,250	1,520	4–6	0.065	0.9	2,021

*The Hollomon equation for flow stress is given by $\sigma = K\varepsilon^n$, where σ is the flow stress, K is the strength coefficient, n is the strain hardening exponent, and ε is the applied strain.

***r-bar* value (also called *Lankford Coefficient*) is the average plastic strain ratio in width and thickness direction in uniaxial tensile testing and indicates an ability of a sheet metal to deform permanently without thinning or fracture. A high *r-bar* value (in the range 1.2–1.9) indicates good drawability and formability of a sheet metal (Ramos et al. 2010). Note that the parameter higher total % elongation (Total EL%) is also a measure of good ductility of the sheet metal.

- *Dispersion strengthening or precipitation hardening*: The steel matrix, ferritic or austenitic, often contains other phases, which may range from fine particles (e.g., cementite particles, islands tempered martensite, or discreet carbide or nitride alloy precipitates) to lamellar sheets (e.g., the ferrite and cementite layers of pearlite). These microstructural features can affect the overall properties of the material considerably and illustrate some of the many ways to increase strength (BH steels).
- *Transformation strengthening*: In the production processing of steel, phase transformations can often occur which enable strengthening by creating microstructures with significant amounts of hard phases, such as martensite or bainite. Such transformations occur in operations like hot rolling, hot-dip galvanizing or continuous annealing where steel can cool from high-temperature austenite and transform to these

harder low-temperature phases. This mechanism is fundamental to the development of advanced high-strength steels and enables dual-phase, TRIP and other AHSSs to be manufactured. For example, in a typical Mo-based TRIP steel, the room temperature microstructure after hot rolling consists of polygonal ferrite, bainite, martensite and a significant amount of retained austenite (RA). The higher level of ductility of this steel is achieved when the RA transforms to martensite during forming operation which causes local strain hardening and thus delays localized deformation or necking (Hore et al. 2015).

Based on the various properties that can be produced in AHSS and other steels, they are utilized for appropriate industrial applications as described in the following section.

9.1.2 Industrial Applications of AHSSs

As mentioned before, the use of HSLA steels and AHSSs in the automotive industry has grown tremendously over the past two decades. Ever increasing demands on automobile design including higher crash resistance, passenger safety, fuel efficiency, vehicle performance, and emission control are pushing new frontiers in steels and other materials research. An example of applications of AHSS in crash resistant design of an automobile is shown schematically in Figure 9.3.

The two generalized areas of the car—the front and the rear end of the vehicle and the passenger compartment have very different safety requirements. The passenger compartment, enclosed in a rigid "safety cage", as shown in Figure 9.3 above is designed to protect the passengers in the event of a low- to high-speed crash; the structure should prevent any deformation or intrusions that would compromise the integrity of the structure and impinge on the space around the passengers. The so-called "crumple zones", (Figure 9.3) located at the front and rear end of the

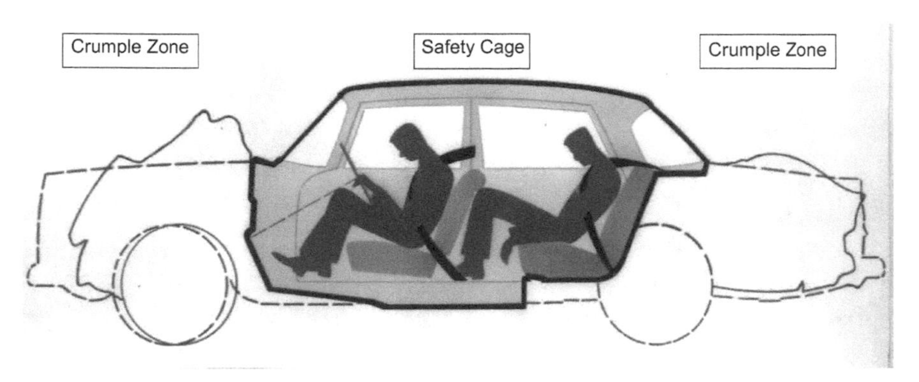

Figure 9.3 The major crash management zones of an automobile (after Tamarelli 2011).

vehicle, are designed to absorb as much energy as possible in the event of a front or rear collision. By absorbing the impact energy over a distance, the crumple zone will cushion the impact and help preserve the structure of the passenger compartment. The general guidelines for materials selection in these zones are outlined below:

Crumple Zone

- Performance requirements: high energy absorption over a distance in a crash event.
- Material properties to meet this need: high work hardening rate, strength, and ductility.
- Evidence of this property: large area under the stress-strain curve.
- Potential steel selection: Duel phase (DP), complex phase (CP), transformation-induced plasticity (TRIP) steels.

Safety Cage—Passenger Compartment

- Performance requirements: no deformation/intrusion during a crash event.
- Material properties to meet this need: high yield strength, high stiffness (high Young's modulus).
- Evidence of this property: highest yield, and ultimate tensile strength of σ-ε curves, high Young's modulus in the elastic rage.
- Potential steel selection: Martensitic, HF, DP (>980 MPa) steels.

The total mass of a typical passenger car is 1,260 kg while the mass of the car body itself (body-in-white–BIW - structure) is 360 kg. By replacing conventional steels with AHSSs in car body, automotive engineers have achieved an overall vehicle mass savings of 9% and in particular BIW mass savings of up to 25% without adding too much cost. The mass savings achieved due to the use of AHSSs has resulted in the lifetime savings of 2,200 kg of CO_2 emissions per vehicle and an increase in fuel efficiency in the range of 2%–8% (World Steel Association 2006).

Going forward, new regulations and new research will spawn new steel grades and their processing technologies will continue to shape the future of AHSS industry. Current research aims to expand the already broad property spectrum of AHSSs even further. One area of particular interest is the "*third generation steels*" (Figure 9.1) that attempts to bridge the gap of strength—ductility balance between conventional and AHSSs at affordable cost (Matlock and Speer 2008). Some steel grades under development in this category include nano-steels that use finely dispersed, nano-scale particles for excellent total and local elongation even at high strengths. Another area of active research is the ultrafine grained steels. Even in plain carbon steels, ultrafine grained steels (grain size of the order 1–2 μm) have been developed that have superior mechanical properties compared to the conventional C-Mn steels (Song et al. 2006, Beladi et al. 2004).

A third area of steel research is the development of Microalloyed and HSLA steels. These steels have found diverse applications over last four decades in areas such as pipes for oil and gas pipe lines (including arctic conditions and sour-gas environments), ship building (tanks and vessels), offshore drilling platforms, mining equipment, power transmission towers, lighting poles, farm machinery, storage tanks, structural engineering such as high rise buildings and bridges (including earthquake resistant ductile RCC constructions) and automotive parts (long members, wheels and axles) (Manohar 1997, Vervynckt et al. 2012). Finally, a recent NSF-funded project is exploring to develop *"fourth generation steels"*, a family of nano-engineered low-alloy steels that surpass the state-of-the-art in coupling high strength (>2,000 MPa tensile strength) with high elongation (>20%) (Nicholas 2014). Considering that mild steels have UTS of the order 270 MPa the advancement on HSLA and AHSS technologies has already seen six fold increase in mechanical properties and it seems certain that there will be further significant improvement in near future through the synergistic effects of the advancement of steel rolling technology, alloy design and post deformation heat treatments.

9.1.3 Primary Manufacturing Routes

The ability to produce useful components from liquid is known for many centuries. Mass production of steel products began in the 1780s and progressed rapidly during the industrial revolution. The key stages in the development of steel manufacturing technologies are presented in Figure 9.4 and explained below.

In days gone by, liquid steel was cast into cast iron molds to produce steel ingots weighing 10–20 tons. The ingots were reheated in soaking pits and then rolled in primary mills to make intermediate products such as slabs, blooms and billets. Continuous casting of steel (hereafter termed Conventional Continuous Casting or CCC) first appeared in the 1950s and quickly gained ascendancy over the ingot casting to develop into a matured industrial technology and over 95% of the world's crude steel production (~1.7 billion tons per year) is now obtained through this casting route (World Steel Association 2015). Following CCC, 200–250 mm thick as-cast slabs are subsequently hot rolled into plate, sheet, strip and various structural shapes. In the last few decades, near net shape casting (NNSC) has attracted worldwide attention as these technologies have the capability of producing low-cost sheet products with desirable mechanical and physical properties (Manohar et al. 2000). NNSC involves the production of steel products to near final shape directly from liquid steel with secondary processing such as rolling reduced to an absolute minimum. The development of thin slab

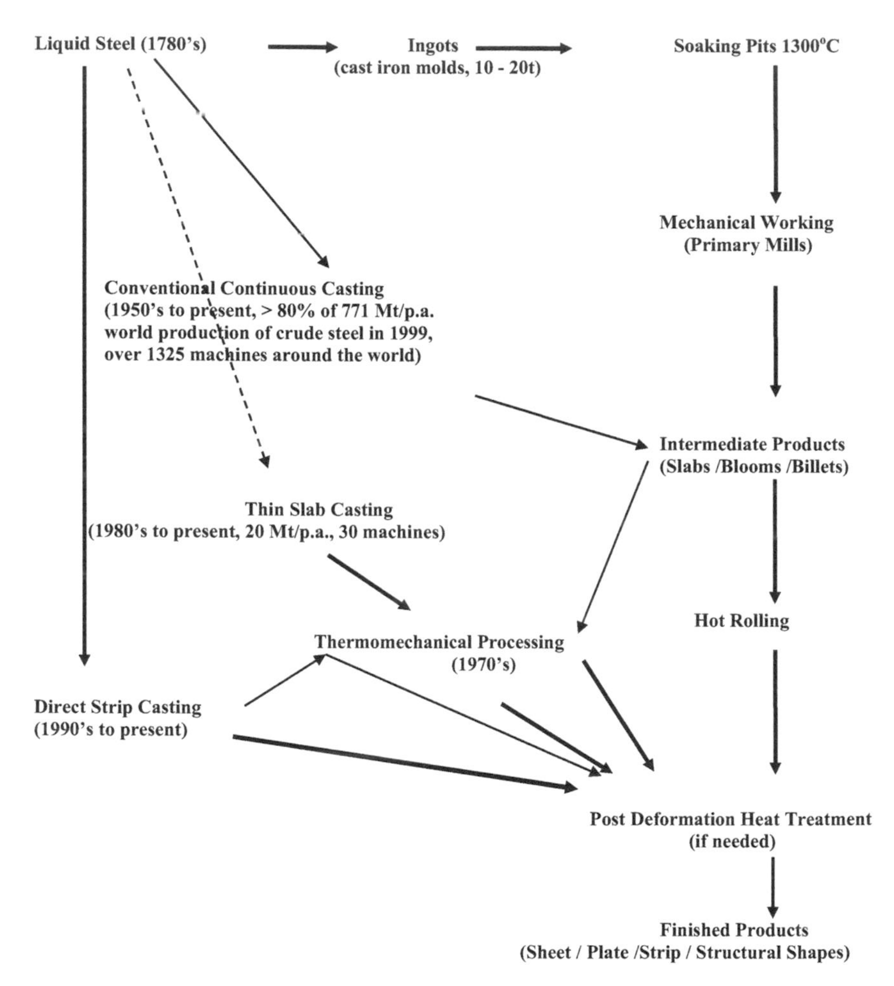

Figure 9.4 Key stages in the development of steel manufacturing technologies (after Manohar et al. 2000).

casting (TSC) which is an integral part of NNSC technology began in 1980s and since then the increasing strategic importance of this technology has been well recognized (Korchynsky 2002). On one hand TSC allows efficient and economic production of flat products (sheets, plates, strips) and on the other hand allows low carbon steel to be replaced by microalloyed, HSLA and AHSSs contributing to weight reduction and thus decreasing the demand for steel and helping the cause of global sustainability for steels. Direct strip casting (DSC) using twin roll casting (TRC) and horizontal single belt casting (HSBC) technologies are key NNSC processes that are now on the verge of full commercialization. They are expected to make a significant impact on steel making operations because DSC will open the flat product (sheet, plate, and strip) market, which is traditionally the forte of integrated steel mills, to the minimills and micromills. NNSC technologies such as TSC and DSC (TRC, HSBC) offer several advantages over CCC (Ge et al. 2012, 2013) because they:

- drastically reduce capital investment costs
- use a much smaller continuous caster
- eliminate either the roughing mill (e.g., in TSC) or almost all of the rolling mill (e.g., in DSC)
- offer a huge process simplification
- minimize scale losses to improve yield
- use recycled scrap or direct reduced iron (DRI) processes that are more "earth friendly" as a source of virgin iron units so that no coke ovens, sintering plants or blast furnaces are needed.

The reduction in energy, labor and capital costs of NNSC technology is anticipated to bring an estimated global savings of 30–60% of the actual costs of strip produced by CCC. Such a huge saving in production costs is a quantum leap forward in the competitive production of sheet steel.

In the context of secondary manufacturing processes, an alternative to conventional hot rolling (CHR) of steel, termed thermomechanical process (TMP), was introduced in 1970 and, since then, progress has been rapid with contemporary TMP routes now replacing almost entirely the CHR process. Some of these secondary manufacturing technologies will now be described.

9.1.4 Important Secondary Manufacturing Processes

Thermomechanical processing is defined as a hot deformation schedule (rolling, forging, extrusion, etc.) designed for the purpose of achieving a predetermined microstructure in austenite prior to transformation to ferrite (DeArdo 1995). The microstructure of austenite can be described by the parameters: grain size, composition, presence or absence of microalloy precipitates, degree of recrystallization and texture. Comparison of conventional and contemporary techniques for hot rolling of steel is shown schematically in Figure 9.5.

Conventional hot rolling involves reheating a continuous cast slab into the austenite phase field followed by soaking, continuous rolling of the slab to the desired thickness and subsequent air cooling to room temperature. In 'conventional controlled rolling' (CCR), the first two stages (reheating and soaking) are similar to CHR, however rolling involves two further stages: *roughing* and *finishing*. In the roughing stage, deformation is carried out *above* a certain critical temperature known as 'No-Recrystallization Temperature' (T_{nr}), and results in austenite grain refinement due to repeated static recrystallization cycles after each rolling pass. In the 'finishing' stage, there is a significant time delay between the roughing and finishing passes

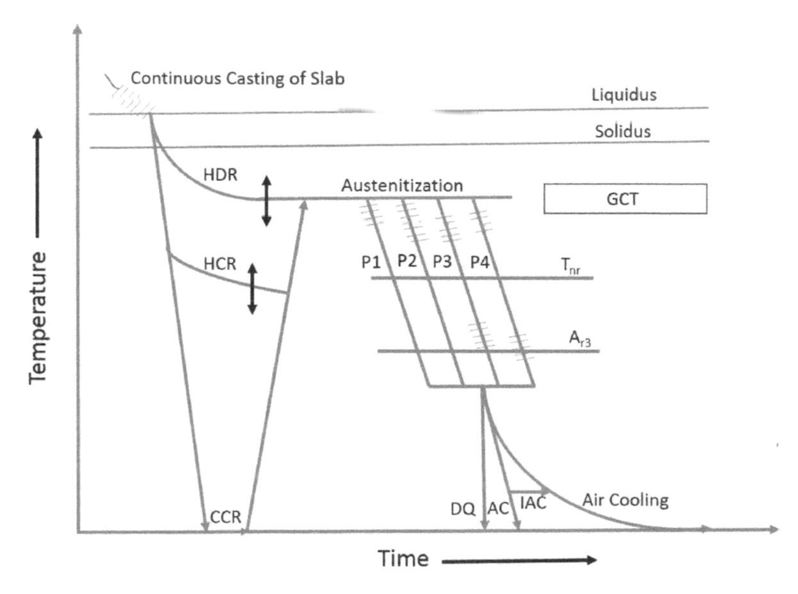

Figure 9.5 Schematic illustration of comparison of conventional and contemporary techniques for hot rolling of advanced steels (after Manohar 1997). P1 through P4 are four different major hot processing routes: P1 is CHR, P2 is RCR, P3 is CCR and P4 is ferrite rolling, a variation of CCR.

to allow deformation of the steel to be carried out below T_{nr}. Austenite does not recrystallize which results in substantial grain flattening to produce a 'pancake' structure with the deformed grains containing a high internal dislocation density as well as deformation bands and twins (Kozasu et al. 1975). The combination of a fine recrystallized austenite grain size after roughing and further deformation of the grains after finish rolling results in an increase in the surface area of austenite grain boundaries per unit volume (S_v). This provides substantial nucleation sites during the subsequent transformation of deformed austenite to ferrite and results in considerable grain refinement.

A process termed 'Recrystallization Controlled Rolling' (RCR) is an alternative means of refining austenite and involves controlled reheating of the as-cast slab to produce a fine austenite grain size prior to roughing which is subsequently refined further by repeated recrystallization cycles during roughing above T_{nr} (Figure 9.5). Some recent TMP technologies, such as hot direct rolling (HDR) and hot charge rolling (HCR), have been developed and these are also shown in Figure 9.5. These processes involve the continuous cast slab being either rolled directly (HDR) or transferred to the hot rolling mill without cooling to room temperature (HCR) and offers several advantages over CHR or CCR, such as significant reduction in energy consumption, efficient furnace utilization, reductions in mill scale

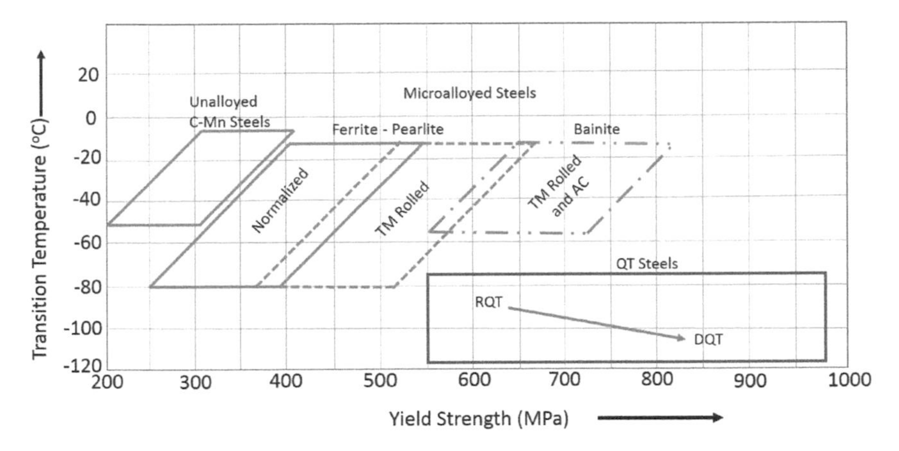

Figure 9.6 Comparison of properties of conventional C-Mn steels and thermomechanically processed microalloyed steels (after Mueschenborn et al. 1995).

loss, improved delivery performance, reduced slab handling and surface and internal quality improvements (Peterson 1994).

The TMP of steel has yielded a two to three fold improvement in the mechanical properties of the steel products compared with CHR, especially when used in combination with microalloying elements to the steel and accelerated cooling (AC). An example of the superior mechanical properties of steel products produced by TMP and AC is shown in Figure 9.6.

9.1.5 Post Deformation Heat Treatments for AHSSs

An area of on going interest is the influence of cooling after TMP. It can be seen in Figure 9.5 that at least four options are possible: air cooling; accelerated cooling (AC); interrupted accelerated cooling (IAC), and direct quenching (DQ). AC was first developed in 1965 for the production of strip and plate steel products with the objective of achieving a low coiling temperature (~600°C) in a hot strip mill to minimize strength variations throughout the coil (Korchynsky 1987). The use of AC was subsequently shown to improve yield and tensile strengths and low temperature Charpy impact energy, without seriously affecting fracture appearance transition temperature (Tamehiro et al. 1987).

The improved mechanical properties of AC-produced steel has been attributed to the refinement of the ferrite grain size and the formation of acicular ferritic, bainitic or martensite-austenite constituent microstructures. The gain in strength via AC allows a reduction in carbon content in the steel, which improves notch toughness and weldability. Microalloyed steels produced by AC have generated properties comparable to and often superior to quench and tempered steels (Korchynsky 1987, Tamehiro et al.

1987). AC steels are also more cost-effective than quench and tempered steels since AC eliminates the need for a separate heat treatment process after rolling.

Interrupted accelerated cooling (IAC) has also been shown to improve significantly the ultimate tensile strength (UTS) of Nb and Nb-Ti microalloyed steel plates (lower IAC temperatures resulted in a higher UTS) (Bognin 1987). A further example of the application of IAC is in the production of low C-Mn-Nb-Ti steels where it was found that this cooling technique has a significant effect on the type of microstructure (ferrite or bainite) and substantial ferrite strengthening by precipitation when the cooling was interrupted between 700–640°C (Pereloma and Boyd 1996).

Some specific heat treatments for different AHSS grades are briefly described as follows.

Bake Hardenable steels: the steels are soft and formable when delivered and car body panels are formed easily. The forming process develops dislocation and these dislocations are pinned down due to interstitial atoms (C, N) that diffuse to the dislocations in the subsequent paint baking process that occurs at 170°C for about 20 min. The material hardens and the strength increases by about 40–60 MPa after heat treatment.

Duel Phase steels: there are two ways of producing DP steels—(i) cold rolling + intercritical annealing in (α + γ two phase field) + quenching ($\gamma \rightarrow$ M) + 400°C aging for a short time allowing precipitation and softening of the steel—a variation of this process is interrupting quenching at 460°C and sent to hot dip galvanizing lines where aging occurs in the zinc bath, or (ii) hot rolling + controlled cooling where the hot rolled steel cooled to intercritical region and then quenched allowing the remaining austenite to transform to martensite, typically about 20% martensite + 80% ferrite at room temperature.

TRIP Steels: there are two ways to develop TRIP microstructures that contain 50–60% ferrite, 25–40% bainite and 5–15% retained austenite. Cold rolled steel can be reheated to intercritical temperatures like DP steels, cooled rapidly at a rate of 15–25°C/s to an isothermal holding temperature in the range 350–500°C to allow bainite form. During bainitic transformation, C, Mn, Al, and Si redistribute and stabilize remaining austenite and depressing M_s temperature so that when the steels are cooled to room temperature the remaining austenite does not transform to martensite. In a second method, after hot rolling is completed in austenite region, the steels are cooled to intercritical temperature and coiled at bainitic temperature of 500°C. The alloy design and TMP is such that pearlite formation or carbide precipitation is suppressed during phase transformations.

CP steels are generated by a combination of controlled cooling and well-designed austenite microstructure subsequent to hot rolling. The steels also have fine dispersion of V and Ti precipitates. Fine austenite and controlled cooling generates a combination of ferrite, bainite and martensite phases at room temperature, they even could have bainitic matrix with good formability.

MART or MS steels are generated by accelerated cooling of the hot rolled steels on the run out table. Here the alloy design ensures good hardenability and a high M_s temperature so that ferrite or bainite formations are avoided.

TWIP steels—these steels contain 18–30% Mn (along with Al, Si, Cr, C and N) and therefore they are mainly austenitic at room temperature and do not need any special heat treatment to stabilize the austenite. The steels deform by various mechanisms such as (i) austenite (fcc) → ε (hcp) martensite phase transformation, or (ii) mechanical twinning or (iii) dislocation guide which causes rapid hardening of the steels. The Mn content has been found to have significant effect on stacking fault energy (SFE) which in turn affects the operating hardening mechanism of these steels (Allain et al. 2004)—for example in Fe+3Al+3Si steels for a Mn content of:

- 22% Mn, the stacking fault energy (SFE) was found to be 15 ± 3 mJ/m^2,
- 25% Mn, SFE was 21 ± 3 mJ/m^2
- 28% Mn, the SFE was determined as 39 ± 5 mJ/m^2 (Pierce et al. 2014).

It has been suggested that $\gamma \rightarrow \varepsilon$ transformation occurs for SFE below 18 mJ/m^2 while mechanical twinning is active for SFE in the range 12 to 35 mJ/m^2. Thus it is clear that for low Mn content when SFE is low, hardening through phase transformation (TRIP) is the thermodynamically preferred mechanism, for intermediate Mn content mechanical twinning (TWIP) is the preferred mechanism while at high Mn content hardening occurs at deformation glide.

PHS—press hardening steels are soft and formable in ferrite—pearlite condition when received. There are austenitized, rapidly formed and quenched by the water-cooled forming die generating martensitic structure in steel after forming. The steels also have 10–30 ppm boron for hardenability and strengthening. Carbon, Mn and Cr additions also control hardenability of steels.

Q&P Steels require carefully controlled three step heat treatment to develop high amounts of retained austenite (~ 50% RA for example) with no need to add high amounts of alloying elements such as Mn. So, the basic idea of this 3rd generation AHSS is to gain the benefits of TRIP or TWIP hardening at a much lower cost. The heat treatment (Speer et al. 2003, 2014) is shown schematically in Figure 9.7.

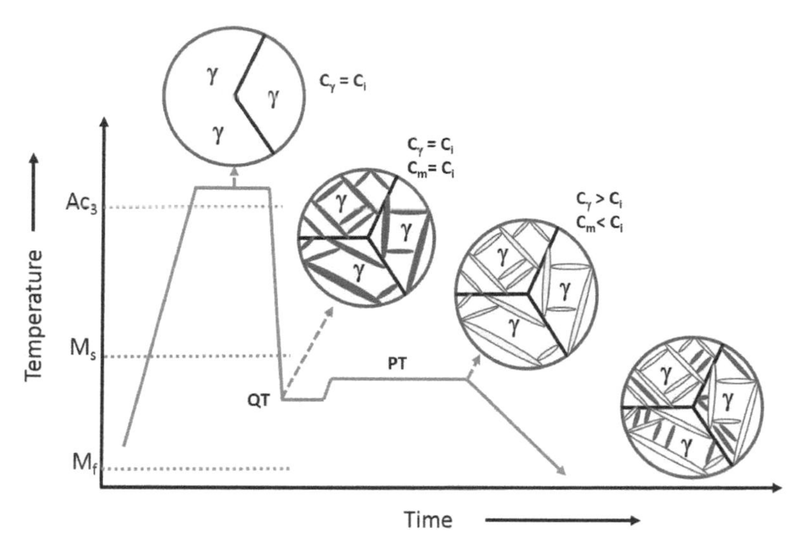

Figure 9.7 Heat treatment steps for developing Q&P AHSS (after Matlock and Speer 2008).

The steel is fully austenitic subsequent to rolling or it has been re-austenitized to an annealing temperature above Ac_3. The carbon content is homogeneously distributed in the microstructure. The steel is then quenched to an intermediate temperature QT, e.g., 240°C (Matlock et al. 2012) between M_s and M_f for this steel to develop specific volume fraction of martensite in the structure. The carbon is still uniformly distributed between austenite and martensite phases, by this metastable arrangement. The steel is then reheated to a partitioning temperature, PT, of the order 350°C (Matlock et al. 2012) where carbon diffuses out of martensite and enriches surrounding austenite and thus makes it stable. In another version of this process, partitioning temperature is the same as quenching temperature so that heat treatment is simplified to a single step process. Many heat treatment schedules, microstructures and resultant properties for several model alloys have shown potential for these Q&P steels to be commercialized (Sun 2013). Alloy design is such that carbide precipitation is suppressed at this stage. Finally, the steel is cooled to room temperature when some additional martensite forms but significant amount of austenite is still retailed due to carbon enrichment and the consequent phase stabilization. Thus at room temperature there is sufficient amount of retained austenite to achieve TRIP effect.

9.2 Fundamentals of Hot Rolling System Design

As the thickness of steel is reduced from 200–250 mm (continuously cast ingot) or 60–80 mm (thin slab) or 5–10 mm (strip) down to any desired thickness from 1 to 10 mm, metal undergoes significant amount of plastic deformation during thermomechanical processing (TMP) of steels. Work hardening occurs during deformation while softening of metals occurs through recovery and recrystallization process subsequent to the deformation. A fraction of the energy of deformation is stored within the metal as strain energy associated with the various types of lattice defects (e.g., dislocations) that are produced in the metal as the deformation progresses. The main consequences of the increase in the number of lattice defects are the increase in strength of the metal, called work hardening, and a reduction in ductility. The ability of the metal to undergo further deformation then depends on its ability to relieve the internal stresses through several competing softening (restoration) processes. The major softening processes that may occur in ferrite and austenite are static and dynamic recovery and recrystallization and metadynamic (postdynamic) recrystallization. There are a number of factors which influence the type and kinetics of the operating softening process in ferrite and austenite. *Material factors* include crystal structure, stacking fault energy (SFE), composition, initial grain size, initial texture and the presence or absence of second phase precipitates. The principle *processing variables* include mode of deformation, strain, strain rate, temperature of deformation and interpass (i.e., inter-deformation treatment) time.

The driving pressure for both recovery and recrystallization is the stored energy of deformation. This is of the order 10–200 $Jmol^{-1}$, generally much lower than the energy associated with phase transformations (Humphreys and Hatherly 2003). The process of recovery reduces the internal energy of the metal through mechanisms such as annihilation of dislocations with opposite signs and/or re-arrangement of dislocations into low angle grain boundaries, a process termed polygonization. When recovery occurs following deformation, it is termed static recovery (SRV). In some cases, the recovery may start while the deformation is still being applied, a process called dynamic recovery (DRV). The mechanism of recovery is complex and is significantly influenced by SFE. In high SFE phases such as body centered cubic (BCC) ferrite, recovery is rapid while low to moderate SFE phases (SFE <80 mJ/m^2) such as face centered cubic (FCC) austenite, recovery is sluggish (Palmiere et al. 1996). There are two main reasons why recovery is predominant in ferrite compared to austenite. First, dislocation climb is easier in materials with high SFE (ferrite) while reverse is true in low to moderate SFE materials (austenite). Secondly, self-diffusivity of iron atoms in ferrite is about 100 times that in austenite at a given temperature. The rate of recovery is very rapid in the initial stage but it falls off significantly as the internal (stored) energy is reduced. If a significant proportion of the stored energy is released by either static or dynamic recovery, then recrystallization may not occur.

However, if the stored energy remains high and there are localized orientation gradients in the microstructure caused by deformation, recrystallization may occur. Recrystallization is initiated by the formation of nuclei of new, strain free grains that subsequently grow into the deformed regions of the microstructure. The process continues until all the original, strained grains are replaced by the new, strain-free grains. The primary mechanism of recrystallization has been suggested to be strain-induced boundary migration (SIBM) especially in microalloyed steels (Rehman and Zurob 2013, Hansen et al. 1980). When recrystallization occurs after deformation, it is termed static recrystallization (SRX). Under certain circumstances, the recrystallization may start during the deformation, a process called dynamic recrystallization (DRX). DRX is the most important softening mechanism in low SFE materials (e.g., austenite, austenitic stainless steel) when DRV is sluggish and the driving force for DRX is maintained (Cahn and Haasen 1996). The phenomenon of DRX is technologically important as it softens the metals during hot deformation thus decreasing the rolling loads needed for hot processing, and can lead to significant grain refinement which in turn improves the mechanical properties and formability of materials after cooling to room temperature (Mandal et al. 2014). In some cases, the dynamic recrystallization may initiate through the nucleation of new grains, but may not proceed to completion during deformation. In these cases, recrystallization is completed after deformation through growth of dynamically nucleated grains. This is known as metadynamic (postdynamic) recrystallization (MDRX).

One of the major challenges in steel processing is to determine which softening mechanism(s) is (are) operative under industrial processing conditions that differ widely depending on the type of steel product being manufactured. For example, the strain rates and interpass times in plate mills are about 1–30 s^{-1} and 8–20 s respectively compared to the high strain rates (10–1000 s^{-1}) and short interpass times (0.015–1.0 s) in bar/wire/rod mills. Both material and processing variables mentioned previously may result in situations where one softening mechanism is predominant or where several softening mechanisms may occur concurrently. Understanding these complex interactions and using them to improve properties of the steel products is a key to the successful design of a thermomechanical processing (TMP) sequence in industry. One way of understanding these softening mechanisms is to study their kinetics. These will be discussed and summarized in the following section.

9.2.1 Mechanisms of Strain Softening

SRX kinetics may be quantified using the classical nucleation and growth theory of phase transformations embodied in the form of the JMAK

(Johnson-Mehl-Avrami-Kolmogorov) equation which relates fraction recrystallized (X_V) to annealing time (t):

$$X_V = 1 - \exp\left[-B(t - t_o)^n\right] \tag{9.1}$$

where t_o is the incubation period needed to form a nucleus, n the JMAK exponent and B is a function of nucleation rate, $\dot{N}$, and growth rate of nuclei, $\dot{G}$. As $\dot{N}$ and $\dot{G}$ increase rapidly with temperature, the kinetics of recrystallization are generally a strong function of annealing temperature. Furthermore, both $\dot{N}$ and $\dot{G}$ are influenced by the stored energy of deformation and the rate of recrystallization is also expected to be dependent on deformation conditions and initial microstructure. Relationships of the form of Eq. (9.1) predict reasonably the static recrystallization kinetics of both cold and hot deformed steel (Humphreys and Hatherly 2003).

In general, the rate of recrystallization after hot working is difficult to predict using physically based models. Empirical models of recrystallization kinetics are frequently used which incorporate both microstructural and processing parameters and have the following form (Sellars and Whiteman 1979):

$$t_x = A\, d_o^a\, \varepsilon^b Z^c \exp\left(Q_{rex}/RT_{rex}\right) \tag{9.2}$$

where t_x is the time for a certain volume fraction to recrystallize during DRX, MDRX or SRX, d_o is the initial grain size, Q_{rex} an activation energy term, R the universal gas constant and T_{rex} the recrystallization temperature. The constants A, a, b and c are derived from experiment. The term Z is the Zener-Hollomon parameter (Zener and Hollomon 1944) that represents the equivalence of the effects of changes in strain rate and in deformation temperature upon the stress-strain relation in metals. Z is given by the following equation:

$$Z = \dot{\varepsilon} \exp(Q_{def}/RT_{def}) \tag{9.3}$$

where $\dot{\varepsilon}$ is the strain rate, Q_{def} is the apparent activation energy for deformation and T_{def} is the deformation temperature. It is noted here that the value of Q_{def} is significantly different from the activation energy for self-diffusion in austenite. This is because of the alloying elements present in the steel composition such as Nb, V, cause drag on grain boundary motion. In addition, in the case of microalloyed steels nano-scale carbo-nitride precipitation may pin the grain boundaries thus reducing the rate of dynamic recovery. Typically the values of Q_{def} have been determined to be in the range 350–550 $kJmol^{-1}$ (Medina and Hernandez 1996, Quelennec and Jonas 2012, Ferdowsi et al. 2014) depending upon steel composition while the activation energy for self-diffusion for austenite in unalloyed iron is determined to be 295 $kJmol^{-1}$ (Evangelista et al. 2004, Vasilyev et al.

2011). The equivalence of the effects of changes in $\dot{\varepsilon}$ and T_{def} on the flow stress of steels was found by Zener and Hollomon in a series of tensile tests conducted in the temperature range –95°C to +20°C using strain rates in the range 10^{-4}–10^{4} s^{-1}. Subsequent research has confirmed this equivalence at high deformation temperatures (e.g., 850–1150°C) as well (Medina and Hernandez 1996). Modeling of austenite flow stress is essential for computing roll separation forces and roll driving torque in rolling mills (Hodgson et al. 2009, Quelennec et al. 2011). Zener–Hollomon parameter has a strong influence on hot ductility behavior as shown in static tension tests of a 8% Cr roller steel (Wang et al. 2015). It is shown in this study that hot ductility as measured by% Reduction of Area (% RA) in a tension test ln(Z) had a parabolic relationship. When the value of ln(Z) were in the range 32–40, the hot ductility was excellent while the steel had low or poor ductility when either $Z < 32$ or when $Z > 40$. The correlation between Z and flow stress of steel (σ) during hot deformation was calculated as follows (Sellars and McTegart 1966):

$$Z = A\,(\sinh \alpha\sigma)^{n} \tag{9.4}$$

where A, α and n are constants that are independent of temperature. Equation (9.4) can be re-written to express the dependence of maximum flow stress (σ_p) on Z during hot deformation of steel (Medina and Hernandez 1996):

$$\sigma_p = \frac{1}{\alpha}\ln\left[\left(\frac{Z}{A}\right)^{\frac{1}{n}} + \left\{\left(\frac{Z}{A}\right)^{\frac{2}{n}} + 1\right\}^{\frac{1}{2}}\right] \tag{9.5}$$

Empirical relations predicting the recrystallized grain size (d_{rex}) during or after hot working of ferrite and austenite are usually of the form:

$$d_{rex} = B\, d_o^{e}\, \varepsilon^{f}\, Z^{g} \tag{9.6}$$

where B, e, f and g are experimentally-determined constants. In case of dynamic recrystallization of austenite, there is little effect of strain or initial grain size on d_{rex}, and the exponents e and f tend to zero. Relationships such as in Eqs. (9.2), (9.5) and (9.6) are necessary components of models that describe industrial process routes for steel manufacturing. A set of hot flow constitutive models for DRV, DRX and SRX have been determined based on the above equations for a boron containing advanced ultra-high strength (A-UHSS) steel and the models validated using experimental results from a torsion test and metallography (Mejia et al. 2014).

9.2.2 Flow Curves, Restoration Processes and Constitutive Relations

Much of the research concerning the mathematical modeling of high temperature mechanical behavior of steels has been carried out by physical simulation in laboratory torsion tests. However, most torsion machines are operated at strain rates ~ 100 s^{-1} or less. Thus, one of the main problems in attempting to model the high strain rate rolling processes, such as wire and rod rolling, based on laboratory torsion test data, is that the operating strain rates in the mills are as high as 3000 s^{-1}. The low strain rate torsion tests have resulted in a constitutive relation of the form shown below (Sellars and McTegart 1966):

$$\dot{\varepsilon}\exp\left(\frac{Q_{def}}{RT}\right) = A_1[\sinh(\alpha\sigma_p)]^{n'} \tag{9.7}$$

It is not clear at present whether the low strain rate models such as Eq. (9.7) could be extrapolated to high strain rate regime to compute the constitutive relations. The problem has been addressed by employing Hopkinson-bar that can deform specimens at high strain rates of the order 4000 s^{-1} (Lim et al. 2003). The stress-strain behavior at various strain rates of a 0.45% C–0.70% Mn steel was investigated in the cited work by means of torsion and Hopkinson-bar compression testing. It was found that at low strain rate regime, a hyperbolic sine law is applicable (as expected), however at high strain rate regime a power law is more accurate in predicting the high temperature—high strain rate flow behavior of austenite.

A typical set of flow curves obtained during low strain rate torsion testing at a test temperature of 1200°C is presented in Figure 9.8.

It can be observed from Figure 9.8 that the flow curves exhibit three distinct stages that reflect the deformation and restoration behavior of austenite. In Stage I, strain hardening is predominant where the stress increases with increasing strain due to storage of dislocations. Stage II is characterized by the decreasing slope of the strain hardening curve until a maximum stress, called the peak stress (σ_p) is reached, the corresponding strain is called the peak strain (ε_p). The decrease in slope of the strain hardening curve is attributed to the dynamic softening. In stage III, the rate of strain hardening is balanced by the rate of dynamic softening so that no increase in stress occurs with increasing strain, thus reaching a steady state. The influence of temperature and strain rate on these key parameters as determined from the torsions tests are summarized in Table 9.4.

It can be seen from the data shown in Table 9.4 that the greatest peak stress is observed at higher strain rate. Also, at higher strain rates and lower temperatures, the initiation of dynamic recrystallization requires more

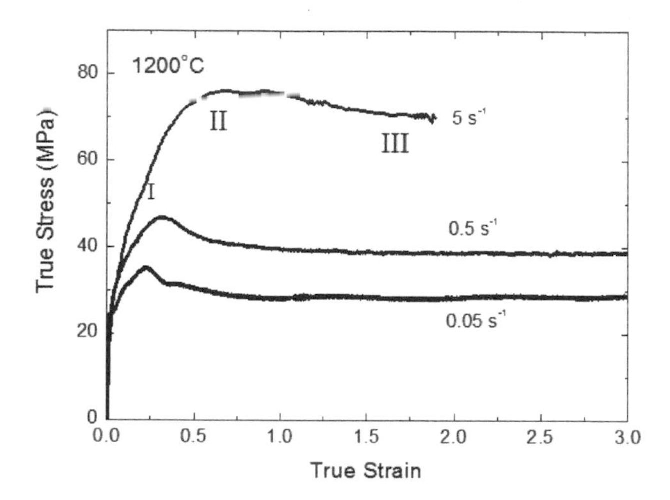

Figure 9.8 Flow curves obtained under continuous torsion testing at 1473 K (after Lim et al. 2003).

strain. In other words, the peak strain and peak stress increase as strain rate increases and temperature decreases.

The hyperbolic sine law (Eq. (9.7)) that indicated the best fit at low strain rate was not adequate to model the high strain rate behavior. In this case, the power law has shown a reasonable fit between the predicted and experimental data. Upon continuous deformation using torsion test and compression test over a wide range of strain rates, 0.05 ~ 4100 s^{-1}, a constitutive equation based on a power law was derived as follows.

$$\dot{\varepsilon} = 6.16\sigma_p^{6.18}\exp(-279140/RT) \tag{9.8}$$

$$Z = 6.16\sigma_p^{6.18} = \dot{\varepsilon}\exp(279140/RT) \tag{9.9}$$

The activation energy of deformation 279 *kJ/mol* is only slightly greater than that obtained for low strain rates (270 *kJ/mol*).

9.2.3 Industrial Exploitation of Various Softening Processes

A brief discussion is now given of the benefits of controlling recrystallization of ferrite and austenite during TMP. As mentioned before, CCR is a term used to describe multi-pass hot rolling of steel. This process was the first widespread industrial application of SRX (Guthrie and Jonas 1990). In CCR, rolling is performed in two stages: (a) "roughing" rolling at high austenite temperatures (1150–1000°C) which results in repeated SRX and refines the initial coarse austenite grain size to ~ 20 μm, and (b) "finishing" rolling where the steel is cooled below T_{nr} and multi-pass rolling is carried out

Table 9.4 Summary of peak stresses and peak strains in torsion tests as a function of deformation temperature (T) and strain rate ($\dot{\varepsilon}$) (Lim et al. 2003).

T (°C)	800			900			1000			1100			1200		
$\dot{\varepsilon}(s^{-1})$	0.05	0.50	5.0	0.05	0.50	5.0	0.05	0.50	5.0	0.05	0.50	5.0	0.05	0.50	5.0
ε_p	0.61	0.68	0.72	0.29	0.68	0.98	-	-	-	-	-	-	0.23	0.30	0.71
σ_p (MPa)	131	169	204	46	123	167	59	88	125	40	60	98	35	47	76

to the final thickness with temperature decreasing to ~800–840°C during the final pass. During deformation below T_{nr}, austenite does not fully recrystallize which results in grain-flattening (pancaking). The deformed structure of austenite offers increased nucleation density for the subsequent transformation to ferrite phase. The final ferrite grain size produced is in the range 5–10 μm, resulting in high-strength steels with excellent toughness and weldability. The CCR process is used extensively for the production of steel plates for oil and gas pipelines and offshore platforms.

An alternative rolling process to CCR, termed the RCR has been developed. In RCR, alloy composition is critically important (DeArdo 1997), and is designed to allow SRX to proceed relatively quickly but minimizing grain growth by particle pinning. Recrystallization between each pass produces fine-grained, recrystallized austenite which subsequently transforms to ferrite to yield a fine grain size (8–10 μm). The RCR process results in hot-rolled products with a good combination of strength and toughness suitable for applications where heavy plate and thick-walled seamless tubes are required (Siwecki and Engberg 1996).

For the production of strip, rod, tubular and bar products, multi-pass deformation of austenite at high strain rates often results in extremely short interpass times (0.03–0.7 s). In this case, SRX is not possible (e.g., Eq. (9.2)) and the accumulation of strain per pass may exceed the critical strain (Figure 9.8), thereby resulting in DRX. Industrial processing by dynamic recrystallization controlled rolling (DRCR) can produce, by transformation of ferrite from dynamically recrystallized austenite, grain sizes in the range 3–8 μm (Samuel et al. 1990). In the processing of some products, longer interpass times can occur (0.3–2.0 s) and may result in MDRX rather than DRX. This process is termed metadynamic recrystallization controlled rolling (MDRCR) (Roucoules et al. 1995). It has been suggested that MDRCR may enable the production of austenite and ferrite grain sizes comparable to those produced by DRCR while not requiring the high strain rates needed for DRCR (Hodgson 1997). A combination of TMP to control recrystallization processes followed by controlled cooling and isothermal transformations has enabled substantial increases in mechanical properties (yield strength, tensile strength), ductility (% elongation, % RA), and toughness (higher shelf energy, lower ductile-to-brittle transition temperature DBTT) in Nb-Ti-V steels (Opiela 2014). Finally, a combination of recrystallization controlled rolling, severe plastic deformation techniques, controlled cooling, transformation control, along with the optimization of chemical composition by the addition of appropriate levels of alloying (Mo, Ni, Cr, Cu, Mn) and microalloying (Ti, Nb, V) is leading pathways to ***ultrafine grained*** steels where ferrite grain sizes of $<=1$ μm are achieved. These steel products possess the desired combination of ultra-high strength and toughness along with formability and weldability at a lower cost (Liu et al. 2009).

Recrystallization of ferrite after cold/warm rolling is carried out by either batch annealing (BA) with low heating rates or continuous annealing (CA) with high heating rates. The choice of annealing method is determined by a number of productivity and quality factors (Humphreys and Hatherly 2003). For example, the major aim in cold rolling and annealing (CRA) of low carbon steel grades is to produce recrystallized strip with a particular preferred orientation of grains (*recrystallization texture*), an important variable directly influencing the formability of these alloys. Warm rolling (WR) of interstitial free (IF) and ultra-low carbon (ULC) steels provides significant metallurgical and economic advantages (Bleck and Langner 1997).

9.3 Hot Rolling Process Design and Evolution of Microstructure

The design of high temperature hot rolling process is of crucial importance that determines the evolution and characteristics of austenite microstructure. Subsequent to the cooling and phase transformations, the resultant room temperature microstructure and hence the properties of steel products are a direct consequence of the high temperature austenite microstructure. It is extremely difficult to experimentally study the high temperature microstructure during industrial processing and hence usually it is done in laboratory on a lab-scale rolling mill or by using advanced TMP simulation equipment. The data obtained in lab scale investigation may then be modeled to generate mathematical models of deformation and phase transformation and then used to predict the properties of the steel product as shown in the following sections.

9.3.1 Kinetics of Microstructure Evolution

The kinetics of SRX may be quantified using the classical nucleation and growth theory of phase transformations embodied in the form of the Johnson-Mehl-Avrami-Kolmogorov (JMAK) equation which relates fraction recrystallized (X_V) to annealing time (t) is given by Eq. (9.1) while the rate of static recrystallization can be determined using Eq. (9.2). The flow strength of austenite is given by Eq. (9.5) while recrystallized grain size is shown in Eq. (9.6). The rate of grain growth depends on temperature, interpass time and precipitation conditions. Precipitation depends on the alloy composition and temperature of processing and significantly affects amount of recrystallization as well and hence the kinetics of precipitation must be taken into account as shown in the following section.

9.3.2 Recrystallization-Precipitation-Time-Temperature (RPTT) Relationships

The complex Interaction of recrystallization and precipitation during TMP of microalloyed steels can be understood with the use of Recrystallization-Precipitation-Temperature-Time (RPTT) diagrams, as initially proposed by Hansen et al. 1980. The RPTT diagram in Figure 9.9 schematically presents the kinetics of precipitation in deformed and undeformed austenite (Kwon and DeArdo 1991).

In this diagram, T_{sol} corresponds to the solution temperature above which Nb-rich precipitates are completely dissolved in austenite, T' is the temperature below which recrystallization and precipitation compete and T_R is the temperature below which precipitation occurs prior to recrystallization. There are three distinct regimes:

Regime I—The steel is initially austenitized above T_{sol}, followed by deformation and holding between T_{sol} and T'. Recrystallization goes to completion (R_f) before the start of precipitation (P_s). This regime corresponds to the roughing stage during TMP.

Regime II—Precipitation occurs after the start of recrystallization (i.e., during the time interval between R_s and R_f) and also results in

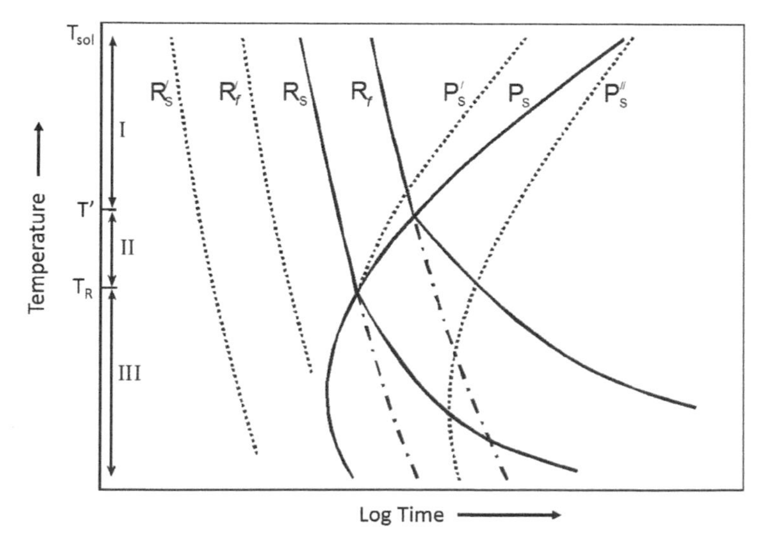

Figure 9.9 Schematic Recrystallization – Precipitation – Time – Temperature (RPTT) diagram showing interaction of precipitation and recrystallization (after Kwon and DeArdo 1991). R_s and R_f refer to the start and finish of recrystallization, respectively, in microalloyed steel; and R_s' and R_f' refer to the start and finish of recrystallization, respectively, in plain carbon steel; P_s' and P_s'' refer to the hypothetical precipitation start times in deformed and undeformed austenite, respectively; and P_s is the actual precipitation start temperature. See text for more details.

accelerated precipitation within the deformation substructure in the partially recrystallized austenite. Deformation within this regime should be avoided due to the undesirable properties associated with a duplex microstructure (Cuddy et al. 1980, Cuddy 1981, Cuddy 1982, Speer et al. 1987). During deformation below T', the significant influence of the interaction of precipitation and recrystallization on microstructure evolution and flow strength of austenite has been demonstrated recently (Gomez et al. 2009) using metallographic technique as well as isothermal and continuous cooling multi-pass hot torsion tests. Deformation within this region may develop duplex and partially recrystallized structure with poor mechanical properties. This regime should therefore be avoided in industrial processing and hence it corresponds to the 'delay' period in the TMP schedule to allow the temperature to drop from T' to T_{nr} (or T_R) which the no recrystallization temperature for steel.

Regime III—In this regime, the precipitation commences prior to recrystallization that generates fine particle dispersion with a large enough Zener pinning force (Zener 1948) to impede recrystallization to such an extent that the deformation substructure may be fully retained after deformation (Palmiere 1995). This regime is below T_{nr} and therefore corresponds to the finishing stage during TMP. T_{nr} is defined as the temperature, below which no complete static recrystallization occurs between two successive rolling passes and is a function of alloy composition, deformation strain, strain rate, and austenite grain size (Vervynckt et al. 2009). In modern thermomechanical processing a lot of deformation is delivered when processing temperature is below T_{nr} to adjust the microstructure of austenite prior to phase transformation during subsequent cooling which has significant influence on the room temperature mechanical properties of advanced high strength steels.

It is clear that an understanding of T_{nr} is of critical importance for microalloyed steels in the design of suitable TMP schedules. Barbosa et al. (1987) used multi-pass torsion to determine experimentally T_{nr} of several steels containing different levels of Nb, V, Ti and Al along with the data reported in literature for 20 other steels. Regression analysis was used to arrive at the following relation (Eq. (9.10)) between T_{nr} and the composition:

$$T_{nr}\,(^{\circ}\text{C}) = 887 + 464\,\text{C} + (6445\,\text{Nb} - 644\sqrt{\text{Nb}}) + (732\,\text{V} - 230\sqrt{\text{V}}) + 890\,\text{Ti} + 363\,\text{Al} - 357\,\text{Si} \qquad (9.10)$$

However, a recent experimental study (Homsher and Van Tyne 2013) of V, Nb and Ti MA steels show that the T_{nr} could be predicted better according to the following Eq. (9.11) given by Bai et al. 1993:

$$T_{nr} = 174\log\left[Nb\left(C + \frac{12}{14}\tilde{N}\right)\right] + 1444 \qquad (9.11)$$

where, *Nb* and *C* are elements in mass % of the steel, and $\tilde{N}$ is free nitrogen remaining in steel after *TiN* precipitation.

Maccagno et al. (1994) have found the relationship satisfactory in the calculation of T_{nr} for Cu-bearing (0.4% Cu) steels as well as Nb and Nb-V steels. They compared this relationship with the experimentally determined T_{nr} that was calculated using the data from rolling mill logs and found good quantitative agreement between T_{nr} values.

9.4 Structure-Property Relationship

9.4.1 Characterization and Parameterization of Microstructures

Microstructures must be characterized and described quantitatively for the purposes of design, specification, comparison, quality control and mathematical modeling of structure/property relationships. The size (e.g., nm, μm), shape (e.g., polygonal, acicular, globular, lamellar, dendritic), type (e.g., ferrite, pearlite, bainite, martensite, retained austenite), amount (i.e., volume fraction) and distribution (e.g., uniform, segregated, banded, clustered, along grain boundaries) of the main phases and the precipitates need to be measured for a detailed description of microstructure. A difficulty often encountered in the characterization of high temperature microstructures of steels is that the prior austenite microstructure is lost at room temperature due the $\gamma \rightarrow \alpha$ transformation during cooling. However, there are special techniques available to reveal the prior austenite microstructure as given in several references (Woodfine 1953, Brownrigg et al. 1975, Riedl 1981, Manohar et al. 1996). The austenite microstructure is an important parameter affecting steel properties and several techniques are available to quantify the mean grain size, size distribution and morphology (Underwood 1968, Leslie 1981, Riedl 1981, ASM Metals Handbook 1985, ASTM E 112–13, 2013, ASTM E 930–99, 2007, ASTM E 1181–02, 2008, ASTM E 1382–97, 2010).

The types of microstructures observed after TMP and cooling in low carbon and microalloyed steels are quite varied and complex and therefore a clear understanding of phase identification, classification and nomenclature is of great importance. Particularly, bainitic microstructures in low carbon PC and MA steels differ in variety and form as compared to the classical bainitic microstructures. In this regard, the publications by Reynolds et al. (1991), Bainite Research Committee of ISIJ, e.g., Araki et al. 1991, Araki and Shibata 1995, and Bramfitt and Speer (1990) provide useful pointers in identifying the various bainitic microstructures in MA steels, along with their properties, conditions of formation and morphology.

9.4.2 Structure-Property Relationships at Room Temperature

The structure of steels, and indeed universally of all materials, exists at different levels of detail (Manohar and Ferry 2005). For example, *nano-scale* structure means atomic bonding, crystal structure, solutes, precipitates and dislocations; *microscopic* structure implies different phases and inclusions; *mesoscopic* structure refers to individual crystals (grains), grain boundaries and interfaces; while *macroscopic* structure corresponds to the structure of bulk materials at polycrystalline level including texture. The word *structure* therefore is a sum total of all structural features mentioned here.

The word *property* also means different things in different contexts. Properties may be broadly classified in to three groups: *intrinsic, functional* and *manufacturing*. The first group (intrinsic properties) includes two sub-groups of properties: the first sub-group covers the *mechanical* properties such as yield and tensile strength, ductility (% elongation and % reduction in area), and toughness while the second subgroup consists of *physical* properties, e.g., electrical, magnetic, thermal, optical, chemical and thermodynamic properties. The intrinsic properties are dependent upon and a consequence of the structural details at all levels. The question mark following toughness needs an explanation here. Toughness may be defined as the ability of a material to resist brittle fracture. This ability depends not only on the structure of material, but also on at least five other external parameters: the presence or absence of tri-axial state of stress, operating temperature, loading rate (impact loads), presence or absence of notches and their sharpness and section thickness. Thus, toughness is strictly not an intrinsic mechanical property. However, traditionally it has been included as a part of mechanical property suite. The second broad group of properties (functional properties) includes wear, corrosion, fatigue, and creep resistance. These properties are the most relevant properties in terms of the performance of a metal product in service. To some extent, these properties can be correlated to the structural details, however to a large extent they also depend on the external service conditions and geometry of a component. In this case, structure—property relationship is extremely difficult to model mathematically. The last broad group of properties (manufacturing properties) expresses the relative ease with which a metallic product could be made. These properties include weldability, machinability, formability, forgeability and castability. Here again, these properties to some extent depend on the structure of material, but also depend significantly on the design of manufacturing tools and components and manufacturing process parameters. In addition, one should not forget the all-pervasive influence of material composition that impacts upon all structural features and all sets of properties.

From the previous discussion, it is clear that it might not be possible to have a global model that can relate all structural details to all sets of properties through some mathematical functional relationship. With this in mind, it should be noted that what is about to be presented is limited to the models appropriate for correlating microstructural details to the mechanical properties in plain carbon steels or C-Mn steels that exhibit a predominantly ferrite + pearlite microstructure.

The yield (or 0.2% proof) strength of steels (σ) may be factorized into components consisting of intrinsic strength of pure iron, contributions by substitutional and interstitial solid solution hardening, dislocation and textural strengthening, strengthening due to dispersed particles or phase refinement strengthening and grain refinement strengthening as follows (Bhadeshia 2001):

$$\sigma = \sigma_o + \Delta\sigma_{comp} + \Delta\sigma_{tex,\rho} + \Delta\sigma_{micro} + \Delta\sigma_{gs} \tag{9.12}$$

where σ_o is the intrinsic strength of pure iron, $\Delta\sigma_{comp}$ is the contribution due to composition, $\Delta\sigma_{tex,\rho}$ is the strength differential due to texture and dislocation density, $\Delta\sigma_{micro}$ is the strengthening due to dispersion and refinement of microconstituents such as particles or inter-lamellar spacing of pearlite and $\Delta\sigma_{gs}$ is strength increase due grain refinement.

The first two factors in Eq. 9.12 are given as follows (Rodriguez and Gutierrez 2003):

$$\sigma_o + \Delta\sigma_{comp} \text{ (MPa)} = 77 + 80\,[\%Mn] + 750\,[\%P] + 60\,[\%Si] + 80\,[\%Cu] + 45\,[\%Ni] + 60\,[\%Cr] + 11\,[\%Mo] + 5000\,[\%C_{ss}] + 5000\,[\%N_{ss}] \tag{9.13}$$

where C_{ss} and N_{ss} are the mass percentages of C and N, respectively. All other elemental compositions are also expressed in mass percent.

The contributions due to texture and dislocation density can be expressed according to the following relation (Gil-Sevillano 1993):

$$\Delta\sigma_{tex,\rho} = \alpha M \mu b \rho^{1/2} \tag{9.14}$$

where α is a constant (~ 0.33), M is the Taylor factor (average value 3) that correlates macroscopic flow stress to the critical resolved shear stress, μ is the shear modulus of ferrite (80 GPa), b is the Burgers vector (0.25 nm) and ρ is the dislocation density (numbers/m^2).

In Eq. (9.14), the dislocation density is a function of strain during cold working: for annealed iron ρ~10^{12} m^{-2} with the maximum limit proposed to be ~10^{16} m^{-2} (Takaki 2003) at true strains >2 (equivalent to a cold reduction of ~90%). The relation between stored strain and dislocation density is highly non-linear in the true strain range 0–2.

The influence of a dispersion of second phase particles or interlamellar spacing of pearlite is accounted for according to the following model:

$$\Delta\sigma_{micro} = k'/\overline{L} \tag{9.15}$$

where k' is a constant and $\overline{L}$ is the mean interparticle or interlamellar spacing.

For strengthening due to a given volume fraction of globular cementite particles (V_θ), the following constant has been proposed (Bhadeshia 2001):

$$k' = 0.52V_\theta \tag{9.16}$$

Finally, the strengthening due to ferrite grain size is expressed by the famous Hall-Petch relation (Hall 1951, Petch 1953):

$$\Delta\sigma_{gs} = kd^{-1/2} \tag{9.17}$$

where k is ~ 0.6 MPa·m$^{1/2}$ (Seto and Sakata 2003) and d is the mean ferrite grain size. It has been recently shown that the Hall-Petch relationship (Eq. 9.17) still holds good for ultrafine grained steels with ferrite grain size of the order 0.2–0.4 μm, see Figure 9.10.

Current literature on ultra-fine grained (UFG) steels has reported a very fine ferrite grain size of the order 0.5 μm or less is obtainable via various high strain thermomechanical processes such as severe plastic deformation (SPD), equal channel angular processing (ECAP) and accumulative roll bonding (ARB). Such a fine grain size results in substantial and simultaneous increase of both strength and toughness of the steel. In fact, a very fine grain size was also found to be suitable for superplastic forming of high C steels at high temperatures where more than 500% uniform elongation was obtained (Sherby et al. 1975).

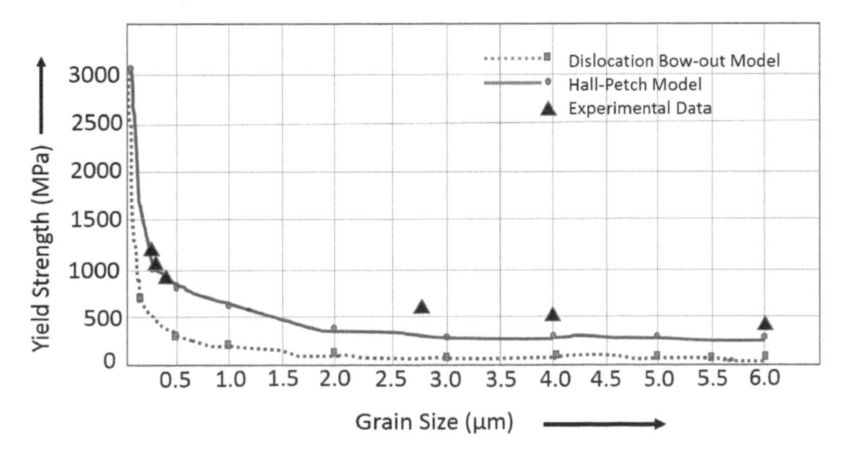

Figure 9.10 Comparison of Hall-Petch model with experimental data for predicting yield strength of ultra-fine grained steels (after Liu et al. 2009).

The influence of compositional and structural parameters on the toughness of ferrite—pearlite steels is now considered. One way of quantifying the toughness of steels is to determine its Ductile to Brittle Transition Temperature (DBTT) by conducting Charpy impact tests over a range of temperatures. The lower the DBTT, the better the toughness of steel. The DBTT is modeled as follows (Pickering 1978):

$$DBTT(^{o}C) = -19 + 44[\%Si] + 700[\%N_{ss}] + 2.2[\%Pearlite] - 11.5d^{-1/2} \quad (9.18)$$

Summarizing, structure—property relationships are vital for the design, product development and control in steel processing. However, care must be exercised in defining the scope of structure—property modeling because both structure and property may include a variety of multi-scale parameters. In the following two sections two industrial-scale hot rolling process design case studies are presented that discuss the manufacturing of plate and hot strip products made from AHSS.

9.5 Process Design Case Study: AHSS/HSLA Steel Plate Rolling

9.5.1 Process Description

Industrial hot rolling processes utilize the softening processes mentioned in Section 9.2.1 to refine the structure at high temperatures. On the other hand, the processes of microalloy precipitation (Section 9.3.2) bring about composition changes as the deformation temperature is lowered. Precipitation may interact with recrystallization, and in some cases may prevent the recrystallization from taking place. Although the interaction between the processing variables and the microstructural evolution during hot working is complex, several distinct steel processing routes have been developed in industry. These processing routes were introduced in Section 9.1.4 and they are briefly described as follows.

Conventional Controlled Rolling (CCR)

It is the first industrial application that utilizes static recrystallization phenomenon. About 8–10% of the total steel tonnage rolled annually is produced this way (Guthrie and Jonas 1990). The process is used for the production of steel plates in applications for oil and gas pipelines and offshore platforms. Rolling is performed in two stages in CCR. The first stage, roughing rolling, is carried out at high temperatures (1150–1000°C) involving repeated static recrystallization of austenite, which reduces austenite grain size from about 100 μm down to about 20 μm. The rolling stock is then cooled to a temperature below T_{nr} where the second stage, finishing rolling, begins which ends at ~800°C. In finish rolling, the austenite is unable to recrystallize (due to low temperatures and microalloy precipitates) so that austenite grains are flattened (pancaked) during rolling. This process increases the nucleation sites for ferrite transformation so that final ferrite grain size of about 5–10 μm is obtained. Ferrite grain refinement, solid solution hardening through alloying (Mn, Mo), precipitation strengthening through microalloying (Nb, V, Ti, Al) in combination with on-line accelerated cooling generating multiphase steels enables CCR to increase the yield strength of the base steel from 200 MPa to over 700 MPa while simultaneously increasing the toughness and weldability.

Recrystallization Controlled Rolling (RCR)

In some cases, CCR is not an appropriate choice because rolling loads increase significantly during finishing passes as austenite strengthens due to the lack of recrystallization. In such cases an alternative method called Recrystallization Controlled Rolling is used. In this case, the alloy composition design is of crucial importance (DeArdo 1997). One way of obtaining RCR is to design alloy system such that it allows recrystallization to proceed relatively quickly (e.g., using V instead of Nb) and to reduce or prevent austenite grain growth that occurs during interpass times (e.g., via the addition of 0.01–0.015% Ti). The fine austenite grain size obtained from recrystallization cycles carried out in RCR at relatively high temperatures (~ 1000°C) result in ferrite grain sizes of about 8–10 μm, leading to mechanical properties in the hot-rolled product that are acceptable for many purposes (heavy plates and thick walled seamless tubes). For example, RCR has been used for industrial production of 20 and 25 mm thick, Ti - V - N steel plates with excellent strength and toughness combination (Siwecki and Engberg 1996).

Dynamic Recrystallization Controlled Rolling (DRCR)

When the interpass times are short and strain rates are high such as in strip, rod, tube and bar rolling, SRX is not possible and therefore strain accumulation occurs which leads to DRX. The process that utilizes the DRX phenomenon is called Dynamic Recrystallization Controlled Rolling. It is possible to obtain a ferrite grain size of about 3–6 μm via DRCR using low finishing temperatures while higher finishing temperatures enable ferrite grain sizes of about to 5–8 μm (Samuel et al. 1990, Roucoules et al. 1994).

Metadynamic Recrystallization Controlled Rolling (MDRCR)

Processing conditions of rod, bar and strip are more suitable for Metadynamic Recrystallization Controlled Rolling rather than DRCR where interpass times are relatively long (0.3–2.0 s) compared to 0.03–0.74 s required for pure DRCR (Roucoules et al. 1994). It has been suggested that industrial realization of MDRCR in these applications may enable the production of austenite and ferrite grain sizes comparable to DRCR while not requiring to give high pass strains that are required for DRCR (Hodgson 1997).

Steel strip produced by TMP may be further rolled in the ferrite phase field to produce thin gauge sheet for a range of applications. Ferrite rolling is not expected to result in DRX but substantial DRV will occur if deformation is carried out at an elevated temperature. In fact, warm rolling of interstitial free and ultra-low carbon steels is a recently developed processing strategy as it provides significant metallurgical and economic advantages (Bleck and Langner 1997). Two important commercial processes for annealing cold/warm-rolled steel are batch annealing low heating rates and continuous annealing with high heating rates. The choice of annealing method is controlled by various productivity and quality factors (Humphreys and Hatherly 2003). For example, the major aim in cold rolling and annealing of a range of carbon steel grades is to produce recrystallized strip with a particular preferred orientation of grains (*recrystallization texture*); an important variable that strongly influences the formability of these alloys.

An example of an industrial-scale recrystallization controlled rolling process for C-Mn as well as Nb-added high strength steel is presented in the following section.

9.5.2 Process Modeling

In the example given here, a new integrated approach is proposed for steel processing design. An expert system flow chart shown in Figure 9.11 is developed to design, evaluate and combine elements from processing routes and different steel compositions in order to maximize the properties of the processed steel products (Manohar 2008, Manohar 2010).

The expert system uses both mathematical (iterative) and knowledge-based approaches. The mathematical models incorporated into the expert system calculate the dynamic microstructural evolution as a function of steel composition and known values of process variables such as pass temperature, strain, strain rate, interpass time and plate cooling rate during the hot rolling of C-Mn and Nb- and Ti-microalloyed steels. The predicted microstructure is used as a basis for the subsequent estimation of the mechanical properties of the steel products using empirical relationships, thus enabling more realistic assessment of the designed process routes. The expert system is expected to assist the product and process development metallurgists in the selection of appropriate process route for a given steel composition and for hot rolling process optimization and development.

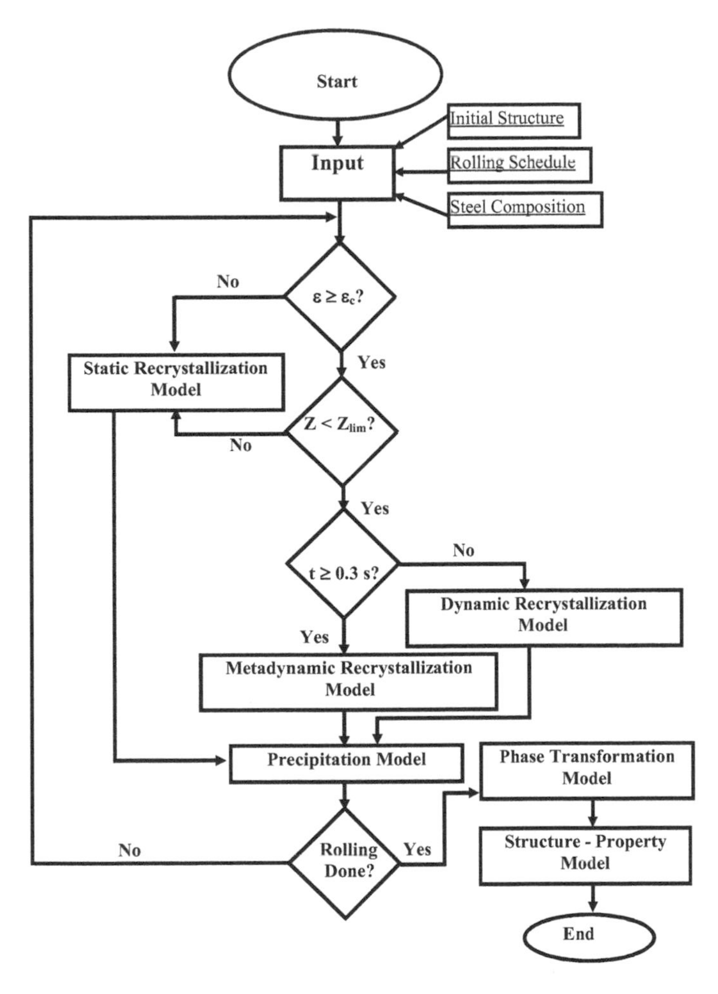

Figure 9.11 Flow chart of an expert system to design the industrial processing of AHSS (after Manohar 2008, 2010).

9.5.3 Microstructural Evolution in Plate Rolling

Mathematical modeling of microstructural evolution during hot rolling of steels has received a great deal of attention over the past two decades and a number of models which describe metallurgical phenomena during steel processing have been published for different steel compositions (e.g., C-Mn, Nb-/Ti-/Nb-Ti/Nb-V microalloyed steels) and a variety of steel processing routes (e.g., conventional, conventional controlled rolling, recrystallization controlled rolling, hot direct rolling, etc.). These models have been reviewed in (Kwon 1992, Sellars 1990). The mathematical models employed in the current work for calculating the microstructural evolution in Nb-microalloyed steels are given in Table 9.5 (Manohar et al. 2000a, Manohar 2008). Macro-scale computer simulation and mathematical modeling of microstructure evolution in industrial-scale production process of hot rolling and cooling of AHSS such as TRIP steels has also been conducted and efforts continue to develop meso-scale models to compute microstructure evolution more accurately that will facilitate process optimization and property prediction (Militzer 2007).

9.5.4 Process Optimization

A sample output of the program for medium C-Mn steel is shown in Figure 9.12 for a full-scale industrial rod rolling process (Manohar et al. 2003, 2014, 2010).

The figure shows the evolution of austenite grain size as a function of process step. By changing the strain, strain rate and rolling temperature along with cooling rate, it is possible to develop process maps. The process maps may then be used to figure out the best combination of rolling variables to develop the desired microstructure that leads to an optimum combination of properties such as yield and tensile strength, toughness, formability and ductility, a low ductile-to-brittle transition temperature along with superior weldability. In case of AHSS-TRIP steels (Fe-0.2C-2Si-3Mn and Fe-0.4C-2Si-4Mn), alloy design and post deformation heat treatment design allows optimization of material properties especially by suppressing carbide (cementite) precipitation during phase transformation of bainite so that at room temperature the microstructure consists of bainitic ferrite and retained austenite that is enriched in carbon (Caballero et al. 2006). Yield strength in the range 1000–1400 MPa and fracture toughness (K_{Ic}) in excess of 125 MPa $m^{0.5}$ was obtained in these steels through alloy design and isothermal heat treatments. Hot rolling scheduled can be designed and optimized using physical simulation in laboratory such as torsion testing (Calvo et al. 2010). Industrial-scale experiments are expensive but necessary to optimize and validate complex processing routes. For example, 45–50 mm thick heavy plates of 0.06C-Ti-Nb-Mn-V-Mn-Cu-Ni microalloyed steels for offshore structures and ship building were processed such that the yield strength was greater than 460 MPa and a ductile fracture was obtained when testing at test temperature of –80°C with Charpy V Notch impact energy in excess of 150 J. These properties were achieved by rolling reductions of 63% below T_{nr}, followed by accelerated cooling that was interrupted at 460°C to develop a well-designed combination of microconstituents consisting of fine-grained quasi-polygonal ferrite (QF), acicular ferrite (AF), Lower Bainite (LB) and Lath Martensite (LM) (Liu et al. 2011, 2011a). A combination of severe plastic deformation, alloy design

Table 9.5 Mathematical models for computing microstructural evolution during the hot rolling of Nb-microalloyed steels (Manohar et al. 2000a, 2003, 2008, 2010).

Parameter	Model	Reference
pass strain ε	$1.155 \ln(h_o/h_f)$	Beynon and Sellars 1992
Critical pass strain for the onset of DRX: ε_c	$\varepsilon_c = 0.83\varepsilon_p$; and $\varepsilon_p = 1.32\text{x}10^{-2}\text{x}d_o^{0.174} \text{ x } \dot{\varepsilon}^{0.165}\text{xexp}(2930/T)$	Sun and Hawbolt 1997
pass strain rate $\dot{\varepsilon}$	$\varepsilon V_R / \sqrt{RR(h_o - h_f)}$	Beynon and Sellars 1992
time for 5% recrystallization $t_{0.05}$	$6.75\text{x}10^{-20}\text{x}d_c^{2}\text{x}\varepsilon^{-4} \text{ x exp}(300000/RT)\text{xexp}\{((2.75\text{x}10^5/T)-185)\text{x}[Nb]\}$	Dutta and Sellars 1987, Kundu et al. 2010
time for 5% precipitation of Nb(C, N) $t_{0.05p}$	$3\text{x}10^{-6}\text{x}[Nb]^{-1}\text{x}\varepsilon^{-1}\text{x } Z^{-0.5} \text{ x exp}(270000/RT)\text{x} \exp\{(2.5\text{x}10^{10})/(T^3(\ln K_s)^2)\}$	Dutta and Sellars 1987, Medina et al. 2014
Nb supersaturation ratio K_s	$\{[Nb]+([C]+0.86[N])\}/10^{(2.26-6770/T)}$	Dutta and Sellars 1987
time for 50% recrystallization $t_{0.5}$	$4.92\text{x}10^{-17}\text{x } \varepsilon^{-2}\text{x } \dot{\varepsilon}^{-0.33} \text{ x } d_o \text{ x exp}(338000/RT)$	Sun et al. 1997
volume fraction recrystallized X	$1\text{-exp}(-0.693(t/t_{0.5})^2)$	Bai et al. 1993
recrystallized grain size d_{rex}	$1.1\text{x}d_o^{0.67}/\varepsilon^{0.67}$ (X≥0.9) $0.5\text{x}d_o^{0.67}/\varepsilon^{0.67}$ (X<0.9)	Williams et al. 1988, Laasraoui et al. 1991
Zener-Hollomon parameter Z	$\dot{\varepsilon} \times \exp(334000/RT)$	Dutta and Sellars 1987

Table 9.5 contd. ...

... Table 9.5 contd.

Parameter	Model	Reference
Zener-Hollomon parameter Z_{lim} (For DRX to occur Z needs to be $< Z_{lim}$ and $\varepsilon > \varepsilon_c$)	$Z_{lim} = 5x10^{15} \text{ x } \exp(-0.0155xd_o)$	Sun and Hawbolt 1997
time for 95% recrystallization $t_{0.95}$	$7.64 \text{ x } t_{0.05}$	Bai et al. 1993
grain growth during interpass time 't' d_f	$d_f^{8.75} - d_o^{8.75} = 2.6x10^{28}x \exp(-437000/RT)xt_{eff}$; $t_{eff} = t - t_{0.95}$	Manohar et al. 1995, 1996
average austenite grain size when $X \leq 0.9$ (partially recrystallized austenite) $\bar{d}$	$\varepsilon_{eff} = \varepsilon_{pass} + \Delta\varepsilon$ $\Delta\varepsilon = \text{const. x } \varepsilon_{previous} x(1-X)$ const. = 1 if $X < 0.1$; 0.5 if $X \geq 0.1$; $\bar{d} = X^{1.33} x d_{rex} + (1-X)^2 d_o$	Laasraoui et al. 1991 Beynon and Sellars 1992

h_o = original slab thickness (mm), h_f = final slab thickness (mm), T = pass temperature (K), t = interpass time (s), V_R = peripheral roll speed (mm/s), RR = roll radius (mm), R = gas constant (8.31 J/mol·K), d_o = initial grain size (μm).

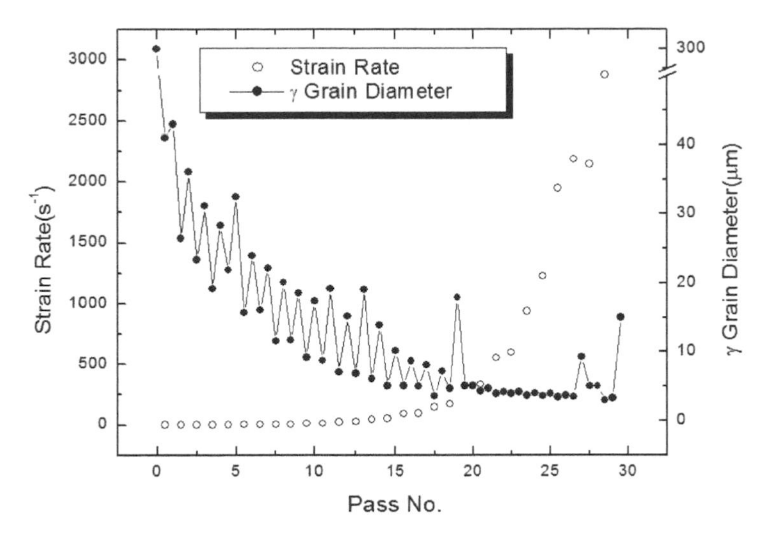

Figure 9.12 Predicted austenite grain evolution during rod rolling of a medium C-Mn steel (after Manohar et al. 2003, 2014, Manohar 2010).

and isothermal heat treatments has resulted in steel with a microstructure of nano-scale lower bainite and Martensite—retained austenite islands that has a yield strength of > 1640 MPa and a yield ratio of 0.84 (Liu et al. 2009). The two-step rolling in a thermomechanical process has also been closely looked at to increase the efficiency of the process. For example, it was realized that a thick slab of 230–250 mm is reduced in multiple passes down to 40 mm or less in roughing rolling. This process occurs at high temperature, consumes lot of energy and is not efficient. Therefore, a high-speed large-reduction semi-continuous hot forging is being investigated to replace the front end of the TMP (Kiuchi 2008). In this process the slab could be reduced from 250 mm thickness straight down to 25 mm in one cycle in a *forging* machine and then the slab could then be fed to the finishing rolling thus making the hot rolling process extremely efficient. Process optimization and innovation is an active research topic. High speed, short interval hot tandem mills, improve lubrication are other areas of research that potentially could affect process optimization in hot rolling of all types of steels.

Mechanical property anisotropy is a problem for high strength linepipe steels (Joo et al. 2013). The property anisotropy is related to chemical segregation, mechanical banding and elongation of grains due to thermomechanical working and anisotropy due to crystal texture. Properties along rolling direction can be significantly different than in transverse direction. For example toughness property of steels as measured by Charpy V Notch impact energy for an American Petroleum Institute (API) X-80 grade linepipe steel (Min. YS 550). In rolling direction is was found to be 300 J while it dropped to only 100 J for samples oriented at 45%

to rolling direction. Therefore mechanical property anisotropy of AHSS/HSLA is an area that needs to be addressed in future via process design and optimization.

For ultra-fine grained carbon, microalloyed and advanced high strength steels there are many ways to optimize the processes and thus to achieve the desired combination of strength, toughness and formability (Song et al. 2006). In addition to severe plastic deformation (SPD) methods such as ECAP, ARB, high pressure torsion (HPT) and bi-directional large strain deformation there are many other methods to achieve ferrite grain refinement. These methods include strain induced ferrite transformation, intercritical hot rolling. DRX of ferrite during warm rolling and cold rolling and annealing of martensitic steels leads to extremely fine ferrite grain size that is of the order ***10 nanometers.*** Some of these methods are still laboratory based and awaiting further development to be elevated to the status of industrial scale production processes but the proofs of concepts are definitely there for these processes to be realized in near future.

Computer based simulations such as those shown in Figure 9.12 has also been constructed by other researchers for hot strip mill for X-70 and X-80 linepipe steel for process optimization via scenario analysis (e.g., Agarwal et al. 2013). The model was constructed and subsequently validated using plant data and laboratory testing. Finally, some scenarios were evaluated, based on the model. These scenarios included questions like: for X80 steel, (i) what is the effect of increasing finish rolling temperature (843°C) by 30°C on ferrite grain size?, (ii) what is the effect of increasing coiling temperature (571°C) by 30°C on ferrite grain size (softer cooling rate)?, (iii) what is the effect of decreasing coiling temperature (571°C) by 30°C on strength (potential for precipitation or second phase hardening)?, (iv) what is the effect of increasing the strip thickness by 10 mm after roughing rolling (more finish rolling) on ferrite grain size and strength; base case: 220 mm slab → 50.8 mm after roughing rolling in 7 passes → 15 mm final coil thickness after 6 finishing rolling passes?, (v) what is the effect of increasing coiling speed by 355 (base case 1.85 m/s) on ferrite grain size and strength (this has an effect of increasing cooling rate)? Scenario analysis like these results in fine tuning and optimization of complex industrial processes.

9.6 Process Design Case Study: AHSS Hot Strip Rolling

High demands on steel quality in automotive industry along with high productivity drive in steel mills using a greater degree of automation has triggered a detailed study and analysis of AHSS manufacturing process. Hot strip mills have been technically sophisticated systems for a long time to accommodate the ever changing needs of manufacturing conditions thus allowing for flexible manufacturing to accommodate a wide range of steels, different quantities, schedules, and quality levels (Senuma et al. 1992, Ataka 2015). There is a significant difference in which the metal is processed in hot conditions for different steel products based on the amount of deformation or roll bite (strain), rate at which strain is delivered or processing speed (strain rate), processing temperature and the number of passes. There are three main types of processing mills for different steel products as follows: plate rolling (covered in the case study presented in Section 9.5), hot strip rolling (covered in this section) and rod rolling (not covered in this chapter but reported elsewhere, see Manohar et al. 2003). Typical strain and interpass time for the three processes is given in Table 9.6.

A detailed history of the 100 years of hot strip mills has been compiled recently (Ataka 2015) that outlines the key changes in hot strip rolling mill theory and technology.

Table 9.6 Typical ranges for strain rates and interpass times in rolling operations (Jonas 2005, Militzer 2007).

Rolling Operation	Strain Rate (s^{-1})	Interpass Time (s)
Plate Rolling	01–30	08–20
Hot Strip Rolling	10–200	0.4–4.0
Rod Rolling	10–1000	0.015–1.0

9.6.1 Process Description

The staring condition for hot strip mill can be diverse. Around 1980s, when thermomechanical processing was well established, rough rolling of a continuously cast slab of thickness 220–280 mm led to an intermediate metal thickness in the range 40–60 mm. This metal (slab) was then hot rolled in a hot strip mill (HSM) leading to a finishing thickness in the range of 1 mm—to a maximum thickness of 25 mm. Subsequent inventions in 1990s such as compact strip processing (CSP) and thin slab casting (TSC) and hot direct rolling (HDR) eliminated the need for roughing rolling when slabs in the thickness range of 40–60 mm were produced directly from liquid steel and processed in the HSM for finishing. Further developments in 1990s have led to the current latest technology of direct strip casting (DSC) which results in steels strips of the order 1–2 mm thickness cast directly from liquid steel with very limited amount of hot rolling needed to adjust the mechanical properties and thickness of steels. DSC technology has been established in low and medium carbon steels and certain grades of stainless steels but not yet proven in AHSS grades. AHSS grades require careful control of cooling and coiling conditions, and sometimes need further cold rolling and annealing heat treatment or some special heat treatments as described in Section 9.1.5 for microstructure and property control. A typical processing schedule for X70 linepipe steel in a roughing and hot strip mill is given in Table 9.7 (Agarwal et al. 2013).

Table 9.7 Hot processing schedule in roughing and hot strip mill for X70 HSLA steel (Agarwal et al. 2013).

Processing Parameter	Roughing Mill	Finishing HSM
Entry Temperature	1478 K	1175
Exit Temperature	1323 K	1093
Number of Passes	7	7
Entry Thickness	220 mm	48 mm
Exit Thickness	48 mm	10.5 mm
Initial Grain Size (μm)	500	160
Exit Grain Size (μm)	150	100
Strain Per Pass	0.17, 0.19, 0.22, 0.23, 0.25, 0.28, 0.36	0.32, 0.33, 0.25, 0.20, 0.15, 0.10, 0.0
Austenite Microstructure	Fully Recrystallized (Retained Strain = 0)	Partially Recrystallized (pancaked structure) (Retained Strain = 1.48)

9.6.2 Process Modeling

The static recrystallization kinetics of austenite during hot strip rolling of AHSS follows the same theory mentioned in Section 9.5.3 for plate rolling. All the mathematical models listed in Table 9.5 are applicable for hot strip processing as well. Material composition specific modeling parameters differ however and they have been reported in literature (historical development of HSM models has been summarized by Senuma et al. 1992, and Agarwal and Shivpuri 2012). Austenite recrystallization and grain growth in interpass time are important parameters that describe high temperature microstructure of hot band. An example of the austenite grain size evolution in hot strip rolling for Nb-bearing steels is presented in Figure 9.13.

Modeling for phase transformation requires the determination of $\gamma \rightarrow \alpha$ transformation start temperature (Ar_3). This temperature depends on steel composition (empirical models have been summarized recently

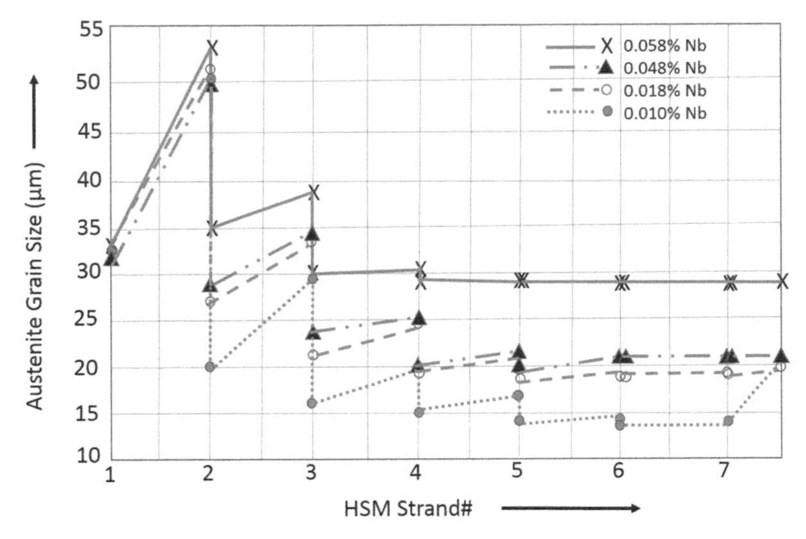

Figure 9.13 Austenite grain size evolution in HSM of Nb-bearing steels (after Pietrzyk 2002).

by Pawlowski 2011), cooling rate, retained strain after HSM rolling, and austenite grain size. A model to calculate Ar_3 based on these parameters is proposed in the following equation (Santos and Barbosa 2010):

$$Ar_3 = Ae_3 - 22.2405(CR)^{0.1677}(\varepsilon_r)^{-0.1027}(d\gamma)^{0.3103} \tag{9.19}$$

where Ae_3 is the equilibrium $\gamma \rightarrow \alpha$ transformation start temperature, CR is the cooling rate subsequent to processing, ε_r is the retained strain in austenite and $d\gamma$ is the austenite grain size. Continuous cooling $\gamma \rightarrow \alpha$ phase transformation during the coiling process is shown in Figure 9.14 below.

After the hot strip rolling is completed, cold rolling and annealing is usually employed for AHSSs for adjustment of thickness, surface finish and mechanical properties. In this case, it is important to know the amount of softening that occurs in annealing. An example of the recrystallization kinetics of 0.06C-0.155Mn-1.86Mo DP steel during continuous heating conditions of a cold rolled steel is shown in Figure 9.15.

Simulation and prediction of cold rolling and annealing leads to many metallurgical reactions such as recovery, recrystallization, phase transformations, precipitation, precipitate growth and dissolution depending on chemical composition of AHSS, straining and processing temperature. These models have been recently summarized (Senuma 2012) for IF, BH, TRIP, DP steels.

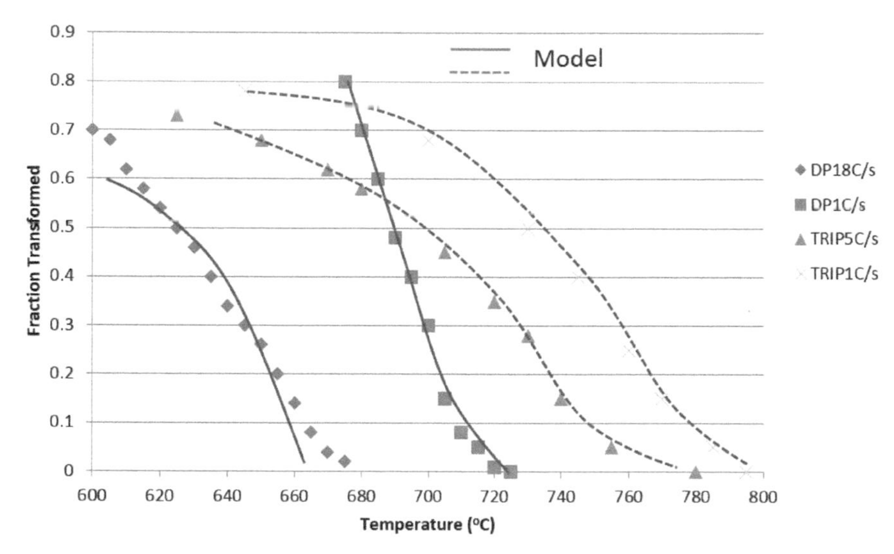

Figure 9.14 Model and experimental data of continuous cooling transformation of austenite → ferrite phase transformation in the coiling process for DP and TRIP steels (after Militzer 2007).

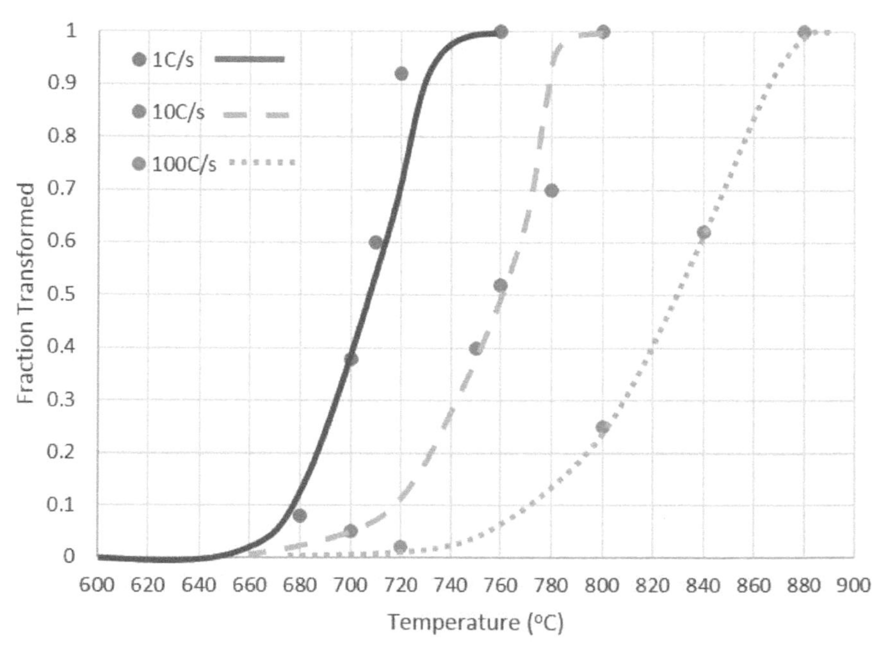

Figure 9.15 Recrystallization kinetics of a cold rolled DP steel during continuous heating (after Militzer 2007).

9.6.3 Microstructural Evolution in Hot Strip Rolling

Ferrite grain size as a function of cooling rate at the runout table is shown in Figure 9.16. Note that cooling rates at or above 100°C/s require water quenching.

Microstructure evolution using Monte Carlo simulation has been also successfully demonstrated for AHSS TRIP steel (Hore et al. 2015). The recrystallized grain size and the evolution of microstructure as a function of strain and stored energy has been validated with published data for Mo-TRIP steel AHSS and a good agreement between model prediction and experiment has been reported in this study. Structure—property modeling during hot stamping process has also been done for ultrahigh strength carbon steels for automotive components (Hidaka et al. 2012).

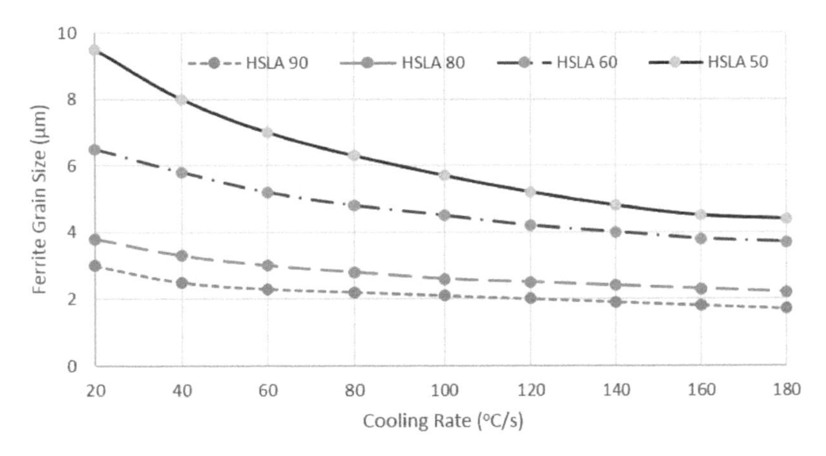

Figure 9.16 Modelled ferrite grain size in HSLA steels as a function of cooling rate (after Militzer 2007).

9.6.4 Process Optimization

The strength increase of AHSS due to precipitation of microalloying elements such as V, Nb, Ti and Cu during cooling and coiling in ferritic matrix is an important parameter in process optimization. There is an optimum window of temperature for coiling for each kind of steel that maximizes the precipitation strengthening effect. This window depends on the precipitation kinetics as described by the precipitation equation given in Table 9.5. Optimum coiling temperatures for Nb and V steels are shown in Figure 9.17 below.

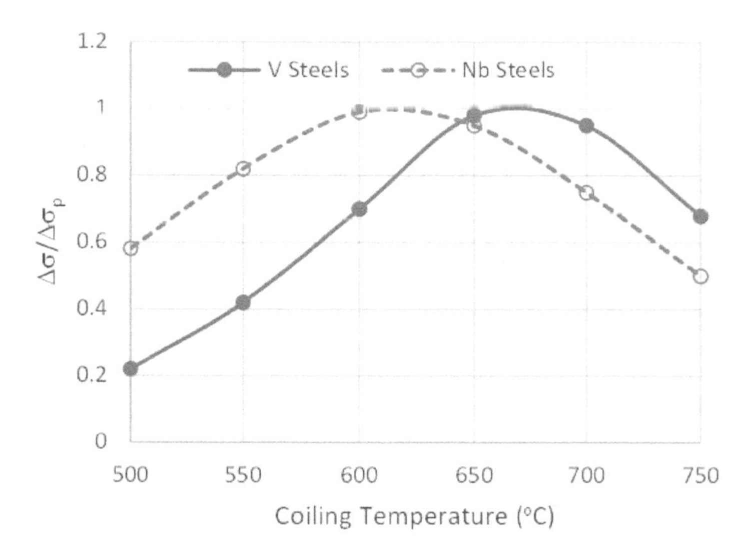

Figure 9.17 Optimum coiling temperature for Nb and V bearing steels at an assumed cooling rate of 30°C/s within the coil. Y-axis is the ratio of actual strength increase due to precipitation hardening ($\Delta\sigma$) divided by the maximum possible strength increase ($\Delta\sigma_p$) due to precipitation hardening. Maximum ratio for full benefits is $\Delta\sigma/\Delta\sigma_p$ = 1.0 (after Militzer et al. 2000).

9.7 Summary

Greater demands placed on the properties, quality, competitiveness and cost is driving innovative breakthroughs in contemporary steel compositions and processing technologies. As a consequence, it has become imperative for downstream process design to utilize a range of thermomechanical processing technologies to optimize steel properties. A combination of fundamental research and industrial trials focusing on mathematically modeling the complex problems of metal deformation and restoration processes at high temperature is facilitating further development. The insights gained through the synergistic application of the various research tools are helping to develop and control new manufacturing technologies such as dynamic recrystallization controlled rolling, metadynamic recrystallization controlled rolling and warm rolling.

As the degree of secondary processing available to adjust and control material properties decreases in step with emerging near-net-shape casting technologies, the design of alloy composition becomes even more critical. High temperature processability of a steel grade is related in a non-linear way to the compositional parameters. The addition of microalloying elements (up to 0.1 mass %) makes significant difference to the mechanical behavior of steels. Niobium, either in solid solution (at T_{def} >1000°C) or precipitated (T_{def} <950°C) retards recrystallization kinetics by an order of magnitude compared to the recrystallization kinetics of steels with no Nb addition. Addition of Ti refines austenite grain size, modifies the shape of MnS inclusions, increases flow strength and retards recrystallization kinetics (although not as much as Nb). The room temperature strength of steels increases substantially with V addition. However, delaying or stopping recrystallization does have an undesirable side effect of increasing the flow strength of austenite so that greater rolling loads are required for continued deformation. This means that not all rolling mills are capable of delivering modern thermomechanical processes.

Properties of the steel grades can be predicted using mathematical modeling and laboratory testing. Microstructural parameterization is achieved through phase identification, grain sizes, grain size distributions and precipitate types and their size distribution. Microstructural features can then be converted to yield strength, tensile strength, hardness, ductility, and toughness values using empirical formulae. Structure-property models may be then be used to optimize process and product design.

Finally, two case studies describing the full scale plate rolling and hot strip rolling process of AHSS and HSLA have been presented. The industrial scale plate and strip rolling processes are described and mathematically modeled. Plate rolling process is characterized by multi-pass deformation sequence at widely varying strain rates (0.4–50 s^{-1}), deformation geometry (square, round, oval) and interpass times (5–20 s). For hot strip mill, the important processing parameters are listed in Table 9.7 and data presented in Figure 9.13. The importance of understanding the mechanisms and kinetics of competing softening processes such as static, dynamic and metadynamic recrystallization is demonstrated by detailed computation of microstructure in each deformation pass shown in Table 9.5 for either of the hot rolling case studies. It is paramount that appropriate mathematical models are developed to compute recrystallization and precipitation kinetics, fraction recrystallized, austenite grain size and structure, and retained strain and subsequently integrated to obtain a comprehensive understanding of the entire processes. Such an approach yields insights into, not only the optimization of existing process sequences, but also the development new ones.

In both cases studies of plate rolling as well as hot strip mill, hot rolling system design aspects have been emphasized. Mathematical models that combine processing parameters such as rolling schedule, roll bite, strain, strain rate, temperature, interpass times and cooling rates are used to model microstructure evolution such as grain size and recrystallization of austenite and precipitation conditions. This high temperature microstructure has profound influence on the room temperature microstructure in combination with steel composition and cooling rate through austenite → ferrite transformation range. Optimization of the processing is achieved through the manipulation of process parameters to carefully control the microstructural evolution during processing and cooling. Some of the AHSS grades subsequent to hot rolling require cold rolling, annealing and specialized heat treatments to obtain the designed microstructure as described in Section 9.1.5. It has been demonstrated in this chapter that careful design and execution of the high and low temperature processes lead to the desired room temperature microstructure which in turn determines all the desirable combination of properties of AHSSs such as strength, toughness, formability and weldability.

Keywords: Advanced high strength steels, thermomechanical processing, recrystallization mechanisms, grain refinement, quench and partition, heat treatment, phase transformations, mathematical modeling, though process modeling, microstructural evolution, structure-property relationships, process design, strip rolling, plate rolling

List of Abbreviations

AC	Accelerated Cooling
AGG	Abnormal Grain Growth
AHSS(s)	Advanced High Strength Steel(s)
API	American Petroleum Institute
Ar_3	Austenite (γ) → ferrite (α) Transformation Start Temperature (during cooling)
ARB	Accumulative Roll Bonding
Aust SS	Austenitic Stainless Steel
A-UHSS	Advanced Ultra-High Strength Steels
BA	Batch Annealing
BH	Bake Hardenable (steels)
BIW	Body In White (auto vehicle body)
CA	Continuous Annealing
CCC	Conventional Continuous Casting
CCR	Conventional Controlled Rolling/Cold Charge Rolling
CE	Carbon Equivalent
CHR	Conventional Hot Rolling
CP	Complex Phase (steels)
CRA	Cold Rolling and Annealing
CST	Cooling Stop Temperature
DRCR	Dynamic Recrystallization Controlled Rolling
DRV	Dynamic Recovery
DRX	Dynamic Recrystallization
DQ	Direct Quenching
DQT	Direct Quenched and Tempered

DSC	Direct Strip Casting
ECAP	Equal Channel Angular Processing
FB	Ferrite–Bainite (steel)
FRT	Finish Rolling Temperature
GCT	Grain Coarsening Temperature
HCR	Hot Charge Rolling
HDR	Hot Direct Rolling
HF	Hot Formed (steel)
HPT	High Pressure Torsion
HSBC	Horizontal Single Belt Casting
IAC	Interrupted Accelerated Cooling
IF	Interstitial Free (steel)
IFHS	Interstitial Free High Strength (steel)
IFLS	Interstitial Free Low Strength (steel)
IS/ISO/IQ	Isotropic or Isotropic Quality (steel)
LC	Low Carbon (steel)
L-IP	Lightweight with Induced Plasticity (steel)
LYS	Lower Yield Strength
MA	Micro alloyed (steel)
MAE	Micro-Alloying Elements
MDRCR	Metadynamic Recrystallization Controlled Rolling
MDRX	Metadynamic (post-dynamic) Recrystallization
NNSC	Near Net Shape Casting
PC	Plain Carbon (steel)
PFHT	Post Forming Heat Treatable (steel)
PHS	Press Hardened Steels (boron based)
ppm	Parts per Million
Q&P	Quench and Partitioning (heat treatment)
QT	Quenched and Tempered (steel)
RCR	Recrystallization Controlled Rolling
RPTT	Recrystallization-Precipitation-Time-Temperature (diagram)
RQT	Reheated, Quenched and Tempered (steel)
SF	Stretch Flangeable (steel)
SFE	Stacking Fault Energy
SIP	Shear band formation Induced Plasticity (steels)
SPD	Severe Plastic Deformation
SRV	Static Recovery
SRX	Static Recrystallization
TMP	Thermo-Mechanical Process(ing)
T_{nr}/T_R	No-Recrystallization Temperature
TRC	Twin Roll Casting
TSC	Thin Slab Casting
UFGS	Ultra-Fine Grained Steels
UHSS	Ultra-High Strength Steels

ULC	Ultra-Low Carbon (steel)
UTS	Ultimate Tensile Strength
UYS	Upper Yield Strength
WR	Warm Rolling

References

Agarwal, P., R. Sah, R.G. Madhusudhan, S. Manjini and A. Chandra. 2013. Optimization in rolling and cooling process of line pipe steels. Steel Res. Int. 84(10): 991–998.

Agarwal, K. and R. Shivpuri. 2012. An on-line hierarchical decomposition based Bayesian model for quality prediction during hot strip rolling. ISIJ Int. 52(10): 1862–1871.

Allain, S., J.P. Chateau, O. Bouaziz, S. Mogot and N. Guelton. 2004. Correlations between the calculated stacking fault energy and the plasticity mechanisms in Fe–Mn–C alloys. Mater. Sci. and Engr. A 387-389: 158–162.

Araki, T. and K. Shibata. 1995. Microstructures and their characterization of modern very low carbon HSLA steels. Proc. Int. Symp. HSLA '95, Guoxun, L. et al. (eds.). CSM, Beijing, China, pp. 13–21.

Araki, T., M. Enomoto and K. Shibata. 1991. Bainite Research Committee of ISIJ Microstructural aspects of bainitic and bainite-like ferritic structures of continuously cooled low carbon (<0.1%) HSLA steels. Mater. Trans. JIM. 32: 729–736.

ASM Metals Handbook. 1985. Metallography and Microstructures. 9th ed., ASM: 170.

ASTM E 112-13. 2013. Standard test methods for determining average grain sizes. ASTM Standards.

ASTM E 930-99 (2007). Standard test methods for estimating the largest grain observed in a metallographic sample (ALA grain size). ASTM Standards.

ASTM E 1181-02. 2008. Standard test methods for characterizing duplex grain size. ASTM Standards.

ASTM E 1382-97. 2010. Standard test methods for determining average grain size using semiautomatic and automatic image analysis. ASTM Standards.

Ataka, M. 2015. Rolling technology and theory for the last 100 years: the contribution of theory to innovation in strip rolling technology. ISIJ Int. 55(1): 89–102.

Bai, D.Q., S. Yue, W.P. Sun and J.J. Jonas. 1993. Effect of deformation parameters on the no-recrystallization temperature in Nb-bearing steels. Metall. Trans. A 24: 2151–2159.

Barbosa, R., F. Boratto, S. Yue and J.J. Jonas. 1987. The influence of chemical composition on the recrystallization behavior of microalloyed steels. Proc. Int. Conf. on Processing, Microstructure and Properties of HSLA Steels, DeArdo, A.J. (ed.). TMS, Pittsburgh, USA, pp. 51–61.

Beladi, H., G.L. Kelly and P.D. Hodgson. 2004. Formation of Ultrafine Grained Structure in Plain Carbon Steels Through Thermomechanical Processing. Mater. Trans. JIM. 45(7): 2214–2218.

Beynon, J.H. and C.M. Sellars. 1992. Modeling microstructure and its effects during multipass hot rolling. ISIJ Int. 32: 359–367.

Bhadeshia, H.K.D.H. 2001. Bainite in Steels. 2nd ed., IOM Communications Inc., London, UK. 286.

Bhattacharya, D. 2006. Overview of advanced high strength steels. In the workshop—Advanced High Strength Steel. Wagoner, R.H. (organizer) and Smith G.R. (chair.), The Ohio State University, Oct. 22–23, 2006, downloaded from http://li.mit.edu/Stuff/RHW/Upload/AHSSDRAFTReport10-29-06.pdf on 08/05/2015.

Billur, E. and T. Altan. 2012. Challenges in forming advanced high strength steels. Engineering Research Center for Net Shape Manufacturing (ERC/NSM), The Ohio State University, www.ercnsm.org: 285–304. downloaded August 08, 2015.

Bognin, G., V. M. Contursi, G. Tanzi, A. Aprile, G. DeFlorio, A. Liguori, P. Borsi and M. Ghersi. 1987. Water blade jet for accelerated cooling of plates. Proc. Int. Symp. on Accelerated Cooling of Steel, Southwick, P.D. (ed.). TMS, Pittsburgh, USA, pp. 69–81.

Bleck, W. and H. Langner. 1997. Softening behaviors of mild steels during hot deformation in the austenite and ferrite range. Proc. Int. Conf. on Thermomechanical Processing of Steels and Other Materials—Thermec '97, Chandra, T. and Sakai, T. (eds.). TMS, Wollongong, Australia, pp. 611–619.

Bramfitt, B.L. and J.G. Speer. 1990. A perspective on the morphology of bainite. Met. Trans. ASM. 21A: 817–829.

Brownrigg, A., P. Curcio and R. Boelen. 1975. Etching of prior austenite grain boundaries in martensite. Metallography. 8: 529–533.

Caballero, F.G., M.J. Santofimia, C. Capdevila, C. Garcia-Mateo and C. Garcia DeAndres. 2006. Design of advanced bainitic steels by optimization of TTT diagrams and T_0 curves. ISIJ Int. 46(10): 1479–1488.

Cahn, R.W. and P. Haasen. 1996. Physical Metallurgy. 4th ed., Vol. III, Cambridge University Press, UK.

Calvo, J., L. Collins and S. Yue. 2010. Design of (Ti- and Nb-) microalloyed steel hot rolling schedules by torsion testing: average schedule vs. real schedule. ISIJ Int. 50(8): 1193–1199.

Cuddy, L.J. 1982. The effect of microalloy concentration on the recrystallization of austenite during hot deformation. Proc. Int. Conf. on Thermomechanical Processing of Microalloyed Austenite, DeArdo, A.J. et al. (eds.). AIME, USA, pp. 129–139.

Cuddy, L.J. 1981. Microstructures developed during thermomechanical treatment of HSLA steels. Metall. Trans. ASM. 12A: 1313–1320.

Cuddy, L.J., J.J. Bauwin and J.C. Raley. 1980. Recrystallization of austenite. Metall. Trans. ASM. 11A: 381–385.

DeArdo, A.J. 1997. The physical metallurgy of thermomechanical processing of microalloyed steels. Proc. Int. Conf. Thermomechanical Processing of Steels and Other Materials—THERMEC 97, Chandra, T. and Sakai, T. (eds.), TMS, Wollongong, Australia, pp. 13–29.

DeArdo, A. J. 1995. Modern thermomechanical processing of microalloyed steel: a physical metallurgy perspective. Proc. Int. Conf. on Microalloying '95, M. Korchynsky et al. (eds.). ISS, Pittsburgh, USA, pp. 15–33.

Dutta, B. and C.M. Sellars. 1987. Effect of composition and process variables on Nb(C, N) precipitation in niobium microalloyed austenite. Mater. Sci. Technol. 3: 197–206.

Evangelista, E., M. Masini, M. El Mehtedi and S. Spigarelli. 2004. Hot working and multipass deformation of a 41Cr4 steel. J. Alloys Compounds. 378: 151–154.

Ferdowsi, M., D. Nakhaie, P. Benhangi and G. Ebrahimi. 2014. Modeling the high temperature flow behavior and dynamic recrystallization kinetics of a medium carbon microalloyed steel. J. Mater. Engr. Performance. 23(3): 1077–1087.

Ge, S., M. Isac and R.I.L. Guthrie. 2013. Progress in strip casting technologies for steel: technical developments. ISIJ Int. 53(5): 729–742.

Ge, S., M. Isac and R.I.L. Guthrie. 2012. Progress of strip casting technologies for steel: historical developments, ISIJ Int. 52(12): 2109–2122.

Gil-Sevillano, J. 1993. Materials Science and Technology, Cahn, R.W. et al. (eds.)., pp. 6: 19–88.

Gomez, M., L. Rancel and S.F. Medina. 2009. Assessment of austenite recrystallization and grain size evolution during multipass hot trolling of a niobium microalloyed steel. Met. Mater. Inter. 15(4): 689–699.

Guthrie, R.I.L. and J.J. Jonas. 1990. Steel processing technology. Metals Handbook. 10th ed., ASM, USA, Properties and Selection: Irons, Steels and High Performance Alloys. 1: 107–125.

Hall, E.O. 1951. The deformation and ageing of mild steel: III discussion of results. Proc. Phys. Soc., Series B. 64: 747–753.

Hansen, S.S., J.B. Vander Sande and M. Cohen. 1980. Niobium carbonitride precipitation and austenite recrystallization in hot-rolled microalloyed steels. Metall. Trans. ASM. 11A: 387–402.

Hidaka, K., Y. Takemoto and T. Senuma. 2012. Microstructural evolution of carbon steels in hot stamping processes. ISIJ Int. 52(4): 688–696.

Hodgson, P.D. 1997. The metadynamic recrystallization of steels. Proc. Int. Conf. Thermomechanical Processing of Steels and other Materials—THERMEC 97, Chandra, T. and Sakai, T (eds.). TMS, Wollongong, Australia, pp. 121–131.
Hodgson, P.D., J.J. Jonas and C.H.J. Davies. 2009. Handbook of Thermal Process Modeling of Steels. Hakan Gur, C. and Pan, J. (eds.). CRC Press, Taylor & Frances Group, London & NY, ISBN 13: 978-0-84935019-1, Chapter 6—Modeling of hot and warm working of steels, pp. 225–264.
Homsher, C.M. and C.J. Van Tyne. 2013. Empirical equations for the no-recrystallization temperature in hot rolled steel plates, Proc. Materials Science & technology Conference, Oct. 27–31, Montreal, Canada. 3: 1815–1822.
Hore, S., S.K. Das, S. Banerjee and S. Mukherjee. 2015. Computational modelling of static recrystallization and two dimensional microstructure evolution during hot rolling of advanced high strength steels. J. of Manuf. Procs. 17: 78–87.
Horvath, C.D. 2004. The Future Revolution in Automotive High Strength Steel Usage, General Motors, Feb. 18, 2004, downloaded from www.autosteel.com on August 05, 2015.
Humphreys, F.J. and M. Hatherly. 2003. Recrystallization and Related Annealing Phenomena. 2nd ed., Elsevier Science, Oxford, UK.
Jonas, J.J. 2005. Effect of quench and interpass time on dynamic and static softening during hot rolling. Steel Res. Int. 76(5): 392–398.
Joo, M.S., D.W. Suh and H.K.D.H. Bhadeshia. 2013. Mechanical anisotropy in steels for pipelines. ISIJ Int. 53(8): 1305–1314.
Keeler, S. 2004. Forming characteristics of advanced high-strength steels. International Iron and Steel Institute. downloaded from www.autosteel.org on August 05, 2015.
Keeler, S., M. Kimchi (eds.). 2014. Advanced High Strength Steels Application Guidelines. Version 5.0, World Auto Steel.
Kiuchi, M. 2008. Integrated production technologies for ultra-fine grained steel sheets. ISIJ Int. 48(8): 1133–1141.
Korchynsky, M. 2002. Strategic importance of thin slab casting technology. The International Symposium on Thin Slab Casting and Rolling, Dec. 3–5, Guangzhou, China, pp. 1–6.
Korchynsky, M. 1987. Development of controlled cooling practice (a historical overview). Proc. Int. Symp. on Accelerated Cooling of Steel, Southwick, P.D. (ed.). Pittsburgh, USA, pp. 3–13.
Kozasu, I., C. Ouchi, T. Sampei and T. Okita. 1975. Hot rolling as a high temperature thermo-mechanical process. Proc. Int. Symp. Microalloying '75, Union Carbide Corp., Washington D.C., pp. 120–135.
Krupitzer, R. 2010. NSF 3rd generation advanced high strength steels. American Iron and Steel Institute, downloaded from www.a-sp.org on August 05, 2015.
Kundu, A., C. Davis and M. Strangwood. 2010. Modeling of grain size distributions during single hit deformation of a Nb-containing steel. Metall. and Mater. Trans A. 41A: 994–1002.
Kwon, O. and A.J. DeArdo. 1991. Interactions between recrystallization and precipitation in hot-deformed microalloyed steels. Acta Metall. 39: 529–538.
Kwon O. 1992. A technology for the prediction and control of microstructural changes and mechanical properties in steel. ISIJ Int. 32: 350–358.
Laasraoui, A. and J.J. Jonas. 1991. Prediction of temperature distribution, flow stress and microstructure during multipass hot rolling of steel plate and strip. ISIJ Int. 31: 95–105.
Leslie, W.C. 1981. The Physical Metallurgy of Steels. 1st ed., McGraw Hill, London: 251.
Lim, K.-H., P.A. Manohar, D. Lee, Y.C. Yoo, C.M. Cady, G.T. Gray III and A.D. Rollett. 2003. Constitutive modelling of high temperature mechanical behavior of a medium C-Mn steel. Mater. Sci. Forum. 426–432: 3903–3908.
Liu, D., B. Cheng and M. Luo. 2011. F460 heavy steel plates for offshore structures and shipbuilding produced by thermomechanical control process. ISIJ Int. 51(4): 603–611.
Liu, D., Q. Li and T. Emi. 2011a. Microstructure and mechanical properties in hot-rolled extra high-yield-strength steel plates for offshore structure and ship building. Metall. and Mater. Trans. 42A: 1349–1361.

Liu, X., H. Lan, L. Du and W. Liu. 2009. High performance low cost steels with ultrafine grained and multi-phased microstructure. Science in China Series E: Technological Sciences. 52(8): 2245–2254.

Maccagno, T.M., J.J. Jonas, S. Yue, B.J. McCrady, R. Slobodian and D. Deeks. 1994. Determination of recrystallization stop temperature from rolling mill logs and comparison with laboratory simulation results. ISIJ Int. 34: 917–922.

Mandal, S., M. Jayalakshmi, A.K. Bhaduri and V.S. Sarma. 2014. Effect of strain rate on the dynamic recrystallization behavior in a nitrogen-enhanced 316L(N). Metall. and Mater. Trans. ASM. 45A: 5645–5656.

Manohar, P.A., P. Wu and S. Acharya. 2014. Enhancing Manufacturing Process Education via Computer Simulation and Visualization. 121st ASEE Conference Indianapolis, June 15–18.

Manohar, P.A. 2010. Thermomechanical Process Innovation and Optimization via Computer Modeling and Simulation. Mater. Sci. Forum. 638-642: 3883–3888.

Manohar, P.A. 2008. The Development of Flexible Hot Rolling Technology Based on Through-Process Modeling. 137th TMS Annual Meeting, Frontiers in Process Modeling Symposium, New Orleans, March 9–13.

Manohar, P. and M. Ferry. 2005. Thermomechanical Processing of Ferrous Alloys. In the book The Deformation and Processing of Structural Materials. Guo, Z.X. (ed.). Woodhead Publishing CRC Press, UK, pp. 76–125.

Manohar, P.A., K. Lim, A.D. Rollett and Y. Lee. 2003. Computational Exploration of Microstructural Evolution in a Medium C-Mn Steel and Applications to Rod Mill. ISIJ Int. 43: 1421–1430.

Manohar, P.A., M. Ferry and A. Hunter. 2000. Direct strip casting of steel—historical perspective and future direction. Mater. Forum. 24: 15–32.

Manohar, P.A., S.S. Shivathaya and R.J. Dippenaar. 2000a. A Hybrid System for the Design and Processing of C–Mn and Microalloyed Steels. Proc. International conference Mathematical Modeling of Metal Technologies (MMT 2000), Zinigrad, M. (ed.). Nov. 12–15, Ariel, Israel, pp. 57–66.

Manohar, P.A. 1997. Grain Growth and Continuous Cooling Transformation Behavior of Austenite in Ti-Nb-Mn-Mo Microalloyed Steels. Ph.D. Thesis, University of Wollongong, Australia.

Manohar, P.A., D.P. Dunne, T. Chandra and C.R. Killmore. 1996. Grain growth predictions in microalloyed steels. ISIJ Inter. 36: 194–200.

Manohar, P.A., D.P. Dunne and T. Chandra. 1995. Grain coarsening behavior of a microalloyed steel containing Ti, Nb and Mo. Proc. 4th Japan Int. SAMPE Symp. Maekawa, Z. et al. (eds.). Sept. 25–28, Japan, pp. 1431–1436.

Matlock, D.K. 2006. Microstructural aspects of advanced high strength sheet steels. In the workshop—Advanced High Strength Steel. Wagoner, R.H. (organizer) and Smith, G.R. (chair.), The Ohio State Univ., Oct. 22–23, downloaded from http://li.mit.edu/Stuff/RHW/Upload/AHSSDRAFTReport10-29-06.pdf on 08/05/2015.

Matlock, D.K. and J.G. Speer. 2008. Chapter 11: Third Generation of AHSS: Microstructure Design Concepts. Included in the Proceedings of the International Conference on Microstructure and Texture in Steels and Other Materials. Feb. 05–07, Halder, A., Suwas, S. and Bhattacharjee, D. (eds.). Springer, Jamshedpur, India, pp. 185–205.

Matlock, D.K., J.G. Speer and E. De Moor. 2012. Recent AHSS developments for automotive applications: processing, microstructure and properties. Workshop, Feb. 9–10, USCAR Offices, Southfield, MI, USA.

Medina, S.F., A. Quispe and M. Gomez. 2014. Model for strain-induced precipitation kinetics in microalloyed steels. Metall. and Mater. Trans A. 45A: 1524–1539.

Medina, S.F. and C.A. Hernandez. 1996. General expression of the Zener—Hollomon parameter as a function of the chemical composition of low alloy and microalloyed steels. Acta Mater. 44: 137–148.

Mejia, I., G. Altamirano, A. Bedolla-Jacuinde and J.M. Cabrera. 2014. Modeling of the hot flow behavior of advanced ultra-high strength steels (A-UHSS) microalloyed with boron. Mater. Sci. and Engr. A. 610: 116–125.

Militzer, M. 2007. Computer simulation of microstructure evolution in low carbon sheet steels. ISIJ Int. 47(1): 1–15.

Militzer, M., E.B. Hawbolt and T.R. Meadowcroft. 2000. Microstructural model for hot strip rolling of high-strength low-alloy steels. Metall. Mater. Trans. A. 31A: 1247–1259.

Mueschenborn, W., K.-P. Imlau, L. Meyer and U. Schriver. 1995. Recent developments in physical metallurgy and processing technology of microalloyed flat rolled steels. Proc. Int. Conf. on Microalloying '95, Korchynsky, M. et al. (eds.). ISS, Pittsburgh, USA, pp. 35–48.

Nicholas, M. 2014. SBIR Phase I: 4th generation advanced high strength steel (AHSS) produced *via* novel high-temperature deformation manufacturing. Wayne Steel Technology, Detroit, MI, USA, downloaded from https://www.sbir.gov/node/704517 on August 08, 2015.

Opiela, M. 2014. Effect of thermomechanical processing on the microstructure and mechanical properties of Nb-Ti-V steels. J. Mater. Engr. Perfor. ASM. 23(9): 3379–3388.

Palmiere, E.J. 1995. Precipitation phenomena in microalloyed steels, Proc. Int. Conf. on Microalloying '95, Korchynsky, M. et al. (eds.). ISS, Pittsburgh, USA, pp. 307–320.

Palmiere, E.J., C.I. Garcia and A.J. DeArdo. 1996. The influence of Niobium supersaturation in austenite on the static recrystallization behavior of low carbon microalloyed steels. Metall. Mater. Trans. ASM. 27A: 951–960.

Pawlowski, B. 2011. Critical points of hypoeutectoid steel—prediction of the pearlite dissolution finish temperature Ac_{1f}. J. of Achievements in Mater. and Manuf. Engr. 49(2): 331–337.

Petch, N.J. 1953. The cleavage strength of polycrystals. J. Iron Steel Inst. 174: 25–28.

Pereloma, E.V. and J.D. Boyd. 1996. Effects of simulated online accelerated cooling processing on transformation temperatures and microstructures in microalloyed steels. Part I. strip processing, Mater. Sci. Technol. 12: 808–817.

Peterson, R.C. 1994. Titanium precipitation in hot charged and hot direct rolled microalloyed steels, Ph.D. Thesis, Monash University, Australia.

Pickering, F.B. 1978. Physical Metallurgy and the Design of Steels, 1st ed., Applied Science, London, UK.

Pierce, D.T., J.A. Jimenez, J. Bentley, D. Raabe, C. Oskay and J.E. Witting. 2014. The influence of manganese content on the stacking fault and austenite/ε-martensite interfacial energies in Fe-Mn-(Al-Si) steels investigated by experiment and theory. Acta Mater. 68: 238–253.

Pietrzyk, M. 2002. Through-process modeling of microstructure evolution in hot forming of steels. J. of Mater. Process. Technol. 125-126: 53–62.

Quelennec, X. and J.J. Jonas. 2012. Simulation of austenite flow curves under industrial rolling conditions using a physical dynamic recrystallization model. ISIJ Int. 52: 1145–1152.

Quelennec, X., N. Bozzolo, J.J. Jonas and R. Loge. 2011. A new approach to modeling the flow curve of hot deformed austenite. ISIJ Int. 51: 945–950.

Ramos, G.C., M. Stout, R.E. Bolmaro, J.W. Signorelli and P. Turner. 2010. Study of drawing-quality sheet steel. I: stress/strain behaviors and Lankford coefficients by experiments and micromechanical simulations. Inter. J. of Solids and Structures. 47: 2285–2293.

Rehman, Md., K. and H. Zurob. 2013. A novel approach to model static recrystallization of austenite during hot rolling on Nb microalloyed steel. Metall. and Mater. Trans. ASM. 44A: 1862–1871.

Reynolds, W.T., Jr., H.I. Aaronson and G. Spanos. 1991. A summary of the present diffusionist views on bainite. Mater. Trans. JIM. 32: 737–746.

Riedl, R. 1981. The determination of austenite grain size in ferrous metals. Metallography. 14: 119–129.

Rodriguez, R. and I. Gutierrez. 2003. Unified formulation to predict the tensile curves of steels with different microstructures. Mater. Sci. Forum. 426-432: 4525–4530.

Roucoules, C., S. Yue and J.J. Jonas. 1995. Effect of alloying elements on metadynamic recrystallization in HSLA steels. Metall. Mater. Trans. ASM. 26A: 181–190.

Roucoules, C., P.D. Hodgson, S. Yue and J.J. Jonas. 1994. Softening and microstructural change following the dynamic recrystallization of austenite. Metall. and Mater. Trans. ASM. 25A: 389–400.

Samuel, F.H., S. Yue, J.J. Jonas and K.R. Barnes. 1990. Effect of dynamic recrystallization on microstructural evolution during strip rolling. ISIJ Int. 30: 216–225.

Santos dos, A.A. and R. Barbosa. 2010. Model for microstructure prediction in hot rolled strip steels. Steel Res. Int. 81(1): 55–63.

Schaeffler, D.J. 2005. Introduction to advanced high-strength steels—Part I. downloaded from http://www.thefabricator.com/article/metalsmaterials/introduction-to-advanced-high-strength-steels---part-i on Aug. 08, 2015.

Sellars, C.M. 1990. Modeling—an interdisciplinary activity, Proc. Int. Symp. on 'Mathematical Modeling of Hot Rolling of Steel', Yue, S. (ed.). Hamilton, Canada, pp. 1–18.

Sellars, C.M. and J.A. Whiteman. 1979. Recrystallization and grain growth in hot rolling. Met. Sci. 13: 187–194.

Sellars, C.M. and W.J. McTegart. 1966. On the mechanism of hot deformation. Acta Met. 14: 1136–1138.

Senuma, T. 2012. Present status and future prospects of simulation models for predicting the microstructure of cold rolled steel sheets. ISIJ Int. 52(4): 679–687.

Senuma, T., M. Suehiro and H. Yada. 1992. Mathematical models for predicting microstructural evolution and mechanical properties of hot strips. ISIJ Int. 32(3): 423–432.

Sherby, O.D., B. Walser, C.M. Young and E.M. Cady. 1975. Superplastic ultra-high-C steels. Scripta Metall. 9: 569–573.

Seto, K. and K. Sakata. 2003. Effect of grain refinement and its application to commercial steels. Mater. Sci. Forum. 426-432: 1207–1212.

Siwecki, T. and G. Engberg. 1996. Recrystallization controlled rolling of steel. Proc. Int. Conf. on Thermomechanical Processing in Theory, Modeling and Practice, Hutchinson, W.B. et al. (eds.). The Swedish Society for Materials Technology, Stockholm, Sweden, pp. 121–144.

Speer, J.G., D.K. Matlock, L. Wang and D.V. Edmonds. 2014. Quenched and Partitioned Steels. Chapter 11, titled Materials Testing and Specialized Materials, in the book Comprehensive Materials Processing, Vol. 1, Van Tyne, C.J. (vol. ed.). Hashmi, S. (ed.-in-chief), Elsevier, pp. 217–225.

Speer, J.G., D.K. Matlock, B.C. De Cooman and J.G. Schroth. 2003. Carbon partitioning into austenite after martensite transformation. Acta Mater. 51: 2611–2622.

Speer, J.G., J.R. Michael and S.S. Hansen. 1987. Carbonitride precipitation in Nb/V microalloyed steels. Metall. Trans. ASM. 18A: 211–222.

Song, R., D. Ponge, D. Raabe, J.G. Speer and D.K. Matlock. 2006. Overview of processing, microstructure and mechanical properties of ultrafine grained bcc steels. Mater. Sci. and Engr. A 441(1-2): 1–17.

Sun, X. 2013. Development of 3rd generation advanced high strength steels (AHSS) with an integrated experimental and simulation approach, Pacific Northwest National Laboratory, Richland, WA, USA, A report 2013 DOE vehicle technology program ID# LM082, downloaded August 05, 2015.

Sun, W.P. and E.B. Hawbolt. 1997. Comparison between static and metadynamic recrystallization—an application to the hot rolling of steel. ISIJ Int. 37: 1000–1009.

Sun, W.P., M. Militzer, E.B. Hawbolt and T.R. Meadowcroft. 1997. Microstructural evolution during thermomechanical processing of V-containing and Nb-containing HSLA steels. Proc. Int. Conf. on 'Thermomechanical Processing of Steels and Other Materials (Thermec'97)', Chandra, T. and Sakai, T. (eds.). TMS, Wollongong, Australia, pp. 685–691.

Takaki, S. 2003. Limit of dislocation density and ultra-grain-refining on severe deformation in iron. Mater. Sci. Forum. 426-432: 215–222.

Tamarelli, C.M. 2011. The Evolving Use of Advanced High Strength Steels for Automotive Applications. AISI, downloaded from www.autosteel.org on August 05, 2015.

Tamehiro, H., R. Habu, N. Yamada, H. Matsuda and M. Nagumo. 1987. Properties of large diameter linepipe steel produced by accelerated cooling after controlled rolling. Proc. Int. Symp. on Accelerated Cooling of Steel, Southwick, P.D. (ed.). TMS, Pittsburgh, USA, pp. 401–413.

Underwood, E.E. 1968. Quantitative Stereology. Addison-Wesley, USA.

Vasilyev, A.A., F.S. Sokolov, N.G. Kolbasnikov and D.F. Sokolov. 2011. Effect of alloying on the self-diffusion activation energy in γ iron. Phys. Of the Solid State. 53(11): 2194–2200.

Vervynckt, S., K. Verbeken, B. Lopez and J.J. Jonas. 2012. Modern HSLA steels and role of non-recrystallization temperature. Int. Mater. Rev. 57(4): 187–207.

Vervynckt, S., K. Verbeken, T. Thibaux, M. Liebeherr and Y. Houbaert. 2009. Austenite recrystallization—precipitation interaction in niobium microalloyed steels. ISIJ Int. 49(6): 911–920.

Wang, Z., S. Sun, Z. Shi and B. Wang. 2015. Hot ductility behavior of an 8 Pct Cr roller steel. Metall. and Mater. Trans. ASM, 46A, published online Jan 23, 2015.

Williams, J.G., C.R. Killmore and G.R. Harris. 1988. Recrystallization behavior of fine grained Nb-Ti austenite at low rolling reductions. Proc. Int. Conf. on 'Physical Metallurgy and Thermomechanical Processing of Steels and Other Metals—THERMEC '88', Tamura, I. (ed.). ISIJ, Tokyo, pp. 224–231.

Woodfine, B.C. 1953. Some aspects of temper-brittleness. J. Iron Steel Inst. 173: 240–255.

World Steel Association. 2006. Environmental case study: Automotive. Downloaded from https://www.worldsteel.org/dms/internetDocumentList/case-studies/Automotive-case study/document/Automotive%20case%20study.pdf on Aug. 08, 2015.

World Steel Association. 2015. World Steel production in figures downloaded from https://www.worldsteel.org/dms/internetDocumentList/bookshop/World-Steel-in-Figures 2014/document/World%20Steel%20in%20Figures%202014%20Final.pdf downloaded on Aug. 08, 2015.

Zener, C. 1948. Referenced in: Grains, Phases and Interfaces: An interpretation of microstructure. by Smith, C. S., Trans. AIME. 175: 15–49.

Zener, C. and J.H. Hollomon. 1944. Effect of strain rate upon plastic flow of steel. J. Appl. Phys. 15: 22–32.

10

Cold Rolling of TWIP Steels

*Xianglong Yu** and *Ji Zhou*

ABSTRACT

This chapter is a review of the current understanding of twinning induced plasticity (TWIP) steels in cold rolling. Two main parts are covered in terms of the TWIP steel, substrate itself and corresponding processing parameters. Some fundamentals have also been introduced such as stacking fault energy and the combined processing route of TWIP steels. The former part of this chapter includes alloying elements, phase composition, grain size, and texture evolution during cold rolling and annealing processes. The latter one involves deformation temperature, reduction, strain rate and pre-deformation before rolling. Both of these focus on the mechanical properties and advantages of TWIP steels relevant to automotive applications. Finally, a clear snapshot has been drawn associated with the upcoming challenges in the research on austenitic TWIP steels.

10.1 Introduction

In the automotive and mechanical engineering industry, the material types and portions by weight for the current vehicle are seven steel grades (91%), two aluminium grades (3%), and thermoplastics and other materials (6%) (De Cooman et al. 2009). For instance, Figure 10.1 (Asghari et al. 2013) illustrates potential applications of advanced steels in a modern car body structure. The development of these advanced steels for a variety of

School of Materials Science and Engineering, Tsinghua University, Beijing 100084, China.
* Corresponding author: xly991@mail.tsinghua.edu.cn

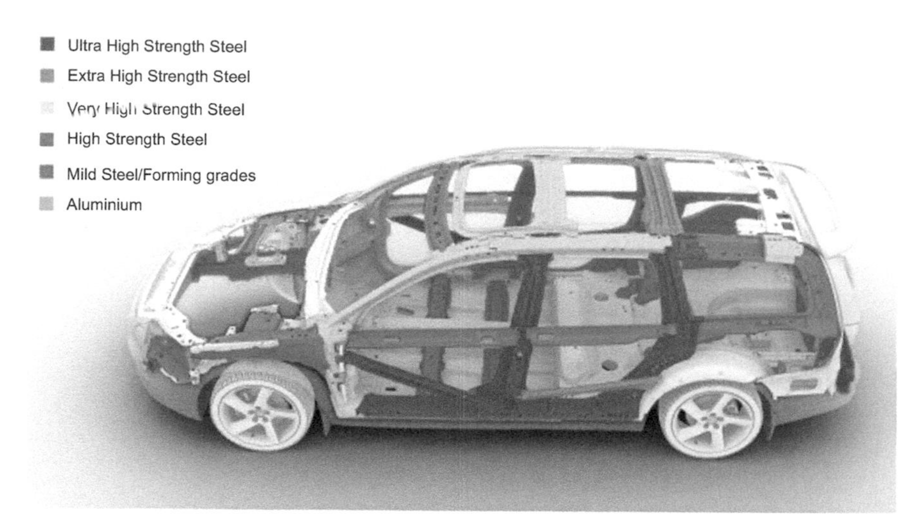

Figure 10.1 Advanced steel applications in a modern car body structure (Asghari et al. 2013).

automotive applications is focused on an increase of strength combined with the preservation or improvement of their formability (Vercammen et al. 2004). The increase of strength enables car manufacturers to reduce automotive body-in-white weight, whereas the increase of ductility and formability allows for more complex car design. Here twinning induced plasticity (TWIP) steels have good combinations of high strength, concurrently high ductility and damage tolerance which satisfy the recent requirements of automotive industries. The TWIP steel is believed to lead to high flow stresses (600–1100 MPa) and exceptional elongations (60–95%) (De Cooman et al. 2009). In addition, Al addition to the TWIP steel could lead to an effective reduction of the specific weight of these steels (6.8–7.3 g/cm^3, depending on Al content) (Chen et al. 2013), which is significant to the automobile industry and makes less carbon emission and fuel consumption possible.

The prototype of TWIP steels comes from Hadfield steels, the high manganese (in 12% wt.% concentration (Adler et al. 1986)) face-centred cubic (fcc) austenitic steel with a low stacking fault energy (SFE) ~23 mJ/m^2, which was firstly discovered by Sir Robert Hadfield (1888). Since then, several alloying concepts for TWIP steels have been developed and reported. Three most popular types are based on the Fe-Mn-C (Bouaziz et al. 2011), Fe-Mn-C-Al (Han et al. 2015, Song et al. 2014), Fe-Mn-Al-Si (Frommeyer et al. 2003, Grässel et al. 2000) systems, and later on the Fe-Mn-Cr-C-N system (Roncery et al. 2010) for corrosion resistant. As there is no standard for the definition of Fe-Mn-C TWIP steels regarding the chemical composition, it is reasonable, according to extensive research, to propose a rough range of various alloying elements in TWIP steels and to exhibit a combination between carbon and manganese respecting the Schumann's equation (Schumann 1974, 1972), as listed in Table 10.1. In this case, carbon contents higher than 1.2 wt.% are not useful due to the detrimental occurrence of cementite precipitation in austenite, whereas the manganese content without carbon has to be higher than 30 wt.% to have only glide (Bouaziz et al. 2011, Frommeyer et al. 2003).

Depending on the above-mentioned chemical composition, the deformation mechanism and stress/strain-induced martensitic phase transformation are closely related to the austenite stability and the SFE of the austenitic TWIP steels. In view of their proper SFE (15~40 mJ/m^2) (Prakash et al. 2008), the austenitic TWIP steels can generate mechanical twins during plastic deformation (Allain et al. 2004a). The important feature of the deformation twinning is the change of shape resulting from the simple shear, e.g., a thickness ranging from a few to several hundred nanometers generally across the whole austenite grain (Christian 2002). The twinning deformation can pose obstacles for the movement of slip dislocation, and thereby lead to a pronounced hardenability effect. Note that the difference between transformation-induced plasticity (TRIP) and TWIP effect lies in that TRIP effect is due to martensitic phase transformation from the metastable fcc γ-austenite into the body-centred cubic (bcc) α'-martensite, whereas TWIP effect is based on mechanical twinning leading to a low work hardening rate and large elongation to fracture. The TWIP effect (i.e., no induced martensite) can therefore be regarded as a dynamic Hall-Petch strengthening effect as the twinned crystal planes act as strong obstacles (Petch 1953).

In the present Chapter, two main Sections are covered with respect to the TWIP steel substrate itself (alloying elements associated with SFE, phase composition, grain size, and texture evolution during cold rolling and annealing processes), and related processing parameters of cold rolling (deformation temperature, reduction, strain rate, and pre-deformation before rolling). All of these cause some variation of the strain hardening of TWIP steels, with special emphasis on the properties and key advantages of TWIP steel products relevant to automotive applications. Finally, a clear snapshot has been drawn associated with the upcoming challenges in the research on austenitic TWIP steels.

Table 10.1 The range of chemical compositions in TWIP steels.

Elements	C	Mn	Al	Si	N	Ti, V, Cu, Nb, Cr
wt.%	0.5–1.2	15–30	2.0–3.0	0–3	<0.21	<0.1

10.2 Fundamentals of TWIP Steels

As mentioned earlier, the main reasons of the industry's interest in austenitic TWIP steels are their ability to achieve high strength and good plasticity during mechanical deformation. These mechanical properties are derived from a low-to-moderate SFE. Hence the TWIP steels can undergo extensive mechanical twinning during deformation, which in turn leads to a good combination of both strength and ductility. To elucidate the mechanical characteristics of the TWIP steels, this Section will address some fundamentals of SFE together with the related processing route of TWIP steels, a combination of pre-straining in the form of cold rolling and recovery annealing.

10.2.1 Stacking Fault Energy

A predominated mechanism of plastic deformation in these steels is related to the SFE of the system. The SFE required for the TWIP effect is generally 20–50 mJ/m^2, which appears to be related to the suppression of the athermal austenite to ε-martensitic transformation (De Cooman et al. 2012). SFE initiates from an interruption process of the normal stacking sequence of an atomic plane in a crystal structure, and thus SFE controls the energetic cost for creating such a defect and the dissociation distance between partial dislocations bounding the defect (Byun 2003, Karaman et al. 2000). The equilibrium width of stacking fault is also partially determined by SFE. When the SFE is low, dislocation glide and particularly cross-slip become more difficult and mechanical twinning is favoured due to the increasing width of stacking faults (Liss and Yan 2010, Vercammen et al. 2004). Remy and Pineau (1977) indicated that the deformation substructure changes from a dislocation cell to ε martensite with decreasing SFE, leading to a work hardening rate increase. The estimation of deformation-dependent SFE is therefore significant in TWIP steels since the SFE can control both final morphology and thickness of mechanical twins, e.g., a linear relationship between SFE and twin thickness (Allain et al. 2004c).

The TEM observation of extended dislocation nodes and split partials, X-ray diffraction line profile analysis, and thermodynamic calculation have been used to determine the SFE of austenitic TWIP steel. For comparison, several studies relating the SFE values in TWIP steels are listed in Table 10.2. The critical value of SFE to obtain the TWIP effect is reported to be in the range of 10–50 mJ/m^2. Specifically, Allain et al. (2004a) reported that strain-induced ε martensite forms when the SFE is lower than 18 mJ/m^2, while mechanical twinning occurs when the SFE value ranges from 12 to 35 mJ/m^2. In medium SFE (20–40 mJ/m^2) TWIP steels, the formation of nanometer-thick deformation twins has been characterised (Gutierrez-Urrutia and Raabe 2012). In practice, for instance, the product of ultimate

Table 10.2 Review of several reported SFE values of TWIP type steels.

Composition (wt.%)	SFE mJ/m^2	Method	Reference
Fe-20Mn-1.2C	15	TEM	(Idrissi et al. 2010)
Fe-22Mn-3Si-3Al	15	TEM	(Pierce et al. 2014)
Fe-25Mn-3Si-3Al	21		
Fe-28Mn-3Si-3Al	39		
Fe-18Mn-3Al-0.6C	45	XRD	(Jeong et al. 2012)
Fe-18Mn-1.5Si-0.6C	14		
Fe-18Mn-1.5Al-0.6C	29	XRD	(Jin and Lee 2012)
	30	TEM	(Kim et al. 2011)
Fe-19Mn-5Cr-0.25C-1Al	20.9	TEM node	(Oh et al. 1995)
Fe-19Mn-5Cr-0.25C-2.5Al	30.5		
Fe-19Mn-5Cr-0.25C-3.5Al	39.4		
Fe-19Mn-5Cr-0.25C-4Al	47.5		
Fe-22Mn-0.6C (T=400°C)	80.0	Thermodynamic calculations	(Allain et al. 2004b)
Fe-22Mn-0.6C (T=20°C)	19.0		
Fe-22Mn-0.6C (T=–196°C)	10.0		
Fe-25Mn-0.15C-0.6Al	7.75	XRD line profile analysis	(Tian et al. 2008)
Fe-25Mn-0.15C-1.5Al	10.67		
Fe-25Mn-0.15C-2.2Al	15.12		
Fe-25Mn-0.15C-3.1Al	14.95		
Fe-25Mn-0.15C-4.8Al	54.74		
Fe-31Mn-0.17C	17.53		(Tian and Zhang 2009)
Fe-18Mn-0.6C-1.5Al	26.4	TEM node	(Kim and De Cooman 2011)
	30.0	TEM WBDF	

WBDF: Weak Beam Dark Field

tensile strength and total elongation (PSE) as a function of Mn content for Fe–xMn–3Al–3Si alloys is shown in Figure 10.2 (Pierce et al. 2015). The SFE range from 15 to 39 mJ/m^2 results in excellent strength and ductility and can serve as guidelines for the design of high-Mn austenitic steel (Dini et al. 2010, Grässel et al. 2000), but the SFE above 50 mJ/m^2 can deteriorate mechanical properties of steel products (Pierce et al. 2014, Saeed-Akbari et al. 2012). SFEs associated with chemical composition, heat treatment and deformation will be introduced in the following sections.

Alternatively, indirect approach to calculate SFE (Γ) on a thermodynamic basis with direct observations of deformation mechanism by TEM (Bouaziz et al. 2011) is carried out in term of a relationship that exists between the SFE and the driving force for ε-martensite formation first established by Hirth

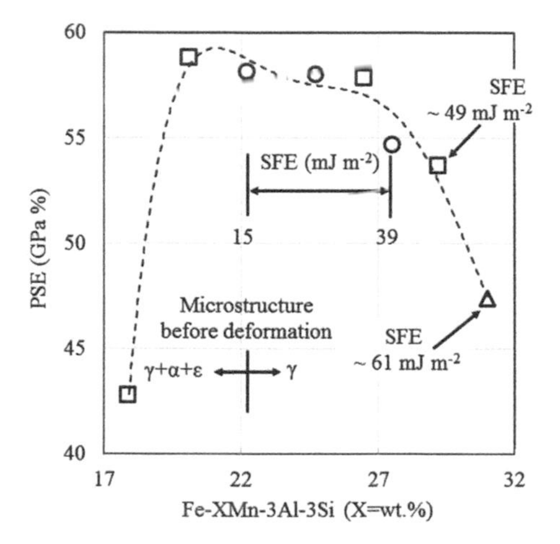

Figure 10.2 The product of ultimate tensile strength and total elongation (PSE) for Fe–*x*Mn–3Al–3Si alloys (Pierce et al. 2015).

(1970) and lately popularised by Olson and Cohen (1976). The approach considers that an intrinsic stacking fault is in fact equivalent to a platelet of ε-martensite of a thickness of only two atomic layers creating two new γ/ε interfaces. It follows that:

$$\Gamma = 2\rho\Delta G^{\gamma\rightarrow\varepsilon} + 2\sigma^{\gamma/\varepsilon} \quad (10.1)$$

where ρ is the molar surface density along {111} planes, $\Delta G^{\gamma\rightarrow\varepsilon}$ the molar Gibbs energy of the transformation $\gamma\rightarrow\varepsilon$, and $\sigma^{\gamma/\varepsilon}$ the surface energy of the interface $\gamma\rightarrow\varepsilon$. The last term is generally taken between 5 and 15 mJ/m^2. The molar surface density is geometrically determined by introducing that lattice parameter a of the alloy:

$$\rho = \frac{4}{\sqrt{3}}\frac{1}{a^2 N} \quad (10.2)$$

where N is the Avogadro number. The estimation of the Gibbs energy $\Delta G^{\gamma\rightarrow\varepsilon}$ for the bulk $\gamma\rightarrow\varepsilon$ transformation can be referred to in previous study (Allain et al. 2004b).

Two reasons can be responsible for the deformation mode in TWIP steel: one is the Dynamic Hall-Petch strengthening effect (Allain et al. 2004a, Gutierrez-Urrutia and Raabe 2012, Petch 1953), another is the dynamic strain aging (DSA) (Hutchinson and Ridley 2006). Nonetheless, the later DSA mechanism cannot explain the high internal back-stresses (Bauschinger effects), which impede the progress of similar dislocations (Bouaziz et al. 2011), although the occurrence of this "jerky flow" in austenitic high Mn steels in terms of Portevin-Le Chatelier (PLC) effect was thought to be related to DSA mechanism (Lebedkina et al. 2009). Further, the details of the respective mechanisms are presented in Table 10.3 (Koyama et al. 2015, Mahato et al. 2015). In general, low SFE is necessary for all the twinning mechanisms. Additionally, multiple slip is required for the pole mechanism and Lomer-Cottrel (LC)-lock mechanism to produce a superjog (Venables 1964) and a LC barrier (Karaman et al. 2000), respectively. In addition, stacking faults in polycrystals showing the TWIP effect are mainly intrinsic and extrinsic in the alloys showing a ε-martensite transformation (Idrissi et al. 2009). The crystallographic aspects of the twinning process and dislocation based nucleation and growth models have been reviewed extensively by Christian and Mahajan (1995).

All these different deformation mechanisms are strongly dependent on the chemical composition, deformation temperature and strain rate. The activation of these secondary deformation mechanisms is controlled in part by the temperature- and composition-dependent SFE. With decreasing SFE, the plasticity mechanisms change from (i) dislocation glide to (ii) dislocation glide and mechanical twinning to (iii) dislocation glide and $\gamma_{fcc} \rightarrow \varepsilon_{hcp}$ martensitic transformations (Curtze and Kuokkala 2010, Dumay et al. 2008, Pierce et al. 2014).

Table 10.3 Details of deformation twinning mechanisms (Koyama et al. 2015, Mahato et al. 2015).

Twinning mechanism	Twin nuclear	Growth	Activation of multi-slip	Interaction of dislocations	Reference
Pole mechanism	Super jog	Motion of sweep dislocation	Necessary	Necessary	(Venables 1964)
Cross-slip mechanism	Extrinsic SF	Sequent stacking from the twin nuclear	Necessary	Necessary	(Karaman et al. 2000)
Three layer mechanism	Intrinsic SF	Sequent stacking from the twin nuclear	Unnecessary	Necessary	(Mahajan and Chin 1974)
Infinite separation mechanism	Infinitely extended intrinsic SF	Random stacking of infinitely extended dislocations	Unnecessary	Unnecessary	(Byun 2003, Lee et al. 2010)

10.2.2 Cold Rolling

Cold rolling is a metal forming process in which the temperature of the metal is below its recrystallization temperature (Roberts 1978). Nevertheless, the cold rolling of TWIP steel has been introduced to obtain high strength with the consequent lowering in ductility. In addition to cold rolling, recovery annealing with different holding times (Nezhadfar et al. 2015) was also introduced as a promising production chain for TWIP steels (Bouaziz et al. 2011, 2012). A combination of pre-straining in the form of cold rolling and recovery annealing was therefore proven to be a potential method to obtain significantly increased YS along with appreciable elongation (Haase et al. 2013a, Santos et al. 2011). During cold rolling both the dislocation density and deformation twin density are increased significantly resulting in an ultra-fine grained microstructure and high yield strength. A subsequent recovery annealing reduces the dislocation density due to dislocation annihilation that leads to regained ductility and work-hardening capacity (Haase et al. 2016). Pre-straining by rolling could further strengthen TWIP steels while still retaining good formability. As such, the combination of mechanical properties can be adjusted over a wide range by varying the processing parameters, i.e., rolling degree and annealing parameters, depending on the final properties required (Haase et al. 2014, 2016). Superior strength-ductility combinations were attained via recovery treatment of cold-rolled TWIP steel by maintaining the deformed nano-twinned structure while reducing the total dislocation density (Bouaziz et al. 2009).

10.3 TWIP Steel Substrate

In the substrate, the mechanical properties of a TWIP steel product depend strongly on chemical composition of alloying elements, phase composition and microstructure, i.e., grain features, texture, etc. To supress martensite transformation of SFEs and to further improve the steel plasticity we need to know under what conditions TWIP affect happens and how twinning during deformation is induced.

10.3.1 Alloying Elements

TWIP steels composed of single austenite phase or multi-phase with high fractions of austenite phase can be alloyed with large amount of alloying elements. Effect of alloying element on properties of high manganese steels is shown in Table 10.4.

High amounts of carbon and manganese can stabilise austenite in the microstructure of TWIP steels. Carbon can improve the stability of austenite and strength of the steels and inhibits the formation of hcp ε-martensite

Table 10.4 Effect of alloying elements on properties of TWIP steels.

Element	C	Mn	Si	B	Ti	N	Al
γ-stabiliser	√	√				√	√
Solid solution strengthening austenite	√		√			√	√
ε-martensite refinement			√				√
Hot ductility				√	√		

by increasing the SFE. The carbon content in TWIP steels is typically in the range of 0.5–1.0 wt.% (De Cooman et al. 2008). Manganese can stabilise austenite, however, if its content is less than 15 wt.%, α'-martensite is formed, which aggravates the formability (Huang et al. 2011). If the content of Mn exceeds 30–32 wt.%, there will be a brittle phase β-Mn to form in the microstructure. Thus, the Mn content is normally in the range of 15–30 wt.%.

Figure 10.3 presents a traditional Schumann diagram (Schumann 1972) for steels with different contents of C and Mn. This diagram summarises the metastable phases vs. C and Mn contents after rapid cooling and additional plastic deformation (Hofer et al. 2011). The straight lines in Figure 10.3a, although give a good indication for the different fields of metastable phases, lead to an oversimplification in some areas, especially in the ranges of 0.6 to 0.8 wt.% C and of 15 to 20 wt.% Mn, which are interesting ranges for steels with a TWIP effect. More recently, compositions in the range from 0.6 to 1.2 wt.% C and from 12 to 22 wt.% Mn have been the focus of several studies due to TWIP effect. It turns out that the austenite field can be expanded, as shown in Figure 10.3b. To confirm this hypothesis, new compositions were produced (large dots in Figure 10.3b), which remained austenitic at room temperature and after a plastic deformation of more than 50% in tensile testing (Mesquita et al. 2013). Figure 10.4a illustrates that with increasing Mn content the ultimate tensile strength R_m decreases from about 930 ± 160 to 630 ± 100 MPa, whereas, the total elongation ε_f in Figure 10.4b increases from 43 ± 4 to 80 ± 10% with increasing Mn content (Grässel et al. 2000). The steels with 25 to 30 wt.% Mn do not show changes in the stress-strain curve. X-ray diffraction analyses revealed that neither a α- nor

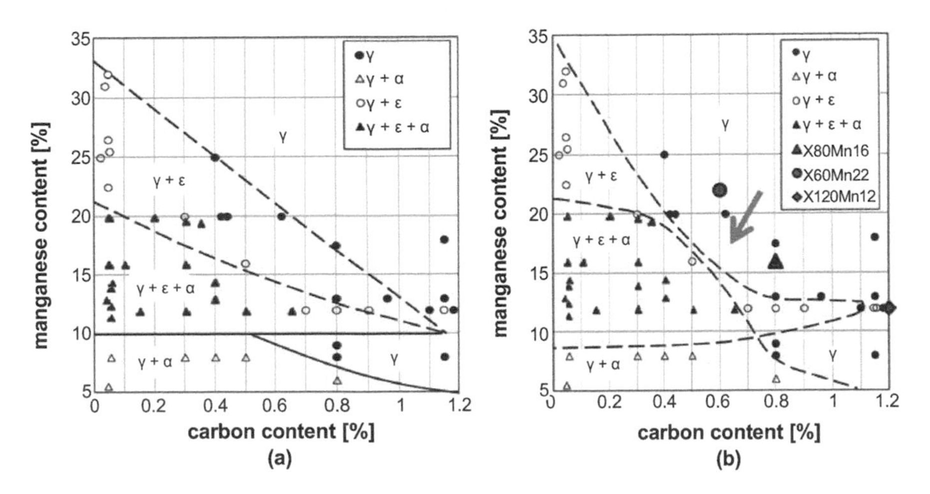

Figure 10.3 (a) Schumann diagram for austenitic ranges for Fe-C-Mn steels, (b) Newly interpreted fields for metastable austenite, enabling the use of alloys with lower Mn; note the arrow in this expanded austenite field (Hofer et al. 2011).

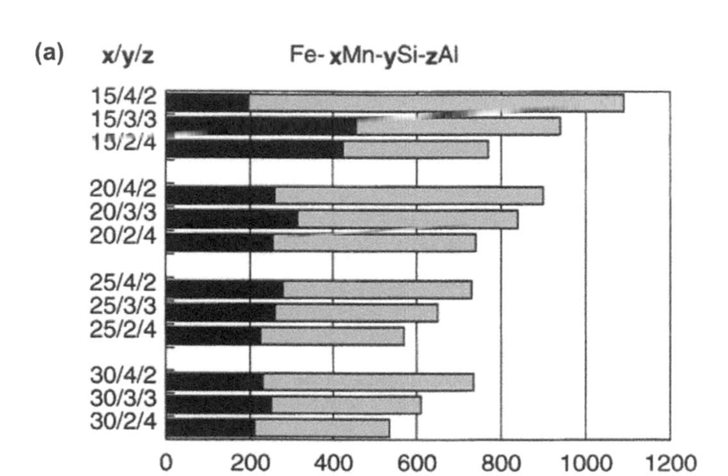

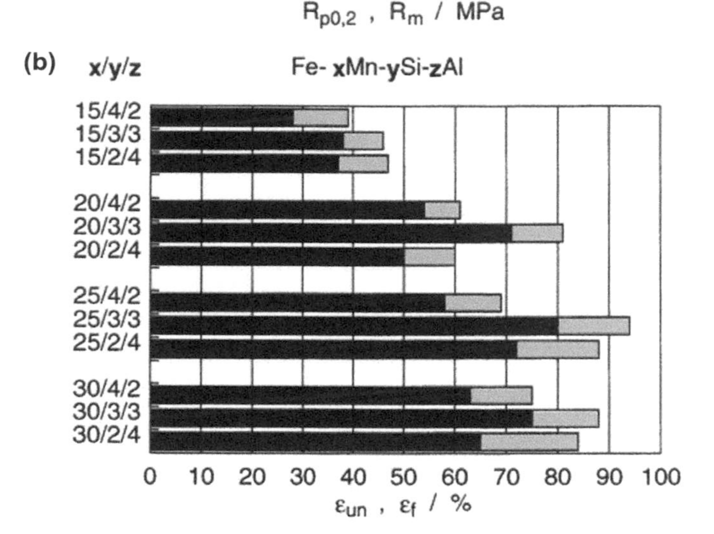

Figure 10.4 Mechanical properties of diverse ultra high manganese steels with additions of aluminium and silicon tested in tension at room temperature, strain rate $\dot{\varepsilon} = 10^{-4}s^{-1}$, (a) yield stress $R_{p0.2}$ (black marked bars) and ultimate tensile strength R_m (grey marked bars), (b) uniform elongation ε_{un} (black marked bars) and total elongation ε_f (grey marked bars) (Grässel et al. 2000).

ε-martensite was formed. These steels deform by strain-induced twinning. Mechanical twinning (TWIP-effect) or a martensitic phase transformation (TRIP-effect) under mechanical load in high manganese steels with additions of silicon and aluminium causes extraordinary mechanical properties even at extremely high strain rates.

The addition of aluminium, silicon, or other elements control the SFE which determines the balance between martensite and twinning formation caused by plastic deformation (De Cooman et al. 2008). Some effects associated with SFE will be introduced in detail in Section 10.3.1.1. The high Al content in TWIP steels increases the SFE of austenite while supresses the formation of ε-martensite, as well as improves low-temperature toughness. In comparison, silicon is known to improve strength by solid solution strengthening, and is also effective for refining ε-martensite plates and increasing fracture strength, although it does not improve ductility. The Al addition, typically less than 3 wt.%, also results in a slight reduction in density. Nevertheless, the presence of aluminium less than 2 wt.% can retard the diffusion of hydrogen in TWIP steel via the atomic trapping mechanism, and thus reduce hydrogen-induced mechanical degradation (Han et al. 2015, Song et al. 2014).

Nitrogen has been found to have similar strengthening effects on aluminium in austenite, e.g., adding nitrogen to the Fe-16.5Mn alloy can decrease the martensite start temperature and also reduce the volume fraction of the ε-martensite (Bliznuk et al. 2002). Other alloying available elements, particularly niobium, can enhance YS via the hardening of dislocations in non-recrystallised grains in the Fe-18Mn-0.6C-1.5Al TWIP steel (Kang et al. 2012).

Microalloying elements, such as Ti, Nb and V, are characterised by small additions <0.1 wt.% and their ability to form carbides or nitrides. They can improve the YS by grain refinement and carbide precipitation, retard or accelerate transformations and affect the diffusion kinetics (Dumay et al. 2008). To improve the oxidation and corrosion resistance, Cr is a usual element to be used in TWIP steels along with Al. Adding small amounts of boron or zirconium can also enhance the hot ductility of the TWIP steels (Kim et al. 1997).

10.3.1.1 Effect of Alloying Elements on SFE

The addition of alloying elements controls the SFE which determines the balance between martensite and twinning formation upon plastic deformation. To induce twining and supress martensite transformation, i.e., TWIP effect, the design for alloying elements is based on SFE. A low SFE (≤ 20 mJ/m^2) enhances $\gamma \rightarrow \varepsilon$ transformation, whereas a high SFE (>20 mJ/m^2), supresses this transformation. In general, increasing Al and Cu contents raise the SFE of austenite, while increasing C and Cr lowers it. Particularly, as silicon content increases, SFE first increases and then decreases. The composition-dependent SFE of Fe-18Mn-0.6C steel was measured to be 13 ± 3 mJ/m^2 and there is considerably increase in SFE due to adding 1 wt.% Al measured to be approximately +11.3 mJ/m^2 (Kim et al. 2011). Nevertheless, the carbon concentration dependence of the deformation twinning behaviour was confirmed to be the same as the conventional SFE dependence (Koyama et al. 2015). If the equation suggested by Schramm and Reed (1975) could be applied, the addition of 1% C could increase the SFE to 410 mJ/m^2 (Tsukahara et al. 2015). As for the Mn effect, the addition of 15% Mn resulted in a monotonously decreasing SFE (Schumann 1974), while the SFE increases by systemically increasing the Mn content from 22 to 28 wt.% (Pierce et al. 2015). If the relationship between the Mn content and the SFE could be extrapolated to a lower Mn range, the 5% Mn steel could have a high SFE which is more than 30 mJ/m^2 (Tsukahara et al. 2015). The SFE value increased almost linearly with increasing N concentration up to 0.21 wt.% N in Fe-15Mn-2Cr-0.6C-xN TWIP steels (Lee et al. 2014). This implies both the retardation of secondary mechanical twinning and the reduction in twin thickness are probably due to the increase in SFE with increasing N concentration in N-added TWIP steels (Jung et al. 2013).

10.3.2 Phase Composition

High manganese austenitic steels are composed of single austenite phase or multi-phase with high content of austenite phase with other alloying elements. Austenite phase (γ) with a relatively low stability can transform to α'-martensite by means of the γ (fcc) $\rightarrow$ ε (hcp) $\rightarrow$ α' (bcc) sequence of martensitic transformations. Suppression of the strain-induced $\gamma\rightarrow\varepsilon$ martensite transformation is therefore considered to imply stability against α'-martensite formation (Bouaziz et al. 2011). Consequently, an efficient TWIP effect are sufficiently stable versus strain-induced ε-martensitic transformation, and also stable versus the α'-martensitic transformation. The resulting martensite platelets and mechanical twins act as planar obstacles and reduce the mean free path of dislocation glide. For instance, Figure 10.5 shows the typical TEM image on the nucleation of α' martensite and mechanical twinning during deformation in the Fe-10Mn-0.45C-1Al steel (He et al. 2016). In addition, after cold rolling of a Fe-17Mn-2Si-3Al-1Ni-0.06C TWIP steel to 42% thickness reduction, the steel comprised of predominant fraction of α'-martensite, a small fraction of blocky ε-martensite and a trace fraction of retained austenite (Pramanik et al. 2015).

Additionally, both mechanisms, twinning and ε martensite transformation, are strongly linked to the SFE, whereas, no direct link from a thermodynamical viewpoint should be drawn between the SFE and the α'-martensite transformation (Ofei et al. 2013). This particular $\gamma\rightarrow\alpha'$ transformation occurs in steels whose compositions are in the bottom-left corner of Schumann diagram (Schumann 1972), below the strain induced ε-martensite boundary, or at low temperature (Hofer et al. 2011). In the particular case of deep cryogenic treatment process prior to intercritical annealing, Figure 10.6 (He et al. 2016) presents the EBSD phase and orientation maps of the samples after this combined treatment. The large austenite grains (>20 μm) are those resisting martensitic transformation during quenching process and remained after intercritical annealing, whereas the small austenite grains are the remaining part of partially transformed prior austenite grain during quenching and thus have the same orientation with the prior austenite grain in Figure 10.6b.

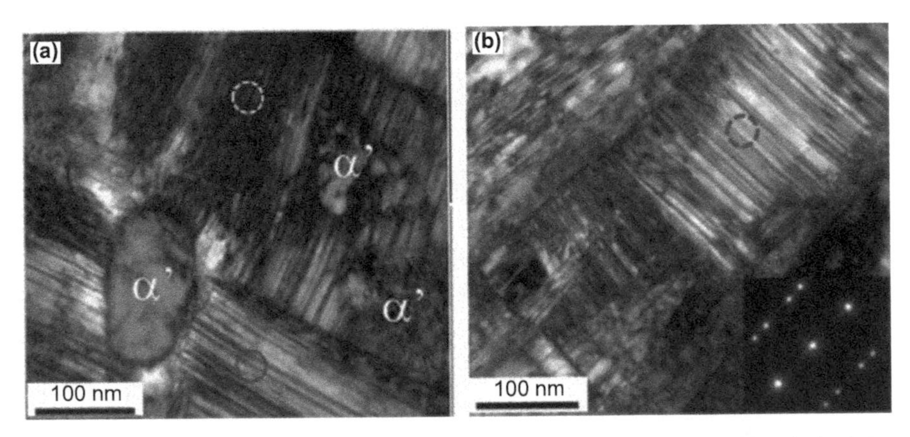

Figure 10.5 Both transformation-induced plasticity (TRIP) and twinning-induced plasticity (TWIP) effects take place in the Fe-10Mn-0.45C-1Al (in wt.%) steel. (a) Nucleation of α' martensite during deformation (b) formation of deformation twins during deformation (He et al. 2016).

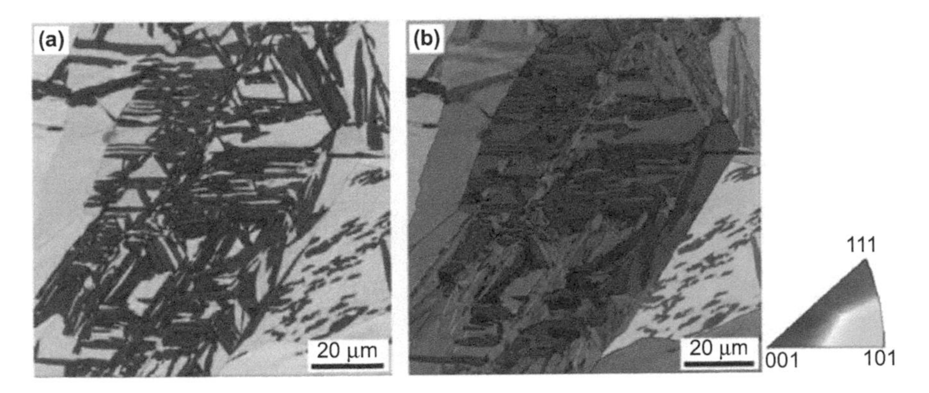

Figure 10.6 (a) EBSD phase map (yellow colour: austenite, red colour: martensite), (b) EBSD orientation image, of the sample after a combined treatment: water quenching, deep cryogenically treatment, intercritical annealing (He et al. 2016).

10.3.3 Grain Size

Grain refinement is a suitable means of the microstructure control for strengthening a material without changing its chemical composition (Ueji et al. 2008). Ultra-fine grained metallic materials with average grain size down to 1 μm or even to the 0.5 μm have become possible by severe plastic deformation techniques (Razmpoosh et al. 2015, Tsuji et al. 2003). Concurrently, grain size affects the degree of strengthening (Hall-Petch effect), while, on the other, a decrease in the grain size diminishes the propensity to ε-martensitic transformation and twinning activity (Gutierrez-Urrutia et al. 2010, Saeed-Akbari et al. 2009), affecting the strain-hardening rate and ductility.

For example, grain refinement suppresses the mechanical twinning triggered by the pile-ups of planar dislocations in TWIP steel with the chemical composition of Fe-31Mn-3Al-3Si and average grain sizes in the range of 2.1–72.6 μm (Dini et al. 2010). In the planar dislocation structure (PDS) (Figure 10.7a), dislocations with straight configuration are piled up

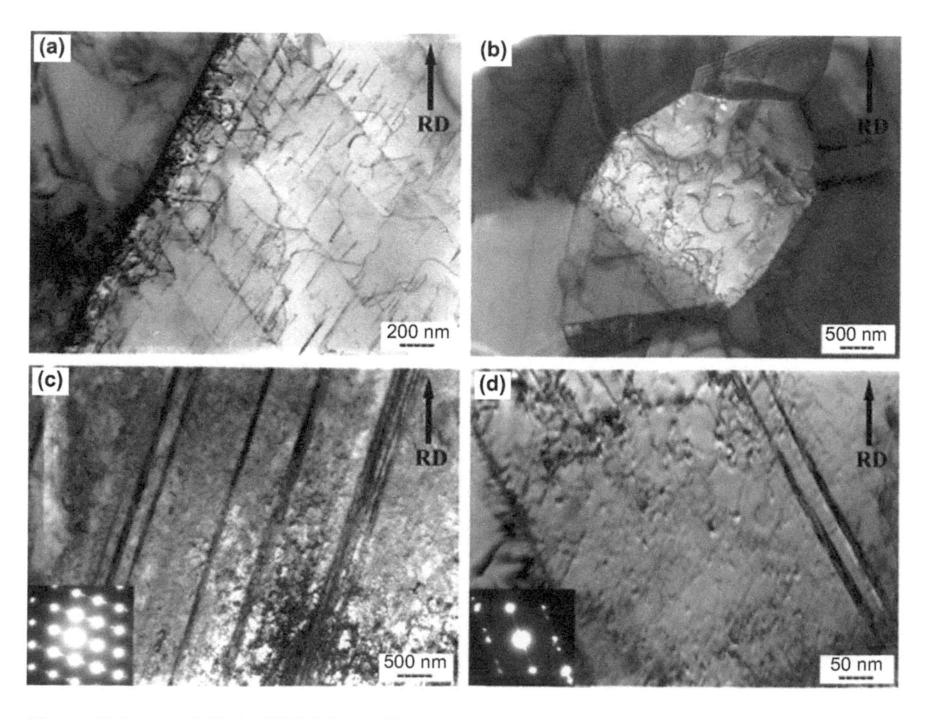

Figure 10.7 (a) and (b) the PDS (planar dislocation structure) and NPDS (non-planar dislocation structure) in coarse and fine grained samples respectively, both at a plastic strain of 0.02, and (c) and (d) the occurrence of mechanical twins in coarse and fine grained samples respectively, both at a plastic strain of 0.11, in Fe–31Mn–3Al–3Si TWIP steel (Dini et al. 2010).

in slip planes while there is no regular arrangement of dislocations in the non-planar dislocation structure (NPDS) (Figure 10.7b). The micro twins in fine grained steel are much shorter and thinner (around 10 nm in thickness in Figure 10.7d) than that of in the coarse grained steel at the same level of strain (around 60 nm in Figure 10.7c) (Dini et al. 2010). Additionally, the number of mechanical twins in a Fe-22Mn-0.6C TWIP specimen with 3 μm grains was much lower than that in a specimen with 50 μm grains after tensile deformation at room temperature (Gutierrez-Urrutia et al. 2010). Hence, although the grain refinement generally restricts mechanical twinning in many materials (Meyers et al. 2001, Ueji et al. 2008), the reason for this is so far unclear.

Generally, the composition-dependent SFE maps (Saeed-Akbari et al. 2009) show a rough estimation of the possible deformation mechanisms within high-manganese steels. These maps can be used to evaluate the influence of grain refinement on the martensitic transformation. This is because twinning stress during phase transformation is also strongly dependent on the SFE (Guitieerez-Urrutia et al. 2010). There are two factors responsible for the increase in the apparent SFE in fine-grained TWIP steels. First, fine-grained steels annealed at relatively low temperatures can result in C segregation near the grain boundary regions where the SFE-induced mechanical twinning occurs (Abe et al. 1990). Another, the restriction of mechanical twinning in fine-grained steels can be from the back-stress of dislocations (Sinclair et al. 2006). Nevertheless, SFE in TWIP steels decrease slightly as austenite grain size increase.

The influence of grain size on the twinning stress within the micrometer size range in TWIP steel can be explained in terms of a Hall-Petch relation:

$$YS = \sigma_0 + \frac{k_y}{\sqrt{d}} \tag{10.3}$$

where σ_0 is the lattice friction stress, k_y is the strengthening coefficient and d is the grain size. The fitted values for σ_0 and k_y are respectively 132 MPa and 449 MPa μm$^{1/2}$ for the alloy Fe-22Mn-0.6C (Bouaziz et al. 2011). Industrial process limitations in conventional cold rolling and annealing steps limit the minimum achievable grain size to ~2.5 μm for this grade. Thus the maximum practical YS for fully recrystallised cold rolled strips is of the order of 450 MPa. For instance, fine grain sizes of around 3 μm can be produced by simple cold rolling and subsequent recrystallization treatment in Fe-17Mn-0.6C (Koyama et al. 2012) and Fe-18Mn-0.6C-1.5Al (Kang et al. 2010) austenitic steels. Grain refinement either by increasing the rolling reduction or decreasing the annealing temperature leads to higher strengths while preserving ductility (Ueji et al. 2008).

10.3.4 Cold Rolling and Recrystallization Textures

A combination of cold rolling and recovery annealing is a promising production chain for TWIP steels. During cold rolling both the dislocation density and deformation twin density are increased significantly resulting in an ultra-fine grained microstructure and high yield strength. A subsequent recovery annealing reduces the dislocation density due to dislocation annihilation that leads to regained ductility and work-hardening capacity (Haase et al. 2016). Both processing are related to the evolution of texture in the TWIP steel substrate.

In general usage, the term "texture" refers to the feel of a material or object, due to some sort of a pattern within the material. In crystallographic orientation, "texture", is said to exist in a polycrystalline material when the distribution of crystal orientations is not random relative to some frame of reference. An understanding of texture can help relate single-crystal properties to those of aggregates of crystals, which influence the isotropic behaviour of a material and may result in enormous cost and waste savings, such as in the drink-can production by mastering the deep-drawing process (Randle and Engler 2000).

In TWIP steels during austenite (γ cubic-close packed crystal structure) to martensite (α' body-centred cubic or ε body-centred tetragonal) transformation, the pattern in which the atoms in the parent crystal are arranged is deformed into that appropriate for martensite, such as, the grain-scale microstructure evolution during this processing procedure is shown schematically in Figure 10.8 (Haase et al. 2014). Consequently, this Section deals with how these atomic planes in TWIP steels during cold rolling are positioned relative to a fixed reference, i.e., the crystallographic texture. Sketch of the position of α and β fibres in Euler space, as well as the ideal orientations of texture components in fcc materials are shown schematically

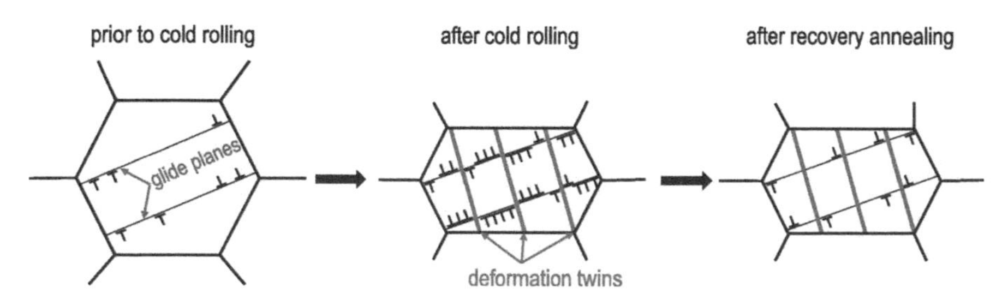

Figure 10.8 Schematic diagram of the grain-scale microstructure evolution during the processing procedure applied (Haase et al. 2014).

in Figure 10.9 (Yan et al. 2014). Through the orientation space, orientation distribution functions (ODFs) in the form of the $\phi_2 = 0°, 45°$ and 65° sections have been given in Figure 10.10 and Table 10.5 (Saleh et al. 2014a).

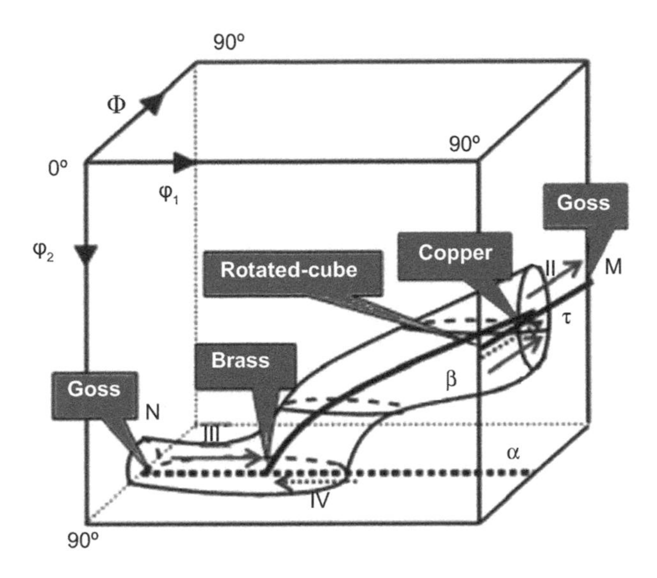

Figure 10.9 Diagram showing ideal positions of α and β fibres and some typical orientations of fcc materials in Cartesian coordinate. The axes are labelled in Bunge convention (Yan et al. 2014).

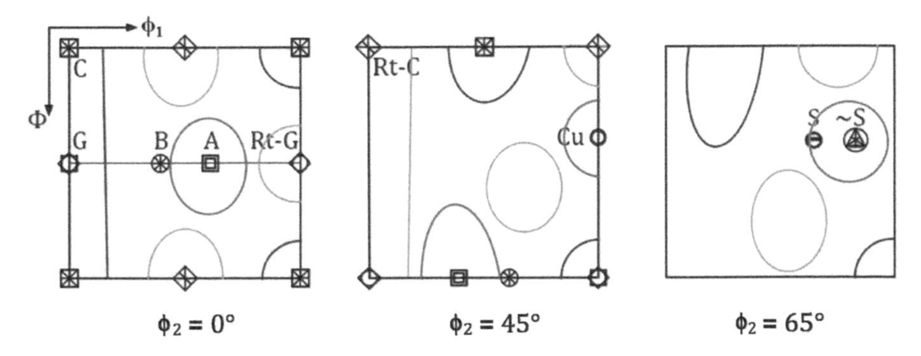

Figure 10.10 A schematic representation of the important texture components in fcc materials. <111> = red, <100> = blue, <110> = green (Saleh et al. 2014a).

Table 10.5 Euler angles and Miller indices for common texture components in fcc metals and alloys (Saleh et al. 2014a).

Texture component	Symbol	Euler angles			Miller indices	Fibre
		ϕ_1	Φ	ϕ_2		
Cube (C)		45	0	45	{001}⟨100⟩	<100>
Cross (G)		90	90	45	{110}⟨001⟩	<100>
Brass (B)		55	90	45	{110}⟨112⟩	-
A		35	90	45	{110}⟨111⟩	<111>
Rotated Goss (Rt-G)		0	90	45	{011}⟨011⟩	<110>
Rotated Cube (Rt-C)		0/90	0	45	{001}⟨110⟩	<110>
Copper (Cu)		90	35	45	{112}⟨111⟩	<111>
S		59	37	63	{123}⟨634⟩	-
~S		75	37	63	{123}⟨111⟩	<111>

10.3.4.1 Cold Rolled Microstructure and Texture

The deformation textures during cold rolling of fcc metals can generally be categorised by the value of the SFE such that: materials with high and medium SFE develop a Copper (Cu)-type texture, which is characterised by the β-fibre that extends from Cu (112<111>) to Brass (B, 110<112>) through S (123<634>) orientations, whereas materials with low SFE, such as austenitic TWIP steel here, develop a Brass (B)-type texture (Tewary et al. 2014, Wenk and Van Houtte 2004), which is characterised by the α-fibre that extends from the Goss (G, 110<001>) to the B orientations (Leffers and Ray 2009). With reducing SFE, cold rolling textures transform from Copper (Cu) to Brass (B)-types, accompanied by deformation twinning and shear banding (Saleh et al. 2014a). For instance, deformation texture in Fe-30Mn-2.7Al-3Si-0.5C TWIP steels with 30% cold rolling is a moderate Copper-type texture, 42% cold-rolled Fe–24Mn–3Al–2Si–1Ni–0.06 C TWIP steel is with higher intensities for Goss ({110}⟨112⟩) compared to Brass ({110}⟨001⟩) (Gazder et al. 2011), whereas a stronger Brass-type texture develops in Fe-30Mn-2.7Al-3Si-0.5C TWIP steels with heavy deformation after 80% cold rolling (Bhattacharya and Ray 2015). Nevertheless, in a Fe-28Mn-0.28C TWIP steel with cold-rolling reductions ranging from 0 to 80%, the evolution of the texture in Figure 10.11 is the {552}⟨115⟩ Copper twin (CuT) component and

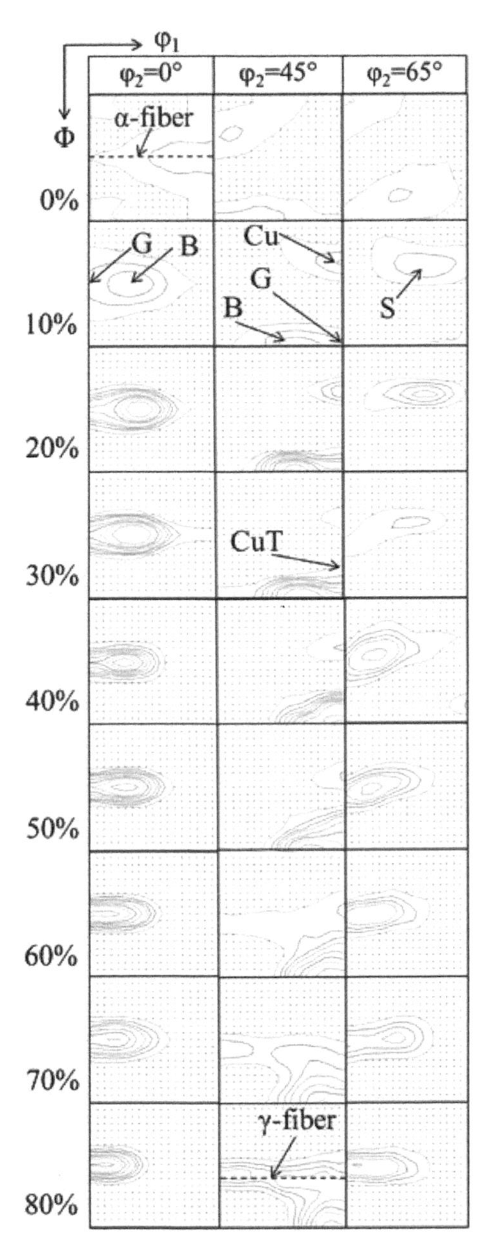

Figure 10.11 Texture evolution during cold rolling in the range between 0 and 80 pct rolling reduction, $\varphi_2 = 0°$, $\varphi_2 = 45°$, $\varphi_2 = 65°$ sections of the ODF (levels: 1, 2, 3, 4, 6, 8, 10, 13, 16, 20) (Haase et al. 2013b).

the γ-fiber (<111>//ND) after 80% cold rolling (Haase et al. 2014, 2013b). Cold rolling textures of TWIP steels are therefore strongly dependent on the composition and temperature of deformation.

10.3.4.2 Recrystallization Texture

Recrystallization of deformed metallic materials is accomplished by the formation of new undeformed grains in the as-deformed microstructures—which is commonly denoted as 'nucleation' and their subsequent growth into the neighbouring deformed matrix (Rollett et al. 2004). The recrystallization texture of deformed TWIP steels can be described in two ways: (i) the oriented nucleation mechanism, where the recrystallised texture depends on the new nuclei orientation, and (ii) the selective growth mechanism, where the final recrystallised texture is not determined by the first nuclei orientation. This effect, combined with the fast recrystallization kinetics and the orientated nucleation mechanism, leads to a fine homogenous microstructure, with a rather weak texture (Barbier et al. 2009).

The typical microstructure evolution in a Fe-24Mn-3Al-2Si-1Ni-0.06C TWIP steel is charted as a function of the annealing temperature in Figure 10.12 (Saleh et al. 2011). The fast progress of recrystallization occurs at 700°C, whereas, recrystallization proceeds even further with annealing up to 850°C, such that the evolution of a high density of annealing twins becomes very apparent in the growing, equiaxed grains (Saleh et al. 2011). Another austenitic Fe-22Mn alloy presents the partially recrystallised microstructure at 650°C (923 K), whereas, the fully recrystallised microstructure after annealing 700°C (973 K) (Bracke et al. 2009). The temperature for partially and fully recrystallization textures in low SFE materials are controversial, partly due to the SFE value of the steel substrate.

The SFE is also closely related to annealing textures via its effect on the microstructure and texture of the deformed state. As such, recrystallization textures in low SFE materials are even more controversial than rolling textures. In general, Cu-type rolling textures typically result in a cube recrystallization texture while B-type rolling textures usually lead to the Recrystallised Brass orientation. However, for the same material the resulting texture may vary with cold-working strain level, annealing temperature and influence the grain growth (Saleh et al. 2011). For instance, recrystallization texture of a cold deformed Fe-30Mn-2.7Al-3Si-0.5C found a strong Bs/Goss texture component located at (ϕ_1 = 74°, Φ = 90°, ϕ_2 = 45°) instead of exact Goss location, formed at the recovery stage (Bhattacharya and Ray 2015, Gazder et al. 2011). Furthermore, Figure 10.13 shows the evolution of partially recrystallised textures in a Fe-24Mn-3Al-2Si-1Ni-0.06C TWIP steel with increasing annealing temperature, as the evolution of the Rotated

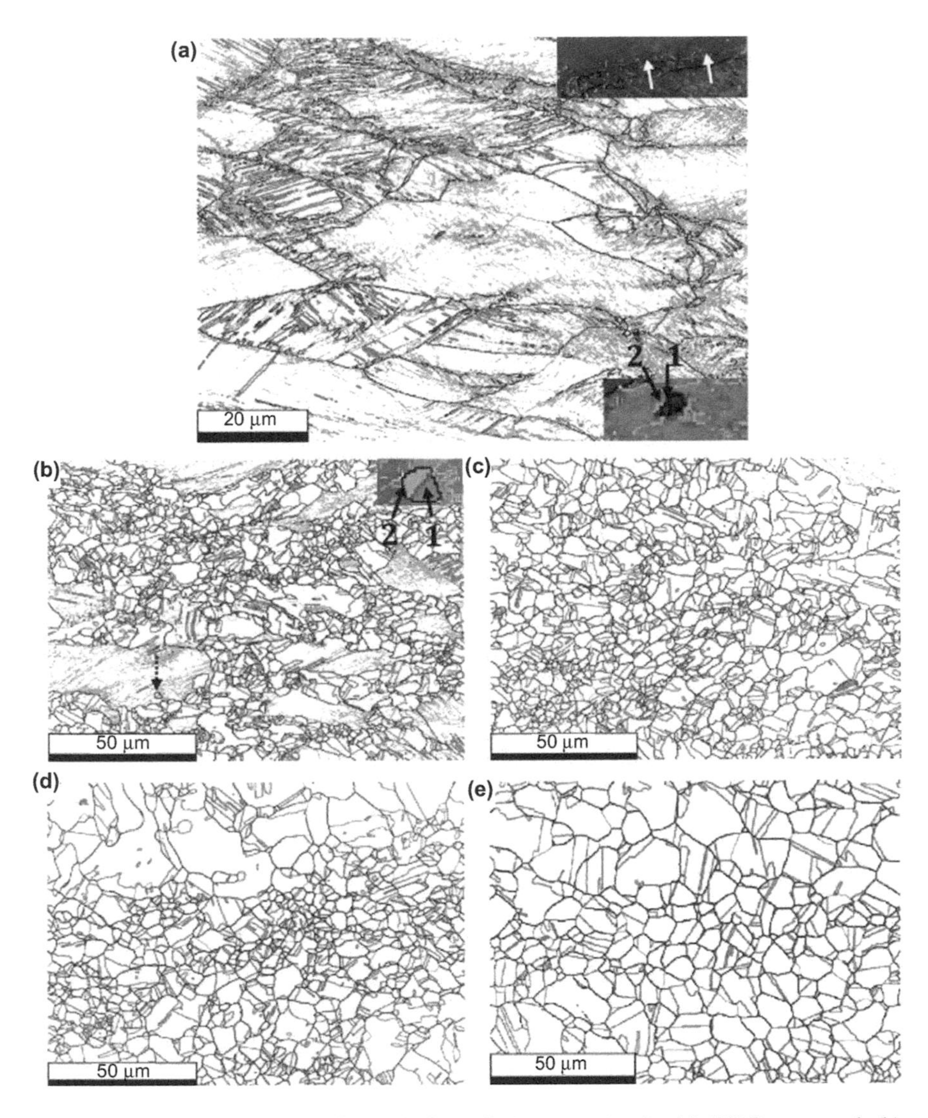

Figure 10.12 Representative EBSD grain boundaries maps for the (a) 600°C recovered, (b) 700°C, (c) 750°C, (d) 775°C and, (e) 850°C partially recrystallised samples. In (a), the top and bottom insets are zoomed-in views of nucleation sites. The top inset in (b) is a zoomed in-view for the recrystallised grain indicated by the dashed arrow. 38.9° ⟨101⟩, $\Sigma9$ GBs = blue lines and RD = horizontal (Saleh et al. 2011).

Copper (Rt-Cu) component increases. Figures 10.13b–e have a $\Sigma3$ relation with the S/B orientation, and the evolution of F component is apparent in the $\varphi_2 = 45°$ section in Figure 10.13e (Saleh et al. 2011). Additionally, the Elasto-Plastic Self-Consistent (EPSC) polycrystal plasticity model can

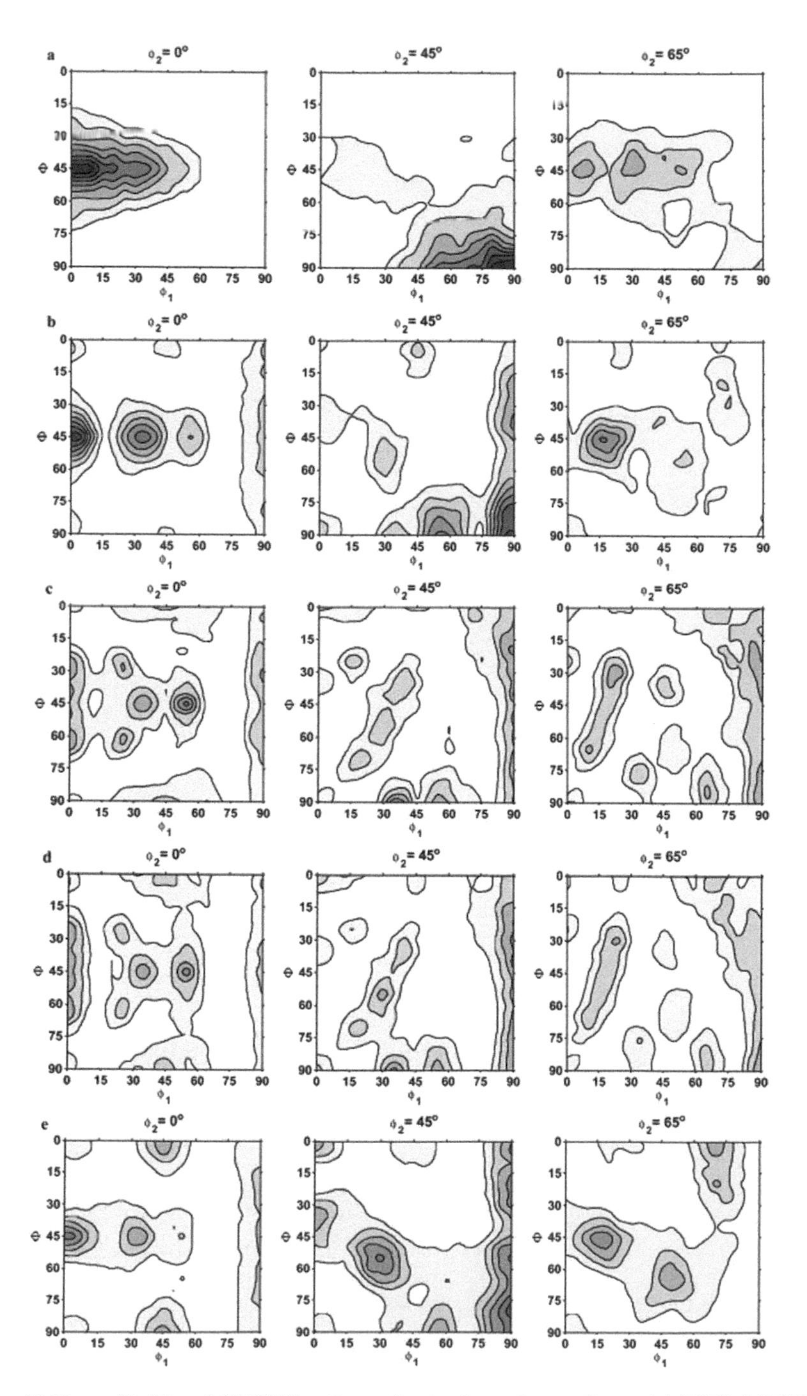

Figure 10.13 $\varphi_2 = 0°$, 45° and 65° ODF sections after isochronal annealing at (a) 600°C, (b) 700°C, (c) 750°C, (d) 775°C and, (e) 850°C. Contour levels = 0.5 × (Saleh et al. 2011).

be used to interpret the *in situ* diffraction measurements, to simulate the deformation behaviour of an fcc material by considering both, slip and twinning mechanisms (Saleh et al. 2014b).

10.4 Effect of Processing Parameters on TWIP Steel

As above Section 10.3 mentioned, there are several methods available to increase mechanical properties of TWIP steels, such as microalloying, austenite stability versus strain-induced ε-martensitic transformation, grain refinement, or bimodal microstructures consisting of deformed and recrystallised grains in partially recrystallised materials. All of these methods are related to the processing routes, in case of cold rolling, i.e., deformation temperature, reduction, strain rate, and pre-staining. This Section deals with these processing parameters during cold rolling on microstructure, texture and corresponding mechanical properties of the TWIP steels.

10.4.1 Effect of Temperature on SFE

As we have known, YS is temperature dependent, namely there is an increase in the YS of high-Mn TWIP steel with a decrease of the temperature below room temperature (Allain et al. 2010), whereas a decrease in YS is reported at higher temperature by the thermally activated dislocation glide (Shterner et al. 2014). The SFE is also temperature dependent, as is the deformation twinning kinetics. Indeed, all these assessments rely on the measured martensite start temperatures and austenite start temperature (Cotes et al. 1998). Consequently, it is believed that there is a transition in twin formation with increasing temperature (Lebedkina et al. 2009). TWIP effect appeared at high temperatures up to 350°C in TWIP steels (Meyers et al. 2001). The temperature increases as SFE decreases (Eskandari et al. 2012).

Deformation temperature has a substantial effect on the SFE, which results in a variation of the relative contributions of twinning and dislocation glide to plastic strain with temperature, thus affecting the strain hardening behaviour (Saeed-Akbari et al. 2009, 2012). The SFE increases with increasing deformation temperature (Remy 1977), and the critical stress and strain for twinning increase with increasing SFE (Remy 1978, Steinmetz et al. 2013). The increasing temperature can suppress the density of deformation twins (Xiong et al. 2015). Temperature effects are usually correlated with strain rate effects through some kind of chemical rate theory equation.

Deformation mechanism and microstructure will change with the deformation temperature, and their exact calculation showed that displacive transformation can occur during plasticity depending on the SFE (Γ), for $\Gamma < 18$ mJ/m^2, the ε-martensitic transformation occurs, while for 12 mJ/m$^2 < \Gamma < 35$ mJ/m^2 mechanical twinning takes place (Allain et al. 2004a). These estimates revealed that the SFE does increase with temperature. In the case of 300°C (Shterner et al. 2014), the magnitude of SFE exceeds the range of SFE values, i.e., 20–60 mJ/m^2, for which mechanical twinning is the deformation-controlling mechanism (Dumay et al. 2008, Saeed-Akbari et al. 2009).

Specifically, Allain et al. (2004b) compared the thermodynamically calculated SFE, deformation mechanisms, and tensile properties of an Fe-22Mn-0.6C wt.% steel at –196°C, at room temperature and 400°C. They reported the deformation mechanism changed with increasing temperature and SFE changes from dislocation glide and ε-martensite formation (SFE ~10 mJ/m^2), to dislocation glide and mechanical twinning (SFE ~19 mJ/m^2) to only dislocation glide (SFE ~80 mJ/m^2). Other temperature-sensitive phenomenons such as thermally activated dislocation dynamics (Allain et al. 2010) and DSA (De Cooman et al. 2012) may obscure the influence of a change in SFE on the microstructure and mechanical properties.

10.4.2 Effect of Reduction

During cold rolling the orientation of the grains turned to the direction of deformation which favours formation of deformation twins and grain refinement of austenite can thus be obtained. As such, this can enhance the strength properties in TWIP steels. Moreover, with the increasing amount of cold deformation, high dislocation density and more deformation twins are evident along with some amount of noticeable micro shear bands, as shown in Figure 10.14, in Fe-21Mn-3Si-3Al-0.06C TWIP steel (Tewary et al. 2014). The cold deformation results in the evolution of sub-structure, comprising dislocations, twins as well as interaction of dislocations and twins. In particular, on Fe-29Mn-5Al dual-phase TWIP steel, the rolling reduction in area increased the density of mechanical twins enhanced (Torabinejad et al. 2011). Therefore, the hardness, YS and tensile strength of the TWIP steel increase, whereas percentage elongation decreases as the amount of cold rolling reduction increases. For instance, in the Fe-23Mn-0.3C-1.5Al TWIP steels (Kusakin et al. 2014), the deformation strengthening and grain boundary strengthening leads to an increase in the YS from 235 MPa in the initial state to 1400 MPa after 80% rolling.

The ductility parameters decrease with the increasing amount of cold rolling deformation which is expected as strength and ductility are inversely related. Table 10.6 (Tewary et al. 2014) illustrates that the maximum strength values are obtained at 50% cold deformation but with lowest ductility. It means that energy absorption capacity decreases with the increase of cold rolling. In practice, a small cold rolling reduction (10%) is favourable for the Fe-21Mn-3Si-4Al-0.06C TWIP steel so as to achieve a YS of 600–700 MPa as potentials for automotive manufacturing (Bouaziz et al. 2011).

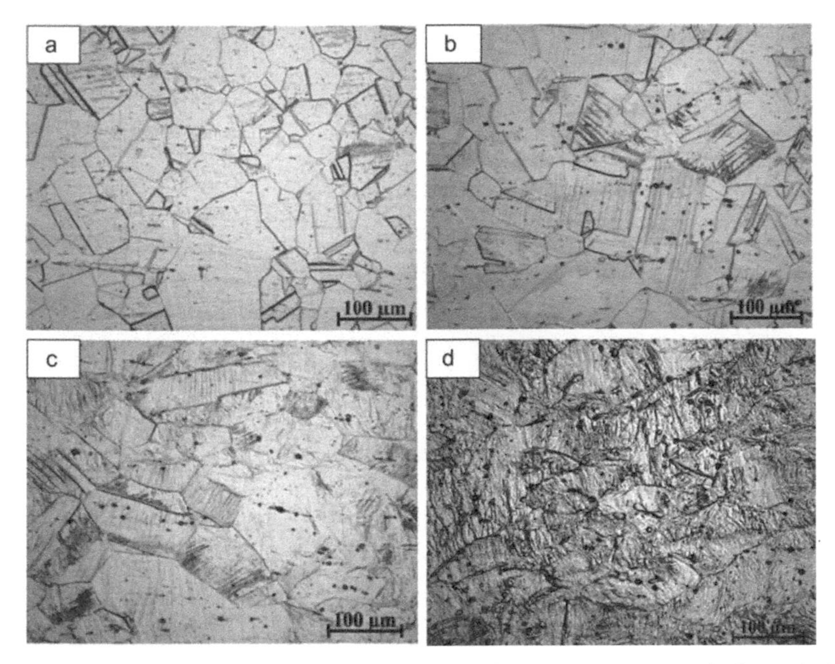

Figure 10.14 Optical micrographs of the Fe-21Mn-3.5Al-3Si-0.06C TWIP sample obtained after (a) 0%, (b) 10%, (c) 30%, and (d) 50%, cold deformation (Tewary et al. 2014).

Table 10.6 Test results of cold deformed Fe-21Mn-3Si-4Al-0.06C TWIP steels (Tewary et al. 2014).

Cold deformed specimen	**YS (MPa)**	**UTS (MPa)**	**Total elongation (%)**	**Tensile toughness ($J/m^3 \times 10^3$)**
0%	275	655	53.85	296.2
10%	744	865	34.17	254.8
30%	837	926	29.05	207.7
50%	1058	1202	19.93	164.4

YS: Yield Strength; UTS: Ultimate Tensile Strength

Different deformation mechanisms become active with increasing cold rolling reduction. After deformation twinning at low strains $\varepsilon < 0.1$ in the Fe-30Mn-3Al-3Si TWIP steel, as the strain increases, the volume fraction of twins increases, while at higher strain levels, non-homogeneous deformation mechanisms such as shear band formation become active (Vercammen et al. 2004). The ease of cross-slip controlled by SFE in the austenitic phase is responsible for different deformation mechanism at different stages (Christian and Mahajan 1995). Table 10.7 shows dislocation density plays a dominant role in strain hardening at lower degree of reduction (at 10% reduction) according to the calculated

Table 10.7 Dislocation density and planar fault (twin) probability for the Fe-21Mn-3Si-4Al-0.06C TWIP steels with different cold rolling reduction (Tewary et al. 2014).

Cold deformed specimen	Dislocation density (m^{-2})	Planar fault (twin) probability
10%	4.93×10^{15}	0.2889
30%	1.51×10^{16}	0.2291
50%	3.67×10^{15}	0.5950
0%	7.24×10^{15}	Very low

dislocation density and planar fault probability (twinning) values in the Fe-21Mn-3Si-4Al-0.06C TWIP steels with cold deformation (Tewary et al. 2014).

Specifically, Figure 10.15 schematically illustrates deformation microstructures evolved in the Fe-28Mn-0.28C TWIP steel with 10 and 80% reductions (Kusakin et al. 2014). With respect to the operation of deformation

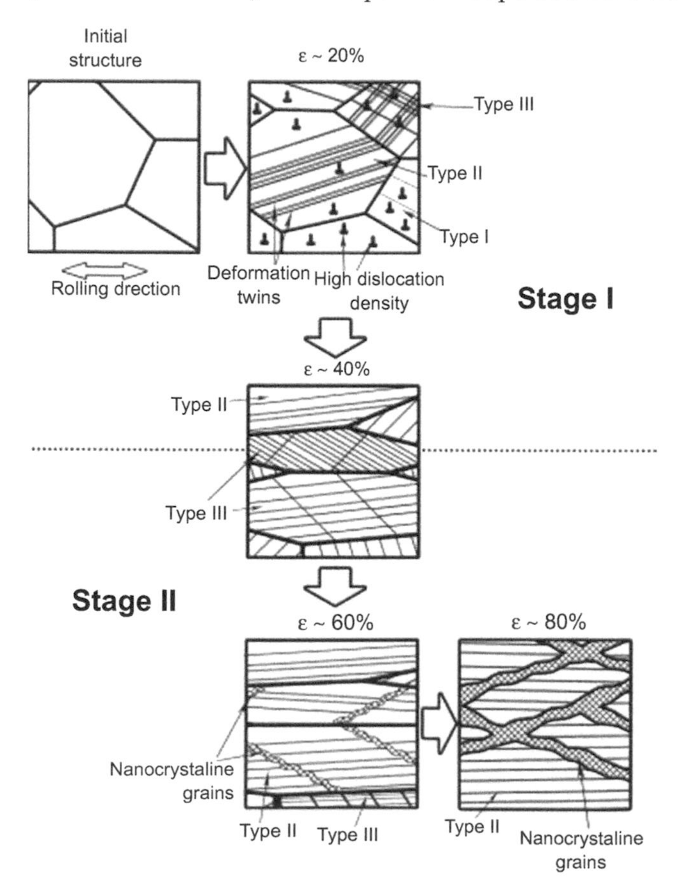

Figure 10.15 Schematic representation of the evolution of deformation structures during rolling (Kusakin et al. 2014).

twinning, three types of grains can be distinguished: Type I grains are almost free of mechanical twins, type II grains contain one active primary twinning system, and type III grains were characterised by more than one primary twinning system or secondary twinning system (Kusakin et al. 2014). In general, at the rolling reduction less than 20% (Figure 10.15), microstructure shows many slip lines, grain elongation and few twin-matrix lamellae. When the rolling reduction is around 40% (Figure 10.15), microstructure shows severe increase of the twins, alignment of twins with the rolling plane and microshear bands. When the rolling reduction is in the range of 60–80% (Figure 10.15), alignment twins, a herring bone structure and macroshear bands can be observed (Haase et al. 2013).

10.4.3 Effect of Strain Rate

Strain rate is the rate of change in strain (deformation) of a material with respect to time, particularly at which it is being deformed by progressive shearing without changing its volume (shear rate). Twinning rate, the rate of twinning induced hardening, is also sensitive to the deformation temperature and strain rate. Generally, the relative contribution of twinning to the overall deformation increases as the strain rate is increased (Christian and Mahajan 1995).

Similar to the effect of deformation temperature, strain rate also has influence on the mechanical behaviour of TWIP steels. An increase in strain rate gave a rise to extensive twinning and results in higher strength values (Curtze et al. 2009), particularly pronounced in UTS than in YS, as shown in Figure 10.16 (Curtze and Kuokkala 2010). Further, the YS increases from 500 to 727 MPa with the strain rate varying from 5.7×10^{-4} to 3000 s^{-1} in a Fe-18Mn-0.6C-1.5Al-0.8Si TWIP steel (Liang et al. 2015b). Nevertheless, the effect of strain rate on the elongation values seems to be more complex. In essence, the temperature and strain rate dependence of twinning will be masked by that of slip. This also means that increasing strain rate extends the value range of SFE for twinning (Xiong et al. 2015). Thus, extensive twinning is expected to occur especially at high strain rates and low temperatures, where the dislocation slip becomes more and more difficult while the twinning stress remains almost the same. At high strain rates an increase in temperature due to adiabatic deformation heating also contributes to the SFE, shifting the SFE of the austenite phase either towards or away from the optimum value for twinning (Curtze and Kuokkala 2010).

The strain-rate sensitivity (Liang et al. 2015b) is defined as (Shen et al. 2006):

$$m = \left.\frac{\partial ln\sigma}{\partial ln\dot{\varepsilon}}\right|_{\varepsilon,T} = \left.\frac{log\sigma_2/\sigma_1}{log\dot{\varepsilon}_2/\dot{\varepsilon}_1}\right|_{\varepsilon,T} \tag{10.4}$$

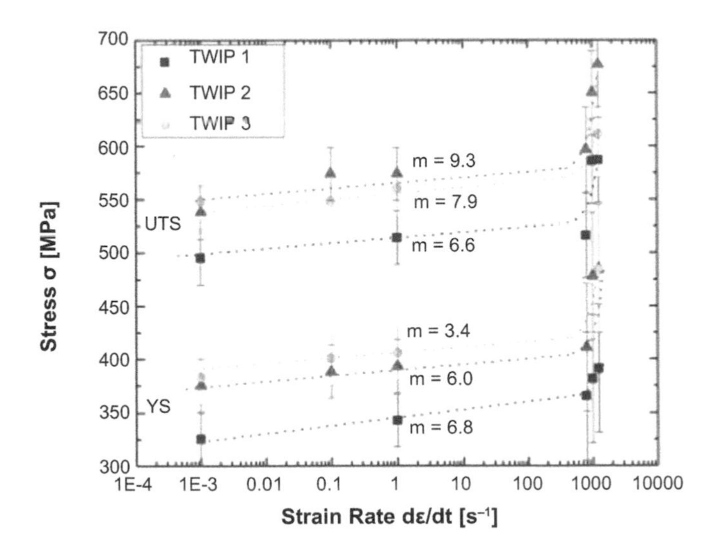

Figure 10.16 YS and UTS of the TWIP steels as a function of strain rate. Slope $m = d\sigma/d\ln\dot{\varepsilon}$ is the strain rate sensitivity factor, TWIP 1: Fe-28Mn-1.6Al-0.28Si-0.08C; TWIP 2: Fe-25Mn-1.6Al-0.24Si-0.08C; TWIP 3: Fe-27Mn-4.1Al-0.52Si-0.08C (Curtze and Kuokkala 2010).

where σ is the flow stress and $\dot{\varepsilon}$ is the strain rate. As the value of m increases, the material is expected to exhibit great resistance to necking, delaying fracture. In addition, they exhibit an opposite behaviour, i.e., negative strain-rate-sensitivity (Shen et al. 2016), or work hardening exponents decrease with increasing strain rate (Yang et al. 2015). As such, a negative strain rate sensitivity of work-hardening, i.e., lower work-hardening rate at higher strain rate, due to the suppression of the dislocations and deformation twins caused by the adiabatic heating associated with high strain rate deformation (Liang et al. 2015a).

10.4.4 Effect of Pre-straining by Rolling

There are several ways available to strengthen TWIP steels, one of which is pre-straining by rolling as the extremely high initial ductility can accommodate a significant pre-deformation step while still retaining good formability. Significant increases in the YS can be achieved: in case of Fe-22Mn-0.6C-0.2V TWIP steel, after 10% reduction the YS is 1000 MPa and the uniform elongation is 25% (Bouaziz et al. 2011). In addition, the pre-strain method was developed to improve the low-cycle fatigue properties of Fe-22Mn-0.5C TWIP steel (Niendorf et al. 2009), which is attributed to the enhanced interaction of glide dislocations with twins of increased density owing to the monotonic pre-deformation. For instance, pre-tension created mainly single variant twins, while pre-drawing generated multi-variant twins in Fe-17Mn-0.8C TWIP steel (Guo et al. 2014). The pre-strained Fe-30Mn-0.9C TWIP steel can improve the YS and introduce high density of fine deformation twins induced by pre-straining. In case of cyclic pre-straining on monotonic flow in Fe-0.6C-22Mn-0.0.3V TWIP steel, the YS increased by 28% by pre-straining at the strain amplitude of 0.5% for 250 cycles, while the plastic strain amplitude was only 0.15%, the monotonic YS increased by 28%, from 500 to 640 MPa (Hamada et al. 2014).

Nevertheless, pre-straining with medium or high degrees of plastic deformation can still be used to enhance the YS but this process reduces the remaining ductility below the critical value that is required for the final shape forming (Haase et al. 2014). The amount of pre-strain required increases with the decreasing temperature of the test and also depends on the final strain rate. To deal with this low yield stress issue, one of the widely accepted methods is applying pre-strain process such as uniaxial tensile, sheet or caliber rolling and wire rod drawing (Chun et al. 2012, Guo et al. 2014). Furthermore, a combination of pre-straining in the form of cold rolling and recovery annealing has proven to be a promising method to obtain significantly increased YS along with appreciable elongation (Santos et al. 2011), so as to reduce the dislocation density without modifying the twinning nanostructure. In doing so, the fatigue lives of the TWIP steel can be effectively improved through pre-straining, since the deformation twins induced by pre-straining lead to the improved YS and the homogenised deformation. Pre-straining by rolling provides possible ways for improving the high-cycle fatigue properties of TWIP steels while still retaining good formability.

10.5 Conclusions

This Chapter has provided a comprehensive overview of development and recent progress in austenitic TWIP steels with the low or moderate SFE (15 ~ 40 mJ/m^2) from two perspective, one from the TWIP steel substrate itself, the other from the processing parameters during cold rolling. A special emphasis has been made on the three main types of Fe-Mn-C, Fe-Mn-C-Al, Fe-Mn-Al-Si alloying system. On the basis of SFE fundamentals, a TWIP steel product has been characterised by its alloying elements, phase composition, grain features, and texture. Depending on tunable alloying design, it is also necessary to obtain the refined grained TWIP steels with a proper texture during cold rolling and annealing. In a combined processing route of cold rolling and recovery annealing, a small reduction (~ 10%) is favourable for the TWIP steel to achieve a potential YS for automotive manufacturing.

A pre-straining by rolling can be used to enhance the high-cycle fatigue properties of TWIP steels while still retaining good formability. In any case, the development of austenitic TWIP steel faces a number of significant technical challenges related to cold rolling, before TWIP sheet steel can be made widely available to the automotive industry.

Acknowledgements

This work was supported by the China Postdoctoral Science Foundation under Grant No. 2015M580094, and the National Natural Science Foundation of China under Grant Nos. 11274198 and 51532004.

Keywords: Twinning-induced plasticity steel, cold rolling, annealing, mechanical twinning, stacking fault energy, recrystallization, texture, grain refinement

References

Abe, T., K. Tsukada, H. Tagawa and I. Kozasu. 1990. Grain boundary segregation behaviour of phosphorus and carbon under equilibrium and non-equilibrium conditions in austenitic region of steels. ISIJ Int. 30: 444–450.

Adler, P.H., G.B. Olson and W.S. Owen. 1986. Strain hardening of Hadfield manganese steel. Metall. Mater. Trans. A 17: 1725–1737.

Allain, S., J.P. Chateau and O. Bouaziz. 2004a. A physical model of the twinning-induced plasticity effect in a high manganese austenitic steel. Mat. Sci. Eng. A 387: 143–147.

Allain, S., J.P. Chateau, O. Bouaziz, S. Migot and N. Guelton. 2004b. Correlations between the calculated stacking fault energy and the plasticity mechanisms in Fe–Mn–C alloys. Mat. Sci. Eng. A 387: 158–162.

Allain, S., J.P. Chateau, D. Dahmoun and O. Bouaziz. 2004c. Modeling of mechanical twinning in a high manganese content austenitic steel. Mat. Sci. Eng. A 387-389: 272–276.

Allain, S., O. Bouaziz and J.P. Chateau. 2010. Thermally activated dislocation dynamics in austenitic FeMnC steels at low homologous temperature. Scripta Mater. 62: 500–503.

Asghari, A., A. Zarei-Hanzaki and M. Eskandari. 2013. Temperature dependence of plastic deformation mechanisms in a modified transformation-twinning induced plasticity steel. Mat. Sci. Eng. A 579: 150–156.

Barbier, D., N. Gey, S. Allain, N. Bozzolo and M. Humbert. 2009. Analysis of the tensile behavior of a TWIP steel based on the texture and microstructure evolutions. Mat. Sci. Eng. A 500: 196–206.

Bhattacharya, B. and R.K. Ray. 2015. Annealing texture of a cold-rolled Fe-Mn-Al-Si-C alloy. Philos. Mag. 95: 3002–3013.

Bliznuk, V.V., N.I. Glavatska, O. Söderberg and V.K. Lindroos. 2002. Effect of nitrogen on damping, mechanical and corrosive properties of Fe–Mn alloys. Mat. Sci. Eng. A 338: 213–218.

Bouaziz, O., C.P. Scott and G. Petitgand. 2009. Nanostructured steel with high work-hardening by the exploitation of the thermal stability of mechanically induced twins. Scripta Mater. 60: 714–716.

Bouaziz, O., S. Allain, C.P. Scott, P. Cugy and D. Barbier. 2011. High manganese austenitic twinning induced plasticity steels: A review of the microstructure properties relationships. Curr. Opin. Solid St. M. 15: 141–168.

Bouaziz, O., D. Barbier, P. Cugy and G. Petigand. 2012. Effect of process parameters on a metallurgical route providing nano-structured single phase steel with high work-hardening. Adv. Eng. Mater. 14: 49–51.

Bracke, L., K. Verbeken, L. Kestens and J. Penning. 2009. Microstructure and texture evolution during cold rolling and annealing of a high Mn TWIP steel. Acta Mater. 57: 1512–1524.

Byun, T.S. 2003. On the stress dependence of partial dislocation separation and deformation microstructure in austenitic stainless steels. Acta Mater. 51: 3063–3071.

Chen, L., Y. Zhao and X. Qin. 2013. Some aspects of high manganese twinning-induced plasticity (TWIP) steel, a review. Acta Metall. Sin. 26: 1–15.

Christian, J.W. and S. Mahajan. 1995. Deformation twinning. Prog. Mater. Sci. 39: 1–157.

Christian, J.W. 2002. The theory of transformations in metals and alloys. Elsevier, Boston.

Chun, Y.S., J. Lee, C.M. Bae, K.T. Park and C.S. Lee. 2012. Caliber-rolled TWIP steel for high-strength wire rods with enhanced hydrogen-delayed fracture resistance. Scripta Mater. 67: 681–684.

Cotes, S., A.F. Guillermet and M. Sade. 1998. Phase stability and fcc/hcp martensitic transformation in Fe–Mn–Si alloys: Part I. Experimental study and systematics of the MS and AS temperatures. J. Alloy. Compd. 278: 231–238.

Curtze, S., V.T. Kuokkala, M. Hokka and T. Saarinen. 2009. Microstructure and texture evolution in high manganese TWIP steels. Dymat: 9th International Conference on the Mechanical and Physical Behaviour of Materials under Dynamic Loading, Belgium. 2: 1007–1013.

Curtze, S. and V.T. Kuokkala. 2010. Dependence of tensile deformation behavior of TWIP steels on stacking fault energy, temperature and strain rate. Acta Mater. 58: 5129–5141.

De Cooman, B.C., K.G. Chin and J. Kim. 2008. High-Mn Twip steels for automotive applications. Proceedings of New Developments on Metallurgy and Applications of High Strength Steels, Buenos Aires. 1-2: 69–83.

De Cooman, B.C., L. Chen, H.S. Kim, Y. Estrin, S.K. Kim and H. Voswinckel. 2009. State-of-the-science of high manganese TWIP steels for automotive applications. pp. 165–183. *In*: A. Haldar, S. Suwas and D. Bhattacharjee (eds.). Microstructure and Texture in Steels and Other Materials. Springer, London.

De Cooman, B.C., O. Kwon and K.G. Chin. 2012. State-of-the-knowledge on TWIP steel. Mater. Sci. Tech. 28: 513–527.

Dini, G., A. Najafizadeh, R. Ueji and S.M. Monir-Vaghefi. 2010. Tensile deformation behavior of high manganese austenitic steel: the role of grain size. Mater. Design. 31: 3395–3402.

Dumay, A., J.P. Chateau, S. Allain, S. Migot and O. Bouaziz. 2008. Influence of addition elements on the stacking-fault energy and mechanical properties of an austenitic Fe–Mn–C steel. Mat. Sci. Eng. A 483: 184–187.

Eskandari, M., A. Zarei-Hanzaki and A. Marandi. 2012. An investigation into the mechanical behavior of a new transformation-twinning induced plasticity steel. Mater. Design. 39: 279–284.

Frommeyer, G., U. Brüx and P. Neumann. 2003. Supra-ductile and high-strength manganese-TRIP/TWIP steels for high energy absorption purposes. ISIJ Int. 43: 438–446.

Gazder, A.A., A.A. Saleh and E.V. Pereloma. 2011. Microtexture analysis of cold-rolled and annealed twinning-induced plasticity steel. Scripta Mater. 65: 560–563.

Gazder, A.A., A.A. Saleh, M.J.B. Nancarrow, D.R.G. Mitchell and E.V. Pereloma. 2015. A transmission kikuchi diffraction study of a cold-rolled and annealed Fe-17Mn-2Si-3Al-1Ni-0.06C wt% steel. Steel Res. Int. 86: 1204–1214.

Grässel, O., L. Krüger, G. Frommeyer and L.W. Meyer. 2000. High strength Fe-Mn-(Al, Si) TRIP/TWIP steels development-properties-application. Int. J. Plasticity. 16: 1391–1409.

Guo, Q., Y.S. Chun, J.H. Lee, Y.U. Heo and C.S. Lee. 2014. Enhanced low-cycle fatigue life by pre-straining in an Fe-17Mn-0.8C twinning induced plasticity steel. Met. Mater. Int. 20: 1043–1051.

Gutierrez-Urrutia, I., S. Zaefferer and D. Raabe. 2010. The effect of grain size and grain orientation on deformation twinning in a Fe–22 wt.% Mn–0.6 wt.% C TWIP steel. Mat. Sci. Eng. A 527: 3552–3560.

Gutierrez-Urrutia, I. and D. Raabe. 2012. Grain size effect on strain hardening in twinning-induced plasticity steels. Scripta Mater. 66: 992–996.
Haase, C., L.A. Barrales, D.A. Molodov and G. Gottstein. 2013a. Tailoring the mechanical properties of a twinning-induced plasticity steel by retention of deformation twins during heat treatment. Metall. Mater. Trans. A 44: 4445–4449.
Haase, C., S.G. Chowdhury, L.A. Barrales-Mora, D.A. Molodov and G. Gottstein. 2013b. On the relation of microstructure and texture evolution in an austenitic Fe-28Mn-0.28C TWIP steel during cold rolling. Metall. Mater. Trans. A 44: 911–922.
Haase, C., L.A. Barrales-Mora, F. Roters, D.A. Molodov and G. Gottstein. 2014. Applying the texture analysis for optimizing thermomechanical treatment of high manganese twinning-induced plasticity steel. Acta Mater. 80: 327–340.
Haase, C., T. Ingendahl, O. Güvenç, M. Bambach, W. Bleck, D.A. Molodov and L.A. Barrales-Mora. 2016. On the applicability of recovery-annealed twinning-induced plasticity steels: potential and limitations. Mat. Sci. Eng. A 649: 74–84.
Hadfield, R.A. 1888. Hadfield's manganese steel. Science. 12: 284–286.
Hamada, A.S., A. Järvenpää, M. Honkanen, M. Jaskari, D.A. Porter and L.P. Karjalainen. 2014. Effects of cyclic pre-straining on mechanical properties of an austenitic microalloyed high-Mn twinning-induced plasticity steel. Procedia Eng. 74: 47–52.
Han, D.K., S.K. Lee, S.J. Noh, S.K. Kim and D.W. Suh. 2015. Effect of aluminium on hydrogen permeation of high-manganese twinning-induced plasticity steel. Scripta Mater. 99: 45–48.
He, B.B., H.W. Luo and M.X. Huang. 2016. Experimental investigation on a novel medium Mn steel combining transformation-induced plasticity and twinning-induced plasticity effects. Int. J. Plasticity 78: 173–186.
Hirth, J.P. 1970. Thermodynamics of stacking faults. Metall. Trans. 1: 2367–2374.
Hofer, S., M. Hartl, G. Schestak, R. Schneider, E. Arenholz and L. Samek. 2011. Comparison of austenitic high-Mn-Steels with different Mn-and C-contents regarding their processing properties. BHM Berg-und Hüttenmän. Monatsh. 156: 99–104.
Huang, M.X., O. Bouaziz, D. Barbier and S. Allain. 2011. Modelling the effect of carbon on deformation behaviour of twinning induced plasticity steels. J. Mater. Sci. 46: 7410–7414.
Hutchinson, B. and N. Ridley. 2006. On dislocation accumulation and work hardening in Hadfield steel. Scripta Mater. 55: 299–302.
Idrissi, H., L. Ryelandt, M. Veron, D. Schryvers and P.J. Jacques. 2009. Is there a relationship between the stacking fault character and the activated mode of plasticity of Fe–Mn-based austenitic steels? Scripta Mater. 60: 941–944.
Idrissi, H., K. Renard, L. Ryelandt, D. Schryvers and P.J. Jacques. 2010. On the mechanism of twin formation in Fe–Mn–C TWIP steels. Acta Mater. 58: 2464–2476.
Jeong, J.S., W. Woo, K.H. Oh, S.K. Kwon and Y.M. Koo. 2012. *In situ* neutron diffraction study of the microstructure and tensile deformation behavior in Al-added high manganese austenitic steels. Acta Mater. 60: 2290–2299.
Jin, J.E. and Y.K. Lee. 2012. Effects of Al on microstructure and tensile properties of C-bearing high Mn TWIP steel. Acta Mater. 60: 1680–1688.
Jung, Y.S., S. Kang, K. Jeong, J.G. Jung and Y.K. Lee. 2013. The effects of N on the microstructures and tensile properties of Fe-15Mn-0.6C-2Cr-xN twinning-induced plasticity steels. Acta Mater. 61: 6541–6548.
Kang, S., Y.S. Jung, J.H. Jun and Y.K. Lee. 2010. Effects of recrystallization annealing temperature on carbide precipitation, microstructure, and mechanical properties in Fe–18Mn–0.6 C–1.5 Al TWIP steel. Mat. Sci. Eng. A 527: 745–751.
Kang, S., J.G. Jung and Y.K. Lee. 2012. Effects of niobium on mechanical twinning and tensile properties of a high Mn twinning-induced plasticity steel. Mater. Trans. 53: 2187–2190.
Karaman, I., H. Sehitoglu, K. Gall, Y.I. Chumlyakov and H.J. Maier. 2000. Deformation of single crystal Hadfield steel by twinning and slip. Acta Mater. 48: 1345–1359.
Kim, J. and B.C. De Cooman. 2011. On the stacking fault energy of Fe-18 pct Mn-0.6 pct C-1.5 pct Al twinning-induced plasticity steel. Metall. Mater. Trans. A 42: 932–936.

Kim, J., S.J. Lee and B.C. De Cooman. 2011. Effect of Al on the stacking fault energy of Fe-18Mn-0.6C twinning-induced plasticity. Scripta Mater. 65: 363–366.

Kim, T.W., Y.G. Kim and S.H. Park. 1997. Process for manufacturing high manganese hot rolled steel sheet without any crack. U.S. Patents# 5647922.

Koyama, M., T. Sawaguchi and K. Tsuzaki. 2012. Inverse grain size dependence of critical strain for serrated flow in a Fe-Mn-C twinning-induced plasticity steel. Philos. Mag. Lett. 92: 145–152.

Koyama, M., T. Sawaguchi and K. Tsuzaki. 2015. Deformation twinning behavior of twinning-induced plasticity steels with different carbon concentrations-Part 2: proposal of dynamic-strain-aging-assisted deformation twinning. ISIJ Int. 55: 1754–1761.

Kusakin, P., A. Belyakov, C. Haase, R. Kaibyshev and D.A. Molodov. 2014. Microstructure evolution and strengthening mechanisms of Fe-23Mn-0.3C-1.5Al TWIP steel during cold rolling. Mat. Sci. Eng. A 617: 52–60.

Lebedkina, T.A., M.A. Lebyodkin, J.Ph. Chateau, A. Jacques and S. Allain. 2009. On the mechanism of unstable plastic flow in an austenitic FeMnC TWIP steel. Mat. Sci. Eng. A 519: 147–154.

Lee, S.J., Y.S. Jung, S.I. Baik, Y.W. Kim, M. Kang, W. Woo and Y.K. Lee. 2014. The effect of nitrogen on the stacking fault energy in Fe-15Mn-2Cr-0.6C-xN twinning-induced plasticity steels. Scripta Mater. 92: 23–26.

Lee, T.H., E. Shin, C.S. Oh, H.Y. Ha and S.J. Kim. 2010. Correlation between stacking fault energy and deformation microstructure in high-interstitial-alloyed austenitic steels. Acta Mater. 58: 3173–3186.

Leffers, T. and R.K. Ray. 2009. The brass-type texture and its deviation from the copper-type texture. Prog. Mater. Sci. 54: 351–396.

Liang, Z.Y., W. Huang and M.X. Huang. 2015a. Suppression of dislocations at high strain rate deformation in a twinning-induced plasticity steel. Mat. Sci. Eng. A 628: 84–88.

Liang, Z.Y., X. Wang, W. Huang and M.X. Huang. 2015b. Strain rate sensitivity and evolution of dislocations and twins in a twinning-induced plasticity steel. Acta Mater. 88: 170–179.

Liss, K.D. and K. Yan. 2010. Thermo-mechanical processing in a synchrotron beam. Mat. Sci. Eng. A 528: 11–27.

Mahajan, S. and G.Y. Chin. 1974. The interaction of twins with existing substructure and twins in cobalt-iron alloys. Acta Metall. Mater. 22: 1113–1119.

Mahato, B., S.K. Shee, T. Sahu, S.G. Chowdhury, P. Sahu, D.A. Porter and L.P. Karjalainen. 2015. An effective stacking fault energy viewpoint on the formation of extended defects and their contribution to strain hardening in a Fe–Mn–Si–Al twinning-induced plasticity steel. Acta Mater. 86: 69–79.

Mesquita, R.A., R. Schneider, K. Steineder, L. Samek and E. Arenholz. 2013. On the austenite stability of a new quality of twinning induced plasticity steel, exploring new ranges of Mn and C. Metall. Mater. Trans. A 44: 4015–4019.

Meyers, M.A., O. Vöhringer and V.A. Lubarda. 2001. The onset of twinning in metals: a constitutive description. Acta Mater. 49: 4025–4039.

Nezhadfar, P.D., A. Rezaeian and M.S. Papkiadeh. 2015. Softening behavior of a cold rolled high-Mn twinning-induced plasticity steel. J. Mater. Eng. Perform. 24: 3820–3825.

Niendorf, T., C. Lotze, D. Canadinc, A. Frehn and H.J. Maier. 2009. The role of monotonic pre-deformation on the fatigue performance of a high-manganese austenitic TWIP steel. Mat. Sci. Eng. A 499: 518–524.

Ofei, K.A., L. Zhao and J. Sietsma. 2013. Microstructural development and deformation mechanisms during cold rolling of a medium stacking fault energy TWIP steel. J. Mater. Sci. Technol. 29: 161–167.

Oh, B.W., S.J. Cho, Y.G. Kim, Y.P. Kim, W.S. Kim and S.H. Hong. 1995. Effect of aluminium on deformation mode and mechanical properties of austenitic Fe-Mn-Cr-Al-C alloys. Mat. Sci. Eng. A 197: 147–156.

Olson, G.B. and M. Cohen. 1976. A general mechanism of martensitic nucleation: Part I. General concepts and the FCC→HCP transformation. Metall. Trans. A 7: 1897–1904.

Petch, N.J. 1953. The cleavage strength of polycrystals. J. Iron Steel Inst. 174: 25–28.
Pierce, D.T., J.A. Jimenez, J. Bentley, D. Raabe, C. Oskay and J.E. Wittig. 2014. The influence of manganese content on the stacking fault and austenite/ε-martensite interfacial energies in Fe-Mn-(Al-Si) steels investigated by experiment and theory. Acta Mater. 68: 238–253.
Pierce, D.T., J.A. Jiménez, J. Bentley, D. Raabe and J.E. Wittig. 2015. The influence of stacking fault energy on the microstructural and strain-hardening evolution of Fe–Mn–Al–Si steels during tensile deformation. Acta Mater. 100: 178–190.
Prakash, A., T. Hochrainer, E. Reisacher and H. Riedel. 2008. Twinning models in self-consistent texture simulations of TWIP steels. Steel Res. Int. 79: 645–652.
Pramanik, S., A.A. Saleh, D.B. Santos, E.V. Pereloma and A.A. Gazder. 2015. Microstructure evolution during isochronal annealing of a 42% cold rolled TRIP-TWIP steel. IOP Conf. Series: Mater. Sci. Eng. 89: 012042.
Randle, V. and O. Engler. 2000. Introduction to texture analysis: macrotexture, microtexture and orientation mapping. CRC Press, Boca Raton.
Razmpoosh, M.H., A. Zarei-Hanzaki, N. Haghdadi, J.H. Cho, W.J. Kim and S. Heshmati-Manesh. 2015. Thermal stability of an ultrafine-grained dual phase TWIP steel. Mat. Sci. Eng. A 638: 5–14.
Remy, L. 1977. Temperature variation of the intrinsic stacking fault energy of a high manganese austenitic steel. Acta Metall. Mater. 25: 173–179.
Remy, L. and A. Pineau. 1977. Twinning and strain-induced FCC→HCP transformation in the Fe-Mn-Cr-C system. Mat. Sci. Eng. A 28: 99–107.
Remy, L. 1978. Kinetics of fcc deformation twinning and its relationship to stress-strain behaviour. Acta Metall. Mater. 26: 443–451.
Roberts, W.L. 1978. Cold rolling of steel. CRC Press, New York.
Rollett, A.D., F.J. Humphreys, G.S. Rohrer and M. Hatherly. 2004. Recrystallization and related annealing phenomena. Elsevier, Boston.
Roncery, L.M., S. Weber and W. Theisen. 2010. Development of Mn-Cr-(C-N) corrosion resistant twinning induced plasticity steels: thermodynamic and diffusion calculations, production, and characterization. Metall. Mater. Trans. A 41: 2471–2479.
Saeed-Akbari, A., J. Imlau, U. Prahl and W. Bleck. 2009. Derivation and variation in composition-dependent stacking fault energy maps based on subregular solution model in high-manganese steels. Metall. Mater. Trans. A 40: 3076–3090.
Saeed-Akbari, A., L. Mosecker, A. Schwedt and W. Bleck. 2012. Characterization and prediction of flow behavior in high-manganese twinning induced plasticity steels: Part I. mechanism maps and work-hardening behavior. Metall. Mater. Trans. A 43: 1688–1704.
Saleh, A.A., E.V. Pereloma and A.A. Gazder. 2011. Texture evolution of cold rolled and annealed Fe–24Mn–3Al–2Si–1Ni–0.06 C TWIP steel. Mat. Sci. Eng. A 528: 4537–4549.
Saleh, A.A., C. Haase, E.V. Pereloma, D.A. Molodov and A.A. Gazder. 2014a. On the evolution and modelling of brass-type texture in cold-rolled twinning-induced plasticity steel. Acta Mater. 70: 259–271.
Saleh, A.A., E.V. Pereloma, B. Clausen, D.W. Brown, C.N. Tome and A.A. Gazder. 2014b. Self-consistent modelling of lattice strains during the *in-situ* tensile loading of twinning induced plasticity steel. Mat. Sci. Eng. A 589: 66–75.
Santos, D.B., A.A. Saleh, A.A. Gazder, A.D. Carman, M.R. Dayanna, A.S. Érica, B.M. Gonzalez and E.V. Pereloma. 2011. Effect of annealing on the microstructure and mechanical properties of cold rolled Fe–24Mn–3Al–2Si–1Ni–0.06 C TWIP steel. Mat. Sci. Eng. A 528: 3545–3555.
Schramm, R.E. and R.P. Reed. 1975. Stacking fault energies of seven commercial austenitic stainless steels. Metall. Trans. A 6: 1345–1351.
Schumann, H. 1974. Influence of stacking fault energy on the crystal lattice change mechanisms in the γ/α phase transformation in high alloy steels. Kristall Technik. 9: 1141.
Schumann, V.H. 1972. Martensitische Umwandlung in austenitischen mangan-kohlenstoff-stählen. Neue Hütte. 17: 605–609.

Shen, Y.F., L. Lu, M. Dao and S. Suresh. 2006. Strain rate sensitivity of Cu with nanoscale twins. Scripta Mater. 55: 319–322.

Shen, Y.F., N. Jia, R.D.K. Misra and L. Zuo. 2016. Softening behavior by excessive twinning and adiabatic heating at high strain rate in a Fe–20Mn–0.6C TWIP steel. Acta Mater. 103: 229–242.

Shterner, V., A. Molotnikov, I. Timokhina, Y. Estrin and H. Beladi. 2014. A constitutive model of the deformation behaviour of twinning induced plasticity (TWIP) steel at different temperatures. Mat. Sci. Eng. A 613: 224–231.

Sinclair, C.W., W.J. Poole and Y. Bréchet. 2006. A model for the grain size dependent work hardening of copper. Scripta Mater. 55: 739–742.

Song, E.J., H. Bhadeshia and D.W. Suh. 2014. Interaction of aluminium with hydrogen in twinning-induced plasticity steel. Scripta Mater. 87: 9–12.

Steinmetz, D.R., T. Japel, B. Wietbrock, P. Eisenlohr, I. Gutierrez-Urrutia, A. Saeed-Akbari, T. Hickel, F. Roters and D. Raabe. 2013. Revealing the strain-hardening behavior of twinning-induced plasticity steels: theory, simulations, experiments. Acta Mater. 61: 494–510.

Tewary, N.K., S.K. Ghosh, S. Bera, D. Chakrabarti and S. Chatterjee. 2014. Influence of cold rolling on microstructure, texture and mechanical properties of low carbon high Mn TWIP steel. Mat. Sci. Eng. A 615: 405–415.

Tian, X., H. Li and Y. Zhang. 2008. Effect of Al content on stacking fault energy in austenitic Fe–Mn–Al–C alloys. J. Mater. Sci. 43: 6214–6222.

Tian, X. and Y. Zhang. 2009. Effect of Si content on the stacking fault energy in γ-Fe–Mn–Si–C alloys: Part I. X-ray diffraction line profile analysis. Mat. Sci. Eng. A 516: 73–77.

Torabinejad, V., A. Zarei-Hanzaki, S. Moemeni and A. Imandoust. 2011. An investigation to the microstructural evolution of Fe-29Mn-5Al dual-phase twinning-induced plasticity steel through annealing. Mater. Design. 32: 5015–5021.

Tsuji, N., Y. Saito, S.H. Lee and Y. Minamino. 2003. ARB (accumulative roll-bonding) and other new techniques to produce bulk ultrafine grained materials. Adv. Eng. Mater. 5: 338–344.

Tsukahara, H., T. Masumura, T. Tsuchiyama, S. Takaki, K. Nakashima, K. Hase and S. Endo. 2015. Design of alloy composition in 5%Mn-Cr-C austenitic steels. ISIJ Int. 55: 312–318.

Ueji, R., N. Tsuchida, D. Terada, N. Tsuji, Y. Tanaka, A. Takemura and K. Kunishige. 2008. Tensile properties and twinning behavior of high manganese austenitic steel with fine-grained structure. Scripta Mater. 59: 963–966.

Venables, J.A. 1964. Deformation twinning in fcc metals. pp. 77–116. *In*: R.E. Reed-Hill, J.P. Hirth and H.C. Rogers (eds.). Deformation Twinning. Gordon and Breach, New York.

Vercammen, S., B. Blanpain, B.C. De Cooman and P. Wollants. 2004. Cold rolling behaviour of an austenitic, Fe-30Mn-3Al-3Si TWIP-steel: the importance of deformation twinning. Acta Mater. 52: 2005–2012.

Wang, B., Z.J. Zhang, C.W. Shao, Q.Q. Duan, J.C. Pang, H.J. Yang, X.W. Li and Z.F. Zhang. 2015. Improving the high-cycle fatigue lives of Fe-30Mn-0.9C twinning-induced plasticity steel through pre-straining. Metall. Mater. Trans. A 46: 3317–3322.

Wenk, H.R. and P. Van Houtte. 2004. Texture and anisotropy. Rep. Prog. Phys. 67: 1367–1428.

Xiong, Z.P., X.P. Ren, S. Jian, Z.L. Wang, W.P. Bao and S.X. Li. 2015. Effect of temperature on microstructure and deformation mechanism of Fe-30Mn-3Si-4Al TWIP steel at strain rate of 700s−1. J. Iron Steel Res. Int. 22: 179–184.

Yan, H.L., X. Zhao, N. Jia, Y.R. Zheng and T. He. 2014. Influence of shear banding on the formation of brass-type textures in polycrystalline fcc metals with low stacking fault energy. J. Mater. Sci. Technol. 30: 408–416.

Yang, H.K., Z.J. Zhang and Z.F. Zhang. 2015. Comparison of twinning evolution with work hardening ability in twinning-induced plasticity steel under different strain rates. Mat. Sci. Eng. A 622: 184–188.

11

Cold Rolling Practice of Martensitic Steel

Màhmoud Nili-Ahmadabadi,[1,2,*] *Hamidreza Koohdar*[1] and *Mohammad Habibi-Parsa*[1]

ABSTRACT

In this chapter, the effects of cold rolling and subsequent annealing on the microstructure and mechanical properties of lath martensitic steels have been studied. These steels are an important group of advanced high strength steels. Investigations show the sequence of microstructural evolution during cold rolling of lath martensite including vanishing of primary lath boundaries, formation of slip bands, microbands and shear bands and elongated dislocation cell blocks, consecutively. Furthermore, monotonous increasing of dislocation density, increasing ratio of screw:edge type dislocations and decreasing of apparent grain size have been found by the analysis of deformed lath martensite. Subsequent annealing after cold rolling of lath martensite steels can produce an ultrafine/nano-grained structure. The formation of fine-grained structure during cold rolling and subsequent annealing leads to gain adequate uniform elongation as well as high tensile strength.

[1] School of Metallurgy and Materials Engineering, University of Tehran, P. O. Box: 14395-731, Tehran, Iran.
[2] Center of Excellence for High Performance Materials, School of Metallurgy and Materials Engineering, College of Engineering, University of Tehran, Tehran, Iran.
* Corresponding author: nili@ut.ac.ir

11.1 Introduction

Advanced high strength steels (AHSSs) are complex and sophisticated materials with carefully selected chemical compositions and multiphase microstructures resulting from precisely controlled heating and cooling processes. Various strengthening mechanisms are employed to achieve a range of strength, ductility, toughness, and fatigue properties. The AHSSs family include dual phase (DP), complex-phase (CP), ferritic-bainitic (FB), martensitic, transformation-induced plasticity (TRIP), hot-formed (HF), and twinning-induced plasticity (TWIP). These 1st and 2nd Generation AHSSs grades are uniquely qualified to meet the functional performance demands of certain parts in automotive industry. Recently, there has been increased funding and research for the development of the "*3rd Generation*" of AHSSs. These are steels with improved strength-ductility combinations compared to the present grades and with potential for more efficient joining capabilities at lower costs. These grades reflect unique alloys and microstructures to achieve the desired properties (Keeler and Kimchi 2014). In this chapter, the study is only focused on the martensitic steels as an important group of AHSSs.

There are three kinds of martensite in ferrous alloys with different crystal structures depending on alloying compositions, i.e., α' martensite with a crystal structure of body-centered cubic (bcc) or body-centered tetragonal (bct), ε martensite with a crystal structure of hexagonal close-packed (hcp) and fct martensite with a crystal structure of face-centered tetragonal (Maki 1990, Maki and Tamura 1986). The morphology of ε martensite and fct martensite is a parallel-sided thin plate type with planar interfaces. On the other hand, in the case of α' martensite, five types of morphologies consisting of lath, butterfly, $(225)_A$ type plate, lenticular, and thin plate have been observed (Davies and Magee 1971, Krauss and Marder 1971, Maki et al. 1972, Thomas 1971, Umemoto et al. 1983).

Researches performed on the cold rolling of martensitic steels are mostly focused on the rolling of steels with an original microstructure of lath martensite. This is related to the fine structure and excellent formability of lath martensite (Ghasemi-Nanesa et al. 2010, Maki et al. 1980, Marder and Krauss 1969, Marder and Marder 1969, Morito et al. 2003, Nili-Ahmadabadi et al. 2011). The sequence of microstructural evolution during the cold rolling of lath martensite encompasses vanishing of primary lath boundaries, formation of slip bands, microbands and shear bands and elongated dislocation cell blocks, consecutively (Huang et al. 2012, Morito et al. 2003, Takaki et al. 1991, Ueji et al. 2002). Furthermore, monotonous increase of dislocation density, increased ratio of screw:edge type dislocations and apparent decrease in grain size have been found by the analysis of deformed lath martensite (Hossein Nedjad and Movaghar Gharabagh 2008, Movaghar Gharabagh 2008, Takaki et al. 1991).

Subsequent annealing after cold rolling of lath martensite can produce an ultrafine/nano-grained structure with large misorientations in low carbon steels (Tianfu et al. 2006, Tsuji et al. 2002, Ueji et al. 2002). It has been reported that very high strain more than 4 is necessary to form ultrafine/nano-grained structure by severe plastic deformation (SPD) (Hansen and Jensen 1999, Ito et al. 2000, Tsuji et al. 2000a,b, Tsuji et al. 1999). The key of the easy formation of the ultrafine/nano-grains in cold rolling and subsequent annealing with a small amount of strain must be in the nature of the starting structure (Tsuji et al. 2002, Tianfu et al. 2006, Ueji et al. 2002). The main reason for the grain refinement under small strain is that the martensite starting structure is itself a fine-grained structure (Tsuji et al. 2002). This structure causes a constraint effect on plastic deformation to enhance inhomogeneous deformation. The high dislocation density as a result of cold rolling in martensite is also expected to play an important role to facilitate grain sub-division (Tsuji et al. 2002, Ueji et al. 2002). The formation of fine-grained structure during cold rolling and subsequent annealing leads to gain adequate uniform elongation as well as high tensile strength (Tsuji et al. 2002, Ueji et al. 2002). The aim of present chapter is to comprehensively investigate the effects of the cold rolling and subsequent annealing on the microstructure and mechanical properties of lath martensitic as an important group of AHSSs.

11.2 Martensitic Transformation

The martensitic transformation is a diffusionless phase transformation in solids in which atoms move cooperatively, often by a shear-like mechanism. Usually, the parent phase (a high temperature phase) is cubic, and the martensite (a lower temperature phase) has a lower symmetry. The transformation is schematically shown in Figure 11.1. When the temperature is lowered below the formation temperature of martensite (M_S), martensitic transformation starts with a shear-like mechanism. According to the figure, martensites in region A and in region B have the same structure, but the orientations are different. These are called the correspondence variants of martensite. Since martensite has a lower symmetry, many variants can be formed from the same parent phase. The above example clearly shows that the characteristics of martensitic transformation lie in the cooperative atoms movement. Therefore, this transformation is sometimes called the displacive or military transformations, which are equivalent in usage to martensitic transformation (Otsuka and Wayman 1998).

11.2.1 Martensite in Ferrous Alloys

As explained, martensite could be formed in ferrous alloys with different structures depend on chemical composition, stacking fault energy and processing conditions, e.g., deformation process. α′(bcc or bct) martensite

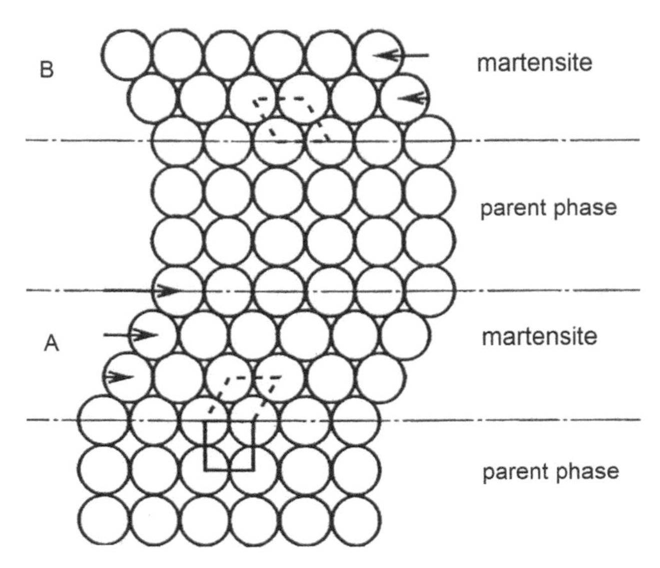

Figure 11.1 A simplified model of martensitic transformation (Otsuka and Wayman 1998).

is reported in Fe-C, Fe-Ni alloys, etc., while ε (hcp) martensite is formed in ferrous alloys with a low stacking fault energy of austenite such as Fe-Cr-Ni and Fe-high Mn alloys. Fct martensite has been found in Fe-pd and Fe-pt alloys (Koohdar et al. 2015, Maki 1990, Maki and Tamura 1986).

11.2.1.1 Morphology and Substructure of Ferrous Martensite

The morphology of ε martensite and fct martensite is a parallel-sided thin plate type with planar interfaces. On the other hand, in the case of α′ martensite, five types of morphologies have been observed so far. Figure 11.2 shows various morphologies of α′ martensite. Various types of α′ martensites shown in the figure are distinguished not only morphologically but also crystallographically. Table 11.1 summarizes the substructure, habit plane, and orientation relationship of these α′ martensites. Among these martensites, lath martensite forms at the highest temperatures and the thin plate martensite forms at the lowest temperatures. With a decrease in M_S temperature, the substructure of martensite changes from dislocated (lath martensite) to twinned (thin plate martensite). The factors that determine the morphology and substructure of martensite remain poorly defined. However, the M_S temperature, the relative strength of austenite and martensite, the critical resolved shear stress for slip and twinned in martensite, and the stacking fault energy of austenite are considered to be the important factors (Davies and Magee 1971, Krauss and Marder 1971, Maki et al. 1972, Thomas 1971, Umemoto et al. 1983).

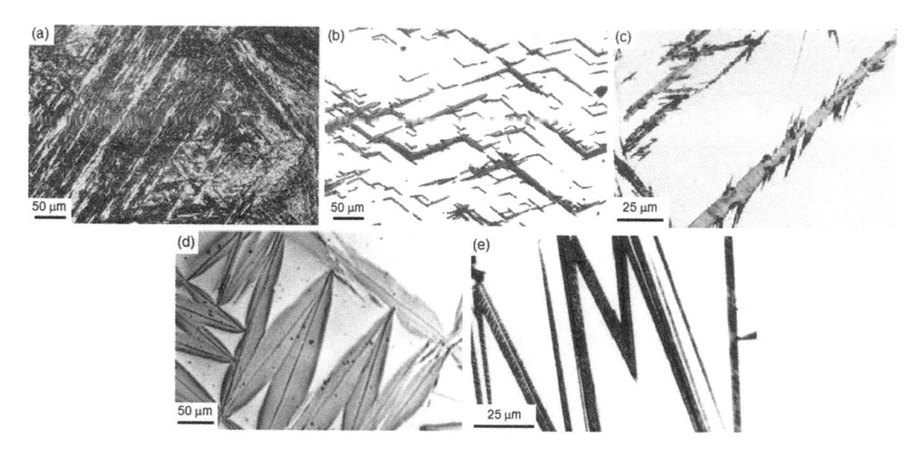

Figure 11.2 Optical micrographs of various types of α′ martensites in ferrous alloys: (a) lath, (b) butterfly, (c) (225)A type plate, (d) lenticular, and (e) thin plate α′ (Maki 1990).

Table 11.1 Morphology, substructure, habit plane (H.P.) and orientation relationship (O.R.) of five types of α′ martensite (Maki 1990).

	Morphology	Substructure	H.P.	O.R.	M_S	Alloy
1	Lath	Dislocations	$(111)_A$ or $(557)_A$	K-S	High	
2	Butterfly	Dislocations + Twins	$(525)_A$	K-S	↑	Fe-Ni-(C) Fe-Ni-Cr-(C)
3	$(225)_A$ type plate	Dislocations + Twins	$(525)_A$	K-S	↕	Fe-7Cr-1C Fe-5Mn-1C Fe-3Mn-3Cr-1C
4	Lenticular	Dislocations + Twins (Mid-rib)	$(259)_A$ or $(3\ 10\ 15)_A$	N G-T	↓	
5	Thin-plate	Twins	$(3\ 10\ 15)_A$	G-T	Low	Fe-high Ni-C Fe-7Al-2C Fe-25at%Pt Fe-Ni-Co-Ti

K-S: Kurdjumov-Sachs relationship, N: Nishiyama relationship
G-T: Grininger-Trolano relationship

Lath and lenticular are the two major types of α martensite (Krauss and Marder 1971, Maki and Tamura 1986). Lath martensite is formed in Fe-C (< 0.6 percentage of C) and Fe-Ni (< 28 percentage of Ni) alloys while lenticular martensite appears in Fe-high C (0.8–1.8 percentage of C) and Fe-high Ni (29–33 percentage of Ni) alloys. Ferrous lath martensite has overwhelming industrial significance because it appears in most of heat-treatable commercial steels.

11.3 Rolling as a Tool of Microstructure Control

Rolling is perhaps the most important metal working process, because a greater volume of material is worked by rolling than by any other deformation process. A significant portion (about ~90%) of steel, aluminum, and copper products may go through the rolling process at least one time during their production. The principal advantage of rolling lies in its ability to manufacture products from relatively large pieces of metals at very high speeds in a somewhat continuous manner (Dieter et al. 2003). The primary objectives of the rolling process are to reduce the cross section of the incoming material; to improve its properties; and to obtain the desired section at the exit from the rolls. The process can be carried out hot, warm, or cold; thus, rolling processes are often classified as hot or cold. From a fundamental point of view, however, it is more appropriate to classify rolling processes on the bases of the complexity of metal flow during the process and the geometry of the rolled product (Lahoti and Semiatin 1988, Schey 1983).

11.3.1 Types of Rolling

11.3.1.1 Hot Rolling

Hot rolling is a metal working process that is normally carried out above the re-crystallization temperature of the material. After the grains deform during processing, they re-crystallize, which maintains an equiaxed microstructure and prevents the metal from hardening. The starting material is usually a large semi-finished metal casting products such as slabs, blooms, and billets. If these products are introduced from a continuous casting operation, they are usually fed directly into the rolling mills at the proper temperature. In small scale operations, the material starts at room temperature and must be heated. During hot rolling, the temperature must be monitored to make sure it remains above the re-crystallization temperature (Degarmo et al. 2003).

Hot rolled metals generally have little directionality in their mechanical properties and deformation induced residual stresses. However, in certain instances non-metallic inclusions will impart some directionality and workpieces less than 20 mm thick often have some directional properties. Also, non-uniform cooling will induce a lot of residual stresses; which usually occur in shapes that have a non-uniform cross-section, such as I-beams. While the finished product is of good quality, the surface is covered in mill scale, which is an oxide that forms at high temperatures. It is usually removed by pickling or the smooth clean surface process, which reveals a smooth surface. Dimensional tolerances are usually 2 to 5% of the overall dimension (Degarmo et al. 2003).

11.3.1.2 Cold Rolling

Cold rolling is usually performed below the material's recrystallization temperature (usually at room temperature), which increases the strength by means of strain hardening. It also improves the surface finish and holds tighter tolerances. Commonly cold-rolled products include sheets, strips, bars, and rods. These products are usually smaller than when they are hot rolled. Because of the smaller size of the workpieces and their greater strength, as compared to hot rolled stock, four-high or cluster mills are used (Degarmo et al. 2003).

Cold-rolled sheets and strips come in various conditions: full-hard, half-hard, quarter-hard, and skin-rolled. Full-hard rolling reduces the thickness by 50%, while the others involve less of a reduction. Skin-rolling, also known as a skin-pass, involves the least reduction amount of 0.5–1%. It is used to produce a smooth surface, a uniform thickness, and reduce the yield point phenomenon (by preventing Luders bands from forming in later processing). It pins the dislocations at the surface and thereby reduces the possibility of the formation of Luders bands. To avoid the formation of Luders bands, it is necessary to create substantial density of un-pinned dislocations in ferrite matrix. Skin-rolled stock is usually used in subsequent cold-working processes where good ductility is required (Degarmo et al. 2003).

11.4 Cold Rolling of Martensitic Steels

Researches performed on the cold rolling of martensitic steels are mostly focused on the rolling of steels with an original microstructure of lath martensite. According to Figure 11.3, lath martensite has a three level hierarchy in its morphology, i.e., packet, block, and lath. The austenite grain is subdivided into several packets (the group of parallel laths with the same habit plane) and blocks (the group of laths of the same orientation) in martensitic transformation. The packet and the block boundaries are high-angle boundaries whereas lath boundaries are rotational dislocation boundaries with low misorientation angles (Marder and Krauss 1969, Marder and Marder 1969, Maki et al. 1980, Morito et al. 2003). Therefore, the lath martensite has a kind of fine structure in as-transformed state. Also, lath martensitic steels show excellent formability in as-solution annealed condition which makes them suitable for rolling process (Ghasemi-Nanesa et al. 2010, Nili-Ahmadabadi et al. 2011).

In the following section, the effects of cold rolling and subsequent annealing processes on the mechanical properties and microstructure of lath martensitic steels will be explored and explained.

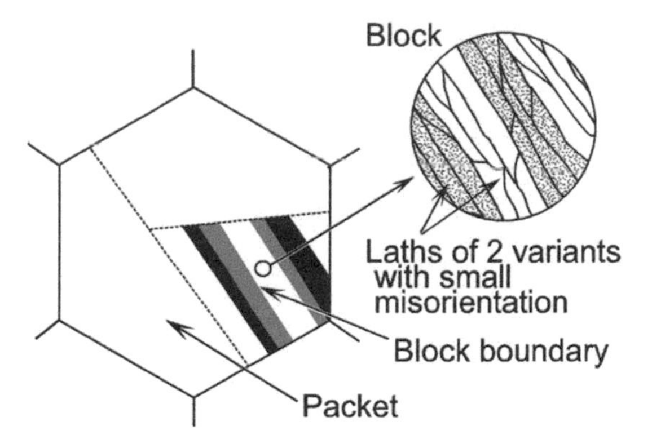

Figure 11.3 Schematic illustration showing the lath martensite structure in a low carbon steel (Morito et al. 2003).

11.4.1 Microstructural Evolution During Cold Rolling of Martensitic Steels

In practice, performing cold rolling on lath martensitic steels leads to a decrease in the average width of laths (Hossein Nedjad et al. 2008, Tianfu et al. 2006, Ueji et al. 2002, Umemoto et al. 2002, Zhao et al. 2001). Figures 11.4a and b illustrate the typical microstructure of lath martensite in a low-carbon steel (0.17 weight percentage of C) before and after cold rolling, respectively. It is clear that the average width of laths in solution-treated steel is close to 300 nm that reduces to about 20 nm after 93 percentage of the cold rolling. Figure 11.4b indicates that most martensite laths are thinned and damaged during cold rolling. Also, according to the corresponding diffraction pattern taken from the cold-rolled sample, although some diffraction spots and some arcs can still be seen in the pattern, the dominant diffraction ring pattern clearly indicates laths after rolling possess very small size and very large misorientation (Tianfu et al. 2006, Umemoto et al. 2002, Zhao et al. 2001).

Detailed microstructural evolution during cold rolling process of low carbon lath martensitic steels indicates that the structure of the cold-rolled steel could be classified into three kinds of microstructures, represented by the alphabetical characters of A, B, and C in Figure 11.5 (Tsuji et al. 2002, Ueji et al. 2002). These microstructures consist of (A) very fine lamellar structure mainly elongated parallel to the rolling direction (RD); (B) irregularly bent lamellar structure; and (C) lump of martensite laths with shear bands. The broken lines in the figure are the boundaries between the different structures. These three different microstructures (A, B, and C) could also be distinguished in TEM observations as shown in Figures 11.6a–c including selected area diffraction (SAD) patterns taken from the circled areas. Figure 11.6a corresponds to the microstructure A in Figure 11.5

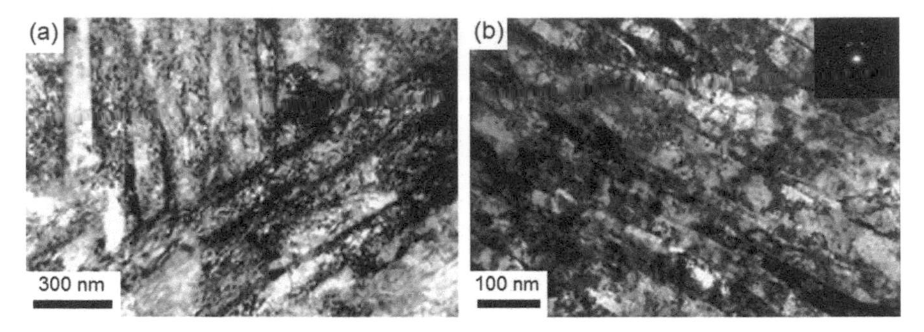

Figure 11.4 Bright field TEM images of the lath martensite in 0.17 wt% C steel (a) before; and (b) after 93% cold rolling (Tianfu et al. 2006).

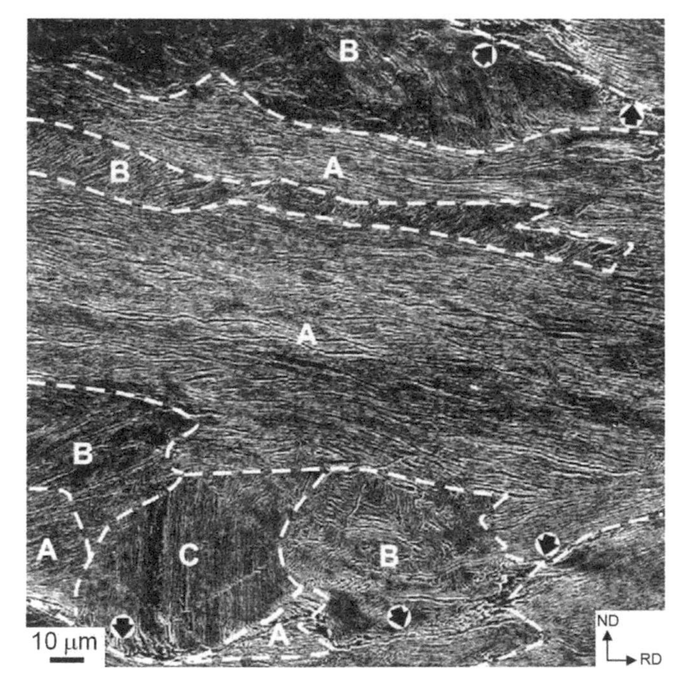

Figure 11.5 SEM image of the 0.13 wt% C martensitic steel 50% cold-rolled consisting of A (lamellar dislocation cells; LDC), B (irregularly bent laths; IBL) and C (kinked laths; KL) microstructures (Ueji et al. 2002).

which is composed of the ultrafine lamella dislocation cells (LDCs) with mean interval lamellar boundaries of only 60 nm. The SAD patterns of the LDC structure are arc-like, which suggests that a number of different orientations exist within the selected small areas (Tsuji et al. 2002, Ueji et al. 2002). Similar lamellar structures have also been observed in materials intensely strained by accumulative roll-bonding (ARB) (Huang et al. 2001, Huang et al. 2002, Ito et al. 2000, Tsuji et al. 1999, Tsuji et al. 2000a, 2000b).

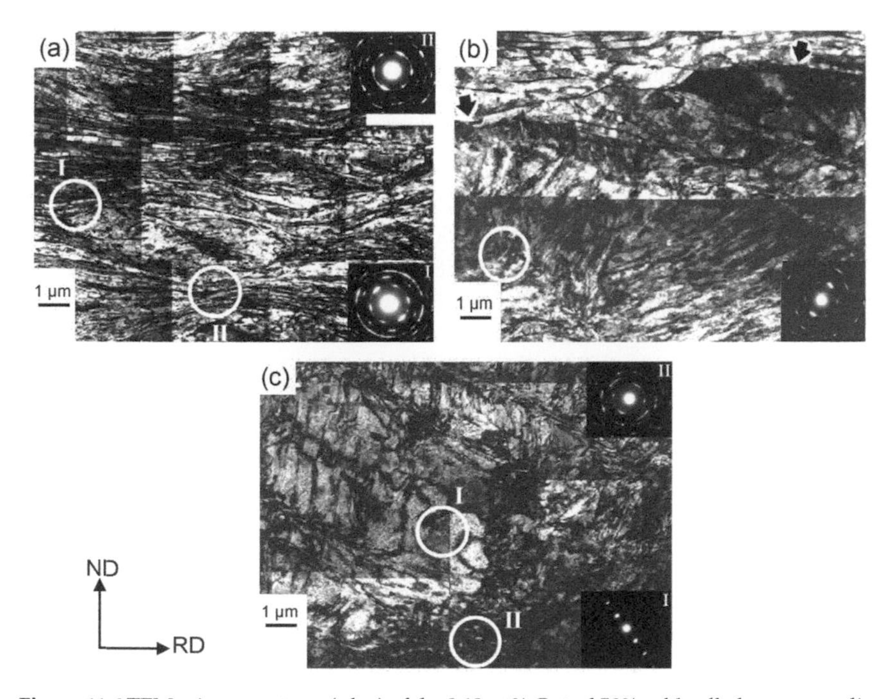

Figure 11.6 TEM microstructures (a,b,c) of the 0.13 wt% C steel 50% cold-rolled, corresponding to the microstructure A (lamellar dislocation cells; LDC), B (irregularly bent laths; IBL) and C (kinked laths; KL) in Figure 11.5, respectively. The starting microstructure was martensite (Tsuji et al. 2002, Ueji et al. 2002).

In Figure 11.6b, the narrow region just below the prior austenite grain boundary (indicated by the arrow) reveals the complicated morphology and darker contrast which suggests high dislocation density. These regions are severely and inhomogeneously deformed due to the constraint imposed by pre-existing grain boundaries. Also, the soft phase (proeutectoid ferrite) sometimes existing at prior austenite grain boundaries would also enhance such concentration of severe deformation. LDC structure is seen above the arrowed boundary in Figure 11.6b, while the lower region mainly showed the irregularly bent lamella (IBL) structure. In the IBL structure, a bunch of laths is bent to orient in various directions. The SAD patterns taken from the IBL structure in Figure 11.6b are ring-like, indicating that large local misorientations also exist in the IBL. Figure 11.6c is a TEM micrograph of the kinked lath (KL) structure. The KL is a martensite block kinked by micro shear bands. The directions of the laths showing the KL structure tend to be nearly parallel to the normal direction (ND) and the shear bands in KL structure are inclined at about 30° to RD. This suggests that the martensite blocks composed of the laths parallel to ND are hard to deform during rolling. The SAD pattern of the KL structure (I) in Figure 11.6c is nearly a single net pattern, reflecting that the KL maintains the martensite lath morphology. It is noteworthy, that the KL structure naturally has a zig-zag shaped outline (prior block boundary). The vicinity of the outline must suffer complicated plastic deformation to satisfy compatibility. Actually, the lower-right region in Figure 11.6c suggests the presence of a very high density of dislocations and the SAD pattern of the region (II) is accordingly ring-like (Tsuji et al. 2002, Ueji et al. 2002).

It is likely that during the cold rolling of the lath martensitic steels, at low to medium strains, lath martensite transforms into a cell block structure composed of cell block boundaries with a high dislocation density and short interconnecting cell boundaries (Huang et al. 2012). The driving force for the transition is suggested to be a decrease in energy per unit length of dislocation line possibly supplemented by a reduction in stored energy. At medium to high strains, cell block structures refine and a typical lamellar structure evolves with extended boundaries almost parallel to the rolling plane. Cold rolling of lath martensitic steels could cause an intrusion of slip bands into the matrix resulting in the destruction of lath martensitic structure and the formation of dislocation cell structure around slip bands (Takaki et al. 1991). The volume fraction of such a damaged martensite, increases with an increase in deformation. In the specimens subjected to heavy cold rolling with more than 80% reduction, the un-damaged lath martensitic structure could rarely be observed and dislocation density becomes one order of magnitude higher.

11.4.2 Effect of Prior Cold Rolling on the re-crystallization and Austenite Formation in Martensitic Steels

The re-crystallization of lath martensite is not observed in as-quenched martensite on tempering even at high temperatures just below A_1 and just above A_3. However, the deformed lath martensite easily re-crystallized on subsequent tempering (Takaki et al. 1991, Tokizane et al. 1982). It is generally accepted that despite the fact that the lath martensite contains a high density of dislocations (e.g., 10^{10}–10^{11}/cm^2) (Kehoe and Kelly 1970, Speich 1969), the re-crystallization is difficult to occur in as-quenched lath martensite compared with cold-rolled ferrite. This phenomenon has been considered to be attributed to the existence of precipitates of fine carbides which suppress the boundary migration and retard the re-crystallization (Caron and Krauss 1972, Lement et al. 1955, Maki and Tamura 1981). However, in addition to such a retarding effect of carbides, the difference in the arrangement of dislocations between martensite and deformed ferrite must be essentially related to the difficulty of re-crystallization of lath martensite. Namely, the deformation on martensitic transformation is uniform, resulting in the fairly uniform distribution of dislocations. On the other hand, the plastic deformation such as cold rolling is usually inhomogeneous, resulting in the formation of deformation bands and cell structure of dislocations. There might be a possibility that the driving force for the re-crystallization in lath martensite structure is somewhat lower than that in deformed ferrite because of such a difference in dislocation arrangement even in the same dislocation density. Moreover, the inhomogeneity of deformation by cold rolling provides effective nucleation sites of re-crystallization (Maki and Tamura 1981, Takaki et al. 1991, Tokizane et al. 1982).

During the annealing of cold rolled lath martensitic steel below A_1 temperature, there is a large difference in the recovery and re-crystallization behavior between the damaged and un-damaged areas (Takaki et al. 1991). Re-crystallized ferrite grains preferentially nucleate within the damaged area and grow to encroach into the un-damaged area where the recovery rate is slower than that in the damaged area. With increasing prior deformation corresponding to an increase in the volume fraction of the damaged martensite and in the density of dislocations, the recovery and re-crystallization of the matrix are markedly prompted.

Similar to re-crystallization, the behavior of austenite formation is changed by the deformation of lath martensite as schematically summarized in Figure 11.7 (Ameyama et al. 1988, Tokizane et al. 1982). In the case of non-deformed lath martensite, re-crystallization does not occur and austenite forms directly from the tempered lath martensite. Austenite particles

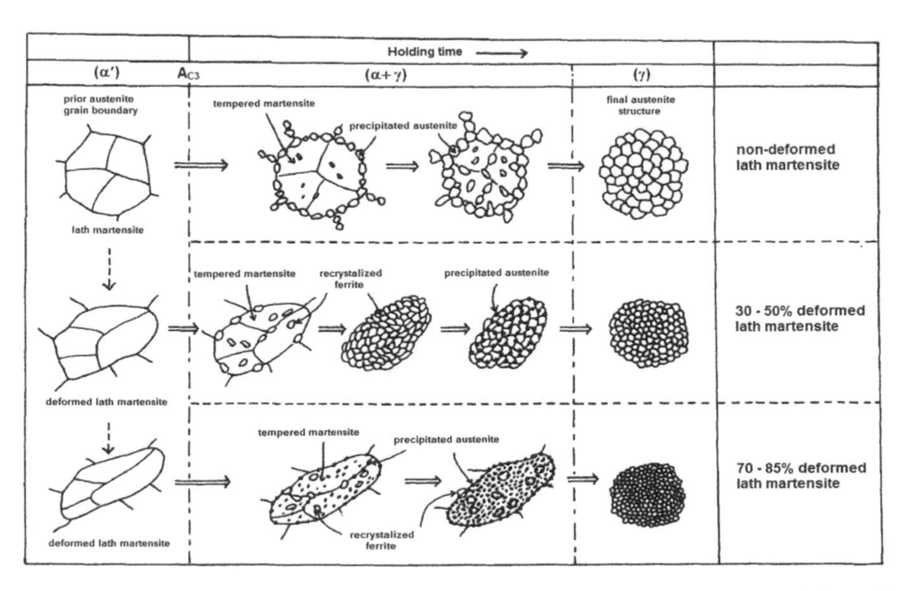

Figure 11.7 Schematic drawing showing the formation process of austenite from non-deformed or deformed lath martensite in 0.2 wt% C steels (Tokizane et al. 1982).

precipitate at first preferentially at prior austenite grain boundaries, and then form within the prior austenite grains mainly along the packet, block, and lath boundaries. Since the austenite particles (acicular type) formed along the parallel block or lath boundaries have almost the same orientation to each other, some of them coalesce to somewhat large massive grains (Kinoshita and Ueda 1974, Law and Edmonds 1980, Matsuda and Okamura 1974a,b, Watanabe and Kunitake 1976).

In deformed lath martensite, the occurrence of the re-crystallization and austenite formation is dependent on the heating rate. In low heating rates, the re-crystallization occurs prior to or simultaneously with the austenite formation. Therefore, the behavior of austenite formation in deformed specimens becomes different from that in non-deformed specimens. Although the re-crystallization is slightly accelerated with an increase in the amount of deformation, the formation rate of austenite from lath martensite is much more accelerated with an increase in the amount of deformation. Consequently, the austenitizing behavior (and thus the austenite grain size) is controlled by the competition between the re-crystallization of martensite and the austenite formation.

Tokizane et al. 1982 indicated that in the case of about 30 to 50% cold rolling, the re-crystallization of deformed lath martensite occurs first and is almost completed prior to the beginning of austenite formation. Thus, a high density of austenite particles is nucleated at the boundaries of re-crystallized fine ferrites. In this case, the austenite particles may have different orientations from each other, and the coalescence is hard to occur, then the final austenite grain size becomes fairly fine.

When the lath martensite is heavily cold-rolled by 70 to 85%, the nucleation of austenite starts almost simultaneously with the re-crystallization of deformed lath martensite at the very early stages of holding. Then, austenite grains are precipitated densely in the deformed (unre-crystallized) regions as well as at the grain boundaries of fine re-crystallized ferrites. Thus, the nucleation sites of austenite are markedly increased in the heavily deformed specimen, and consequently the final austenite grain becomes very fine (Tokizane et al. 1982). More finer austenite grains <5 μm, could be obtained in the presence of precipitates and reheating a cold-rolled tempered martensite in a Nb microalloyed steel (0.1 weight percentage of C) (Priestner and Ibraheem 2000).

In high heating rates, the re-crystallization of deformed lath martensite is suppressed since there is insufficient time for diffusional process and austenite formation is derived by diffusionless (displacive) mechanism (Apple and Krauss 1972, Koohdar et al. 2015).

11.4.3 Pressure-induced Austenite Formation during Cold Rolling of Martensitic Steels

During severe deformation of martensitic steels at room temperature, austenite can be formed from martensite by displacive mechanism that holds Kurdjumov-Sachs (K-S) orientation relationship with martensite (Allain et al. 2004, Ghasemi-Nanesa et al. 2010, Koohdar et al. 2014, Koohdar et al. 2016, Nili-Ahmadabadi et al. 2011). The reverse transformation of martensite to austenite is possible even at room temperature in a low carbon martensitic steel if the deformation energy stored in microstructure during cold rolling could provide the required driving force for reverse transformation (Ghasemi-Nanesa et al. 2014). It was shown that 60 percentage of the cold rolling provided the required driving force for reversed austenite formation in a Fe-Ni-Mn lath martensitic steel (Ghasemi-Nanesa et al. 2014). The martensitic steels with lath-like morphology have a high dislocation density and cold rolling process could introduce a non-uniform dislocation distribution in their microstructure (Kozlov et al. 1991, Nili-Ahmadabadi et al. 2011). Also, the localized highly strained part of the specimen during cold rolling could increase the internal friction which in turn leads to the formation of localized heat generation in the specimen. Locally increased temperature may trigger the reverse transformation of martensite to austenite. The same concept is reported during high-pressure torsion (HPT) process (Litovchenko et al. 2012). During SPD not only direct martensite transformation ($\gamma \rightarrow \alpha'$) but also reverse transformation ($\alpha' \rightarrow \gamma$) may occur. It is likely that the reverse transformation is attributed to the local increase in temperature and high hydrostatic pressure during the deformation process.

11.4.4 Severe Cold Rolling of Martensitic Steels and Resulted Free Volume

11.4.4.1 Severe Cold Rolling of Martensitic Steels

Microstructure of the lath martensitic steels could be affected by severe cold rolling methods such as repetitive corrugation and straightening by rolling (RCSR) technique which is one of the new methods of severe cold rolling (Mirsepasi et al. 2012). It has been reported that in the microstructure of Fe-10Ni-7Mn martensitic steel after 50 cycles of RCSR at room temperature, in addition to the compressed and bent laths, there are severely distorted regions between the laths as a consequence of strain accumulation (Mirsepasi et al. 2012). The microstructure of Fe-10Ni-7Mn martensitic steel during severe cold rolling (80 percent) and 50 cycles of RCSR with a final strain close to 4 has been shown in Figure 11.8. According to the figure, the ultrafine structure consists of lamellar dislocation cell blocks (Ghasemi-Nanesa et al. 2014). The bent contours are easily shown by arrows. Some

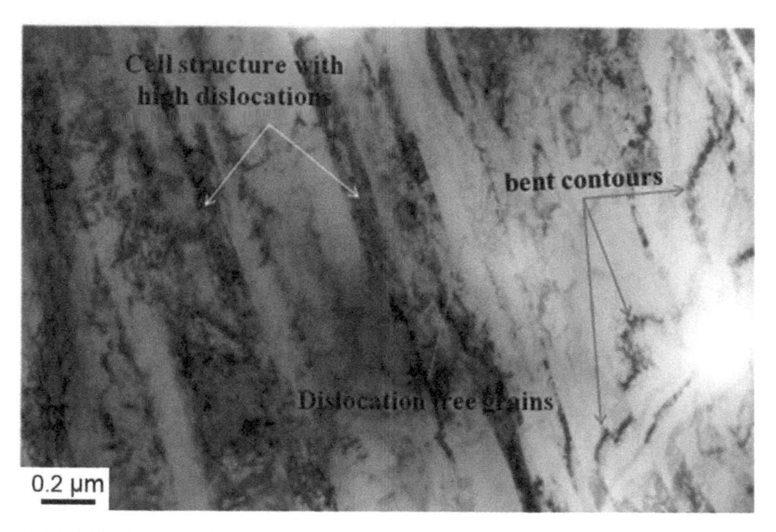

Figure 11.8 TEM micrograph, severe cold-rolled Fe-10Ni-7Mn martensitic steel, bent contours, cell structure with high dislocations; and dislocation free grains (Ghasemi-Nanesa et al. 2014).

of them are extended along the lath and some cross over the lath along the kink at angles close to 90°. The contours can bend more when the strain is increased and also they can be close to the sub-boundary or become self-closed (Kozlov et al. 1992). From the three possible types of grains and sub-grains commonly formed after SPD processes, elongated grains with high dislocation density and dislocation free grains are observed as shown in Figure 11.8. The cell structure with high dislocation density in Figure 11.8 has a bcc crystal structure and is related to the martensite and the dislocation free grains have fcc crystal structure, related to the reverted martensite to austenite during severe deformation (Ghasemi-Nanesa et al. 2014) as reported earlier.

11.4.4.2 Free Volume and Voids Formation

Plastic deformation introduces excess free volume to the severely cold-rolled materials in the accumulated high energy points of microstructure. The "*non-equilibrium*" grain boundaries (Grabski and Korski 1970, Pumphrey and Gleiter 1974) in the severely deformed poly-crystals are assumed to contain a higher excess free volume or non-equilibrium vacancies (Setman et al. 2008). The coalescence of excess free volume may lead to small and large vacancy clusters, nano-pores, and even submicron voids and cracks. Figure 11.9 presents an example of micro-voids with an average size of 100–150 nm and spherical morphology which mostly took place along the lath martensite severely deformed by RCSR (Mirsepasi et al. 2012).

The formation mechanism of micro-voids is explained in terms of disclination reactions, which represent triple junction defects in nanocrystalline materials (Nazarov et al. 1993). A triple junction disclination becomes unstable upon reaching a critical strain, giving rise to micro-void or crack formation. This cracking occurs at grain boundaries near triple junctions (Zhou et al. 2008).

The studies performed on severely cold-rolled Fe-10Ni-7Mn martensitic steel indicated that misfits and transformation induced contractions at the interfaces of martensite and reverted austenite could act as nucleation sites for the formation of voids (Ghasemi-Nanesa et al. 2014). HRTEM images of interface in severe cold-rolled Fe-10Ni-7Mn martensitic steel reveal a series of small voids at the scale of nanometer as shown in Figure 11.10a. It is supposed that one of the sources of nano-void and micro-void formation can be at the interface of austenite and martensite, because it is known that the reverse transformation of martensite to austenite is accompanied by crystal volume contraction. As shown in Figure 11.10b, this contraction is revealed by small detachment at the interface. The void nucleation at the

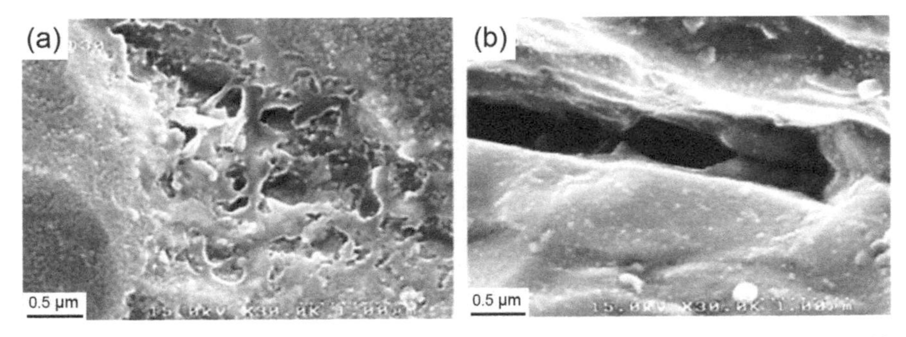

Figure 11.9 FESEM micrographs of a Fe-10Ni-7Mn steel after RCSR processing showing (a) the formation and evolution of the micro-voids in the microstructure; and (b) coalescence of at least two microvoids (Mirsepasi et al. 2012).

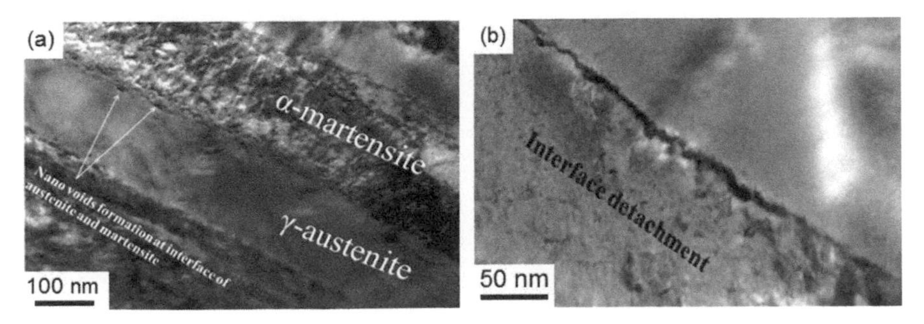

Figure 11.10 HRTEM micrographs of (a) nano-voids at the interface of austenite (γ) and martensite (α); and (b) interface detachment in severe cold-rolled Fe-10Ni-7Mn martensitic steel (Ghasemi-Nanesa et al. 2014).

interface can be attributed to the misfit strain at the interface of austenite and martensite and also to molar volume contraction between bcc and fcc crystals. Since the mentioned material is severely deformed, further strain results from the misfit and the molar volume contraction provides the required stress for void nucleation. With further deformation, these nano-voids are connected to each other to form sub-micron cracks and with further straining at the final stage, the micro-cracks start at the interface from sub-micron voids coalescence and easily propagate along the interface (Ghasemi-Nanesa et al. 2014).

11.4.5 Cold Rolling of Dual Phase Microstructure Composed of Ferrite and Martensite

It is important to note that starting structures for plastic deformation do not have to be uniform. During cold rolling of low-carbon steels with a mixed structure of ferrite and martensite phases, a larger strain is introduced in the soft phase (ferrite) (Okitsu et al. 2009). Figure 11.11a shows the TEM microstructure of the 91 percent cold-rolled specimen with a dual-phase microstructure composed of ferrite and martensite in a low-carbon steel. The "F" and "M" signs in the figure represent ferrite and martensite, respectively. In the ferrite region neighboring to the martensite phase, a fine lamellar structure with a mean spacing of 140 nm was observed. Figure 11.11b shows SAD pattern taken from the ferrite region illustrated by the dotted circle (b) in Figure 11.11a. The SAD pattern shows ring-like spots indicating that various orientations exist within the selected area (Okitsu et

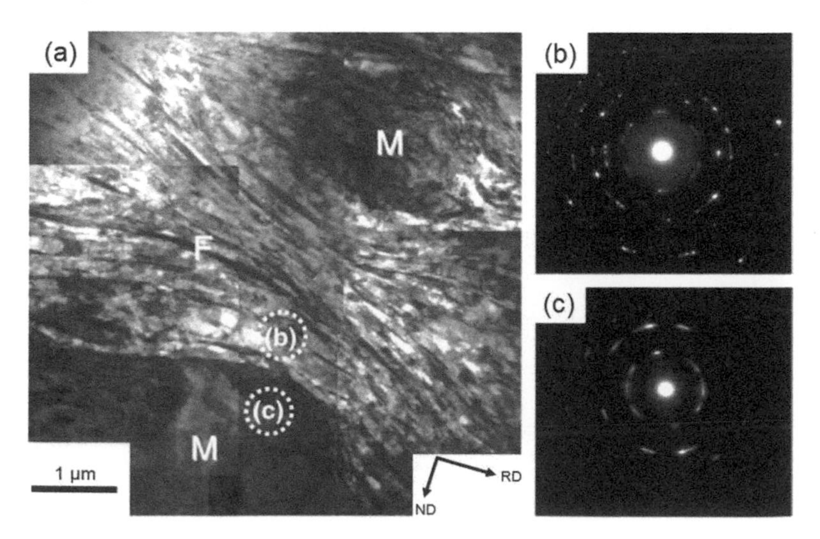

Figure 11.11 (a) TEM microstructure of the low carbon steel cold-rolled to 91% reduction. The starting microstructure is a duplex structure composed of ferrite and martensite. (b) and (c) SAD pattern taken from ferrite and martensite, respectively (Okitsu et al. 2009).

al. 2009). Similar lamellar structures with a mean spacing of about 300 nm in a commercial purity aluminum which is heavily cold-rolled to a strain of 5 have also been reported (Hansen and Jensen 1999). However, it should be emphasized that in the mentioned work the ultrafine lamellar structure of ferrite is obtained only after conventional 91 percentage of the rolling (equivalent strain of 2.4). This fact indicates that the hard martensite phase caused strain concentration in the ferrite matrix, resulting in the formation of the fine lamellar structure with large misorientations. Figure 11.11c shows a SAD pattern taken from the martensite region illustrated by the dotted circle (c) in Figure 11.11a. The pattern is again ring-like suggesting large local misorientations in the martensite regions.

11.4.6 Subsequent Annealing of Cold-Rolled Martensitic Steels

Subsequent annealing after cold rolling can lead to the formation of equiaxed ultrafine/nano-grain structure in martensitic steels. An equiaxed UFG structure with the mean grain size of 180 nm is reported after subsequent annealing of 0.13 weight percentage of carbon cold-rolled martensitic steel (Tsuji et al. 2002, Ueji et al. 2002). The final grain size is dictated by cold rolling severity, chemical composition and time and annealing temperature of martensitic steel. For example, TEM microstructures of the specimens 50 percent cold-rolled and annealed at various temperatures for 1.8 ks are depicted in Figure 11.12. The microstructure of the specimen annealed at 673 K, as shown in Figure 11.12a, is not so different from the as-deformed one (Figure 11.5). The LDC structure can be observed in this figure. In the specimen annealed at 773 K (Figure 11.12b), the equiaxed ultrafine ferrite grains with the mean diameter of 180 nm are formed in most of the areas. The majority of grains are ultrafine and equiaxed with clear high angle grain boundary, but some minor elongated grains are also observed. On the other hand, some of the regions do not show ultrafine grains like the upper-right area of Figure 11.12b. The blocky region without ultrafine grains corresponds to the KL structure. Ultrafine grains exist only along the line with the slope of about 45° to RD, which is a micro-shear band in the as-deformed state. The KL structure becomes tempered martensite by annealing. It can be concluded that the LDC and the IBL structures with large local misorientations turn into the ultrafine ferrite grains. In addition to the ultrafine ferrite grains, nano-carbides precipitate uniformly in Figure 11.12b because martensite (the starting microstructure) in 0.13 weight percentage of C steel is a supersaturated solid solution of carbon. The multi-phased nanostructure composed of ultrafine ferrite grains, nano-carbides, and tempered martensite blocks are also observed in the specimen annealed at 823 K (Figure 11.12c). Annealing at and above 873 K causes grain growth to result in coarse ferrite grains with spheroidized coarse carbides (Figure 11.12d).

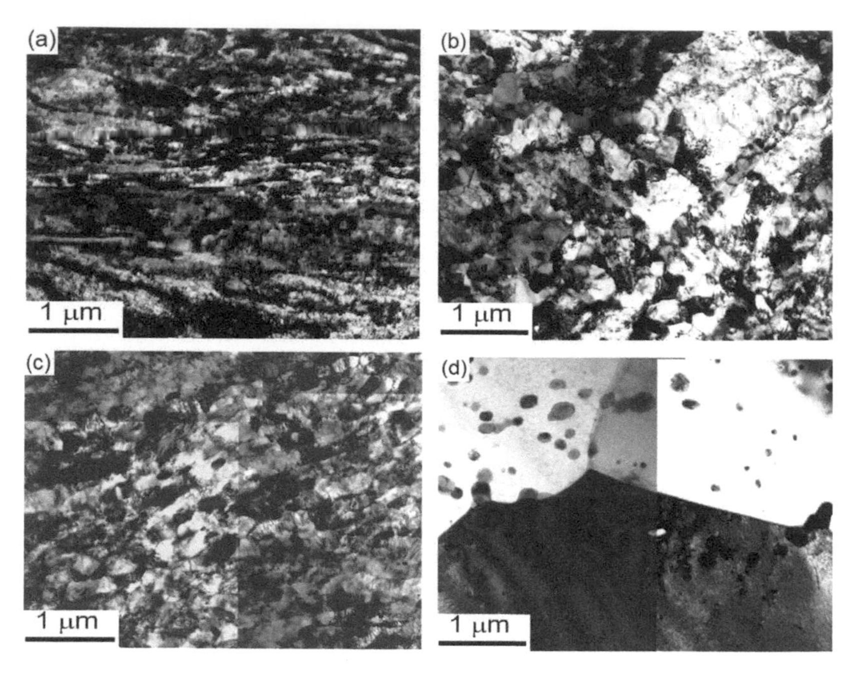

Figure 11.12 TEM microstructures of the 0.13 wt% C steel 50% cold-rolled and subsequently annealed at (a) 673 K, (b) 773 K, (c) 823 K; and (d) 873 K for 1.8 ks. Starting microstructure was martensite (Ueji et al. 2002).

Martensite grain size and rolling reduction or the value of induced strain are the key issues to form UFG structure by SPD as reported by different researcher (Hansen and Jensen 1999, Ito et al. 2000, Iwahashi et al. 1997, Saito et al. 1998, Tsuji et al. 2000a,b, Tsuji et al. 1999). In fact, martensite with fine-grained structure is the main reason for the grain refinement in a small strain (Tsuji et al. 2002). The fine grained structure causes constraint effect of plastic deformation to enhance inhomogeneous deformation (Tsuji et al. 2002, Ueji et al. 2002). High dislocation density and a number of solute carbon atoms in martensite are also expected to play an important role to improve grain subdivision (Tsuji et al. 2002, Ueji et al. 2002).

An ultrafine and nanocrystalline steel with an average grain size of 20 to 300 nm has been introduced by using severe cold rolling (93 percentage of the rolling) and subsequent annealing at different temperature of a low carbon lath martensitic steel (0.17 mass percentage of carbon) (Tianfu et al. 2006). The results of XRD and TEM observation and the corresponding annealing regime are given in Table 11.2. Table 11.2 indicates that after annealing below 623 K, regardless of the annealing time, the average crystal size is about 20 nm, however, as the annealing temperature is increased, crystal size increases drastically (Tianfu et al. 2006).

Figure 11.13 shows TEM images and the corresponding SAD pattern of a 93 percent cold-rolled steel annealed at 623 K for 28.8 ks. Equiaxed crystals and diffraction ring pattern clearly indicate the existence of very small crystals with random orientations. This means that a larger misorientation between laths is obtained when the lath martensite structure is subjected to SPD and subsequent annealing. It is possible that these low angle boundaries in the same packet rotated and evolved into a kind of fine non-equilibrium structure with curved and serrated boundaries, which have a large local misorientation caused by dislocation accumulation, interaction, tangling, and spatial rearrangement during heavy cold rolling (Tianfu et al. 2006, Zhao et al. 2005). As mentioned earlier, similar to other SPD methods such as equal channel angular press (ECAP), repetitive corrugation and straightening (RCS), high pressure torsion (HPT), multiaxial compressions

Table 11.2 Crystal sizes (nm) determined by XRD Scherrer method and TEM dark images along with the corresponding processing performed on 93% cold-rolled low carbon steel with lath martensite microstructure (Tianfu et al. 2006).

XRD				TEM average value of about 250 grains (nm)	Processes
011	200	211	Average (nm)		
20.0	11.9	13.9	17.4	–19	At room temperature
23.1	16.6	14.8	18.2	-	573 K × 3.6 ks
25.8	11.1	13.6	17.0	-	573 K × 7.2 ks
31.4	21.0	22.4	24.9	-	573 K × 28.8 ks
28.1	13.8	17.5	20.1	-	623 K × 3.6 ks
29.2	15.5	19.1	21.3	-	623 K × 7.2 ks
28.3	17.1	18.8	21.4	22.0 ± 9.1	623 K × 28.8 ks
72.3	34.5	43.7	50.2	52.0 ± 16.8	673 K × 3.6 ks
				316.0 ± 24.6	773 K × 3.6 ks
				~4700	873 K × 3.6 ks

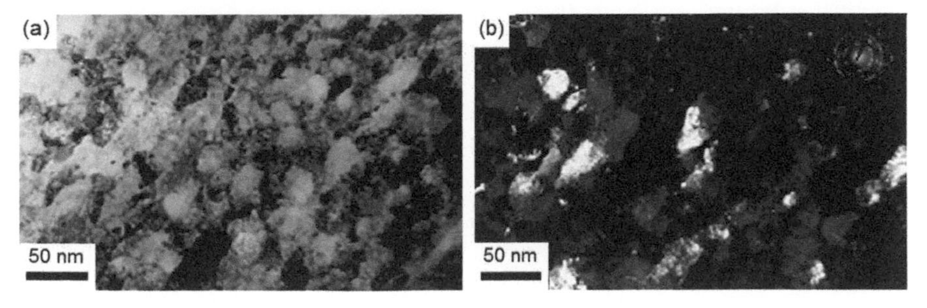

Figure 11.13 (a) Bright field; and (b) dark field TEM images of 93% cold-rolled steel annealed at 623 K for 28.8 ks, the inset shows the corresponding diffraction pattern (Tianfu et al. 2006).

and lattice rotation occurred during rolling deformation (Belyakov et al. 2002, Valiev et al. 2000).

11.4.7 Aging Treatment after Cold Rolling of Martensitic Steels

Mechanism of ultrafine grain (UFG) formation during cold rolling and isothermal aging treatment in Fe-10Ni-7Mn (weight percent) martensitic steel has been studied and the results are shown in Figure 11.14 (Hossein Nedjad et al. 2008). The figure shows bright-field transmission electron micrographs of 85 percent cold-rolled steel along with steels cold-rolled and aged for 0.36, 3.6 and 86.4 ks at 753 K. In the cold-rolled steel, as shown in Figure 11.14a, a typical deformed structure consisting of lamellar dislocation boundaries and shear bands is identified. After aging for 0.36 ks (Figure 11.14b), a microstructure consisting of UFGs and elongated cells is observed. Figure 11.14c shows microstructure of the steel aged for 3.6 ks after cold rolling. Two distinct regions are present; (i) a grain-refined region in the lower part of the micrograph and (ii) a lamellar region, in the upper part of the micrograph, composed of precipitation hardened laths and ferrite bands within which coarse precipitates have been embedded. It should be noted that θ-NiMn precipitates with face centered tetragonal (fct) crystal structure could be formed in Fe-Ni-Mn steel during aging heat treatment (Hossein

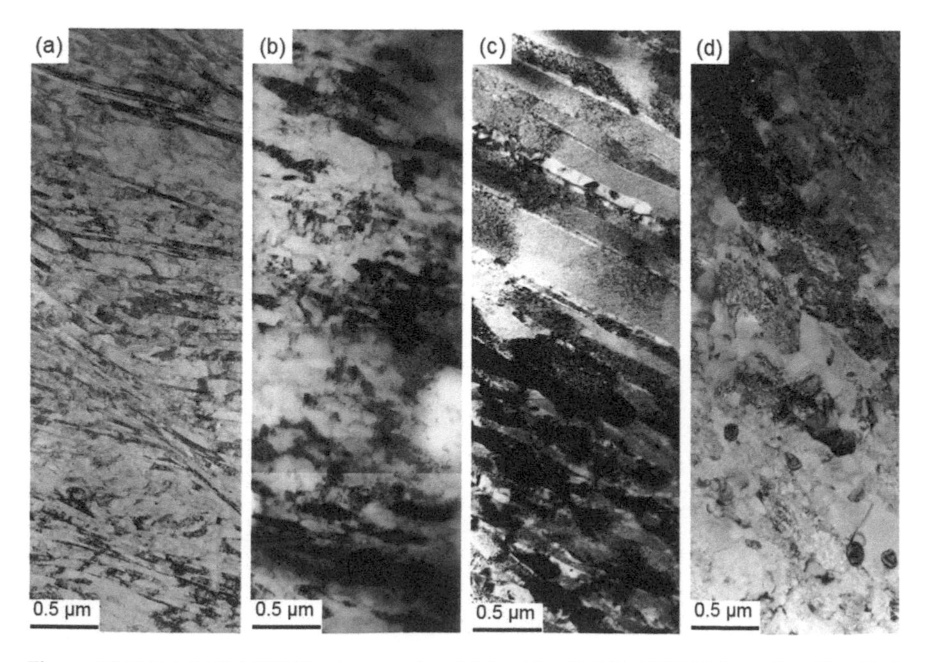

Figure 11.14 Bright field TEM micrographs of (a) cold-rolled Fe-10Ni-7Mn steel, cold-rolled steels aged for (b) 0.36 ks, (c) 3.6 ks; and (d) 86.4 ks (Hossein Nedjad et al. 2008).

Nedjad et al. 2006, Hossein Nedjad et al. 2009, Yodogawa 1976, Yodogawa and Tanaka 1978). After aging for 86.4 ks (Figure 11.14d), a multi-phase structure consisting of equiaxed ferrite grains with a size of about 100 nm, coarse precipitates and overaged laths has been observed.

For better understanding of the microstructural evolutions during isothermal aging of cold rolled steel, the sequence of UFG formation in the SPD nanomaterials is considered accordingly. It has been well established that during SPD, deformed structure undergoes repetitive grain sub-divisions along with dislocation cell rotations and consequently, a thoroughly refined structure with large fraction of high angle grain boundaries is realized at large strains. Further, recovery, re-crystallization, and grain growth occur successively during subsequent annealing of the severely deformed structures (Takaki et al. 1991, Tsuji et al. 2000b, Tokizane et al. 1982, Zhao et al. 2001). However, normal re-crystallization of deformed structures is suppressed in the presence of second-phase particles. Consequently, recovery process is not interrupted by the onset of re-crystallization and is to be extended at later stages of annealing, named as *"extended recovery"*. In fact, extended recovery continues by *virtue* of sub-grain growth which is correlated with coarsening of second-phase particles at sub-grain boundaries (Humphreys and Hatherly 1996). Extended recovery of severely deformed structures with large local misorientation has been coined *"continuous re-crystallization"* (Belyakov et al. 2001) or *"in situ crystallization"* (Tsuji et al. 2000b). Accordingly, a gradual evolution of the low angle boundaries into high angle boundaries during intense straining in SPD has been coined *"continuous dynamic re-crystallization"* which is proposed as the main mechanism of UFG formation in the SPD nanomaterials (Kaibyshev et al. 2005). Similarly, continuous re-crystallization has been assumed as a mechanism of UFG formation in the cold-rolled and warm annealed iron-carbon lath martensite (Tsuji et al. 2002). As mentioned earlier, the efficacy of cold rolling in producing severely deformed structures of large local misorientations is attributed to microstructural characteristics of lath martensite. In addition, the interrelation of aging and recovery should be taking into account. For example in the case of Fe-10Ni-7Mn martensitic steel recovery of deformed structure and precipitation of fct-NiMn intermetallic particles competitively proceeds during the initial stages of isothermal aging (Hossein Nedjad et al. 2008). Hence, it is likely to assume that unconstrained recovery in terms of dislocation annihilation and conversion of dislocation cells into sub-grains may occur at the initial stages of aging, before nucleation and growth of the precipitates. Then, all restoration processes would be hindered with effective particle pinning during increased aging times. Subsequently, the extension of recovery and re-crystallization should be correlated with coarsening of precipitates at later stages of aging. The qualitative increase of local misorientation, deduced from sharpening of grain contrast in bright field transmission electron micrographs (Figure 11.14) is attributed to preliminary recovery processes. However, the effect of recovery on the mean misorientation among sub-grains needs to be understood. It has been indicated that mean misorientation decreases upon sub-grain growth in the absence of orientation gradient (Humphreys 1992). In contrast, mean misorientation increases upon sub-grain growth in the presence of orientation gradient (Furu and Nes 1992). Huang et al. 1995 demonstrated that mean misorientation remains constant during sub-grain growth in a wide range. Nevertheless, they suggested that misorientation could increase during sub-grain growth, if the extent of sub-grain growth was less than the half wave length of a periodic orientation gradient. Therefore, it is proposed that a preliminary sub-grain growth in a long-range orientation gradient increase means misorientation at initial stages of aging in the mentioned martensitic steel, aimed at the enhanced illumination of ultrafine grains after short aging. Upon further aging, ferrite bands were found along lamellar dislocation boundaries within which coarse precipitates had been embedded. It is assumed that a re-crystallization assisted coarsening of precipitates occurs by lateral migration of lamellar dislocation boundaries. Consequently, the coarsening reaction is a "*discontinuous coarsening*" phenomenon which, in this case, combines driving forces of normal re-crystallization and precipitate coarsening (Williams and Butler 1981). Alternatively, it is proposed that discontinuous re-crystallization proceeds at later stages of aging by *virtue* of discontinuous coarsening reaction.

11.4.8 Mechanical Properties of the Martensitic Steels after Cold Rolling and Subsequent Annealing

Cold rolling and subsequent annealing processes can affect the mechanical properties of the martensitic steels. Engineering stress-strain curves of 0.13 weight percentage of C steel started from as-quenched martensite, cold-rolled by 50 percent, and then annealed at various temperatures for 1.8 ks are shown in Figure 11.15 (Tsuji et al. 2002, Tsuji et al. 2008). The as-rolled specimen shows very high strength over 1500 MPa, but it has limited uniform elongation similar to the SPD material. The flow stress decreases with an increase in annealing temperature. However, the specimens annealed at 773 K or 823 K, which show multi-phased UFG structures of ferrite and cementite like Figure 11.12, produce obvious strain-hardening after macroscopic yielding (Tsuji et al. 2002, Ueji et al. 2002). As a result, these specimens illustrate adequate uniform elongation as well as high strength. Especially, the 823 K annealed specimen shows a (0.2 precent) proof stress of 710 MPa, tensile strength of 870 MPa, uniform elongation of 8 percent and total elongation of 20 percent. Since the starting steel is a 400 MPa class

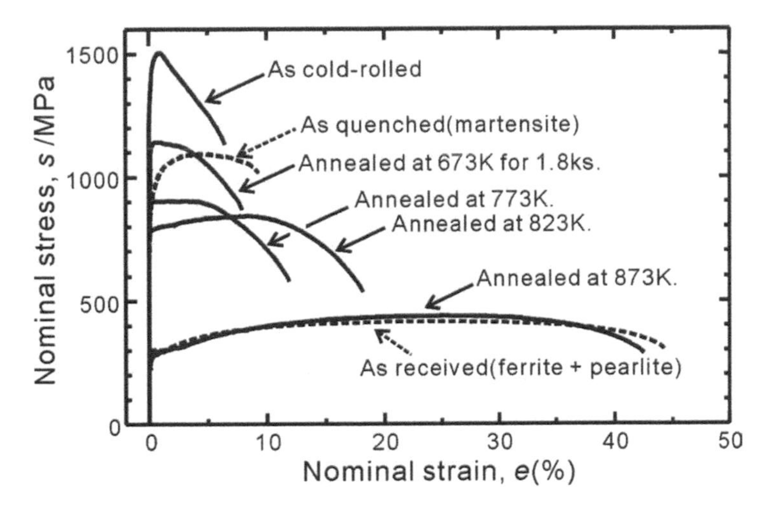

Figure 11.15 Engineering stress-strain curves of the 0.13 wt% C steel started from as-quenched martensite, cold-rolled by 50% and then annealed at various temperatures for 1.8 ks (Tsuji et al. 2002, Tsuji et al. 2008).

steel, the multi-phased UFG specimen obtained has strength more than two times higher than that of the starting material with ferrite-pearlite structure (Tsuji et al. 2002, Tsuji et al. 2008, Ueji et al. 2002).

The rolling reduction ranging from 25 to 70 percent (equivalent strains of 0.3–1.5) of plain low-carbon martensitic steel shows that in the as-deformed specimen, the strength increases with an increase in the rolling reduction (Ueji et al. 2004). Tensile test exhibits that the specimen rolled to the intermediate reduction (50 percent) performs the best strength-ductility balance after annealing at intermediate temperature. The reason for the good strength-ductility balance of the specimen rolled to the intermediate reduction is related to the obtained proper microstructure. It has been reported that at higher temperatures, conventional re-crystallization takes place, and the re-crystallization temperature become lower with increase in the reduction since the stored energy (the driving force for re-crystallization) becomes larger (Hossein Nedjad et al. 2008). Because of the enhanced re-crystallization, the specimen cold-rolled to higher reduction (70 percent) cannot keep high strength until adequate ductility is realized. On the other hand, the 25 percentage of the cold rolling is not enough to obtain significant high strength.

Tensile stress-strain curve of 0.17 weight percentage of C martensitic steel after cold rolling and subsequent annealing at different temperatures is shown in Figure 11.16. It is clear that the strength of martensitic steel subjected to cold rolling is about 2110 MPa. After annealing, the typical feature of the tensile curves of the samples annealed below 673 K is entirely elastic, Curves 1–6. The plasticity increases with annealing temperature

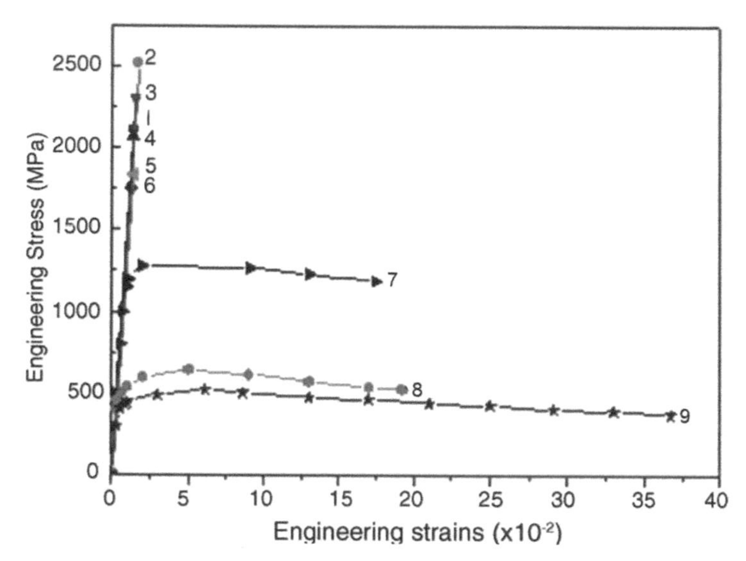

Figure 11.16 Stress-strain curves of the 0.17 wt% C martensitic steel in (1) Cold-rolled 93% without annealing, (2–8) cold-rolled 93% and then respectively annealed at 373, 473, 573, 623, 673, 773, 873 K; and (9) as-received conditions (Tianfu et al. 2006).

above 773 K, Curves 7 and 8. The Curves 1–6, the corresponding crystal size ranges from 18 to 50 nm (Table 11.2), demonstrate pure elastic deformation. It is reported that the entirely elastic behavior of the samples cold-rolled and annealed below 673 K may be caused by the integral effect of very heavy work-hardening and intensive strain-aging which occurred during severe rolling and subsequent low temperature annealing processes accompanied by the recovery and low temperature re-crystallization (Tianfu et al. 2006, Zhao et al. 2005).

The detailed mechanism for the pure elastic deformation phenomenon is so far not very clear. It is possible that the interaction intensity between solute atoms and dislocations, grain boundaries and dislocations, and dislocations and dislocation forest are so strong that the dislocations cannot move at all during tensile straining. In contrast, as shown in Figure 11.16, the samples annealed at 773 and 873 K show considerable plasticity. This means that the de-pinning and glide of dislocations occur during the tensile deformation and the increase in plasticity with corresponding decrease in strength from the conventional re-crystallization (Tianfu et al. 2006, Zhao et al. 2005).

Figure 11.17 shows the SEM images of the tensile specimen fracture surface in 0.17 weight percentage of C martensitic steel after cold rolling and subsequent annealing at different temperatures (Zhao et al. 2005). The phenomenon of fracture delamination is observed from the specimens, which were cold-rolled and annealed up to 773 K. The delamination plane

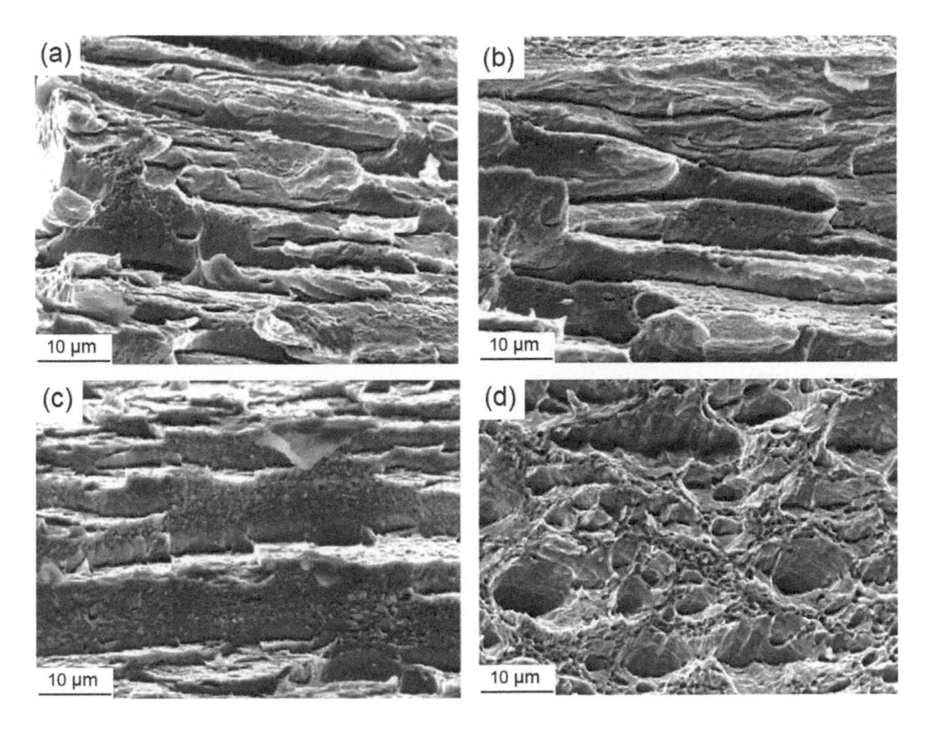

Figure 11.17 SEM images of tensile fracture surface of (a) cold-rolled martensitic steel, and subsequent annealing at (b) 673 K, (c) 773 K; and (d) 873 K for 3.6 ks (Zhao et al. 2005).

is parallel to the rolling plane. The phenomenon of fracture delamination disappears in the specimens annealed at 873 K as shown in Figure 11.17d. It is the heavy cold rolling process that leads to the nano-layered structure and makes the {110} planes parallel to the rolling plane at the same time. Researches indicate that the potential {100} cleavage planes form the delaminating crack during the necking stage of the tensile test. The delamination-toughening effect is beneficial to the cryogenic toughness of steel plate (Cai et al. 2001). Thus, fracture delamination is an expected phenomenon in the mentioned thermo-mechanical treatment. Zhao et al. 2005 reported that during the annealing treatments, the {100} texture is strong enough up to 773 K. So, the weak interfaces of potential {100} cleavage planes are stable up to 773 K. The phenomenon of fracture delamination has also been realized in 90 percent cold-rolled Fe-10Ni-7Mn martensitic steel (Ghasemi-Nanesa et al. 2012).

Mirsepasi et al. 2012 reported that the RCSR process increases the ultimate tensile strength value of solution treated Fe-10Ni-7Mn martensitic steel from 830 MPa to 910 MPa. Higher strength of RCSR steel sheets is attributed to the effect of SPD which causes high density of dislocations leading to high density shear bands and very fine grains. According to their study, the solution treated steel shows an elongation of 14 percent, due to the good strain hardening capacity of the starting material that decreases to 8.6 percent after RCSR. This reduction may be a result of dislocation generation and accumulation in the shear bands, grain boundaries, and the enhancement of strain hardening in the micro-voids during the deformation process.

The fracture surface of the solution treated exhibits features indicating ductile fracture through dimple nucleation (Figure 11.18a). On the other hand, fracture surface of the RCSR specimen shows both dimples and cleavage planes. The fracture mechanism is transformation from ductile to brittle (Figure 11.18b). High energy and distorted microstructure of the steel after RCSR is observed resulting in reduction of the dimples (Mirsepasi et al. 2012).

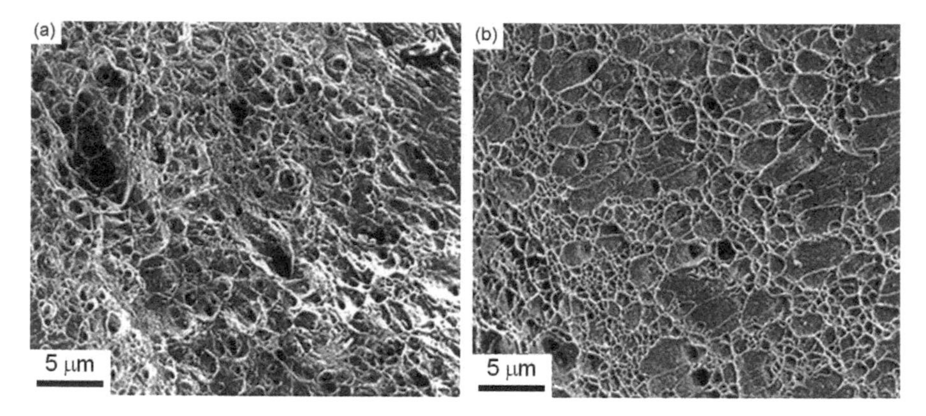

Figure 11.18 SEM images of fracture surface of (a) solution treated; and (b) RCSR Fe-10Ni-7Mn specimens (Mirsepasi et al. 2012).

11.5 Summary

In this chapter, the effects of the cold rolling practice and subsequent annealing on the microstructure and mechanical properties of lath martensitic steels as an important group of the AHSSs have been studied. The main points of the conclusion are as follow.

Researches performed on the cold rolling practice of the martensitic steels are mostly focused on the rolling of steels with an original microstructure of lath martensite due to their fine structure and excellent formability.

The sequence of microstructural evolution during cold rolling of lath martensite encompasses vanishing of primary lath boundaries, formation of slip bands, microbands and shears bands and elongated dislocation cell blocks. Furthermore, monotonous increase of dislocation density, increased ratio of screw:edge type dislocations, and apparent decrease in grain size have been found by the analysis of the deformed lath martensite.

Subsequent annealing after cold rolling of lath martensite can produce an ultrafine/nano-grained structure with large misorientations in low carbon steels. The main reason for the quick grain subdivision is that the lath martensite starting structure is itself a fine grained structure. The fine-grained structure causes the constraint effect of plastic deformation to enhance inhomogeneous deformation, in other words, grain subdivision. The formation of fine-grained structure leads to adequate gain in uniform elongation as well as high tensile strength.

Acknowledgment

The authors gratefully acknowledge Dr. Seyamak Hossein Nedjad at Sahand University of Technology and Dr. Hassan Shirazi at University of Tehran for reviewing and giving their valuable comments.

Keywords: Martensitic steels, cold rolling, lath martensite, dislocation, misorientation, cell structure, subsequent annealing, austenite formation, ultrafine-grained structure, re-crystalization, mechanical properties

References

Allain, S., J.P. Chateau, O. Bouaziz, S. Migot and N. Guelton. 2004. Correlations between the calculated stacking fault energy and the plasticity mechanisms in Fe-Mn-C alloys. Mater. Sci. Eng. A 387-389: 158–162.

Ameyama, K., N. Matsumura and M. Tokizane. 1988. Ultrafine austenite grains obtained by thermomechanical processing in low and medium carbon steels. J. Jpn. Soc. Heat Treat. 28: 233–240.

Apple, C.A. and G. Krauss. 1972. The effect of heating rate on the martensite to austenite transformation in Fe-Ni-C alloys. Acta Metall. 20: 849–856.

Belyakov, A., T. Sakai, H. Mina, R. Kaibyshew and K. Tsuzaki. 2002. Continuous re-crystallization in austenitic stainless steel after large strain deformation. Acta Mater. 50: 1547–1557.

Belyakov, A., T. Sakai, H. Miura and K. Tsuzaki. 2001. Grain refinement in copper under large strain deformation. Phil. Mag. A 81: 2629–2643.

Cai, D., C. Zhang, Y. Gao, Y. Hu and T. Jing. 2001. Study on the *in-situ* composite effect induced by plastic deformation of lath martensite in CrMoV steel. Trans. Mater. Heat Treat. 22: 43–47.

Caron, R.N. and G. Krauss. 1972. The tempering of Fe-C lath martensite. Metall. Trans. 3: 2381–2389.

Davies, R.G. and C.L. Magee. 1971. Influence of austenite and martensite strength on martensite morphology. Met. Trans. 2: 1939–1947.

Degarmo, E.P., J.T. Black and R.A. Kohser. 2003. Materials and processes in manufacturing. Wiley, USA.

Dieter, G.E., H.A. Kuhn and S.L. Semiatin. 2003. Handbook of Workability and Process Design. ASM International, pp. 232–242.

Furu, T. and E. Nes. 1992. Recrystallization'92. pp. 311. *In*: M. Fuentes and G. Sevillano (eds.). Trans Tech Publications. San Sebastian, Spain.

Ghasemi-Nanesa, H., M. Nili-Ahmadabadi and H. Shirazi. 2010. Mechanical properties of Fe-10Ni-7Mn martensitic steel subjected to severe plastic deformation via cold rolling and wire drawing. J. Phys.: Conference Series. 240: 1–4.

Ghasemi-Nanesa, H., M. Nili-Ahmadabadi, H. Shirazi, S. Hossein Nedjad and S.H. Pishbin. 2010. Ductility enhancement in ultrafine-grained Fe-Ni-Mn martensitic steel by stress-induced reverse transformation. Mater. Sci. Eng. A 527: 7552–7556.

Ghasemi-Nanesa, H., M. Nili-Ahmadabadi, H.R. Koohdar, M. Habibi-Parsa, S. Hossein Nedjad, S.A. Alidokht and T.G. Langdon. 2014. Strain-induced martensite to austenite reverse transformation in an ultrafine-grained Fe–Ni–Mn martensitic steel. Phil. Mag. 94: 1493–1507.

Ghasemi-Nanesa, H., M. Nili-Ahmadabadi, A. Mirsepasi and C. Zamani. 2014. Nano and micro-void formation in ultrafine-grained martensitic Fe-Ni-Mn steel after severe cold rolling. Met. Mater. Int. 20: 201–205.

Ghasemi-Nanesa, H., M. Nili-Ahmadabadi, H. Shirazi and S. Hossein Nedjad. 2012. Observation of reverse transformation of α-ε-γ in ultrafine-grained Fe-Ni-Mn age hardenable martensitic steel. Int. J. Mod. Phys. 5: 9–17.

Grabski, M.W. and R. Korski. 1970. Grain boundaries as sinks for dislocations. Philos. Mag. 22: 707–715.

Hansen, N. and D.J. Jensen. 1999. Development of microstructure in FCC metals during cold work. Phil. Trans. R. Soc. Lond. A 357: 1447–1469.

Hossein Nedjad, S. and M.R. Movaghar Gharabagh. 2008. Dislocation structure and crystallite size distribution in lath martensite determined by X-ray diffraction peak profile analysis. Int. J. Mater. Res. 99: 1248–1255.

Hossein Nedjad, S., M. Nili Ahmadabadi and T. Furuhara. 2008. The extent and mechanism of nanostructure formation during cold rolling and aging of lath martensite in alloy steel. Mater. Sci. Eng. A 485: 544–549.

Hossein Nedjad, S., M. Nili-Ahmadabadi, R. Mahmudi, T. Furuhara and T. Maki. 2006. Analytical transmission electron microscopy study of grain boundary precipitates in an Fe–Ni–Mn maraging alloy. Mate. Sci. Eng. A 438: 288–291.

Hossein Nedjad, S., M. Nili-Ahmadabadi and T. Furuhara. 2009. Annealing behavior of an ultrafine-grained Fe-Ni-Mn steel during isothermal aging. Mater. Sci. Eng. A 503: 156–159.

Hossein Nedjad, S., J. Teimouri, A. Tahmasebifar, H. Shirazi and M. Nili-Ahmadabadi. 2009. A new concept in further alloying of Fe-Ni-Mn maraging steels. Scripta Mater. 60: 528–531.

Huang, X., N. Tsuji, Y. Minamino and N. Hansen. 2001. Characterisation of ultrafine microstructures in aluminium heavily deformed by accumulative roll-bonding (ARB). Proc. of the 22nd Risø Int. Symp. On Mater. Sci. Denmark, pp. 255–262.

Huang, X., N. Tsuji, N. Hansen and Y. Minamino. 2002. Microtexture of lamellar structures in Al heavily deformed by accumulative roll-bonding (ARB). Mater. Sci. Forum. 408: 715–720.

Huang, X., S. Morito, N. Hansen and T. Maki. 2012. Ultrafine structure and high strength in cold-rolled martensite. Metal. Mater. Trans. A 43: 3517–3531.

Huang, X., K. Tsuzaki and T. Maki. 1995. Subgrain growth and misorientation of the α matrix in an (α+ γ) microduplex stainless steel. Acta Metall. Mater. 43: 3375–3384.

Humphreys, F.J. 1992. A network model for recovery and recrystallisation. Scripta Mater. 27: 1557–1562.

Humphreys, F.J. and M. Hatherly. 1996. Re-crystallization and related annealing phenomena. Pergamon Press, Great Britain.

Ito, Y., N. Tsuji, Y. Saito, H. Utsunomiya and T. Sakai. 2000. Change in microstructure and mechanical properties of ultrafine-grained aluminum during annealing. J. Jpn. Inst. Met. 64: 429–437.

Iwahashi, Y., Z. Horita, M. Nemoto and T.G. Langdon. 1997. An investigation of microstructural evolution during equal-channel angular pressing. Acta Mater. 45: 4733–4741.

Kaibyshev, R., K. Shipilova, F. Musin and Y. Motohashi. 2005. Continuous dynamic recrystallization in an Al–Li–Mg–Sc alloy during equal-channel angular extrusion. Mater. Sci. Eng. A 396: 341–351.

Keeler, S. and M. Kimchi. 2014. Advanced high strength steels-application guidelines version 5.0. WordAutoSteel, Middletown, Ohio.

Kehoe, M. and P.M. Kelly. 1970. The role of carbon in the strength of ferrous martensite. Scripta Met. 4: 473–476.

Kinoshita, S. and R. Ueda. 1974. Some observations on formation of austenite grains. Trans. ISIJ 14: 411–418.

Koohdar, H.R., M. Nili-Ahmadabadi, M. Habibi-Parsa and H. Ghasemi-Nanesa. 2014. Investigating on the reverse transformation of martensite to austenite and pseudoelastic behavior in ultrafine-Grained Fe-10Ni-7Mn (wt%) steel processed by heavy cold rolling. Adv. Mater. Res. 829: 25–29.

Koohdar, H.R., M. Nili-Ahmadabadi, M. Habibi-Parsa and H.R. Jafarian. 2015. Development of pseudoelasticity in Fe-10Ni-7Mn (wt.%) high strength martensitic steel by intercritical heat treatment and subsequent ageing. Mater. Sci. Eng. A 621: 52–60.

Koohdar, H.R., M. Nili-Ahmadabadi, M. Habibi-Parsa, H.R. Jafarian, H. Ghasemi-Nanesa and H. Shirazi. 2016. Observation of pseudoelasticity in a cold rolled Fe-Ni-Mn martensitic steel. Mater. Sci. Eng. A 658: 86–90.

Kozlov, E.V., N.A. Popova, N.A. Grigor'eva, L. No Ignatenko, T.A. Kovalevskaya, L.A. Teplyakova and B.D. Chukhin. 1991. Plastic deformation stage substructure evolution and slip picture in alloys with disperse hardening. Izvest. Vys. Uch. Zav. Fiz. 34: 112–128.

Kozlov, E.V., L.A. Teplyakova, N.A. Popova, Y.F. Ivanov, D.V. Lychagin, L.N. Ignatenko and N.A. Koneva. 1992. Band structure and packet-martensite structure. Comparison of evolution paths. Russ. Phys. J. 35: 906–911.

Krauss, G. and A.R. Marder. 1971. The morphology of martensite in iron alloys. Met. Trans. 2: 2343–2357.

Lahoti, G.D. and S.L. Semiatin.1988. Flat, bar, and shape rolling, forming and forging. ASM Handbook. 14: 343–360.

Lement, B.S., B.L. Averbach and M. Cohen. 1955. Further study of microstructural changes on tempering iron-carbon alloys. Trans. ASM. 47: 291–319.

Law, N.C. and D.V. Edmonds. 1980. The formation of austenite in a low-alloy steel. Metall. Trans. A 11: 33–46.

Litovchenko, L.Y., A.N. Tyumentsev, M.I. Zahozheva and A.V. Korznikov. 2012. Direct and reverse martensitic transformation and formation of nanostructured states during severe plastic deformation of metastable austenitic stainless steel. Rev. Adv. Mater. Sci. 31: 47–53.

Maki, T. 1990. Microstructure and mechanical behavior of ferrous martensite. Mater. Sci. Forum. 56-58: 157–168.

Maki, T., S. Shimooka, M. Umemoto and I. Tamura. 1972. The morphology of strain-induced martensite and thermally transformed martensite in Fe–Ni–C alloys. Trans. JIM. 13: 400–407.

Maki, T., K. Tsuzaki and I. Tamura. 1980. The morphology of microstructure composed of lath martensites in steels. Trans. ISIJ. 20: 207–214.

Maki, T. and I. Tamura. 1986. Shape memory effect in ferrous alloys. Proc. ICOMAT-86. Nara, pp. 963–970.

Maki, T. and I. Tamura. 1981. Morphology and substructure of lath martensite in steels. Trans. ISIJ. 67: 852–866.

Marder, A.R. and J.M. Marder. 1969. The morphology of iron-nickel massive martensite (Structural features of Fe-Ni massive martensite observed by light, electron and hot stage microscopies showing parallel block packet. ASM Trans. Quart. 62: 1–10.

Marder, A.R. and G. Krauss. 1969. Formation of low-carbon martensite in Fe-C alloys. ASM Trans. Quart. 62: 957–964.

Matsuda, S. and Y. Okamura. 1974a. Microstructural and Kinetic Studies of Reverse Transformation in a Low-C Low-Alloy Steel. Trans. ISIJ. 14: 363–368.

Matsuda, S. and Y. Okamura. 1974b. The Later Stage of Reverse Transformation in Low-C Low-Alloy Steel. Trans. ISIJ. 14: 444–449.

Mirsepasi, A., M. Nili-Ahmadabadi, M. Habibi-Parsa, H. Ghasemi-Nanesa and A.F. Dizaji. 2012. Microstructure and mechanical behavior of martensitic steel severely deformed by the novel technique of repetitive corrugation and straightening by rolling. Mater. Sci. Eng. A 551: 32–39.

Morito, S., S. Iwamoto and T. Maki. 2003. International forum for the properties and applications of IF steels. pp. 365–368. *In*: H. Takeuchi (ed.). The Iron Steel Institute of Japan. Tokyo.

Morito, S., H. Tanaka, R. Konishi, T. Furuhara and T. Maki. 2003. The morphology and crystallography of lath martensite in Fe-C alloys. Acta Mater. 51: 1789–1799.

Movaghar Garabagh, M.R., S. Hossein Nedjad and M. Nili Ahmadabadi. 2008. X-ray diffraction study on a nanostructured 18Ni maraging steel prepared by equal-channel angular pressing. J. Mater. Sci. 43: 6840–6847.

Nazarov, A., A.E. Romanov and R.Z. Valiev. 1993. On the structure, stress fields and energy of nonequilibrium grain boundaries. Acta Metall. Mater. 41: 1033–1040.

Nili-Ahmadabadi, M., H. Shirazi, H. Ghasemi-Nanesa, S. Hossein Nedjad, B. Poorganji and T. Furuhara. 2011. Role of severe plastic deformation on the formation of nanograins and nano-sized precipitates in Fe-Ni-Mn steel. Mater. Des. 32: 3526–3531.

Okitsu, Y., N. Takata and N. Tsuji. 2009. A new route to fabricate ultrafine-grained structures in carbon steels without severe plastic deformation. Scripta Mater. 60: 76–79.

Otsuka, K. and C.M. Wayman. 1998. Shape memory materials. Cambridge university press.

Priestner, R. and A.K. Ibraheem. 2000. Processing of steel for ultrafine ferrite grain structures. Mate. Sci. Technol. 16: 1267–1272.

Pumphrey, P.H. and H. Gleiter. 1974. The annealing of dislocations in high-angle grain boundaries. Philos. Mag. 30: 593–602.

Saito, Y., N. Tsuji, H. Utsunomiya, T. Sakai and R.G. Hong. 1998. Ultrafine-grained bulk aluminum produced by accumulative roll-bonding (ARB) process. Scripta Mater. 39: 1221–1227.

Schey, J. 1983. Rolling, Tribology in Metalworking. American Society for Metals, pp. 249.

Setman, D., E. Schafler, E. Korznikova and M.J. Zehetbauer. 2008. The presence and nature of vacancy type defects in nanometals detained by severe plastic deformation. Mater. Sci. Eng. A 493: 116–122.

Speich, G.R. 1969. Tempering of low-carbon martensite. Trans. Met. Soc. AIME. 245: 2553–2564.

Takaki, S., S. Iizuka, K. Tomimura and Y. Tokunaga. 1991. Influence of cold working on recovery and re-crystallization of lath martensite in 0.2% C steel. J. Jpn. Inst. Metals. 55: 1151–1158.

Thomas, G. 1971. Electron microscopy investigations of ferrous martensites. Met. Trans. 2: 2373–2385.

Tianfu, J., G. Yuwei, Q. Guiying, L. Quan, W. Tiansheng, W. Wei, X. Furen, C. Dayong, S. Xinyu and Z. Xin. 2006. Nano-crystalline steel processed by severe rolling of lath martensite. Mater. Sci. Eng. A 432: 216–220.

Tokizane, M., N. Matsumura, K. Tsuzaki, T. Maki and I. Tamura. 1982. Recrystallization and formation of austenite in deformed lath martensitic structure of low carbon steels. Metal. Trans. A 13: 1379–1388.

Tsuji, N., Y. Saito, Y. Utsunomiya and S. Tanigawa. 1999. Ultrafine-grained bulk steel produced by accumulative roll-bonding (ARB) process. Scripta Mater. 40: 795–800.

Tsuji, N., Y. Saito, Y. Ito, H. Utsunomiya and T. Sakai. 2000a. Ultrafine-grained ferrous and aluminum alloys produced by accumulative roll-bonding. Ultrafine-Grained Mater. TMS. USA, pp. 207–218.

Tsuji, N., R. Ueji, Y. Ito and Y. Saito. 2000b. *In situ* recrystallization of ultrafine grains in highly strained metallic materials. Proc. of the 21st Risø Int. Symp. on Mater. Sci. Denmark. 607–616.

Tsuji, N., K. Shiotsuki and Y. Saito. 1999. Superplasticity of ultrafine-grained Al-Mg alloy produced by accumulative roll-bonding. Mater. Trans. JIM. 40: 765–771.

Tsuji, N., R. Ueji,Y. Minamino and Y. Saito. 2002. A new and simple process to obtain nanostructured bulk low-carbon steel with superior mechanical property. Scripta Mater. 46: 305–310.

Tsuji, N., N. Kamikawa, R. Ueji, N.Takata, H. Koyama and D. Terada. 2008. Managing both strength and ductility in ultrafine-grained steels. ISIJ Int. 48: 1114–1121.

Ueji, R., N. Tsuji, Y. Minamino and Y. Koizumi. 2002. Ultragrain refinement of plain low carbon steel by cold-rolling and annealing of martensite. Acta Mater. 50: 4177–4189.

Ueji, R., N. Tsuji, Y. Minamino and Y. Koizumi. 2004. Effect of rolling reduction on ultrafine-grained structure and mechanical properties of low-carbon steel thermomechanically processed from martensite starting structure. Sci. Technol. Adv. Mater. 5: 153–162.

Umemoto, M., E. Yoshitaka and I. Tamura. 1983. The morphology of martensite in Fe-C, Fe-Ni-C and Fe-Cr-C alloys. J. Mater. Sci. 18: 2893–2904.

Umemoto, M., B. Huang, K. Tsuchiya and N. Suzuki. 2002. Formation of nano-crystalline structure in steels by ball drop test. Scripta Mater. 46: 383–388.

Valiev, R.Z., R.K. Islamgaliev and I.V. Alexandrov. 2000. Bulk nanostructured materials from severe plastic deformation. Prog. Mater. Sci. 45: 103–189.

Watanabe, S. and T. Kunitake. 1976. Formation of austenite grains from prior martensitic structure. Trans. ISIJ 16: 28–35.

Williams, D.B. and E.P. Butler. 1981. Grain boundary discontinuous precipitation reactions. Int. Matal. Rev. 26: 153–183.

Yodogawa, M. 1976. Precipitation behaviour in Fe-Ni-Mn martensitic alloys. Trans. JIM. 17: 799–808.

Yodogawa, M. and M. Tanaka. 1978. Effect of molybdenum additions on the age-hardening behavior and mechanical properties of Fe–9%Ni–4.5%Mn martensitic alloy. Trans. ISIJ. 18: 295–303.

Zhao, Y.H., H.W. Sheng and K. Lu. 2001. Microstructure evolution and thermal properties in nano-crystalline Fe during mechanical attrition. Acta Mater. 49: 365–375.

Zhao, X., T.F. Jing, Y.W. Gao, G.Y. Qiao, J.F. Zhou and W. Wang. 2005. Annealing behavior of nano-layered steel produced by heavy cold-rolling of lath martensite. Mater. Sci. Eng. A 397: 117–121.

Zhou, K., M.S. Wu and A.A. Nazarov. 2008. Relaxation of a disclinated tricrystalline nano-wire. Acta Mater. 56: 5828–5836.

12

Cold Rolling and Annealing of Advanced High Strength Steels

Rosalia Rementeria and *Francisca G. Caballero**

ABSTRACT

Advanced High Strength Steels (AHSSs) are materials with carefully designed chemical compositions and multi-phase microstructures resulting from precisely controlled cold rolling and annealing processes. Microstructures can be tailored to meet specific requirements of automotive parts, giving rise to the different AHSSs generations. The *first generation* of AHSSs includes steels with ferrite-based microstructures, i.e., dual phase (DP), transformation induced plasticity (TRIP), complex-phase (CP) and martensitic (MS) steels. The *second generation* of AHSSs are high-manganese austenitic steels, namely twinning-induced plasticity (TWIP) steels and Al-added lightweight steels with induced plasticity (L-IP). The *third generation* of AHSSs combines microstructures consisting of a high strength phase and a significant amount of ductile austenite, specifically quenching and partitioning (Q&P) and carbide-free bainite (CFB) steels. This chapter will give an overview of the alloy design process and the cold rolling and annealing practices needed to produce the microstructures from the first to the third generation of AHSSs.

Department of Physical Metallurgy, Spanish National Center for Metallurgical Research (CENIM-CSIC), Avda. Gregorio del Amo 8, E-28040 Madrid, Spain.
E-mail: rosalia.rementeria@cenim.csic.es
* Corresponding author: fgc@cenim.csic.es

12.1 Introduction

Global regulations of greenhouse gases are evolving and calling for stricter laws considering a variety of regulatory measures, such as CO_2 emissions or emission-based taxation of vehicles. Responding to these regulations, automakers have drawn up an approach based on the weight reduction of vehicles and thereby request an equitable contribution to the required increase in fuel efficiency by all manufacturers. To reduce weight, light-weight materials and/or thinner structures can be used in the auto body without compromising the strength levels required in an affordable manner.

Advanced High Strength Steels (AHSSs) were developed by the Ultra-Light Steel Auto Body (ULSAB) Consortium whose aim was the design of a lightweight steel auto body structure that would meet a wide range of safety and performance targets in an executable and affordable manner (Consortium ULSAB 1998). Afterwards, several projects emerged following this trail in order to refine the fabrication methods, achieve an even greater mass reduction and further stretch the envelope for strength/ductility levels. Nowadays, the Future Steels Vehicle (FSV) project brings a portfolio including more than 20 different AHSSs grades from the *first* and the *second generations* that represent materials expected to be commercially available in the 2015–2020 technology horizon.

Terminology has evolved to categorize AHSSs in "generations" or "steel families". The *first generation* of AHSSs encompasses steels that possess primarily ferrite-based microstructures, specifically dual phase (DP), transformation induced plasticity (TRIP), complex-phase (CP) and martensitic (MS) steels. The need for higher strength, supra-ductile steels led to the *second generation* of AHSSs, which are essentially high-manganese austenitic steels, usually referred to as twinning-induced plasticity (TWIP) steels, or Al-added lightweight steels with induced plasticity (L-IP), both closely related to conventional austenitic stainless steels (AUST SS). The mechanical properties of the different families of AHSSs are usually visualized in the strength/ductility plot, also known as the Global Formability Diagram shown in Figure 12.1. This figure stands today as the frame of reference for assessing new AHSSs developments. The lowest property band (or the classical "banana plot") includes the *first generation* of AHSSs (DP, TRIP, CP and MS) along with other more conventional steels with ferrite-based microstructures, such as interstitial free (IF), CMn and high-strength low-alloy (HSLA) steels. The evolution of this figure has unseated the steels from the lowest property band towards the higher strength/ductility combinations exhibited by the second generation of AHSSs (TWIP and L-IP), which have an exceptional formability. However, the production of these austenitic steels requires high alloying additions and challenging processing routes which make them cost-ineffective.

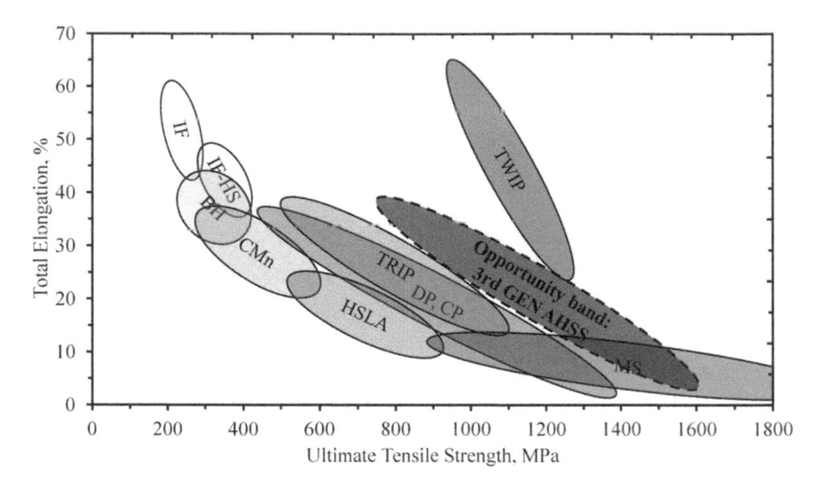

Figure 12.1 Tensile strength and total elongation data for various classes of conventional and AHSSs. Reproduced from (Keeler and Kimchi 2015).

The *third generation* of AHSSs arises to fill the "opportunity band" in between the *first*, and the *second generation* of AHSSs, allowing strengths over 1 GPa in combination with 20–30% of total elongation but without high levels of alloying elements, resulting in steel grades economically viable for the automotive industry. The basis of these new steels lies on microstructures consisting of a high strength phase (e.g., ultra-fine grained ferrite, martensite, or bainite) and a significant amount of ductile austenite that improves the work-hardening of the composite-like structure mainly by TRIP effect. The latest approaches to the development of the *third generation* of AHSSs focus in quenching and partitioning (Q&P) steels and carbide-free bainite (CFB).

The identification standard, first adopted by the ULSAB Consortium, specifies both yield strength (YS) and ultimate tensile strength (UTS). Accordingly, AHSSs are identified by the system:

$$\text{XX aaa/bbb} \tag{12.1}$$

where XX = type of steel, aaa = minimum YS in MPa, and bbb = minimum UTS in MPa. In the following, commercial steel grades will be referred to according to this standard.

Each grade of the first and second generation of AHSSs is specifically qualified to meet the functional performance demands for certain parts. Thus, these grades are utilized differently depending on the requirements, e.g., crash zones need steels with high energy absorption while structural elements need extremely high strength steels. Contrastingly, the third generation of AHSSs reflects unique alloys and microstructures to achieve the desired properties through effective processing of carefully designed steel compositions. Indeed, AHSSs provide a wide range of properties with the same chemical composition solely by adjusting the volume fraction of the second phases. Concerning individual AHSSs steels, these can be considered as composite materials from the mechanical point of view, so that the overall mechanical properties of the material are related to the behavior of each of the phases within the microstructure and the synergic interaction between them (De Cooman 2004). In the end, the alloy and process design allows the control of the mechanical properties through the microstructure.

AHSSs steels used for automotive industry (body work and structural parts) are generally delivered as coiled sheet metal (thin strips) to the customer in thicknesses from 0.4 to 3 mm and in widths of 1900 mm. Cold rolling enables the attainment of such thicknesses, which cannot be obtained by hot rolling due to excessive cooling. In addition, the appearance and properties of the final surface will be defined during the cold rolling stage by eliminating oxidized layers and inducing a controlled roughness (Béranger et al. 1996).

The cold rolling mill is fed with hot rolled strips of thickness between 2 and 6 mm. The mill transmits a very high pressure to the strip via the rolls, thereby reducing the strip thickness and driving the strip. During this process the sheet undergoes cold working, leading to strain hardening accompanied by a loss of ductility. The subsequent annealing step is introduced to produce an optimized microstructure in order to control the mechanical properties.

12.2 Alloy and Process Design of AHSSs

To achieve the desired microstructures, mechanical properties and final thickness and flatness of the thin strips, AHSSs need to be processed through different stages which include hot rolling, pickling, cold rolling and/or annealing. The microstructural developments at each of these stages will determine the final performance of the product, so that the chemical composition, the temperatures through the process and the heating and cooling rates must be carefully selected. Since every steel producer has different melting, rolling, annealing and cooling facilities, the choice of the steel composition best suited to the existing production capabilities is mandatory. Thus, a single widely accepted composition for each grade of AHSSs steel family is out of consideration.

Today it is not possible to physically follow the phase transformations in-line during processing. For the success of the process, it is of great importance to determine accurately, continuous cooling transformation (CCT) or deformation continuous cooling transformation (DCCT) diagrams, where the influence of composition and heat treatment conditions on the desired microstructure and properties can be tracked (Takahashi and

Bhadeshia 1989). CCT and DCCT diagrams are based on dilatometric analyses, which are quite suitable for the simulation of the whole production cycle (Zhao et al. 2002). Both diagrams provide information of the stability of the phases during continuous cooling of the austenite, but in the DCCT diagrams the austenite is strained to simulate the real process conditions. The effect of plastic deformation on the austenite to ferrite transformation kinetics is generally recognized as an accelerating effect. Specifically, in cooling experiments, both—the transformation start temperatures and the transformation finish temperatures increase, the temperature range of the transformation contracts, the final fraction of proeutectoid ferrite increases and the final ferrite grain size decreases (Crooks et al. 1982, Grajcar and Opiela 2008). Prior austenite deformation affects the transformation principally via a reduction in the undercooling required for nucleation rather than via an acceleration of the growth kinetics (Hanlon et al. 2001). Figure 12.2 illustrates the CCT and DCCT diagrams obtained in a

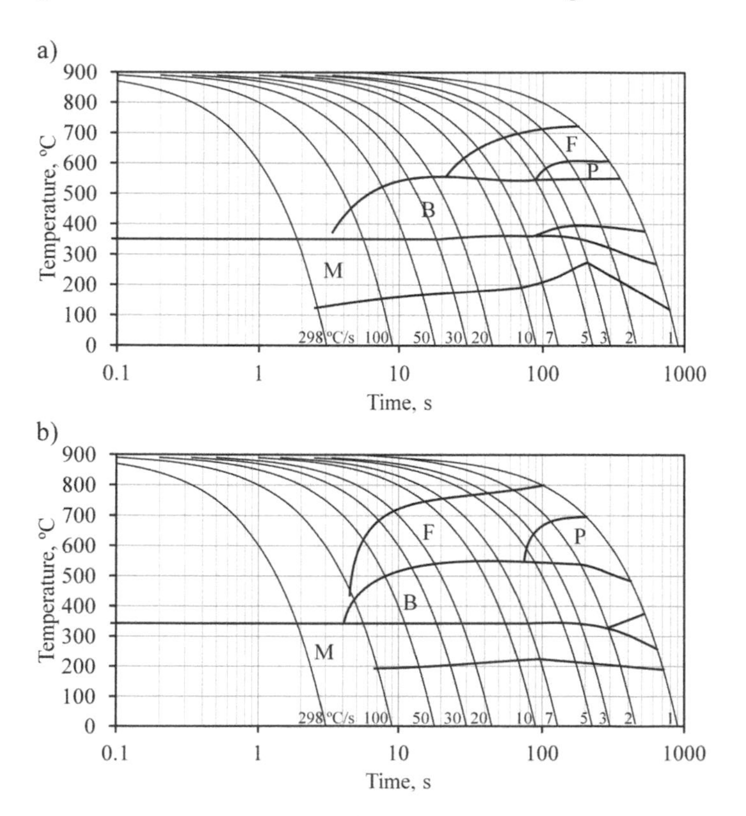

Figure 12.2 (a) CCT and (b) DCCT diagrams of a Fe—0.24 C—1.55 Mn—0.87 Si—0.40 Al (wt.%) TRIP assisted steel after cooling from 900°C at different cooling rates. The DCCT diagram was obtained after prior 50% deformation of the sample with a strain rate of 1 s^{-1} at 900°C (adapted from (Grajcar and Opiela 2008)).

Fe—0.24 C—1.55 Mn—0.87 Si—0.40 Al (wt.%) TRIP assisted steel, where the DCCT diagram was obtained after prior 50% deformation of the sample with a strain rate of 1 s^{-1} at 900°C.

12.3 First Generation of AHSSs

The metallurgical concepts of the first generation of AHSSs originated in the 1970s and the mechanical properties and applications are well established now. These are multiphase steels with a microstructure consisting of hard islands of martensite, bainite and/or retained austenite dispersed in a soft ferritic matrix. In the following sections, the main features regarding chemical composition, microstructure, processing and mechanical properties of DP, TRIP, CP and MS steels will be outlined.

12.3.1 Dual Phase Steels

DP steels are produced from lean elemental compositions, e.g., Fe—(0.06–0.2) C—(0.8–1.5) Mn—(0.2–0.5) Si (wt.%). The microstructure usually consists of a soft ferrite matrix and a hard martensitic second phase in the form of embedded islands whose volume fraction is around 15–25% (Davies 1978, Rashid 1981, Speich et al. 1981). The term "dual-phase" was first adopted by Hayami and Furukawa (Hayami and Furukawa 1975) and thereafter adopted by the steel community to refer to the presence of those two phases, though small amounts of retained austenite, pearlite or bainite may also be present.

Regardless of the chemical composition of the steel, current commercial production of DP steels is carried out by intercritical annealing after hot or cold rolling through two different routes: (a) by holding the material in the austenite-ferrite two-phase region followed by quenching or (b) by adjusting the quenching rate from the fully austenitic regime to the ferritic region so that most of the austenite transforms into ferrite and the rest into martensite. Figure 12.3 schematizes these annealing treatments used to obtain DP microstructures. The thermomechanical heat-treatment, and also the initial microstructure must be optimized to control the mechanical properties through the grain size of the ferrite matrix and the volume fraction, size and morphology of the martensite.

DP steels exhibit a continuous yielding behavior, low yield strength, high work hardening rate, high tensile strength and remarkably high uniform and tensile elongations. Both yield and tensile strength increase with the increase of the martensite volume fraction (for a fixed C content) and C content (for a fixed martensite volume fraction), though this is followed by a loss in ductility. For further strengthening without compromising ductility, decreasing ferrite grain size and increasing martensite dispersion are effective methods (Tasan et al. 2015). Furthermore, equiaxed microstructures result in higher strength and lower ductility compared to specimens with finely dispersed elongated particles (Pierman et al. 2014). The continuous yielding and the high work hardening behavior are assumed to be a consequence of the high dislocation density and residual stress arising in ferrite as a consequence of the volume expansion associated with the austenite-to-martensite transformation.

Current production of DP steels includes several grades regarding tensile strength and formability in order to meet the different design requirements of individual components. DP commercial grades range from DP 210/440 to DP 1150/1270, whose strength/ductility combinations are represented in Figure 12.4 together with literature data (Ahmad et al. 2000, Bag et al. 1999, Bergstrom et al. 2010, Jiang et al. 1995, Kim and Thomas 1981, Zhang et al. 2014).

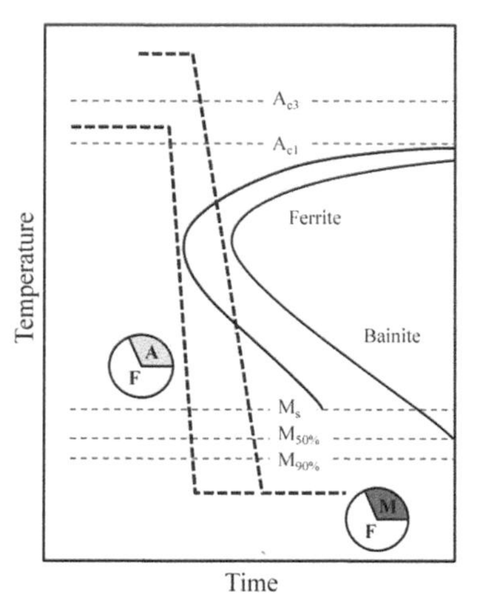

Figure 12.3 Scheme indicating the annealing treatments used to obtain a DP microstructure.

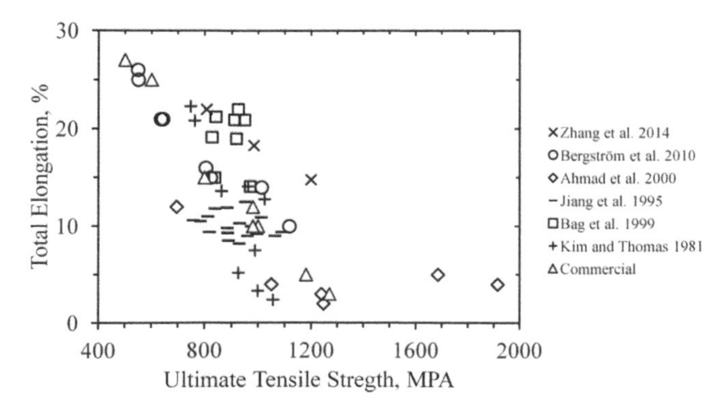

Figure 12.4 Strength/ductility data of current DP steels on varying specimen geometries.

Advances in the experimental and simulation techniques in the last decades have significantly improved the understanding of DP steels. Now it is feasible to precisely predict the microstructure evolution during processing, as well as to simulate the macromechanical behavior from the experimental characterization of micromechanical behavior (Tasan et al. 2015).

12.3.2 Transformation Induced Plasticity Steels

Conventional TRIP steel compositions are usually based on the original Fe—(0.10–0.45) C—(0.5–1.8) Si—(0.5–3.0) Mn (wt.%) concept proposed by Takechi et al. (Takechi et al. 1987a,b). However, current TRIP steel compositions have moved to lower contents of alloying elements to improve weldability, with typical compositions of Fe—(0.20–0.25) C—(1.3–1.8) S—1.5 Mn (wt.%) (Tsukatani et al. 1991), where Si can be partially replaced by Al (max. 1 wt.%). TRIP steels are processed by intercritical annealing after hot or cold rolling followed by isothermal transformation in the bainitic range, as illustrated in Figure 12.5.

During the intercritical annealing the austenite C content is raised to ~ 0.3–0.4 wt.% as a result of C partitioning from ferrite to the adjacent austenite. After the intercritical annealing, the steel is rapidly cooled down (> 30°C/s) to a temperature in the bainitic transformation range and isothermally held for several minutes. During the bainite transformation stage, retained austenite grains decrease in size and become further enriched with C (~ 1–1.5 wt.%), as the C excess is expelled from the bainitic ferrite into the austenite. The final microstructure consists of a matrix of proeutectoid ferrite, baintic ferrite and retained austenite. Minor phases,

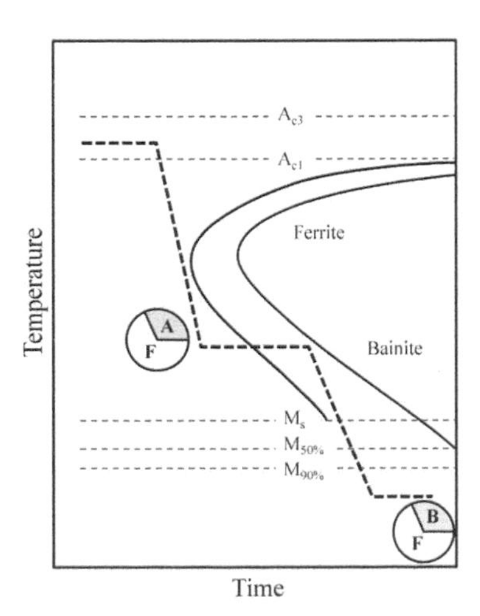

Figure 12.5 Scheme indicating the annealing treatment used to obtain TRIP microstructures.

such as martensite or carbides, may also be present depending on the alloy composition and the processing route. Suppressing the carbide precipitation during bainitic transformation appears to be crucial for TRIP steels, and this is accomplished by Si and Al additions (Kozeschnik and Bhadeshia 2008).

In low alloy TRIP steels proeutectoid ferrite constitutes the soft matrix phase, whose high intrinsic stacking fault energy (SFE) results in a low strain hardening. Bainitic ferrite is the strongest phase as a result of a very small grain size, a high dislocation density and C-supersaturation. Retained austenite is a low SFE phase, which results in a high strain hardening. The presence of austenite and its transformation into martensite by TRIP effect during straining is the key factor in the improved ductility and toughness of these steels. The stabilization of austenite is controlled by its C content, and it has to be well balanced in order to prevent the formation of martensite during cooling and to allow the continuous martensite formation over a large straining range. The imperative point in the alloying concept and the heat treatment procedure of TRIP steels is the attainment of a proper, relatively high C content in the austenite (Tsukatani et al. 1991). This latter is achieved by an alloy design adjusted to the characteristic "process windows" of the industrial continuous annealing lines or continuous galvanizing lines.

Current commercially available TRIP steels are employed for manufacturing complex automotive parts due to their excellent formability, and for parts susceptible to impact due to the high work hardening exhibited during crash. TRIP commercial grades range from TRIP 350/600 to TRIP 600/980, whose strength/ductility combinations are represented in Figure 12.6 together with literature data (Baik et al. 2001, De Cooman 2004, Kim et al. 2003, Li and Wu 2006, Mahieu et al. 2002, Meyer et al. 1999, Zaefferer et al. 2004).

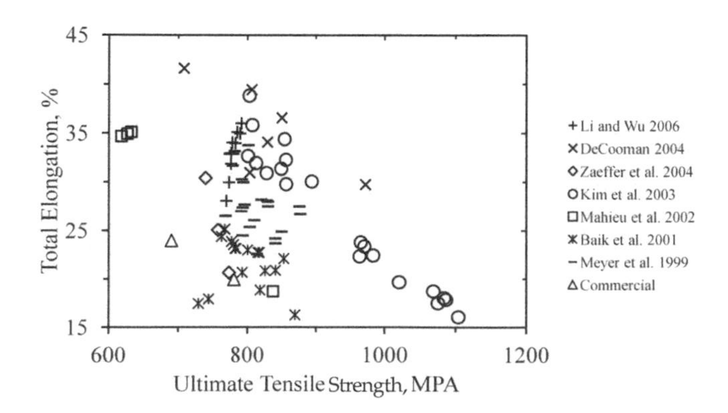

Figure 12.6 Strength/ductility data of current TRIP steels on varying specimen geometries.

12.3.3 Complex-phase Steels

The demand for higher strength of DP and TRIP grades (800–1200 MPa) for automotive applications could be met by increasing C content to ~ 0.4 wt.%. However, higher C levels introduce weldability problems and may cause hot-rolling difficulties, particularly for wide formats. Thus, CP steels emerged as a microalloyed TRIP concept in order to increase strength but keeping a low C content (Scott et al. 2004). The chemical composition and processing route of CP steels, also referred to as multiphase steels, are based on that of TRIP and DP steels but additions of small amounts of Nb, V and or Ti are employed to provoke the precipitation strengthening effect and grain refinement.

Nb combines with C and N to form precipitates that retard recrystallization and grain growth, resulting in precipitation strengthening (Bleck et al. 1998). Nb also increases the ferrite start transformation temperature, refines the ferrite microstructure, and stimulates the formation of acicular ferrite (Timokhina et al. 2001b), though it retards the kinetics of the austenite to ferrite transformation (Onink et al. 1998). Under bainite transformation conditions, Nb promotes the stabilization of retained austenite, which influences favorably the mechanical properties of CP steels (Bai et al. 1998, Timokhina et al. 2001b).

The solubility of V in both ferrite and austenite is much greater than that of Nb and Ti, and it is used to control the transformation behavior of CP steels during the last stages of the annealing process (Scott et al. 2004). By choosing the appropriate V additions and thermomechanical process path, a high fraction of vanadium carbides and nitrides V(C,N) precipitate in ferrite during the continuous annealing step, providing a large strengthening effect. However, if the intercritical annealing temperature is too large, a decrease in the UTS might occur due to solubilization of small V(C,N) precipitates in the austenite and coarsening of the V(C,N) precipitates in ferrite (Scott et al. 2004).

Ti is added to steels to induce multiple effects. Ti readily stabilizes nitrogen during continuous casting by forming Ti nitrides, which help to inhibit the austenite grain coarsening during reheating. Ti carbides formed during hot rolling by strain-induced precipitation retard static recrystallization causing the pancaking of the austenite grains and the refinement of the final ferrite microstructure (Panigrahi 2001). Moreover, the strengthening due to the precipitation of Ti carbides is more effective than the grain refinement. In solid solution, Ti retards the transformation of austenite into ferrite.

The microstructure of CP steels usually consists of a ferrite–bainite matrix containing small amounts of martensite, retained austenite and pearlite. The final phase fractions vary with the cooling rate from austenite, i.e., high cooling rates suppress the formation of proeutectoid ferrite and pearlite favoring the transformation to bainite or martensite. The precipitation behavior of Nb, V and Ti compounds is complex and non-intuitive, as precipitation takes place inside a continuously evolving matrix. Also, precipitates themselves have time varying solubilities due to their changing C/N ratio (Perrard and Scott 2007), and their presence is highly dependent on the thermomechanical processing.

Figure 12.7 shows current strength/ductility data for various CP steels as reported in (Hanzaki et al. 1995, Krizan and De Cooman 2014, Scott et al. 2007, Shi et al. 2006, Timokhina et al. 2001a, Zhang et al. 2006). The mechanical properties of these steels depend on the volume fraction, size and morphology of the microconstituents, but also on the amount, size and distribution of the precipitates. These can act as obstacles against the motion of mobile dislocations, but they also serve as nucleation sites for void formation. Smaller void spacing results in an easier void linking, which in turn results in a decrease in elongation (Krizan and De Cooman 2014).

Therefore, small additions of Nb, V or Ti have a strong influence on the transformation and precipitation history during the cold rolling and annealing schedule, which ultimately controls the final mechanical properties of the microstructure. These additions must be carefully designed and controlled in order to obtain the desired property balance.

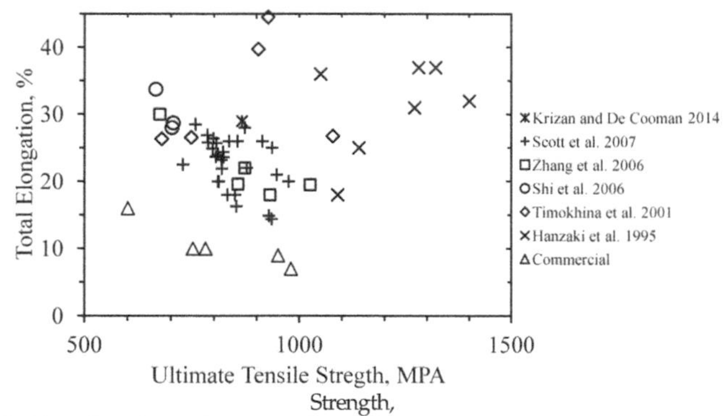

Figure 12.7 Strength/ductility data of current CP steels on varying specimen geometries.

12.3.4 Martensitic Steels

MS steels have been present in the humankind even before the 13th century B.C., but it was not until 1890 that Adolf Martens gave a name to this microstructure. Chemical compositions of MS steels currently used for automotive applications are Fe—(0.10–0.25) C—(0.30–0.50) Si—(1.50–2.40) Mn (wt.%), but elements such as Cr, Mo, V and Ni are also used individually or in combination to increase hardenability. These steels can be produced directly at the steel mill by quenching after hot rolling or cold rolling and annealing and/or via post-forming heat treatment, as schematized in

Figure 12.8. The austenite composition and grain size, which play important roles in determining the ultimate structure and properties, are controlled largely by the austenitizing temperature (Taylor and Hansen 1991). Regardless of the production process, the microstructure obtained consists of a martensitic matrix containing small amounts of ferrite and/or bainite.

Martensite inherits the C content of the parent austenite, which results in a supersaturated solution of C in a tetragonal lattice. The high hardness of martensitic steels is due to solid solution strengthening of C atoms which impede the slip of dislocations in the distorted lattice, and leads to a general increase in plastic deformation resistance. In addition to solid solution strengthening, the effective grain size plays an important role in the mechanical properties of MS steels. The smallest microstructural unit separated by high-angle boundaries in martensite is the crystallographic block, and it is accepted to be the effective grain size in a Hall-Petch type relationship (Morito et al. 2006a, 2006b). The martensite laths or plates formed within a parent austenite grain, group themselves into several packets that are further subdivided into blocks. The laths/plates within a packet have similar habit planes but different crystallographic orientations, resulting in a grouping of the laths into different blocks of which each contains laths of similar orientations, as schematized in Figure 12.9. There is a high angle misorientation between both the packet and block boundaries, while the misorientation across lath boundaries are low angles that vary within a few degrees.

MS steels have the highest tensile strength level of the first generation of AHSSs. Tensile strength is extremely high, with values up to 1900 MPa, but due to the large number of internal stresses associated during transformation ductility is low, ranging from 3 to 15%. As formed, martensite is of little

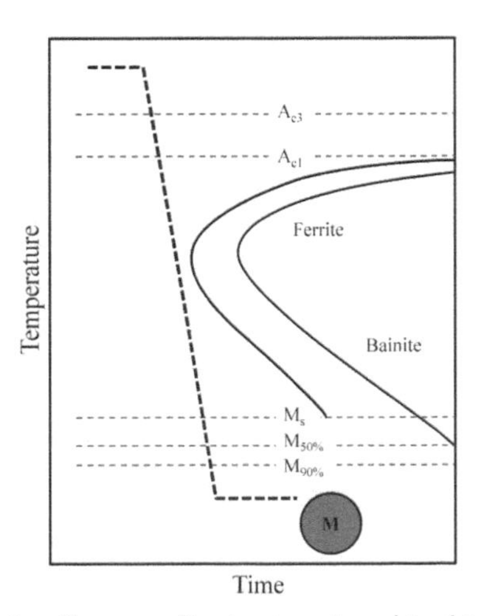

Figure 12.8 Scheme indicating the annealing treatment used to obtain an MS microstructure.

use and must be softened for metal working, and either heat treatment or hot forming is applied to overcome their formability issues. Softening is accomplished by tempering martensite at temperatures below the eutectoid temperature. Properties achieved in tempered martensite depend on the tempering temperature and the holding time at that temperature, making tempering a simple widespread technique to tailor the microstructure according to the requirements. Figure 12.10 collects strength/ductility data for commercial and MS steels found in the literature in the as-quenched and tempered condition (Chang 2002, Duang et al. 2012, Lee and Su 1999, Meysami et al. 2010, Mohrbacher 2015, Muckelroy et al. 2013, Taylor and Hansen 1991).

Typical applications for MS steels are those that require high strength and good fatigue resistance, with relatively simple shapes for traditionally stamped parts and more complex shapes for hot formed parts. Applications of automotive MS steels include bumper reinforcements, door intrusion beams, side sill reinforcements, springs, clips, and more recently, roll formed rocker panel inners and reinforcements.

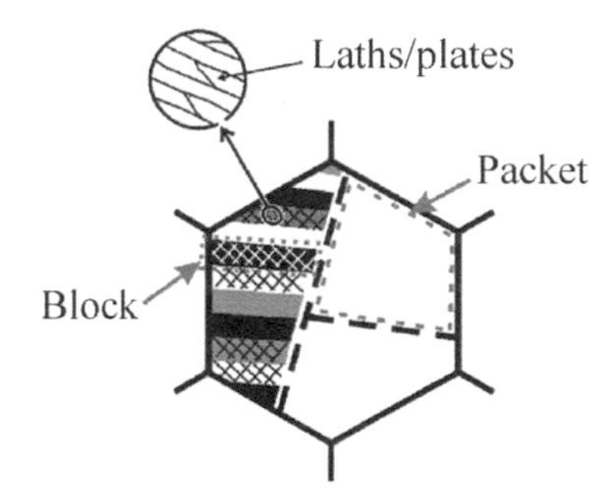

Figure 12.9 Scheme of the martensitic substructures formed at a single prior austenite grain.

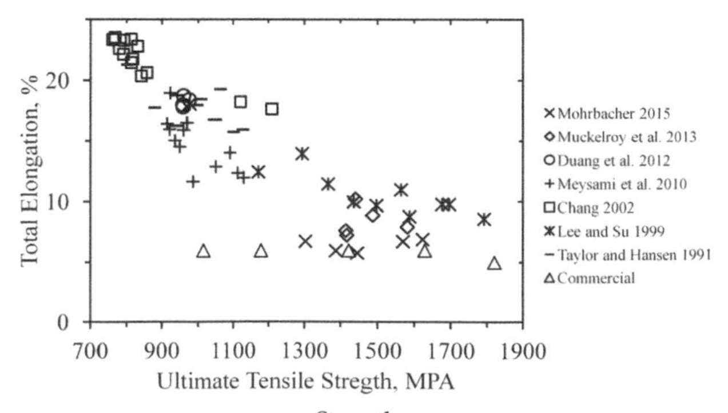

Figure 12.10 Strength/ductility data of current MS steels on varying specimen geometries.

12.3.5 Cold rolling and annealing of the first generation of AHSSs

12.3.5.1 Microstructure after hot and cold rolling in the first generation of AHSSs

The hot rolling (outlined in Figure 12.11) and cold rolling routes to obtain the required thicknesses before annealing in the *first generation* of AHSSs are essentially similar. The finishing rolling temperature (FRT), the cooling rate (CR) and the coiling temperature (CT) will determine the microstructure before annealing, and can be optimized to facilitate cold rolling and eliminate microstructural banding. As reference parameters, semi rolled slabs 30 mm thick are soaked at 1200°C during 45 min and hot rolled to about 3 mm in several passes finishing at 900°C. Slow CRs after finishing, lead to ferrite and pearlite microstructures with an increasing amount of pearlite with increasing CT. Figures 12.12a to d illustrate the microstructures obtained in a Fe—0.15 C—0.19 Mn—0.20 Si—0.20 Cr—0.03 Al (wt.%) steel with a FRT of 900°C and a CR of 7°C/s. A higher amount of pearlite arranged in layers is revealed for the highest CT (dark-etched areas in Figures 12.12c and d). On the other hand, hot rolled sheets more rapidly cooled after finishing consist mainly of martensite and bainite, as illustrated in Figures 12.12e to h for a CR of 60°C/s.

The effect of 68% cold rolling on the latter microstructures is shown in Figure 12.13. Whatever the microstructure is after hot rolling, the grains become elongated and the amount of phases remains roughly inalterable after cold rolling. For the case of the microstructure consisting of ferrite and pearlite (Figures 12.12c and d), the pearlite colonies become smaller exhibiting a finer interlamellar spacing (see Figures 12.13c and d).

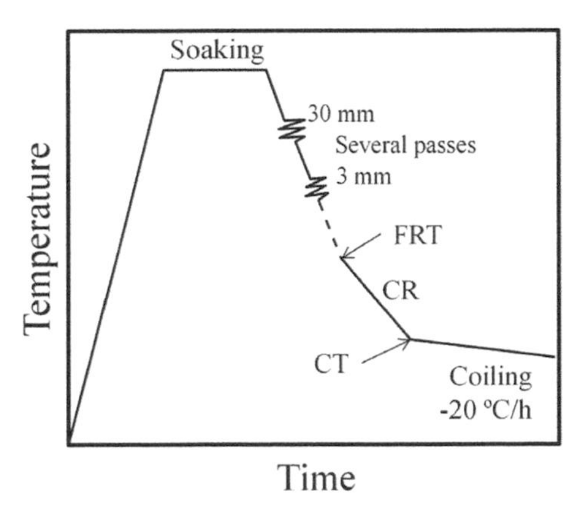

Figure 12.11 Scheme of temperature–time profile used for the hot rolling and processes of the *first generation* of AHSSs. FRT is finishing rolling temperature, CR is cooling rate and CT is coiling temperature.

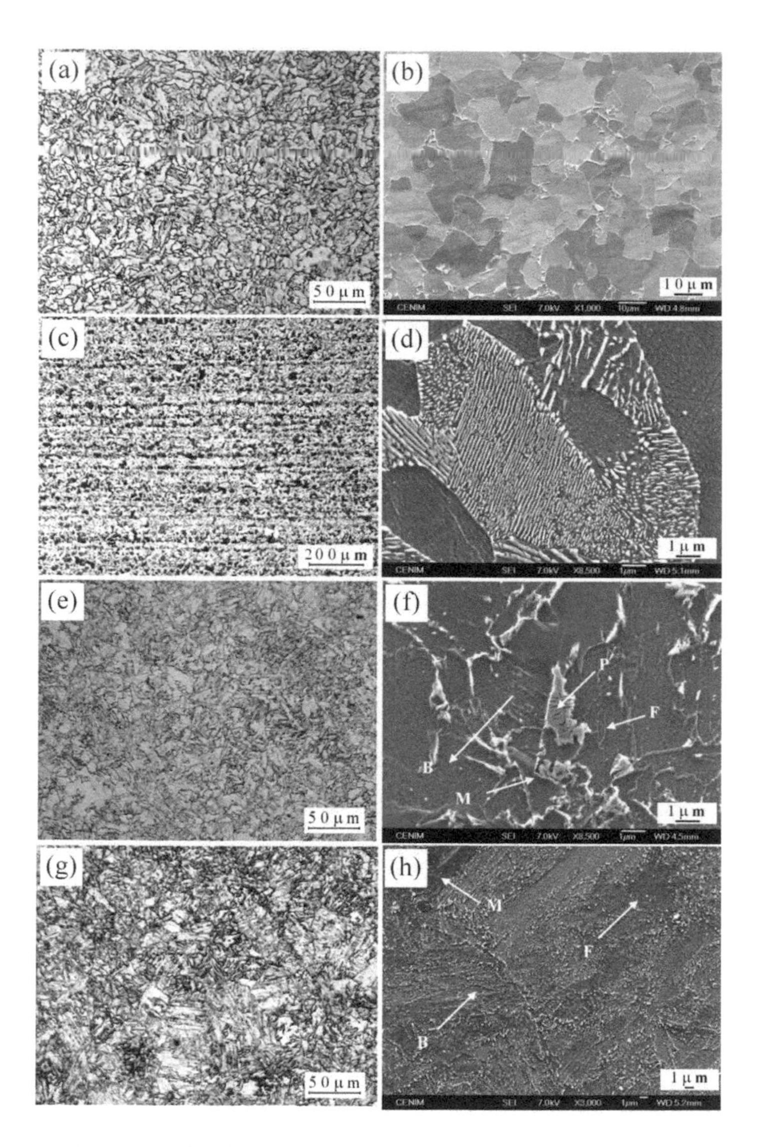

Figure 12.12 Light optical micrographs and scanning electron micrographs of hot rolled samples parallel to the rolling direction in a Fe—0.15 C—0.19 Mn—0.20 Si—0.20 Cr—0.03 Al (wt.%) steel with FRT = 900°C and different CR and CT: (a)(b) CR = 7°C/s, CT = 500°C; (c)(d) CR = 7°C/s, CT = 650°C; (e)(f) CR = 60°C/s, CT = 500°C; (g)(h) CR = 60°C/s, CT = 650°C. B is bainite, M is martensite, F is ferrite, P is pearlite.

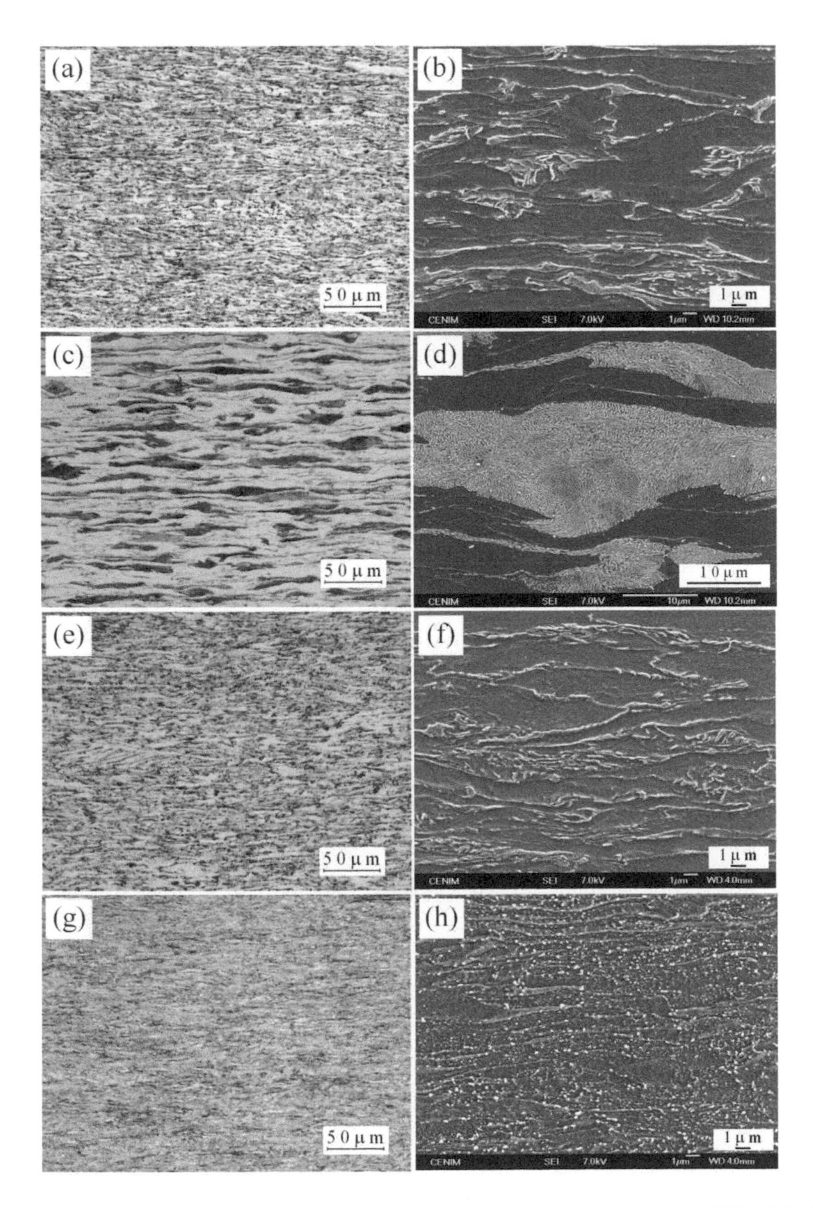

Figure 12.13 Light optical micrographs and scanning electron micrographs of 68% cold rolled samples parallel to the rolling direction in a Fe—0.15 C—0.19 Mn—0.20 Si—0.20 Cr—0.03 Al (wt.%) steel with FRT = 900°C and different CR and CT: (a)(b) CR = 7°C/s, CT = 500°C; (c)(d) CR = 7°C/s, CT = 650°C; (e)(f) CR = 60°C/s, CT = 500°C; (g)(h) CR = 60°C/s, CT = 650°C.

12.3.5.2 Microstructure after Intercritical Annealing in the First Generation of AHSSs

The formation of austenite from ferrite is a diffusion-controlled phase transformation. Thus, the volume fraction of austenite and its coarseness increase with increasing the soaking temperature and time, and results in lower amounts of ferrite in the final microstructure. Figure 12.14 shows microscopic evidences of how austenite formation occurs in the cold rolled sample (CR = 7°C/s, CT = 650°C) shown in Figures 12.13c and d throughout optical micrographs from intercritical annealing specimens at different temperatures and times. LePera's reagent reveals pearlite and ferrite as darker phases in the microstructure, whereas martensite formed during quenching appears as lighter regions. Microstructure in Figure 12.14a annealed at 750°C for 1 s is formed mainly of ferrite, pearlite and some grains of martensite. At this quench-out time, the pearlite-to-austenite transformation has already started. Once pearlite dissolution has finished, the annealed microstructure consists of a mixture of ferrite and martensite (Figures 12.14b and c). Increasing the soaking temperature

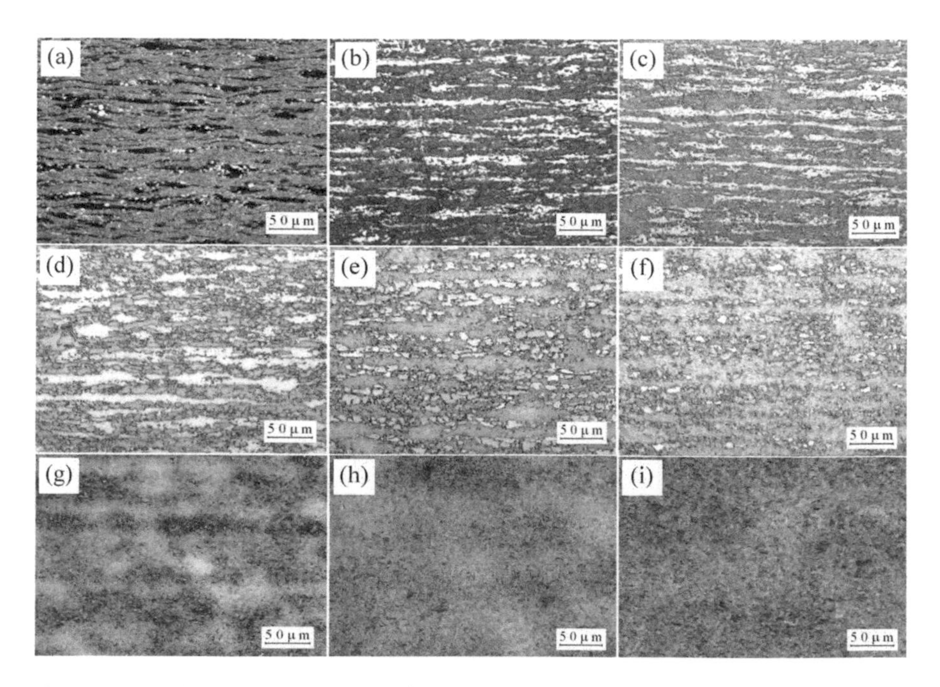

Figure 12.14 Light optical micrographs from intercritical annealed samples of 68% cold rolled samples parallel to the rolling direction in a Fe—0.15 C—0.19 Mn—0.20 Si—0.20 Cr—0.03 Al (wt.%) with FRT = 900°C, CR = 7°C/s and CT = 650°C: (a) at 750°C for 1 s; (b) at 750°C for 20 s; (c) at 750°C for 100 s; (d) at 800°C for 1 s; (e) at 800°C for 20 s; (f) at 800°C for 100 s; (g) at 850°C for 1 s; (h) at 850°C for 20 s; (i) at 850°C for 100 s. LePera reagent.

from 750 to 800°C results in lower amounts of ferrite in the microstructure (Figures 12.14d to f). This is due to the larger amount of austenite formed at higher temperatures, which transforms into martensite on quenching. At 850°C, when the intercritical soaking is reached, a few ferrite grains remain untransformed (Figure 12.14g), but it takes less than 20 s to complete the transformation to austenite, which is responsible for the fully martensitic microstructure formed on quenching (Figures 12.14h and i).

The initial microstructure slightly affects the kinetics of austenite formation and its coarsening (Caballero et al. 2006b). Lower CTs in the hot rolling stage lead to slightly finer austenite grains, while higher CTs result in a more sluggish formation of austenite associated with a higher amount of dispersed carbides in the cold rolled microstructure. At 750°C and 1 s of soaking time, austenitization process has already started for all the initial microstructures given in Figure 12.13. Figure 12.15 shows electron micrographs corresponding to the beginning of the transformation in annealed samples. In samples consisting of ferrite and pearlite (CR = 7°C/s) in Figures 12.15a and b, the nucleation of austenite takes place inside pearlite preferentially at the points of intersection of cementite with the edges of the pearlite colony. Likewise, in samples mainly formed of

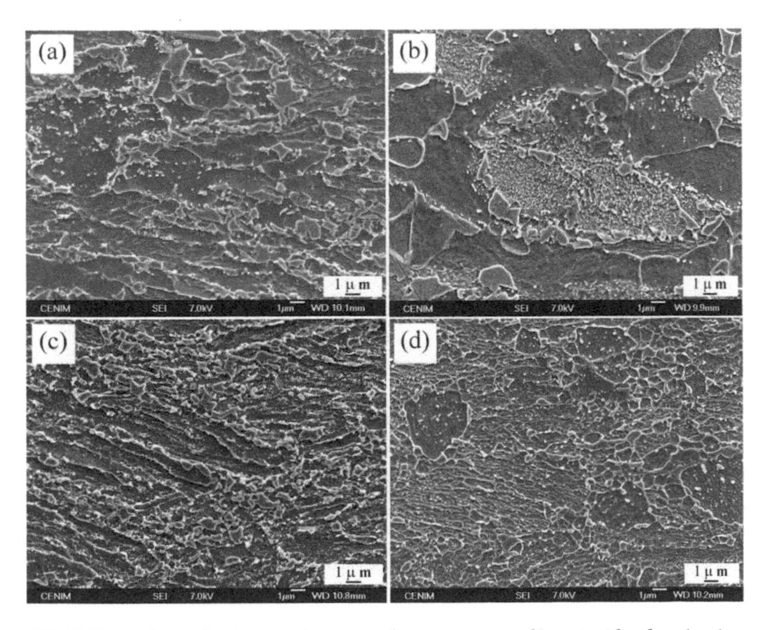

Figure 12.15 Scanning electron micrographs corresponding to the beginning of the austenitization process in annealed samples at 750°C for 1 s of 68% cold rolled sheets in a Fe—0.15 C—0.19 Mn—0.20 Si—0.20 Cr—0.03 Al (wt.%) steel with FRT = 900°C and: (a) CR = 7°C/s, CT = 500°C; (b) CR = 7°C/s, CT = 650°C; (c) CR = 60°C/s, CT = 500°C; (d) CR = 60°C/s, CT = 650°C.

bainite (CR = 60°C/s), austenite nucleates at the interface between the plates of ferrite in the sheaves of bainite (Figures 12.15c and d). Moreover, carbides at grain boundaries and inside ferritic grains are important nucleation sites for austenite in cold rolled samples.

When the intercritical soaking stage is reached, ferrite grains are fully recrystallized (Caballero et al. 2006b, Geib et al. 1980, Rocha et al. 2005, Speich et al. 1981). On the other hand, Figure 12.15b suggests that pearlite spherodizes before austenite formation. Spherodization of the deformed pearlite occurs concurrently with ferrite recrystallization during annealing of the cold rolled samples (Yang et al. 1985).

12.3.5.3 Microstructural Banding Control in the First Generation of AHSSs

Upon hot and cold rolling of steels, substantial microstructure and texture inhomogeneity can occur owing to the through-thickness gradients in shear, total deformation, and temperature (Hölscher et al. 1994). Microstructure inhomogeneity is often characterized by a continuous change in the initial microstructures from a banded morphology (Speich et al. 1981) in the sheet center to a heterogeneous distribution near the surface. Therefore, the mechanical properties of the sheets before and after annealing can be anisotropic and dependent on the through-thickness position.

Microstructural banding after hot rolling is mainly due to the segregation of substitutional alloying elements during dendritic solidification. Several investigations have shown manganese to be the alloying element most responsible for the development of microstructural banding in low alloy steels (Grange 1971, Grossterlinden et al. 1992). In hot rolled low alloy steels, pearlite and ferrite are, as a rule, arranged in layers visible as a banded structure in the longitudinal section (Krauss 2003), as seen in Figures 12.12c and d.

Hot rolled bands remain inalterable after cold rolling and continuous annealing of AHSSs, since during the intercritical heat treatment austenite formation takes place only in the C-rich regions featuring pearlite, while the low-C regions remain ferritic (Grossterlinden et al. 1992). When rapid cooling, martensite will then form in the regions previously occupied by pearlite, as shown in Figure 12.14. Major factors affecting the severity of microstructural banding are the austenitization temperature, the prior austenite grain size, the cooling rate and the coiling temperature (Majka et al. 2002).

The banded microstructure affects mainly the ductility and the impact energy of the steel (Grange 1971), which are critically dependent on the morphology of the band, as well as on the mechanical behavior of the phase that composes the band. In microstructures where there is a continuous microstructural band, shear bands develop through the band and percolate through its narrowest section forcing the banded phase to deform beyond its plastic limit. This effect is especially noticeable if there is a significant difference in the ultimate strains of the phases composing the banded microstructure (e.g., the case of DP steels). For discontinuous microstructural bands, shear bands naturally cross at the gaps within the band, thereby delaying early damage initiation (Tasan et al. 2010).

Microstructural banding can be first eliminated by increasing the cooling rate from the austenitic condition during hot rolling as Ar_3 temperature differences of the segregated bands are reduced (Thompson and Howell 1992). However, banding might appear after intercritical annealing at high temperatures resembling the original chemical segregation (Caballero et al. 2006a). In the case of cold rolled DP and TRIP steels, increasing the cooling rate during hot rolling, and using low intercritical annealing temperatures and long soaking times (e.g., 750°C and 100 s) permanently eliminates microstructural banding (Caballero et al. 2006a, Rivera-Díaz-Del-Castillo and Van Der Zwaag 2004).

From an industrial point of view, the complete removal of banding may not always be economically feasible. In those cases, the detrimental influence of a banded microstructure can be significantly reduced by altering the morphology of the band in order to avoid microstructures with continuous bands and decrease the thickness variation of the band (Tasan et al. 2010).

12.4 Second Generation of AHSSs: Twinning-induced Plasticity Steels

The composition of TWIP steels is based in the Fe–Mn–C system, with usual compositions of Fe—(18–30) Mn—(0.10–0.6) C—(3 Si—1.5 Al) (wt.%). The role of such amounts of manganese in the chemical composition is first, the stabilization of the austenite at ambient temperature and second, the control of the SFE of the austenite. The SFE determines the occurrence of the TWIP effect; low SFE promotes the formation of mechanical twins that gradually reduce the effective glide distance of dislocations, which results in the "Dynamical Hall-Petch effect". As the formation of mechanical twins involve the creation of new crystal orientations, twins progressively reduce the effective mean free path of dislocations (see Figure 12.16) and increase the flow stress, resulting in a high strain hardening behavior.

It is essential for the occurrence of the strain-induced twinning that the SFE lies within a very specific range for mechanical twin formation. A low SFE, i.e., less than <20 mJ/m^2, favors the strain induced transformation of austenite into martensite (TRIP effect), while moderate SFE favors the mechanical twinning–induced plasticity (Sato et al. 1989). C additions are also required to obtain a low SFE, but they are limited by the formation

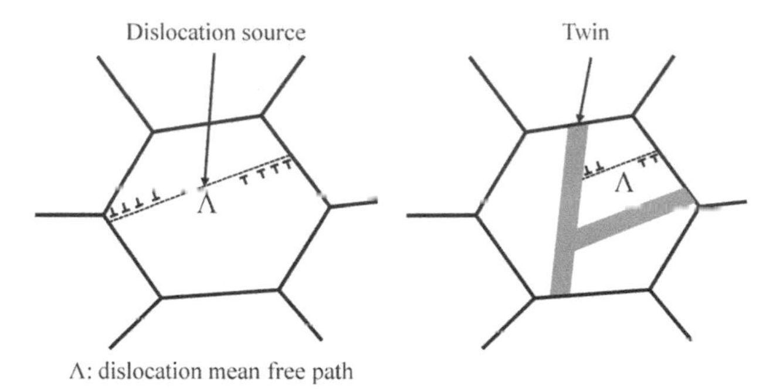

Figure 12.16 Scheme diagram of the dynamical Hall-Petch effect. Mechanical twins gradually reduce the effective glide distance of dislocations, resulting in the ultra-high strain hardening observed in TWIP steels.

of carbides. Stabilizing the austenite at room temperature requires Mn contents in excess of 27 wt.% in the binary Fe-Mn alloy system. However, C additions of approximately 0.6 wt.% allow to obtain fully austenitic microstructures and avoid the formation of martensite in alloys with less than 25 wt.% Mn (De Cooman et al. 2011). An alternative approach to this TWIP steel composition concept requires Si and Al additions to control the stacking fault energy without the addition of any C. This alternative approach is what L-IP steels are based on. Small additions of Al result in much improved TWIP properties and low sensitivity to delayed fracture. Al effectively suppresses the martensitic transformation, increases the SFE and lowers the strain hardening resulting in TWIP steels with slightly lower tensile strengths (Ishida and Nishizawa 1974). Contrastingly, Si decreases the austenite stability by sustaining the martensitic transformation during cooling and deformation (Schramm and Reed 1975). The combination of alloying elements will determine the SFE of the steels and thus, the operating deformation mechanism (TRIP/TWIP) (Frommeyer et al. 2003), as illustrated in Figure 12.17.

The high rate of strain hardening associated with the deformation twinning phenomenon allows for the combination of higher strengths and higher uniform elongations of TWIP steels. Tensile strength in commercial grades ranges from 900 to 1100 MPa with total elongation ranging from 55 to 70%. Strength/ductility combinations of current TWIP steels are represented in Figure 12.18; further details of individual data can be found in De Cooman et al. 2011, Ding et al. 2006, Dini et al. 2010, Grässel et al. 2000.

Typical applications of TWIP steels include automotive parts related to passenger safety, such as the B-pillar and the door impact beam, which are essential elements for passenger protection in side impact collisions.

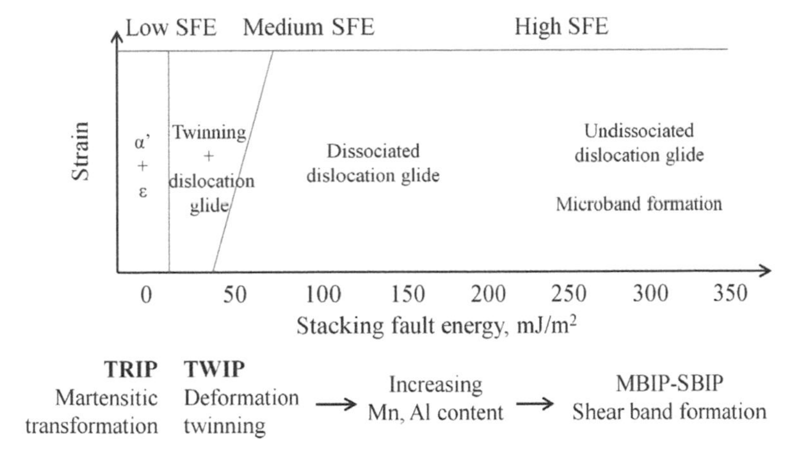

Figure 12.17 Scheme showing the relation between SFE and the operating deformation mechanism in FCC metals and alloys.

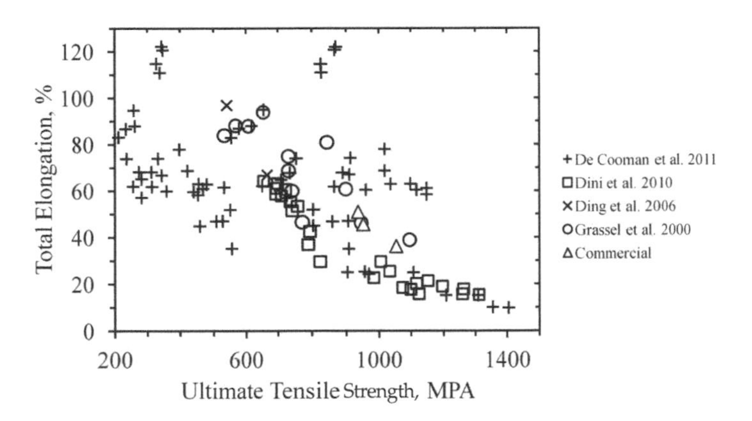

Figure 12.18 Strength/ductility data of current TWIP steels on varying specimen geometries.

12.4.1 Cold Rolling and Annealing of the Second Generation of AHSSs

Despite the attractive mechanical properties of TWIP steels, the cost issues related to both the alloy cost and the manufacturing process are of prime importance. The industrialization of TWIP steels has been primarily focused on strip products. At present, the economically attainable hot strip thickness by conventional continuous casting routes is too large to allow the weight reduction potential due to processing difficulties. The conventional route for producing hot strips in TWIP steels faces the following problems:

a) Ductility of TWIP steels is very low in the 900°C to 1300°C temperature range (Kim et al. 1997). As a result, casting and hot rolling have to be performed within a very narrow process window in order to avoid cracking (Akhlaghi and Yue 2001, Zaitsev and Mogutnov 1995).
b) The addition of Si and Al results in the formation of reactive slags, which can attack refractories during casting (Rivlin 1983). The slag can subsequently agglomerate with particles which have come away from the refractory lining, resulting in clustering or "*clogging*", that decreases both product quality and productivity (Jacobi and Wünnenberg 1999).
c) The macroscopic and microscopic segregation of such a high amount of alloying elements during solidification has a strong impact in the mechanical properties of TWIP steels (Kim et al. 1997), as the operating deformation mechanism varies with the local composition leading to a heterogeneous mechanical behavior.

The direct strip casting (DSC) process has arisen as an alternative economically viable route to produce TWIP steels by significantly decreasing the hot rolling effort by reducing the as-cast thickness to about 3 mm (Daamen et al. 2014, 2015). Additionally, the rapid solidification achieved during DSC could reduce segregation problems and result in a more homogeneous microstructure (Senk et al. 2000). The high strip temperature combined with the thin strip gauge would also offer the possibility of performing thermomechanical post-treatment, perhaps even in-line, to achieve the desired material properties (Ferry 2006). However, DSC process is still on the first steps of development and issues related to the stability of the process and inhomogeneity of element and grain size distributions have been encountered (Daamen et al. 2011, 2014, 2015).

On the other hand, cold rolling reductions are slightly limited due to the high strain hardening behavior of TWIP steels. Nevertheless, cold rolling of TWIP steels has been the object of active research in order to establish the influence of the alloying elements, C, Si, and Al on the constitutive flow and the transformation behavior during processing. During cold rolling with different degrees of deformation, mechanical twinning and pronounced textures are developed (Vercammen et al. 2004). After subsequent recrystallization annealing, the rolling texture is retained, which is believed to be a consequence of the energetically homogeneous deformation structure, which results in the absence of preferred nucleation sites (Bracke et al. 2009). Besides, homogenized cold strips can be produced by adding an additional annealing prior to the cold rolling to dissolve microsegregations.

12.5 Third Generation of AHSSs

To improve the strength/ductility combinations observed for the first generation AHSSs without the increased cost of the highly alloyed steels of the second generation, new complex microstructures and processing routes have been integrated. The microstructural designs must contain constituents to increase strength (e.g., martensite) and enhance strain hardening (e.g., austenite).

The production of the third generation of AHSSs requires a systematic design methodology to identify and predict the combinations of microstructural constituents that lead to the properties within the opportunity band shown in Figure 12.1. Composite models developed by Matlock et al. (Matlock and Speer 2009) evaluated the resulting strength/elongation combinations as a function of the austenite stability by considering an austenite/martensite mixture according to the continuous composite model developed by Mileiko (Mileiko 1969). Results predict that the opportunity band can be reached by an austenite/martensite mixture, as long as the austenite stability and volume fraction are high. Thus, the next generation of AHSSs steels must incorporate complex microstructures consisting of significant fractions of high strength phases which may be martensite, bainite or ultra fine grained ferrite, in combination with highly-ductile austenite with controlled stability against transformation of austenite to martensite with strain.

12.5.1 Quenching and Partitioning Steels

Quench and partitioning has been shown to be an effective way to produce high strength steels with good ductility containing stabilized retained austenite in a martensitic microstructure. Tensile strength/total elongation combinations range from 1030 MPa/>30% to 1500 MPa/>9%, providing great flexibility for Q&P product applications that can be controlled through carefully designed processing parameters and alloy compositions.

12.5.1.1 Processing and Alloy Design of Q&P Steels

In the Q&P process, steels are rapidly cooled to a specific quench temperature between M_S and M_F to create controlled fractions of martensite and austenite. This step is followed by a thermal treatment at a specific partitioning temperature, at which C migrates from martensite to austenite to increase the austenite stability, resulting in higher austenite fraction at room temperature after cooling. In 1-step processing, partitioning is carried out at the quenching temperature, while 2-step processing involves reheating to a higher partitioning temperature than that of the quench temperature. Figure 12.19 schematizes evolution of the microstructure at the different stages in the 2-step Q&P process. One-step processing might apply to the

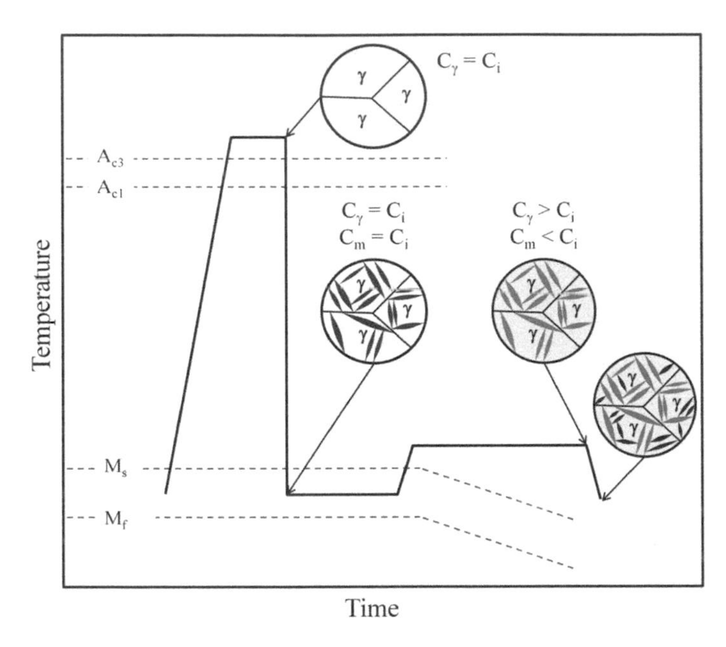

Figure 12.19 Scheme of the Q&P process used to obtain austenite-containing microstructures. C_i, C_γ and C_m represent the carbon concentrations in the initial alloy, austenite and martensite, respectively.

production of hot-rolled sheet steel using low coiling temperatures, while 2-step treatment is envisioned for cold-rolled gages (Speer et al. 2004).

The final volume fraction of martensite can be controlled through the stabilization of the austenite during partitioning. To stabilize the austenite, carbide precipitation must be suppressed (Bhadeshia and Edmonds 1979), as for that, Si and/or Al additions are included in the alloy composition. In addition, austenite stability is assumed to be dependent on its C concentration, and it is not influenced by either the martensite volume fraction or the bulk C content, but by the partitioning of substitutional elements between ferrite and carbides (Toji et al. 2015).

Models have been developed for the prediction of the final microstructure through alloy design (Speer et al. 2003a, Toji et al. 2015). The C partitioning from martensite into austenite was first suggested to be controlled by the constrained C equilibrium (CCE) criterion (Clarke et al. 2008, 2009, Speer et al. 2003a, 2004, 2005). Here, the C concentration in austenite is predicted under the condition where competing reactions, such as carbide formation or bainite transformation, are suppressed. However, carbide precipitation in martensite is often observed during the partitioning step even if the alloys contain a high amount of Si (Edmonds et al. 2006, Santofimia et al. 2009, Toji et al. 2014). If carbides precipitate, the remaining amount of C in martensite that can partition into the austenite is reduced. Thus, the austenite C concentration after the partitioning step, is presumed to be lower than that predicted under the CCE conditions (Toji et al. 2015). A new model has been recently developed to predict the austenite C concentration after the Q&P process considering carbide precipitation (Toji et al. 2015), giving C content values closer to the experimental values than that predicted by the original CCE model.

Compositions considered for the production of Q&P sheet products were first focused in the Fe—0.19 C—1.96 Al—1.46 Mn (wt.%) TRIP sheet steel (Speer et al. 2003b). Due to concerns related to uncertainty in the effects of Al on the Ms temperature, and its influence on the overlapping of the C partitioning and bainite formation, following studies were conducted using Fe—0.20 C—(1.50–1.60) Si—(1.60–1.90) Mn (wt.%) TRIP sheet steels—(Streicher et al. 2004, Sun and Yu 2013, Wang and Feng 2011).

12.5.1.2 Microstructure and Mechanical Properties of Q&P Steels

Q&P microstructures extend the strength/ductility combination levels of the first generation of AHSSs, as shown in the results of Figure 12.20 for current Q&P steels (Clarke 2006, De Moor et al. 2008, Streicher et al. 2004, Sun et al. 2014, Wang and Feng 2011, Yan et al. 2015, Zhang et al. 2015).

Tensile strengths in the 800 to 1500 MPa range can be obtained with ductilities from 8% to 30%. UTS values decrease and elongation values increase with increasing partitioning temperature and time, i.e., with increasing volume fraction of austenite (De Moor et al. 2008). A maximum volume fraction of austenite can be observed as function of partitioning time if cementite formation, bainite transformation or austenite/martensite interface movement occurs.

Austenite stability and volume fraction provide an important contribution to the strength *via* TRIP effect. But, to fully understand and

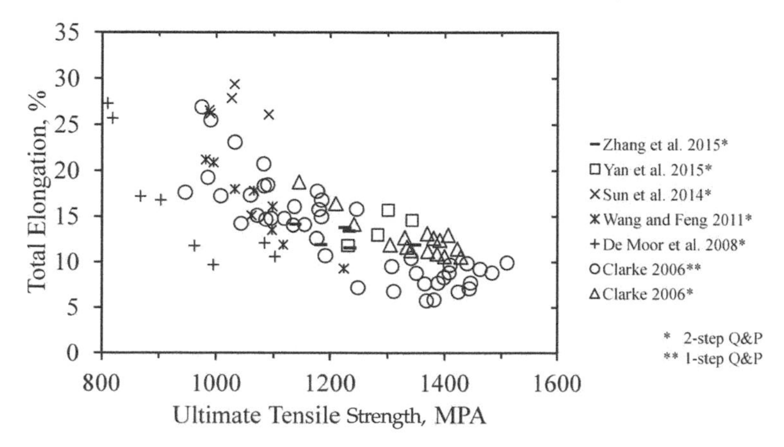

Figure 12.20 Strength/ductility data of current Q&P steels on varying specimen geometries.

exploit the observed strength levels and work hardening behavior, other microstructural combinations including the substructures, dislocations, carbides, solute, and the C depleted martensite need to be considered (Matlock and Speer 2009).

12.5.1.3 Future Perspective of Cold-rolled and Annealed Q&P Steels

Q&P steels can be easily cold-rolled in conventional mills following the strategies used for the first generation of AHSSs. Q&P microstructures are achieved after the subsequent annealing process. However, nowadays it is not feasible to produce Q&P steels in conventional continuous annealing lines. The reasons for this are: (a) the partitioning times required in the 1-step process are too long for current industrial capabilities, and (b) there are no systems available to reheat the coils up to the partitioning temperature in the 2-step process.

Nevertheless, in late 2010 a special ultra-high strength steels (UHSSs) line was developed by Baosteel, which is equipped with a shear line and a slitting line with maximum tensile strength of 1500 MPa. The first full-scale production coils were reported in 2011 (Wang and Feng 2011) for a cold rolled TRIP steel sheet annealed by the Q&P process in that line, achieving a good property balance (tensile strength of 1000–1200 MPa with 9–21% total elongations).

On the other hand, as the implementation of UHSSs lines is not likely to immediately and widely happen, other alternatives as press hardening (PH) have been explored (Seo et al. 2014). The Q&P processing of a compatible Si- and Cr-added PH steel resulted in a considerable improvement of ductility (17% total elongation) without degradation of strength (1320 MPa) (Seo et al. 2014). Thus, the Q&P-processed PH steel might be applied to parts where additional in-service bending performance is required to avoid catastrophic brittle fracture in passenger car collisions.

12.5.2 Carbide-free Bainite Steels

Bainitic AHSSs have been designed on the basis of diffusionless bainite transformation theory (Allain and Iung 2008, Caballero et al. 2006c, 2008, 2009, Hell et al. 2011). According to thermodynamic and kinetic models, CFB steels with compositions Fe—(0.2–0.3) C—1.5 Si—(1.0–2.3) Mn—(0–1.5 Cr)—(0–1.5) Ni (wt.%) were designed and successfully manufactured, first following conventional hot rolling practices (Allain and Iung 2008, Caballero et al. 2006c, 2008, 2009) and later following cold rolling and continuous annealing (Caballero et al. 2013a). As aforementioned, Si additions suppress the precipitation of brittle cementite during bainite formation, and hence should lead to an improvement in toughness. The resulting microstructure consists of a mixture of bainitic ferrite, retained austenite, and some martensite (Bhadeshia 1983, Bhadeshia and Edmonds 1983). These steels present significant combinations of strength and ductility, with tensile strengths ranging from 1300 to 1800 MPa and total elongations over 14%, and are good candidates in the future prospect of the automotive sector.

12.5.2.1 Theoretical Calculations and Alloy Design of CFB Steels

Bainite reaction takes place by displacive transformation of the austenite within the bainitic range. The bainite transformation window is delimited by the martensite start temperature (M_S) and the bainite start temperature (*Bs*), which is defined as the highest temperature at which austenite starts to transform to martensite and bainite respectively. Nevertheless, a more rapid onset of bainite reaction can be achieved at temperatures below the M_s due to the accelerating effect of the presence of martensite plates on the nucleation of bainite in adjacent regions of untransformed austenite (Radcliffe and Rollason 1959).

The bainite transformation progresses by the diffusionless growth of platelets whose C excess partitions into the residual austenite immediately after the growth event (Bhadeshia and Edmonds 1980). This diffusionless growth can only occur if the C concentration of the residual austenite is below that given by the T_0 curve. The T_0 curve is the locus of all points, on a temperature versus C concentration plot, where austenite and ferrite of the same chemical composition have the same free energy (Bhadeshia and Edmonds 1980). The construction of the T_0 curve is illustrated in Figure 12.21a. It follows that the maximum amount of bainite that can be obtained at any temperature is limited by the fact that the C content of the residual austenite must not exceed the T_0 curve of the phase diagram.

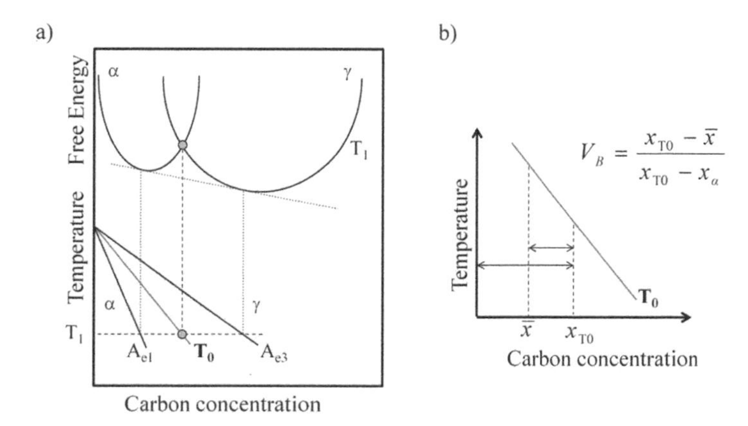

Figure 12.21 (a) Definition of the T_0 curve. (b) Lever rule applied to the T_0 curve to determine the maximum volume fraction of bainite formed at a given temperature.

The maximum volume fraction of bainite, V_{Bmax}, which can be formed at a given temperature is directly related to the T_0 curve and can be calculated as follows (see Figure 12.21b):

$$V_{Bmax} = \frac{(x_\gamma - \bar{x})}{(x_\gamma - s)} \tag{12.2}$$

where x_γ is the C content of austenite according to the T_0 curve, $\bar{x}$ is the average C content of the alloy, and s is the amount of C trapped in the bainitic ferrite in solid solution ($s = 0.03$ wt.%). In spite of the independence of the T_0 curve on the average C content of the steel (Bhadeshia 1981), the maximum volume fraction of bainite formed at a given transformation temperature will increase as the C content of the steel is decreased since the critical C concentration for displacive transformation is reached at a later stage (Bhadeshia 1983, Bhadeshia and Edmonds 1983).

The design procedure maximizes the volume fraction of bainite formed in two ways: by adjusting the T_0 curve to greater C concentrations with the use of substitutional solutes such as Mn and Cr and by controlling the mean C concentration (Bhadeshia 1983, Bhadeshia and Edmonds 1983).

Apart from controlling the T_0 curve and B_S temperature, substitutional solutes also affect hardenability, which is an important design parameter to avoid transformations such as proeutectoid ferrite and pearlite. To predict the effect of the alloy composition on the isothermal and continuous transformation diagrams, thermodynamic and kinetic models are used in the alloy design process (Bhadeshia 1982, Caballero et al. 2002, Jones and Bhadeshia 1997). There are other output parameters, such as the martensite and Widmanstätten start temperatures. An extensive description of all the models currently used on the design procedure of CFB steels can be found elsewhere (Caballero et al. 2006c). It must be always borne in mind that for commercial manufacturing of steels in continuous annealing lines, the chemical composition of the alloys has to be adapted in order to ensure a low Ac_3 temperature to guarantee full austenitization during annealing and avoid proeutectoid ferrite upon cooling.

12.5.2.2 Cold Rolling and Annealing of CFB Steels

For commercial manufacturing of steel sheets in conventional continuous annealing lines, a soft microstructure must be first achieved after hot rolling to allow cold rolling of the material without any softening treatment. Bainitic microstructures are achieved after the subsequent annealing process. Hence, ingots are hot rolled to a thickness of ~3 mm following the thermomechanical control process of commercial SiMn TRIP steels described in Figure 12.22.

A high coiling temperature (over 550°C) must be selected to trigger the ferrite–pearlite transformation region, and thus, to achieve soft hot

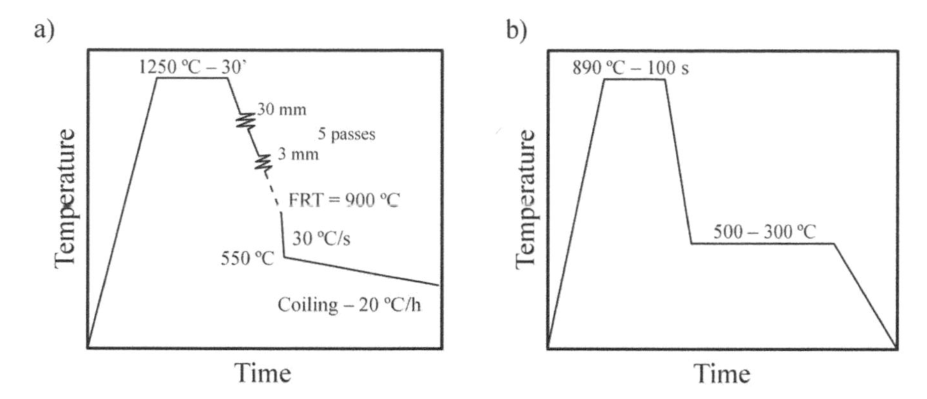

Figure 12.22 Scheme of temperature–time profile used for (a) hot rolling and (b) continuous annealing processes of CFB steels. FRT stands for finishing rolling temperature.

rolled strips and to enable the cold rolling without limiting the accessible gauge thickness for the final product. Considering the high Si content of the designed steels, higher coiling temperatures are not recommended for preventing inter-granular oxidation, which could cause the deterioration of product performance (Wise et al. 2001). The low hardness of the hot rolled strips (below 400 HV) enables the subsequent cold rolling of the material without any further softening treatment. The sheets can be subsequently cold-rolled to a total reduction rate of 50% in multiple passes, without any cracks (Caballero et al. 2013a).

The described process has been recently applied to annealed cold rolled bainitic steels designed for uncoated (i.e., bare) or electrogalvanized (EG) products and manufactured by a continuous annealing line. The microstructures achieved showed outstanding mechanical performances compared to present-day very high strength offers (e.g., DP750/980 and MS1400) (Caballero et al. 2013a).

However, tensile properties of CFB steels produced by conventional hot dip galvanizing (HDG) annealing are disappointing and comparable to those of high-Si DP steels (Caballero et al. 2013b). This process cannot be recommended to obtain fully CFB steels (i.e., without the presence of equiaxed ferrite in the microstructure), unless some improvements on soaking and cooling capabilities are explored. In this sense, a HDG annealing with a low end-cooling strategy, consisting of an over ageing at temperatures lower than 460°C and followed by an induction heating, could be an option to achieve CFB microstructures with properties comparable to those obtained with continuous annealing lines capabilities (Wang and Feng 2010).

12.5.2.3 Microstructure and Mechanical Properties of CFB Steels

Figure 12.23 shows an example of the microstructure obtained by bainitic transformation at 400°C in a Fe—0.2 C—1.5 Mn—1.5 Si—1.5 Cr (wt.%) steel sheet of 1.0 mm thickness. Microstructures in CFB steels consist of a lath-like matrix composed of bainitic ferrite and/or a-thermal martensite. Bainite is formed during the isothermal holding stage and martensite is formed during quenching before the holding stage, if this is below the M_s temperature.

Figure 12.24 shows the strength/ductility combinations of current CFB steels as reported in (Alvarez et al. 2014, Caballero et al. 2001, 2006c, 2008, 2009, 2013a). Unprecedented strength/ductility combinations of 1945 MPa/7% and 1525 MPa/25% have been achieved in these steels, making them attractive candidates for the future automotive industry. The main microstructural contribution to the strength of CFB microstructures is from the extremely fine grain size of lath-like ferrite. It is known that the effective lath size defined by low angle boundaries explains yield strength since both lath boundaries and dislocation cell boundaries have the similar capability of being dislocation obstacles.

Moreover, it is difficult to separate the effect of retained austenite on strength in these steels from other factors (see Figure 12.25a). Qualitatively, austenite can affect strength in two ways: (i) since austenite is a soft and ductile phase, this alone could cause a decrease in the yield strength; and (ii) retained austenite could increase the tensile strength by transforming into martensite during testing, similar to the behavior of TRIP steels. On

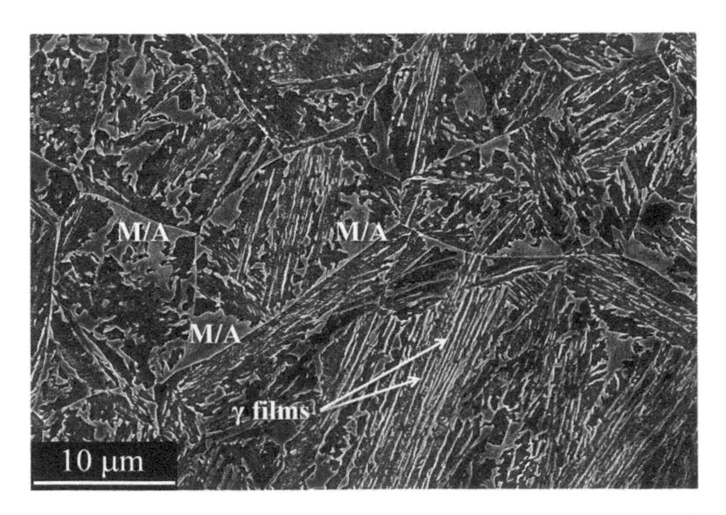

Figure 12.23 Scanning electron micrographs of bainitic microstructure obtained in a Fe—0.2 C—1.5 Mn—1.5 Si—1.5 Cr (wt.%) steel sheet (1.0 mm thickness) by transformation at 400°C. M/A is martensite/austenite constituent; γ is austenite.

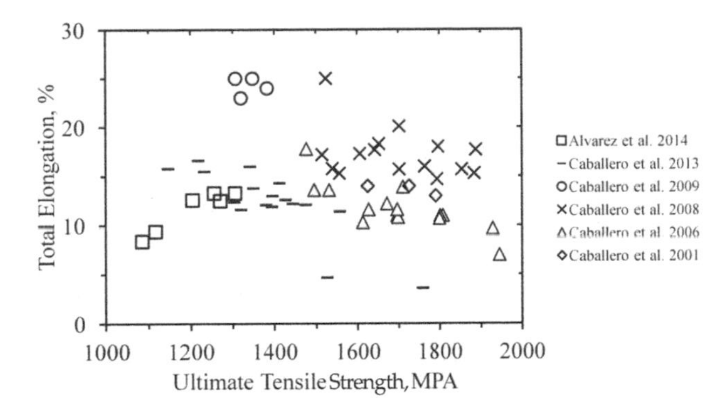

Figure 12.24 Strength/ductility data of current CFB steels on varying specimen geometries.

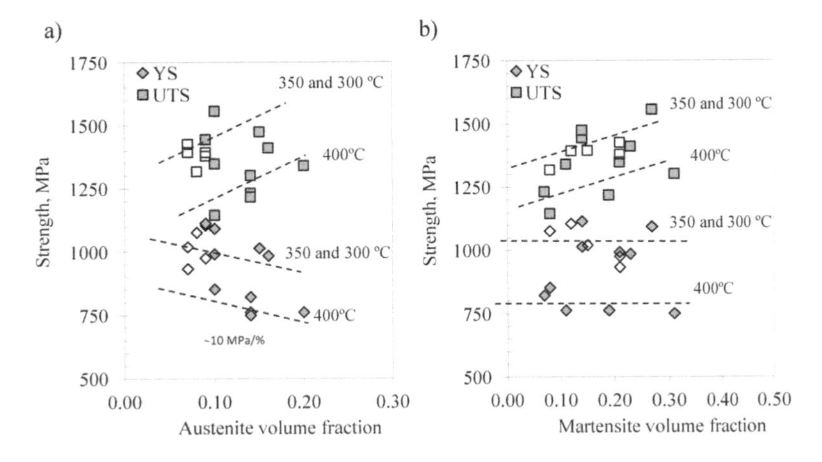

Figure 12.25 Effect of (a) retained austenite and (b) martensite on strength of annealed cold rolled CFB steels. Hollow symbols correspond to samples transformed below M_S temperature. YS is yield strength and UTS is ultimate tensile strength.

the other hand, the amount of martensite in the microstructure plays an important role on the tensile strength, as illustrated in Figure 12.25b.

A general observation in these steels is that yield strength seems to increase as the microstructure is generated at the lower temperature. In addition to the ferrite lath thickness, other strengthening mechanisms are active, such as precipitation strengthening, dislocation density, and C trapping at crystal defects.

12.5.2.4 Future Perspective for Cold-rolled and Annealed CFB Steels

CFB steels for continuous annealing lines are very promising. They present high tensile performances and their properties can be adjusted for cold-stamping applications as a result of an adequate choice of the bainitic transformation temperature. However, the following constraints remain to be overcome (Caballero et al. 2013a):

a) In order to achieve a fully austenitic soaking during annealing, the Ac_3 temperature of the alloy should be below the highest temperature accessible by plants.
b) The critical cooling rate of the alloy should be designed sufficiently high to avoid ferrite upon cooling.
c) The bainite holding temperature is a relevant parameter to optimize the mechanical properties of the final products. Temperatures above M_S should be selected to reach optimized elongation levels, whereas low temperature bainitic holding should be used to produce high strength products. In both cases, the expected performances are higher than those of conventional very high strength steels (i.e., DP steels). However, bainitic holding at too high temperature could lead to insufficient transformed fractions and should be avoided.
d) Acceleration of bainitic transformation is a key issue for the design of industrial products. Most of the durations used for laboratory sample production are for the moment incompatible with industrial capability (too long a bainitic plateau). Too short bainitic holding would result in low damaging performance of the microstructure (too high fraction of fresh martensite).
e) Austenite stabilizing elements, such as C, Mn, Cu, and Ni significantly improve the mechanical performance of carbide-free bainitic steels. Cu and Ni seem to have the same effect as Mn and C. Nevertheless, all these elements slow down the bainitic transformation kinetics.

Summary

The development of AHSSs has been driven by the necessity of vehicle weight reduction for fuel efficiency and materials with enhanced crash performance and passenger safety. Current production of AHSSs involves ferrite-based multiphase steels with lean compositions and different cold rolling and annealing routes that lead to specific strength/ductility combinations within the lowest property band in Figure 12.1. The properties of these steels, namely DP, TRIP, CP and MS, are derived from suitable combinations of strengthening mechanisms, the most attractive of them being the TRIP effect. To take advantage of the TRIP effect while increasing strength, developments for new AHSSs are based in lath-like matrices (either martensitic or bainitic) with retained austenite. New Q&P and CFB steels show strength/ductility combinations towards the opportunity band that meet the objectives of the latest generation of AHSSs. An overview of total elongation and strength combinations of current first and third generation of AHSSs is given in Figure 12.26.

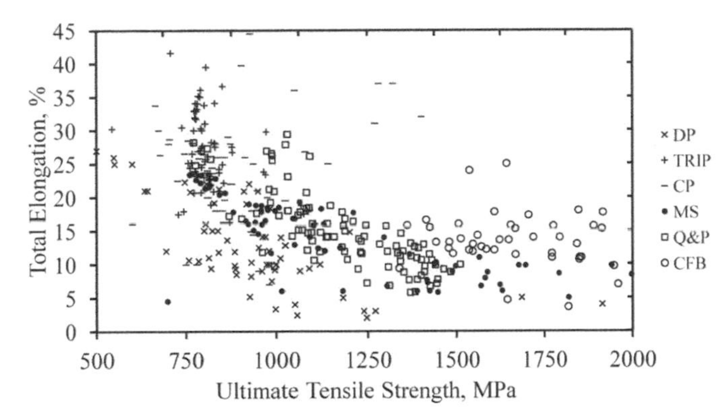

Figure 12.26 Strength/ductility data of current first and third generation of AHSSs on varying specimen geometries.

Keywords: Cold rolling, annealing, processing, dual-phase steels, complex-phase steels, TRIP-assisted steels, martensitic steels, quenching and partitioning steels, bainitic steels, microstructure, alloy design, mechanical properties

References

Ahmad, E., T. Manzoor, K.L. Ali and J.I. Akhter. 2000. Effect of microvoid formation on the tensile properties of dual-phase steel. J. Mater. Eng. Perform. 9: 306–310.

Akhlaghi, S. and S. Yue. 2001. Effect of thermomechanical processing on the hot ductility of a Nb-Ti microalloyed steel. ISIJ Int. 41: 1350–1356.

Alvarez, D., J. Ferreño and J.M. Artimez. 2014. Development of low Silicon carbide free bainitic steel for automotive industry, Proc. Transport Research Arena (TRA) 2014, Paris.

Allain, S. and T. Iung. 2008. Development of hot rolled copper/nickel alloyed TRIP steels with carbide-free bainitic matrix Rev. Metall./Cah. Inf. Tech. 10: 520–530.

Bag, A., K.K. Ray and E.S. Dwarakadasa. 1999. Influence of martensite content and morphology on tensile and impact properties of high-martensite dual-phase steels. Metall. Mater. Trans. A 30: 1193–1202.

Bai, D.Q., A. Di Chiro and S. Yue. 1998. Stability of retained austenite in a Nb microalloyed Mn-Si TRIP steel. Mater. Sci. Forum. 284-286: 253–260.

Baik, S.C., S. Kim, Y.S. Jin and O. Kwon. 2001. Effects of alloying elements on mechanical properties and phase transformation of cold rolled TRIP steel sheets. ISIJ Int. 41: 290–297.

Béranger, G., G. Henry and G. Sanz. 1996. The Book of Steel. Lavoisier Publishing, Paris.

Bergstrom, Y., Y. Granbom and D. Sterkenburg. 2010. A dislocation-based theory for the deformation hardening behavior of DP steels: Impact of martensite content and ferrite grain size. J. Metall, pp. 1–16.

Bhadeshia, H.K.D.H. and D.V. Edmonds. 1979. The bainite transformation in a silicon steel. Metall. Trans. A 10: 895–907.

Bhadeshia, H.K.D.H. and D.V. Edmonds. 1980. Mechanism of bainite formation in steels. Acta Metall. 28: 1265–1273.

Bhadeshia, H.K.D.H. 1981. A rationalisation of shear transformations in steels. Acta Metall. 29: 1117–1130.

Bhadeshia, H.K.D.H. 1982. Thermodynamic analysis of isothermal transformation diagrams. Met. Sci. 16: 159–165.

Bhadeshia, H.K.D.H. 1983. Bainite in silicon steels: New composition-property approach. Part 2. Met. Sci. 17: 420–425.

Bhadeshia, H.K.D.H. and D.V. Edmonds. 1983. Bainite in silicon steels: New composition-property approach. Part 1. Met. Sci. 17: 411–419.

Bleck, W., K. Hulka and K. Papamentellos. 1998. Effect of niobium on the mechanical properties of TRIP steels. Mater. Sci. Forum. 284-286: 327–334.

Bracke, L., K. Verbeken, L. Kestens and J. Penning. 2009. Microstructure and texture evolution during cold rolling and annealing of a high Mn TWIP steel. Acta Mater. 57: 1512–1524.

Caballero, F.G., H.K.D.H. Bhadeshia, K.J.A. Mawella, D.G. Jones and P. Brown. 2001. Design of novel high strength bainitic steels: Part 2. Mater. Sci. Technol. 17: 517–522.

Caballero, F.G., C. Capdevila and C. Garcia de Andrés. 2002. Evaluation and review of simultaneous transformation model in high strength low alloy steels. Mater. Sci. Technol. 18: 534–540.

Caballero, F.G., A. García-Junceda, C. Capdevila and C. García de Andrés. 2006a. Evolution of microstructural banding during the manufacturing process of dual phase steels. Mater. Trans. JIM 47: 2269–2276.

Caballero, F.G., A. García-Junceda, C. Capdevila and C.G. De Andrés. 2006b. Evolution of microstructural banding during the manufacturing process of dual phase steels. Mater. Trans. JIM. 47: 2269–2276.

Caballero, F.G., M.J. Santofimia, C. Capdevila, C. García-Mateo and C. De García Andrés. 2006c. Design of advanced bainitic steels by optimisation of TTT diagrams and T0 curves. ISIJ Int. 46: 1479–1488.

Caballero, F.G., C. García-Mateo, J. Chao, M.J. Santofimia, C. Capdevila and C.G. De Andrés. 2008. Effects of morphology and stability of retained austenite on the ductility of TRIP-aided bainitic steels. ISIJ Int. 48: 1256–1262.

Caballero, F.G., M.J. Santofimia, C. García-Mateo, J. Chao and C.G. de Andrés. 2009. Theoretical design and advanced microstructure in super high strength steels. Mater. Des. 30: 2077–2083.

Caballero, F.G., S. Allain, J. Cornide, J.D. Puerta Velásquez, C. Garcia-Mateo and M.K. Miller. 2013a. Design of cold rolled and continuous annealed carbide-free bainitic steels for automotive application. Mater. Des. 49: 667–680.

Caballero, F.G., S. Allain, J.D. Puerta-velásquez and C. Garcia-mateo. 2013b. Exploring carbide-free bainitic structures for hot dip galvanizing products. ISIJ Int. 53: 1253–1259.

Clarke, A. 2006. Carbon Partitioning into Austenite from Martensite in A Silicon-Containing High Strength Sheet Steel. Ph.D. Thesis, Colorado School of Mines, Golde, Colorado.

Clarke, A.J., J.G. Speer, M.K. Miller, R.E. Hackenberg, D.V. Edmonds, D.K. Matlock, F.C. Rizzo, K.D. Clarke and E. De Moor. 2008. Carbon partitioning to austenite from martensite or bainite during the quench and partition (Q&P) process: A critical assessment. Acta Mater. 56: 16–22.

Clarke, A.J., J.G. Speer, D.K. Matlock, F.C. Rizzo, D.V. Edmonds and M.J. Santofimia. 2009. Influence of carbon partitioning kinetics on final austenite fraction during quenching and partitioning. Scr. Mater. 61: 149–152.

Consortium, ULSAB. 1998. UltraLight Steel Auto Body Final Report, Washington D.C.

Crooks, M.J., A.J. Garratt-Reed, J.B.V. Sande and W.S. Owen. 1982. The isothermal austenite-ferrite transformation in some deformed vanadium steels. Metall. Trans. A 13: 1347–1353.

Chang, W.S. 2002. Microstructure and mechanical properties of 780 MPa high strength steels produced by direct-quenching and tempering process. J. Mater. Sci. 37: 1973–1979.

Daamen, M., B. Wietbrock, S. Richter and G. Hirt. 2011. Strip casting of a high-manganese steel (FeMn22C0. 6) compared with a process chain consisting of ingot casting and hot forming. Steel Res. Int. 82: 70–75.

Daamen, M., O. Güvenç, M. Bambach and G. Hirt. 2014. Development of efficient production routes based on strip casting for advanced high strength steels for crash-relevant parts. CIRP Annals—Manuf. Technol. 63: 265–268.

Daamen, M., C. Haase, J. Dierdorf, D.A. Molodov and G. Hirt. 2015. Twin-roll strip casting: A competitive alternative for the production of high-manganese steels with advanced mechanical properties. Mater. Sci. Eng. A 627: 72–81.

Davies, R.G. 1978. Influence of martensite composition and content on the properties of dual phase steels. Metall. Trans. A 9: 671–679.

De Cooman, B.C. 2004. Structure–properties relationship in TRIP steels containing carbide-free bainite. Curr. Opin. Solid State Mat. Sci. 8: 285–303.

De Cooman, B.C., K.G. Chin and J. Kim. 2011. High Mn TWIP steels for automotive applications. pp. 101–128. *In*: M. Chiaberge (ed.). New Trends and Developments in Automotive System Engineering, InTech Open Access Publisher.

De Moor, E., S. Lacroix, A.J. Clarke, J. Penning and J.G. Speer. 2008. Effect of retained austenite stabilized via quench and partitioning on the strain hardening of martensitic steels. Metall. Mater. Trans. A 39: 2586–2595.

Ding, H., Z.Y. Tang, W. Li, M. Wang and D. Song. 2006. Microstructures and Mechanical Properties of Fe-Mn-(Al, Si) TRIP/TWIP Steels. J. Iron. Steel Res. Int. 13: 66–70.

Dini, G., A. Najafizadeh, R. Ueji and S.M. Monir-Vaghefi. 2010. Improved tensile properties of partially recrystallized submicron grained TWIP steel. Mater. Lett. 64: 15–18.

Duang, Z., Y. Li, M. Zhang, M. Shi, F. Zhu and S. Zhang. 2012. Effects of quenching process on mechanical properties and microstructure of high strength steel. Journal of Wuhan University of Technology-Mater. Sci. Ed. 27: 1024–1028.

Edmonds, D.V., K. He, F.C. Rizzo, B.C. De Cooman, D.K. Matlock and J.G. Speer. 2006. Quenching and partitioning martensite-A novel steel heat treatment. Mater. Sci. Eng. A 438-440: 25–34.

Ferry, M. 2006. Direct Strip Casting of Metals and Alloys. Woodhead Publishing Limited, Cambridge.

Frommeyer, G., U. Brüx and P. Neumann. 2003. Supra-ductile and high-strength manganese-TRIP/TWIP steels for high energy absorption purposes. ISIJ Int. 43: 438–446.

Geib, M.D., D.K. Matlock and G. Krauss. 1980. The effect of intercritical annealing temperature on the structure of niobium microalloyed dual phase steel. Metall. Trans. A 11: 1683–1689.

Grajcar, A. and M. Opiela. 2008. Influence of plastic deformation on CCT-diagrams of low-carbon and medium-carbon TRIP steels. JAMME. 29: 71–78.

Grange, R.A. 1971. Effect of microstructural banding in steel. Metall. Trans. A 2: 417–426.

Grässel, O., L. Krüger, G. Frommeyer and L.W. Meyer. 2000. High strength Fe–Mn–(Al, Si) TRIP/TWIP steels development—properties—application. Int. J. Plast. 16: 1391–1409.

Grossterlinden, R., R. Kawalla, U. Lotter and H. Pircher. 1992. Formation of pearlitic banded structures in ferritic-pearlitic steels. Steel Res. Int. 63: 331–336.

Hanlon, D.N., J. Sietsma and S.v.d. Zwaag. 2001. The Effect of Plastic Deformation of Austenite on the Kinetics of Subsequent Ferrite Formation. ISIJ Int. 41: 1028–1036.

Hanzaki, A.Z., P.D. Hodgson and S. Yue. 1995. Hot deformation characteristics of Si-Mn TRIP steels with and without Nb microalloy additions. ISIJ Int. 35: 324–331.

Hayami, S. and T. Furukawa. 1975. A family of high strength cold rolled steels. Microalloying. 75: 78–87.

Hell, J.C., M. Dehmas, S. Allain, J.M. Prado, A. Hazotte and J.P. Chateau. 2011. Microstructure —Properties relationships in carbide-free bainitic steels. ISIJ Int. 51: 1724–1732.

Hölscher, M., D. Raabe and K. Lücke. 1994. Relationship between rolling textures and shear textures in f.c.c. and b.c.c. metals. Acta Metall. Mater. 42: 879–886.

Ishida, K. and T. Nishizawa. 1974. Effect of alloying elements on stability of epsilon iron. Trans. Jap. Inst. Met. 15: 225–231.

Jacobi, H. and K. Wünnenberg. 1999. Solidification structure and micro-segregation of unidirectionally solidified steels. Steel Res. Int. 70: 362–367.

Jiang, Z., Z. Guan and J. Lian. 1995. Effects of microstructural variables on the deformation behaviour of dual-phase steel. Mater. Sci. Eng. A 190: 55–64.

Jones, S.J. and H.K.D.H. Bhadeshia. 1997. Kinetics of the simultaneous decomposition of austenite into several transformation products. Acta Mater. 45: 2911–2920.

Keeler, S. and M. Kimchi. 2015. Advanced High-Strength Steels Application Guidelines V5, WorldAutoSteel.

Kim, N.J. and G. Thomas. 1981. Effects of morphology on the mechanical behavior of a dual phase Fe/2Si/0.1C steel. Metall. Trans. A 12: 483–489.

Kim, S.J., C. Gil Lee, T.-H. Lee and C.-S. Oh. 2003. Effect of Cu, Cr and Ni on mechanical properties of 0.15 wt.% C TRIP-aided cold rolled steels. Sci. Mater. 48: 539–544.

Kim, T.W., Y.G. Kim and S.H. Park. 1997. Process for manufacturing high manganese hot rolled steel sheet without any crack. Process for Manufacturing High Manganese Hot Rolled Steel Sheet Without Any Crack. US Patent #US5647922 A.

Kozeschnik, E. and H.K.D.H. Bhadeshia. 2008. Influence of silicon on cementite precipitation in steels. Mater. Sci. Technol. 24: 343–347.

Krauss, G. 2003. Solidification, Segregation, and banding in carbon and alloy steels. Metall. Mater. Trans. B 34: 781–792.

Krizan, D. and B. De Cooman. 2014. Mechanical properties of trip steel microalloyed with Ti. Metall. Mater. Trans. A 45: 3481–3492.

Lee, W.S. and T.T. Su. 1999. Mechanical properties and microstructural features of AISI 4340 high-strength alloy steel under quenched and tempered conditions. J. Mater. Process. Technol. 87: 198–206.

Li, Z. and D. Wu. 2006. Effects of Hot Deformation and Subsequent Austempering on the Mechanical Properties of Si-Mn TRIP Steels. ISIJ Int. 46: 121–128.

Mahieu, J., J. Maki, B.C. De Cooman and S. Claessens. 2002. Phase transformation and mechanical properties of Si-free CMnAl transformation-induced plasticity-aided steel. Metall. Mater. Trans. A 33: 2573–2580.

Majka, T.F., D.K. Matlock and G. Krauss. 2002. Development of microstructural banding in low-alloy steel with simulated Mn segregation. Metall. Mater. Trans. A 33: 1627–1637.

Matlock, D.K. and J.G. Speer. 2009. Third Generation of AHSS: Microstructure Design Concepts. pp. 185–205. *In*: A. Haldar, S. Suwas and D. Bhattacharjee (eds.). Microstructure and Texture in Steels, Springer, London.

Meyer, M.D., D. Vanderschueren and B.C.D. Cooman. 1999. The influence of the substitution of Si by Al on the properties of cold rolled C-Mn-Si TRIP steels. ISIJ Int. 39: 813–822.

Meysami, A.H., R. Ghasemzadeh, S.H. Seyedein and M.R. Aboutalebi. 2010. An investigation on the microstructure and mechanical properties of direct-quenched and tempered AISI 4140 steel. Mater. Des. 31: 1570–1575.

Mileiko, S.T. 1969. The tensile strength and ductility of continuous fibre composites. J. Mater. Sci. 4: 974–977.

Mohrbacher, H. 2015. Martensitic Automotive Steel Sheet-Fundamentals and Metallurgical Optimization Strategies. Adv. Mater. Res. 1063: 130–142.

Morito, S., X. Huang, T. Furuhara, T. Maki and N. Hansen. 2006a. The morphology and crystallography of lath martensite in alloy steels. Acta Mater. 54: 5323–5331.

Morito, S., H. Yoshida, T. Maki and X. Huang. 2006b. Effect of block size on the strength of lath martensite in low carbon steels. Mater. Sci. Eng. A 438-440: 237–240.

Muckelroy, N.C., K.O. Findley and R.L. Bodnar. 2013. Microstructure and Mechanical Properties of Direct Quenched Versus Conventional Reaustenitized and Quenched Plate. J. Mater. Eng. Perform. 22: 512–522.

Onink, M., T.M. Hoogendoorn and J. Colijn. 1998. Nb and the transformation from deformed austenite. Mater. Sci. Forum 284-286: 185–192.

Panigrahi, B.K. 2001. Processing of low carbon steel plate and hot strip-an overview. Bull. Mat. Sci. 24: 361–371.

Perrard, F. and C. Scott. 2007. Vanadium precipitation during intercritical annealing in cold rolled TRIP steels. ISIJ Int. 47: 1168–1177.

Pierman, A.P., O. Bouaziz, T. Pardoen, P.J. Jacques and L. Brassart. 2014. The influence of microstructure and composition on the plastic behaviour of dual-phase steels. Acta Mater. 73: 298–311.

Radcliffe, S.V. and E.C. Rollason. 1959. The kinetics of the formation of bainite in high-purity iron-carbon alloys. J. Iron Steel Inst. 191: 56–65.

Rashid, M.S. 1981. Dual phase steels. Ann. Rev. Mater. Sci. 11: 245–266.

Rivera-Díaz-Del-Castillo, P.E.J. and S. Van Der Zwaag. 2004. Assuring microstructural homogeniety in dual phase and trip steels. Steel Res. Int. 75: 711–715.

Rivlin, V.G. 1983. Phase equilibria in iron ternary alloys - 12: Critical review of constitution of aluminium-iron-manganese and iron-manganese-silicon systems. Int. Met. Rev. 28: 309–337.

Rocha, R.O., T.M.F. Melo, E.V. Pereloma and D.B. Santos. 2005. Microstructural evolution at the initial stages of continuous annealing of cold rolled dual-phase steel. Mater. Sci. Eng. A 391: 296–304.

Santofimia, M.J., L. Zhao and J. Sietsma. 2009. Microstructural evolution of a low-carbon steel during application of quenching and partitioning heat treatments after partial austenitization. Metall. Mater. Trans. A 40: 46–57.

Sato, K., M. Ichinose, Y. Hirotsu and Y. Inoue. 1989. Effects of deformation induced phase transformation and twinning on the mechanical properties of austenitic Fe-Mn-Al alloys. ISIJ Int. 29: 868–877.

Scott, C., P. Maugis, P. Barges and M. Gouné. 2004. Microalloying with vanadium in TRIP steels. Proc. Int. Conf. Adv. High Strength Sheet Steels Auto. App., pp. 181–193.

Scott, C., L. Gavard, A. Dero, T. Ewertz and T. Maiwald. 2007. New metallurgy for microalloyed TRIP steels. Final Report No. EUR 22391, Research Found for Coal and Steel.

Schramm, R.E. and R.P. Reed. 1975. Stacking fault energies of seven commercial austenitic stainless steels. Metall. Trans. A 6: 1345–1351.

Senk, D., F. Hagemann, B. Hammer, R. Kopp, H.P. Schmitz and W. Schmitz. 2000. In-line rolling and cooling of direct cast steel strip. Stahl und Eisen. 120: 65–69.

Seo, E., L. Cho and B. De Cooman. 2014. Application of Quenching and Partitioning (Q&P) Processing to Press Hardening Steel. Metall. Mater. Trans. A 45: 4022–4037.

Shi, W., L. Li, C.-X. Yang, R.-Y. Fu, L. Wang and P. Wollants. 2006. Strain-induced transformation of retained austenite in low-carbon low-silicon TRIP steel containing aluminum and vanadium. Mater. Sci. Eng. A 429: 247–251.

Speer, J., D.K. Matlock, B.C. De Cooman and J.G. Schroth. 2003a. Carbon partitioning into austenite after martensite transformation. Acta Mater. 51: 2611–2622.

Speer, J.G., A.M. Streicher, D.K. Matlock, F.C. Rizzo and G. Krauss. 2003b. Quenching and partitioning: A fundamentally new process to create high strength TRIP sheet microstructures. pp. 505–522. *In*: E.B. Damm and M. Merwin (eds.). Austenite Formation and Decomposition, Warrendale.

Speer, J.G., D.V. Edmonds, F.C. Rizzo and D.K. Matlock. 2004. Partitioning of carbon from supersaturated plates of ferrite, with application to steel processing and fundamentals of the bainite transformation. Curr. Opin. Solid State Mat. Sci. 8: 219–237.

Speer, J.G., F.C. Rizzo **Assunção, D.K. Matlock and** D.V. Edmonds. 2005. The "quenching and partitioning" process: Background and recent progress. Mater. Res. 8: 417–423.

Speich, G.R., V.A. Demarest and R.L. Miller. 1981. Formation of austenite during intercritical annealing of dual-phase steels. Metall. Trans. A 12: 1419–1428.

Streicher, A.M., J.G. Speer, D.K. Matlock and B.C. De Cooman. 2004. Quenching and partitioning response of a Si-added TRIP sheet steel. Proc. Int. Conf. Adv. High Strength Sheet Steels Auto. App., pp. 51–62.

Sun, J. and H. Yu. 2013. Microstructure development and mechanical properties of quenching and partitioning (Q&P) steel and an incorporation of hot-dipping galvanization during Q&P process. Mater. Sci. Eng. A 586: 100–107.

Sun, J., H. Yu, S. Wang and Y. Fan. 2014. Study of microstructural evolution, microstructure-mechanical properties correlation and collaborative deformation-transformation behavior of quenching and partitioning (Q&P) steel. Mater. Sci. Eng. A 596: 89–97.

Takahashi, M. and H.K.D. Bhadeshia. 1989. The Interpretation of dilatometric data for transformations in steels. J. Mater. Sci. Lett. 8: 477–478.

Takechi, H., O. Matsumura and K. Sakuma. 1987a. Production Of High-Strength Steel Sheet Having Excellent Ductility. Japan Patent, JP19860023642 19860205.

Takechi, H., O. Matsumura and K. Sakuma. 1987b. Production Of High-Strength Steel Sheet Having Good Ductility. Japan Patent, JP19860023643 19860205.

Tasan, C.C., J.P.M. Hoefnagels and M.G.D. Geers. 2010. Microstructural banding effects clarified through micrographic digital image correlation. Scr. Mater. 62: 835–838.

Tasan, C.C., M. Diehl, D. Yan, M. Bechtold, F. Roters, L. Schemmann, C. Zheng, N. Peranio, D. Ponge, M. Koyama, K. Tsuzaki and D. Raabe. 2015. An overview of dual-phase steels: Advances in microstructure-oriented processing and micromechanically guided design. Annual Review of Materials Research. 45: 391–431.

Taylor, K.A. and S.S. Hansen. 1991. Effects of vanadium and processing parameters on the structures and properties of a direct-quenched low-carbon Mo-B steel. Metall. Trans. A 22: 2359–2374.

Thompson, S.W. and P.R. Howell. 1992. Factors influencing ferrite/pearlite banding and origin of large pearlite nodules in a hypoeutectoid plate steel. Mater. Sci. Technol. 8: 777–784.

Timokhina, I.B., E.V. Pereloma and P.D. Hodgson. 2001a. Microstructure and mechanical properties of C-Si-Mn(-Nb) TRIP steels after simulated thermomechanical processing. Mater. Sci. Technol. 17: 135–140.

Timokhina, I.B., E.V. Pereloma and P.D. Hodgson. 2001b. Microstructure and mechanical properties of C–Si–Mn (–Nb) TRIP steels after simulated thermomechanical processing. Mater. Sci. Technol. 17: 135–140.

Toji, Y., H. Matsuda, M. Herbig, P.P. Choi and D. Raabe. 2014. Atomic-scale analysis of carbon partitioning between martensite and austenite by atom probe tomography and correlative transmission electron microscopy. Acta Mater. 65: 215–228.

Toji, Y., G. Miyamoto and D. Raabe. 2015. Carbon partitioning during quenching and partitioning heat treatment accompanied by carbide precipitation. Acta Mater. 86: 137–147.

Tsukatani, I., S.i. Hashimoto and T. Inoue. 1991. Effects of silicon and manganese addition on mechanical properties of high-strength hot-rolled sheet steel containing retained austenite. ISIJ Int. 31: 992–1000.

Vercammen, S., B. Blanpain, B.C. De Cooman and P. Wollants. 2004. Cold rolling behaviour of an austenitic Fe–30Mn–3Al–3Si TWIP-steel: the importance of deformation twinning. Acta Mater. 52: 2005–2012.

Wang, L. and W. Feng. 2010. Industry Trials of C-Si-Mn Steel Treated by Q&P Concept in Baosteel. SAE Technical Paper 2010-01-0439.

Wang, L. and W. Feng. 2011. Development and application of Q&P sheet steels. pp. 255–258. *In*: Y. Weng, H. Dong and Y. Gan (eds.). Advanced Steels: The Recent Scenario in Steel Science and Technology. Springer, Berlin.

Wise, J.P., G. Krauss and D.K. Matlock. 2001. Microstructure and fatigue resistance of carburized steels. Vol. 2, pp. 1152–1161. *In*: K. Funatani and G.E. Totten (eds.). Heat Treating: Proceedings of the 20th Conference. ASM International, Materials Park, Ohio.

Yan, S., X. Liu, W.J. Liu, H. Lan and H. Wu. 2015. Comparison on mechanical properties and microstructure of a C–Mn–Si steel treated by quenching and partitioning (Q&P) and quenching and tempering (Q&T) processes. Mater. Sci. Eng. A 620: 58–66.

Yang, D.Z., E.L. Brown, D.K. Matlock and G. Krauss. 1985. Ferrite recrystallization and austenite formation in cold-rolled intercritically annealed steel. Metall. Trans. A 16: 1385–1392.

Zaefferer, S., J. Ohlert and W. Bleck. 2004. A study of microstructure, transformation mechanisms and correlation between microstructure and mechanical properties of a low alloyed TRIP steel. Acta Mater. 52: 2765–2778.

Zaitsev, A.I. and B.M. Mogutnov. 1995. Thermodynamic properties and phase equilibria in the MnO-SiO2 system. J. Mater. Chem. 5: 1063–1073.

Zhang, J., H. Ding and J.W. Zhao. 2015. Effect of pre-quenching process on microstructure and mechanical properties in a Nb-microalloyed low carbon Q-P steel. Mater. Sci. Forum 816: 729–735.

Zhang, M., L. Li, R.Y. Fu, D. Krizan and B.C. De Cooman. 2006. Continuous cooling transformation diagrams and properties of micro-alloyed TRIP steels. Mater. Sci. Eng. A 438-440: 296–299.
Zhang, M.D., J. Hu, W.Q. Cao and H. Dong. 2014. Microstructure and mechanical properties of high strength and high toughness micro-laminated dual phase steels. Mater. Sci. Eng. A 618: 168–175.
Zhao, J.Z., C. Mesplont and B.C. De Cooman. 2002. Model for extracting phase transformation kinetics from dilatometry measurements for multistep transformations. Mater. Sci. Technol. 18: 1115–1120.

13

Chattering in Rolling of Advanced High Strength Steels

Maria Cristina Valigi[1,*] and *Mirko Rinchi*

ABSTRACT

The Advanced high-strength steels (AHSSs) are a newer generation of steel grades that at the same time provide extremely high-strength and maintain the high formability required for manufacturing. The AHSSs continue to grow and evolve in particular for the automotive sheet product because materials in automotive applications meet the requirement of light weight solutions, and key criteria including crash performance, stiffness, and forming requisites. For this reason the rolling mill production requires a fundamental evolution in order to obtain the thickness and the high surface quality of the sheet. This chapter is about the chatter in rolling mill process: an undesired phenomenon generating detrimental conditions for the plant and the quality of rolled products. Since the improvement of manufacturing processes are, in many cases, key contributors to the implementation of AHSS technologies, models of chatter phenomena in rolling mill process are showed.

[1] Dipartimento di Ingegneria, Università degli Studi di Perugia, Via Goffredo Duranti, 93, 06125 Perugia, Italy.

[2] Dipartimento di Ingegneria Industriale, Università degli Studi di Firenze, Viale Morgagni 40/44, 50134 Firenze (FI), Italy.
E-mail: mirko.rinchi@unifi.it

* Corresponding author: mariacristina.valigi@unipg.it

13.1 Introduction

Recently, the market of steel has been characterized by an increasing demand for a higher product quality level, both in terms of surface quality and mechanical properties. In fact, using stronger steel enables engineers to use thinner steels or reduced gauge, in order to produce lighter-weight parts, while maintaining or improving the strength and other performance properties. To meet this challenge the steel industry has developed the family of advanced high strength steels (AHSSs). The range of AHSSs continues to grow and evolve in many applications and in particular in the vehicle industry, where the achievement of lighter, safer, greener vehicles turns out to be an important issue. Thus, lighter steels are more sustainable for the environment, thanks to reduced fuel consumption, lower CO_2 emissions and lower aerodynamic resistances. In this scenario, the improvement of the rolling mill process, associated with the steel production chain, is essential for a higher product quality in determining the thickness and the high surface quality of the sheet. In particular, detecting and avoiding the insurgence of undesired vibrations during the working is fundamental. The rolling mills, as other rotating machines, are vulnerable to many different vibrations during the working, but even if most of them do not cause failures, there are several types of rolling mill vibrations that could have a significant impact on the quality and/or the productivity of the rolling process.

This chapter deals with the undesired mechanical vibrations commonly known as "chatter", often observed during the rolling mill, and adversely affecting the quality of the rolled product. This undesirable vibration is observed in most of the rolling mills operating at high speed and rolling thin metal sheet. The chatter is a particular phenomenon of self-excited vibration, which arises in some processes as a result of the interaction between the structural dynamics of the machine and the dynamics of the machining process. It is more frequent in rolling mill processes with a multi-stand rolling mill. Couples and dynamic forces in the process distort the structure; these deformations in turn cause a further variation of the rolling actions. Under certain conditions, the interaction between the structure and the process ensures that the energy released by the mill motors is captured by the process itself and transformed into energy of the structure; the result is the creation of instability phenomena, both in the stand and in the transmission system. Thus, chatter is a very important issue to model and analyze, having major impact on machine operability and sheet product quality. It is considered to be one of the most important directions in developing cold rolling technology for thin sheet in the field of advanced high-strength steels, which in comparison with traditional production, AHSS rolled products are much thinner and much harder sheet materials.

At the beginning of this chapter, a classification of the chatter is presented, together with the main consequences and causes of this phenomenon as reported in literature. Then, understanding the basic mechanisms of the chatter is fundamental in order to investigate the instability conditions and to limit the problem. It is for this reason that some models of the rolling mill structural dynamics and models of the rolling process are described. In the second part of this chapter, a record of cases is presented regarding rolling mills used in the production of AHSS.

13.2 Consequences, Causes and Classification

The origin of the vibrations can be self-excited or forced (for instance, due to the inlet transients of the steel in the rolling stands). The chatter occurs more frequently in the early stands, for high values of the rolling speed and high values of the reduction ratio and especially in the cold-rolling of special steels. One of the classical consequences of chatter is the manifestation of regular parallel marking across the width of the metal sheet, called "chatter marks" or "skid marks". The presence of these signs compromise the quality of the steel and must be absolutely avoided in the production of the AHSSs. Figure 13.1 shows an example of marks on the steel sheet.

Chatter marks are periodical faults in the thickness on the surface of the sheet, transverse to the rolling direction. When the chatter marks occur, undesired rumble from stands of rolling mill is detected audibly. When chatter marks appear, heavy vibrations of the roll stand may occur and even cause ruptures of the sheet generating the instability of process, oscillation of rolling speed, and in some cases, structural damage to the rolling mill

Figure 13.1 Skid marks (Petrucci 2012).

itself. The structural damage is due to dynamic phenomena that overload the mechanical components of the rolling mill structure. The amplitude and wavelength of chatter marks depend on the vibration system and on the vibration frequency. The chatter marks impressed on the sheet are equally spaced according to a well-determined wavelength λ that is correlated to the frequency (f) of the vibration source and the speed of the sheet (v):

$$\lambda = \frac{v}{f}. \tag{13.1}$$

Figure 13.2 shows chatter marks due to vibration phenomena at constant frequency. By varying the speed and measuring the wavelength of the chatter marks on the metal sheet, the source of vibrations excitation becomes distinguishable. In fact, if the frequency is variable, the source will be an element of the transmission system, otherwise the cause will be found elsewhere.

In particular, the sources of vibration at a constant frequency can be due to: marks already present on the work rolls or back-up rolls due to the inadequate grinding; sheet in entrance already having marks; and forced vibration due to roll bending or elasticity of stand.

In the last case, the rolling process leads the rolling rolls in resonance, and chatter marks may occur on the rolls, as well as on the sheet, if the circumference of the back-up rolls is an integer multiple of the wavelength of the defect. The relation describing this condition is:

$$\lambda = \frac{\pi D}{N}; \tag{13.2}$$

where N is the *critical speed* (that is the speed in which marks on rolls are generating) and D is the roll diameter (Rinchi 1999). Regarding the marks

Figure 13.2 Chatter marks due to vibration phenomena at constant frequency.

in the rolls due to grinding operation, they occur because the oscillation in the motion of the working tool is combined with the motion of the roll to be grinded. The roll behaves as a rigid body during the grinding operation, so that the marks will be uniform along the roll and the grinding operation can generate micro-grooves of uniform intensity on the rolls in a wide range of frequencies (150–1200 Hz and above). If the marks on the sheet are due to the rigid motion of the rolls or due to grinding defects, they will be uniform along the width of the sheet, but in the case of non-uniformity they are due to the bending modes of the rolls.

The phenomenon of chatter in rolling mills is distinguishable in respect to the resonant frequency, as Roberts classified, giving an overview of chatter phenomena and summarizing many of the studies of chatter (Roberts 1978, Yun et al. 1998).

The chatter phenomena has been characterized as follows: torsional chatter occurs in a range of about 5–25 Hz, third-octave-mode chatter in the range of 125–240 Hz and fifth-octave-mode chatter in the 500–1000 Hz range (Roberts 1988, 1983, Valigi et al. 2014, Valigi and Papini 2013, Wu et al. 2015, Chen et al. 2002, Joseph and Gallenstein 1981).

Torsional chatter has little effect, causing very slight change in gauge. The main causes of insurgence of torsional chatter are: oscillations due to an electrical problem and oscillations attributed to the variations of coefficient of friction with the rolling mill speed. One of the first papers regarding the torsional chatter was published by Moller and Hoggart (1967), where they concluded that the chatter persists if the gradient of the friction-velocity curve is negative (i.e., when friction coefficient decrease whilst mill speed increases). In particular, they attributed that the use of lubricants made coefficients of friction decrease with the speed, causing a decrease of the required rolling torque with increasing speeds. The decrease in torque, as speed increases could create an instability in the rolling process inducing speed changes. These speed changes would then cause the torque variations due to gauge variations or other causes, finally generating a steady-state oscillation (Joseph and Gallenstein 1981).

Very often, the torsional chatter can be due to the malfunctioning of motor speed-control circuits. Torsional oscillations of the vertical stand of a bloom and billet mill were studied, and it was observed that a small electrical oscillation drove the entire system to resonance, when a low range of critical speeds was reached. In this case, the chatter appears as the resonance vibration of the mill drive train: the rolls acting as rotational inertia and the spindles acting as torsional springs. Since motor itself has a relatively high, inertia and stiffness, its contribution in torsional dynamics may be neglected if compared to the transmission system's one. For this reason, Roberts modelled the motor as a rigid body, and studied torsional chatter through a simple one degree-of freedom torsional vibration model (roll inertia and spindle stiffness are the only model's parameters) (Roberts 1983). The magnitude of torsional chatter can be characterized by the magnitude of oscillations of the spindle torque and by the magnitude of the gauge variations on the sheet, and Gallenstein, showed their dependence to rolling speed in a 4-high stands cold rolling mill (Joseph and Gallenstein 1981). Ubici et al. (2001) contended that chatter is caused by the negative gradient of the friction–velocity curve for the roll. The investigation conducted by Joseph and Gallenstein (1981) suggests that the frictional conditions in the roll gap are the principal causes of chatter in the cold rolling mill; Chen et al. (2002) observed that the frictional conditions appear to be associated with the thickness and properties of oxide formed on rolls. The literature survey suggests that chatter in hot rolling is often associated with torsional vibrations, and this is consistent with the calculated natural frequency of the roll-spindle system in hot rolling mills.

The third-octave-mode chatter is considered to be the most critical, because it generates large gauge variations, which produce large thickness variations and sheet rupture. It is observed to happen very suddenly and it reaches its maximum amplitude in seconds. It is more prevalent and it can cause significant surface quality issues and high tension variation through the stands that sometimes lead to the rupture of the sheet in the cold rolling mill. Usually it begins with a low-frequency rumble, recognizable by ear. In order to avoid the incipient phenomena, the rolling speeds at the entry of stand have to be lower. Thus, it poses a difficult problem for the industry, causing significant loss of productivity. The third-octave-mode chatter appears to be self-excited vibration, and in the 80's Tlusty et al. (1982) had already developed a theory that explains that third octave chatter is due to the phase delay. These studies found a criterion for determining the instability of the system and the influence of rolling parameters on it. In addition, a model that was able to predict the instability of the rolling process was developed by calculating the reactions on a stand when a disturbance is introduced. Tlusty et al. (1982), established that the effect of inter-stand tensions on rolling force induced self-excitation as the main reason of chatter. Paton and Critchley (1985) have developed theoretical methods to increase the damping in the case of chatter of the third octave, relying on the fact that the vibrations of a vibrating system can be eliminated if there is a sufficient damping coefficient. This solution was implemented by placing calibrated dampers on the upper surface of the backup rolls. Chefneux et al. (1984) studied the third-octave chatter considering that the sheet is rolled in subsequent stands; they developed a numerical model and numerical simulations showed that chatter tends to occur when the mill sensitivity to chatter is high and there is a sudden change in rolling force due to, for example, the presence of a weld line or sudden change in lubrication. They concluded that chatter is a self-excited vibration, caused by inter-stand tension.

The literature describes three main mechanisms considered responsible for third-octave chatter: negative damping, mode coupling and regeneration (Yang and Tong 2012, Meehan 2002, Zhao and Ehmann 2013, Hu and Ehmann 2001, Zhong et al. 2002).

Chatter, in rolling, can also be due to variations of the entry tension. The corresponding force component on the rolls associated with this tension can be modelled as a negative damping force: in specific working conditions this may lead to the increase of roll vibrations even without external forces' contribution (self-excitation phenomenon). The existence of the negative damping effect can no longer maintain the system stable with the increase of rolling speed: it is a phenomenon of tension caused by roll vibrations generating roll force variations, which in turn, induced further vibrations. In relation to the regeneration mechanism, the influence of time delays in a multi-stand system causes chatter, so that a tandem rolling mill may also become unstable even though the individual stands are stable. In fact, the variation of tension between consecutive stands has an immediate consequence on the roll forces and torque. Hu and Ehmann stated that the regenerative effect in sheet rolling is the result of a sheet thickness variation generated through inter-stand interactions at a prior time, when the stand was vibrating and influences the rolling process dynamics at the current time (Hu and Ehmann 2001). Another mechanism for dynamic instability of the rolling mill is the coupling of two or more of the principal modes of vibration of the mill, excited by the rolling process and Paton and Critchley (1985) experimentally showed that the work rolls do indeed vibrate in the horizontal direction, as well as in the vertical direction.

The fifth-octave-mode chatter is characterized by a very gradual occurrence until it reaches a very high level of undesired acoustic emissions also with the manifestation of chatter marks on the rolls (Figure 13.3) (William 1978, Lin et al. 2003, Lin et al. 2002, Xie 2013, Kimura et al. 2003).

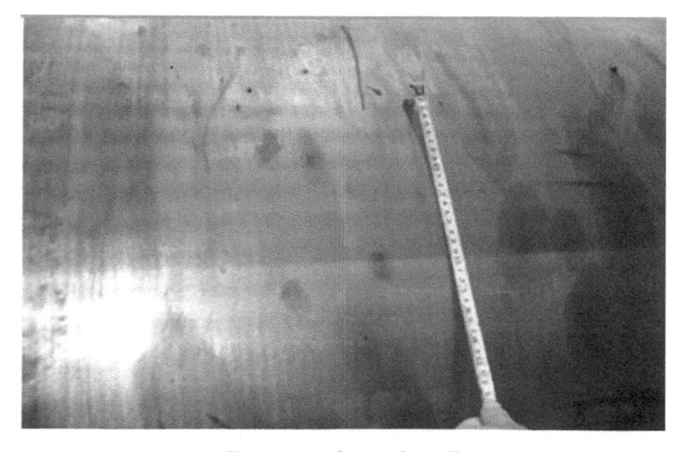

Figure 13.3 Chatter marks on the rolls (Xie 2013).

It results in transverse banding of the back-up and work rolls and matching transverse surface marking of the sheet. Thus, when the chatter marks have begun to form on rolls, the marks themselves can excite the vibration mode at certain roll speeds and the problem becomes self-exciting. The fifth-octave-mode chatter can occur at any rolling speed and even if the thickness of the rolling sheet is not influenced, it is harmful for the superficial quality of the product and for the process. In fact, in the rolling mills suffering from this problem, rolls must be changed frequently; however changing rolls frequently causes economic losses in terms of productivity, excessive cost and the shortened useful life of rolls. Roberts has developed a simple mathematical model to explain the chatter frequency, and highlighted some conditions in which it appears as: rolls operating at speeds close to the critical values for long periods; form irregularities or surface irregularities of the backup rolls; incorrect positioning of the rolling rolls; work parameters; streaks sheet in input to the stand. Then, he proposed solutions for the suppression of the fifth-octave-mode chatter such as: adequate modulation of rolling speed; careful selection of the backup rolls according to their diameter; the use of back-up rolls with diameters as large as possible in the last stand; checking the surface wear of the back-up rolls; special attention to the grinding rolls and isolation from external vibrations.

According to literature, chatter is also influenced by the friction conditions of rolls. For example, Kimura et al. (2003) discovered the existence of an optimum range for friction coefficient, in which the vibrations are damped and the rolling mill is stable despite the disturbances (Kimura et al. 2003). Many experimental investigations show that lubrication is one of the main factors chatter depends on (Meehan 2002, William 1978, Johmson and Qi 1994, Arnken et al. 2007). The practical investigation of Yarita et al. (1978) led to the point that chatter occurs at reductions close to the limit for rolling, when coupled with inadequate lubrication. They suggested that lubricant film strength, emulsion stability, and particle size distribution should be taken into consideration in selecting the rolling lubricants, in order to prevent chatter. Due to the complexity of the problem, the majority of authors have used simple friction models in their analysis considering friction at the roll-sheet interface as a function of relative velocity between the rolls and sheet (Yarita et al. 1978). Also, they concluded that chatter occurs at reductions close to the limit for rolling, when coupled with inadequate lubrication (Yarita et al. 1978), and only a very few researches have described how chatter occurs through unsteady lubrication phenomenon (Kimura et al. 2003, Heidari et al. 2014).

13.3 Modeling

The central problem of the analysis and prevention of chatter is to identify the conditions leading to the dynamic instability of the process. Therefore, the interaction between the structural dynamics of the mill and the rolling deformation process must be investigated (Misonoh 1980, Swiatonoswki and Bar 2003). The general block diagram of chatter is displayed in Figure 13.4 and describes the closed loop diagram schematically representing the interaction between the structure and the process (Valigi et al. 2014, Malvezzi and Valigi 2008). The dynamic actions of the deformation process deform the structure and in turn, these deformations cause a further variation of the rolling actions. According to a closed loop diagram and under certain conditions, the interaction between the structure and the process generates instability phenomena.

In the following sections, some models of the stand, of the drive train system, and of the deformation process are described. They are useful and provide insights into interactions between structure and the process and to simulate, prevent and study vibration and chatter. Obviously, as for all the mathematical models, validation of the numerical results by experimental tests is a crucial step for their application.

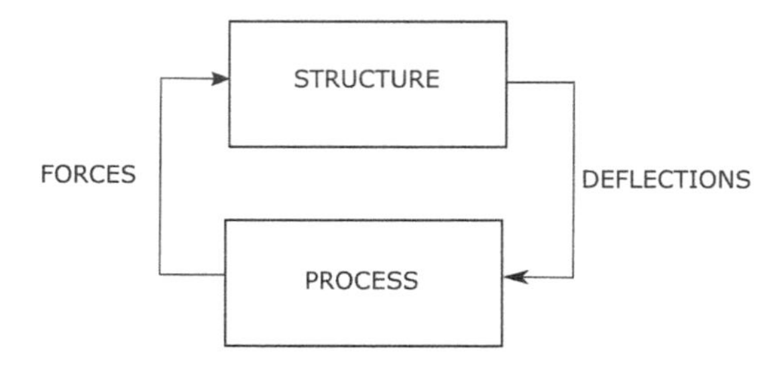

Figure 13.4 Closed loop diagram schematically representing the chatter.

13.3.1 Models of Stands

Lumped parameter models are useful to understand the structural dynamic of rolling mills (Rinchi 1999, Benvenuti 1999, Casini 1999, Yun et al. 1998). The analysis of all single rolling mill stand components, and the estimation of the interactions between them are fundamental to obtain dynamic characteristics, create a satisfactory model and evaluate the natural frequency and the vibration modes of the stand. Considering, for instance, that a 4-high stand rolling mill, and a lumped parameter model with four degrees of freedom, can describe just the movement of the rolls, in respects to the stand and between the rolls (as Figure 13.5 shows). In this model, the shoulders are assumed to be rigid, if compared to the other elements of the stand. The forces due to the stand deformation and the contact force between the rolls, are represented by means of linearly deformable and damping elements (characterized by stiffness and viscous damping coefficients k_1, k_2, k_3, k_4, c_1 and c_2).

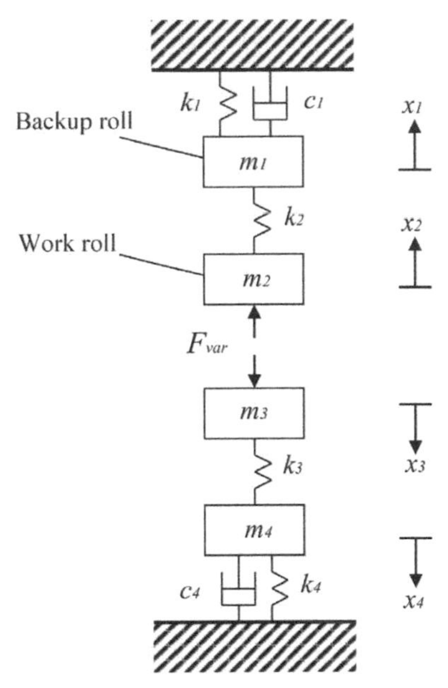

Figure 13.5 4-DOFs model.

x_i being the vertical displacements (with respect to the static equilibrium position) of the various DOF and m_i the related inertial mass, the equations of the system are:

$$\begin{cases} m_1\ddot{x}_1 = -c_1\dot{x}_1 - k_1x_1 - k_2(x_1 - x_2) \\ m_2\ddot{x}_2 = k_2(x_1 - x_2) - F_{var} \\ m_3\ddot{x}_3 = k_3(x_4 - x_3) - F_{var} \\ m_4\ddot{x}_4 = -c_2\dot{x}_4 - k_4x_4 - k_3(x_4 - x_3) \end{cases}; \tag{13.3}$$

where the dynamic component of rolling force F_{var} is applied to the system as an external force.

Usually, the dynamic analysis of the model is simplified, assuming that the system is symmetrical in respect to the steel sheet, so that the number of degrees of freedom are halved. Even if the work roll masses are assumed to be smaller in respect to masses of the back-up rolls, the simplification in two degrees of freedom can be realized by neglecting the work roll masses. In this case, the equations of the system become:

$$\begin{cases} m_1\ddot{x}_1 = -c_1\dot{x}_1 - k_1x_1 - F_{var} \\ m_4\ddot{x}_4 = -c_2\dot{x}_4 - k_4x_4 - F_{var} \\ 0 = k_2(x_1 - x_2) - F_{var} \\ 0 = k_3(x_4 - x_3) - F_{var} \end{cases}. \tag{13.4}$$

From the above equations, the relationship between the gauge variation $h_{g\ var}$ and the force is:

$$h_{g\ var} = (x_1 + x_4) - F_{var}\left(\frac{k_2+k_3}{k_2k_3}\right); \quad (13.5)$$

and, with the hypothesis of symmetry ($x_1 = x_4 = x$, $k_2 = k_3 = k$), it becomes:

$$h_{g\ var} = 2\left(x - \frac{1}{k}F_{var}\right); \quad (13.6)$$

Associated to the frequency of the thickness variation of sheet, the presented model can accurately describe the problem of chatter in many situations at the frequency of the principal vibrational mode, and it is sufficient for investigating chatter due to negative damping. Usually, the experimental evidence shows that the work rolls vibrate both in the horizontal direction, as well as in the vertical direction, so that multi-directional models have to be considered.

Multi-directional models can explain chatter due to model-coupling. For example, Figure 13.6 shows the 2-DOF model, for a two-high rolling mill stand vibrating along two orthogonal principal directions, assuming that the mass of the work roll is negligible, and that the system is symmetrical in respect to the rolled sheet.

Considering a fixed reference frame (X,Y), being $X_1(t)$ and $X_2(t)$ the roll displacement components along two orthogonal directions X_1, X_2, the equations of motion are:

$$\begin{cases} m_1\ddot{X}_1(t) + c_1\dot{X}_1(t) + k_1X_1(t) = -F_{var\ 1}(t) \\ m_1\ddot{X}_2(t) + c_2\dot{X}_2(t) + k_2X_2(t) = -F_{var\ 2}(t) \\ F_{var\ 1}(t) = F_{var}(t)\cos(\beta - \alpha) \\ F_{var\ 2}(t) = F_{var}(t)\cos\left(\alpha + \frac{\pi}{2} - \beta\right) \end{cases}; \quad (13.7)$$

where α and β are the angles respectively determining the direction of X_1 and F_{var} vector with respect to Y axis. In many cases, when the effect of roll bending and house torsional modes of the stand have to be analyzed, the described models are not sufficient and the development of a 3D finite element model of the mill stand is more convenient. Figure 13.7 shows a 3D finite element model able to analyze those effects.

An improvement of the previous lumped parameter model, with four degrees of freedom is shown in Figure 13.8. It also considers the mass and the stiffness of shoulders and the stiffness connection with the backup roll. The mass m_s represents the portion of the shoulders mass and the upper crosspiece mass participating in the vertical vibrations. The stiffness k_s models the shoulders connected to the ground by means of base plates. The stiffness k_{1s} models the upper backup roll connection with the stand (including stiffness of the screw, the chock and the bushing); in addition,

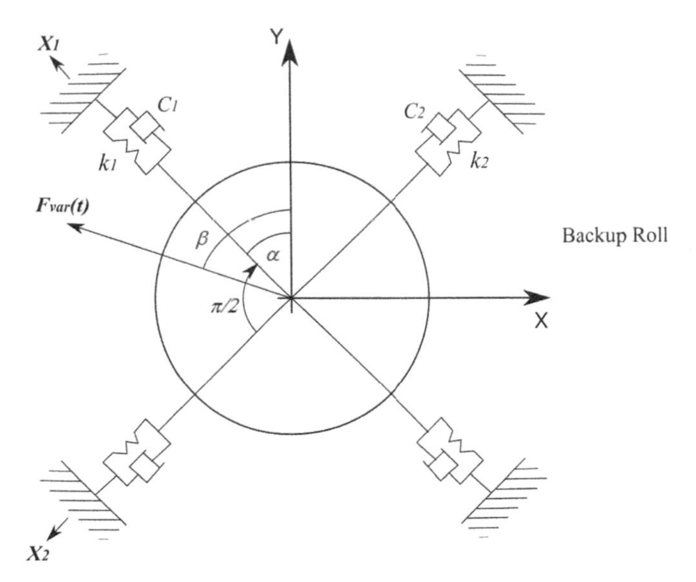

Figure 13.6 Multi-directional model.

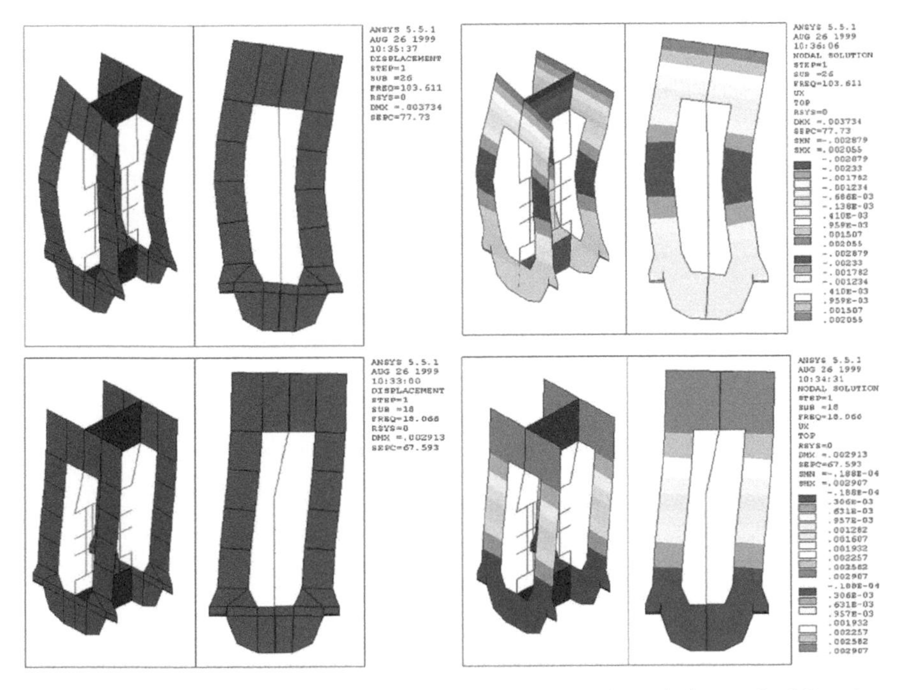

Figure 13.7 Deformation model of a stand in the plane perpendicular to the longitudinal direction.

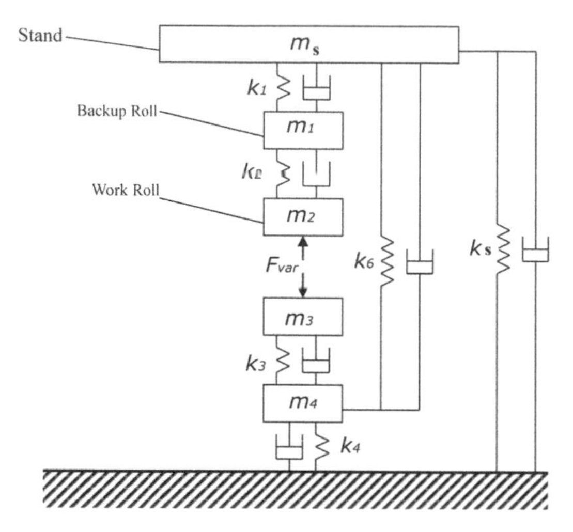

Figure 13.8 5-DOFs model.

the model also highlights the different stiffness frame connections of upper and lower backup rolls (k_6 k_4). Dynamic equations of this model are not reported here.

Another improvement of the model is the 9 degrees of freedom model (Figure 13.9), where the shoulders are considered independent (connected elastically to each other by means of a crosspiece) and the movement of the work rolls is considered with 2 degrees of freedom (the vertical oscillation and the rotation around the barycenter, assumed positioned in the center line). The contact forces between the work rolls and the backup rolls are modeled as a distributed stiffness, according to the model of the Winkler springs bed.

The equations of the model are:

$$\begin{cases} m_5\ddot{x}_5 + k_{57}(x_5 - x_7) + k_{56}(x_5 - x_6) = 0 \quad \text{(crosspiece)} \\ m_7\ddot{x}_7 + k_{75}(x_7 - x_5) + k_{71}(x_7 - x_1) + k_{74}(x_7 - x_4) + k_{70}x_7 = 0 \text{ (shoulder 7)} \\ m_6\ddot{x}_6 + k_{65}(x_6 - x_5) + k_{61}(x_6 - x_1) + k_{64}(x_6 - x_4) + k_{60}x_6 = 0 \text{ (shoulder 6)} \\ m_1\ddot{x}_1 + k_{17}(x_1 - x_7) + k_{16}(x_1 - x_6) = -F_{sbu} \quad \text{(upper back-up roll)} \\ m_2\ddot{x}_2 - F_{sb} = -F_{var} \quad \text{(upper work roll)} \\ J_2\ddot{\vartheta}_2 + M_{sbu} = 0 \quad \text{(rotation upper work roll)} \\ m_3\ddot{x}_3 - F_{sbl} = -F_{var} \quad \text{(lower work roll)} \\ J_3\ddot{\vartheta}_3 - M_{sbl} = 0 \quad \text{(rotation lower lower roll)} \\ m_4\ddot{x}_4 + k_{43}(x_4 - x_3) + k_{47}(x_4 - x_7) + k_{46}(x_4 - x_6) + k_{40}x_4 = -F_{sb} \text{ (lower back-up roll)} \end{cases} \tag{13.8}$$

where M_{sbu}, F_{sbu}, M_{sbu} and F_{sbl} are the moments and the contact forces generated by the "spring beds" (respectively on upper and lower work rolls). In the model, the horizontal displacements of the work rolls are neglected, and also the presence of the spindle is neglected together with its eventual effects on the eccentricity movement of the work roll.

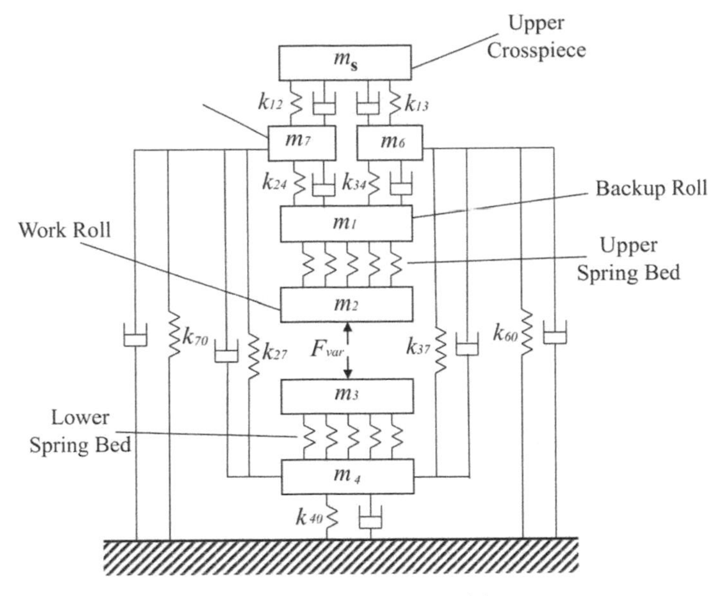

Figure 13.9 9-DOFs model.

Considering *A* to be the generic point of the contact surface between the (upper) work roll and the bearing surface. Assuming R to be work roll radius, its coordinates in static condition are (R, y_A). When the work roll moves from its static condition, the generic point moves to *A*′. If ϑ is the rotation angle of the work roll with respect to its (static equilibrium) horizontal position and $x_{G'}$ the vertical displacement of roll barycenter with respect to its equilibrium position, the coordinates of *A*′ (see Figure 13.10) may be expressed by:

$$A' \begin{cases} y_{A'} = y_A \cos(\vartheta) + R\sin(\vartheta) \\ x_{A'} = x_{G'} - y_A \sin(\vartheta) + R\cos(\vartheta) \end{cases} \tag{13.9}$$

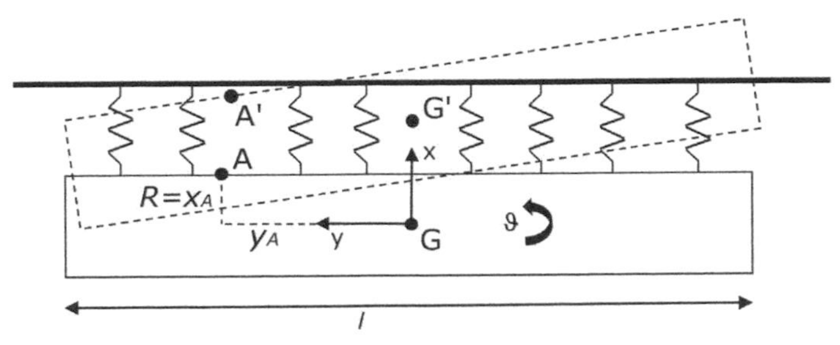

Figure 13.10 Work roll model.

Considering Eq. (13.9) it is possible to calculate the elastic force per unit length f_{sb} and, taking into account all points of contact surface it is possible to evaluate the expression of the global spring bed force F_{sb} and moment M_{sb}. By assuming K_d is the value of the distributed stiffness of the spring bed f_{sb} is:

$$f_{sb} = -K_d(x_{A'} - x_A) \quad \rightarrow \quad f_{sb} = -K_d(x_{G'} - y_A \sin(\vartheta) + R\cos(\vartheta) - R); \tag{13.10}$$

l being the work roll length:

$$\begin{aligned} F_{sb} &= \int_{-\frac{l}{2}}^{\frac{l}{2}} f_{sb}\, dy_A = -\int_{-\frac{l}{2}}^{\frac{l}{2}} K_d(x_{G'} - y_A \sin(\vartheta) + R\cos(\vartheta) - R)dy_A = \\ &= -K_d l[x_{G'} + R(\cos(\vartheta) - 1)]; \end{aligned} \tag{13.11}$$

$$\begin{aligned} M_{sb} &= \int_{-\frac{l}{2}}^{\frac{l}{2}} (f_{sb} y_A) dy_A = -\int_{-\frac{l}{2}}^{\frac{l}{2}} [K_d(x_{G'} - y_A \sin(\vartheta) + R\cos(\vartheta) - \\ & R)y_A]dy_A = -K_d \frac{l^3}{12}\sin(\vartheta). \end{aligned} \tag{13.12}$$

These equations can be simplified assuming small rotations around the barycenter, so that they become:

$$\begin{cases} F_{sb} = -K_d l x_{G'} \\ M_{sb} = -K_d \dfrac{l^3}{12}\vartheta \end{cases} . \tag{13.13}$$

The general expressions in Eq. (13.13) may be specialized in order to model spring bed actions on both upper and lower work rolls, and more specifically to calculate the mathematical model of M_{sbu}, F_{sbu}, M_{sbl} and F_{sbl}, introduced in Eq. (13.8).

13.3.2 Models of Driving Train System

The model of the driving train system is also useful to investigate chatter (in particular torsional chatter) and vibration in a rolling mill plant, and to develop solutions or effective control systems able to minimize the inevitable vibratory phenomena, improving finished product quality and increasing plant productivity. Usually, for each rolling stand there is a power driving train system connecting motor and work rolls. For economic reasons, the motors of the same rolling mill line are of the same type and for reason of alignment, maintenance and cleaning; the lengths of the kinematic chains are equal. A generic driving system is shown in Figure 13.11; where one

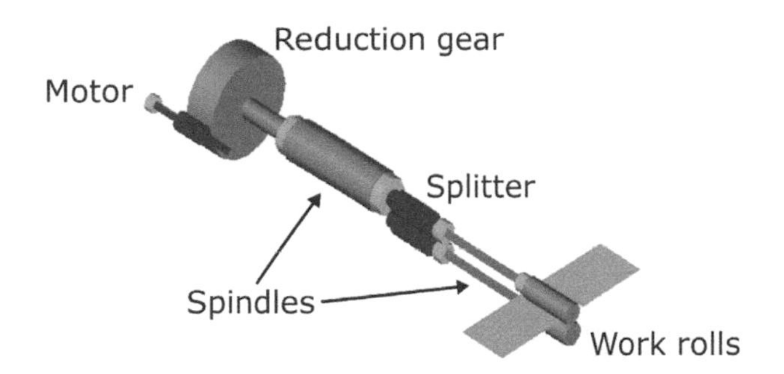

Figure 13.11 Sketch of the driving train system.

or two dc motors, in series, move a reduction gear connected to the pinion gear through a shaft. The pinion gear transmits the driving torque to the work rolls through two quasi-parallel shafts.

Obviously, different stands have different gearboxes because variations of the transmission ratio are necessary to obtain different rolling torques and different angular speed, in order to assure the continuity of the rolled material flow. In particular, the transmission system of the first stand must generate higher rolling torques so that the shafts, gears and joints of the first stand are different in respect to the following stands.

Figure 13.12 shows a common drive train and the corresponding scheme of a lumped parameter model with 14 degrees of freedom.

After analyzing the dynamic characteristics of the single mechanical components, 14 equations can be written in order to describe the model according to the matrix form:

$$[M]\{\ddot{\Theta}(t)\} + [C]\{\dot{\Theta}(t)\} + [K]\{\Theta(t)\} = \{M_{ext}(t)\}; \tag{13.14}$$

where $[M]$, $[C]$ and $[K]$ are respectively system mass, viscous damping and stiffness matrixes, $\{M_{ext}(t)\}$ represents the (vector of the) external torques acting on the system and $\{\Theta(t)\}$ the vector of the angular position of the various DOF of the model.

More complex models with higher numbers of degrees of freedom can be created to study torsional vibrations of drive train, but usually they do not supply better results on the dynamic behavior of the system, especially for low frequencies.

On the other hand, reducing the model can be more convenient in order to manage and carry out numerical simulations.

Thus, a reduced 4-DOF model is considered and shown in Figure 13.13, where:

- the first DOF is the rotation of mass including the motor, the gear unit, the connections between them, and half the spindle between gear unit and splitter;

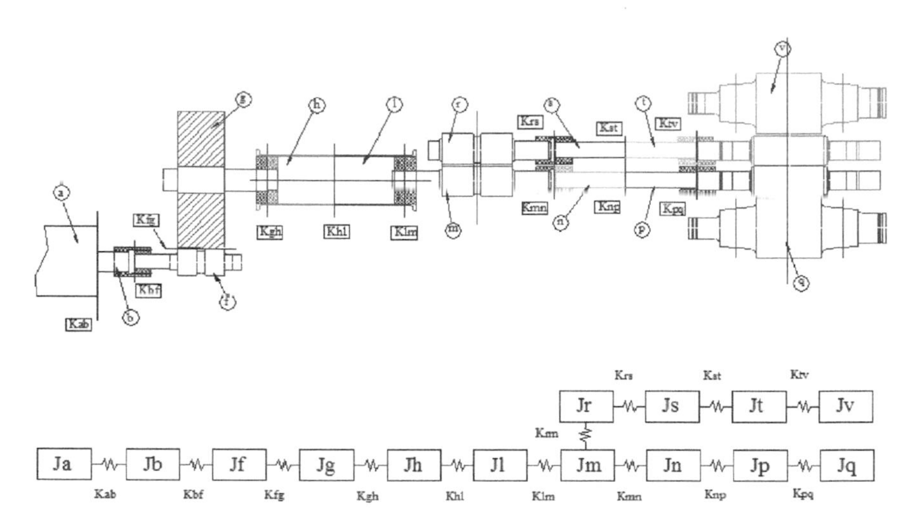

Figure 13.12 Typical drive train with its 14-DOFs model (Benvenuti 1999).

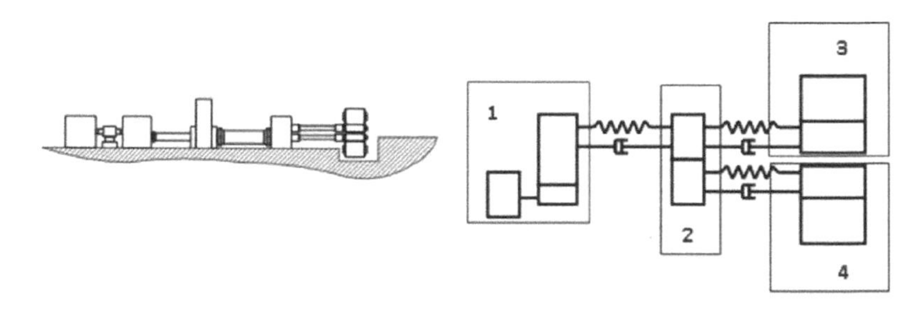

Figure 13.13 Typical drive train system and its 4-DOFs model.

- the second DOF is the rotation of mass including the splitter, the other half of the spindle between gear unit and splitter, and the half spindle between the two splitter and work rolls;
- the third DOF is the movement of the mass including the upper work roll together with the upper back up roll (since its mass contributes to the motion) and the half spindle of the upper work roll;
- similarly, the fourth DOF is the movement of the lower work roll mass, together with the lower back up roll and the half spindle connecting the upper pinion and the work roll.

One can also consider a more complex model, with 5-DOFs as Figure 13.14 shows. In this case, the only variation, in respect to the 4-DOFs model, is that the connection between the motor and the gear unit is not considered as rigid.

In addition, the multibody dynamic analysis is a useful tool to investigate the chatter in a rolling mill. For example, Kim et al. (2012)

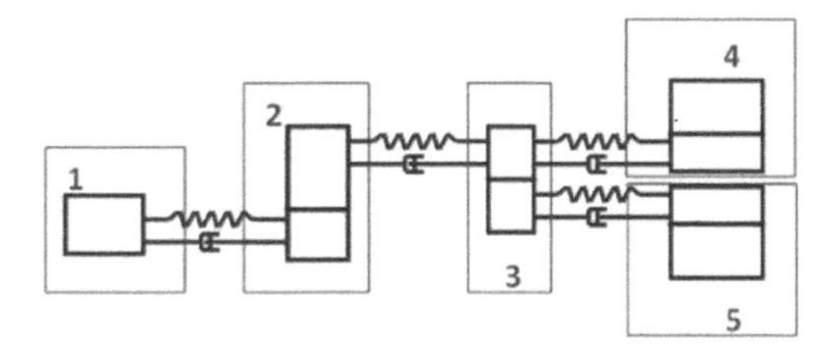

Figure 13.14 5-DOFs model of a drive train system.

developed a numerical model of a 6-high cold rolling mill including the driving system. This model was built through connections of rigid bodies to perform the multibody dynamic analysis. As the rolling mill is structurally symmetrical, only the upper part of the rolling mill was investigated. The backup, intermediate and work rolls were included in the model and, in addition, also the drive train system, including the helical gears that transmit the torque into the spindle and the spindle that transmits the torque into the work roll were considered. The spindle consists of three components that are connected by two universal joints.

The model took into account the transmission of torque by friction contact between the rolls. Roller bearings that support the axle of helical gears and both ends of the rolls were modeled by spring-damper elements.

13.3.3 Models of the Rolling Process (Rolling Theory)

In the literature, a number of analytical models of the plastic deformation are developed. In this paragraph, some plastic deformation theories, useful to model rolling actions on the stand, are described. The aim of these theories is to compute the distribution of roll pressure and shear stress, due to friction over the contact arc and to evaluate roll force and torque, acting between sheet and work rolls, through the integration of these distributions. One of the most used models of the rolling deformation process is the slab theory. This theory is applicable when the length of the roll bite is several times larger than its thickness, and assuming the radii of the undeformed rolls, about hundred times greater than the sheet thickness and the ratio between sheet width and sheet thickness greater than ten. Thus, the rolling takes place in plane strain conditions and the problem can be considered bi-dimensional. Under these conditions and by hypothesizing the material as homogenous and isotropous, the horizontal stresses and velocities along any vertical slice of the roll gap may be assumed constant (this condition is generally called 'homogeneous deformation').

Figure 13.15 is a schematic diagram of the stresses acting on a vertical slice of dx thickness, where the roll curvature is neglected.

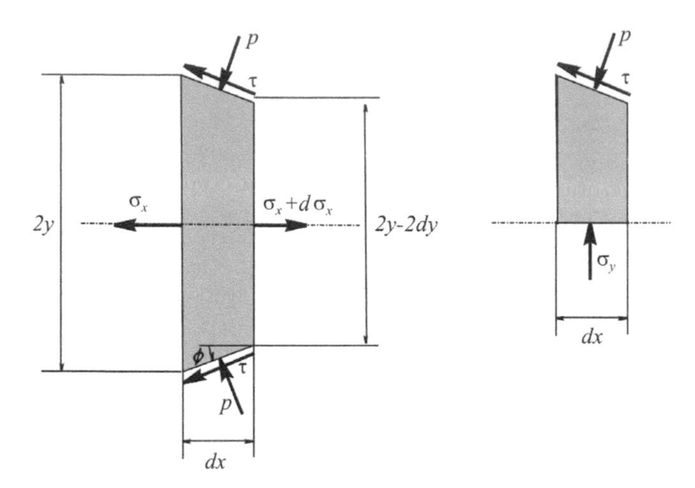

Figure 13.15 Reference scheme for slab method theory.

Thus, supposing that the infinitesimal segments in deformation are delimited by two surfaces that remain flat during the process, starting at the classic Orowan's equilibrium equation, assuming the shear stress $\tau = \mu p$ and the sheet thickness h=2y, the following equation is obtained:

$$\frac{dp}{dx} = 2\frac{K}{y}\frac{dy}{dx} \pm \mu\frac{p}{y} + 2\frac{dK}{dx}; \tag{13.15}$$

where p is the rolling pressure, μ is the friction coefficient and $K = K(x)$ is the mean yield stress in conditions of plane strain ($2\,K = \sigma_x + p$ Huber-Mises criterion—being σ_x the stress in the direction of rolling). The obtained differential equation has to be integrated from the roll gap entrance to the exit. The boundary conditions are the horizontal tensile stress at entry and at exit of the stand σ_{in} and σ_{out}, respectively. According to a yield criterion, typically used in most of the equilibrium equations, it is assumed that the sheet material is deformed homogeneously, therefore the roll pressure at entry and exit are respectively:

$$\begin{cases} p_{in} = 2K - \sigma_{in} & \text{(at the entrance)} \\ p_{out} = 2K - \sigma_{out} & \text{(at the exit)} \end{cases} . \tag{13.16}$$

The model determines the rolling force F_{roll} and the rolling torque T_{roll}, assuming roll radius R to be constant, i.e., work rolls are assumed to remain rigid, and may be implicitly written as a function of several variables, such as the half thickness of rolled sheet at the entry and at the exit of the stand (y_{in}, y_{out}), the horizontal tensile stress at the entry and exit of the stand and the deformation resistance K_f dependent on the strain hardening characteristics:

$$F_{roll} = F_{roll}\left(y_{in}, y_{out}, \sigma_{xin}, \sigma_{xout}, \mu, K_f\right); \tag{13.17}$$

$$T_{roll} = T_{roll}\left(y_{in}, y_{out}, \sigma_{xin}, \sigma_{xout}, \mu, K_f\right). \tag{13.18}$$

Orowan's inhomogeneous theory allows accurate analysis of the interaction between work rolls and steel, even in the presence of vertical and torsional vibrations of work rolls and of drive train components (Orowan 1943). The presence of regions steel slips on work rolls and on some others, and adhesion (stick) between rolls and steel, are taken into account. Internal friction of the material is generally calculated as the product between the normal rolling pressure and the friction coefficient. However, if the friction coefficient is too high, the shear stress could exceed the maximum value allowed by material plasticity conditions. When the product between the rolling pressure and the friction coefficient exceeds the yield stress, friction actions between the surfaces of the rolls and the sheet cannot further increase and the steel plasticizes: in such zones the steel sticks to the rolls. The plastic deformation region can be basically subdivided in 4 zones. In the first zone (first slip zone), starting from the entry section, the speed of the steel is lower than the horizontal component of the rolls peripheral speed. Therefore, local friction actions favour the advancement of the steel into the plastic region. In the zone (second slip zone), preceding the exit section, steel is faster than the rolls, and therefore the friction actions change direction, hindering the steel flow. In the intermediate zones (stick zones), there is adherence between the work roll surface and the rolled material. Friction actions act in such a way as to prevent local slipping between rolls and steel. Such actions, vanish in correspondence with the neutral plane, where the speed of the sheet equals the horizontal component of roll peripheral speed. In the rolling direction the rolled sheet is divided into segments of material constituted by a portion of material between two cylindrical surfaces A and A′, as shown in Figure 13.16. The rolling distribution pressure p, as a function of angle ϕ, has different expressions inside the plastic deformation region. In slip and stick zones, the relationships taken from Orowan's inhomogeneous theory are respectively:

$$p = \frac{F}{h(y_{out})} + \sigma_s\left[w(\phi) \pm \frac{1}{2}\left(\frac{1}{\phi} - \frac{1}{\tan(\phi)}\right)\right]; \tag{13.19}$$

$$p = \left(\frac{F}{h(y_{out})} + \sigma_s w\right) + \left[1 \pm \mu\left(\frac{1}{\phi} - \frac{1}{\tan(\phi)}\right)\right]^{-1}; \tag{13.20}$$

where

- σ_s is the yield stress of the material subjected to homogeneous compression;
- F is the roll tension per unit width of the strip.

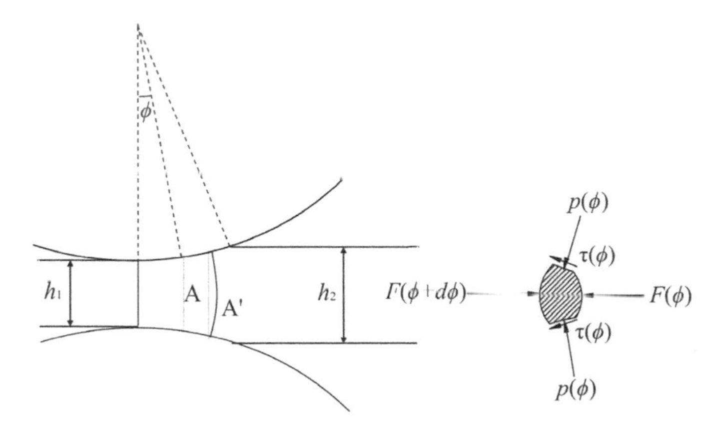

Figure 13.16 Non-homogeneous theory: division of strip and segment of material.

The nonlinear function $w(\phi)$ takes into account the inhomogeneity of plastic deformations inside the bite zone (see Orowan 1943):

$$w = \frac{1}{\sin(\phi)} \int_0^{\varphi} \sqrt{1 - a^2 \left(\frac{\varphi}{\phi}\right)^2} \cos(\varphi)\, d\varphi. \tag{13.21}$$

Assuming $D = 2R$ is the work roll diameter, such functions can be calculated through the following relationships, so that:

$$\frac{dF}{d\phi} = 2R\cos(\phi)(\tan(\phi) \pm \mu). \tag{13.22}$$

The rolling force F_{roll} and the rolling torque T_{roll} can be expressed through the following integral formulas:

$$F_{roll} = \int_{\varphi_{in}}^{\varphi_{out}} p d\phi; \tag{13.23}$$

$$T_{roll} = \frac{D^2}{2}\left[\mu\left(\int_{\varphi_{t1}}^{\varphi_{in}} p d\phi - \int_{\varphi_{out}}^{\varphi_{t2}} p d\phi\right) + \frac{1}{2}\left(\int_{\varphi_n}^{\varphi_{t1}} \sigma_s d\phi - \int_{\varphi_{t2}}^{\varphi_n} \sigma_s d\phi\right)\right]; \tag{13.24}$$

where φ_{in}, φ_{out} and φ_n represent the angles, respectively relative to the entry section and to the exit one and to the neutral plane, and φ_{t1} and φ_{t2} are the angles relative to the transition sections between stick and slip zones, respectively, before and after the neutral plane.

Another model, for the calculation of the rolling force is described by Kimura et al. (2003). This is a simplified Pawelsky's model and it describes the dynamic response against the periodical change of the roll bite. In order to model the work roll vertical dynamics, the roll bite geometry represented in Figure 13.17 is considered. In order to preserve the Authors' original formulation, some of the constants and variables differ from those previously adopted in this chapter. The nomenclature introduced in Figure 13.17 and the relative mathematical models (Eqs. 13.25–13.32) is the following:

h, h_{in}, h_{out}—strip thickness, at entry and exit (m);
h_a, h_n—thickness at minimum gap and neutral point (m);
k_f—deformation resistance (MPa);
R_w, R_b, R'—Work roll radius, back-up roll radius, flattened work roll radius (m);
u, u_{in}, u_{out}—strip speed, at entry and exit (m/s);
x_{in}, x_{out}—position of roll bite, entrance and exit (m);
t_b, t_f—back and front tension (MPa).

When the gap changes with rather high speed in the vertical direction, the exit position of the bite moves forward or backward. Specifically, if the change of the gap is quick or the speed of the sheet is slow, the extension or reduction of the bite length is significant. As a consequence, a large change in the rolling force occurs. Therefore, the response of the rolling force is determined not only by static conditions but also by the relationship between the vibrating rolls vertical speed and the material travelling speed, which means that the bite has a damping effect in the system.

Thus, the contour of the roll gap is assumed as parabolic and the condition of vanishing the strain rate gives the exit position of the gap as the solution of the following quadratic equation:

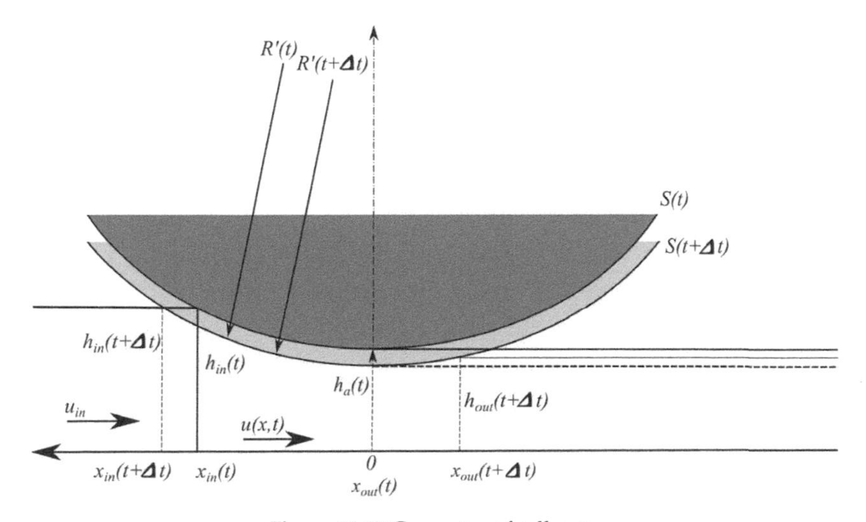

Figure 13.17 Geometry of roll gap.

$$x^2 + 2\left(\frac{u_{in}h_{in}}{(\partial h_a/\partial t)} - x_{in}\right) x - h_a R' = 0; \tag{13.25}$$

where the entry position x_{in} is:

$$x_{in} = \sqrt{(h_{in}(t) - h_a(t))R'(t)}; \tag{13.26}$$

and from the momentum equation, the following expression is obtained:

$$\frac{\partial p}{\partial x} \mp 2\mu\frac{p}{h} - \frac{1}{h}\frac{\partial(hK_f)}{\partial x} = 0. \tag{13.27}$$

which is similar to the equation obtained by the slab method, but differs for the partial equation. Thus, an approximate expression for the pressure is derived, and the pressure distribution in the forward and backward slip region can be written as:

$$p^+ = \left(\frac{k_f h}{h_{out}}\right)\left(1 - \frac{t_f}{k_f^{out}}\right) exp(\mu(H - H_{out})); \tag{13.28}$$

$$p^- = \left(\frac{k_f h}{h_{in}}\right)\left(1 - \frac{t_f}{k_f^{in}}\right) exp(\mu(H_{out} - H)); \tag{13.29}$$

where

$$H = 2\sqrt{\frac{R'}{h_a}}\tan^{-1}\left(\phi\sqrt{\frac{R'}{h_a}}\right); \tag{13.30}$$

and where the position of the neutral point can be calculated by:

$$\varphi_n = \sqrt{\frac{h_a}{R'}}\tan\left(\frac{H_n}{2}\sqrt{\frac{h_a}{R'}}\right); \tag{13.31}$$

and where H_n is defined as follows:

$$H_n = \frac{H_{in} + H_{out}}{2} - \frac{1}{2\mu} ln\left(\frac{h_{in}}{h_{out}}\left(\frac{1 - t_f/k_f^{out}}{1 - t_b/k_f^{in}}\right)\right). \tag{13.32}$$

The rolling force can be calculated by numerical integration of the pressure along the roll bite described by (13.26) (13.27), and also the inlet and outlet velocity of the sheet can be calculated by Eq. (13.32), together with a continuity law of the material.

13.3.4 The Rolling Process Model in Full Film Regime

In the cold rolling mill, in order both to improve the surface quality of the sheet and to dispose the heat produced during the deformation process, it is usual to spray proper lubricant between the working rolls and sheet, especially in the production of thin sheets, where the friction plays a crucial role.

Specifically, to precisely control the friction, it is crucial when the AHSS products are rolled. Compared with traditional productions, AHSS rolled products are much thinner and much harder sheets and they must be rolled at much lower friction levels, and controlled in a much narrower range of friction characteristics. Therefore, the effect of the inadequate lubrication has to be taken into account in the analysis of chatter phenomena (Heidari et al. 2014, Laugier et al. 2015).

In order to focus the effect of inadequate lubrication on the chatter, a rolling mill process model for cold rolling mill, using lubricant emulsions sprayed on the entrance of the sheet is necessary.

Such a model is described below, and Figure 13.18 illustrates the coordinate system and geometry of the rolling process in the full-film regime. The lubricant film separating the rolls from the sheet may be divided, as usual, into three zones: the inlet zone, the work zone and the outlet zone (Malvezzi and Valigi 2008).

- *Inlet zone*: within the lubricant film, between the rolls and the rigid sheet. In this zone the pressure rises rapidly until the deformation of the sheet starts at the inlet edge of the work zone;
- *Work zone*: the material has a significant deformation from an initial thickness S_1 to a final thickness S_2 and the lubricant film is carried along by the motion of the sheet and the rolls;

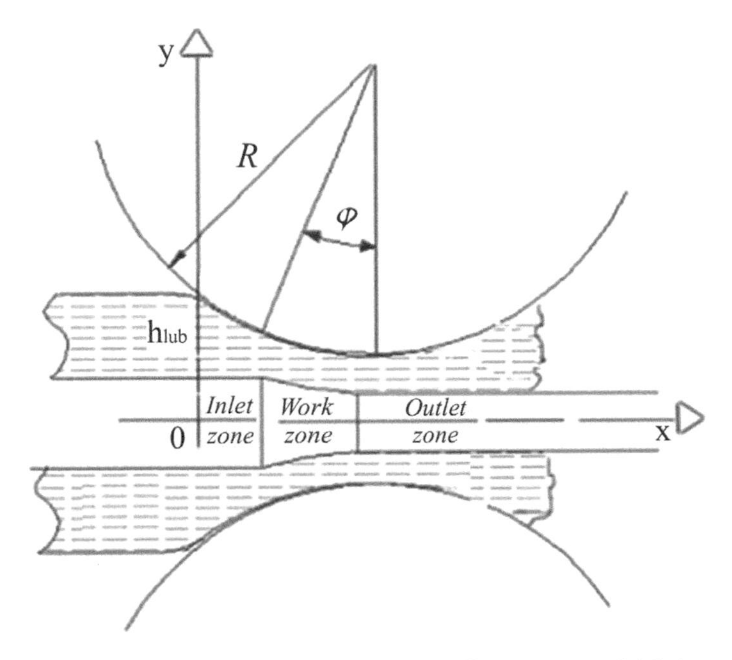

Figure 13.18 Schematic representation of the rolling process with lubricant.

- *Outlet zone*: in this zone the sheet can be considered as rigid again and the lubricant pressure falls back to the atmospheric value.

By knowing the flow rate of the lubricant Q sprayed on the entrance and the speed of the sheet $u_{1,0}$ before the yield point, it is possible to evaluate the inlet zone thickness of lubricant film $h_{lub}(0) = h_0$ at the origin of the x axis. In the inlet zone the film shape $h_{lub}(x)$ represents the region between the roll and the undeformed sheet so that the Reynolds equation can be written as follows, where ω is roll angular speed:

$$\frac{d}{dx}\left(\frac{{h_{lub}}^3}{e^{\gamma \cdot p_{lub}}}\frac{\partial p_{lub}}{\partial x}\right) = -6\mu_0\left(R\omega + u_{1,0}\right)\frac{dh_{lub}}{dx}. \tag{13.33}$$

In the work zone, the pressure $p_{lub}(x)$ has to assure the deformation of sheet at each point, so a combination between the equation of the rolling process (one of the models previously described) and the Reynolds equation is used.

$$\frac{d}{dx}\left(\frac{{h_{lub}}^3}{e^{\gamma \cdot p_{lub}}}\frac{\partial p_{lub}}{\partial x}\right) = 12\mu_0(v_2 - v_1) + 6\mu_0 h_{lub}\frac{d(u_1+u_2)}{dx}$$
$$- 6\mu_0(u_2 - u_1)\frac{d(y_1+y_2)}{dx}. \tag{13.34}$$

Thus, the governing equations in the work zone are, the deformation equation and the Reynolds equation, together with the equation of the material mass conservation. The thickness variation $y_1(x)$, is defined on the basis of the interaction between the plastic deformation of the sheet and the dynamics of the fluid film, together with $u_1(x)$, being the value defined by the material mass conservation; $v_1(x)$ calculated as a function of $u_{1,0}$ and $y_{1,0}$.

The governing equation of the outlet zone is analogous to that of the inlet zone, so that:

$$\frac{d}{dx}\left(\frac{{h_{lub}}^3}{e^{\gamma \cdot p}}\frac{\partial p}{\partial x}\right) = -6\mu\left(R\omega + u_{1,f}\right)\frac{dh_{lub}}{dx}. \tag{13.35}$$

When the chatter occurs, the models evaluate the instability conditions of the process-structure system. In particular, by properly arranging the models of stands with the models of the deformation process, the formulation of the closed loop diagram (described in Figure 13.4) can be achieved, in order to simulate the conditions in which chatter occurs.

13.4 Chatter in Rolling Mills: Case Studies

In this section three examples of studies on chatter in rolling mills are presented.

13.4.1 S6-High Rolling Mill

The first case analyses a chatter in a S6-High Rolling Mill, it is a rolling technology for highest quality standards. This type of rolling mill is mainly used in rolling mills for cold rolling for a wide range of strip widths.

The utilization range is even wider when using the S6-high solution in combination with a standard 4-high mill. The S6-high rolling mill is able to carry out high strip reduction (until 1:12) in just one pass: it allows the use of very small work rolls laterally guided by individually adjustable side support rolls, which are supported by two rows of roller bearings mounted in cassettes. The cassettes have the function of creating a packing condition of the rolls, which is helpful to provide a sufficient compression inside the stand. The rows of roll bearing axes are parallel to the side support roll, thus having the function of reducing the strokes originated by work rolls during the process. During the rolling process the WR is pushed onto SSR due to the horizontal force and transmitted to the roll bearings with the aim of restraining the force. Little fluctuations on the process parameters generate little fluctuations on the force values so that WR is pushed towards SSR with a vibration mode. When the WR pushes SSR towards its roll bearings with little force, SSR must react to follow the WR contact in order to damp the stroke effects of the following load increase. This is an important role played by the springs located on the extremities of the SSRs necks within the cassette. Valigi et al. (2014) studied an S6-high rolling mill in which a series of chatter marks, perpendicular to the rolling direction, appeared prematurely thus compromising the aesthetic quality of the product, and also created a self-exciting behavior causing a shorter grinding life of side support rolls Valigi et al. 2014.

The frequency of marks identified the problem as a third-octave-mode chatter. In order to investigate the reason of chatter-marks a vibration analysis was carried out, to identify parameters involved in the self-exiting behavior and to understand how a rolling mill can be adjusted to ensure maximum productivity and highest quality. Experimental measures showed that the occurrence of chatter marks followed a periodic trend of rolling mill campaigns with to the replacement of SSR cassettes. Also chatter marks were noted on SSRs' surfaces (called "facets") and the problem disappeared with the replacement of the cassettes. The immediate relationship between the age of the SSR rolls and the chatter marks on the strip was deduced. Vibration measurements made directly on the rolling stand before and after the change of the cassettes, showed a critical value (Figure 13.19). This result showed the origin of the chatter inside the stand and specifically in the cassettes.

A linearized lumped parameter model of the rolling mill having ten degrees of freedom describes the stand where the masses are reduced to the ten rolls involved in the process. Considering that the stand is symmetrical

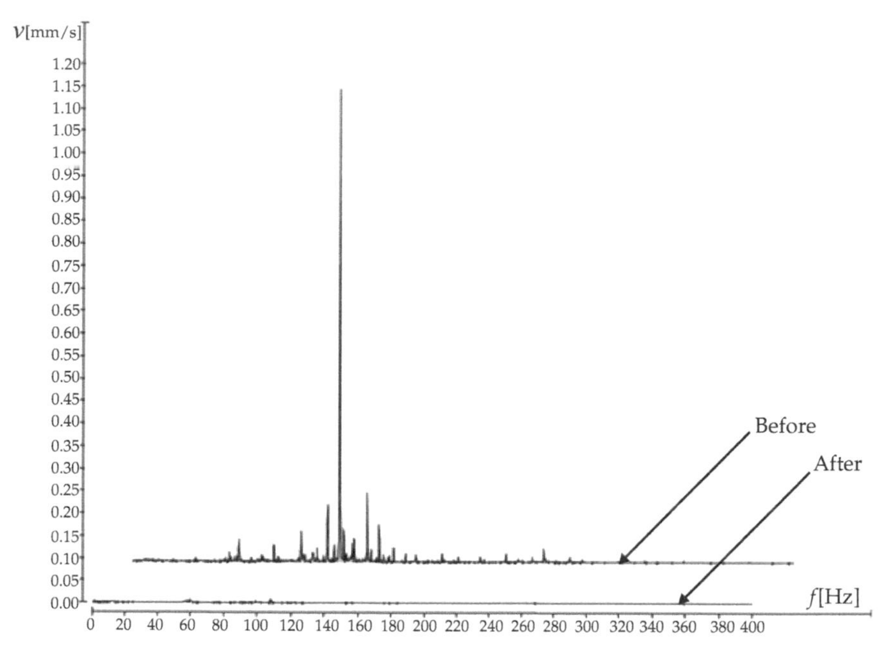

Figure 13.19 Vibration in the stand before end after the replacement of the side support rolls.

in relation to the rolled strip, as well as to the vertical axis of the stand, the model has been reduced into the simplest system, with two degrees of freedom (Figure 13.20). This model takes into account the equivalent mass m_1^* of the working rolls, the intermediate rolls and the back-up rolls and the equivalent mass m_2^* of the side support rolls.

The stiffness and the damping of the mass frame connection are represented by k_1^* and c_1^*, the contact stiffness between working rolls and side support rolls k_2^*, the stiffness of spring in the cassettes k_3^*. Then the vertical component F_{var} of the rolling force acting between strip and working rolls can be evaluated by the slab theory previously described.

$$\begin{bmatrix} m_1^* & 0 \\ 0 & m_2^* \end{bmatrix} \begin{Bmatrix} \ddot{x}_1 \\ \ddot{x}_2 \end{Bmatrix} + \begin{bmatrix} c_1^* & 0 \\ 0 & 0 \end{bmatrix} \begin{Bmatrix} \dot{x}_1 \\ \dot{x}_2 \end{Bmatrix} + \begin{bmatrix} k_1^* + k_2^* & -k_2^* \\ -k_2^* & k_2^* + k_3^* \end{bmatrix} \begin{Bmatrix} x_1 \\ x_2 \end{Bmatrix} = \begin{Bmatrix} F_{var} \\ 0 \end{Bmatrix}. \quad (13.36)$$

The model can be considered as a mass damper, and the problem of the short grinding life of SSRs is solved by properly changing the springs located on the extremities of the SSRs' necks in the cassettes, in order to have anti-resonance at the chatter frequency. Thus, the result of this proposed solution is, doubling of the SSRs' life. In the work zone, the geometry of the fluid film is not defined as a priority, since it depends on the deformed sheet

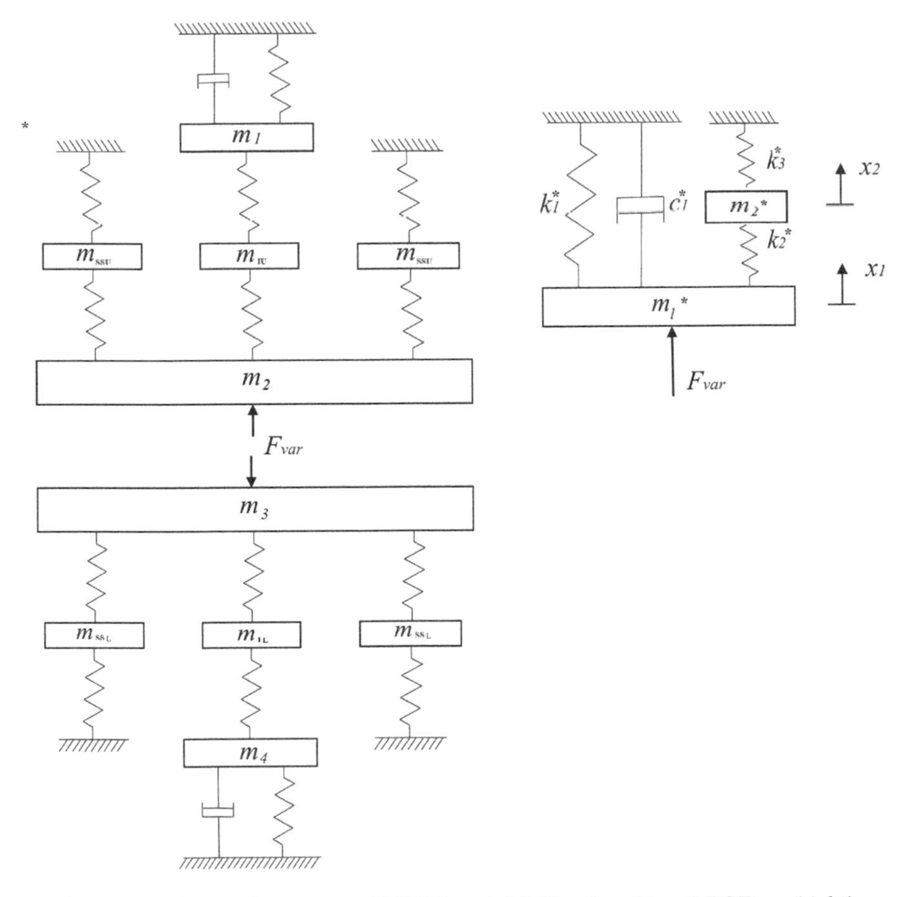

Figure 13.20 Lumped parameter 10-DOF model (left) reduced to a 2-DOF one (right).

profile; in this phase the equation describing the lubricant flow interacts with the plastic deformation.

13.4.2 Strip Tension Control in a Multi Stand Finishing Rolling Mill

The second presented case, analyses and studies a 4-h multi-stand finishing rolling mill and in particular the control techniques of the tension between the various stands of rolling mills in order to reduce the vibratory phenomena. This case highlights the importance of effective control systems to prevent strip fluctuations (Malvezzi and Rinchi 2000).

According to the general scheme showed in Figure 13.21 (where S_1 and S_2 are the input and output strip thickness in the stand), the complete model of the stand is subdivided into the following subsystems:

- the stand (vertical vibrations of work rolls);
- drive train (fluctuations of rolling speed);
- plastic deformation region (roll vertical force and torque);
- control systems in the inter-stand region (regulating rolling speed and strip tension).

In the subsystem regarding the rolling stand, the vertical vibrations of the stand are analyzed in order to determine variations in the strip thickness. Thus, a 5-DOF lumped parameter model (as described in the previous section) and a FEM model of the stand has been studied and compared. The results are presented in Table 13.1.

The subsystem of plastic deformation region has been modeled on the basis of the analytical model of the plastic deformation region derived from Orowan's inhomogeneous theory described previously. The evaluation of rolling pressure distribution in the plastic deformation region can be evaluated as function of the entry tension force as Figure 13.22 shows. Figures 13.23 and 13.24 represent respectively the vertical rolling force and the rolling torque versus entry tension.

The simulations, carried out on the model obtained by assembling the described subsystems, were performed in order to verify what was already known in the literature (Tamiya et al. 1980), that fluctuations of rolling tension can make the system unstable. In fact, Figure 13.25 shows that strip thickness oscillations gradually increase in time when the stand is subjected to a forced harmonic fluctuation of the entry tension force at the frequency of 170 Hz (inside the third octave chatter range). After about

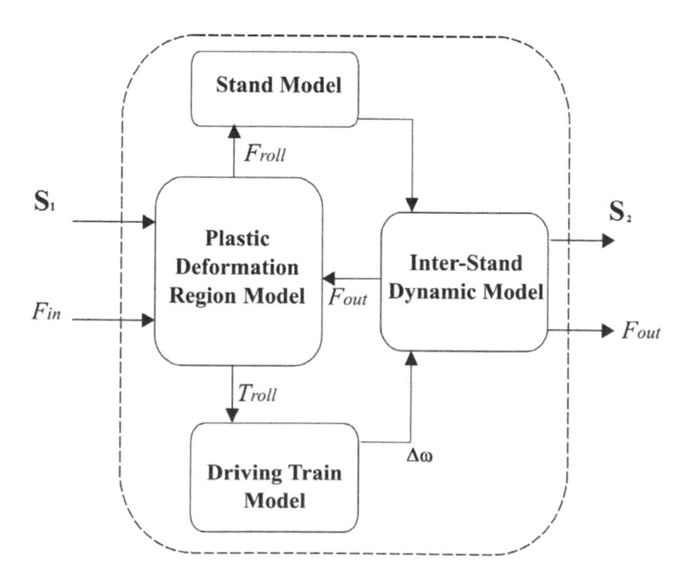

Figure 13.21 Block diagram of a single rolling mill stand.

Table 13.1 Comparison between natural frequencies of the FEM model and the lumped parameter model.

Model	I freq. (Hz)	II freq. (Hz)	III freq. (Hz)	IV freq. (Hz)	V freq. (Hz)
5-DOF	85.11	128.03	270.04	430.7	531.36
FEM	79.74	120.14	250.05	314.05	537.94

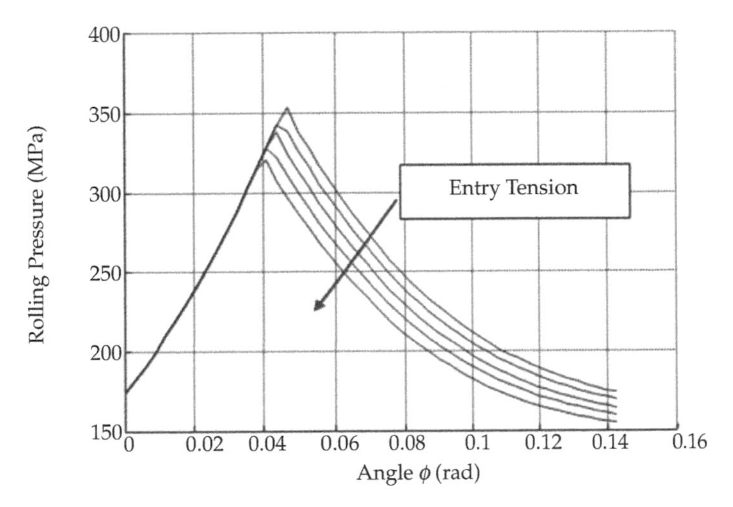

Figure 13.22 Rolling pressure.

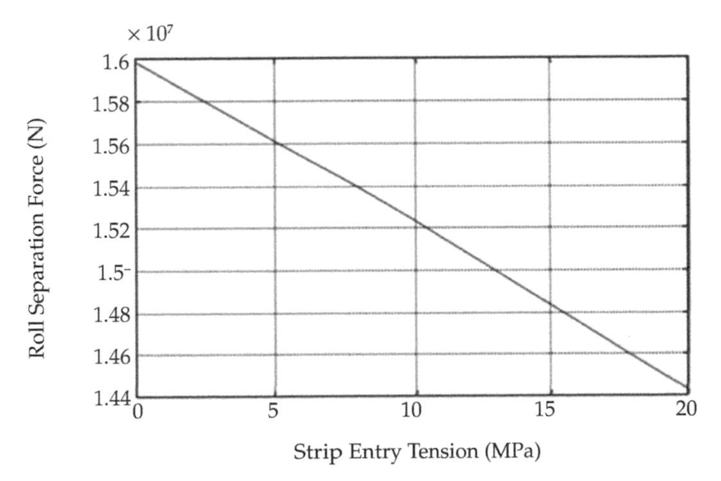

Figure 13.23 Roll separation force.

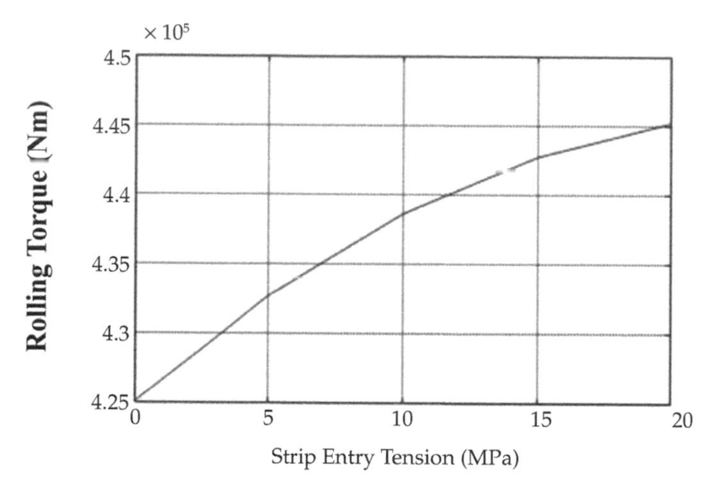

Figure 13.24 Rolling torque.

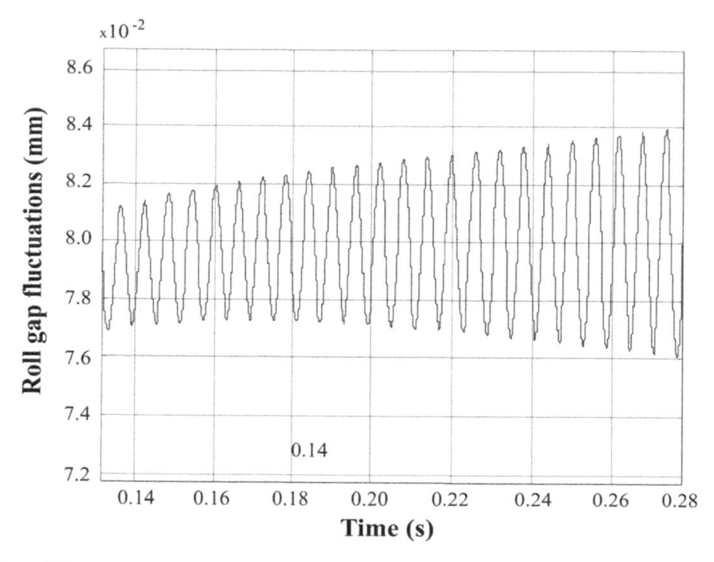

Figure 13.25 Vibrations of the upper work roll in the presence of varying entry tension force.

one second, the amplitude of these work roll vibrations reaches the value of about 0.1 mm (about 2–3% of the nominal thickness). Such vibrations are sufficient to significantly lower the quality of the finished product. In such simulations the control system regulating the tension force in the inter-stand region has not been introduced yet. This control system is generally present in the plants in order to limit chatter phenomena.

Then, the drive train subsystem was studied. Analyzing the main characteristics of the single mechanical components, the parameters of two different lumped parameter models described in the previous sections (with 14 and 5-DOF) were identified and their dynamic behavior analyzed and compared. Table 13.2 shows the results where the lumped parameter model with 5-DOF, is substantially equivalent to the 14-DOF one in the same frequency range and it is more easily managed numerically.

The control subsystems in the inter-stand region is studied considering that a looper is positioned in the zone between two successive stands, whose function is to regulate strip tension and to decouple each stand

Table 13.2 Natural frequencies of the 14 and the 5-DOF lumped parameter model of drive train system.

Model	I freq. (Hz)	II freq. (Hz)	III freq. (Hz)	IV freq. (Hz)
14-DOF	13.1	21.9	44.1	48.3
5-DOF	13.0	21.8	43.4	47.8

from the adjacent ones. The strip tension control system is essential in order to assure high quality strip. The looper is able to compensate possible rolling speed fluctuations that could lead to defects or accumulations of mass in the inter-stand region. The loopers are basically constituted by a looper arm, hinged at one end and put into rotation by electric motors (Figure 13.26). In order to improve finished product quality strip thickness closed loop control systems are usually introduced into each stand. They are needed to limit the amplitude of thickness defects, yet they may also be sources of severe disturbances of strip tension. It is for this reason that strip tension control systems have to be fast enough to effectively balance these undesirable disturbances.

Referring to the scheme in Figure 13.27, strip tension in the inter-stand region depends on the variation of strip length L in the inter-stand induced by arm rotation. Strip tension T_{st} fluctuations can be expressed as a function of the angular displacement θ of looper arm and roll speed by the following non-linear differential equation with time-varying coefficients:

$$\frac{dT_{st}}{dt} = \frac{Ebh}{L} v_{i,i+1} = \frac{Ebh}{L}\left(v_{in}^{i+1} - v_{out}^{i} + \frac{dL}{dt}\right) = \frac{Ebh}{L}\left(v_{in}^{i+1} - v_{out}^{i} + \frac{dL}{d\theta}\frac{d\theta}{dt}\right); \tag{13.37}$$

where E is the Young modulus, b is strip width, v_{in}^{i+1} and v_{out}^{i} are strip speed at entry to the (i+1)-th stand (downstream) and at exit of the i-th one (upstream) respectively. The term $dL/d\theta$ can be expressed as a function of looper and inter-stand region geometry while the term $d\theta/dt$ can be derived from the looper and its driving motor dynamic equations.

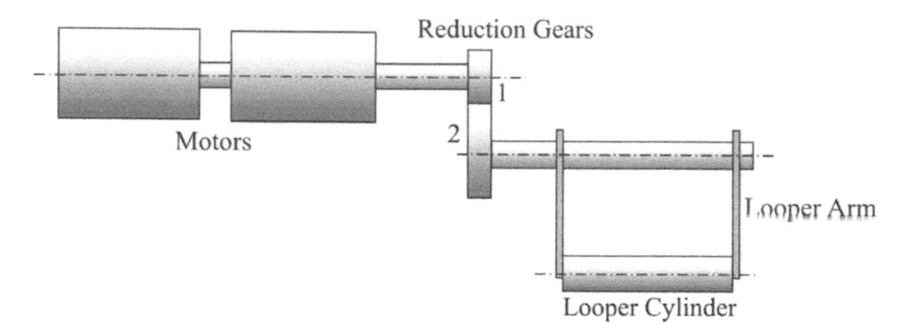

Figure 13.26 Mechanical configuration of the looper.

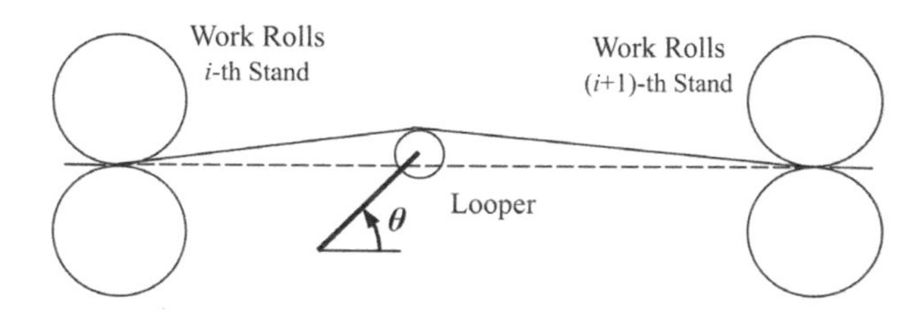

Figure 13.27 Geometry of the inter-stand region with the looper arm.

The looper's motors are adjusted by acting on the supply current, with a good approximation, proportional to the torque. Normally there is also a speed feedback using the angular velocity signal, measured by the tachometer on the motor shaft. However, the speed feedback is not generally sufficient to ensure that the system has the desired dynamic behavior. For this reason, an additional control loop is employed by using the strip tension as a feedback: the reference angular speed of the looper results from processing such a signal, typically through a simple PID.

The reference scheme is shown in Figure 13.28.

The looper arm is able to compensate small strip tension fluctuations around its reference value but it cannot regulate extremely high tension variations nor very low frequency fluctuations. In such cases the looper arms run the risk of reaching the end limit positions. Control systems in the inter-stand region simultaneously regulate strip tension and looper position, varying rolling reference speed of the upstream stand. The control system regulates the speed of work rolls through current feeding of motor.

Control actions however generate very slow roll speed variations, due to the very high inertial properties of the mechanical group composed by dc driving motors, drive train and work and backup rolls. The design of control system was carried out considering the linearization of the models.

In particular the looper dynamic equations are considered linear, assuming the term $dL/d\theta$ constant and the behavior of the material perfectly elastic (in this way the term regarding the material can be neglected).

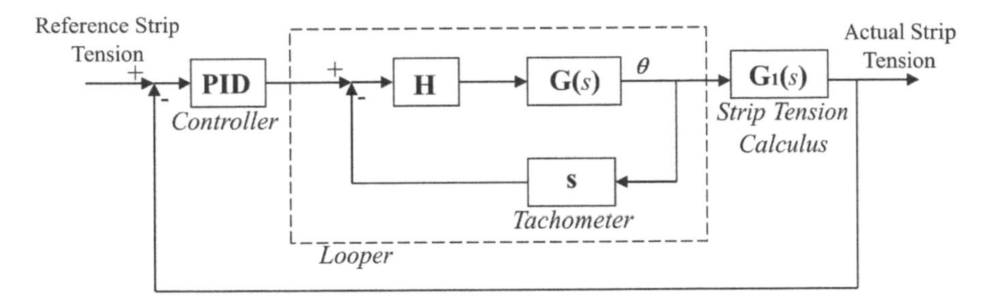

Figure 13.28 Scheme of the tension control system.

Nevertheless, in order for the model to be representative of the phenomenon, it is suggested that a nonlinear element of the "saturation" be included in the simulation. A comparison of the two different tension control systems (looper) and the work rolls speed (that are inherently interacting), has been carried out. The two control laws can be established by considering the two controllers separately, as conventional, or considering the controllers interacting between control loops (control system NIC) (Bryant and Higham 1973). The only scheme in Figure 13.29 is enough to represent both the NIC and traditional control system. Both constants $\mathbf{K}_{12}$ and $\mathbf{K}_{21}$ differ from zero and refer to the NIC control system, whereas the traditional control system is obtained by setting both $\mathbf{K}_{12}$ and $\mathbf{K}_{21}$ to zero.

A number of dynamic simulations were carried out on the linearized model, previously described, subjected to a series of typical inputs and external disturbances. In particular, step response rise time, steady state error and maximum overshoot were analyzed. Controller parameters were chosen in order to optimize system dynamic response and their performance was compared. The same linear model equipped with different control system was subjected to the same input signals and disturbances (strip tension reference value and strip speed variations). The control system performance, when the system is subjected to a step strip speed input, is summarized in Figure 13.30.

As can be observed, NIC performance is substantially better than that of the usual decoupled control system. Also a number of numeric simulations were carried out with complete non-linear model in order to analyze the effects of external disturbances such as strip thickness variations. The full model is extremely non-linear: principal non-linearity sources are the plastic deformation region model, the strip tension calculation formulas and the looper dynamic equations (moment of inertia and resistant roll torque are trigonometric functions of looper displacement).

An example of simulation results obtained by a model built by coupling three rolling stands (two interstands, each with a looper) is shown in Figure 13.31. In this diagram, the strip tension in the first interstand is shown, with

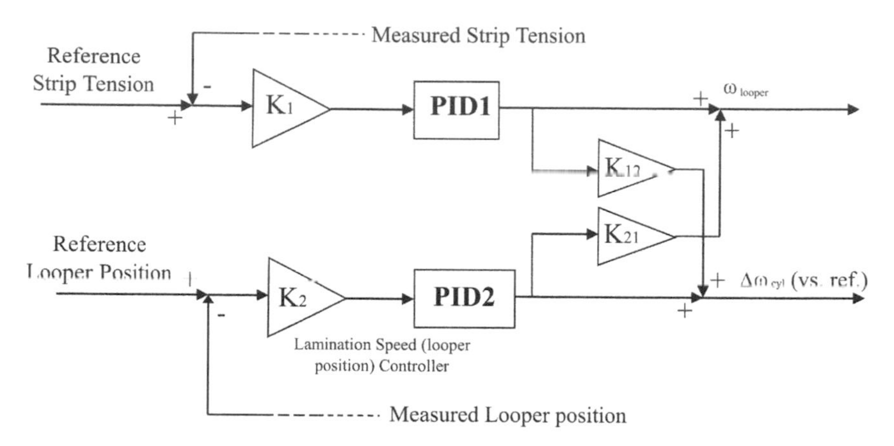

Figure 13.29 Controller: general scheme.

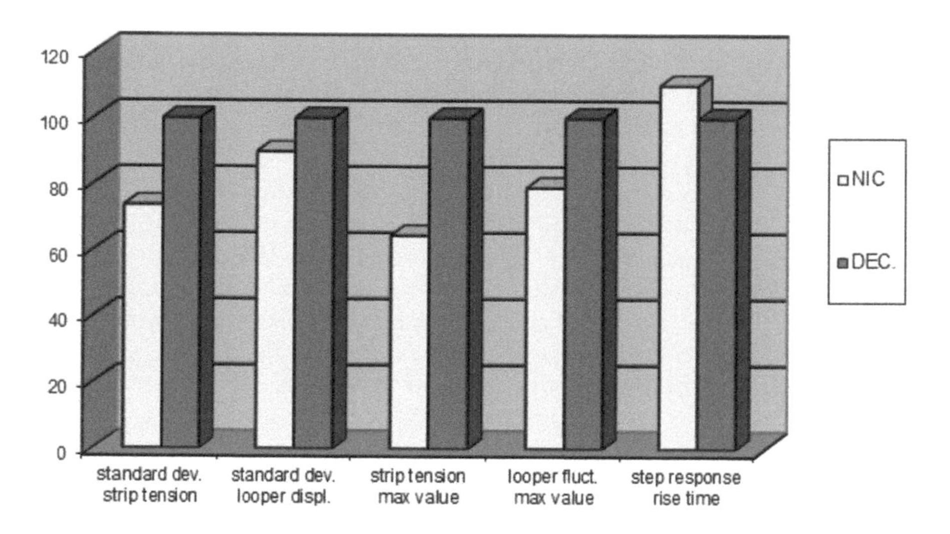

Figure 13.30 Comparison between usual decopuled and NIC control system.

the intent of comparing the performance of the NIC and traditional control system, if applied to a more realistic (non-linear) model of the system.

From the beginning of the simulation (when the rolling starts in the first stand) to the moment when the strip enters in the second stand (approximately t = 0.7 s), the tension in the first interstand is zero. When the material is positioned in the second stand, the tension increases rapidly and the looper adjusts on the reference value (50 kN). Significant differences in the performance of the two system controls are observed at the entrance of the material into the following stand (about t = 1.4 s).

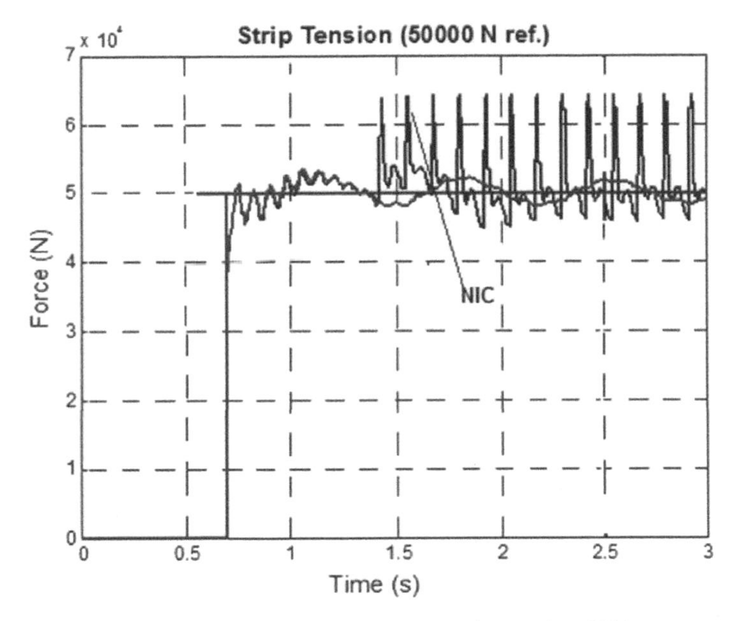

Figure 13.31 Strip tension: conventional control vs. NIC.

When the material is rolled in the first stand, the performance of the two control systems is very similar and vice-versa, when the material is simultaneously at the outlet of the two stands, the performance of the traditional control systems is markedly better than the more complex control system (NIC). This obviously does not mean that this control system is completely unsuitable for the system, but that the usual techniques of parameterization of the control systems (based on linearization of the models and optimization of the relative performance) have proven to be unsuitable for such a complex non-linear system.

Experimental evidences show that the synthesized control systems (usually simple PID), starting from a linearized model of a single stand and implemented in the plants, perform effectively even in presence of limited parametric variations of the dynamic characteristics of the different sub-models. On the other hand, in the case of models of rolling mill stands in series (dynamically interacting), the control systems determined through the previously described procedure, may be inadequate in controlling the non-linear model of the entire system, and are very sensitive to possible variations in the mode. In such cases, the problem of the parameterization of the control systems is resolved at the level of individual plants through a consistent series of experimental tests. Alternatively, robust control techniques (such as neural networks or control $H\infty$) (Lisini et al. 2000) appear to be more suitable.

13.4.3 Rolling Chatter Model with Lubrication

Another reported case is on third octave chatter of a tandem 4-h stand with tandem rolling mill with full film lubrication (Heidari et al. 2014).

The thickness reductions are at 60 to 90 percent with an incoming strip of 2 mm and pass through 4-high stands 2 to 3 times. Chatter usually occurs in the second stand and during the third pass.

The model of the rolling process assumes full film lubrication regime and the domain is divided as previously described, in the inlet, work and outlet zones (Figure 13.32).

The inlet zone is described by Reynolds equation including also the squeeze action and Barus equation for lubricant viscosity. The evaluation of the variation pressure is obtained by applying the boundary conditions in the inlet zone considering the entry pressure equal to the difference of the yield stress in the plane strain condition and the entry strip tensile stress. Film thickness $h_0(t)$ at the end of the inlet zone and at the entry of the work zone may be calculated by numerical integration of the system (differential) equations.

In the work zone the strip model describes the deformation process where the friction stress τ_f can be expressed as:

$$\tau_f = \frac{\eta\,(V_s - V_r)}{h_{lub}}; \tag{13.38}$$

where h_{lub} (x,t) is the film thickness distribution, v_s is the strip speed, V_r the roll peripheral. Since the high pressure in the work zone, the dynamic viscosity η is assumed according to the Roelands equation and also the behavior of lubricant is considered viscoplastic. Thus, the friction stress is

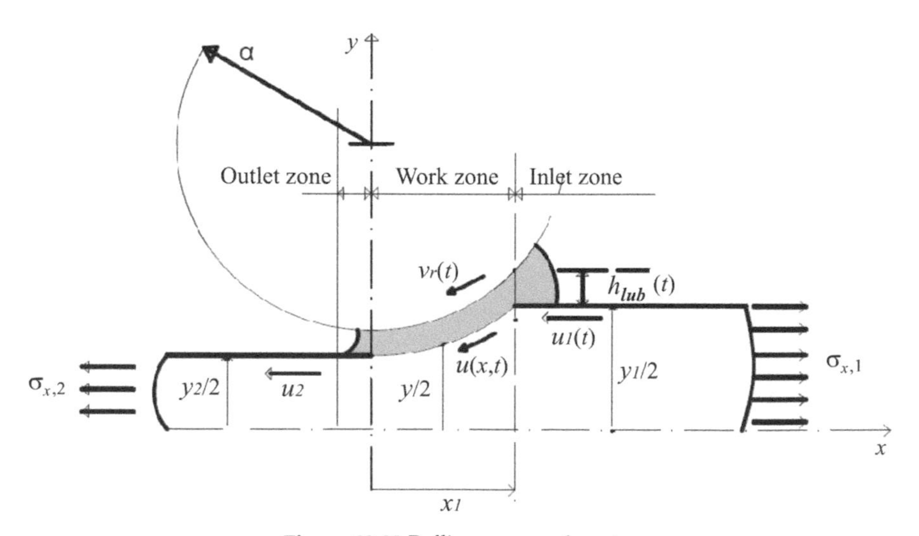

Figure 13.32 Rolling process domain.

set equal to the limiting shear stress when the friction stress passes the shear stress limit. The h_{lub} (x,t) is obtained from the work zone analysis using the continuity equation.

In turn h_{lub} (x,t) is the input at the "slab model" so that the pressure distribution and an estimation of rolling force in the work zone are evaluated. Also, considering the strip speed at neutral point equal to the roll peripheral velocity, the strip speed at entry and exit have been evaluated together with the exit strip thickness by means the mass conservation equation. The rolling mill structural model Figure 13.33 (a 2-DOF lumped parameter model) has to be properly coupled with the described model of the rolling process in which the inputs are the strip thickness, the strip tensile stress at entry and the exit; and the outputs are the strip thickness at exit, the strip speed at the entry and exit and the rolling force. In this way the chatter model for a single stand is achieved.

The output speeds are used to calculate the strip tensile stress useful to model the inter-stand zone which deals with the transportation time delay and the inter-stand tension. In regards to the model of the stand, the modulus of rolling mill is evaluated experimentally and linearized around the steady state conditions, a FEM method calculates the stiffness of the contact between rolls.

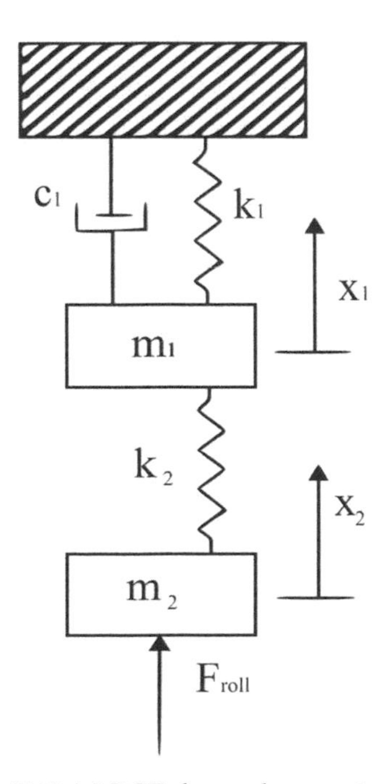

Figure 13.33 A 2-DOFs lumped parameter model.

No damping is assumed for work rolls, but a damping coefficient is considered for back-up rolls, which are included in order that the simulated results of the oscillation domain of the rolling force is comparable to experimental data.

On the basis of the single second stand model, simulations during the third pass for different speeds have reproduced work roll fluctuations: for low speed vibrations result damped and the system stable but with the increasing speed vibration becomes instable. The obtained results have shown that the undamped vibrations make the system unstable within the frequency range of third octave chatter. The results of a series of simulations have shown that the critical speeds are, on an average, about 40 percent higher than the critical speeds. It can be connected to the fact that the rolling chatter, in the simulations, regards the single stand rolling, while the experiments regard, double stand rolling. In the single stand rolling, the negative damping mechanism is activated but the regenerative chatter also occurs in the two rolling stands. It is known that the critical speed is much higher in the negative damping mechanism than in the regenerative chatter and, as expected, the experimental results show that chatter occurs much faster than that in the simulation.

In addition, the effect of some parameters on chatter's critical speed such as viscosity pressure coefficient, lubricant viscosity and limiting shear stress are investigated by simulations. The results of this research show that just limiting shear stress has an important effect on the chatter's critical speed.

Summary

Avoiding the insurgence of chatter vibration is a very important goal in the rolling of the production of advanced high strength steel, due to the fact that the quality of steel is strongly influenced by such vibrations. In this chapter, a description of the undesired vibration phenomenon, known as chatter, has been given and models of the deformation process and the plant have been studied, in order to describe and simulate the dynamical behavior of rolling and to detect the insurgence of chatter. In addition, some examples of chatter in the rolling mill process have been described and some solutions proposed.

Acknowledgements

Authors warmly acknowledge the Italo-American Dott.Elisa Tacchi for her support to the English review of this chapter.

Keywords: Chatter in rolling mill, AHSS, deformation process, models of stand, models of process, lubrication in rolling mill process, models of driving train system, inter-stand control system

Key of Abbreviations

λ	chatter marks wavelength (m)
f	vibration frequency (Hz)
v	sheet speed (m/s)
N	sheet critical speed (m/s)
D	roll diameter (m)
R	roll radius (m)
m_i	inertial mass coefficient (kg)
k_i and kij	stiffness coefficients (N/m)
c_i	viscous damping coefficient (Ns/m)
x_i	vertical displacement of i-th inertial mass (m)
m_1	upper backup roll inertial mass (kg)
m_2	upper work roll inertial mass (kg)
m_3	lower work roll inertial mass (kg)
m_4	lower backup roll inertial mass (kg)
m_5	upper crosspiece inertial mass (kg)
m_6	shoulder inertial mass (kg)
m_7	shoulder inertial mass (kg)
m_s	inertial mass of stand (kg)
J_2	inertial moment of upper work roll (kg m^2)
J_3	inertial moment of lower work roll (kg m^2)
F_{var}	dynamic component of rolling force (roll separating force) (N)
$h_{g\,var}$	gauge variation (m)
$X_1(t)$ and $X_2(t)$	roll displacement components in multidirectional model (m)
α	angle between X_1 direction and Y axis (rad)
β	angle between F_{var} and Y axis (rad)
ϑ	angular deflection of work roll with respect to its static equilibrium (rad)
$X_{G'}$	vertical displacement of roll barycenter with respect to its equilibrium position
M_{sbu} and M_{sbl}	moments generated by the upper and lower "spring bed" (Nm)
F_{sbu} and F_{sbl}	forces generated by the upper and lower "spring bed" (N)
M_{sb} and F_{sb}	general expression of moment and force generated by "spring bed"
f_{sb}	elastic force per unit length (N/m)
K_d	distributed stiffness of the "spring bed" (N/m)
l	work roll length (m)
$[M]$, $[C]$ and $[K]$	system mass, viscous damping and stiffness matrixes

$\{M_{ext}(t)\}$	vector of the external torques acting on the system (Nm)
$\{\Theta(t)\}$	vector of angular positions (rad)
x_A and $y_{A'}$	coordinates of A′
$x_{A,}$ and y_A	coordinates of A
τ	shear stress (MPa)
$h = 2y$	strip thickness (m)
y	half thickness of strip (m)
y_{in}	half thickness of strip at the entry of the stand (m)
y_{out}	half thickness of strip at the exit of the stand (m)
p	rolling pressure (MPa)
μ	friction coefficient
$K = K(x)$	mean yield stress in conditions of plane strain (MPa)
σ_x	stress in the rolling direction (MPa)
σ_{xin}	horizontal tensile stress at entry of the stand (MPa)
σ_{xout}	horizontal tensile stress at exit of the stand (MPa)
p_{in}	roll pressure at entry of the stand (MPa)
p_{out}	roll pressure at exit of the stand (MPa)
K_f	deformation resistance (MPa)
a	Orowan's coefficient
F_{roll}	rolling force (N)
T_{roll}	rolling torque (Nm)
ϕ	angle identifying the various sections in the bite zone (rad)
F	the roll tension per unit width of the strip (N/m)
σ_s	yield stress of the material subjected to homogeneous compression (MPa)
φ_{in}, φ_{out} and φ_n	angles, respectively relative to the entry section, to the exit one and to the neutral plane (rad)
φ_{t1} and φ_{t2}	angles relative to the transition sections between stick and slip zones before and after the neutral plane (rad)
$w(\varphi)$	non-linear function taking into account the inhomogeneity of plastic deformations
h_{in}	strip thickness at the entry section (m)
h_{out}	strip thickness at the exit section (m)
$h_a - h_n$	strip thickness at minimum gap and neutral point (m)
k_f	deformation resistance (MPa)
k_f^{out}	deformation resistance at the exit section (MPa)
k_f^{in}	deformation resistance at the entry section (MPa)
R'	flattened work roll radius (m)
R_w,	work roll radius (m)
R_b	back-up roll radius (m)
u	strip speed (m/s)
u_{in}	strip speed at entry section (m/s)
u_{out}	strip speed at exit section (m/s)

x_{in}	position of roll bite at the entrance (m)
t_b and t_f	back and front tension (MPa)
x	coordinate identifying the sections in the bite zone (m)
p^+ and p^-	pressure distribution in forward sleep region (MPa)
u_1	horizontal component of strip speed (m/s)
$u_{1,0}$ and $u_{1,f}$	horizontal component of strip speed in the inlet and in the outlet zone (m/s)
Q	flow rate of the lubricant sprayed on the entrance (m^3/s)
ω	roll angular speed (rad/s)
u_2	horizontal component of roll speed (m/s)
v_1 and v_2	vertical component of strip speed and roll speed (m/s)
$y_1(x)$	function of strip profile (m)
$y_2(x)$	function of roll profile (m)
$h_{lub}(x)$	thickness of lubricant film $[y_2(x)-y_1(x)]$ (m)
h_0	thickness of lubricant film in the inlet zone (m)
S_1	initial thickness of the strip (m)
S_2	final thickness of strip (m)
$y_{1,0}$	half final thickness of strip in the inlet zone (m)
$p_{lub}(x)$	lubricant pressure (MPa)
γ	Viscosity coefficient in Barus law (MPa^{-1})
μ, μ_0	lubricant viscosity, when $p = 0$ (m^2/s)
θ	angular displacement of looper arm (rad)
L	strip length in the inter-stand (m)
T_{st}	Strip tension (N)
E	Young modulus (MPa)
b	strip width (m)
v_{in}^{i+1}, v_{out}^{i}	strip speed at entry to the (i+1)-th stand, at exit of the i-th stand (m/s)
$\mathbf{K}_{12}$ and $\mathbf{K}_{21}$	constant of the control system
τ_f	friction stress (MPa)
$h_{lub}(x,t)$	film thickness distribution (m)
V_s	strip speed (m/s)
$V_r = R\omega$	roll peripheral velocity (m/s)
η	dynamic viscosity coefficient in accordance to the Reolands' equation (MPa s)
m_1^*	equivalent mass of work roll, intermediate roll and back-up rolls (kg)
m_2^*	equivalent mass of the side support rolls (kg)
k_1^*	stiffness of mass frame connection (N/m)
c_1^*	damping of mass frame connection (Ns/m)
k_2^*	contact stiffness between work rolls and side support rolls (N/m)
k_3^*	stiffness of spring in the cassettes (N/m)

References

Arnken, R., C. Bilgen and W. Hennig. 2007. Commissioning and optimization of the roll gap lubrication system at the ANSDK CSP plant. Iron Steel Technol. 4(8): 81–86.

Benvenuti, S. 1999. Treni di laminazione a caldo delle lamiere: analisi dei fenomeni di instabilità nelle linee di trasmissione di potenza. Degree Thesis, University of Florence, Italy.

Bryant, G.F. and J.D. Higham. 1973. A method for realizable non-interactive control design for a five stand cold rolling mill. Automatica. 9(4): 453–466.

Casini, J. 1999. Treni di laminazione a caldo delle lamiere: analisi dei fenomeni di chatter dovuti all'elasticità della gabbia. Degree Thesis, University of Florence, Italy.

Chefneux, L., J. Fischbach and J. Gouzou. 1984. Study and industrial control of chatter in cold rolling. Iron Steel Eng. 61(11): 17–26.

Chen, Y., S. Liu, T. Shi, S. Yang and G. Liao. 2002. Stability analysis of the rolling process and regenerative chatter on 2030 tandem mills. P. I. Mech. Eng. C-J Mec. 216(12): 1224–1225.

Heidari, A., M.R. Forouzan and S. Akbarzadeh. 2014. Development of a rolling chatter model considering unsteady lubrication. ISIJ International. 54(1): 165–170.

Hu, P. and K.F. Ehmann. 2001. Regenerative effect in rolling chatter. J. Manuf. Proc. 3(2): 82–93.

Johnson, R.E. and Q. Qi. 1994. Chatter dynamics in sheet rolling. Int. J. Mech. Sci. 36(7): 617–630.

Joseph, D. and H. Gallenstein. 1981. Torsional chatter on a 4-h cold mill. Iron Steel Eng. 58(1): 52–57.

Kim, Y., C.W. Kim, S.J. Lee and H. Park. 2012. Experimental and numerical investigation of the vibration characteristics in a cold rolling mill using multibody dynamics. ISIJ International. 52(11): 2042–2047.

Kimura, Y., Y. Sodani, N. Nishiura, N. Ikeuchi and Y. Mihara. 2003. Analysis of chatter in tandem cold rolling mills. ISIJ International. 43(1): 77–84.

Laugier, M., M. Tornicelli, J. Cebey, L. Schiavone, L. Peris, A. Devolder, R. Guillard and F. Kop. 2015. Flexible lubrication for controlling friction in cold rolling, crucial to be successful for the AHSS challenge. Proc. 2nd Metec & Estad, Düsseldorf: 1–8.

Lin, Y.J., C.S. Suh, R. Langari and S.T. Noah. 2003. On the Characteristics and mechanism of rolling instability and chatter. J. Manuf. Sci. Eng. 125: 778–786.

Lin, Y.J., C.S. Suh and S.T. Noah. 2002. Dynamic characteristics and sheet rolling instability. Int. J. Struct. Stab. Dy. 2(03): 375–394.

Lisini, G.G., P. Toni and M.C. Valigi. 2000. H∞ control system for a four-high stand rolling mill. P. I. Mech. Eng. I-J. Sys. 214(2): 79–86.

Malvezzi, M. and M. Rinchi. 2000. Control systems of strip tension in hot steel rolling mills trains. Proc. of Nineteenth IASTED International Conference—MIC 2000—Modeling, Identification and Control, Innsbruck, Austria.

Malvezzi, M. and M.C. Valigi. 2008. Cold rolling mill process: A numerical procedure for industrial applications. Meccanica. 43(1): 1–9.

Meehan, P.A. 2002. Vibration instability in rolling mills: Modelling and experimental results. J. Vib. Acoust. 124(2): 221–228.

Misonoh, K.1980. Analysis of chattering in cold rolling of steel strip. Journal of the JSTP. 21(238): 1006–1010.

Moller, R.H. and J.S. Hoggart. 1967. Periodic surface finish and torque effects during cold strip rolling. J. Aust. Inst. Met. 12: 155–165.

Orowan, E. 1943. The calculation of roll pressure in hot and cold flat rolling. P. I. Mech. Eng. 150: 1847–1982.

Paton, D.L. and S. Critchley. 1985. Tandem mill vibration: its cause and control. I & SM. 37–43.

Petrucci, A. 2012. Problemi vibrazionali nella laminazione a freddo mediante tecnologia s6-high. Degree Thesis, University of Perugia, Terni, Italy.

Rinchi, M. 1999. Controllo della tensione della sulla lamiera nei treni di laminazione a caldo, Ph.D. Thesis, University of Bologna, Firenze, Italy.

Roberts, W.L. 1978. Four-high mill-stand chatter of the fifth-octave mode. Iron. Steel Eng. 55: 41–47.

Roberts, W.L. 1978. Cold Rolling of Steel. Marcel Dekker Inc. New York.

Roberts, W.L. 1983. A review of hot rolling of steel. Marcel Dekker Inc. New York.
Roberts, W.L. 1988. Flat processing of steel. Marcel Dekker Inc. New York.
Swiatonoswki, A. and A. Bar. 2003. Parametrical Excitement vibration in tandem mills-mathematical model and its analysis. J. Mater Process Tech. 134: 214–224.
Tamiya, T., K. Furui and H. Hida. 1980. Analysis of chattering phenomenon in cold rolling. Proc. of Int. Conf. on Steel Rolling. 1191–1207.
Tlusty, J., G. Chandra, S. Critchley and D. Paton. 1982. Chatter in cold rolling. CIRP Ann-Manuf. Tech. 31(1): 195–199.
Ubici, E., M. Borda, A. Klempnow and J. Pineyro. 2001. Identification and countermeasures to resolve hot strip mill chatter. AISE Steel Tech. 78(6): 48–52.
Valigi, M.C., S. Cervo and A. Petrucci. 2014. Chatter marks and vibration analysis in a S6-high cold rolling mill. Lecture Notes in Mechanical Engineering. 5: 567–575.
Valigi, M.C. and S. Papini. 2013. Analysis of chattering phenomenon in industrial S6-high rolling mill. Diagnostyka. 14(3): 3–8.
Wu, S., Y. Shao, L. Wang, Y. Yuan and C.K. Mechefske. 2015. Relationship between chatter marks and rolling force fluctuation for twenty-high roll mill. Eng. Fail. Anal. 55: 87–99.
Xie, Y. 2013. Vibration Analysis and Test of Backup Roll in Temper Mill. Sensors & Transducers. 21(5): 105–110.
Yang, X. and C. Tong. 2012. Coupling dynamic model and control of chatter in cold rolling. J. Dyn. Syst-T. ASME. 134(4): 041001/1-041001/8.
Yarita, I., K. Furukawa, Y. Seino, T. Takimoto, Y. Nakazato and K. Nakagawa. 1978. Analysis of chattering in cold rolling for ultrathin gauge steel strip. Trans. Iron Steel Inst. Jpn. 18(1): 1–10.
Yun, I.S., W.R.D. Wilson and K.F. Ehmann. 1998. Review of chatter studies in cold rolling. Int. J. Mach. Tool Manu. 38(12): 1499–1530.
Zhao, H. and K.F. Ehmann. 2013. Stability analysis of chatter in tandem rolling mills—Part 1: Single and multi-stand negative damping effect. J. Manuf. Sci. E-T ASME. 135: 031001/1-031001/8.
Zhao, H. and K.F. Ehmann. 2013. Stability analysis of chatter in tandem rolling mills—Part 2: The regenerative effect. J. Manuf. Sci. E-T ASME. 135: 031002/1-031002/11.
Zhong, J., H. Yan, J. Duan, L. Xu, W. Wang and P. Chen. 2002. Industrial experiments and findings on temper rolling chatter. J. Mater Process Tech. 120(1-3): 275–280.

14

Hot-dip Galvanizing of Advanced High Strength Steels

Martin Arndt

ABSTRACT

Hot-dip galvanized second generation advanced high strength steel (AHSS), 15.8 wt.% Mn, 0.79 wt.% C, was analyzed at the interface between steel and zinc by scanning Auger electron spectroscopy (AES) and microscopy in the nanometer range in order to confirm and improve an existing model of an additional pre-oxidation treatment step before annealing and immersing into the zinc bath. Furthermore these steel samples were fractured in the analysis chamber of the Auger system and analyzed without breaking vacuum. In these measurements the results of an aluminothermic reduction of the manganese and iron surface oxides on the steel could be confirmed by AES. Additionally, it could be found that no inhibition layer was formed between the manganese and iron oxides and the zinc layer. The result of the additional pre-oxidation is an excellent wetting behavior but the adhesion of the zinc layer is weak because of the brittle underlying oxide layer.

14.1 Introduction

Improving passenger safety and minimizing fuel consumption due to weight reduction of the vehicles are the two contradicting requirements

ZONA, CDL-MS-MACH, Johannes Kepler Universität, Altenbergerstraße 69, 4040 Linz, Austria, voestalpine Stahl GmbH, voestalpine-Straße 3, 4020 Linz, Austria.
E-mail: martin.arndt@jku.at

the modern automotive industry has as a challenge. Therefore advanced high strength steels (AHSSs) are used as materials for structural body applications (De Cooman et al. 2009, 2011), because these steel grades exhibit excellent mechanical properties.

Second generation AHSSs consist primarily of an austenitic microstructure stabilized by high amounts of Mn and C. They are typically designed with a Mn content ranging from 10 to 30 wt.%. High-Mn steels are classified according to their characteristic deformation mechanisms which occur during plastic deformation (Samek et al. 2012a, 2012b). Depending on their composition and deformation mechanism, they are often referred in literature as TRIP/TWIP (transformation induced plasticity/twinning induced plasticity), TWIP and Nano-TWIP steels (Grässel et al. 2000, Frommeyer et al. 2003, Krüger et al. 2003, De Cooman et al. 2011, Samek et al. 2012b). The so-called TWIP effect (Cornette et al. 2005, Samek et al. 2012b), in which twins are formed during tensile deformation, is responsible for the remarkable material characteristics of TWIP steels. The so-called nano-sized TWIP steels have a reduced Mn content of about 14–16 wt.%. Nano-TWIP steels offer the outstanding mechanical properties of 1000 MPa tensile strength by ~100% total elongation, due to the intense refinement of the microstructure during deformation (Samek et al. 2012a, 2012b). Regarding automotive applications, the simple and cost effective formability by cold rolling instead of expensive and cumbersome hot forming operations makes this steel even more attractive.

On the other hand, fully austenitic steels are known to be liable to aqueous corrosion (Shih et al. 1993, Zhu and Zhang 1998, Dieudonné et al. 2014a, b), and thus a suitable corrosion protection is often necessary. In this context, hot-dip galvanizing of stated steels with standard Zn or some more advanced Zn alloyed coatings (e.g., Zn-Al, Zn-Mg-Al, etc.) may provide sufficient and also cost efficient corrosion protection (Rodnyansky et al. 2000, Feliu et al. 2003, Volovitch et al. 2011, Schuerz et al. 2009, LeBozec et al. 2013, Persson et al. 2013, Diler et al. 2014, Salgueiro Azevedo et al. 2015). Nevertheless, the high Mn content of conventional TWIP steels may lead to difficulties during the galvanizing process. These difficulties are linked to the formation of manganese oxides on the surface during the technologically necessary annealing step in production. These Mn-based oxides were found to hinder the growth of a continuous Zn layer on the steel surface during galvanizing (Arndt et al. 2013).

However, there are some approaches to improve the wettability of high Mn steel with Zn. Blumenau et al. (2011a) suggested an additional pre-oxidation step in order to create a Mn/Fe mixed oxide on the steel surface. The Fe in the mixed oxide should then be chemically reduced during the annealing process by an extremely dry atmosphere and form an Fe_2Al_5 inhibition layer with the dissolved Al from the liquid Zn bath on top of the MnO, allowing the growth of the Zn layer. This approach was successfully applied on simulator samples. They also reported an improvement in the wettability by changing the Al concentration in the bath, the bath temperature and the temperature of the steel (Blumenau et al. 2011b). Cho and De Cooman (2012) suggested two additional methods: one involving annealing at high dew points to provoke internal Mn oxidation, and the other involving flash coating of the surface with pure Fe. Furthermore, Kavitha and McDermid (2011) had observed that longer dwell times in the Zn bath could reduce the thickness of a MnO layer due to an aluminothermic reduction of the MnO to metallic Mn, which was brought about by the oxidation of metallic Al to Al_2O_3.

Therefore, the main part of this chapter deals with recent results of AES analyses of simulator hot-dip galvanized Nano-TWIP steels in which the pre-oxidation approach proposed by Blumenau et al. (2011b) was repeated (Arndt et al. 2015). The motivation for this repetition was the lack of information from the nanometer scale because experience has shown that chemical analyses of surfaces using coarse methods such as X-ray photoelectron spectroscopy (XPS), X-ray diffraction (XRD), glow discharge optical emission spectroscopy (GDOES), energy dispersive X-ray spectroscopy (EDX) and even transmission electron microscopy (TEM) combined with electron energy loss spectroscopy (EELS) may yield misleading results, because these techniques have a high volume interaction (Blumenau et al. 2011a, b, Cho and De Cooman 2012, Blumenau et al. 2010a, b). Based on the results of these AES measurements a new model for the Zn wetting behavior could be proposed.

14.1.1 Inhibition Layer

Hot-dip galvanizing of steel sheets takes place normally in a liquid Zn bath at temperatures of about 460°C for a few seconds. The Zn bath contains about 0.2 wt.% Al which not only creates a thin Al_2O_3 layer on top of the final Zn layer brightening its surface but also creates an inhibition layer between the steel and the Zn layer (Marder 1990, Baril and L'Espérance 1999). This inhibition layer consists of a Fe-Al alloy, $FeAl_3$ or Fe_2Al_5, sometimes combined with Zn to ternary alloy or a mixture of these. The inhibition layer separates the iron from the Zn and hinders the formation of unwanted Zn-Fe phases, the monoclinic ζ-phase ($FeZn_{13}$), the hexagonal δ-phase ($FeZn_7$), the face centered cubic (FCC) Γ_1-phase ($FeZn_4$) or the body centered cubic (BCC) Γ-phase (Fe_3Zn_{10}). It is reported that the formation of a defect free inhibition layer is essential for galvanizability of steel (Sagl 2013).

14.1.2 Aluminothermic Reduction

Thin Mn oxide layers at the surface of steel can be reduced and dissolved during hot-dip galvanizing by an aluminothermic reduction (Khondker et al. 2007). The Mn oxide reacts with the dissolved Al in the Zn bath via the following equation:

$$3MnO_{(solid)} + 2Al_{(solution)} \rightarrow Al_2O_{3(solid)} + 3Mn_{(solution)} \quad (14.1)$$

Kavitha and McDermid (2012) found that the reduction of the Mn oxide at the interface between the steel and the Zn bath depends, linearly on the dwell time of the steel in the bath. Transmission electron microscopy (TEM) images have shown a 3–4 nm thick Al_2O_3 layer between the Zn and the remaining MnO independent of the dwell time. The appearance of ζ-$FeZn_{13}$ crystals at the Zn Fe interface is often interpreted as an indirect proof for the aluminothermic reduction because the consumption of metallic Al according to equation (14.1) leads to a depletion of Al in the Zn bath near the steel surface which hinders the formation of the Fe_2Al_5 inhibition layer. Without an inhibition layer, Fe and Zn react to ζ-$FeZn_{13}$ (Alibeigi et al. 2011, Kavitha and McDermid 2012).

14.2 Sample Material

The steel used in this study is a second generation AHSS taken from industrial production. Its chemical composition (wt.%) is: 0.79 C, 15.8 Mn, 0.05 Si, <0.05 Al, 0.03 P, 0.002 Ti, 0.022 Nb, 0.036 N, and 0.03 Cr. The steel is fully austenitic and its strengthening mechanism is nano twinning therefore it is called Nano-TWIP. The advantage of the material is its chemical simplicity. The only relevant expected oxide at the surface is MnO because the amount of Si, Al and Cr in the bulk is low.

The sample material was cut out from the steel band into smaller sheets (see Figure 14.1) and subsequently hot-dip galvanized in a galvanizing simulator under controlled temperature and atmospheric conditions. The Zn bath contains 0.2 wt.% Al and is heated to 460°C. Two kinds of heat treatments were applied to the steel sheets (see Figure 14.2). In the first the steel was annealed in 100% H_2 atmosphere at a dew point of –45°C and afterwards immersed into the hot Zn bath. In the second case, an additional pre-oxidation step in a 1% O_2 atmosphere was inserted before the annealing.

14.3 Analytical Methods and Sample Preparation

The analytical work in this study was performed with the aid of a scanning Auger electron microscope JAMP 9500F (from JEOL). It is equipped with a field emission electron gun which allows a lateral resolution in elemental mappings down to less than 10 nm. The base pressure in the analysis chamber was in the ultra-high vacuum region below 10^{-7} Pa. There is also an argon sputter gun at the chamber which was used for cleaning the surface of the samples, as well as for performing sputter depth profiles using ion energies between 0.2 and 3 keV. The sputter gun can also be used for

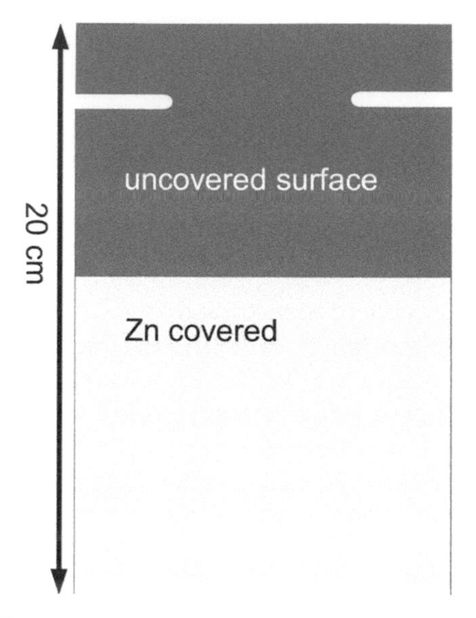

Figure 14.1 The steel sheet is cut into pieces of 20 cm length suitable for the galvanizing simulator and heat treated under controlled atmospheric conditions. Subsequently the lower part of the piece is immersed into a liquid Zn bath and thus covered by a Zn layer, as illustrated in the figure.

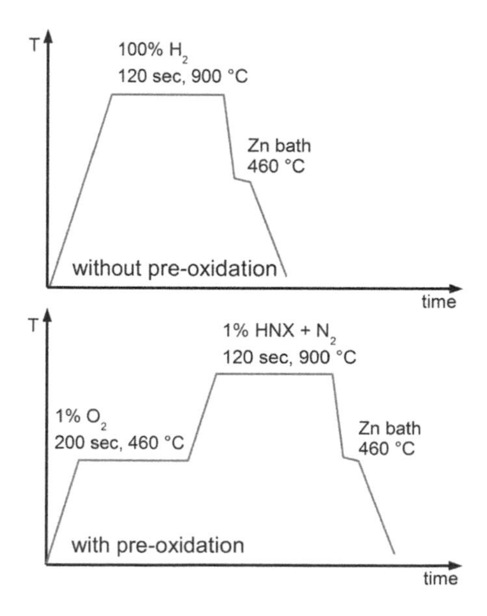

Figure 14.2 The upper figure shows the standard heat treatment. The steel is first annealed at 900°C and subsequently immersed into a liquid Zn bath. The annealing leads to the formation of MnO at the steel surface and a bad coatability with Zn. In the lower figure an additional pre-oxidation step is implemented before the annealing.

charge neutralization with ion energies between 10 and 50 eV. The energy of the electrons which leave the sample surface is analyzed by a concentric hemispherical analyzer (CHA) equipped with 7 channeltrons in parallel, which allows the measurement of an energy range from 0 to 2500 eV. In addition, a fracture stage is mounted on the analysis chamber; inside it, samples can be cooled via liquid nitrogen, and the top of the sample can be chopped off by a hammer in order to prepare a fresh surface for Auger analysis without breaking the vacuum. A Panalytical X'Pert Pro device was used for XRD analysis. It is equipped with a Co-Kα source.

For the evaluation of the chemical state of the involved elements at the interface between Zn and steel, namely Fe, Mn, Al and Zn, it is necessary to measure high energy resolution Auger spectra. Therefore a constant retard ratio (CRR) of 0.05 in the CHA was chosen, which resulted in an energy resolution of $\Delta E/E = 0.15\%$. The obtained spectra were then compared with reference spectra of these elements in the metallic and oxidic state: Fe, FeO, Mn, MnO_2, Al, Al_2O_3, Zn and ZnO. No other oxides of Fe and Mn appear in this list because it is extremely difficult to distinguish between them. Therefore distinction distinguished is made only between metallic Fe and Mn on the one hand and oxidic ones on the other.

Sample pieces with the dimension of 8×8 mm^2 were cut out from the galvanized steel sheet and cleaned in an ultrasonic bath to remove adventitious carbon based surface contaminations, which originate from sample handling and preparation in the galvanizing simulator. These contaminations especially accumulate in trenches and craters on the surface and make Auger spectroscopy in these areas impossible, as experience has shown (Arndt et al. 2012). As a first clean up step, the glass beakers used for cleaning in the ultrasonic bath were washed with a mixture of 100 ml sulphuric acid (Merck, 95–97%, p.A.) and 100 ml hydrogen peroxide (Merck, 30%, stabilized for synthesis). For the following cleaning steps, the samples were consecutively placed for 15 minutes in organic solvents, namely tetrahydrofuran (Sigma-Aldrich, anhydrous, ≥99.9%, inhibitor free), isopropanol (VWR, 99.9%) and ethanol (VWR, 99.9%), with each step carried out twice and always using fresh organic solvent.

For the preparation of cross sections of the interface between the steel and the zinc coating the samples were cast in synthetic resin and subsequently, mechanically polished with both SiC grinding paper and diamond particles of a size 1 µm, or less. Because the Zn is softer than the oxides and the steel and the Zn is smeared over the whole polished surface, the surface was cleaned via 30 seconds 3 keV argon ion (4.5 µA) sputtering in the scanning Auger or polished by a grazing incidence 30 keV gallium focused ion beam (FIB) in a Zeiss 1540 XB FIB-SEM. Cross sections of sample areas without Zn coating were prepared by cutting out a lamella of the approximate dimension of 10 µm × 10 µm × 1 µm via FIB. Beforehand, the sample surface was covered by a platinum deposit to prevent curtaining effects from the gallium beam.

14.4 Analytical Results (without pre-oxidation)

14.4.1 AES Surface Overview

Due to the reduced wettability of the TWIP steel, as indicated above, the surface is not fully covered with Zn. In general, there are surface regions exhibiting Zn droplets extending some millimeters in diameter, while other areas appear to be fully uncovered. Figure 14.3 shows such an uncovered area between two Zn droplets on microscopic scale. The sample was cleaned ultrasonically as described before but no Ar^+-sputtering was applied. Nonetheless, the surface is very clean and is not C contaminated. The spectra were in each case taken out of an area as marked in the corresponding image with a rectangle.

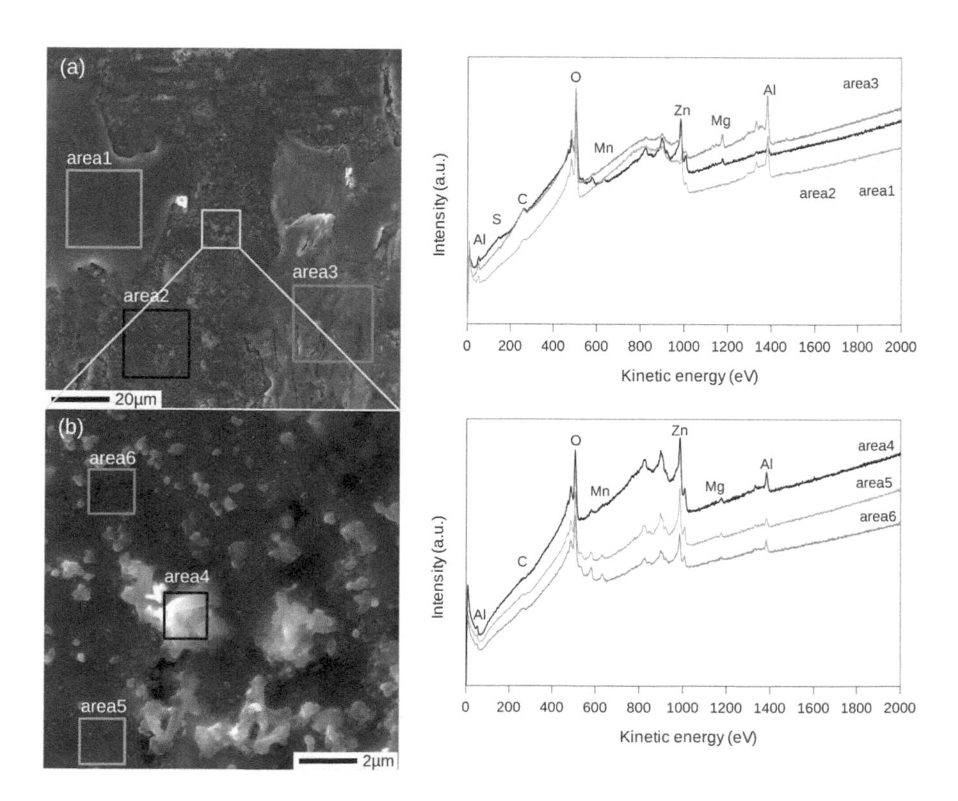

Figure 14.3 SEM images and corresponding AES surface spectra from non-pre-oxidized hot-dip galvanized surface. Area (b) is a magnification of the middle of area (a). The surface is covered with Al_2O_3. Zn droplets are found on some places but no continuous Zn wetting is found.

From the upper image on the left (area1) it can be seen that corresponding spectrum shows the typical characteristics of a surface of hot-dip galvanized steel, consisting mainly of a homogeneous Al_2O_3 layer with an approximate thickness of 5 nm. This layer thickness is an estimation because the spectra look similar to AES spectra taken from industrial produced hot-dip galvanized steel samples, where the layer thickness was calculated to be 5 nm using the inelastic mean free path of Zn $L_3M_{45}M_{45}$ electrons in Al_2O_3 and the Beer-Lambert law (results not shown here). Below this Al oxide layer metallic bulk Zn can be found, as derived from the spectrum considering two effects. At first, the Zn peak in the area1 spectrum—at an energy of 990 eV—is not as well developed as the ones in the other spectra, especially when compared to the Zn signal of the area4 spectrum from the plot in the bottom of Figure 14.3b. At the same time, a broad hump is visible at lower energies between 700 and 950 eV, proving that a large reservoir of Zn lies below the surface in which Zn Auger electrons are generated and inelastically scattered when passing through the overlying Al oxide layer. The second effect is the clear appearance of the Al_{LVV} signal in the energy region below 100 eV. These electrons have an extreme low inelastic mean free path in solid material and are therefore only visible when they are originating from the top most surface.

The spectrum from area2 is rather similar to the spectrum from area1: there is also an Al_2O_3 layer on top of metallic Zn. However, one difference is the additional appearance of Mg in the spectrum. The shape of the Mg peak also shows that Mg is located on the surface most probably in the form of MgO, as shown by Arndt et al. (2012). Mg might in this case come from minor contaminations of the Zn bath in the simulator, since no Mg was added to the steel. Due to its affinity to oxygen, Mg is most probably located on the topmost surface.

Area3 shows a slightly different spectrum than the other two: not only the Zn signal is more clearly developed and does not exhibit the broad hump in the lower energy range of the Zn region as the two earlier mentioned spectra, hinting at the fact that the Zn signal at area3 originates more from the surface, but additionally Mn (besides traces of S) appears in the spectrum.

The lower picture b now shows a magnification of the area between the two larger Zn droplets. The three spectra taken from this area are quite similar to area3, with the Zn signal being clearly developed and with missing signs of inelastic scattering from an overlayer. Furthermore, Al, Mn, Mg, and O appear in all spectra and there is surprisingly not much difference between the spectrum from area4 taken from a crystal and the other ones taken from the darker regions. One can see in Figure 14.3 that the hot-dip galvanized surface is covered by an Al_2O_3 layer because all six spectra show clearly the low energy Al_{LVV} Auger line. It is also worth while to mention that no Fe could be found at the surface.

14.4.2 AES Mappings on a FIB-lamella, Uncoated Surface

A cross section of the uncoated non-pre-oxidized material was prepared by FIB in form of a lamella, covered by Pt deposition as a protective layer, and subsequently analyzed by Auger mappings (see Figure 14.4). The surface was slightly sputtered with Ar^+ ions in order to remove the native oxide layer which grow during the transport from the SEM to Auger microscope. One can see pure Mn oxide at the surface without any Fe in it. Obtaining the O signal was difficult because the electron beam changes the surface chemistry. It removes O from oxides and oxidizes metal. Therefore, the oxygen mappings are sometimes blurry compared to the other element mappings because only the first O scan contributes to a good signal to noise ratio. There are also some inner oxides visible in the image which contains pure Mn oxide.

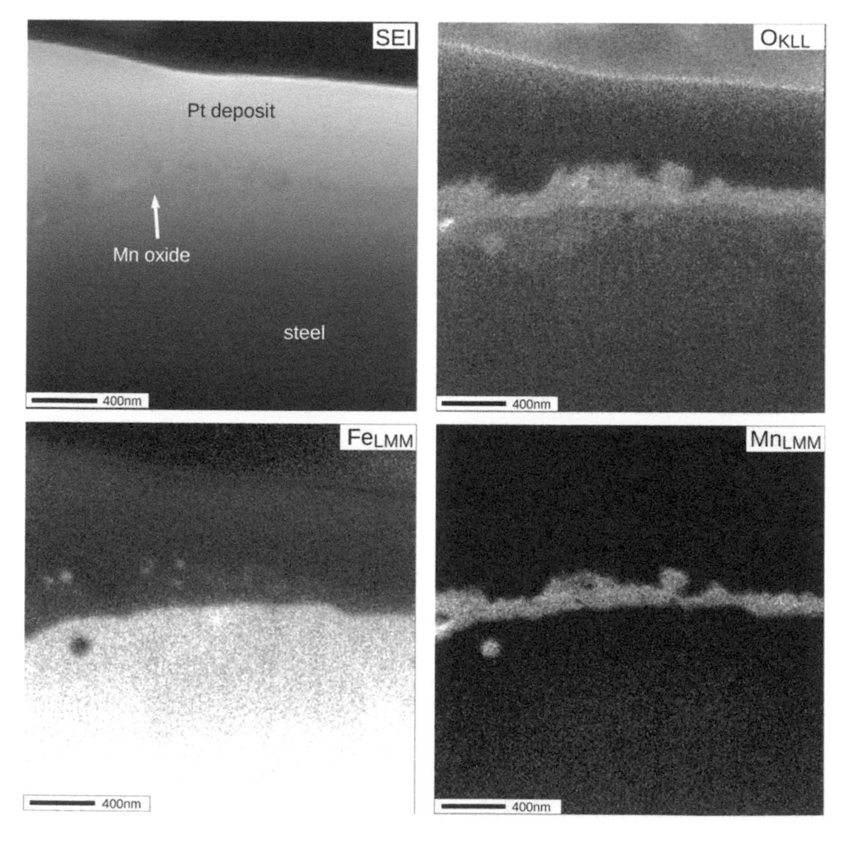

Figure 14.4 SEM images and corresponding Auger mappings of FIB cross sections from the uncoated surface of the non-pre-oxidized sample. The surface is covered by a Mn oxide layer. Also inner Mn oxides are visible.

14.4.3 AES Mappings on FIB-lamellas, Zn-coated Surface

On the lamella presented in Figure 14.5 several AES mappings were obtained. Evidently, a continuous MnO layer is formed on top of the steel substrate with a film thickness between 100 and 200 nm. In addition, it is also visible that MnO layers are even situated below the surface. It is also clearly visible that no continuous Zn coating is present at the surface. Instead, one can see an Fe_2Al_5 inhibition layer crystal as well as an $FeZn_{13}$ ζ-phase crystal. They appear in regions where the MnO layer is interrupted and metallic Fe reaches the surface. The growth of ZnFe crystals indicates an aluminothermic reduction (see Equation 14.1).

Figure 14.6 shows AES element mappings from an area below a Zn droplet. It is clearly visible that there is a gap between the MnO, which covers the steel surface, and the Zn droplet. This is the main reason why the non-pre-oxidized AHSS surface cannot be coated with Zn. There is no chemical bonding between the MnO and the Zn. The only reason why Zn droplets stick to the surface is because the MnO layer is not continuous. There are crevasses and interruptions of the MnO layer in which the Zn has direct contact to the steel. In these areas the Fe forms AlFe and ZnFe phases which adhere to the metallic Zn.

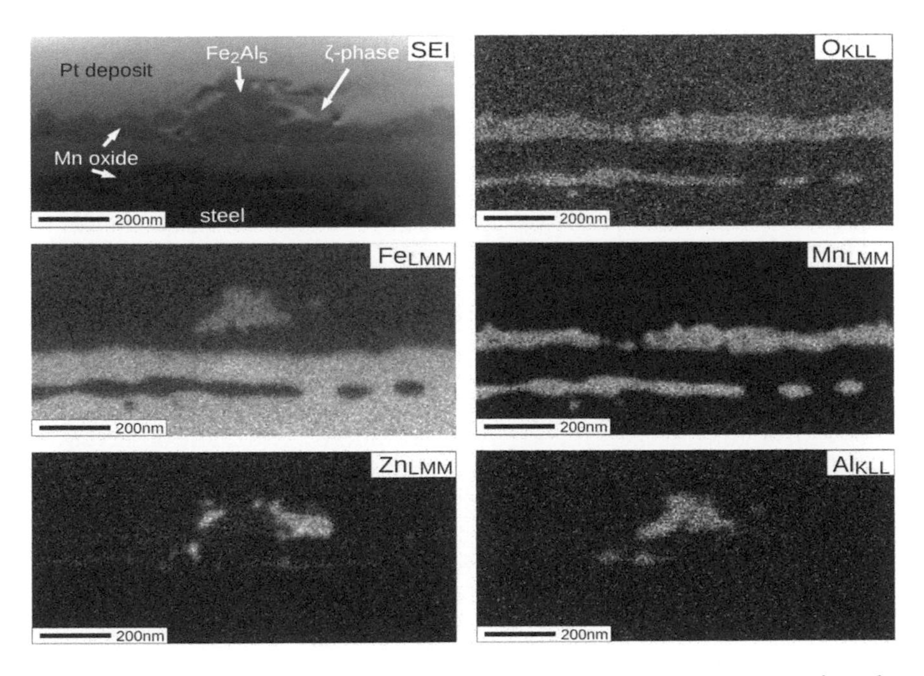

Figure 14.5 SEM images and corresponding Auger mappings of FIB cross sections from the Zn-coated surface of the non-pre-oxidized sample. The surface is mostly covered by a Mn oxide layer. There are Fe_2Al_5 inhibition layer crystals and $FeZn_{13}$ ζ-phase crystals visible.

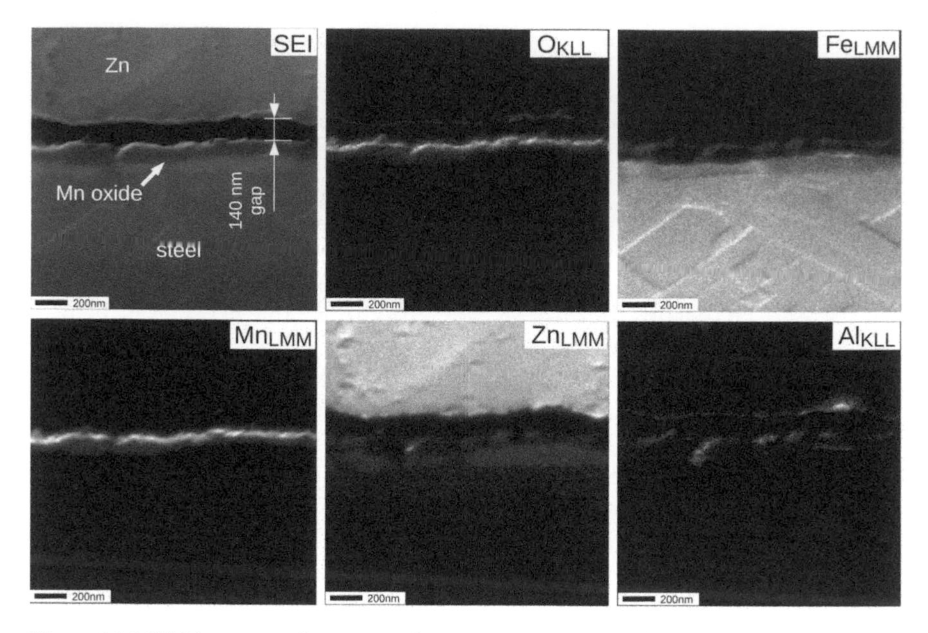

Figure 14.6 SEM images and corresponding Auger mappings of FIB cross sections from the Zn-coated surface of the non-pre-oxidized sample below a Zn droplet. The steel at interface is covered by a MnO layer similar to Figures 14.4 and 14.5. There is a gap between the MnO surface and the Zn droplet.

14.4.4 AES Mappings on Zn-coated Surface

Figure 14.7 shows high quality Auger mappings of a non-pre-oxidized Zn-coated surface after the removal of the outermost aluminium oxide layer by argon ion sputtering. The AES mappings of Zn and Fe demonstrate that the signal is sufficiently clean in order to distinguish between $FeZn_{13}$ and pure Zn in the crystalline area in the lower left part of the image. Furthermore, the Al signal is not distributed over the whole area but shows sharply separated regions with high concentrations corresponding to regions exhibiting a high Fe concentration. Mn is found outside these areas with augmented Fe, Zn, and Al concentrations. A puzzling observation can be made for the O signal: the highest amounts are neither found in the Mn rich areas nor at some round bright structures visible in the secondary electron image which could be oxides. Instead, the highest oxygen signal is located within the AlFe phases. This is, as already indicated above, a measuring artefact due to a reaction of small amounts of O from the residual gas with the surface after a long measurement of time, even when working in the UHV pressure range. This effect is even amplified by the heat applied by the primary electron beam and the high oxygen affinity of Al. On the other

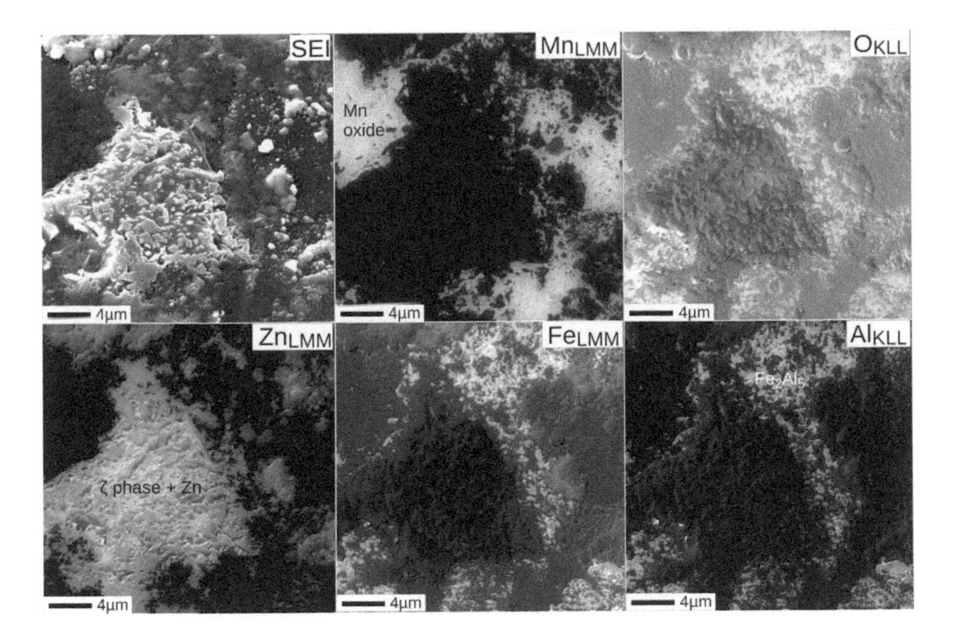

Figure 14.7 SEM overview image and AES mappings of the sputter cleaned surface in a region between Zn droplets of a non-pre-oxidized sample.

hand, MnO is also being reduced by the electron beam and loses O after long time measurement. This effect was tested and proven by taking Auger spectra from a MnO region after sputtering a second time (not shown here).

14.5 Analytical Results (with pre-oxidation)

14.5.1 Surface Overview, Uncoated Surface

The main result of the two different heating treatments under controlled atmospheric conditions in the galvanizing simulation oven is that the sample with pre-oxidation could be homogeneously covered in Zn while the sample with the standard treatment could not.

Since the pre-oxidized sample behaves differently, the first analysis was conducted on the oxides of the uncoated surface. That is on the part of the sample which was treated by pre-oxidation and annealing but had no contact with the Zn bath. Figure 14.8 shows an XRD measurement on this surface. The 2θ-plot reveals ferrite, austenite and Mn/Fe mixed oxides in different compositions. The appearance of ferrite can easily be explained by a preferred segregation of Mn out of the austenitic steel matrix to the surface during heating. This is in agreement with the observations of Cho and De Cooman (2012) and the model of Blumenau et al. (2011b). Furthermore, these findings make it clear that the simultaneous occurrence of Mn/Fe

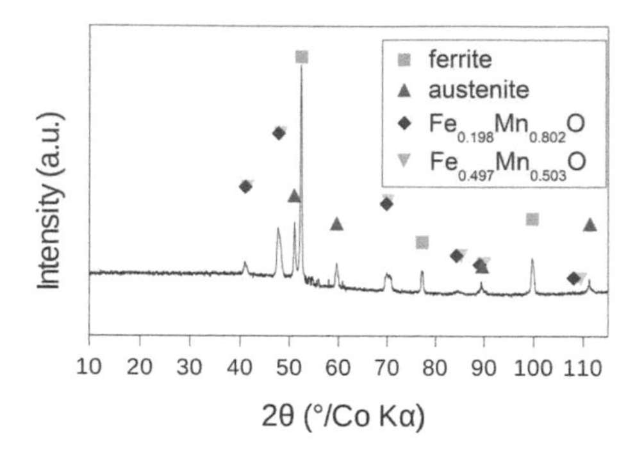

Figure 14.8 XRD 2θ-plot from the surface of the pre-oxidized, uncoated steel. The surface region of the steel is mostly ferritic with small amounts of austenite. The oxides found on the surface are a mixture of iron and manganese oxides.

mixed oxides and ferrite is not a proof of the existence or non-existence of Mn oxide with metallic Fe. No pure MnO could be found in the 2θ-plot but its lines overlap with the Fe/Mn mixed oxides. The shown diagram is a result of a best fit calculation based on a database. Depending solely on that XRD result it is impossible to confirm the model of Blumenau et al. (2011b) where the reduction of Fe in the mixed oxide layer is proposed.

In order to get further information from the sample surface detailed Auger measurements were performed (see Figure 14.9). The secondary electron image (Figure 14.9a) shows a region of the surface with a higher magnification in Figure 14.9b. The surface is covered by small crystallites with an approximate lateral size of 1 μm. The surface is not homogeneous, which is shown by the Auger survey spectra in Figure 14.9c. Note that these spectra are presented in an integrated form, although it is usual to display Auger spectra in a differential form as is the case with all high energy resolution spectra in this work. The reason for this is the shape of the background of the spectra in both positions 1 and 2 and Figure 14.9c makes it immediately obvious that C and Zn are surface contaminations. The amount of C is very low due to the ultrasonic bath cleaning procedure. The Zn stems from the gas phase over the liquid Zn bath, and it covers the whole surface. In contrast, the Fe and Mn concentrations are completely different at both positions. The lateral distribution of Mn and Fe is mapped on the right-hand side of Figures 14.9a and 14.9b. The reason for these different distributions is the fact that in some areas the steel reaches the surface directly without being covered by an oxide. This will be confirmed later by means of cross sections.

A sputter depth profile was made at position 2 in order to measure the chemical composition of an oxide crystal (see Figure 14.9d). The depth was calibrated by sputtering a 100 nm thick thermal SiO_2 on Si(100). For the depth profile a series of spectra were taken with a kinetic energy between 450 and

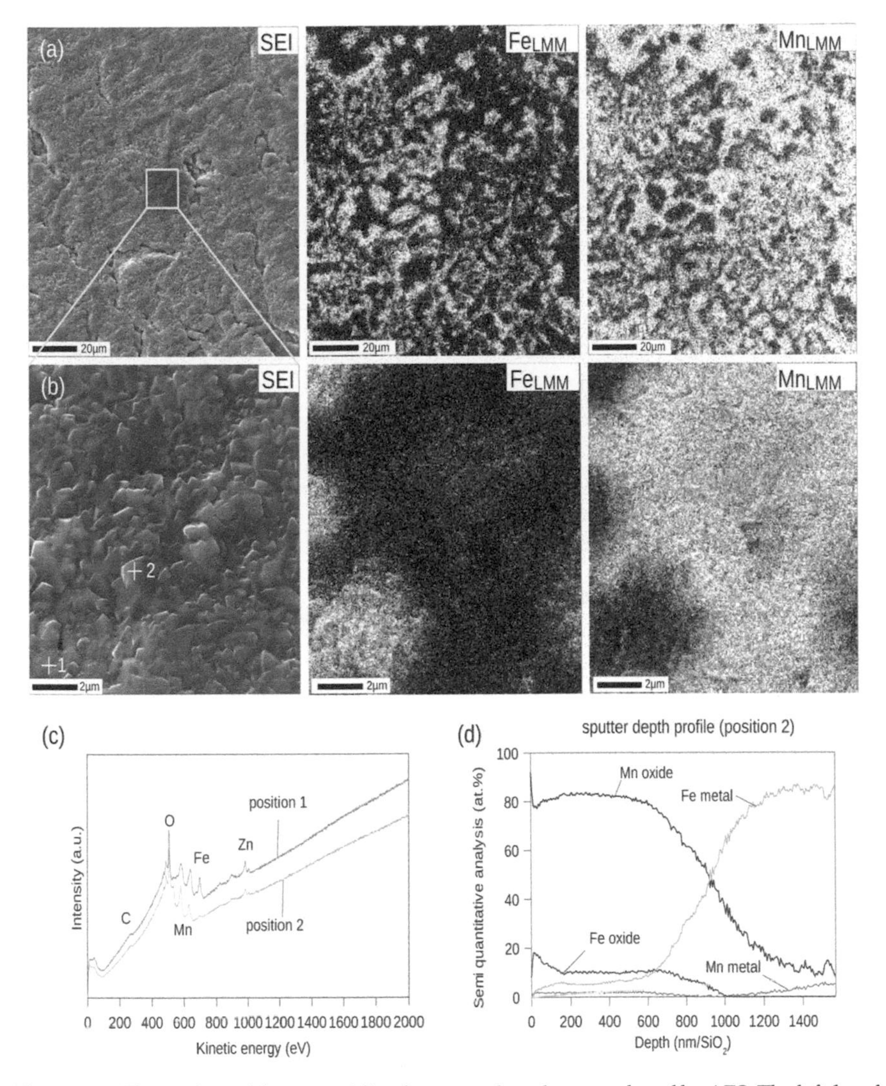

Figure 14.9 The surface of the pre-oxidized uncoated steel was analyzed by AES. The left-hand side image (a) shows the secondary electron image and (b) its magnification of a surface area while the two images on the right-hand side show the corresponding Auger mappings from the same regions. The two Auger survey spectra (c) were taken at the marked positions in image (b). Diagram (d) shows a sputter depth profile taken on position 2.

750 eV in high energy resolution mode ($\Delta E/E = 0.15\%$). Then reference spectra from a database of the Auger microscope manufacturer Jeol were linear least square fitted on the first derivative of the measured spectra. The relative amounts of the reference materials were normalized and plotted as a depth profile, Figure 14.9d. C and Zn were neglected since they appear only at the

surface. It becomes obvious that the surface is mostly covered by oxidic Mn, which is in agreement with the survey spectrum shown in Figure 14.9c. After the first sputter cycle the Fe oxide concentration increases to nearly 20 wt%. At the same time metallic Fe and Mn appear. At a depth of approximately 200 nm the composition of the four components becomes stable over a larger depth range. At the lower end of the Fe/Mn oxide crystallite at a nominal depth of 700 nm the Fe and Mn oxide concentration decreases. At a depth of 1000 nm the steel bulk appears in the profile because the Fe oxide vanishes completely and the metallic Fe and Mn signals increase.

There are two findings which can unquestionably be derived from this depth profile. Firstly, there are crystallites which are composed of Mn/Fe oxide. Secondly, the thickness of these crystallites is in the order of 1 μm. The question whether metallic Fe or Mn exists in the crystals cannot be answered because of possible damage and reduction caused by the Ar^+ ions during the sputtering process. This will be discussed later in Section 14.5.4.

14.5.2 Cross Section, Uncoated Surface

Figure 14.10 shows a FIB cross section of the pre-oxidized uncoated surface. The crystals at the surface consist of Fe, Mn and O in agreement with the XRD measurements and the surface mappings (Figures 14.8 and 14.9). Below the surface, inner oxides in form of MnO can be found. It is also noteworthy that the inner oxides have a round shape while the surface oxides are angular. The gap between the Mn/Fe oxides in middle of the image proves the assumption that the steel appears in some areas directly at the surface without an oxide cover and that the surface is therefore not homogeneous.

14.5.3 Cross Section, Zn-coated Surface

The Zn layer on the coated surface of the pre-oxidized sample has a homogeneous thickness of approximately 30 μm. The interface between the Zn and the steel of the Zn-coated pre-oxidized steel surface was analyzed by Auger mappings after the sample was first conventionally polished, subsequently milled with a focused 30 kV Ga^+ ion beam at a shallow angle and finally 3 kV Ar sputtered for 30 s to remove the Zn from both the steel surface and the interface (see Figure 14.11). On the left-hand side of the image the same Mn/Fe mixed oxides are found at the interface as on top of the uncoated sample. The Zn layer is directly formed on top of the oxide layer without an Fe_2Al_5 inhibition layer. On the right-hand side of the image the Fe_2Al_5 inhibition layer is present at regions where the Zn had direct contact with the steel. It is also noteworthy that no gap between the oxide and the Zn was found in contrast to the non-pre-oxidized sample (see Figure 14.6). It is hitherto not clear if the distribution of Al in the Zn layer is real or

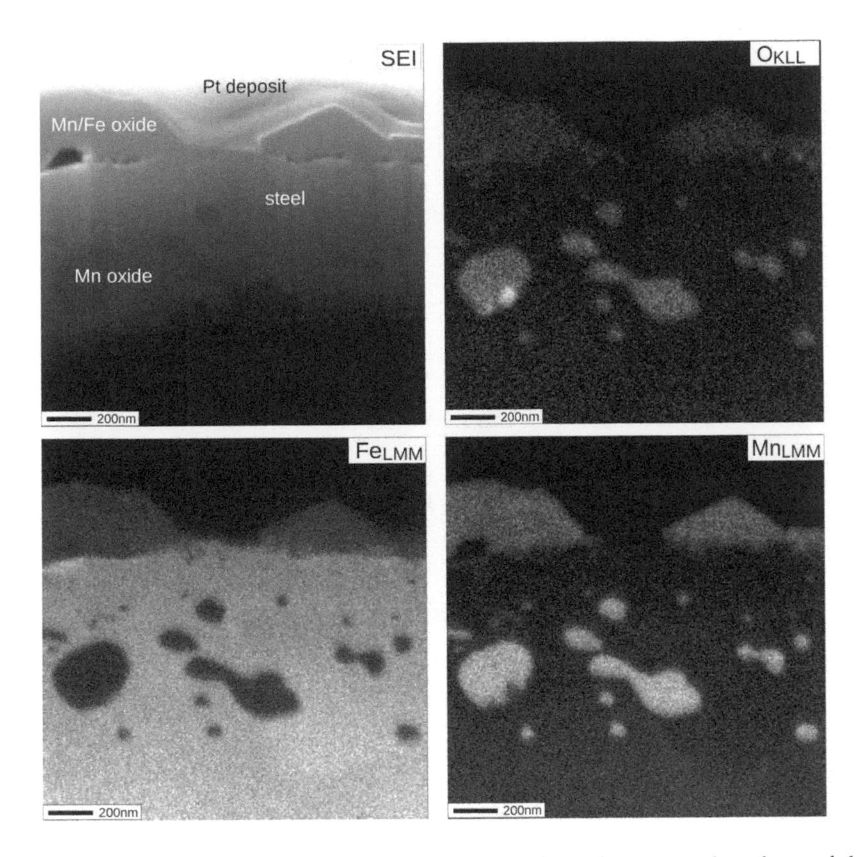

Figure 14.10 Auger mappings of a FIB cross section from the uncoated surfaces of the pre-oxidized sample. The surface is covered by an Mn/Fe mixed oxide while pure Mn oxide is found as an inner oxide below the surface. One can also see in the middle of the image that the steel partially appears directly at the surface.

mostly a preparation or measurement artefact. It seems that Al is following the Zn grain boundaries as shown in the upper right corner of Figure 14.11. There is no depletion of Al towards the interface visible in comparison to the Al mapping in Figure 14.6. It could mean that Al is not consumed in an aluminothermic reduction and therefore is still present in the Zn layer.

14.5.4 Analysis of the Oxide Layer, Zn-coated Surface

The oxide layer was further investigated by high energy resolution Auger electron spectroscopy in order to find an explanation for the different wetting behavior of the Zn on the non-pre-oxidized and the pre-oxidized surface (see Figure 14.12). Firstly, the polished interface was sputtered by 3 keV Ar^{+} ions for 30 seconds in order to remove the Zn from the surface and to expose the oxide. Secondly, the Auger spectrum in the energy range

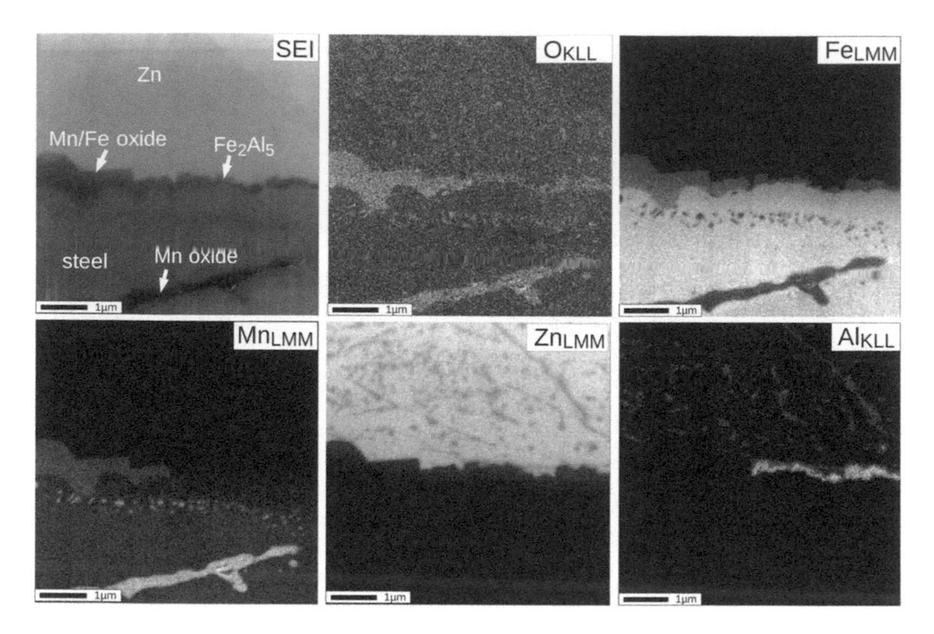

Figure 14.11 A larger Auger element mapping of a FIB cross section of the interface between the steel and the Zn coating of the pre-oxidized sample. The left-hand side of the interface consists of an Mn/Fe mixed oxide without an Fe_2Al_5 inhibition layer on top while the right hand side consists of a pure Fe_2Al_5 inhibition layer without oxides, as the liquid Zn had direct contact to the steel at that area.

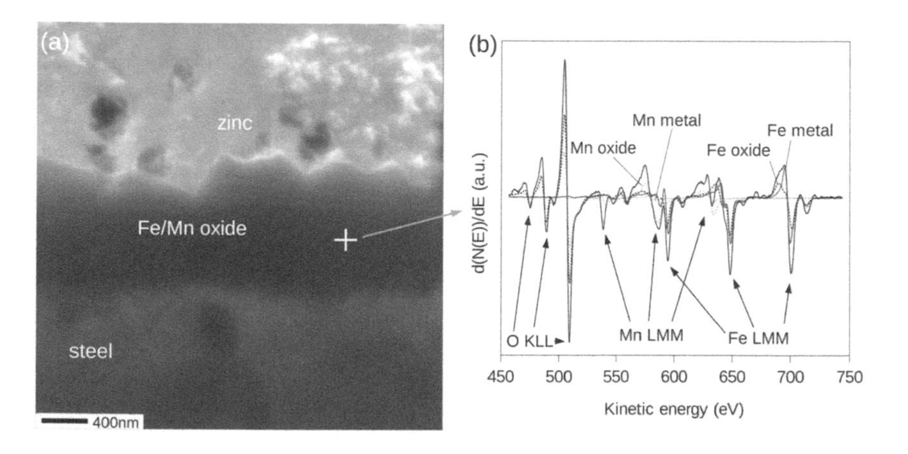

Figure 14.12 The interface oxide between the steel and the Zn layer, see cross section image (a), is analyzed by high energy resolution Auger spectroscopy after the surface was prepared by 30 s, 3 keV Ar^+ sputtering. The oxide consists of a combination of Mn, Mn oxide, Fe and Fe oxide, as shown in spectrum (b).

between 450 and 750 eV was taken, differentiated and linear least square fitted with four reference spectra from our database, these are metallic Fe, FeO, metallic Mn and MnO_2. The cross in the SE image, Figure 14.12a, shows the position of the Auger measurement and Figure 14.12b shows the corresponding spectrum in the first derivative representation. One can see that all four fitted components are present in the oxide layer. The oxide on the non-pre-oxidized surface contains only Mn and O but no Fe.

Steinberger et al. (2014) have shown via XPS that different iron oxide powders, namely FeO, Fe_2O_3 and Fe_3O_4, are reduced to metallic Fe during long time sputter depth profiling. Therefore, it is not clear if the metallic iron found in the mixed oxide layer does really exist.

14.5.5 Reverse Depth Profile, Zn-coated Surface

In order to find out if there exists at least a very thin inhibition layer between the Fe/Mn mixed oxides and the Zn coating, which cannot be seen in a cross section because it is too thin, the peeled off Zn coating created in a ball impact test was taken and analyzed from the back side. Figure 14.13 shows the results of an Auger depth profile taken on a Fe/Mn mixed oxide. The Auger spectra were separated into oxidic and metallic parts. For better clarity only the most important components were plotted into the graph of Figure 14.13. Important point to note is the progression of the Fe oxide signal which vanishes at a nominal depth between 1200 and 1400 nm. At that depth no increase of the Al metal signal can be observed. This shows clearly that no Fe_2Al_5 inhibition layer exists between the Fe/Mn oxides and the Zn coating.

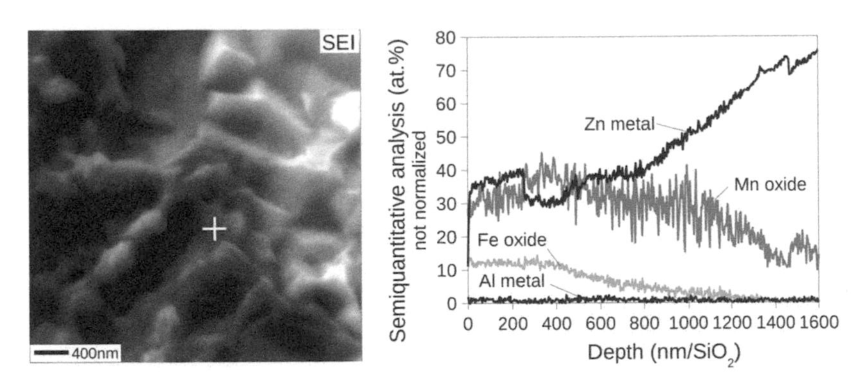

Figure 14.13 The image shows the backside of peeled off Zn coating from a pre-oxidized sample. The cross marks an Mn/Fe mixed oxide crystal where the depth profile shown at the right-hand side was taken. The Al metal signal shows no increase at the point where the Fe oxide signal disappears, between 1200 and 1400 nm depth. This is a proof that no Fe_2Al_5 inhibition layer lies on top of the mixed oxide.

14.5.6 Analysis of the Zn-coated Interface Layer Prepared by Fracture Stage

One idea to avoid the damage induced by Ar^+ ions is the preparation of the interface between Zn and the steel via the fracture stage. In this case the new surface is created by breaking a sample in the analysis chamber of the Auger device with a hammer, after the sample was cooled with liquid nitrogen. By this procedure the fractured surface was never exposed to the oxygen and water in the atmosphere. Also chemical changes due to sputtering do not occur by fracturing. But there are some difficulties one has to face. Firstly, the sample heats up from the liquid nitrogen temperature to room temperature during the measurement. This results in a large drift of the sample stage, which has to be compensated by a software drift correction. Secondly, it is difficult to identify the interface because the sample breaks most probably along the grain boundaries, which results in a rough topography. The SE image shows therefore more topography than chemical contrast. Therefore it is not possible to obtain Auger mappings from the fractured surface. And finally, although the chemistry is unchanged after fracturing, it could change during electron bombardment. One has to make a compromise between fast measurement in order to let the chemistry intact and longtime measurement in order to obtain a better signal to noise ratio.

Results of high energy resolution Auger electron spectroscopy measurements are shown in Figure 14.14. Three different places were analyzed as shown in Figure 14.14a. The corresponding differentiated spectra are shown in Figure 14.14c. One can see immediately that the interface crystals contain not only Mn, Fe and O but also Zn and Al in different chemical states. The interface breaks most probably along grain boundaries and these boundaries were covered by Zn and Al during galvanizing in the hot Zn bath. One reason could be that diffusion of Zn and Al along grain boundaries, in orders of magnitude, higher than through grains. A more obvious reason could be that the grain surface was directly attached to the Zn coating. Bar chart in Figure 14.14b shows a semi-quantitative evaluation of the linear least squares fitted reference spectra from Figure 14.14c, in which the total amount is normalized to 100 at.%. The used reference materials were MnO_2, Mn, FeO, Fe, ZnO, Zn, Al_2O_3 and Al.

For the understanding and evaluation of the three different positions one should look at position 3 first. The Al_{KLL} spectrum in Figure 14.14c shows a combination of metallic and oxidic Al. But the marked plasmon peak, visible in the Al metal reference spectrum, cannot be fitted. This means one can distinguish between pure metallic Al which is present in the Zn layer and the metallic Al of the inhibition layer by the shape of the Auger spectrum of Al. It is also noticeable that at that position only metallic Fe is present. From this we conclude that the crystal at position 3 is an Fe_2Al_5 inhibition layer crystal. In contrast, the crystals at positions 1 and 2 are Fe/

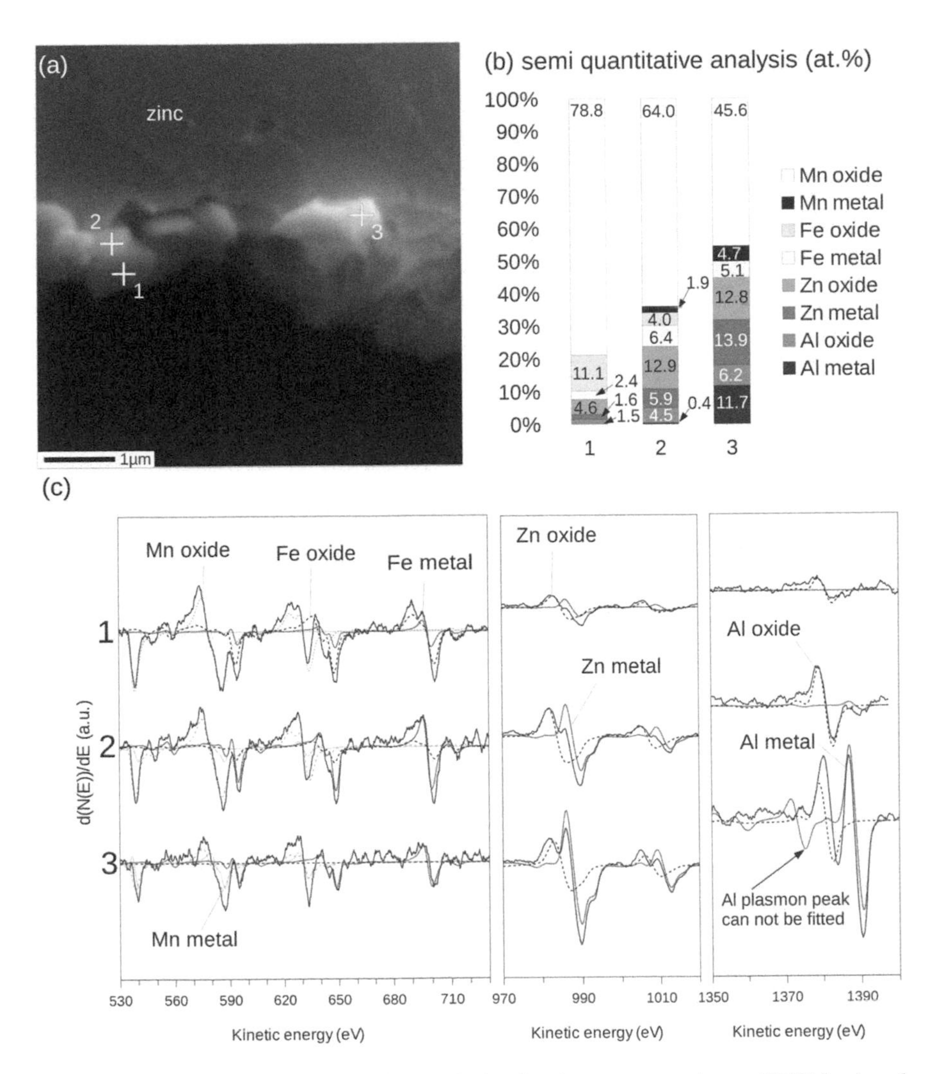

Figure 14.14 Three high energy resolution derivative Auger spectra from a UHV fractured surface of a pre-oxidized Zn coated sample. The secondary electron image (a) shows the position and bar chart (b) the semi quantitative evaluation of the spectra (c). At positions 1 and 2 are Mn/Fe oxide crystals while at position 3 is an Fe_2Al_5 inhibition layer crystal.

Mn mixed oxide crystals. One can see mostly oxidic Mn and oxidic Fe. Both positions also show almost pure oxidic Al. This is most probably due to an aluminothermic reduction in which the Al picks up the oxygen from the Mn/Fe oxide. The observation of small amounts of metallic Fe at positions 1 and 2 and metallic Mn at position 2 confirm this assumption. The chemical state of Zn in all three positions is metallic and oxidic. Another observation is that the amount of metallic Fe and Mn is higher at position 2 compared to that at position 1, although position 1 lies nearer to the steel. This is also a hint for the aluminothermic reduction, because its effect should be greater near the interface to the Zn.

14.5.7 Analysis of the Uncoated Surface Prepared by Fracture Stage

The same preparation method as for the Zn-coated interface was used to create a chemically unchanged surface between the oxide crystals on top of the uncoated steel surface. Then high energy resolution Auger electron spectra were taken at three positions at the fractured surface (see Figure 14.15). One sees that all three positions are quite similar in their chemical composition, only the intensity at position 3 differs, because its surface is tilted away from the detector. At all three positions, metallic Fe besides oxidic Fe and oxidic Mn is found. That means metallic Fe exists at least at the boundaries between the Mn/Fe mixed oxide crystals.

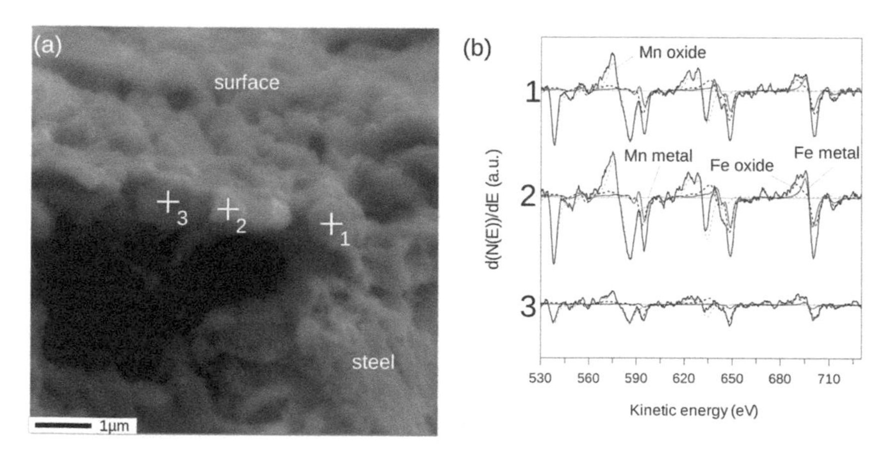

Figure 14.15 Differentiated high energy resolution AES spectra (b) from three different positions, see secondary electron image (a), which are prepared by the fracture stage and broken in the UHV after cooling with liquid nitrogen. At all three positions metallic Fe besides oxidic Mn and oxidic Fe could be found. There is no metallic Mn.

14.6 Discussion

It could be confirmed that pre-oxidation leads to an excellent wetting of high manganese second generation advanced high strength TWIP steel with hot Zn. In contrast, without pre-oxidation the same TWIP steel could not be covered with Zn, because of the formation of a 100–200 nm thick MnO layer at the steel surface (Arndt et al. 2013). Figures 14.9a and 14.9b show, that after pre-oxidation Mn/Fe mixed oxide lie at the surface of the steel, which is lateral not homogeneously distributed. It demonstrates that coarse methods like XPS or XRD cannot be applied for the analysis of the interface region between the steel and the Zn coating. Therefore AES was used as a technique which connects the excellent chemical analysis features of XPS with the nanometer scale. An Auger depth profile at a single Mn/Fe crystal at the uncoated surface (Figure 14.9d) proves the existence of a mixed Mn/Fe oxide. Pure metallic Fe or Mn would lead to a disappearing of the oxide phase as seen at higher depth in the profile shown in Figure 14.9d.

Cross section preparations (Figures 14.11 and 14.12) show that the Mn/Fe mixed oxides are still present, after galvanizing. The predicted Fe_2Al_5 continuous inhibition layer on top of the oxide could not be confirmed (Figure 14.13). Quite the contrary, the Fe_2Al_5 crystals were created at gaps in the oxide layer where the liquid Zn had direct contact to the steel. Exactly the same observations were made at the non-pre-oxidized surface. Hence it can be concluded that the Zn layer adheres directly to the oxides in case of pre-oxidation, in contrast to the non-pre-oxidized sample, where the Zn is repelled from the oxides (Figure 14.6). A new model was developed (see Figure 14.16) from the previous one by Blumenau et al. (2011a) including these new observations.

The question, whether the Mn/Fe mixed oxide contains metallic Fe due to a reduction during annealing in dry atmosphere, cannot be answered. From fracture stage prepared surfaces of the uncoated sample it was found that metallic Fe occurs at the grain boundaries of the oxide crystals while the Mn is completely oxidic (see Figure 14.15). There were also Auger spectra taken from the fractured Zn-coated surface (see Figure 14.14) which show metallic iron on top of the oxide crystals. The results from these measurements are not without ambiguity because of the following reason. Khondker et al. (2007) introduced a model of an aluminothermic reduction of manganese oxides by Al in the Zn bath. As a hint for this model, Kawano and Renner (2012) found Al_2O_3 on top of MnO after contact with liquid Zn containing 0.2 at.% Al. Kavitha and McDermid (2012) have shown this aluminothermic reduction of MnO on medium Mn AHSS directly via TEM and EELS:

$$3MnO_{(solid)} + 2Al_{(solution)} \rightarrow Al_2O_{3(solid)} + 3Mn_{(solution)} \quad (14.2)$$

The same aluminothermic reduction could be confirmed by Sagl et al. (2013) on dual phase steel with a Mn concentration of 1.82 wt.% via XPS. Pure oxidic Al was observed at the interface (Figure 14.14) but the possible reactions are more complex. From the Ellingham diagram (Pawlek 1983) one can extract the standard free energies ΔG of the formation of oxides at 460°C:

$$4/3Al + O_2 \rightarrow 2/3Al_2O_3 \quad \Delta G \approx -960\ kJmol^{-1} \quad (14.3)$$

$$2Mn + O_2 \rightarrow 2MnO \quad \Delta G \approx -660\ kJmol^{-1} \quad (14.4)$$

$$2Zn + O_2 \rightarrow 2ZnO \quad \Delta G \approx -540\ kJmol^{-1} \quad (14.5)$$

$$2Fe + O_2 \rightarrow 2FeO \quad \Delta G \approx -410\ kJmol^{-1} \quad (14.6)$$

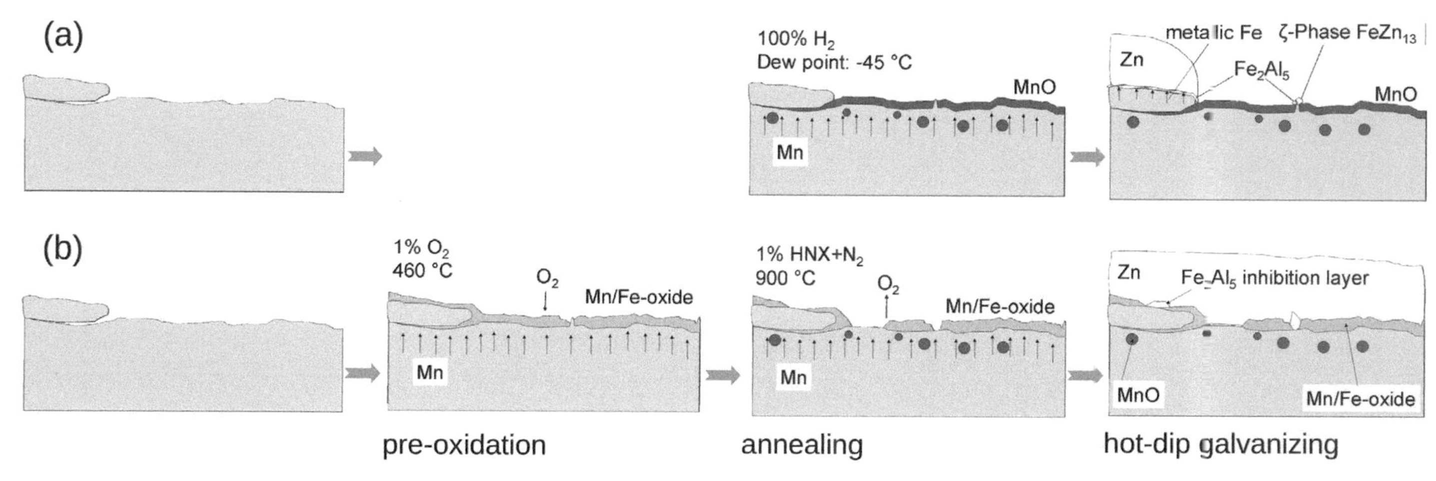

Figure 14.16 Model of the influence of pre-oxidation on the Zn wetting behavior of high Mn steel (b) in contrast to untreated material (a). The oxides after pre-oxidation consist of Mn/Fe oxide while there is only MnO at the untreated surface. The subsequent galvanizing does not form a continuous Fe_2Al_5 inhibition layer at the pre-oxidized sample, instead the Zn adheres directly at the Mn/Fe mixed oxide.

Therefore, these additional reactions are very likely:

$$2Al_{(solution)} + 3FeO_{(solid)} \rightarrow Al_2O_{3(solid)} + 3Fe_{(solution)} \quad (14.7)$$

$$2Al_{(solution)} + 3ZnO_{(solid)} \rightarrow Al_2O_{3(solid)} + 3Zn_{(solution)} \quad (14.8)$$

$$Mn_{(solution)} + FeO_{(solid)} \rightarrow MnO_{(solid)} + Fe_{(solution)} \quad (14.9)$$

$$Mn_{(solution)} + ZnO_{(solid)} \rightarrow MnO_{(solid)} + Zn_{(solution)} \quad (14.10)$$

$$Zn_{(solution)} + FeO_{(solid)} \rightarrow ZnO_{(solid)} + Fe_{(solution)} \quad (14.11)$$

This is in good agreement with the experimental results. Al is completely oxidic on top of the Mn/Fe oxide crystals (positions 1 and 2 in Figure 14.14). Mn is mostly oxidic with small amounts of metal (position 2). Zn is a mixture of oxide and metal as well as Fe. On top of the inhibition layer crystal (position 3) the Fe is completely metallic which shows it is the least oxide affine component of Fe, Mn, Zn and Al. The reason why Fe is not completely metallic is that the underlying Mn/Fe oxide is still contributing to the Auger signal, although AES is a very surface sensitive method where the information depth is approximately 6 nm. Furthermore, the presence of oxidic Zn is an indirect proof for the existence of oxidic Fe at the surface of the oxide crystals, because the reduction of oxidic Fe is the only energetically favored chemical reaction which produces Zn oxide.

14.7 Summary

Hot-dip galvanizing of high manganese TWIP steel is a challenge which is hitherto not solved. The situations are shown as two models in Figure 14.16.

In case (a) one can see that a Mn oxide layer is growing at the surface of the steel during annealing in dry atmosphere. The Mn segregates from the sub surface region to the steel surface which leads to a Mn depletion. Therefore the austenite near the surface is no longer stable and the sub surface area becomes ferritic. When the steel sheet is immersed into the liquid Zn bath the dissolved Al from the bath has no contact to the iron in areas where Mn oxide is present at the steel surface and therefore no Fe_2Al_5 inhibition layer can grow. Furthermore an aluminothermic reaction takes place on top of the Mn oxide consuming the metallic Al from the bath near the surface of the steel. That leads to the growth of brittle $FeZn_{13}$ crystals in areas where the Mn oxide layer is interrupted and the Zn is in direct contact to the Fe. The Zn has bad adhesion to the Mn oxide and therefore the steel cannot be wetted with Zn.

Case (b) illustrated the function of an additional pre-oxidation step before the annealing of high manganese steel. Pre-oxidation creates an Fe/Mn mixed oxide at the steel surface. In the following annealing step the formation of pure Mn oxide is prevented. During the immersion into the Zn bath the metallic Zn reduces the oxidic Fe and a uniform Zn layer forms on top of it. Neither an Fe_2Al_5 inhibition layer nor an $FeZn_{13}$ ζ-phase can be formed because there is not sufficient metallic Fe present.

Further research is necessary since the steel surface can be covered with Zn but the underlying Fe/Mn mixed oxide is brittle. The adhesion between the steel and the oxide layer is weak. It should be investigated if the processing parameters can be optimized to create a much thinner oxide layer which also fulfils the stability requirements of a Zn coating.

Acknowledgements

Thanks to my co-workers: J. Duchoslav, R. Steinberger, G. Hesser, C. Commenda, L. Samek, E. Arenholz and D. Stifter.

The financial support by the Federal Ministry of Economy, Family and Youth and the National Foundation for Research, Technology and Development is gratefully acknowledged.

The financial support of the Austrian Forschungsförderungsgesellschaft GmbH research program, under the project number 823398, is gratefully acknowledged. The authors thank voestalpine Stahl GmbH of Linz in Austria for providing materials and the support of the research center.

Keywords: Pre-oxidation, hot-dip galvanizing, Auger electron spectroscopy, Nano-TWIP, aluminothermic reduction

References

Alibeigi, S., R. Kavitha, R. Meguerian and J. McDermid. 2011. Reactive wetting of high Mn steels during continuous hot-dip galvanizing. Acta Mater. 59: 3537–3549.

Arndt, M., J. Duchoslav, H. Itani, G. Hesser, C. Riener, G. Angeli, K. Preis, D. Stifter and K. Hingerl. 2012. Nanoscale analysis of surface oxides on ZnMgAl hot-dip-coated steel sheets. Anal. Bioanal. Chem. 403: 651–661.

Arndt, M., J. Duchoslav, K. Preis, L. Samek and D. Stifter. 2013. Nanoscale surface analysis on second generation advanced high strength steel after hot dip galvanizing. Anal. Bioanal. Chem. 405: 7119–7132.

Arndt, M., J. Duchoslav, R. Steinberger, G. Hesser, C. Commenda, L. Samek, E. Arenholz and D. Stifter. 2015. Nanoscale analysis of the influence of pre-oxidation on oxide formation and wetting behavior of hot-dip galvanized high strength steel. Corros. Sci. 93: 148–158.

Baril, E. and G. L'Espérance. 1999. Studies of the morphology of the Al-Rich interfacial layer formed during the hot dip galvanizing of steel sheet. Metall. Mater. Trans. A 30: 682–695.

Blumenau, M., M. Norden, F. Friedel and K. Peters. 2010a. Wetting force and contact angle measurements to evaluate the influence of zinc bath metallurgy on the galvanizability of high-manganese-alloyed steel. Surf. Coat. Technol. 205: 828–834.

Blumenau, M., M. Norden, F. Friedel and K. Peters. 2010b. Galvannealing of (High-) Manganese-Alloyed TRIP and X-IP®-Steel. Steel Res. Int. 81: 1125–1136.

Blumenau, M., M. Norden, F. Friedel and K. Peters. 2011a. Reactive wetting during hot-dip galvanizing of high manganese alloyed steel. Surf. Coat. Technol. 205: 3319–3327.

Blumenau, M., M. Norden, F. Friedel and K. Peters. 2011b. Use of pre-oxidation to improve reactive wetting of high manganese alloyed steel during hot-dip galvanizing. Surf. Coat. Technol. 206: 559–567.

Cho, L. and B. De Cooman. 2012. Selective oxidation of TWIP steel during continuous annealing. Steel Res. Int. 83: 391–397.

Cornette, C., A. Hildenbrand, M. Bouzekri and G. Lovato. 2005. Ultra high strength FeMn TWIP steels for automotive safety parts. Metall. Res. Technol. 102: 905–918.

De Cooman, B., L. Chen, H. Kim, Y. Estrin, S. Kim and H. Voswinckel. 2009. *In*: A. Haldar, S. Suwas and D. Bhattacharjee (eds.). Microstructure and Texture in Steels and Other Materials. Springer, London.

De Cooman, B., K. Chin and J. Kim. 2011. pp. 101–128. *In*: M. Chiaberge (ed.). New Trends and Developments in Automotive System Engineering. InTech, Rijeka, Croatia.

Dieudonné, T., L. Marchetti, M. Wery, J. Chêne, C. Allely, P. Cugy and C. Scott. 2014a. Role of copper and aluminum additions on the hydrogen embrittlement susceptibility of austenitic Fe–Mn–C TWIP steels. Corros. Sci. 82: 218–226.

Dieudonné, T., L. Marchetti, M. Wery, F. Miserque, M. Tabarant, J. Chêne, C. Allely, P. Cugy and C. Scott. 2014b. Role of copper and aluminum on the corrosion behavior of austenitic Fe–Mn–C TWIP steels in aqueous solutions and the related hydrogen absorption. Corros. Sci. 83: 234–244.

Diler, E., B. Lescop, S. Rioual, G. Nguyen Vien, D. Thierry and B. Rouvellou. 2014. Initial formation of corrosion products on pure zinc and MgZn2 examined by XPS. Corros. Sci. 79: 83–88.

Feliu, S. and V. Barranco. 2003. XPS study of the surface chemistry of conventional hot-dip galvanized pure Zn, galvanneal and Zn-Al alloy coatings on steel. Acta Mater. 51: 5413–5424.

Frommeyer, G., U. Brux and P. Neumann. 2003. Supra-Ductile and high-strength manganese-TRIP/TWIP steels for high energy absorption purposes. ISIJ Int. 43: 438–446.

Grässel, O., L. Krüger, G. Frommeyer and L. Meyer. 2000. High strength Fe–Mn–(Al, Si) TRIP/TWIP steels development—properties—application. Int. J. Plast. 16: 1391–1409.

Kavitha, R. and J. McDermid. 2011. Aluminothermic Reduction of Manganese oxides in the continuous galvanizing bath. Proc. Galvatech ´11. Genua. Italy.

Kavitha, R. and J. McDermid. 2012. On the *in-situ* aluminothermic reduction of manganese oxides in continuous galvanizing baths. Surf. Coat. Technol. 212: 152–158.

Kawano, T. and F. Renner. 2012. Studies on wetting behaviour of hot-dip galvanizing process by use of model specimens with tailored surface oxides. Surf. Interface Anal. 44: 1009–1012.

Khondker, R., A. Mertens and J. McDermid. 2007. Effect of annealing atmosphere on the galvanizing behaviour of dual-phase steel. Mat. Sci. Eng. A 463: 157–165.

Krüger, L., L. Meyer, U. Brüx, G. Frommeyer and O. Grässel. 2003. Stress-deformation behaviour of high manganese (AI, Si) TRIP and TWIP steels. J. Phys. IV France. 110: 189–194.

LeBozec, N., D. Thierry, M. Rohwerder, D. Persson, G. Luckeneder and L. Luxem. 2013. Effect of carbon dioxide on the atmospheric corrosion of Zn-Mg-Al coated steel. Corros. Sci. 74: 379–386.

Marder, R. 1990. Microstructural Characterization of Zinc Coatings. pp. 55–82. *In*: G. Krauss and K. Matlock (eds.). Zinc-Based Steel Coating Systems: Metallurgy and Performance. The Minerals, Metals & Materials Society, Warrendale, Pennsylvania.

Pawlek, F. 1983. Metallhüttenkunde, de Gruyter Verlag, Berlin-NewYork.

Persson, D., D. Thierry, N. LeBozec and T. Prosek. 2013. *In situ* infrared reflection spectroscopy studies of the initial atmospheric corrosion of Zn-Al-Mg coated steel. Corros. Sci. 72: 54–63.

Rodnyansky, A., Y. Warburton and L. Hanke. 2000. Segregation in hot-dipped galvanized steel. Surf. Interface Anal. 29: 215–220.

Sagl, R., A. Jarosik, D. Stifter and G. Angeli. 2013. The role of surface oxides on annealing high-strength steels in hot dip galvanizing. Corros. Sci. 70: 268–275.

Salgueiro Azevedo, M., C. Allely, K. Ogle and P. Volovitch. 2015. Corrosion mechanism of Zn(Mg, Al) coated steel in accelerated tests and natural exposure: 1. The role of electrolyte composition in the nature of corrosion products and relative corrosion rate. Corros. Sci. 90: 472–481.

Samek, L., E. Arenholz and J. Gentil. 2012a. Extended Tensile Ductility of a Formable High-Performance High-Manganese Steel. BHM. 157: 187–193.

Samek, L., S. Tollabimazraehno, E. Arenholz and K. Hingerl. 2012b. Characterisation of the microstructure of a high-strength manganese steel using different microscopy methods. Proc. 4th TMP Sheffield.

Schuerz, S., M. Fleischanderl, G. Luckeneder, K. Preis, T. Haunschmied, G. Mori and A. Kneissl. 2009. Corrosion behaviour of Zn-Al-Mg coated steel sheet in sodium chloride-containing environment. Corros. Sci. 51: 2355–2363.

Shih, S., C. Tai and T. Perng. 1993. Corrosion behavior of two-phase Fe-Mn-Al alloys in 3.5% NaCl solution. Corros. Sci. 49: 130–134.

Steinberger, R., J. Duchoslav, M. Arndt and D. Stifter. 2014. X-ray photoelectron spectroscopy of the effects of Ar+ ion sputtering on the nature of some standard compounds of Zn, Cr, and Fe. Corros. Sci. 82: 154–164.

Volovitch, P., T. Vu, C. Allely, A. Abdel Aal and K. Ogle. 2011. Understanding corrosion via corrosion product characterization: II. Role of alloying elements in improving the corrosion resistance of Zn-Al-Mg coatings on steel. Corros. Sci. 53: 2437–2445.

Zhu, X. and Y. Zhang. 1998. Investigation of the electrochemical corrosion behavior and passive film on Fe-Mn, Fe-Mn-Al and Fe-Mn-Al-Cr alloys in aqueous solution. Corros. Sci. 54: 3–12.

15

Control Model of Heat Treatment Process in the Rolling of Advanced High Strength Steels

Dadong Zhao,[1] *Hongmei Zhang*[1,*] and *Zhixin Chen*[2]

ABSTRACT

The water cooling (includes lamination cooling and Ultra Fast Cooling) after a hot rolling and the continuous annealing line (CAL) after a cold rolling are two heat treatment methods applied in the production of advanced high strength steels (AHSSs). In the water cooling process after hot rolling, the mathematical models of the convective heat transfer, the radiation heat transfer, the heat transfer between water and the strip, and the heat transfer in the strip are described by using dual phase (DP) steels. In the CAL, the mathematical models of tension setting (uncoiler, looper and coiler) and cooling process (the roller cooling and the air jet cooling) are described, and some methods to improve the profile are discussed.

[1] School of Materials and Metallurgy, University of Science and Technology Liaoning, Anshan, 114051, China.

[2] School of Mechanical, Materials and Mechatronic Engineering, University of Wollongong, NSW, 2522, Australia.

* Corresponding author: lilyzhm68@163.com

15.1 Production Process of AHSSs

Advanced High Strength Steels (AHSSs) refer to the steels reinforced through the phase transition and they include dual phase (DP) steels consisting of martensite, bainite or retained austenite, transformation induced plasticity (TRIP) steels, complex phase (CP) steels, quenching and partitioning (Q&P) steels, twinning induced plasticity (TWIP) steels and press hardening (PH) steels. Compared with conventional strength steels, AHSSs have the advantages of combined high strength, good formability and toughness which are particularly valuable in automobile applications. AHSS products are divided into hot rolled and cold rolled products which are usually accompanied with a heat treatment on a continuous annealing line (CAL). Table 15.1 shows a list of AHSS products of a steel company.

Since the main strengthening of AHSSs is obtained through phase transition hardening the rolling of AHSSs is not much different from rolling of conventional steels. Therefore this chapter will focus on the control model of post rolling heat treatment process. The control model is further divided into two parts, the process control model of cooling after a hot rolling and the process control model of CAL after a cold rolling.

In Section 15.2 of this chapter, a typical layout of a hot rolling factory and a new cooling method, known as ultra fast cooling (UFC) will be described. Then a DP production process is presented as an example to illustrate the requirements for a control model of heat treatment after a hot rolling. Finally, a new control model will be introduced and described in detail. In Section 15.3 four control models in CAL: uncoiler model, looper model, coiler model and cooling model will be described and discussed.

Table 15.1 Product list of steel company.

Type	Grade (*MPa*)	Production
TRIP	590	Cold rolled + CAL
	690	Cold rolled + CAL
	780	Cold rolled + CAL
	980	Cold rolled + CAL
CP	1180	Cold rolled + CAL
TWIP	980	Cold rolled + CAL
DP	590	Hot rolled
	780	Hot rolled
	980	Cold rolled + CAL

15.2 Control Model of Heat Treatment in Hot Rolling

15.2.1 Equipment Layout of Hot Rolled Production Line

Generally speaking, as shown in Figure 15.1, typical hot rolled production line consists of five parts, a walking beam type reheating furnace, one or two roughing mills, five or six continuous finishing mills, a lamination cooling system and two or three coilers. A slab is heated in the austenite region, in the walking beam type reheating furnace, rolled in the roughing mills and continuous finishing mills to the required specifications, then cooled in the laminar cooling system to the required temperature, finally curled into coils by the coilers.

In recent years, a new UFC (Ultra Fast Cooling) process which is based on TMCP (Thermo Mechanical Control Process) is introduced after the last finishing mill or laminar cooling. The strip can be directly cooled to below 200°C by its powerful nozzles and strong water flows. It also greatly expands the temperature range of the cooling process. Consequently, the microstructure of DP steel obtained on this line is more stable than that of the DP steel obtained on the hot rolling line which is not equipped with a UFC. Figure 15.2 shows a typical equipment layout of a hot rolled production line equipped with a UFC.

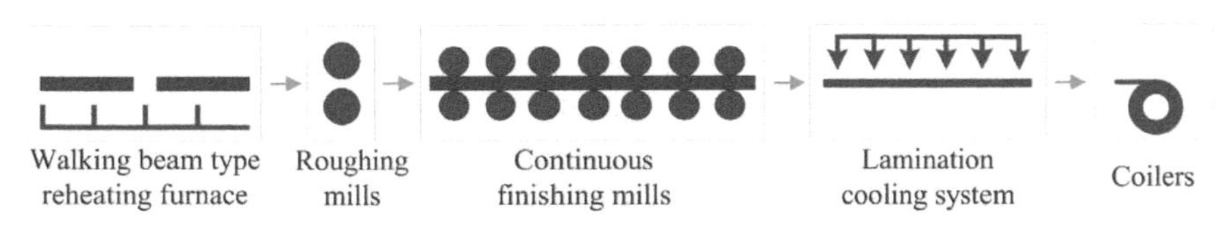

Figure 15.1 Typical layout of equipment in hot rolled production line.

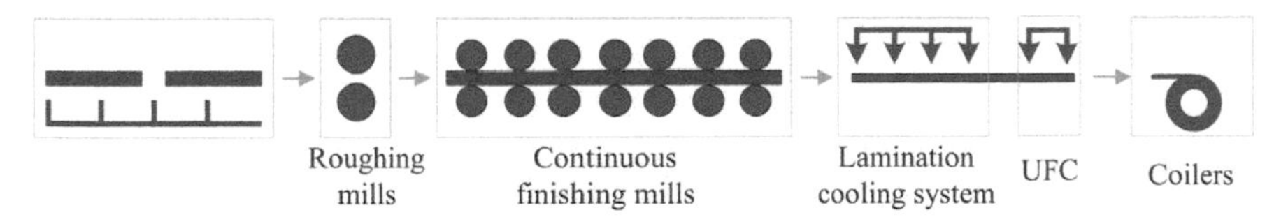

Figure 15.2 Typical equipment layout of Hot rolled production line with UFC.

15.2.2 Production Process and Control Strategy of DP Steel

A typical cooling curve of a hot rolled DP strip is shown in Figure 15.3. In a typical hot rolled production line FT (the temperature after finish mill) is about 850°C, MT (the temperature after lamination cooling) is between 640 and 680°C and CT (coil temperature) is below 200°C. In order to produce DP steel, the strip is rapidly cooled to the austenite (γ) $\rightarrow$ ferrite (α)

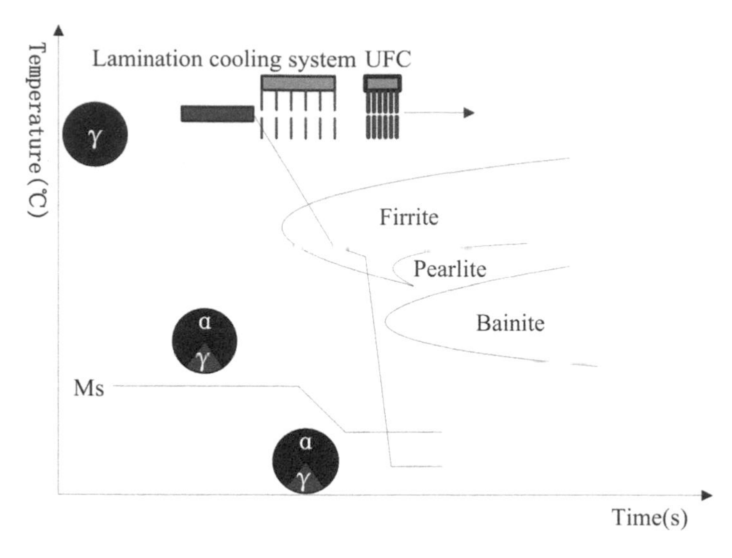

Figure 15.3 Cool curve of hot roll DP steel.

transformation nose temperature and then cooled slowly by the lamination cooling system so that a certain percentage of ferrite is formed. Then the strip is rapidly cooled to below Ms (martensitic transformation temperature). The last rapid cooling is achieved in the UFC. Finally, a DP strip is obtained.

15.2.3 Control Flow of Heat Treat Line in Hot Rolled Production

There are six variables: the hot rolling finishing temperature, the cooling rate of the laminar cooling, the exit temperature of the laminar cooling, the cooling rate of the UFC, the exit temperature of UFC and the coil temperature, that need to be controlled. The control flow chart of the heat treatment line is shown in Figure 15.4. The control system mainly consists of the following components:

1) Temperature monitoring at the exit of the hot rolling.
2) Computation of the temperature at the exit of UFC and the coiler.
3) Self learning of temperature model.
4) Feed forward control of strip travel speed.
5) Feedback control of strip temperature.

The strip is divided into several separated samples of equal length, usually equal with one or several banks along the rolling direction. Each bank contains a row of nozzles along the transverse direction. Before the strip enters the cooling control regions, finishing mill settings are calculated. After the calculation is completed, or after finish mill F2 (second finishing

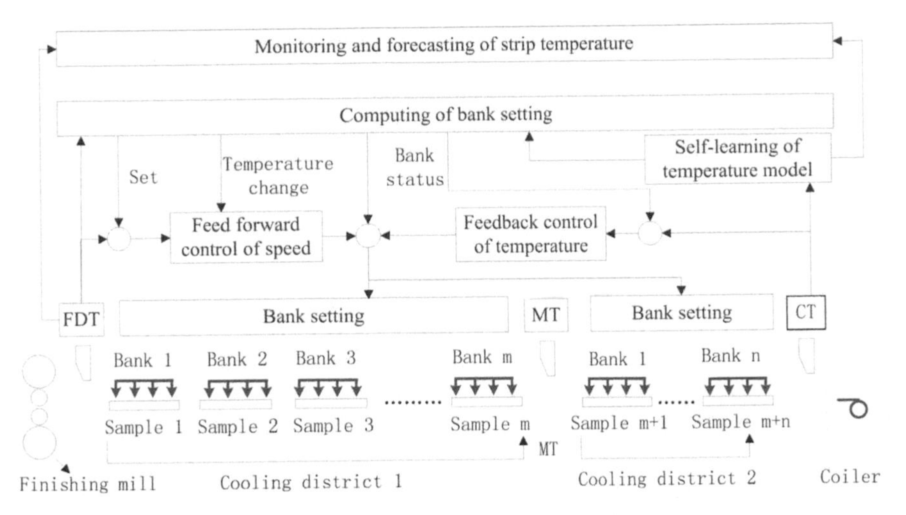

Figure 15.4 Control flow of heat treat line.

mill) biting, the banks in the cooling control regions are configured to satisfy with the temperature requirement of all samples in the case of the strip speed which is used to calculate the cooling time and thickness of the strip have been determined. When each sample pass through the temperature detector in the finishing outlet, the configuration of the banks and the location of the samples in the cooling district are recalculated, in order to meet the temperature requirement of the samples. When there is an obvious difference between the actual speed and predicted speed in the cooling process, the feed forward function will be triggered to modify the configuration of the banks and eliminate the temperature fluctuation which is caused by the change of the strip speed. When each sample passes through the CT (temperature detector before coiler), the temperature feedback function will be triggered to keep the differences between the actual and calculated coiling temperature in a reasonable range. When the study point reaches to the cooling outlet or CT, self learning function will be triggered to modify parameters of the model and the modification will be used on the control of the next strip.

When a strip head reaches FDT, the monitor function of cooling temperature control will be triggered. Based on soft measurement technology and parameterized with FDT, the strip speed and actual open or close status of cooling water are used to get the required temperature anywhere in the cooling district by computing with a mathematical model after the rolling line monitoring. The real-time temperature anywhere along the cooling district can be obtained from the online monitoring module and the value can be used to feed forward control of the strip speed and the choice of the cooling path.

15.2.4 Keep CT When the Rolling Speed Fluctuation

Today, most of hot rolling factories still use the adjustment of the strip travel speed as the most important way to control FDT. Some of them even use the adjustment of the strip travel as the only method to control FDT and ignore the effect of the cooling water between the mills. So, in the most of the cases, the strip speed is always changing and unpredictable. When the difference between actual speed and configured speed of the strip is not big enough and can be ignored, the temperature deviation can be eliminated by the feedback of CT. But if the difference is too large to ignore, the CT feedback cannot eliminate the temperature deviation.

In the CT feedback control, when the actual speed deviates from the scheduled speed, a second compensation is applied to the CT deviation which is caused by speed fluctuation by separately adjusting the bank in the lamination cooling section and UFC section. As shown in Figure 15.5, assuming the bank k is opened last in the lamination cooling section, according to the location X_k of the bank k and the valve response time, the critical point of the lamination cooling section is obtained by the following equation:

$$X = X_k - V \cdot t_k \qquad (15.1)$$

where X is the critical point of the lamination cooling section, X_k is the location of the bank k, V is the speed of the strip and t_k is the response time.

When the actual speed deviates from the scheduled speed, the temperature at the critical point deviates from the scheduled temperature. According to the measured temperature at the critical point and scheduled speed curve between the critical point X and the CT position, based on the temperature math model, the CT temperature is recalculated and compared with the target value of the CT temperature. Then the CT deviation between FDT and the critical point X which is caused by speed fluctuation of the strip, is further eliminated. For the UFC section, repeat the above work to reduce the CT deviation to an acceptable level and meet the production requirement.

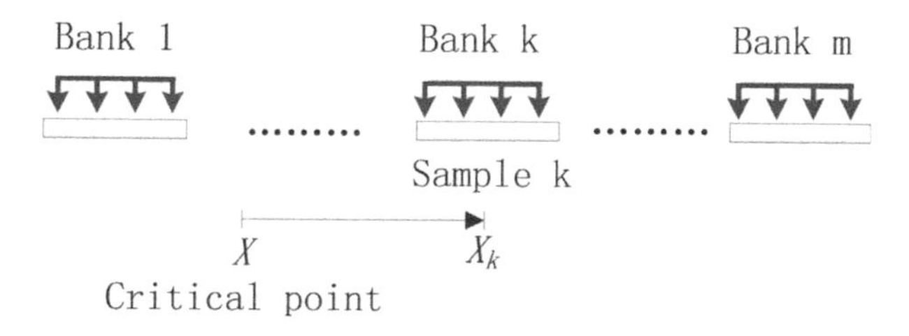

Figure 15.5 Critical point.

15.2.5 The Heat Transfer Process of Strip in Cooling District

The heat transfer process of the strip in the cooling district is very complicated and mainly consists of convective heat transfer, radiation heat transfer and thermal convection of cooling water. When the strip is located in the section between the finishing mill and the lamination cooling, convective heat transfer and radiation heat transfer are two main methods of heat transfer. When the strip is in the lamination cooling or UFC, thermal convection of cooling water is considered as the most important method of heat transfer.

1. Convective heat transfer

The heat flux between the strip surface and the surrounding air is given by the following equation:

$$q_n = h_n \cdot (T_s - T_a) \tag{15.2}$$

where q_n is the heat flux between the strip surface and the surrounding air, h_n is the convective heat transfer coefficient, T_s is the steel surface temperature and T_a is the air temperature.

The convective heat transfer on strip surface reduces to natural convection heat transfer on the horizontal surface of the strip. The flow status is described by the Prandtl number P_r and Grashot number G_r.

The Prandtl number P_r is a dimensionless number, defined as the ratio of momentum diffusivity to thermal diffusivity. Its value is given by following equation:

$$P_r = \frac{c_p \mu}{k} \tag{15.3}$$

where c_p is the specific heat, μ is the dynamic viscosity, k is the thermal conductivity. The Prandtl numbers of the air at different temperatures can be easily found from relevant literature.

The Grashot number G_r is also a dimensionless number in fluid dynamics and heat transfer and it approximates the ratio of the buoyancy to viscous force acting on a fluid. For vertical flat plates, its value is given by following equation:

$$G_r = \frac{g\beta(T_s - T_\infty)L^3}{\nu^2} \tag{15.4}$$

where g is the acceleration due to earth's gravity, β is the coefficient of thermal expansion, T_s is the surface temperature, T_∞ is the bulk temperature, L is the vertical length, v is the kinematic viscosity.

The Nusselt number N_u is the ratio of convective to conductive heat transfer across the boundary. Its value is given by following equation:

$$N_u = \frac{h_n L}{k} \quad (15.5)$$

where h_n is the convective heat transfer coefficient of the flow, L is the characteristic length, k is the thermal conductivity of the fluid.

If $10^4 < G_r \cdot P_r < 10^9$, the air flow is in the laminar flow status and the Nusselt number N_u can also be obtained by the following equation:

$$N_u = 0.59\left(G_r \cdot P_r\right)^{1/4} \quad (15.6)$$

If $10^4 < G_r \cdot P_r < 10^9$, the air flow is in the turbulence status and the N_u can also be obtained by the following equation:

$$N_u = 0.1\left(G_r \cdot P_r\right)^{1/3} \quad (15.7)$$

So the convective heat transfer coefficient h_n is obtained by Eqs. (15.5), (15.6) and (15.7).

If the air flow is in the laminar flow status, h_n can be derived by the following equation:

$$h_n = \frac{0.59 \cdot k \cdot \left(G_r \cdot P_r\right)^{1/4}}{L} \quad (15.8)$$

If the air flow is in the turbulence flow status, h_n is calculated by the following equation:

$$h_n = \frac{0.1 \cdot k \cdot \left(G_r \cdot P_r\right)^{1/3}}{L} \quad (15.9)$$

2. Radiation heat transfer

According to Stefan-Boltzmann law, Heat flux q_r and heat transfer coefficient h_r of the strip surface by the radiation heat transfer are given by the following equations:

$$q_r = \varepsilon \cdot \sigma \cdot \left[\left(\frac{T_s + 273}{100}\right)^4 - \left(\frac{T_a + 273}{100}\right)^4\right] \quad (15.10)$$

$$h_r = \frac{q_r}{T_s - T_a} \quad (15.11)$$

where ε is the blackness value of the strip, σ is the thermal radiation constant, T_s is the surface temperature of strip, T_a is the ambient temperature.

Assuming cooling area and volume of the strip is $2A_r$ and V_r, then the dissipated heat dQ_r within the time $d\tau$ is:

$$dQ_r = 2 \cdot q_r \cdot A_r \cdot d\tau = 2A_r \cdot \varepsilon \cdot \sigma \left[\left(\frac{T_s + 273}{100} \right)^4 - \left(\frac{T_a + 273}{100} \right)^4 \right] d\tau \tag{15.12}$$

Meanwhile, the lost heat dQ_r causes a temperature drop $d\tau$ to the strip, then:

$$dQ_r = \rho(T)c_p(T)V_r dT \tag{15.13}$$

where $\rho(T)$ is the density of the strip, $C_p(T)$ is the specific heat of the strip.

So, the next two equations can be obtained as:

$$2A_r \cdot \varepsilon \cdot \sigma \left[\left(\frac{T_s + 273}{100} \right)^4 - \left(\frac{T_a + 273}{100} \right)^4 \right] d\tau = \rho(T)c_p(T)V_r dT \tag{15.14}$$

$$\frac{dT}{d\tau} = \frac{2A_r \cdot \varepsilon \cdot \sigma}{\rho(T)c_p(T)V} \left[\left(\frac{T_s + 273}{100} \right)^4 - \left(\frac{T_a + 273}{100} \right)^4 \right] \tag{15.15}$$

When the temperature of the strip is between 500 and 900°C, $T_s^4 >> T_a^4$, the item $((T_a + 273)/100)^4$ containing T_a can be ignored. In addition, due to the shape characteristics of the strip, that is the width w and the length $l >>$ the thickness h, the surface temperature T_s equals to the average temperature T of the strip. It is also assumed that the density and specific heat are independent of the temperature. Equation (15.15) can be simplified to:

$$\frac{dT}{d\tau} = \frac{2 \cdot \varepsilon \cdot \sigma}{\rho c_p h} \left(\frac{T + 273}{100} \right)^4 \tag{15.16}$$

During this period of time τ, if the temperature of the strip decreases from T_1 to T_2, the above equation is integrated in the corresponding time interval and temperature range. The following equation thus can be obtained:

$$T_2 = 100((\frac{T_1 + 273}{100})^{-3} + \frac{6\varepsilon\sigma\tau}{100c_p\rho h})^{-1/3} - 273 \tag{15.17}$$

3. Heat transfer for the strip

As shown in Figure 15.6, heat flux densities caused by radiation and convection for the strip decrease with time. When the strip temperature is higher, the heat flux density of the radiation is far larger than that of the convection. When the convection temperature is lower, they become equal.

In engineering applications, to simplify the heat transfer calculation only heat radiation is considered and the heat flux density caused by convection is merged into radiation coefficient. The heat transfer is given by Eq. (15.18).

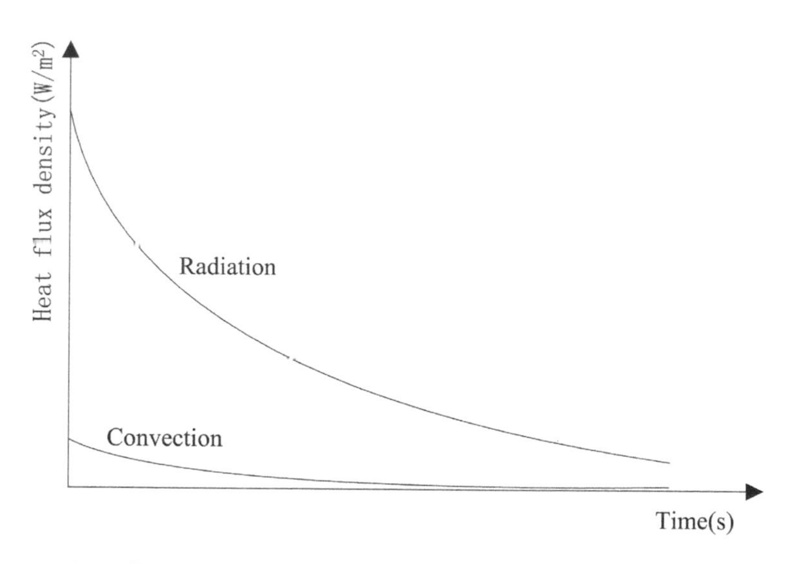

Figure 15.6 Heat flux densities caused by radiation and convection for the strip decrease with time.

$$\frac{dT}{d\tau} = -\frac{2\varepsilon\sigma}{c_p\rho h}\left[(T+273)^4 - (T_a+273)^4\right] - \frac{\alpha_w}{c_p\rho h}(T-T_w) \tag{15.18}$$

where d_T is the strip temperature, $d\tau$ is the cooling time, ε is the thermal radiation coefficient of the strip, σ is the Stefan Boltzmann constant, C_p is the specific heat of the strip, α_w is the convection heat transfer coefficient of the cooling water, ρ is the density of the strip, h is the thickness of the strip, T is the surface temperature of the strip, T_a is the air temperature, T_w is the temperature of the cooling water.

Taking into account the impacts of the strip speed, the flow of the cooling water, water temperature and water pressure, the heat transfer coefficient of the cooling water α_w can be obtained from the following equation:

$$\alpha_w = \frac{Q_U + K_L Q_L}{B \cdot L} A_1 h^{A2} \exp\left(-A_3 \cdot (t - t_w)\right) \left(\frac{t_{w0}}{t_w}\right)^{B1} \left(\frac{V}{V_0}\right)^{B2} \left(\frac{P}{P_0}\right)^{B3} \tag{15.19}$$

where Q_U and Q_L are the upper banks and lower banks of the water flow respectively, K_L is the equivalent coefficient of the lower banks, B is the width of the strip, L is the sample length of the strip, h is the thickness of the strip, A_1, A_2 and A_3, B_1, B_2 and B_3 are model coefficients, t_{w0} is the water reference temperature, V is the actual speed of the strip, V_0 is the reference speed of the strip, P is the actual water pressure, P_0 is the reference water pressure.

Figure 15.7 shows there is a difference in terms of heat transfer mechanism between the lamination cooling and UFC, though they have similar heat transfer coefficients in control models. When the cooling water contacts with the surface of the strip, a vapor film is introduced because the surface has a high temperature (800 to 1000°C). The vapor film significantly reduces the heat transfer coefficient. Ranges in the heat transfer coefficient between the lamination cooling and UFC are different. The cooling water in lamination cooling cannot come in contact with the strip because its pressure is not enough to penetrate the vapor film. Nevertheless, the cooling

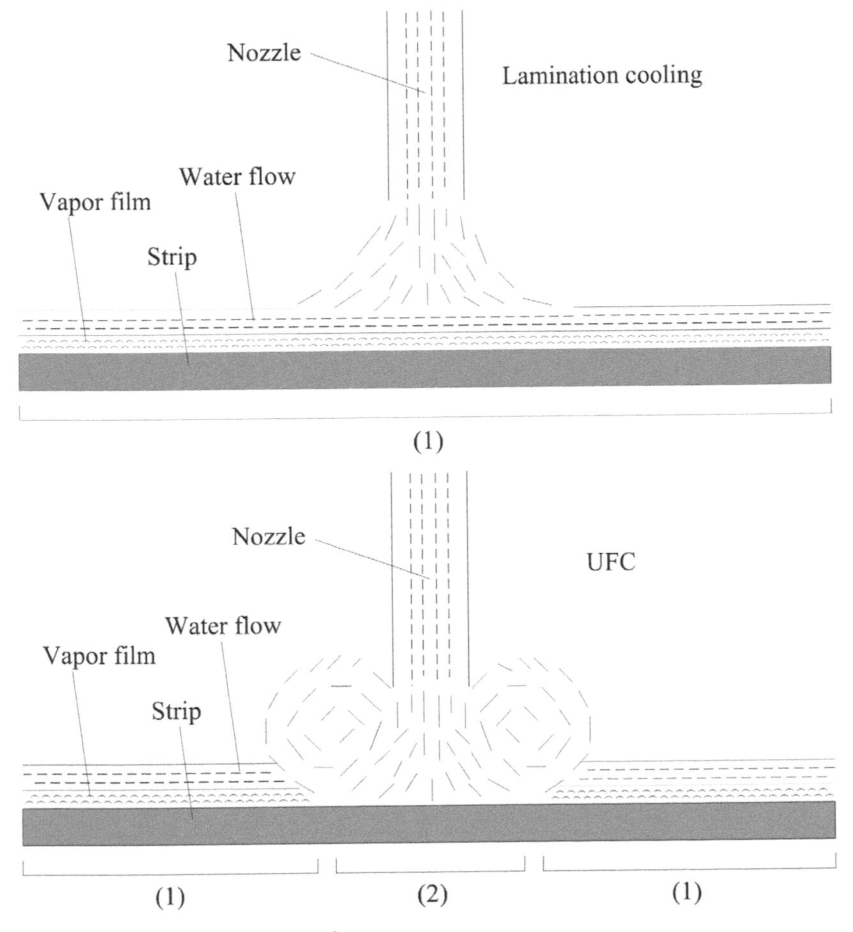

Figure 15.7 Heat transfer mechanism between the lamination cooling and UFC.

water in UFC can directly reach to the surface of the strip, due to the higher pressure (about 0.8 MPa).

So their heat transfer coefficients are different. The heat transfer coefficients of some cooling method are shown in Table 15.2.

Table 15.2 Heat transfer coefficient ($W{\cdot}m^{-2}{\cdot}k^{-1}$) with different strip thickness (*mm*).

Thickness	Air cooling	Lamination cooling	UFC
3.0	50	550	4400
5.0	50	850	5100
8.0	50	900	5900
11.0	50	700	5800
13.0	50	700	5790

15.2.6 Numerical Solution of Two Dimensional Unsteady for Heat Conduction Process

The numerical solution method is based on Discrete Mathematics and Computer Science. Its theoretical basis is not rigorous. However, it shows great adaptability in practical applications. It can solve the problems of variable physical parameters in the cooling and heating process of steel plate or coil. So far, a less complicated heat conduction problem can be solved by using the numerical method.

Finite difference method and finite element method are two main methods for the numerical solution of temperature distribution problem. The mathematical basis of the finite element method is variation principle and subdivision interpolation, suitable for objects with a complex geometry. The physical foundation of the finite difference method is the law of energy conservation. The continuous distribution problem of object temperature in space and time is transformed to solve the problem of temperature values of finite discrete points. The temperature values on these discrete points are shown in Figure 15.8 and used to approximate continuous temperature distribution.

The two-dimensional heat conduction equation for temperature distribution is:

$$\frac{\partial T}{\partial \tau} = \frac{\lambda}{C_p \rho}\left(\frac{\partial^2 T}{\partial x^2} + \frac{\partial^2 T}{\partial y^2}\right) \tag{15.20}$$

Initial condition is:

$$T(x, y, 0) = \phi(T), (t = 0, 0 \le x \le b/2, 0 \le y \le a/2) \tag{15.21}$$

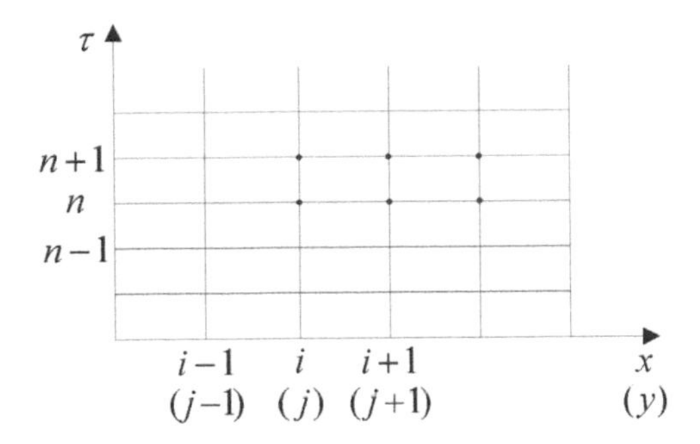

Figure 15.8 Discrete point schemes.

Boundary conditions are:

$$-\lambda\frac{\partial T}{\partial x}=b(T-T_{\infty}),(x=b/2,t\geq 0) \tag{15.22}$$

$$-\lambda\frac{\partial T}{\partial y}=a(T-T_{\infty}),(y=a/2,t\geq 0) \tag{15.23}$$

$$\frac{\partial T}{\partial x}=\frac{\partial T}{\partial y}=0,(t\geq 0,x=0,y=0) \tag{15.24}$$

where $\Phi(T)$ is the initial temperature distribution of steel plate section.

Heat conduction differential equation is applied to node (i, j), and then the next equation can be obtained:

$$\left(\frac{\partial^2 T}{\partial x^2}\right)_{i,j}^{k+1}+\left(\frac{\partial^2 T}{\partial y^2}\right)_{i,j}^{k}=\frac{C_p\rho}{\lambda}\left(\frac{\partial T}{\partial t}\right)_{i,j}^{k} \tag{15.25}$$

The Crank-Nicolson difference method can be used to solve Eq. (15.25) and obtain a stable numerical solution.

$$\frac{T_{i+1,j}^{k+1}-2T_{i,j}^{k+1}+T_{i-1,j}^{k+1}}{(\Delta x)^2}+\frac{T_{i+1,j+1}^{k+1}-2T_{i,j}^{k+1}+T_{i-1,j-1}^{k+1}}{(\Delta y)^2}=\frac{1}{h}\frac{T_{i,j}^{k+1}-T_{i,j}^{k}}{\Delta t} \tag{15.26}$$

The difference of the boundary condition on the surface of the strip is:

$$T_{L+1,j}^{k+1}+\frac{2b\Delta x}{\lambda}T_{L,j}^{K+1}-T_{L-1,j}^{K+1}=\frac{2b\Delta x}{\lambda}T_h \tag{15.27}$$

$$T_{i,M+1}^{k+1}+\frac{2b\Delta x}{\lambda}T_{i,M}^{K+1}-T_{i,M-1}^{K+1}=\frac{2b\Delta x}{\lambda}T_h \tag{15.28}$$

The alternating direction method is applied to solve the problem and then the equations will only consist of three unknown variables. The corresponding equations of the coefficient matrix have the characteristics of tridiagonal matrices thus can be solved by chasing method. The specific formulae are:

$$\frac{h\Delta\tau}{(\Delta x)^2}\left(T_{i+1,j}^{k+1}-2T_{i,j}^{k+1}+T_{i-1,j}^{k+1}\right)+\frac{h\Delta\tau}{(\Delta y)^2}\left(T_{i+1,j+1}^{k}-2T_{i,j}^{k}+T_{i-1,j-1}^{k}\right)=T_{i,j}^{k+1}-T_{i,j}^{k} \quad (15.29)$$

$$\frac{h\Delta\tau}{(\Delta y)^2}\left(T_{i+1,j}^{k+1}-2T_{i,j}^{k+1}+T_{i-1,j}^{k+1}\right)+\frac{h\Delta\tau}{(\Delta x)^2}\left(T_{i,j+1}^{k+2}-2T_{i,j}^{k+2}+T_{i,j-1}^{k+2}\right)=T_{i,j}^{k+2}-T \quad (15.30)$$

The matrix has the form:

$$\begin{bmatrix} \left(2+\frac{h\Delta\tau}{(\Delta x)^2}\right) & -2\frac{h\Delta\tau}{(\Delta x)^2} & & \\ -\frac{h\Delta\tau}{(\Delta x)^2} & \left(2+\frac{h\Delta\tau}{(\Delta x)^2}\right) & -\frac{h\Delta\tau}{(\Delta x)^2} & \\ & \cdots & \cdots & \\ & -\frac{h\Delta\tau}{(\Delta x)^2} & \left(1+2\frac{h\Delta\tau}{(\Delta x)^2}+2\frac{h\Delta\tau}{(\Delta x)^2}\frac{b\Delta x}{\lambda}\right) & 1 \end{bmatrix} \begin{bmatrix} T_{i,j} \\ T_{2,j} \\ \vdots \\ T_{L,j} \end{bmatrix} =$$

$$\begin{bmatrix} \left(1-\frac{h\Delta\tau}{(\Delta y)^2}\right) & -2\frac{h\Delta\tau}{(\Delta y)^2} & & \\ \frac{h\Delta\tau}{(\Delta y)^2} & \left(1-2\frac{h\Delta\tau}{(\Delta y)^2}\right) & -\frac{h\Delta\tau}{(\Delta y)^2} & \\ & \cdots & \cdots & \\ & \frac{h\Delta\tau}{(\Delta y)^2} & \left(1-2\frac{h\Delta\tau}{(\Delta y)^2}\right) & \frac{h\Delta\tau}{(\Delta y)^2} \end{bmatrix} \begin{bmatrix} T_{i,1} \\ T_{i,2} \\ \vdots \\ T_{i,L} \end{bmatrix} + \begin{bmatrix} 0 \\ 0 \\ \vdots \\ 2\frac{h^2\Delta\tau}{(\Delta y)^2}\frac{\Delta x}{\lambda}T_w \end{bmatrix} \quad (15.31)$$

When the heat transfer boundary condition is given, if the temperature of any point in the strip is known at a certain time, the temperature distribution of the strip at any time can be obtained by solving the above difference equations simultaneously.

15.2.7 Determine of Thermal Parameters in Cooling Process

Thermal physical properties are described by a set of thermal physical property parameters which can be divided into two kinds, some of them are not a function of the temperature and other parameters are. The thermal physical parameters involved in the strip cooling process include density, thermal conductivity and specific heat as well as the thermal conductivity coefficient, and the latter three are more dependent on the temperature.

1. Processing variable thermal parameters

When the finite difference method is used to solve the temperature field, how to deal with the variable parameter problem?

The heat flow density from node j to i is given by:

$$q_{ij} = K_{ij}\left(T_j - T_i\right) \tag{15.32}$$

where K_{ij} is the conduction heat between the node j and the node i. It can be obtained by the next equation:

$$K_{ij} = \frac{\lambda \cdot A_{ij}}{L_{ij}} \tag{15.33}$$

where λ is the thermal conductivity of the strip, L_{ij} is the distance between node i and node j, A_{ij} is the average area vertical to the heat flow direction.

The thermal conductivity λ of the strip depends on the temperature and so:

$$K_{ij} = \frac{\left[\lambda(T_i) + \lambda(T_j)\right] \cdot A_{ij}}{2 \cdot L_{ij}} \tag{15.34}$$

where $\lambda(T_i)$ and $\lambda(T_j)$ are the thermal conductivities of the nodes i and j at the temperatures T_i and T_j respectively.

In order to simplify the calculation, the thermal conductivity $\lambda(T)$ is usually arranged in the form of a function of temperature (using the least square method for discrete experimental curve fitting, and then derive the function).

If Eq. (15.34) is used to obtain K_{ij}, according Eqs. (15.26), (15.27) and (15.28), the equations for the internal node and the boundary node can be obtained. The rest of the problem is simplified to calculate temperature under the condition of the same thermal conductivity. The calculation of K_{ij} relies on the values of T_i and T_j to be known but they are unknown, in other words the calculation of the thermal conductivity depends on a unknown temperature field, thus it is impossible to directly obtain K_{ij}. In this case an iterative calculation is needed:

1) The temperature distribution is assumed and a certain value $T_i^{(n)}$ is assigned to the node of the unknown temperature, $n = 0$.

2) For the $T_i^{(n)}$, a distribution of λ is obtained using the experimentally derived relationship between the thermal conductivity and temperature, $n=0$.
3) The conduction heat between node i and node j is obtained by Eq. (15.32).
4) The new temperature distribution $T_i^{(n+1)}$ is calculated using the Eqs. (15.26), (15.27) and (15.28).
5) If the difference between the new value $T_i^{(n+1)}$ and the old value $T_i^{(n)}$ of the temperature distribution is less than a given convergence value ε, then the calculation will stop. The iteration method is used to converge $T_i^{(n)}$, $n = 1, 2, \ldots$.
6) The new distribution $\lambda_i^{(n+1)}$ of λ is obtained by using the $T_i^{(n+1)}$ and repeating step (2), $n=0$.
7) Repeat (3) to (6), until $\lambda_i^{(n+1)}$ and $\lambda_i^{(n)}$ meet the requirements of the convergence value ε, $n=1, 2, \ldots$.

2. Determination of thermal conductivity

Thermal conductivity refers to the rate at which heat passes through a specified material, expressed as the amount of heat that flows per unit time through a unit area with a temperature gradient of one degree per unit distance. It links the temperature gradient and the heat flux vector at every point inside the object. The thermal conductivities of different materials are different, because they relate not only to temperature but also other factors, such as structure and density. Their definitions are given by the mathematical expression of Fourier's law:

$$\lambda = -q / \mathrm{grad}t \tag{15.35}$$

where grad$t$ is the temperature gradient, q is the heat flux vector.

Thermal conductivities at different temperatures used in an engineering numerical calculation for various steels are determined experimentally. Table 15.3 and Figure 15.9 show how some thermal conductivity coefficients of steels change with temperature.

In an actual calculation, the method of interpolation in terms of chemical composition and temperature range is used to determine the thermal conductivity coefficient. The carbon content plays an important role to determine it.

3. Determination of specific heat value

The value of a specific heat of a general solid metal C is composed of three parts:

$$C = C_s + C_e + C_m \tag{15.36}$$

Table 15.3 Thermal conductivity coefficient (W/m·°C).

Temperature (°C)	Rimmed steels	Killed steels	Low-carbon steels	Medium carbon steels	Si-Mn Steels	Stainless steels
0	65.4	59.4	51.7	51.7	25.1	15.9
50	62.8	58.6	51.5	51.5	26.7	15.9
100	60.2	57.8	50.5	50.7	28.5	16.3
150	57.8	55.2	49.8	49.8	29.3	16.7
200	55.7	53.6	48.6	48.1	30.1	17.2
250	53.1	51.5	46.5	46.9	30.9	17.6
300	51.1	49.4	44.4	45.6	30.9	18.4
400	46.5	44.8	42.7	41.7	30.9	20.1
500	41.1	40.2	39.3	38.1	30.9	21.7
600	37.6	36.1	35.6	34	30.1	23.8
700	34	31.9	31.9	30.1	28	25.6
800	30.1	28.5	25.9	24.7	25.1	26.7
850	27.7	27.2	25.9	24.7	25.1	26.4

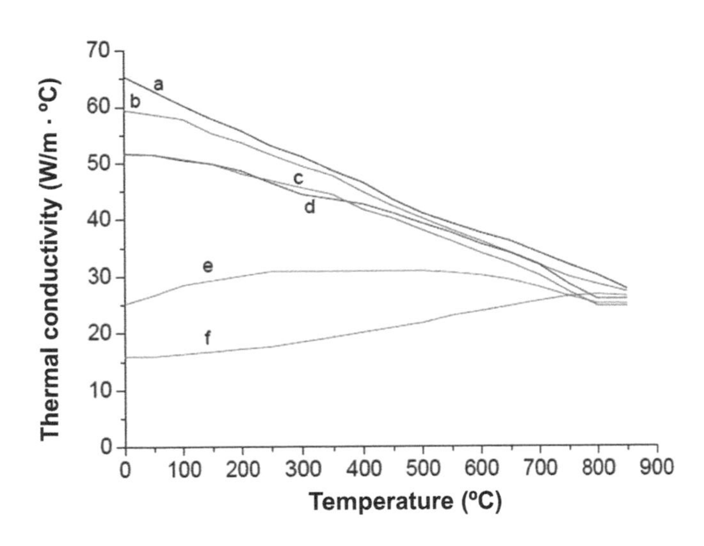

Figure 15.9 Thermal conductivities of steels changes with temperature: (a) rimmed steels, (b) killed steels, (c) low-carbon steels, (d) medium carbon steels, (e) Si-Mn Steels, and (f) stainless steels.

where C_s is the shaking specific heat, C_e is the electronic specific heat, C_m is the magnetic specific heat.

In most cases, although the specific heat increases with increasing temperature, there are some differences in the components of the specific heat for α-Fe, γ-Fe and ε-Fe. For α-Fe, the magnetic transition from a paramagnetic state at high frequencies into a ferromagnetic state will occur at 770°C, the change is usually called A_2 change, and the magnetic transition temperature of iron is called Curie temperature. The heat capacity sharply changes in the A_2 changing process. So the interpolation calculation is used to get the specific heat value between two adjacent points where specific heat values are known. Table 15.4 and Figure 15.10 show some specific heats of steels change with temperature.

In an actual calculation, the specific heat value is determined by the method of linear interpolation in the interval of two chemical composition points and two temperature points.

Assuming the specific heat value on the node x_i is λ_i, the specific heat values between the two nodes follow the function $\lambda(a)$, where $x_i<a<x_{i+1}$. For non-equidistant nodes, the forward difference formula or the post difference formula can be used to get $\lambda(a)$:

$$\lambda(a) = \lambda(x_i) + \frac{(\lambda(x_{i+1}) - \lambda(x_i))(a - x_i)}{x_{i+1} - x_i} \tag{15.37}$$

or

$$\lambda(a) = \lambda(x_{i+1}) - \frac{(\lambda(x_{i+1}) - \lambda(x_i))(x_{i+1} - a)}{x_{i+1} - x_i} \tag{15.38}$$

Table 15.4 Specific heat (W/(m^2·°C)).

Temperature (°C)	Rimmed steels	Killed steels	Low-carbon steels	Medium carbon steels	Si-Mn Steels	Stainless steels
0	0.469	0.469	0.469	0.469	0.502	0.494
50	0.486	0.486	0.486	0.486	0.503	0.511
100	0.502	0.502	0.502	0.502	0.504	0.528
150	0.519	0.519	0.519	0.511	0.528	0.536
200	0.536	0.544	0.536	0.528	0.544	0.537
250	0.553	0.553	0.553	0.553	0.553	0.553
300	0.569	0.569	0.578	0.569	0.578	0.552
400	0.628	0.628	0.628	0.611	0.628	0.586
500	0.703	0.695	0.703	0.687	0.703	0.628
600	0.804	0.787	0.787	0.729	0.779	0.628
700	1.105	1.139	1.431	1.563	0.904	0.62
800	0.804	0.862	0.737	0.511	0.611	0.644
850	0.837	0.812	0.645	0.544	0.628	0.643

15.2.8 Self Adjust Control System in Cooling Process

It is possible that some changes in product specifications or the external environment in the hot rolling strip production process may occur.

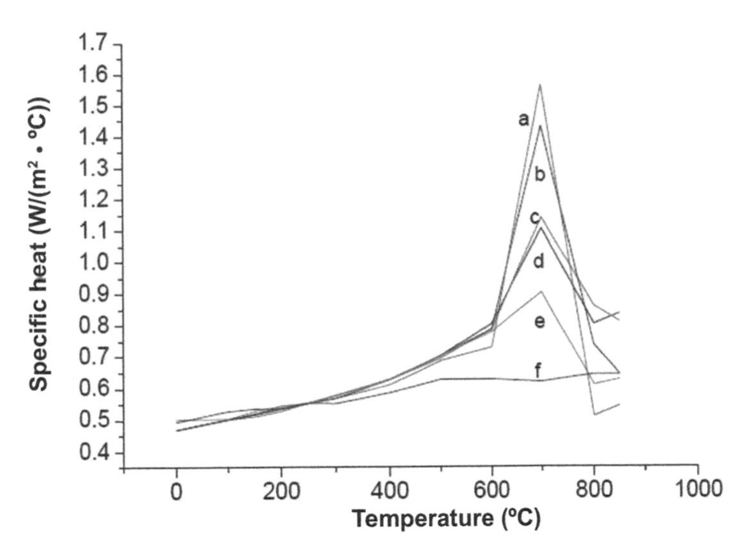

Figure 15.10 Specific heats of steels changes with temperature: (a) rimmed steels, (b) killed steels, (c) low-carbon steels, (d) medium carbon steels, (e) Si-Mn Steels, and (f) stainless steels.

Therefore, adaptive ability of the system is important to ensure the control accuracy and the operation stability of the mathematical model.

1. Self adjust model

Mathematically many unpredictable or incurable factors are grouped into one or more variables in the self adjust model. The strip is divided into three parts: head, middle and tail sections and controlled separately. The strip is divided into n sample sections along the length. According the difference between the actual temperature and calculated temperature of each sample section, the heat transfer coefficient of the water cooling in the control model is recalculated and used to correct the heat transfer coefficient of the water cooling in the control model. Then the corrected heat transfer coefficient is applied to optimize the control of the next strip. The corrected coefficient can be calculated by using the equation below:

$$Z^m = \left(\ln\left(\frac{T_f^a - T_w}{T_0^a - T_w} \right) + \sum_j \alpha_j^a \cdot t_{bj} \right) \cdot \left(\ln\left(\frac{T_f^c - T_w}{T_0^c - T_w} \right) + \sum_j \alpha_j^c \cdot t_{bj} \right)^{-1} \tag{15.39}$$

where Z^m is the calculated self learning coefficient, T_f^a is the actual temperature at the final cooling point, T_0^a is the actual temperature at the start cooling point, T_f^c is the calculated temperature at the final cooling point, T_0^c is the calculated temperature at the start cooling point, α_j^a is the actual temperature declension caused by air cooling in unit time, α_j^c is the calculated temperature declension, t_{bj} is the time in the each cooling section.

In practical engineering applications, according to the size information of the product, such as thickness and steel type, strips are divided into several classes and each class has its own adjust coefficient. The set of classes is called class table. After the strip is produced, a new adjust coefficient for this class is recalculated from the old one and production data. The impact of accidental factors on the adjusted coefficient in class tables is smoothened by weighing instantaneous value of the self learning coefficients. The self learning coefficient for one sample section is obtained by following equation:

$$Z^n = Z_m^0 + \lambda_m \left(\left(\sum_i^n Z_i^m - Z_{\min}^m - Z_{\max}^m \right) \cdot (i-2)^{-1} - Z_m^0 \right) \tag{15.40}$$

where Z_i^m is the self learning coefficient, $Z_{\min}^m$ is the minimum instantaneous value of the calculated self learning coefficient, $Z_{\max}^m$ is the maximum instantaneous value of the calculated self learning coefficient, Z_m^0 is the calculated self learning coefficient in class table, λ_m is the smoothing factor.

2. *Control strategy of self adjust*

Based on the above temperature control model, a precise control on the strip temperature can be achieved by using the control strategy of the self learning coefficient. The temperature drop of the strip is related to the cooling time which is determined by the roll speed or coiler speed. For the head section of the strip, the roll speed is set to a smaller value to improve the strip stability because the strip does not enter the coiler and the tension is not established yet. After being coiled, the middle section of the strip passes through the cooling equipment with a higher and stable speed. Finally, the tail section of strip leaves from the finish mill and its speed is determined by the coiler. As a result, the self learning of the temperature model used the learning control strategy which includes three zones: the head, the middle and the tail section of the strip.

The temperatures of the head, the middle and the tail section of the strip are calculated separately and every zone has its own studying position and self learning coefficient. For sample sections that are not used for self learning, their self learning coefficients can be obtained from adjacent sample sections that are used for self learning by using linear interpolation method. The self learning control strategy is shown in Figure 15.11, where n_h, n_b and n_t are the numbers of the sample section in the head, middle and tail section of the strip respectively, m_h, m_b and m_t are the numbers of the learning sample in the head, middle and tail section of the strip respectively. The above parameters are all classified according to the chemical composition,

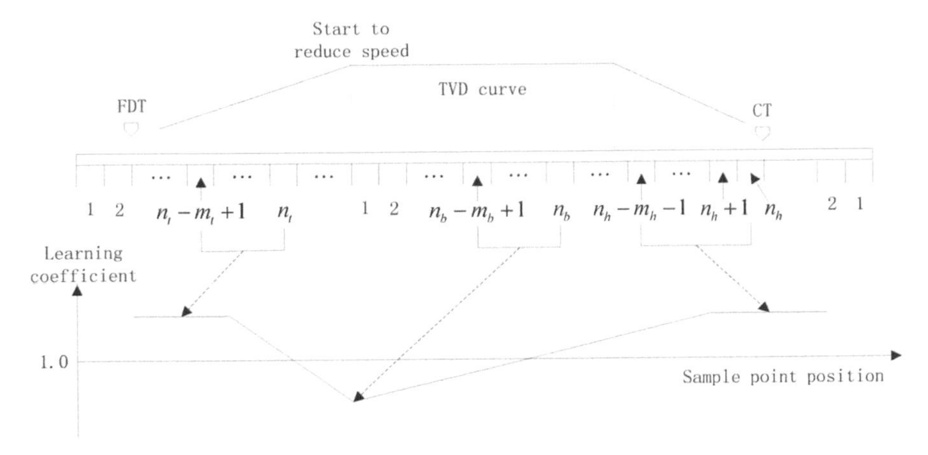

Figure 15.11 Control strategy of self learning.

the target thickness of the strip, the finishing rolling temperature and the finishing cooling temperature.

The self learning is activated when each learning point in the cooling section reaches the thermo detector which is used to detect CT, in order to improve the prediction accuracy and control precision of the temperature model. The calculation of the self learning coefficient are mainly divided into three processes, those are the data preparation, the instantaneous value of the calculated data and the update value of calculated learning coefficients. The main task of the data preparation is to obtain the information of the actual measurements of the starting cooling and finishing cooling temperature of the strip. The validity of the data, the TVD (time, speed and distance) curve which is calculated by the model and the control information of the configuration bank should also be confirmed. In the period of calculating instantaneous value, the control and tracking information of the strip sample which is collected in the data preparation process, the calculated value of the finishing temperature, the instantaneous learning value calculated by the model in each sample section and the average value of the instantaneous learning value are obtained. In the stage of updating the learning coefficients, the effect of causal factors on the learning coefficients are studied using the average weighted smoothing method and the update learning values are computed based on the succession of the same class learning coefficients characteristic. The self learning control flow is shown in Figure 15.12.

15.3 Control Model of CAL

The layout of a typical CAL is shown in Figure 15.13. There are three sections in the CAL, called entry section, process section and exit section. In the

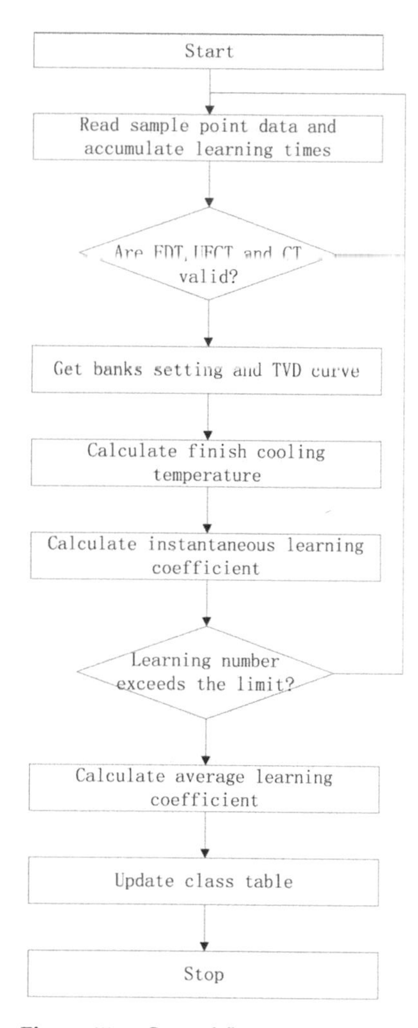

Figure 15.12 Control flow of self adjust.

entry section, a cold rolled coil is transported to the production line and placed on an uncoiler by using a charging car. After being opened, the strip and the previous strip are welded into a new strip and a welding gap is produced. A tension roller is used to measure the tension and line speed of strip. Then the new coil enters the entry looper which is used as a coil buffer and to set up proper tension for the process section. In the process section, when the coil is running in an annealing furnace, it is heated to above Ac_3 (transformation of the ferrite into the austenite ends at this temperature) or Ac_1 (beginning of the formation of the austenite at this temperature), solid treated or quenched in the cooling section to below 200°C and then

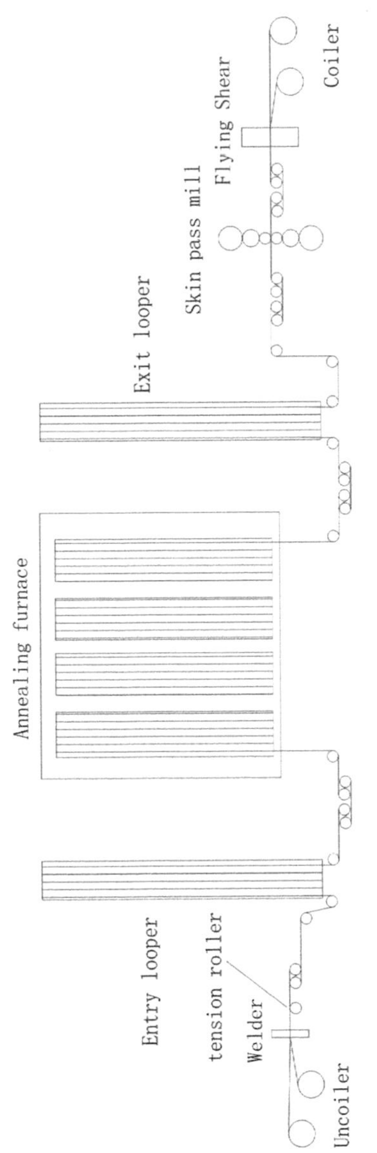

Figure 15.13 Layout of typical continuous annealing line.

required microstructure is obtained. The new coil is also buffered in an exit looper and then flattened by using a skin pass mill. Finally, the new coil is divided into several short strips by using a flying shear and these short strips are separately coiled by using two coilers.

To meet the process requirement, there are some functions of entry section:

1) Moving of the coil.
2) Auto measuring of diameter and width of the coil.
3) Auto inserting the uncoiler of the coil.
4) Auto cutting the head and tail of the coil.
5) Automatic passing through the welder and auto welding.
6) Master speed control.
7) Tension control.
8) Calculating the coil diameter and the remaining length in the uncoiler.
9) Inertia compensation and the constant tension control of the coil.

The functions of middle section include:

10) Tension and position control of entry looper and exit looper.
11) Master speed control.
12) Tension control in the furnace.
13) Tracking weld point of the strip.
14) Heating of the strip in the furnace.
15) Cooling path control of the strip in the cooling district.

The functions of exit section include:

16) Pressure control, gap control, elongation control and shape control of the strip in the skin pass mill.
17) Shearing of the strip.
18) Auto coiling of the strip.

Although many points are mentioned, only seven important points which are (6), (7), (8), (9), (12), (14) and (15) will be discussed in this section.

15.3.1 Mathematical Model of Uncoiler

Actually, the working process of the uncoiler is maintaining a constant tension of the strip. When the coil is loaded onto the uncoiler, there is no tension on the strip. After the coil is loaded onto the uncoiler, the uncoiler begins to rotate clockwise. Then the strip is coiled and tensioned. Finally the uncoiler is stalled an the tension increase to a value needed by technical requirement and then remain the same until the whole strip leave the uncoiler. In this period, the uncoiler works in motor-driven status and a silicon controlled rectifier converter is used to power the uncoiler works in the rectification mode. When the diameter of the coil becomes smaller and smaller, the armature current must be changed to maintain the tension.

1. Calculation of theoretical diameter

When the uncoiler is running, the coil diameter changes as shown in Figure 15.14. Ideally, the thickness of the strip is uniform and the gap between the coiled strips is negligible. The strip is wound on the roll to a ring and its area is filled by the strip. So the relationship between length and diameter of the strip can be obtained:

$$L = \frac{\pi(\frac{d_1}{2})^2 - \pi(\frac{d_2}{2})^2}{T} \tag{15.41}$$

where d_1 is the internal diameter of the strip, d_2 is the outer diameter, T is the thickness of the strip and L is the length of the strip.

After a period of time, the relationship between the coil diameter and the length is:

$$L + \Delta T = \frac{\pi(\frac{d_1}{2})^2 - \pi(\frac{d_2}{2})^2}{T} + \Delta T \tag{15.42}$$

where ΔL is the change of the strip length.

Then, the new diameter is given by:

$$d_2' = \sqrt{d_2^2 + \frac{4 \cdot \Delta L \cdot T}{\pi}} \tag{15.43}$$

The change of the strip length ΔL can be calculated by reading the line speed of the uncoiler from PLC. If the initial diameter d_2 and the strip thickness T are known, the diameter can be calculated.

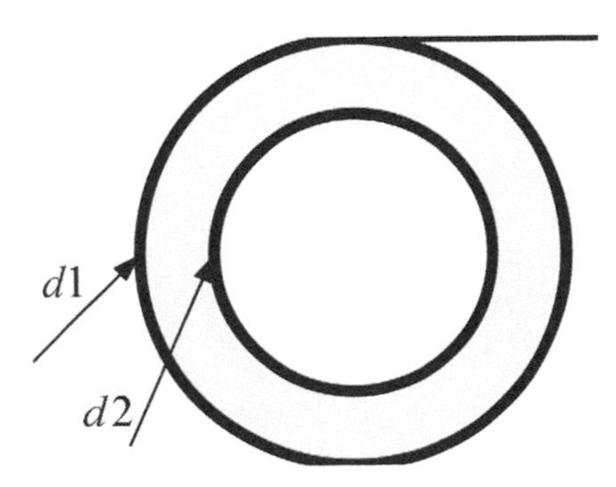

Figure 15.14 Theoretical analysis of the calculation of the coil diameter.

2. Control principle of uncoiler constant tension

The resistance moment M can be obtained from the tension T and the coil diameter D:

$$M = \frac{TD}{2} \tag{15.44}$$

The resistance moment M also can be obtained from following equation.

$$M = C_M \cdot \phi \cdot I \tag{15.45}$$

where C_M is the coefficient of the motor torque and it is a constant, Φ is the excitation flux, I is the current.

The above two equations imply,

$$\frac{TD}{2} = C_M \cdot \phi \cdot I \tag{15.46}$$

$$T = \frac{2C_M \cdot \phi \cdot I}{D} = K \cdot \phi \cdot I \tag{15.47}$$

where K is a constant.

$$K = 2C_M \tag{15.48}$$

Below the base speed, Φ is a constant.

$$\phi = \phi_e \tag{15.49}$$

Then:

$$T = \frac{K \cdot \phi_e \cdot I}{D} = \frac{K_1 \cdot I}{D} \tag{15.50}$$

$$K_1 = K \cdot \phi_e \tag{15.51}$$

Obviously, below the base speed, in order to keep the tension constant, the ratio of the armature current I and the diameter D of the Unwind need to be kept constant. Above the base speed, the back electromotive force E is a constant. According to the electro mechanics, the relationship of Φ and n can be obtained:

$$\phi = \frac{E}{C_E \cdot n} \tag{15.52}$$

where C_E is an electromotive force constant.

Substituting the value of ϕ in Eq. 15.50, the following equation is obtained:

$$T = \frac{K \cdot \phi \cdot I}{D} = \frac{K \cdot E \cdot I}{n \cdot C_E \cdot D} \tag{15.53}$$

The coil diameter D can be obtained from the line speed of the strip V and the angular speed of the uncoiler n:

$$D = \frac{60V}{\pi n} \tag{15.54}$$

Then:

$$T = \frac{K \cdot \pi \cdot E \cdot I}{60 \cdot C_E} = \frac{K_2 \cdot I}{V} \tag{15.55}$$

$$K_2 = \frac{K \cdot \pi \cdot E}{60 \cdot C_E} \tag{15.56}$$

So, above the base speed, the tension can be a constant only if the ratio of the armature current I and the line speed of the strip V is a fixed value.

When the strip line speed V changes, an extra part of the torque is required to overcome the mechanical inertia of the strip.

$$T_C = \left(\frac{C_1}{D^2} + C_2 \cdot D^2 \right) \cdot \frac{dV}{dt} \tag{15.57}$$

where C_1 and C_2 are constants.

3. Tension control model of uncoiler

Theoretically, the coil diameter D can be calculated from above equations. However, the change of the strip thickness and the interlayer slippage should also be considered in the calculation of coil diameter. As a result, the computed coil diameter deviates from the actual coil diameter. Thus the computed coil diameter needs to be adjusted.

When the strip is running with a tension, ignoring tiny elastic elongation caused by the strip tension, the speed is equal at every point. Thus in a given period of time, the length of the strip through the tension roller is equal to the length of the strip through the uncoiler spool.

$$\Delta\omega_m \frac{D_C}{2} = \Delta\omega_n \frac{D_B}{2} \tag{15.58}$$

where $\Delta\omega_m$ is the rotation angle of the coil in the period of time τ, D_C is the effective diameter of the coil, $\Delta\omega_m$ is the rotation angle of the tension roller in the period of time τ, D_B is the diameter of the tension roller.

When the drive motor of the tension roller rotates a circle (2π), the number of pulses is P_{m0} which is transmitted by the photoelectric encoder fixed on the drive motor shaft. Meanwhile, corresponding to the coil turn the angle $\Delta\omega_n$, the number of pulses is P_n. The following equations can be obtained:

$$\frac{P_{m0}}{2\pi} = \frac{P_m}{\Delta\omega_m \cdot i_C} \tag{15.59}$$

$$\frac{P_{n0}}{2\pi} = \frac{P_n}{\Delta\omega_n \cdot i_B} \tag{15.60}$$

where i_B is the transmission ratio of the uncoiler, i_C is the transmission ratio of the bridle.

Respectively, $\Delta\omega_m$ and $\Delta\omega_n$ can be obtained from Eqs. (15.59) and (15.60). Substituting the value of $\Delta\omega_m$ and $\Delta\omega_n$ in Eq. 15.58, the following equation is obtained:

$$D_C = \frac{D_B P_{m0} i_C}{P_{n0} i_B} \frac{P_n}{P_m} \tag{15.61}$$

If $\frac{D_B P_{m0} i_C}{P_{n0} i_B}$ is considered as a constant K, and then the above equation can be written as:

$$D_C = K\frac{P_n}{P_m} \tag{15.62}$$

From the number of pulses which are transmitted by the photoelectric encoder fixed on the drive motor shaft and the bridle shaft, the real-time value of the coil diameter can be obtained. The coil diameter D_C is calculated based on the flow equivalent method. The calculation accuracy is mainly determined by the accuracy of the photoelectric encoder. So the accuracy of the formula can be determined by the actual condition.

Due to the deviation, the calculated D_C cannot be directly used to modify the value D. After being numerically treated, the diameter of the coil D_C can be used to obtain the modifier diameter of the coil by comparing it with D.

Therefore, after every measurement cycle, the values of the coil diameter need to be compared with the diameter value of the former coil.

$$\Delta D_C = D_C(n) - D_C(n-1) \tag{15.63}$$

where ΔD_C is the deviation of the coil diameter and saved after each calculation. Then the arithmetic mean and the variance of deviations can be calculated. Generally, the deviations on the last eight times are analyzed.

$$\Delta D_{CMV} = \frac{1}{8}\sum_{n-1}^{8} \Delta D_C(n) \tag{15.64}$$

$$\Delta D_{CVar} = \frac{1}{8-1}\sum_{n-1}^{8} (\Delta D_C(n) - \Delta D_C(n-1))^2 \tag{15.65}$$

where ΔD_{CMV} is the arithmetic mean of the deviation, ΔD_{CVar} is the variance of the deviation.

If the value ΔD_{CVar} is within the set range then the diameter D_C calculated is credible, $\Delta D_{CMV}/2$ will be as the strip thickness value H. Comparing D_C with D, if the deviation exceeds a certain value, the D need to be modified by using ΔD_{CK}.

$$\Delta D_{CK} = \pm\frac{1}{3}(D - D_C) \tag{15.66}$$

After the D is modified in the next three measurement cycles, the values of D_C and D are compared. If the deviation is within the setting range, ΔD_{CK} is equal to zero. Otherwise the modified coil diameter can be obtained by following equation:

$$D' = D + \Delta D_{CK} \tag{15.67}$$

Since the deviation of the coil diameter calculated is ensured within a certain range, the value of the diameter is more accurate on the above equation. The value ΔD_{CK} contributes to reduce the errors caused by the initial diameter error, the inaccurate strip thickness and the interlayer sliding of the coil. The modified D is used as the initial value of the next cycle.

After the coil diameter D is determined, based on the principle that current I is proportional to the coil diameter D, the current I is determined by the coil diameter D to keep the tension of the uncoiler as a constant value.

15.3.2 Mathematical Model of Looper

As a buffer of the strip, the structure of a looper is shown in Figure 15.15 and consists of four parts: a framework, several fixed rollers, several moving rollers and one or two looper cars. When the looper car moves up, some length of the strip is released to the annealing furnace. When the looper car moves down, some new coming length of the strip is stored to the looper.

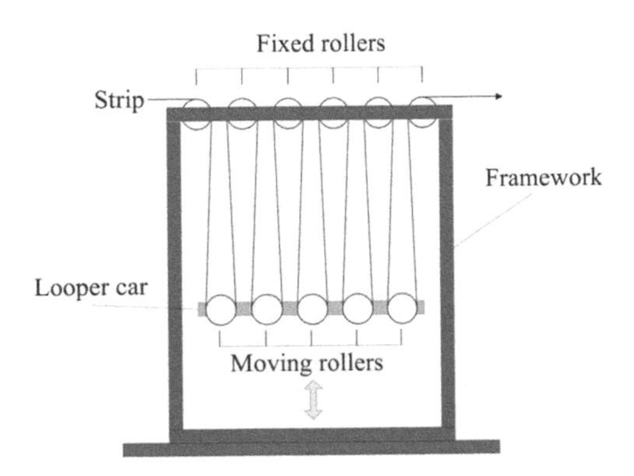

Figure 15.15 Structure of the looper.

There are two looper control modes: tension control mode and torque control mode.

Torque control means controlling strip tension through controlling motor torque. The strip tension increases with the motor torque and it consists of several parts:

$$T_q = T_{qT} + T_{qa} + T_{qF} + T_{qG} + T_{qZ} \tag{15.68}$$

where T_q is the total motor torque, T_{qT} is the setting tension torque, T_{qa} is the acceleration torque, T_{qF} is the friction torque which is related to the speed of the strip, T_{qG} is the self-weight torque of the strip, T_{qZ} is the bend torque of the strip.

PI (Proportional Integral) controller is used to adjust current tension to set tension.

1. Torque control of looper

Tension set torque is calculated as:

$$T_{qT} = \frac{T_{set} \cdot D}{2i} \tag{15.69}$$

where T_{qT} is the set tension torque, T_{set} is the set point for tension, D is the lopper outside diameter and i is the transmission gear ratio.

2. Calculation of acceleration torque

Acceleration torque includes three parts: the motor acceleration, the sleeve acceleration torque and strip acceleration torque.

$$T_{qa} = \frac{2i(A_{in} + A_{out})}{D(J_F + J_V)} \tag{15.70}$$

where T_{qa} is the acceleration torque, J_F is the inertia for the motor and the sleeve, J_V is the inertia for strip, A_{in} is the acceleration of lopper entry, A_{out} is the acceleration of lopper outlet, D is the outer diameter of the sleeve, i is the transmission ratio of gearbox.

3. Calculation of friction torque

The friction torque consists of two parts: one part is produced by the gearbox or drive shaft. The friction torque and motor speed has a non linear relationship and another part of the friction torque is the friction between the roller and strip. This friction torque is related to the car location. Taking into account the error of the friction, inertia compensation, the motor speed regulator has a relatively small amount of the adjustment. Although this

adjustment amount is small, it is meaningful to the stable station, the speed of charging car and dynamic tension.
The friction torque analysis is shown as follows:

$$T_{qF} = \alpha F_t \frac{D}{2} \tag{15.71}$$

where α is the journal friction coefficient, F_t is the tension and the sum of the volume weight, D is the coil diameter.

When the motor speed is less than a certain value, the friction torque and rotational speed is proportional to motor speed. If the speed is greater than this certain value, the friction torque is a constant. The friction torque compensation variable is the motor torque.

4. Calculation of self weight torque

During the vertical movement of strip, its weight G should be compensated for tension calculation and is calculated as follows:

$$G = H \cdot S \cdot L \cdot B \cdot h \cdot \rho \cdot g \tag{15.72}$$

where H is the height of the lopper, S is the layer number of lopper, L is the storage length of strip, B is the width of strip, h is the thickness of strip, ρ is the density of steel and g is the acceleration due to gravity.

The total weight torque:

$$T_{qG} = \frac{2(G_1 + G)}{D} \tag{15.73}$$

where G_1 is the weight of the looper car, D is the diameter of the roll, G is the weight of the strip.

5. Calculation of bending torque

In the case of elastic deformation, the bending moment of the strip can be obtained by the algorithm:

$$T_{qZ} = \frac{B}{4} h^2 \eta \times 10^{-3} \tag{15.74}$$

where η is the yield strength of the steel.

The bending torque of n layers strip is converted to the roll torque:

$$T_{qZn} = \frac{2nT_{qZ}D}{d} \tag{15.75}$$

where d is the steering roller diameter.

6. Calculation of tension setting value

The strip tension is related to the yield strength and calculated as follow:

$$F = \delta_s \cdot B \cdot h \cdot k \tag{15.76}$$

$$k = \beta(0.33 - 0.14h + 0.02h^2) \tag{15.77}$$

where k is the tension coefficient, β is the coefficient and = 0.5.

So, the machine tension set point value is:

$$F_r = F + \frac{2T_q}{Di} \tag{15.78}$$

where D is the windlass diameter, i is the speed ratio of the reduction box which is connected with the looper motor, T_q is the torque of looper motor.

7. Tension control for looper

The main function of the looper is to store strips, synchronize with the movement of the strip at the inlet and outlet and ensuring a continuous production. In a production process, looper tension is only needed to keep the strip in a live kit deviation and the strip can be wrapped tight on the roller without relative sliding and surface scratches which is caused by the sliding. Because the strip in the looper is very long and there might be tension losses in the looper, therefore, various kinds of strip tension loss in the looper must be considered when a reasonable looper tension set point value is set.

1) Tension loss caused by the steering roller

The unit tension caused by steering roller is related to the strip thickness, the yield strength and the diameter of steering roller. It can be calculated by the following equation:

$$F_{in} = \left[C_{h1} + C_{h2}e^{(-h/C_{h3})}\right] \times \left[C_{s1}e^{(\sigma_s/C_{s2})}\right] \times \left[C_{D1}e^{(D/C_{D2})}\right] \tag{15.79}$$

where F_{in} is the steering tension in looper entrance, h is the strip thickness, σ_s is the yield strength, D is the diameter of the steering roller, C_{h1}, C_{h2} and C_{h3} are the steering tension coefficients of strip thickness, C_{s1} and C_{s2} are the steering tension coefficients of yield strength for steel, C_{d1} and C_{d2} are the steering tension coefficients of the diameter of the steering roller. In order to meet the process requirements of the strip and let the strip under a stable operation condition, the tension of the entrance steering roller should be selected as the benchmark tension. Eventually, based on the benchmark tension, the corrective tension can be obtained by overlaying tension losses.

2) Tension loss caused by elastic-plastic bending

The strip which is bent to wrap the looper roller causes some tension loss. When the strip bends 180° on a looper roller, the tension loss is calculated as follows:

$$F_b = \frac{1}{6}\frac{\sigma_s Bh^2(3-e^2)}{D} \tag{15.80}$$

where F_b is the elastic-plastic bending tension loss, σ_s is the yield strength of the strip, B is the strip width, h is the strip thickness, D is the looper roller outer diameter and e is the ratio of the thickness of the elastic deformation zone to strip thickness.

When the strip is under a pure elastic tension, the e is 1, when the strip is under plastic deformation, the e is equal to zero. When the strip bends 180° on a looper roller, the tension loss is calculated as follows:

$$F'_b = F_b / 3 \tag{15.81}$$

where F'_b is the tension loss under a pure elastic tension.

So the elastic-plastic bending tension loss for a strip is a major part in the tension loss when the strip thickness and yield strength is large.

3) Tension loss caused by the bearing of the looper roller

The frictional resistance of the bearing is shown in Figure 15.16. The strip tensions at the entry and the outlet of the looper roller are T_1 and T_2. The tension of the strip on the roller is shown as follows:

The tension is:

$$F_m = C_0 \mu N \frac{d}{D} \tag{15.82}$$

where F_m is the tension loss which overcomes the bearing friction, μ is the friction coefficient, d is the diameter of bearing, D is the diameter of looper roller, N is the surface pressure between the looper roller and the strip, C_0 is the coefficient of friction compensation.

Since the $T_1 \approx T_2$, then:

$$N = 2T_1 \sin\frac{\theta}{2} \tag{15.83}$$

where θ is the strip wrap angle on the looper roller.

So there is:

$$F_m = 2C_0 \mu T_1 \frac{d}{D} \sin\frac{\theta}{2} \tag{15.84}$$

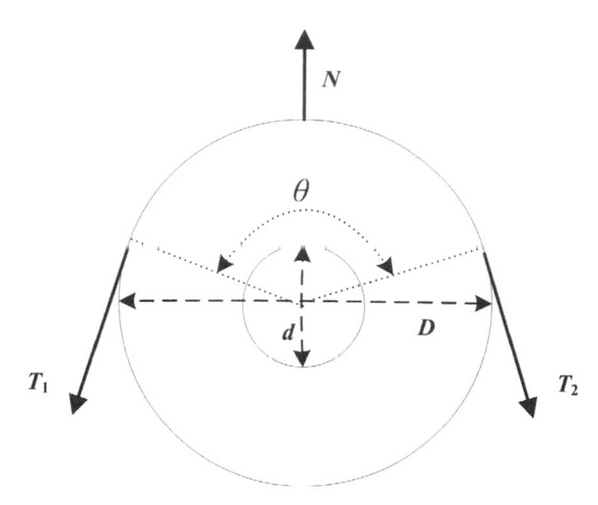

Figure 15.16 Friction resistance of the looper roller bearings.

The tension caused by the bearing friction loss is relatively small, but the friction becomes bigger when the tension is bigger.

4) Tension loss caused by the acceleration and deceleration of the strip

If the strip is in an accelerated state, it requires a forward acceleration forces. If the strip is in a deceleration state, it requires a backward deceleration force:

$$F_{-} = Lqa, \quad F_{+} = -Lqa \tag{15.85}$$

where F_{-} is the strip tension loss at acceleration station, F_{+} is the tension loss when the strip at a deceleration state, a is the acceleration, L is the strip length, q is the strip unit weight.

5) Tension loss caused by the weight of the strip

Due to the height of the vertical looper, the strip weight has a certain impact on the tension. The tension loss which is caused by strip weight is:

$$F_g = G = V\rho g = hBL'\rho g \tag{15.86}$$

where F_g is the tension loss caused by the weight of the strip, G is the weight of the strip, V is the strip volume, ρ is the steel density, g is the gravity acceleration, h is the strip thickness, B is the strip width, L' is the height between looper roller and tension roller.

Based on the model of tension loss, the total tension loss is obtained as the follow:

$$F_s = n \cdot (F_b + F_m + F_- + F_+) + m \cdot \dot{F}_b - F_g \tag{15.87}$$

where n is the number of looper roller wrapped at 180° by the strip, m is the number of the looper roller wrapped at 90° by the strip.

15.3.3 Coiler Mathematical Model

1. Calculation of coil diameter

In order to control the tension of the coiler accurately, it is necessary to measure the diameter of the coiler. The calculation process of the coiler diameter is shown in Figure 15.17 by the speed ratio of the pinch roll N_p to the coiler N_M.

$$D_n = D_{n-1} + \Delta D \tag{15.88}$$

$$\Delta D = \frac{N_P}{N_M} d - D_{n-1} \tag{15.89}$$

$$\Delta D_{\max} = \frac{N_M}{60} \times 2h \times T_S + B_j \tag{15.90}$$

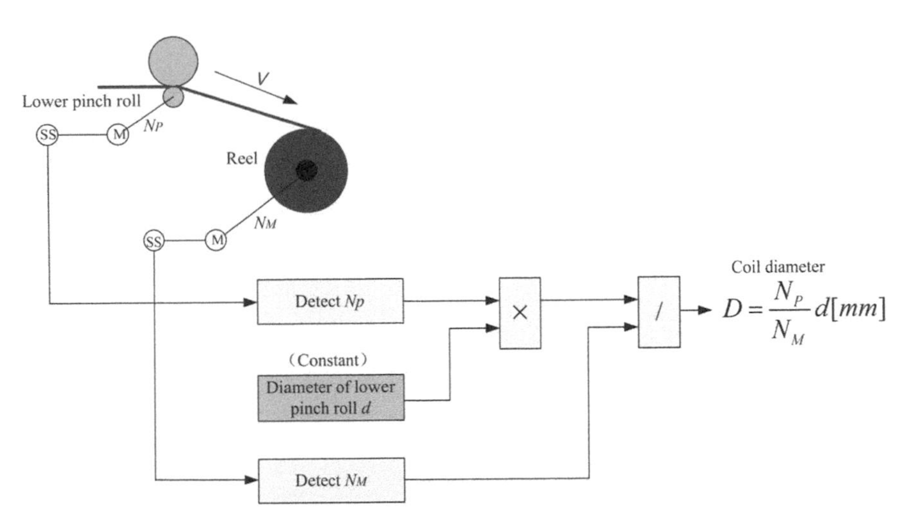

Figure 15.17 The diameter calculation of the coiler.

If $\Delta D \leqslant 0$, then $\Delta D = 0$.

If $\Delta D \geqslant \Delta D_{max}$, then set $\Delta D = \Delta D_{max}$.

2. Calculation of tension torque

In the case of the open loop tension control, the motor torque M_D consists of the following three parts:

$$M_D = M_T + M_d + M_f \tag{15.91}$$

where M_T is the torque caused by the setting tension of the strip, M_d is the dynamic compensation torque, M_f is the friction torque and it is determined by experiment in debug period. The M_T is calculated using the following equation:

$$M_T = U_T \times B \times h \times \frac{D}{2} \tag{15.92}$$

where U_T is the unit tension of the strip, B is the width of the strip, D is the coil diameter and h is the thickness of the strip.

The dynamic compensation torque M_d is determined by the change of line speed and the coil diameter and is calculated as follows:

$$M_d = \frac{GD^2}{375}\frac{dn}{dt} \tag{15.93}$$

where n is the rotation speed of the coiler, $\frac{dn}{dt}$ is the acceleration speed of the coiler, GD^2 is the moment inertia of the whole system of the coiler which includes the moment inertia of the motor and the drive device GD_0^2 and the moment inertia of the strip GD_C^2.

$$GD^2 = GD_0^2 + GD_C^2 \tag{15.94}$$

where GD_C^2 changes with the change in coil diameter:

$$GD_C^2 = \frac{\pi}{8} \times \rho \times B \times (D^4 - D_0^2) \tag{15.95}$$

where ρ is the density of the strip, B is the width of the strip.

The compensation of the friction torque M_f is a nonlinear function of the speed. According to design experience and practice, the function is determined by experiment in debug period.

15.3.4 Mathematical Model of Cooling Process in CAL

Figure 15.18 shows an annealed product's requirements for temperature and there are eight segments in CAL: the preheat segment, the heat segment, the soaking heat segment, the gas jet cooling segment, the roll cooling segment,

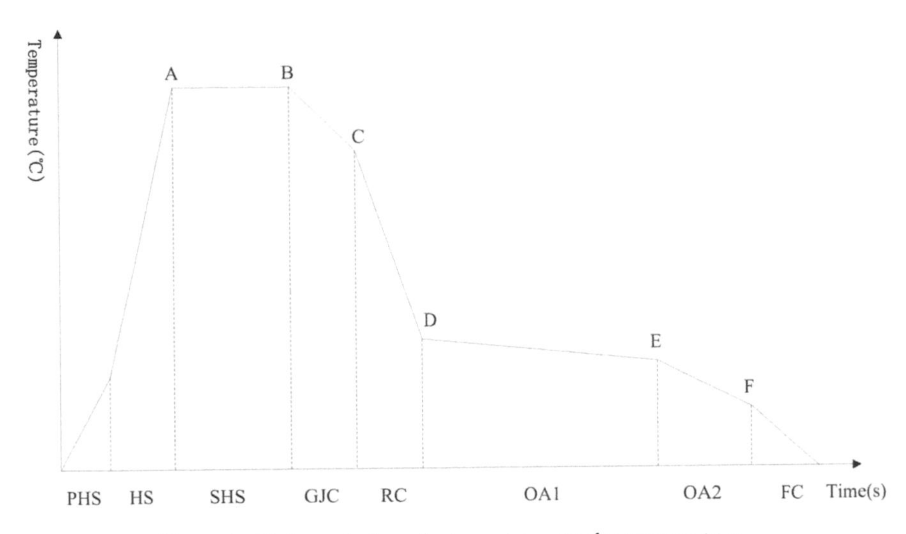

Figure 15.18 Annealed product requirements for temperature.

the over aging segment 1, the over aging segment 2 and the final cooling segment. In this section, an emphasis will be put on the cooling process.

The cooling process of a modern continuous annealing consists of GJC (Gas Jet Cooling), WQ (Water Quench), RQ (Roll Quench) or ACC (Accelerated Control Cooling). The comparison of the four cooling methods is shown in Table 15.5.

The cooling speed of RQ is between that of GJC and WQ, furthermore, the strip temperature is easy to control. The advantages of RQ are:

1) The cooling speed is fast, up to 100~400°C/s, which meets the temperature demand of multiple grade strips.
2) The cooling is well distributed and the strip can have a good performance in flatness.

The temperature control in the RQ section is the most complex control process and is shown in Figure 15.19. There are three basic heat transfer modes in the strip cooling process: transmission, heat convection and radiation heat transfer. The heat transfer of the strip is through the contact with the water cooling rollers, and the cooling by using the opposite jets and the back jets. At the same time, there is also a radiation heat transfer between the strip and furnace wall in this system.

1. *Mathematical model of RQ and opposite jet cooling*

During the cooling process, the heat transfer between the strip and the water cooling roller and the convection between the side blown gas and

Table 15.5 The comparison of GJC, WQ, RQ and ACC in cooling process.

Cooling methods	**Cooling speed (°C/s)**	**Process**	**Applicable material**
GJC	30~50	Gas jet cooling	Cold rolling sheet
WQ	500~2000	Water quench after cooling to 560°C	Cold rolling sheet High strength sheet
RQ	100~400	Roll touch quench by cooling water roll	Cold rolling sheet High strength sheet
ACC	80~300	The mixed medium of protective gas and water jet cooling	Cold rolling sheet High strength sheet

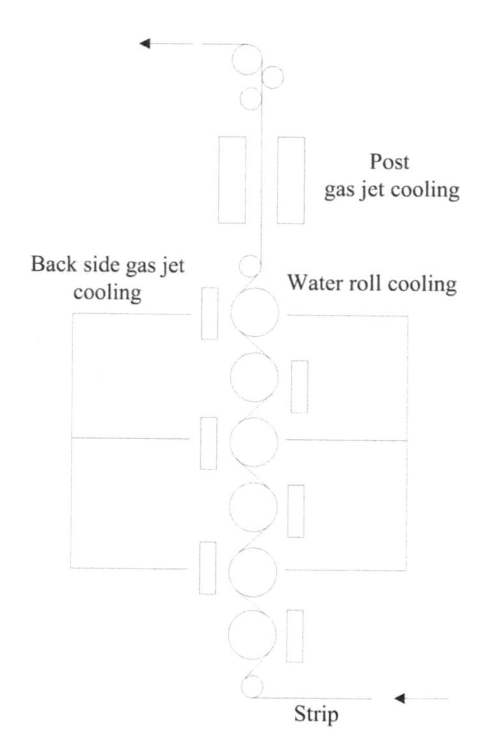

Figure 15.19 The composition of equipments and functions in RQ section.

the strip are the two main parts. The exit temperature of the strip T_{out} can be obtained by following equation:

$$T_{out} = T_c - (T_c - T_{in}) \cdot e^{-\frac{\Delta T \cdot U\left(C_1 \theta^{C_2} + C_3\right)}{C_p h \rho}} - (T_{in} - T_a) - S \tag{15.97}$$

where ρ is the density of the strip, h is the thickness of the strip, U is the self learning coefficient of the model, C_p is the specific heat of the strip, ΔT is the contact time between the strip and water cooling roller, C_1, C_2 and C_3 are constants, θ is the contact angle, T_c is the cooling water temperature, T_a is the cooling gas temperature of opposite jet cooling, T_{out} is the exit temperature of the strip, T_{in} is the entry temperature of the strip, S is a calculation parameter.

2. *Mathematical model of back jet cooling*

For a step time, the transfer heat is obtained by following equation:

$$\Delta q = \frac{\left[F_1\left(T_g - T_s\right) + F_2\left[\left(T_f + 273\right)^4 + \left(T_s + 273\right)^4\right] + F_3\left(T_f - T_s\right)\right]\Delta T_g}{\rho h} \tag{15.98}$$

where F_1, F_2 and F_3 are calculation factors, T_g is the temperature of cooling air, T_f is the temperature of the furnace, T_s is the temperature of the strip, ΔT_g is the step time, Δq is the transfer heat.

3. *Profile control*

Due to many different products, some of the strips may be insufficiently cooled in RQ section. Under the condition that other parameters are constants, increasing the contact angle between the strips and the cooling roller is useful to control strip temperature properly and avoid shape wave.

When the temperature of the furnace is unstable, the strip might be heated inadequately in the heating section and the strip could be harder than it should be. In this case, the strip may be unstably cooled in the RQ section. Decreasing the running speed of the strip in the heating section is usually used to improve its shape quality.

Defects appear in incoming strips (especially in high strength steel strip) and might lead to the insufficient contact between the strip and the cooling roller in the RQ section. So the profile of the strip becomes bad because of uneven temperature distribution caused by the uneven contact between the strip and the cooling roller. The problem can be solved by increasing the tension in order to improve the contact between the strip and the cooling roller.

Middle wave defect will be caused if an unsuitable back jet cooling or no back jet cooling is used. Through a reasonable back jet and opposite jet cooling, a suitable temperature distribution is established in the strip, as a result, the profile of the strip in the exit of the RQ section can be controlled.

Keywords: Lamination cooling, ultra fast cooling, heat transfer coefficient, continuous annealing line, tension control, roller quench, air jet cooling

Index